Organometallic Compounds of Low-Coordinate Si, Ge, Sn and Pb

Organometallic Compounds of Low-Coordinate Si, Ge, Sn and Pb

From Phantom Species to Stable Compounds

VLADIMIR Ya. LEE and AKIRA SEKIGUCHI

Department of Chemistry,
Graduate School of Pure and Applied Sciences,
University of Tsukuba, Tsukuba, Japan

WILEY

A John Wiley and Sons, Ltd., Publication

This edition first published 2010

Registered office
John Wiley & Sons Ltd, The Atrium, Southern Gate, Chichester, West Sussex, PO19 8SQ, United Kingdom

For details of our global editorial offices, for customer services and for information about how to apply for permission to reuse the copyright material in this book please see our website at www.wiley.com.

Library of Congress Cataloging-in-Publication Data

Vladimir Ya. Lee, 1960-
Organometallic compounds of low-coordinate Si, Ge, Sn, and Pb : from phantom species to stable compounds / Vladimir Ya. Lee and Akira Sekiguchi.
p. cm.
Includes bibliographical references and index.
ISBN 978-0-470-72543-6 (cloth)
1. Organometallic compounds. I. Lee, Vladimir Ya. II. Title.
QD411.S45 2010
547′.0568–dc22

2010013108

A catalogue record for this book is available from the British Library.
ISBN: 9780470725436

Typeset in 10/12pt Times by Laserwords Private Limited, Chennai, India.

Contents

Preface

The variety of organometallic compounds based on the group 14 elements heavier than carbon is tremendous, spanning the wide range from low(mono, di, tri)- to normal tetra- to hyper(penta, hexa)-coordinate derivatives. However, in contrast to the two last classes of compounds, the low-valent group 14 organometallics have never appeared in books. The lack of such books, which is particularly surprising given the permanently growing interest in this hot field of modern main group chemistry as reflected by the vast number of outstanding reviews and book chapters, prompted us to undertake an attempt to survey, analyse and summarize the current state of affairs in this area. The fundamental achievements in the field are associated, first of all, with the recent advances in state-of-the-art sophisticated synthetic and spectroscopic techniques, as well as rapid progress in theory and computational methods. This has enabled isolation and structural studies of unique stable low-coordinate species, which only a few decades ago were believed to exist only as fleeting intermediates, not isolable or even detectable. In our book, subtitled *From Phantom Species to Stable Compounds*, we specifically deal with the synthetic accomplishments that have been made in the field of **isolable low-coordinate derivatives of heavy group 14 elements**, giving only short comments regarding their transient congeners. Having specialized in this topic, we do not aim to compete with the previously published books on the heavy group 14 organometallics which cover a broad range of rather different topics (from theoretical insights to experimental achievements and material science applications), such as the brilliant series *The Chemistry of Organic Silicon Compounds* and The *Chemistry of Organic Germanium, Tin and Lead Compounds* (edited by Patai, Rappoport and Apeloig) and *Organosilicon Chemistry: From Molecules to Materials* (edited by Auner and Weis).

Given that the specific field of low-coordinate group 14 organometallics is flourishing with many spectacular achievements that deserve mentioning, it is not realistic to cover all of them in a single volume and we do not intend to do this. Instead, we focus on the most important and most recent (the literature coverage is up to the end of 2009 – beginning of 2010) milestone advances that are crucial for a general understanding of the peculiar structural bonding and chemical properties of the low-coordinate group 14 organometallics, which in many cases are distinctly different from those of their organic analogs. The book is organized into six chapters, each devoted to an independent class of the most fundamental low-valent species: heavy analogs of carbenium ions (Chapter 1), heavy analogs of free radicals (Chapter 2), heavy analogs of carboanions (Chapter 3), heavy analogs of carbenes (Chapter 4), heavy analogs of unsaturated hydrocarbons: alkenes, 1,3-dienes, allenes, and alkynes (Chapter 5), and heavy analogs of aromatic compounds (Chapter 6). Each chapter begins with a review of general synthetic approaches, continues with a consideration of particular structural features and synthetic applications, and concludes with a discussion of the most important recent advances in the field of stable derivatives.

We are particularly grateful to many of our highly talented and greatly motivated students, with whom we have been lucky to work with and who have made numerous outstanding experimental contributions. We would also like to thank our distinguished collaborators from many research groups from all over the world, both experimentalists and theoreticians, with whom we have been very pleased to work with throughout our research careers and whose names are listed in a number of our joint publications.

Regarding the audience of our book, we hope that it will be useful to the entire scientific community; however, first of all, we address it to advanced graduate and postgraduate students, especially to those who intend to specialize in the field of organometallic chemistry. We believe that our contributions will also be of interest and be helpful to those who have already been involved in the fascinating and challenging world of organosilicon, organogermanium, organotin, and organolead chemistry. We hope that the book will find its readers not only among specialists in the field of group 14, but also among others working in the areas of both main group and transition metal chemistry, as well as those from interdisciplinary fields such as polymer, material science, nanotechnology etc. Let us finally hope that our book will serve as a useful guide and reference source to interested specialists and even more importantly, to those inexperienced beginners who are still seeking inspiration.

Vladimir Ya. Lee
Akira Sekiguchi
University of Tsukuba (Tsukuba)

January 2010

Abbreviations

Ad	1-Adamantyl
acac	2,4-Pentanedionate
AIBN	Azobisisobutyronitrile
ASE	Aromatic Stabilization Energy
Bbt	2,6-Bis[bis(trimethylsilyl)methyl]-4-[tris(trimethylsilyl)-methyl]phenyl
BLYP	Becke 1988 Exchange Functional with the Lee–Yang–Parr Correlation Functional
B3LYP	Becke Three-Parameter Hybrid Functional with the Lee–Yang–Parr Correlation Functional
BP86	Becke Exchange Functional and the Perdew Correlation Functional
CC-pVDZ	Correlation-Consistent Polarized Valence Double-Zeta Basis Set
CCSD(T)	Coupled Cluster Method Including Singles, Doubles and Optional Triples terms
CGMT	Carter–Goddard–Malrieu–Trinquier
CI	Configuration Interaction
CIDNP	Chemically Induced Dynamic Nuclear Polarization
CIP	Contact Ion Pair
CISD	Single and Double Excitations, Single Reference CI
Cp	Cyclopentadienyl η^5-C_5H_5
Cp*	Pentamethylcyclopentadienyl η^5-C_5Me_5
CPMAS	Cross Polarization Magic Angle Spinning
CSA	Chemical Shift Anisotropy
CST	Chemical Shift Tensor
CV	Cyclic Voltammetry
CW	Continuous Wave
Cy	Cyclohexyl
DBU	1,8-Diazabicyclo[5.4.0]undec-7-ene
DFT	Density Functional Theory
diglyme	Bis(2-methoxyethyl) ether
diox	1,4-Dioxane
Dip	2,6-Diisopropylphenyl

Dis	Bis(trimethylsilyl)methyl
Ditp	2,6-Bis(2-*iso*-propylphenyl)phenyl
DME	1,2-Dimethoxyethane
dmpe	1,2-Bis(dimethylphosphino)ethane
DMPU	*N*,*N*′-Dimethylpropyleneurea
DZD	Double-Zeta Diffuse Basis Set
DZP	Double-Zeta with Polarization Basis Set
DZVP	Double Zeta Valence Basis Set Augmented with Polarization Functions
EA	Electron Affinity
ECP	Effective Core Potential
EDA	Energy Decomposition Analysis
ENDOR	Electron Nuclear Double Resonance
EPR	Electronic Paramagnetic Resonance
eV	Electron volt
EXAFS	Extended X-Ray Absorption Fine Structure
FT	Fourier Transformation
G2	Gaussian-2
GIAO	Gauge Independent Atomic Orbital
HF	Hartree–Fock
hfcc	Hyperfine Coupling Constant
HMPA	Hexamethylphosphortriamide
HOMO	Highest Occupied Molecular Orbital
IE	Ionization Energy
IGLO	Individual Gauge for Localized Orbital
IR	Infrared
LDMAN	Lithium 1-(dimethylamino)naphthalenide
LUMO	Lowest Unoccupied Molecular Orbital
Λ	Diamagnetic Susceptibility Exaltation
MCSCF	Multiconfigurational SCF
Mes	2,4,6-Trimethylphenyl
Mes*	2,4,6-Tri-*tert*-butylphenyl
2-Me-THF	2-Methyltetrahydrofuran
MO	Molecular Orbital
3-MP	3-Methylpentane
MP2	Second-order Møller–Plesset Perturbation Theory
MP4	Fourth-order Møller–Plesset Perturbation Theory
MPW1PW91	Hybrid Density Functional Employing Modified Perdew–Wang 1991 Exchange and Perdew–Wang 1991 Correlation
NBO	Natural Bond Orbital
NHC	N-Heterocyclic Carbene
NICS	Nucleus Independent Chemical Shift
NMR	Nuclear Magnetic Resonance
NPA	Natural Population Analysis
NL–SCF	Non-Local Self-Consistent Field
OTf^-	$OSO_2CF_3^-$

PBE1PBE	Adiabatic Connection Method Functional Derived from Perdew–Burke–Enzerhof Functional
PES	Potential Energy Surface
PES	Photoelectron Spectroscopy
PMDTA/PMDETA	Pentamethyldiethylenetriamine
PSO	Paramagnetic Nuclear Spin–Electron Orbit
Py	Pyridine
Pz	Pyrazolyl
RE	Resonance Energy
REMPI	Resonance-Enhanced Multiphoton Ionization Spectroscopy
RHF	Restricted HF
SCF	Self-Consistent Field
SDB-cc-pVTZ	Stuttgart–Dresden–Bonn Relativistic Effective Core Potential with the Correlation-Consistent Polarized Valence Triple Zeta Basis Set
SDD	Stuttgart–Dresden Effective Core Potential with the Double Zeta Basis Set
SET	Single Electron Transfer
SINDO1	Intermediate Neglect of Differential Overlap Method Modified on the Basis of Symmetrically Orthogonalized Orbitals and Commutator Relations
S_N2	Substitution Nucleophilic Bimolecular
SOMO	Singly Occupied Molecular Orbital
SSIP	Solvent-Separated Ion Pair
STO-3G	Slater-Type Orbital Minimal Basis Set using Three Gaussians to Fit an Exponential
Tbt	2,4,6-Tris[bis(trimethylsilyl)methyl]phenyl
TCSCF	Two-Configurational SCF
TD-DFT	Time-Dependent Density Functional Theory
TfOH	Trifluoromethanesulfonic acid
$TFPB^-$	Tetrakis[3,5-bis(trifluoromethyl)phenyl]borate
THF	Tetrahydrofuran
THT	Tetrahydrothiophene
Tip	2,4,6-Triisopropylphenyl
Titp	2,6-Bis(2,4-diisopropylphenyl)phenyl
TMDAP	1,3-Bis(dimethylamino)propane
TMEDA	Tetramethylethylenediamine
TPB^-	Tetraphenylborate
$TPFPB^-$	Tetrakis(pentafluorophenyl)borate
$TSFPB^-$	Tetrakis{4-[*tert*-butyl(dimethyl)silyl]-2,3,5,6-tetrafluorophenyl}borate
$TTFPB^-$	Tetrakis(2,3,5,6-tetrafluorophenyl)borate
TZ2P	Triple-Zeta Basis Set with Two Sets of Polarization Functions
TZV	Triple Zeta Valence Basis Set
TZVP	Triple Zeta Valence Basis Set Augmented with Polarization Functions
UB3LYP	Unrestricted B3LYP

UHF	Unrestricted Hartree–Fock
UHF–NO CI	Unrestricted Hartree–Fock Natural Orbitals Configuration Interaction
UMP2	Unrestricted Second-order Møller–Plesset Perturbation Theory
UV	Ultraviolet
UV–PES	Ultraviolet Photoelectron Spectroscopy
VDZ+P	Valence Double Zeta Plus Polarization Functions Basis Set
WBI	Wiberg Bond Index
Xyl	2,6-Dimethylphenyl
6-31G(d)	Valence Double Zeta Basis Set with d-Type Polarization Functions for Heavy Atoms
6-311G(d)	Valence Triple Zeta Basis Set with d-Type Polarization Functions for Heavy Atoms
6-311+G(2d,p)	Valence Triple Zeta Basis Set with s- and p-Type Diffuse Functions for Heavy Atoms, Two d-Type Polarization Functions for Heavy Atoms, and One p-Type Polarization Function for Hydrogen
6-311+G(2df,p)	Valence Triple Zeta Basis Set with s- and p-Type Diffuse Functions for Heavy Atoms, Two d-Type and One f-Type Polarization Functions for Heavy atoms, and One p-Type Polarization Function for Hydrogen

1

Heavy Analogs of Carbenium Ions: Si-, Ge-, Sn- and Pb-Centered Cations

1.1 Introduction

The classical textbook definition of the carbenium ions R_3C^+ (*carbenium* ions are tri-coordinate carbocations, while those with a coordination number of five and above are named *carbonium* ions) describes them as trivalent species with a positively charged central sp^2-hybridized carbon atom, which features planar geometry and R–C–R bond angles close to ideal values of 120°. The unhybridized $2p_z$-orbital on the central carbon is vacant and orthogonal to the R_3C plane: the geometry which has, for example, the simplest methylium ion CH_3^+ isoelectronic to BH_3. Given the intrinsic electron deficiency of the carbenium ions, which have only six valence electrons in their valence shell, one would expect them to possess very high Lewis acidity and extreme electrophilicity. This is indeed the case, and in the early stages the carbenium ions were commonly considered only as short-lived fleeting reactive intermediates of classical electrophilic reactions, such as S_N1 solvolysis, electrophilic addition to alkenes, aromatic substitution, etc.: the pioneering contributions to this field were done by Meerwein (Germany), Ingold (UK) and Whitmore (USA). Accordingly, the existence of the transient (unobservable) carbenium ions was firmly supported by a number of experimental facts, including substituent effects, orientation in electrophilic reactions, solvent effects on the rates of solvolysis, rearrangements, etc. In a limited number of cases carbenium ions have been thermodynamically and kinetically stabilized by appropriate substituents. Thus, the first example of such persistent carbenium ions, namely the triphenylmethylium ion Ph_3C^+

Organometallic Compounds of Low-Coordinate Si, Ge, Sn and Pb: From Phantom Species to Stable Compounds
Vladimir Ya. Lee and Akira Sekiguchi

(otherwise known as the trytil cation), was prepared at the very beginning of the twentieth century due to the seminal works of Norris and Wentzel in 1901 (the crystal structure of its perchlorate salt $Ph_3C^+ \bullet ClO_4^-$ was reported much later, in 1965). The other milestone achievement in the chemistry of carbenium ions is related to the generation and direct NMR spectroscopic observation of the stable long-lived alkyl cations in superacidic media (SbF_5–SO_2, HF–SbF_5, 'magic' acid HSO_3F–SbF_5), developed by the group of Olah and nicely covered in a series of his papers published in the 1950–1960s. The major advantage of using superacids was their extreme acidity allowing the smooth formation of carbocations through halogen abstraction from alkyl halides: $Me_3CF + SbF_5/SO_2 \rightarrow Me_3C^+ \bullet SbF_6^-$. On the other hand, the very low basicity and nucleophilicity of the counteranions (SbF_6^-) prevented their reaction with carbocations, thus promoting the formation of true ion pairs.

The generation of the analogs of carbenium ions of the heavy group 14 elements, that is silylium, germylium, stannylium and plumbylium ions R_3E^+ (E = Si, Ge, Sn, Pb), was one of the most attractive and long-standing goals in contemporary organometallic chemistry, and is still a field of very active investigation. From the early stages of heavy carbenium ion chemistry, it quickly became apparent that there is a huge difference between the carbenium ions R_3C^+ and their heavy analogs R_3E^+ because of the sharply distinctive properties of carbon and its heavy congeners: size, polarizability and electronegativity. Consequently, the synthetic approaches, which were very successfully used for generation of stable carbenium ions in organic chemistry, proved to be rather inefficient in the synthesis of silylium ions, because of the high electrophilicity of the latter species leading to their intrinsic kinetic instability. Another important problem, hampering the generation of heavy group 14 element-centered cations, deals with the degree of 'freedom' of such cations from external nucleophiles, such as counter anions and solvents. It is therefore not surprising that the real nature of the bonding interaction between such cationic species and their counteranions, *ionic* vs *covalent*, has been one of the most important questions to solve in the problem of the true cations of the heavy group 14 elements.

Accordingly, the successful synthesis of silylium, germylium, stannylium and plumbylium ions has required the design of new synthetic strategies based upon the utilization of counterions and solvents of particularly low nucleophilicity to prevent their reaction (or coordination) to the cationic part. The first crystal structures of silylium ion derivatives were reported in the early 1990s; however, their real silylium ion nature has been severely criticized. Meanwhile, taking advantage of the particularly low nucleophilicity of borate and carborane as counteranions and using benzene and toluene as solvents finally enabled the synthesis of true R_3E^+ (E = Si, Ge, Sn, Pb) cations, free from any covalent interactions with either counterion or solvent. Although some of these cations were intramolecularly stabilized by cyclic π-conjugation, the acyclic tricoordinate cations were almost entirely electronically unperturbed, being genuine heavy analogs of the classical carbenium ions.

The chemistry of the heavy analogs of carbenium ions has been repeatedly reviewed during the past several decades, describing both transient and stable representatives.[1] In this chapter, we will briefly overview the whole story of the cations of heavy group 14 elements (generation of cations, their reactions and synthetic applications) with particular emphasis given to the latest progress in the field, which deals with the synthesis and structural characterization of stable free cations of the type R_3E^+.[2]

1.2 Synthesis of RR′R″E$^+$ Cations (E = Si–Pb)

The general synthetic approaches for the preparation of the heavy group 14 element centered cationic species can be classified into several groups based on the starting material used.

1.2.1 From Halides RR′R″EX

Ionization of the carbon–halogen bond is a key step in the monomolecular substitution reaction $R_3C–X \rightarrow R_3C^+ + X^-$ and is the most general method for the generation of stable carbocations in organic chemistry. In a marked contrast, this synthetic approach is definitely not the best choice for the preparation of the heavy analogs of the carbenium ion RR′R″E$^+$, because of the strong E–X bonds of the precursor RR′R″EX on the one hand and great reactivity of the developing cationic species RR′R″E$^+$ towards the halide leaving group X^- on the other hand (much higher halophilicity of Si–Pb compared with that of C). Therefore, cations generated by this method are to be classified as strongly polarized donor–acceptor complexes featuring only a partial positive charge on E, rather than true silylium ions (Scheme 1.1).[3,4]

$$Me_3Si–Br + AlBr_3 \xrightarrow{CH_2Br_2} Me_3Si^{\delta+}–Br \cdots AlBr_3^{\delta-} \quad \textbf{(A)}$$

$$Et_3Si–OTf + BCl_3 \longrightarrow Et_3Si^{\delta+}–OTf \cdots BCl_3^{\delta-} \quad \textbf{(B)}$$

Scheme 1.1

1.2.2 From Hydrides RR′R″EH

This so-called 'hydride-transfer reaction' is the most commonly used and straightforward method for the generation of stable RR′R″E$^+$ cations. The driving force of this process, involving oxidation of the starting hydride RR′R″EH with a powerful Lewis acid (typically, trityluim ion Ph_3C^+), is the relative strength of the breaking and forming bonds: stronger C–H vs weaker E–H. A variety of heavy analogs of carbenium ions, intra- or intermolecularly stabilized by coordination to n/π-donors, counteranions or nucleophilic solvents, can be readily prepared by this route (Scheme 1.2).[5–7] As a drawback of this synthetic approach one should mention the steric bulkiness of the Ph_3C^+ reagent, which may hamper its interaction with hydrides RR′R″EH bearing voluminous substituents necessary for the kinetic stabilization of the resulting cation.

1.2.3 From RR′R″E–R‴ and RR′R″E–ERR′R″

A most impressive example of the generation of R_3E^+ cations by cleavage of R_3E^+–C bonds was reported by Lambert *et al.* They treated allylic derivatives

$$Et_3SiH + Ph_3C^{+}\bullet TPFPB^{-} \xrightarrow{C_6H_6} [Et_3Si(C_6H_6)]^{+}\bullet TPFPB^{-} + Ph_3CH \qquad \textbf{(A)}$$

$$i\text{-}Pr_3SiH + Ph_3C^{+}\bullet[CB_{11}H_6X_6]^{-} \xrightarrow[(X = Cl,\ Br,\ I)]{C_7H_8} i\text{-}Pr_3Si^{+}\bullet[CB_{11}H_6X_6]^{-} + Ph_3CH \qquad \textbf{(B)}$$

$$n\text{-}Bu_3SnH + Ph_3C^{+}\bullet TFPB^{-} \xrightarrow{CD_2Cl_2} n\text{-}Bu_3Sn^{+}\bullet TFPB^{-} + Ph_3CH \qquad \textbf{(C)}$$

Scheme 1.2

$Mes_3E–CH_2–CH{=}CH_2$ (E = Si, Ge, Sn) with $[Et_3Si(C_6H_6)]^{+}\bullet B(C_6F_5)_4^{-}$ to form at first intermediate β-silyl-substituted carbenium ions $Mes_3E–CH_2–CH^{+}–CH_2SiEt_3$, which then undergo E–C bond breaking to produce more favorable Mes_3E^{+} cations and allyltriethylsilane $Et_3Si–CH_2–CH{=}CH_2$ as a side product (Scheme 1.3).[8]

$$Mes_3E–CH_2–CH{=}CH_2 + [Et_3Si(C_6H_6)]^{+}\bullet B(C_6F_5)_4^{-} \xrightarrow{C_6H_6}$$

$$\longrightarrow [Mes_3E–CH_2–CH^{+}–CH_2–SiEt_3]\bullet B(C_6F_5)_4^{-} \longrightarrow$$

$$\longrightarrow Mes_3E^{+}\bullet B(C_6F_5)_4^{-} + H_2C{=}CH–CH_2–SiEt_3$$

Scheme 1.3

The ease of oxidation of hexamethyldistannane $Me_3Sn–SnMe_3$ by one-electron oxidizing reagents in acetonitrile, producing the solvent-coordinated trimethylstannyl cation Me_3Sn^{+}, stems from the low oxidation potential of the Sn–Sn bond.[9] Likewise, heteronuclear compounds $Me_3Sn–EMe_3$ (E = Si, Ge, Sn) can be oxidized (two-electron oxidation) forming acetonitrile-solvated cations Me_3Sn^{+} and Me_3E^{+}, whereas disilane $Me_3Si–SiMe_3$, digermane $Me_3Ge–GeMe_3$ and silagermane $Me_3Si–GeMe_3$ were inert under such oxidation conditions because of the markedly higher oxidation potentials of the Si–Si, Ge–Ge and Si–Ge bonds.[9a] Hexaphenyldiplumbane $Ph_3Pb–PbPh_3$ can also be oxidized by Ag^{+} ions in acetonitrile to generate the solvated cation Ph_3Pb^{+}.[10]

Other examples of R_3E^{+} cations generated by cleavage of the E–E bonds of $R_3E–ER_3$ with a strong Lewis acid include: (1) oxidation of t-$Bu_3E–Et$-Bu_3 (E = Si, Ge, Sn) with $Ph_3C^{+}\bullet TFPB^{-}$ in the presence of nitriles R–C≡N (R = Me, t-Bu) to form nitrilium complexes of t-Bu_3E^{+} cations[11] (Scheme 1.4, A); (2) oxidation of n-$Bu_3Sn–Snn$-Bu_3 with the free radical $CB_{11}Me_{12}\bullet$ to produce a solvent-free n-Bu_3Sn^{+} cation weakly coordinated to the Me groups of two $CB_{11}Me_{12}^{-}$ counteranions[12a] (Scheme 1.4, B). Similarly, $Me_3E^{+}\bullet CB_{11}Me_{12}^{-}$ derivatives (E = Ge, Sn, Pb), lacking solvent coordination, were synthesized by the oxidation of $Me_3Ge–GeMe_3$, $Me_3Sn–SnMe_3$ and Me_4Pb in pentane with the free radical $CB_{11}Me_{12}\bullet$.[12b]

$$t\text{-Bu}_3\text{E–E}t\text{-Bu}_3 + 2\,\text{Ph}_3\text{C}^+\bullet\text{TFPB}^- \xrightarrow{\text{R–C}\equiv\text{N}} 2\,[t\text{-Bu}_3\text{E}\leftarrow\text{:N}\equiv\text{C–R}]^+\bullet\text{TFPB}^- \quad \textbf{(A)}$$

[E = Si, Ge, Sn; R = Me, *t*-Bu]

$$n\text{-Bu}_3\text{Sn–Sn}n\text{-Bu}_3 + 2\,\text{CB}_{11}\text{Me}_{12}\bullet \xrightarrow{\text{pentane}} 2\,n\text{-Bu}_3\text{Sn}^+\bullet\text{CB}_{11}\text{Me}_{12}^- \quad \textbf{(B)}$$

Scheme 1.4

1.2.4 From Heavy Carbene Analogs RR′E:

The oxidative addition of Lewis acids to the heavy analogs of carbenes results in an increase of the central element coordination number from 2 to 3 and formation of element-centered cations, strongly stabilized by intramolecular electron donation. Such a synthetically attractive approach is still not widely developed, and one can mention only a couple of representative examples, namely the reaction of decamethylsilicocene $(\eta^5\text{-Me}_5\text{C}_5)_2\text{Si}$: with catechol producing a silyl cation in the form of protonated decamethylsilicocene[13] (Scheme 1.5, A) and the reaction of the stable Lappert's germylene $[(\text{Me}_3\text{Si})_2\text{CH}]_2\text{Ge}$: with $[(4\text{-}t\text{-Bu–C}_6\text{H}_4)]_3\text{C}^+\bullet\text{TPFPB}^-$ unexpectedly yielding an intramolecularly stabilized germyl cation after a series of consecutive rearrangements[14] (Scheme 1.5, B).

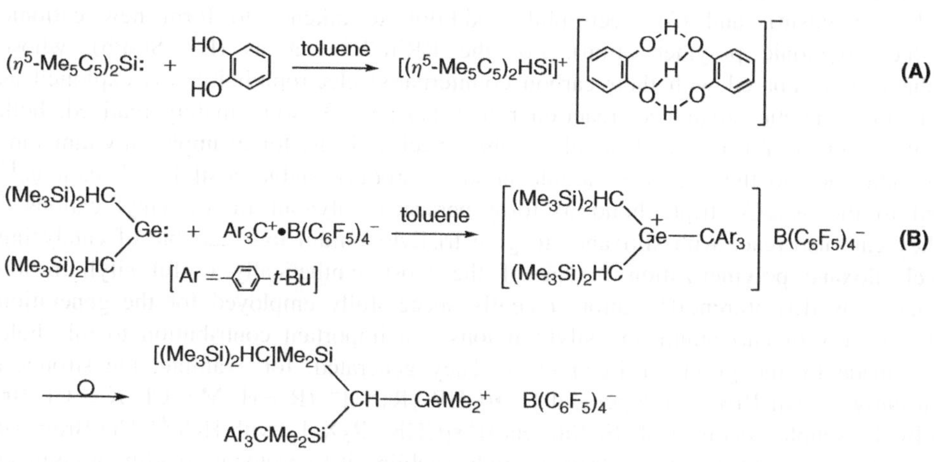

Scheme 1.5

1.2.5 From Free Radicals RR′R″E•

This synthetic route, involving one-electron oxidation of the free radicals RR′R″E• with powerful Lewis acids (such as Ph_3C^+), represents one of the best methods for cleanly forming element-centered cations RR′R″E$^+$ with no formation of any side products, except for the inert Ph_3CH. Although this approach requires isolable radical species as readily available starting materials, the recent discovery of the stable persilyl-substituted radicals of the type $(t\text{-}Bu_2MeSi)_3E\bullet$ (E = Si, Ge, Sn) (see Chapter 2, Section 2.4.1.2) turned this approach into a highly attractive and easily realizable synthetic route for preparation of the stable 'free' $(t\text{-}Bu_2MeSi)_3E^+$ cations (Scheme 1.6).[15]

$$(t\text{-}Bu_2MeSi)_3E\bullet \quad + \quad Ph_3C^+\bullet B(C_6F_5)_4^- \xrightarrow[\text{[E = Ge, Sn]}]{C_6H_6} (t\text{-}Bu_2MeSi)_3E^+\bullet B(C_6F_5)_4^-$$

Scheme 1.6

1.3 Reactions and Synthetic Applications of RR′R″E⁺ Cations[16]

Although reactivity studies and synthetic utilization of the heavy group 14 element analogs of carbenium ions are not sufficiently realized yet, even now it is evident that the major synthetic interest of silylium, germylium, stannylium and plumbylium ion derivatives is parallel to that of the classical carbocations. Thus, among the typical reactions of carbocations in organic chemistry one should mention: (1) reaction with nucleophiles to form substitution products with a novel C–C σ-bond (S_N1 mechanism); (2) removal of a proton to form elimination products with a novel C=C π-bond (E1 mechanism); and (3) electrophilic addition to alkenes to form new cationic adducts (cationic polymerization). For the $RR'R''E^+$ cations (E = Si–Pb), whose enhanced (compared with their carbon counterparts) electrophilicity was exploited as a major synthetic advantage, reaction routes (1) and (3) were mainly realized, both resulting in the formation of novel cationic species. Thus, for example, silylium ions smoothly add to the >C=C< double bond to produce stable β-silyl carbocations,[17] and to the –C≡C– triple bond to form persistent silyl-substituted vinyl cations.[18] They can also react with siloxanes to give trisilyloxonium ions capable of catalysing cyclosiloxane polymerization.[19] One of the most synthetically useful silylium ion reagents is $[Et_3Si(arene)]^+$ cation, recently successfully employed for the generation of a variety of carbenium and silylium ions. An important contribution to this field was made by the group of Reed *et al.* They generated, for example, the strongest currently known Brønsted superacid $H^+\bullet[CHB_{11}R_5X_6]^-$ (R = H, Me, Cl; X = Cl, Br, I) by the simple treatment of $[Et_3Si(arene)]^+\bullet[CHB_{11}R_5X_6]^-$ with HCl.[20] The Brønsted acidicity of this superacid is extremely high, enabling it to protonate readily at ambient temperatures such stable aromatic systems as fullerene C_{60} and Me-substituted benzenes $C_6Me_nH_{6-n}$ (n = 0, 1, 2, 3, 5, 6) generating the fullerene cation $[HC_{60}]^+$[20b] and benzenium ions $[HC_6Me_nH_{6-n}]^+$,[20a–c] respectively. On the other hand, the treatment of $[Et_3Si(arene)]^+\bullet[CHB_{11}Me_5X_6]^-$ (X = Cl, Br) with alkyl triflates ROTf (R = Me, Et) resulted in the formation of alkylium ion derivatives $R^+\bullet[CHB_{11}Me_5X_6]^-$, which are extremely electrophilic alkylating reagents, even stronger than alkyl triflates.[21] Thus, the high electrophilic power of $Me^+\bullet[CHB_{11}Me_5Br_6]^-$ was spectacularly demonstrated by its reactions with benzene C_6H_6 and alkanes R–H (R = C_4H_9, C_5H_{11}, C_6H_{13}), providing access to the corresponding toluenium $[Me(C_6H_6)]^+$ and tertiary carbenium R^+ ions, respectively.[21] Undoubtedly, the extreme reactivity of $R^+\bullet[CHB_{11}Me_5X_6]^-$ exceeds that of the conventional alkyl triflates. Reaction of $[Et_3Si(arene)]^+\bullet[CHB_{11}I_{11}]^-$ with *p*-F-C_6H_4-CF_3 or CH_3CF_3 results in immediate fluorine abstraction to produce intermediate *p*-F-C_6H_4-CF_2^+ or $CH_3CF_2^+$ difluorocations, which subsequently participate in an electrophilic aromatic substitution reaction with the fluorobenzene solvent to form the stable (*p*-F-C_6H_4)$_2$$CF^+$ or (*p*-F-C_6H_4)CH_3CF^+ fluorinated carbocation derivatives.[22] $[Et_3Si(arene)]^+\bullet[CHB_{11}H_5Cl_6]^-$ reagent is able to abstract a chloride ion from the $[IrCl(CO)(PPh_3)_2]$ complex to form a new $[Ir(CHB_{11}H_5Cl_6)(CO)(PPh_3)_2]$ system undergoing an unusually smooth oxidative addition of chlorobenzene to produce the coordinatively unsaturated $[IrCl(C_6H_5)(CO)(PPh_3)_2]^+$ cation.[23] Among other examples of the practical applications of silylium ion derivatives, one can mention silanorbornyl cations, which were shown to be the key intermediates in the metal-free catalytic intramolecular hydrosylilation of C=C double bonds under mild conditions,[24] as well as chiral silyl cation complexes with acetonitrile, claimed to be novel Lewis acid catalysts for Diels–Alder cycloaddition reactions.[25] Readily available cationic complexes $[Me_3Si(arene)]^+\bullet B(C_6F_5)_4^-$ (arene = benzene, toluene) smoothly reacted with persilylated phosphane and arsane $(Me_3Si)_3E$ (E = P, As) to produce the corresponding phosphonium and arsonium salts $[(Me_3Si)_4E]^+\bullet B(C_6F_5)_4^-$.[26]

The reactivity of cations centered on the heavier than silicon group 14 elements is represented mainly by that of stannylium ions. Thus, *n*-$Bu_3Sn^+\bullet[CB_{11}Me_{12}]^-$ readily reacted with PhMgBr to produce *n*-Bu_3SnPh almost quantitatively.[12a] It was found that stannyl cations R_3Sn^+ (R = Me, Bu) can serve as excellent leaving groups in electrophilic aromatic *ipso*-substitution reactions, widening the scope of the Friedel–Crafts acylation, Vilsmeier formylation, sulfinations, and sulfonations.[27] Stannylium ions are also able to promote the cationic polymerization of simple alkenes. For example, the stable *sec*-alkyl β-stannylcarbocation, believed to be formed through the addition of a transient Me_3Sn^+ cation to the C=C double bond, effectively polymerized a number of simple alkenes, such as isobutene, to produce high-molecular weight polymers.[28] The stannylium ion $[n\text{-}Bu_3Sn]^+\bullet TPFPB^-$, generated *in situ* from *n*-Bu_3SnH and $[Ph_3C]^+\bullet TPFPB^-$, may serve as an effective catalyst for allylation of *ortho*-anisaldehyde with *n*-Bu_3Sn–CH_2–CH=CH_2, providing an excellent *ortho–para* regioselectivity.[29] The bis(acetonitrile) complexes of trialkylstannylium ions $[R_3Sn(N{\equiv}CMe)_2]^+\bullet SbF_6^-$ (R = cyclohexyl, *tert*-butyl, neopentyl), prepared from the corresponding bromides R_3SnBr or hydrides R_3SnH, have been shown to be effective Lewis acid catalysts for the Diels–Alder addition of α,β-unsaturated nitriles to furan.[30]

The reactivity of the stable 'free' cations of heavy group 14 elements, such as $(t\text{-}Bu_2MeSi)_3E^+$ (E = Ge, Sn) (see below), is still largely unexplored. One can mention only the pronounced electrophilicity of the germylium derivative $(t\text{-}Bu_2MeSi)_3Ge^+\bullet B(C_6F_5)_4^-$, which readily forms a complex with acetonitrile $[(t\text{-}Bu_2MeSi)_3Ge \leftarrow :N{\equiv}C\text{–}CH_3]^+\bullet B(C_6F_5)_4^-$, can be reduced with $LiAlH_4$ to form the hydride $(t\text{-}Bu_2MeSi)_3GeH$, undergoes one-electron reduction with *t*-BuLi to produce the free radical $(t\text{-}Bu_2MeSi)_3Ge\bullet$ and causes a ring-opening polymerization of THF.[15a]

1.4 Theoretical Studies

The computational accomplishments have been thoroughly discussed in the recent reviews by Apeloig *et al.*,[31] Schleyer *et al*,[11, 32] and Müller,[1p] therefore in this section we will just very briefly overview the most important achievements illuminating the theoretical contribution to the chemistry of heavy analogs of carbenium ions.

1.4.1 Structure of Cations

Two minima structures were located on the PES of EH_3^+ ions (E = group 14 element): a planar D_{3h} form (for E = C–Pb) and C_s side-on complex $HE^+\cdots H_2$ (for E = Si–Pb) (Scheme 1.7).[33] For silylium H_3Si^+ and germylium H_3Ge^+ ions the D_{3h} structure represents a global minimum, D_{3h}/C_s = 0/27.1 and 0/10.0 kcal/mol; whereas for stannylium H_3Sn^+ and plumbylium H_3Pb^+ ions the C_s complex is most favorable, D_{3h}/C_s = 0/−5.2 and 0/−23.3 kcal/mol [calculated at the B3LYP DFT level of theory with the 6-311++G(2d,2p) (for C, Si and Ge) and TZ+2P (for Sn and Pb) basis sets].[33b]

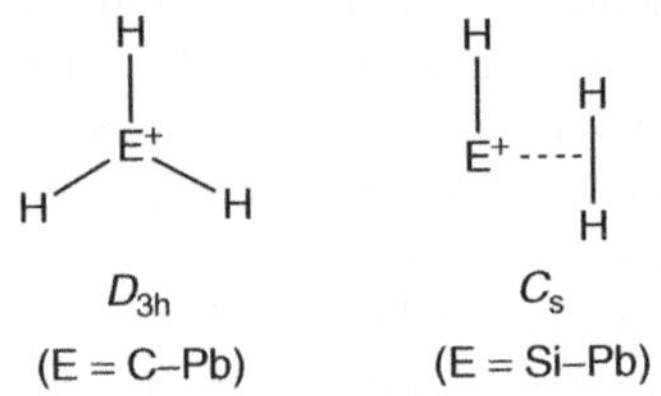

Scheme 1.7

The remarkable preference for the C_s structure over its symmetrical D_{3h} counterpart for H_3Pb^+ can be attributed to relativistic effects, which stabilize the 6s lone pair on the H–Pb^+ fragment.[33b] Such side-on C_s complexes $HE^+\cdots H_2$ are best viewed as donor–acceptor aggregates, in which the HOMO (H–H σ-bond) donates its electron density to the LUMO (empty p-orbital on E of the HE^+ fragment).[33b,c]

1.4.2 Stability of Cations

The stability of the parent cations H_3E^+ steadily increases going down from C to Pb (stabilization energies in kcal/mol calculated at the MP2/VDZ+P level are given in parentheses): $H_3C^+(0.0) < H_3Si^+(58.9) < H_3Ge^+(70.7) < H_3Sn^+(87.5) < H_3Pb^+(97.9)$.[31,34] This trend of increasing thermodynamic stability descending group 14 is evidently due to the changes in intrinsic properties of group 14 elements: decrease of electronegativity and increase of polarizability.

Although the same type of substituents stabilize both carbenium and silylium ions, the extent of such stabilization for the latter class of cations is markedly lower, which leads to an appreciable decrease of the inherent stability of H_3Si^+ vs H_3C^+ in their substituted derivatives.[31a] Thus, while the benefits of the stabilization of heavy group 14 element cations with electropositive silyl substituents are still rather important [$(H_3Si)_3Si^+$ and $(H_3Si)_3Pb^+$ are more stable than $(H_3Si)_3C^+$ by 32.9 and 63.5 kcal/mol, respectively], the stabilization effect of alkyl substituents is markedly smaller (Me_3Si^+ and Me_3Pb^+ are more stable than Me_3C^+ by 12.0 and 35.2 kcal/mol, respectively).[1p] The extent of substituents stabilization further drops in the case of the phenyl group: Ph_3Pb^+ is more stable than Ph_3C^+ by only 10.5 kcal/mol, whereas Ph_3Si^+ is destabilized compared with Ph_3C^+ by 2.7 kcal/mol.[1p]

Moreover, whereas the effect of stabilization of carbenium ions by alkyl substituents is highly pronounced, it is markedly smaller for the heavy analogs. Thus, if Me_3C^+ is more stable than H_3C^+ by 74.8 kcal/mol, the analogous stabilization of Me_3Si^+ and Me_3Pb^+ ions (vs their unsubstituted analogs H_3Si^+ and H_3Pb^+) amounts to only 40.6 and 29.7 kcal/mol, respectively. Even phenyl groups, traditionally commonly used for stabilization of carbenium ions in organic chemistry (Ph_3C^+ is more stable than H_3C^+ by 111.3 kcal/mol), are much less effective in stabilization of the heavier cations (stabilization energies of Ph_3Si^+ and Ph_3Pb^+ ions (vs H_3Si^+ and H_3Pb^+) are only 64.5 and 42.2 kcal/mol, respectively).[1p] On the other hand, the silylium ions were predicted to be stabilized by electropositive substituents such as Li and BeH.[31a,35] Thus, the D_3 tris(dimethylboryl)silylium ion $(Me_2B)_3Si^+$ was stabilized by 61.9 kcal/mol compared with the parent H_3Si^+ (B3LYP/6-31G* level), whereas the Me_3Si^+ ion was more stable than H_3Si^+ by only 43.2 kcal/mol.[36]

In contrast to carbenium ions, which are stabilized by any halogen substituents (this effect increases from F to I), such stabilization is much less pronounced in the case of the cations of the heavy group 14 elements. Thus, although Br and I stabilize the silylium ion R_3Si^+, the more electronegative F and Cl destabilize it. Only the most electropositive I stabilizes germylium R_3Ge^+ and stannylium R_3Sn^+ ions, whereas all other halogens destabilize them. For the most electropositive Pb atom, all halogens destabilize its cation R_3Pb^+.[31b,34]

Amino groups are also capable of stabilizing the silylium ions, although the degree of such stabilization is smaller than that of carbon analogs. Thus, the D_3 tris(amino)silylium ion $(H_2N)_3Si^+$ can benefit from ca. 40% of the stabilization energy of the corresponding carbenium ion $(H_2N)_3C^+$.[37] It was therefore concluded that amino groups are significantly more effective than methyl groups in the stabilization of silylium ions.

Overall, it can be concluded that the substituent effects for the heavy analogs of carbenium ions do not play such a decisive role in their thermodynamic stabilization as they play in the chemistry of organic carbocations.

1.4.3 Calculation of the NMR Chemical Shift of Cations

The central element E of the cationic tricoordinate derivatives of group 14 elements R_3E^+ is diagnostically strongly deshielded with respect to neutral tetracoordinate counterparts R_4E. It is therefore evident that NMR chemical shift calculations (for E = C, Si, Sn, Pb) represent a very powerful tool for straightforward identification of cationic species in the condensed phase and estimation of their degree of ionicity. Below, the major conclusions drawn from the ^{29}Si and ^{119}Sn NMR chemical shift calculations of R_3Si^+ and R_3Sn^+ cations will be discussed. Neither good empirical estimates nor reliable ^{207}Pb NMR chemical shift calculations are available for plumbylium ion derivatives. Because ^{13}C NMR chemical shift calculations of the heavy group 14 element centered cations are only of very limited value, they will not be discussed in the present chapter. Sometimes the ^{13}C NMR computational data are useful in identification of the germylium ions R_3Ge^+, because the direct NMR spectroscopic observation of germanium centers is precluded by the lack of a convenient and sensitive Ge nuclide.

1.4.3.1 ^{29}Si NMR Chemical Shift Calculations

Reliable NMR chemical shift calculations for organosilicon compounds became available at the beginning of the 1990s. Since then, such computations have been widely used as a major tool for proof (or disproof) of claims on the synthesis of genuine silylium ions.

Similar to their carbon analogs, silylium ion derivatives exhibit characteristic highly deshielded ^{29}Si NMR chemical shifts, a tendency that was nicely supported by theoretical calculations. Thus, the deshielding of R_3Si^+ ions (R = alkyl group) compared with their R_3SiH precursors amounts to ca. 400 ppm.[1p] The chemical shifts of H_3Si^+ and Me_3Si^+ ions in the gas phase were calculated to be 264.7 and 346.7 ppm, respectively.[32] One should note that in solution the extent of NMR deshielding of the silylium ion species strongly correlates with the degree of solvent nucleophilicity, sharply dropping with an increase in the solvent coordinating ability. This tendency was computationally studied in the elaborate work by Cremer *et al.*[38] (see Section 1.5). The predicted region for the tricoordinate silylium ions is very wide, ranging from the rather high-field resonance of $(Me_2N)_3Si^+$ (42 ppm)[37] to the extremely low-field signals of $(Me_2B)_3Si^+$ (572 ppm)[36] and particularly $(Me_3Si)_3Si^+$ (920 ppm).[39] Clearly, the magnitude of the ^{29}Si NMR chemical shifts of the above-mentioned silylium ion derivatives is totally governed by the influence of substituents: strongly π-donating Me_2N groups vs electropositive Me_3Si substituents. This phenomenon is now well-recognized and was realized on the basis of the following considerations.[1p,39] The paramagnetic contribution, which is dominant in the overall NMR chemical shifts of heteronuclei, is directly related to the energy gap between occupied and vacant frontier orbitals. When this gap tends to decrease, the paramagnetic contribution becomes larger and consequently, the nucleus is more deshielded. In tricoordinate cations R_3E^+ such occupied and vacant orbitals are typically represented by the σ(E–R)- and np(E)-orbitals, respectively. When R is electropositive silyl group, the σ(E–R)-orbitals level is raised resulting in a decrease of σ(E–R)–np(E) energy separation and consequently in a strongly deshielding contribution for E. By contrast, electronegative substituents lead to an increase in the energy gap and decrease in the deshielding contribution. The same is true for the π-donating groups R (such as amino groups), which destabilize the vacant np(E)-orbitals through their interaction, resulting in an increase of the energy separation.

The cationic Si centers of the H_3Si^+ and Me_3Si^+ ions were markedly shielded upon the approach of such typically inert molecules as CH_4, He, Ne and Ar. On the basis of this computational result, Schleyer *et al.* concluded that the silylium ions can be coordinated by even such non-nucleophilic media as aliphatic hydrocarbons and noble gases.[11,32] This led them to a rather pessimistic statement: 'Thus, it seems unlikely that free silyl cations can exist in solution, not even in the most non-nucleophilic solvents, unless, perhaps, very bulky substituents hinder coordination. . . . One major conclusion can be drawn: the prospects for obtaining and observing truly "free" silyl cations in condensed phases are very poor.'[32] However, this discouraging conclusion proved to be somewhat exaggerated, at least from the viewpoint of experimental organometallic chemists. Actually, the highly desirable synthetic challenge of the preparation and isolation of tricoordinate silylium, germylium and stannylium ions, truly 'free' in both solid state and in solution, was realized by the groups of Lambert and Sekiguchi in the early 2000s (see Section 1.6.2.2).

1.4.3.2 ^{119}Sn NMR Chemical Shift Calculations

Accurate calculations of the ^{119}Sn chemical shifts, which cover a very broad range from ca. −2500 to +4000 ppm using Me_4Sn as a reference, turned out to be an important computational tool only recently.[1p,40] Before that, estimation of the ^{119}Sn resonances of stannylium ions was made based on the empirical correlation between the ^{29}Si and ^{119}Sn NMR chemical shifts, which was successfully applied for the evaluation of the chemical shifts of isostructural tetracoordinate organosilicon and organotin compounds.[41] Accordingly, the ^{119}Sn chemical shifts of stannylium ion derivatives R_3Sn^+ were predicted to be ca. 1770 ppm (for R = alkyl) and ca. 1250 ppm (for R = aryl).[1p] However, such expectations, based on the empirical ^{29}Si–^{119}Sn chemical shift correlation, overestimated the degree of deshielding of the cationic Sn centers in stannylium ions, as was demonstrated by IGLO calculations giving the Me_3Sn^+ chemical shift estimation as ca. 1075 ppm.[42] Subsequent computations revealed that the ^{119}Sn chemical shifts of the 'free' stannylium ions spread over a wide region, ranging from 596 ppm for H_3Sn^+ [GIAO/HF level with the 6-31G(d) and tzv basis sets] to 3450 ppm for $(Me_3Si)_3Sn^+$ [GIAO/MPW1PW91 level with the 6-31G(d) and tzv basis sets].[1p] The chemical shifts of the Me_3Sn^+ ion were calculated to be in the range of 1075–1466 ppm depending on the theoretical method used, whereas those of the Mes_3Sn^+ and Tip_3Sn^+ ions were estimated as 856 and 763 ppm, respectively.[1p] The extreme deshielding of the persilyl-substituted stannylium ions [3450 ppm for $(Me_3Si)_3Sn^+$ and 2880 ppm for $(H_3Si)_3Sn^+$ vs 1466 ppm for H_3Sn^+ at the same computational level][1p] is explained by the same reasons as those responsible for the deshielding of structurally related tris(silyl)silylium ions (see above); namely, by the very large paramagnetic contribution to the overall NMR chemical shift because of the small energy gap between the occupied σ(Sn–Si)- and vacant 5p(Sn)-orbitals. This agrees well with a recent experimental finding: the resonance of the $(t\text{-}Bu_2MeSi)_3Sn^+$ ion was observed at a record low-field shift of 2653 ppm[15b] (see Section 1.6.2.2).

1.5 Early Studies of RR′R″E$^+$ Cations: Free or Coordinated?

The early belief in the ease of preparation of silylium ions RR′R″Si$^+$ (and other cations of heavier group 14 elements) was based on the higher polarizability and lower electronegativity of silicon (as well as germanium, tin and lead) compared with that of carbon (1.90 for Si vs 2.55 for C, Pauling electronegativity scale).[43] It was, for example, expected that the heterolysis of the R_3Si–X bond would be facilitated by the thermodynamic stabilization of silylium ions R_3Si^+ compared with their carbon analogs R_3C^+. This was indeed the case in the gas phase, where a number of tricoordinate silylium ions have been detected and their reactivity studied by both classical mass spectrometry and special methods, such as ion cyclotron resonance spectroscopy and tandem mass spectrometry techniques.[16a–e] Thus, the recent investigation of the relative hydride affinities for silylium and carbenium ions and equilibrium constants of hydride transfer reactions by FT ion cyclotron resonance spectroscopy clearly demonstrated that the silylium ions in the gas phase are significantly thermodynamically stabilized compared with the corresponding carbenium ions, and the positive charge of the silylium ions is mostly localized on the Si atom.[44] The existence of silylium ions in the gas phase was reliably supported by theoretical calculations, which also confirmed that the planar D_{3h} silylium ion H_3Si^+ is substantially more stable that its carbon analog, methylium ion H_3C^+, at all computational levels.[11,31] However, the generation of silylium ions in condensed media, mostly desired by synthetic organometallic chemists, was a long-standing problem whose solution has required several decades of very intensive research. Given the above-discussed intrinsic thermodynamic stabilization of silylium ions, one should definitely acknowledge the kinetic origin of their overall instability. The extreme electrophilicity of silylium ions, greatly exceeding that of their carbon counterparts, results in the interaction of the former species with a variety of π- and σ-donors, including even such weakly nucleophlic and typically inert solvents as toluene and benzene. This prevented the use of traditional leaving groups (such as tosylates and halides), that have been widely and very efficiently used for the generation of carbenium ions in organic chemistry, due to the extraordinarily high oxo- and halophilicity of the silylium ions. Thus, whereas the *tert*-butylium ion derivative $Me_3C^+{\bullet}Sb_2F_{11}^-$ can be smoothly generated and isolated under superacidic conditions,[45] the corresponding silylium ion derivative did not exist as an ion pair, forming instead a neutral compound with a covalent bond between silicon and oxygen or fluorine atoms.[46] The other problem, greatly contributing to the overall instability of silylium ions, is the significant difference in the size of the silicon and carbon atoms: atomic radii are 117 and 77 pm, respectively.[43] For this reason, the bonds from substituents to silicon are longer than those to carbon, which results in an appreciable decrease in the degree of hyperconjugative stabilization of the cationic center on going from carbon to silicon. On the other hand, the bigger size of silicon is associated with its increased coordination sphere, which is manifested in the general tendency of the silicon compounds (unlike their carbon counterparts) to form hypercoordinate derivatives with the coordination numbers 5 or 6 because of the intra- or intermolecular stabilizing coordination of Lewis bases, which results in a partial or complete loss of the silylium ion character. It is, therefore, not surprising that the story of generation, identification and, at last, isolation of truly ionic silylium ions was neither straightforward nor simple, being full of controversial reports and hot debates concerning the real nature of the 'silylium ion' species, the synthesis of which has been declared from time to time.[1,31,32] It is therefore particularly instructive to follow the progress in the search for tricoordinate silylium ion derivatives.

As the first step towards the synthesis of cations of heavy group 14 elements, several groups in the 1970s tried to prove the existence of silylium ion derivatives by physicochemical methods previously successfully used for the study of carbenium ions (cryoscopic, conductivity, UV and NMR measurements), however, all of these attempts failed to observe silicon centered cationic species.[1a] A number of attempts were made to detect the presence of silylium ions as reactive intermediates in solvolysis reactions (hydrolysis of Ph_3SiF), halogen abstraction from a carbon next to a silicon in $R_3Si–CH_2–X$ by Lewis acids ($AlCl_3$, SbF_5, BF_3), reaction of β-functional silicon compounds $R_3Si–CH_2–CH_2–X$, hydride transfer reactions from the hydrosilane Ph_3SiH to the carbenium ion derivatives $Ph_3C^+{\bullet}X^-$, and reactions accompanied by racemization at the silicon center.[1a] However, in no cases has clear evidence for the formation of silylium ion intermediates been obtained.

In the following decade an important contribution to the problem of silylium ions was made by the group of Lambert, whose work, however, has led sometimes to controversial conclusions.[1d,47] Thus, they presented experimental data on the attempted ionization of simple silyl perchlorates [such as $(i\text{-PrS})_3SiOClO_3$, $Ph_3SiOClO_3$, $Me_3SiOClO_3$] in CH_2Cl_2 and sulfolane, the results of which were interpreted in terms of the formation of stable R_3Si^+ (R = *i*-PrS, Ph, Me) cations in the form of their perchlorate salts as a silicon analog of the trityl cation, Ph_3C^+.[1c,48] However, the subsequent detailed investigation by Olah *et al.* disproved such claims based on a careful investigation of the NMR spectral and X-ray crystal data along with theoretical calculations, clearly demonstrating the covalent, rather than ionic, nature of the bonding between the R_3Si and $OClO_3$ parts and, consequently, absence of the free silylium ion species in solution.[49]

Thus, at the beginning of the 1990s the synthesis of real silylium ions, featuring a positive charge on the Si atom, had not been achieved. It became clear that their successful synthesis required nonclassical approaches greatly distinctive from traditional organic chemistry methods. The numerous unsuccessful attempts described above led to a definite conclusion that the three important factors most responsible for either success or failure in the synthesis of silylium ion derivatives are: the counteranion, solvent and substituents. The first problem to be solved was the right choice of the counteranion and solvent, which was finally overcome at the beginning of the 1990s. The major requirement for counteranions was their minimal nucleophilicity to prevent their close contact with the target silylium ions to form tight ion pairs or, in the extreme case, formation of covalently bonded compounds (such as triphenylsilyl perchlorate).[48b,49b] The requirements for the solvents were the same: as low as possible nucleophilicity to avoid possible coordination to the highly electrophilic silylium ion. In the case of such coordination of either counteranion or solvent, one should expect an appreciable transfer of the positive charge onto the nucleophilic counterpart (counteranion, solvent) and, consequently, significant electronic perturbation around the cationic center. Overall, this will result in a great (or complete) loss of the silylium ion character. The major breakthrough in resolving the silylium ion problem was achieved following the successful introduction of borate and carborane counteranions of particularly low nucleophilicity $B(C_6F_5)_4^-$ and $CB_{11}H_6Br_6^-$, and utilization of nonpolar aromatic hydrocarbons (benzene, toluene) as the solvents of choice. The critical choice of the substituents was determined by two major demands: (1) steric bulkiness necessary for kinetic stabilization of the cationic center to avoid coordination of both anions and solvents; (2) electron donating properties essential for the thermodynamic stabilization of the positive charge.

The first milestone discoveries were accomplished in 1993, when the groups of Lambert[50] and Reed[51] published the crystal structures of their Et_3Si^+ and $i\text{-}Pr_3Si^+$ derivatives. Thus, $[Et_3Si(toluene)]^+$•$TPFPB^-$ (**1$^+$•TPFPB$^-$**) was prepared by Lambert *et al.* by the hydride transfer reaction between Et_3SiH and Ph_3C^+•$TPFPB^-$ in benzene (Scheme 1.8).[50]

The crystal structure analysis of **1$^+$•TPFPB$^-$** revealed no direct cation–anion interaction, however, there was a 'distant' coordination of the Si cationic center to the solvent (toluene) with a long Si–C interatomic distance of 2.18 Å. The geometry of the toluene molecule was almost undistorted and essentially planar, which was realized as an

$$Et_3SiH + Ph_3C^{+}\bullet TPFPB^{-} \xrightarrow{C_6H_6} [Et_3Si(C_6H_6)]^{+}\bullet TPFPB^{-} + Ph_3CH$$

$\mathbf{1^{+}\bullet TPFPB^{-}}$

Scheme 1.8

indication of its very weak bonding interaction with the Si cation, resulting in extraordinary little (if at all) charge transfer from the Si to the C atom. Thus, the authors concluded that **1**$^+$ represents a stable silylium ion lacking coordination to the counteranions and only very weakly coordinated to the toluene solvent. However, two experimental observations were in sharp conflict with such a conclusion: (1) the Si cationic center was pronouncedly pyramidal (the sum of the bond angles around the Si atom was 342°), whereas trigonal-planar geometry (360°) was expected for the real silylium ion; (2) the resonance of the cationic Si atom of **1**$^+$ was observed at 92.3 ppm, a value that was by far high-field shifted compared with the several hundred ppm calculated for the planar noncoordinated silylium ion. These problematic issues provoked very hot debates around the real nature of **1**$^+$, in the course of which Lambert's original claim of the nearly 'free' silylium ion was severely criticized by both experimentalists and theoreticians (Scheme 1.9).

$\mathbf{1^{+}\bullet TPFPB^{-}}$

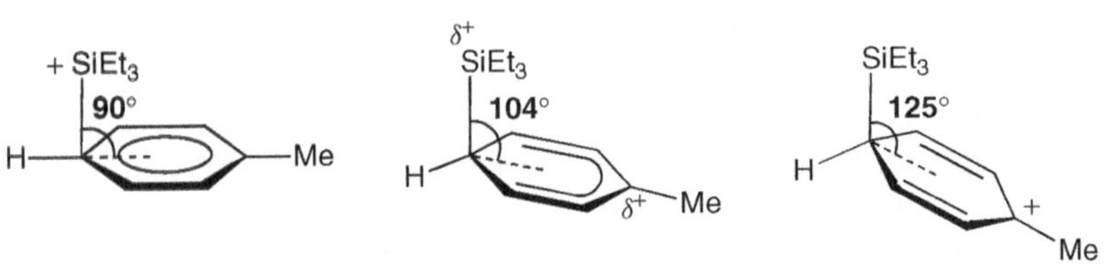

η^1 π-arene complexed *silylium* ion — Experimental structure — σ-complexed *arenium* ion (Wheland intermediate)

Scheme 1.9

Thus, Pauling pointed out that the calculated bond order between the Si and *para*-C of a coordinated toluene molecule in $\mathbf{1^{+}\bullet TPFPB^{-}}$ is 0.35, a value that cannot be neglected.[52] Olah *et al.* calculated that the ^{29}Si NMR resonance of the planar free Et_3Si^+ cation should be expected at a very low field, 354.6 ppm[53a] or even at 371.3 ppm,[53b] whereas the experimentally observed value of 92.3 ppm[50] in $\mathbf{1^{+}\bullet TPFPB^{-}}$ was rather attributed to the covalently bonded compound that can be best described as a Wheland σ-complex (Scheme 1.9).[53] In independent experimental studies, the formation of such a σ-complex in the gas phase was confirmed by radiolytic experiments and FT ion cyclotron resonance mass spectrometry.[54] The comprehensive theoretical insight by Cremer *et al.*[38] agreed well with Olah's conclusions[53] regarding the degree of deshielding of the cationic Si atom: ^{29}Si NMR resonances of R_3Si^+ (R = Me, Et) were calculated to be ca. 400 ppm (in the gas phase, free silylium ions), 370–400 ppm (in noncoordinating solvents), or 200–370 ppm (in weakly coordinating solvents). Such a shift to higher field clearly reflects the progressive decrease in cationic character in line with the increasing coordinating power of the solvents. In the limiting case of nucleophilic solvents, the silylium ion character is totally lost because of the covalent bonding between the cationic center and solvent molecules. Schleyer *et al.* also concluded that the pyramidalization at the silyl cation center and relatively high-field ^{29}Si NMR resonance of $\mathbf{1^+ \bullet TPFPB^-}$ testify to its σ-complex structure.[55] To reach a consensus, Reed *et al.* suggested[56] (and Lambert finally concurred)[1n,5b] that the real structure of $[Et_3Si(toluene)]^+$ represents a hybrid of an η^1 π-complex and σ-complex (Scheme 1.9) with a predominant contribution from the former.

The next milestone contribution to the silylium ions issue was achieved by Reed *et al.*, who synthesized $i\text{-}Pr_3Si^+ \bullet [CB_{11}H_6Br_6]^-$, ($\mathbf{2^+ \bullet [CB_{11}H_6Br_6]^-}$), by the hydride transfer reaction of $i\text{-}Pr_3SiH$ and $Ph_3C^+ \bullet [CB_{11}H_6Br_6]^-$ in toluene,[51a] taking advantage of the very low nucleophilicity of the carborane anion[57] (Scheme 1.10).

$$i\text{-}Pr_3SiH + [Ph_3C]^+ \bullet [CB_{11}H_6Br_6]^- \xrightarrow{\text{toluene}} \underset{\mathbf{2^+ \bullet [CB_{11}H_6Br_6]^-}}{[i\text{-}Pr_3Si]^+ \bullet [CB_{11}H_6Br_6]^-} + Ph_3CH$$

Scheme 1.10

Although exhibiting no interaction with toluene solvent, the $i\text{-}Pr_3Si^+$ cation $\mathbf{2^+}$ was not totally 'free', being weakly bound to a carborane ion through its Br atoms with a long Si–Br distance of 2.479(9) Å. However, $i\text{-}Pr_3Si^+$ $\mathbf{2^+}$[51a] was more planar than Et_3Si^+ $\mathbf{1^+}$[50] (351° vs 342°) and more deshielded (109.8 ppm vs 92.3 ppm), observations that led to the conclusion that $\mathbf{2^+ \bullet [CB_{11}H_6Br_6]^-}$ more closely approached the 'free' silyl cation, possessing the highest degree of silylium ion character yet observed. This claim, however, was doubted by Olah *et al.*,[53] who assigned the structure of $\mathbf{2^+ \bullet [CB_{11}H_6Br_6]^-}$ to a polarized silylbromonium zwitterion rather than the true silylium ion. Again, similar to the above case of $[Et_3Si(toluene)]^+ \bullet TPFPB^-$, Reed has concluded that $\mathbf{2^+ \bullet [CB_{11}H_6Br_6]^-}$ can be best viewed as '. . . lying on a continuum between a bromonium ion and a silylium ion' with the major contribution from the latter (Scheme 1.11).[56]

In due course, Reed *et al.* prepared several other trialkylsilylium ion derivatives $R_3Si^+ \bullet [CB_{11}H_6Br_6]^-$ ($R_3Si^+ = Et_3Si^+, t\text{-}Bu_2MeSi^+$, and $t\text{-}Bu_3Si^+$) employing the same synthetic protocol; namely, hydride abstraction from R_3SiH by

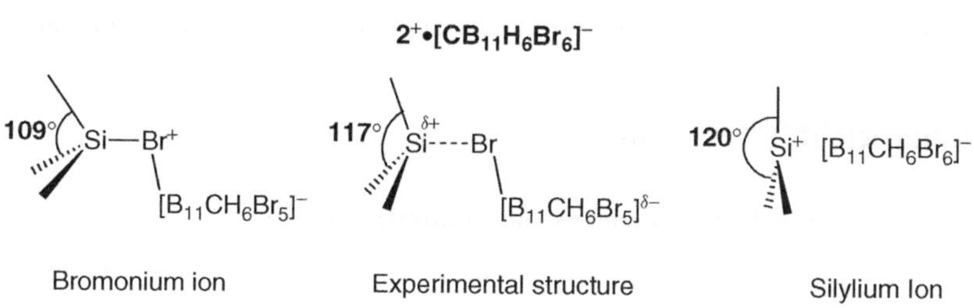

Scheme 1.11

Ph_3C^+•$[CB_{11}H_6Br_6]^-$.[58] All of these compounds featured structural and spectral characteristics very similar to those of **2^+•$[CB_{11}H_6Br_6]^-$**: a long Si–Br interatomic distance (2.43–2.48 Å), tendency to planarization around the Si-cationic center (345–351°), and low-field ^{29}Si NMR resonances (105–115 ppm). The high degree of silylium ion character of these compounds was proved by their smooth reactions with organic halides to form silyl halides R_3SiX and with water to give protonated silanols $R_3Si(OH_2)^+$.[58] In subsequent studies, Reed *et al.* expanded the range of their least-coordinating weakly nucleophilic anions by adding newly synthesized hexachloro- and hexaiodocarboranes $[CB_{11}H_6X_6]^-$ (X = Cl, I) to the original hexabromocarborane (X = Br).[6] Consequently, the novel salts of the i-Pr_3Si^+ cation, i-Pr_3Si^+•$[CB_{11}H_6X_6]^-$ (X = Cl, I), were successfully prepared, of which the hexachloro derivative developed the highest degree of silylium ion character, whereas the hexaiodo derivative manifested the strongest halonium ion character and the most covalent Si–X bond.[6] Thus, the i-Pr_3Si^+•$[CB_{11}H_6Cl_6]^-$ derivative represented the closest approach to a trialkylsilylium ion known at that time.[1k]

The search for stable heavier congeners of silylium ions, namely, germylium and stannylium ions, met with little success in the 1990s. Lambert *et al.* reported the synthesis of the protonated digermyl and distannyl ethers $(R_3E)_2OH^+$•$TPFPB^-$ (R = Me, Et; E = Ge, Sn), which were claimed to possess considerable amounts of germylium and stannylium ions character.[59]

$$R_3E\text{–}(H)O^+\text{–}ER_3 \leftrightarrow R_3E^+\bullet HO\text{–}ER_3$$

The same authors also prepared the cationic derivative n-Bu_3Sn^+•$B(C_6F_5)_3H^-$ by the oxidation of the stannyl hydride n-Bu_3SnH with $B(C_6F_5)_3$ and characterized it as a tricoordinate stannylium ion because of its low-field ^{119}Sn NMR resonance of 360 ppm.[60] The identical n-Bu_3Sn^+ cation with a $TPFPB^-$ counterion revealed a less deshielded value of +263 ppm (room temperature),[17a] which was, however, corrected in subsequent studies to +434 ppm (−60 °C).[29,61] Kira *et al.* also prepared the n-Bu_3Sn^+ derivative as its $TFPB^-$ salt, using the traditional hydride transfer method (Section 1.2.2, Scheme 1.2, C) and reported the low-field resonance of the cationic Sn atom to be 356 ppm.[7] This conclusion was, however, later questioned by Edlund *et al.*,[41] who pointed out that the values of 360[60] and 356[7] ppm assigned to the cationic Sn atom are better attributed to the covalently bound arene complexes, quite similar to the case discussed above of the silylium ions vs Wheland σ-complexes problem. On the basis of an empirical correlation between the ^{29}Si and ^{119}Sn NMR chemical shifts, the resonance of the truly free trigonal-planar Me_3Sn^+ cation was expected to be observed at a much lower field of 1500–2000 ppm, a conclusion that also gained support from theoretical calculations.[41]

1.6 Stable RR′R″E$^+$ Cations

1.6.1 Intramolecularly Stabilized (Coordinated) Cations

Stabilization of the highly electrophilic cations of the heavy group 14 elements can be achieved either intermolecularly (by coordinating solvents, see previous section) or intramolecularly (by n- or π-donating atoms or groups, such as N, O, C=C, C≡C, etc.).

However, such stabilization occurs at the expense of the degree of cationic character at the central atom, thus leading to an increase in its coordination number and formation of cationic species which are not tricoordinate but tetra-, penta- or hexacoordinate instead. Because such species cannot be classified as real low-coordinate derivatives, we will just briefly overview the most important achievements in this field describing the crystal structures of these compounds. Some of them, in which the positive charge is incorporated into the bis- or trishomoaromatic 2π-electron cyclic system, will be discussed in the last chapter devoted to the aromaticity of group 14 organometallics (see Chapter 6, Section 6.3.3).

The first compound of this type, a 1:1 adduct of Me_3SiBr/Me_3SiI and pyridine (Py) $[Me_3Si \leftarrow :Py)]^+ \bullet X^-$ ($X = Br$, I), $\mathbf{3^+ \bullet X^-}$, was described in 1983.[62] In the solid state the $[Me_3Si(py)]^+$ cation of $\mathbf{3^+ \bullet X^-}$ revealed no interaction with the X^- counteranions, featuring a distorted trigonal-pyramidal geometry of the tetracoordinate Si atom and a long Si–N bond distance of 1.858(9) Å.

Reed *et al.* reported a remarkable protonated silanol $[t\text{-}Bu_3Si \leftarrow (:OH_2)]^+ \bullet [CB_{11}H_6Br_6]^-$, $\mathbf{4^+ \bullet [CB_{11}H_6Br_6]^-}$, that according to NMR (strongly low-field shifted resonance of the coordinated water protons at 8.04 ppm) and X-ray [trigonal flattening around the tetracoordinate Si center, long Si–O bond of 1.779(9) Å] data, contains a predominantly silylium ion $t\text{-}Bu_3Si^+ \leftarrow (:OH_2)$ rather than oxonium ion $t\text{-}Bu_3Si\text{–}(OH_2)^+$.[63] In the crystalline form the $[t\text{-}Bu_3Si \leftarrow (:OH_2)]^+$ cation of $\mathbf{4^+ \bullet [CB_{11}H_6Br_6]^-}$ was weakly coordinated to the carborane anion through the hydrogen bonding of its acidic O–H bonds to the carborane Br atoms.

Several silylnitriluim and silyloxonium ion derivatives, $[R_3Si\text{–}^+N{\equiv}C\text{–}CH_3] \bullet TFPB^-$ [$R_3Si = Me_3Si$, Ph_2MeSi, $(2\text{-thienyl})_2MeSi$, $(3,5\text{-}t\text{-}Bu_2\text{-}C_6H_3)_3Si$] and $[R_3Si\text{–}^+OEt_2] \bullet TFPB^-$ [$R_3Si = Me_3Si$, Ph_2MeSi, $(2\text{-thienyl})_2MeSi$], were prepared by standard hydride transfer reactions of the corresponding hydrides R_3SiH with $Ph_3C^+ \bullet TFPB^-$ and were detected by low temperature NMR spectroscopy.[64]

Stabilization of silyl cations by the intramolecular coordination of two n-donating ligands (N, O, S) results in the trigonal-bipyramidal geometry of the pentacoordinate Si atom. Thus, bis{2-[(dimethylamino)methyl]phenyl}silyl triflate $\mathbf{5^+ \bullet [OTf]^-}$, featuring no interaction between the cationic and anionic parts, revealed a perfectly planar Si center having both dimethylamino groups at the apical positions with considerably longer Si–N bond distances (Scheme 1.12).[65]

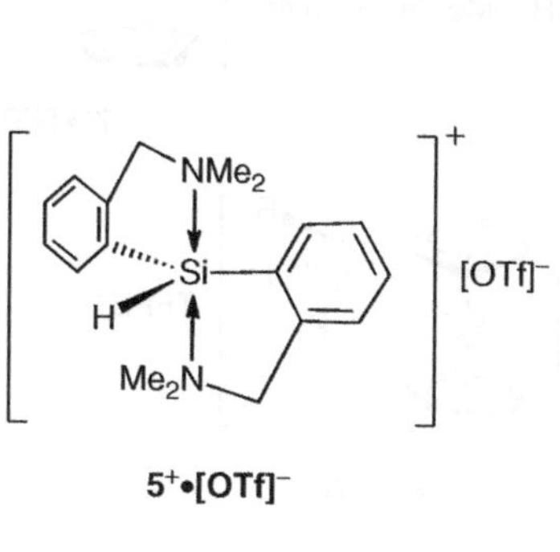

5⁺•[OTf]⁻

Scheme 1.12

Similarly, silyl triflates $\mathbf{6^+\bullet[OTf]^-}$ were prepared by reaction of the corresponding chlorosilane precursors with trimethylsilyl triflate (Scheme 1.13).[66]

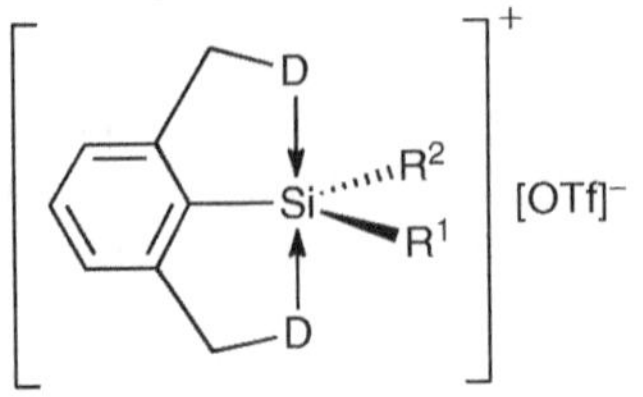

6⁺•[OTf]⁻
[D = OMe, SMe; R¹ = Me, Ph; R² = H, Me]

Scheme 1.13

The cationic portions of $\mathbf{6^+\bullet[OTf]^-}$ were well separated from their anionic parts in the solid state, featuring a slightly distorted trigonal-bipyramidal configuration at the Si center with the n-donating ligands at the apical positions.

An interesting method for the generation of transient THF solvated silylium ion species by the treatment of [1]silaferrocenophanes with $[H(OEt_2)]^+\bullet TFPB^-$ was recently described by Manners *et al.* Being fleeting under ambient conditions, such ferrocenyl-substituted silylium ions $\mathbf{7^+\bullet TFPB^-}$ can be stabilized by coordination with Lewis bases, such as pyridine, to form room temperature stable pyridinium ion derivatives $\mathbf{8^+\bullet TFPB^-}$ (Scheme 1.14).[67] The Si–N bond order in the cation $\mathbf{8^+}$ of 0.61, estimated from its bond length of 1.858(5) Å, is characteristic of a predominantly covalent interaction, which agrees well with the marked pyramidalization about the Si atom (337.7°).

A series of organosilicon compounds featuring pentacoordinate cationic Si centers, stabilized by the intramolecular coordination of the two O-ligands at the apical positions of the trigonal bipyramid, was reported by Baukov *et al.*[68]

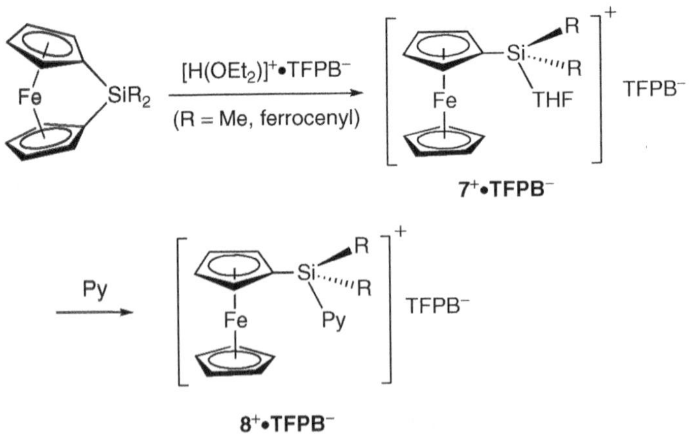

Scheme 1.14

Nitrilium complexes $[t\text{-}Bu_3E^+ \leftarrow :N{\equiv}C\text{–}R]^+ \bullet TFPB^-$ (E = Si, Ge, Sn; R = Me, *t*-Bu) were synthesized by the oxidative cleavage of E–E bonds with the Ph_3C^+ ion in the presence of either acetonitrile or pivalonitrile, as depicted in Scheme 1.4, A (Section 1.2.3). All of these complexes uniformly displayed a tetracoordinate central atom E featuring a greatly distorted tetrahedral geometry as a result of the interaction with the nucleophilic nitrile molecule occupying the fourth coordination site around the E center.[11]

The delocalized halogen-bridged cation $\mathbf{9^+ \bullet TPFPB^-}$ was prepared by the unusual reaction of bromosilirene with $[Et_3Si(C_6H_6)]^+ \bullet TPFPB^-$ (Scheme 1.15).[69] The ^{29}Si NMR resonance of $\mathbf{9^+ \bullet TPFPB^-}$ (X = Br) was observed at 90.8 ppm, that is, far upfield of the free silyl cations range, which testified to the minor extent of its silylium ion character and the major contribution of the bromonium ion character.

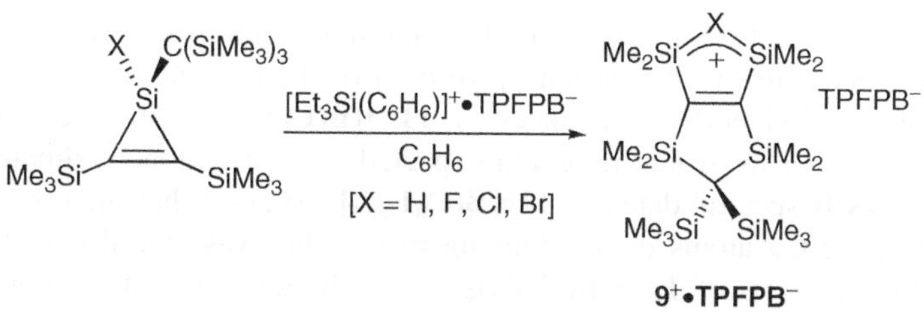

Scheme 1.15

Several important intramolecularly stabilized silyl cation derivatives have been published very recently. Among them one should mention the remarkable trimethylsilylium ion salt $Me_3Si^+ \bullet [RCB_{11}F_{11}]^-$ (R = H, Et) $\mathbf{10^+ \bullet [RCB_{11}F_{11}]^-}$ prepared by the hydride transfer reaction between $Ph_3C^+ \bullet [RCB_{11}F_{11}]^-$ and Me_3SiH.[70] $\mathbf{10^+ \bullet [RCB_{11}F_{11}]^-}$ revealed a conductivity that is typical of ionic liquids; however, in contrast to most of them it was highly reactive. In the crystalline form $\mathbf{10^+ \bullet [RCB_{11}F_{11}]^-}$ revealed coordination of the cationic Me_3Si^+ units to one of the fluorine atoms of the carborane counteranion, which caused a slight pyramidalization (354.4°) at the Si centers and stretching of the coordinating B–F bonds of the carborane cage (compared with the noncoordinating B–F bonds). In the liquid phase without solvent, $\mathbf{10^+ \bullet [RCB_{11}F_{11}]^-}$ was also not entirely 'free' from anion coordination as was manifested by its ^{29}Si NMR resonance observed at 138 ppm. Despite this value being low-field shifted compared with that of $Me_3Si^+ \bullet [B(C_6F_5)_4]^-$ (84.8 ppm),[5b] it was still by far smaller than the value of 400 ppm calculated for the 'free' Me_3Si^+ cation.

The stabilization of silyl cations can also be achieved through intramolecular coordination of π-donors. Thus, the cationic derivatives $\mathbf{11^+ \bullet [B(C_6F_5)_4]^-}$, possessing some degree of silylium ion character, were synthesized by the classical hydride transfer between $Ph_3C^+ \bullet [B(C_6F_5)_4]^-$ and dimethylarylsilyl hydrides (Scheme 1.16).[71] The very bulky aryl substituent, featuring flanking rings at the 2- and 6-positions, provided sufficient kinetic (steric protection) and thermodynamic (p-π interaction) stabilization, thus enabling the isolation of $\mathbf{11^+ \bullet [B(C_6F_5)_4]^-}$ as a room temperature stable derivative.

From the NMR spectral data it was concluded that the interaction of the $\mathbf{11^+}$ cation with either the aromatic solvents (C_6D_6 or C_7D_8) or the $[B(C_6F_5)_4]^-$ counteranion was

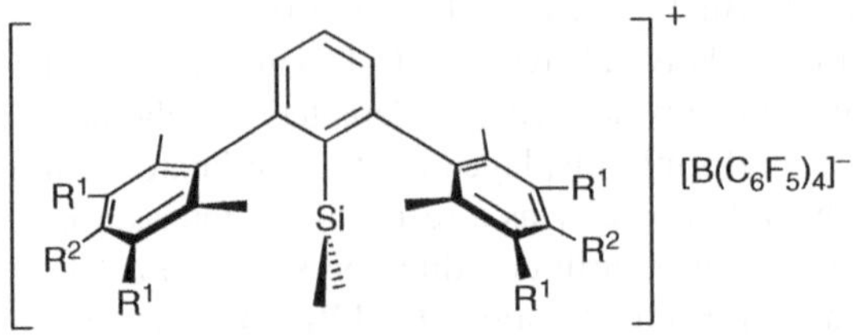

11⁺•[B(C₆F₅)₄]⁻
[R¹= H, Me; R²= H, Me]

Scheme 1.16

very small. However, the importance of the stabilizing intramolecular π-coordination of the flanking aryl rings to the vacant $3p_z$-orbital of the cationic Si center was clearly manifested in the ^{29}Si NMR resonances of $\mathbf{11^+ \bullet [B(C_6F_5)_4]^-}$ observed in the range 58.6–80.1 ppm, rather far from the region expected for 'free' (noncoordinated) silylium ions. The ^{13}C NMR spectral data of $\mathbf{11^+ \bullet [B(C_6F_5)_4]^-}$ suggest that such $Si^+ \leftarrow$Ar interaction involves the C_β atoms of the flanking rings. This was indeed confirmed by the crystal structure analysis of $\mathbf{11^+ \bullet [B(C_6F_5)_4]^-}$, which shows no interaction of the $\mathbf{11^+}$ cation with solvent molecules or counteranions. The Si^+–C_β(aryl) coordinating bond distance of 2.126(1) Å was ca. 16% longer than the average Si–C covalent bond (bond order approximately 0.66). The intramolecular π-coordination resulted in a significant departure of the geometry of the Si center from planarity: the sum of the bond angles around the Si atom was 346.1°, reflecting its marked pyramidality. Overall, the structure of the cationic portion $\mathbf{11^+}$ of the molecule was described on the basis of an η^1 π-coordination of the flanking tetramethylphenyl ring to the cationic Si center.

An interesting silyl cation derivative $\mathbf{12^+ \bullet [B(C_6F_5)_4]^-}$ featuring a two-coordinate Si center was synthesized by Driess *et al.* by the protonation of N-heterocyclic silylene **13** with the Brønsted acid $[H(OEt_2)_2]^+ \bullet [B(C_6F_5)_4]^-$ (Scheme 1.17).[72]

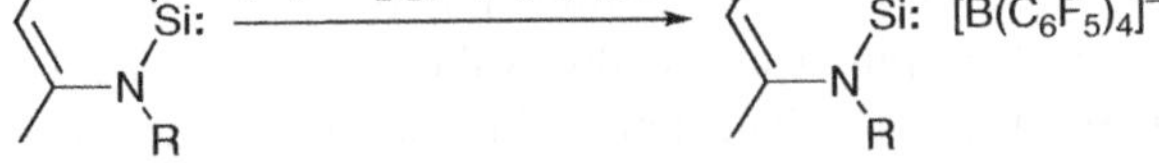

13 **12⁺•[B(C₆F₅)₄]⁻** (R = 2,6-*i*-Pr₂C₆H₃)

Scheme 1.17

$\mathbf{12^+ \bullet [B(C_6F_5)_4]^-}$ exists as a separated ion pair with no cation–anion contacts in the solid state. The ^{1}H NMR chemical shift of the C=C*H* proton of the SiC_3N_2 ring was observed at 6.92 ppm, thus suggesting the presence of aromatic 6π-electron stabilization (the resonance of the Si atom was observed at 69.3 ppm). The cyclic π-delocalization was manifested also by the shortening of the cyclic N–C bonds and stretching of the

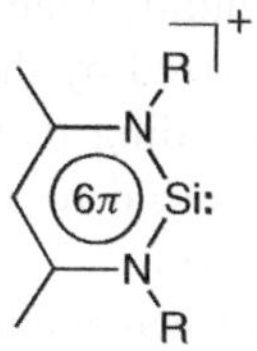

Scheme 1.18

Si–N bonds, taking place upon the protonation of **13** to form $\mathbf{12^{+} \bullet [B(C_6F_5)_4]^-}$. This points to an important contribution of the resonance structure with a delocalized system of the cyclic 6π-electrons (Scheme 1.18), which was corroborated also with the magnetic properties calculation on the appropriate model compound: NICS(1) = −3.9.

The germanium version of $\mathbf{12^+}$ with the $[HO\{B(C_6F_5)_3\}_2]^-$ counteranion was reported earlier by Power *et al.*[73] The cationic part of the molecule was well separated from its anion (closest approach from Ge to the nearest F atom was 3.01 Å), exhibiting structural properties similar to those of its silicon homolog $\mathbf{12^{+} \bullet [B(C_6F_5)_4]^-}$ (an essentially planar GeC_3N_2 ring with the two-coordinate Ge atom, cyclic Ge–N, N–C and C–C bond distances diagnostic of a cyclically delocalized system of 6π-electrons). The structural characteristics of this germyl cation are reminiscent of those of other structurally comparable Ge-centered cations intramolecularly stabilized by adjacent nitrogen ligands. They include an aminotroponiminate salt with a two-coordinate cationic Ge(II) unit featuring weak interactions to two Cl atoms from counteranions,[74] and a poly(pyrazolyl)borate complex in which the cationic Ge(II) center is pyramidally coordinated by the three neighboring nitrogen ligands with the germanium lone pair occupying the fourth tetrahedral site.[75]

A silylium ion $\mathbf{14^{+} \bullet [B(C_6F_5)_4]^-}$, stabilized by polyagostic Si–H···Si interactions, was prepared by the hydride transfer reaction between hexakis(dimethylsilyl)benzene and $Ph_3C^{+} \bullet [B(C_6F_5)_4]^-$ (Scheme 1.19).[76] The manifestation of the hydride transfer between the Si^{α} centers, assisted by the agostic bonding with the Si^{β}–H bonds, was seen in the ^{1}H- and ^{29}Si NMR spectra of $\mathbf{14^{+} \bullet [B(C_6F_5)_4]^-}$ measured at −80 °C and at room temperature. The existence of such a H-bridged silylium ion structure, featuring two $Si^{\alpha} \leftarrow H–Si^{\beta}$ agostic interactions, was further supported by DFT calculations. The

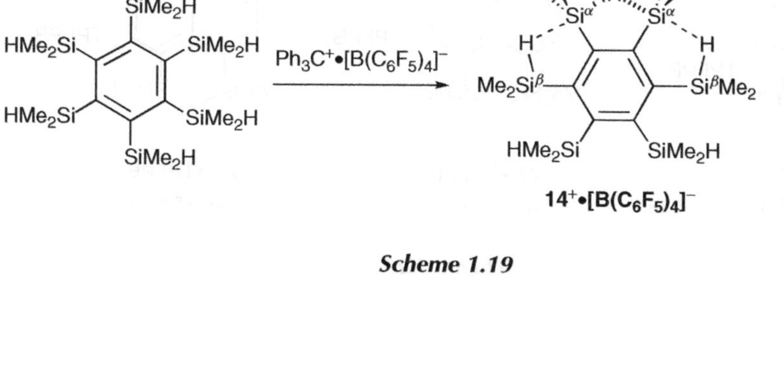

$\mathbf{14^{+} \bullet [B(C_6F_5)_4]^-}$

Scheme 1.19

α-silylium center was found to be essentially planar (358.9°), and the three-center two-electron Si^{α}–H^{α}–Si^{α} bond was characterized by the markedly larger Wiberg bond order (0.426) than the $Si^{\alpha} \leftarrow H$–Si^{β} agostic bonding (0.182). Accordingly, the NBO electron occupancy of the H^{β}–Si^{β} bond was reduced to 1.806 because of the electron density transfer to the adjacent cationic Si^{α} centers.

An important contribution to the field was made by Müller and his group, who published a series of papers describing the synthesis of intramolecularly stabilized cations of the heavy group 14 elements as well as silyl-substituted vinyl cations and arenium ions, prepared by classical hydride transfer reactions with Ph_3C^+•$TPFPB^-$.[77] Thus, the transient 7-silanorbornadien-7-ylium ion **15⁺** was stabilized by coordination to nitrile and isolated as the nitrilium complex **[15←:N≡C–CD₃)]⁺•TPFPB⁻** (Scheme 1.20), whereas the free **15⁺** was unstable at room temperature and rearranged possibly into the highly reactive [$PhSi^+$/tetraphenylnaphthalene] complex.[77a,i] In contrast, 2-silanorbornyl cation **16⁺•TPFPB⁻** (Scheme 1.20) was stable because of the effective internal coordination of its silylium ion center to the C=C double bond and revealed no interaction with either counterion or solvent.[77b,h,i] The intramolecular π-complexation was manifested in the ^{29}Si NMR chemical shift of **16⁺•TPFPB⁻** observed at 87.4 ppm, in the range diagnostic for silyl cation–π arene complexes, as well as by the appreciable

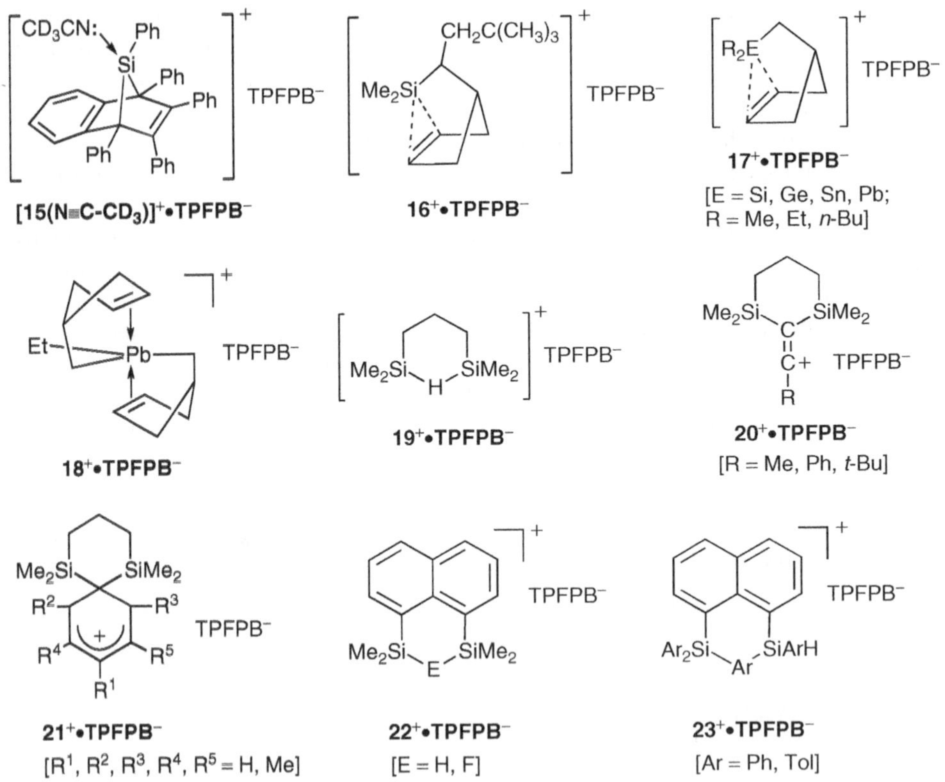

Scheme 1.20

deshielding of the olefinic C atoms (ca. 20 ppm compared with the starting hydrosilane) indicating substantial charge transfer from the Si to the C=C bond. Like the transient **15**$^+$, **16**$^+$ can be readily stabilized by coordination of the highly nucleophilic acetonitrile forming the corresponding silylated nitrilium ion. Certainly, **16**$^+$ possesses some degree of silylium ion character, although it can be alternatively viewed as a bridged β-silyl carbocation. The scope of this synthetic approach was then expanded to produce novel stable norbornyl cations free from aromatic solvent interaction **17**$^+$**•TPFPB**$^-$, based on the heavier group 14 elements (Ge, Sn, Pb) (Scheme 1.20).[77c,i] These cations **17**$^+$**•TPFPB**$^-$ revealed the characteristic low-field resonance of their central nuclei at 80.2–87.2 ppm (E = Si), 334.0 ppm (E = Sn) and 1049 ppm (E = Pb); values that, however, were markedly smaller than those expected for the free tricoordinate ions. Cations of this type can be stabilized by the intramolecular π-coordination of the two C=C double bonds from a pair of cyclopentenyl ligands (Scheme 1.20).[77d] The plumbylium ion **18**$^+$**•TPFPB**$^-$ was stable at room temperature for several weeks, exhibiting a low-field ^{207}Pb NMR resonance at 807 ppm. Its crystal structure analysis revealed well-separated cationic and anionic parts, and a trigonal-bipyramidal coordination geometry around the Pb atom.[77d] The silyl cation **19**$^+$**•TPFPB**$^-$, featuring a three-center two-electron Si–H–Si bond with hydrogen as a bridging atom, exhibited a low-field resonance of the cationic Si atom at 76.7 ppm, was observed as a doublet with a $^1J(^{29}\text{Si}–^1\text{H})$ coupling constant of 39 Hz (Scheme 1.20).[77e,f] Stable β-disilacyclohexenylidene-substituted vinyl cations **20**$^+$**•TPFPB**$^-$, formed upon the intramolecular addition of the transient silylium ion to a C≡C triple bond, were also free from solvent interaction (Scheme 1.20).[18a,77f] Their high stability was reasonably attributed to the hyperconjugative interaction of the vinyl cationic center with the two β-silyl substituents, as well as to electron donation from the α-phenyl group. The crystal structure of **20**$^+$ (R = *t*-Bu) with the hexabromocarborane counterion, **20**$^+$**•[CB$_{11}$H$_6$Br$_6$]**$^-$, disclosed a free vinyl cation with a markedly short C=C double bond length of 1.221 Å, closely approaching the length of a usual C≡C triple bond.[18b,77f] Moreover, the =C–Si bonds were rather long (1.984 and 1.946 Å), thus providing solid support for the existence of β-silyl hyperconjugation, giving the **20**$^+$ cation some extent of silylium ion character. A series of bissilylated arenium ions **21**$^+$**•TPFPB**$^-$ was prepared by the intramolecular complexation of the transient silylium ions with aromatic rings (Scheme 1.20).[77f,g] Similar to the above case of **20**$^+$, the surprising thermodynamic stability of **21**$^+$**•TPFPB**$^-$ was ascribed to the important hyperconjugative effects of the two β-silyl substituents. The hydrogen-bridged bis(silyl)cation **22**$^+$**•TPFPB**$^-$ (E = H) revealed the deshielded resonance of its Si atoms at 54.4 ppm ($^1J_{\text{Si–H}}$ = 46 Hz).[77j] In the solid state cation **22**$^+$, both tetracoordinated Si atoms of which revealed a pyramidal geometry (345.5° and 346.7°), showed no covalent contacts to the TPFPB$^-$ counteranion. Interestingly, a fluoronium ion **22**$^+$**•TPFPB**$^-$ (E = F) was formed upon the reaction of **22**$^+$**•TPFPB**$^-$ (E = H) with alkyl fluorides ($C_{10}H_{21}F$, $PhCF_3$). The ^{29}Si NMR chemical shift of **22**$^+$**•TPFPB**$^-$ (E = F) was observed at 77.2 ppm ($^1J_{\text{Si–F}}$ = 243 Hz). Similar to the case of **22**$^+$ (E = H), **22**$^+$ (E = F) also exists as a 'free' cation, lacking any bonding interaction to either counteranion or solvent, and featuring a tetrahedral coordination of the Si atoms (345.9° and 347.7°). The bonding in the two cations **22**$^+$**•TPFPB**$^-$ (E = H) and **22**$^+$**•TPFPB**$^-$ (E = F) is different: electron-deficient two-electron, three-center Si–H–Si bonding in **22**$^+$**•TPFPB**$^-$ (E = H) and Lewis acid–base interaction in **22**$^+$**•TPFPB**$^-$ (E = F). The aryl-bridged bis(silyl) cations **23^+•TPFPB$^-$** with 1,8-naphthalenediyl backbone exhibited an arenium-type nature, as was evidenced by their characteristic ^{13}C NMR chemical shift pattern (Scheme 1.20).[77k] This conclusion was also supported by the diagnostic bond length alternation in the bridging aryl ring of **23^+** (R = Tol), namely, long–short–intermediate, expected for the bissilylated arenium ion. The X-ray analysis of **23^+•TPFPB$^-$** (R = Tol) revealed also the absence of the cation–anion covalent interactions in the solid state.

Several remarkable examples of stable cations of group 14 elements heavier than silicon have been recently reported. Thus, a series of germyl cations **24^+•I$^-$**, stabilized by the intramolecular coordination of a nitrogen ligand, was synthesized by the alkylation of heteroleptic germylenes with MeI (Scheme 1.21).[78] X-ray diffraction analysis of **24^+•I$^-$** (R = *t*-Bu) revealed no bonding interaction between the cationic and anionic portions of the molecule (separated by at least 4.887 Å).[78b] Interestingly, the geometry around the Ge cationic center was rather planar (351.5°), thus implying the unremarkable influence of the nitrogen coordination on the configuration of the Ge cationic center. On the other hand, the ^{1}H NMR spectrum of **24^+•I$^-$** revealed a large deshielding of the N–*Me* and N–CH_2 protons (comparing with those of the starting germylenes), thus pointing to the importance of the N: → Ge^+ coordination.

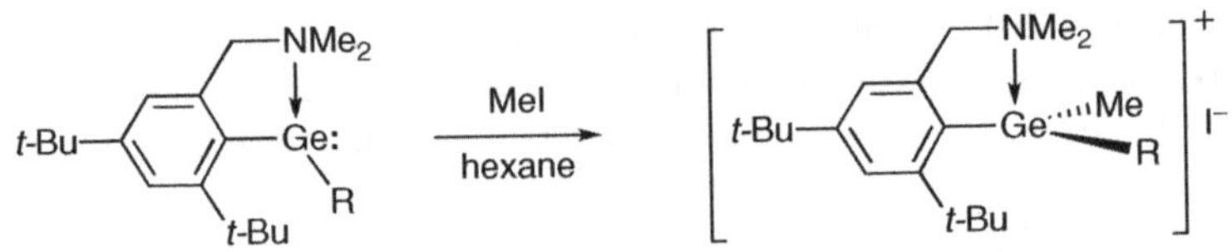

Scheme 1.21

Among other examples of the n-donor stabilized germyl cations are germyl triflate **25^+•[OTf]$^-$**[79] and the recently reported germyl chloride **26^+•Cl$^-$**.[80] The composition of **25^+•[OTf]$^-$** as a hydrated germyl cation was deduced from its crystal structure analysis, which revealed a slightly distorted trigonal-bipyramidal geometry around the Ge center with the apical positions occupied by a coordinated H_2O molecule [long Ge–O bond distance of 1.951(6) Å] and one of the methoxy groups (Scheme 1.22).[79] No direct contacts between the cationic part and triflate anion were detected in the solid state. The strong deshielding of the Ge–H proton (7.76 ppm) is also indicative of the positive charge being predominantly localized on the Ge center. The ionic derivative **26^+•Cl$^-$** contains well-separated cationic and anionic units with the closest Ge···Cl distance between them being 4.668(2) Å (Scheme 1.22).[80] Because of the intramolecular coordination of a pair of β-dimethylaminoethoxy ligands to the cationic center [Ge^+←:N bonds in the cationic portion **26^+** are 2.092(4) Å], the Ge atom is pentacoordinate featuring the geometry of a distorted trigonal bipyramid with both nitrogen ligands occupying apical positions.

As was mentioned in Section 1.5, Lambert *et al.* described their protonated digermyl and distannyl ethers $[(R_3E)_2OH]^+$•TPFPB$^-$ (R = Me, Et; E = Ge, Sn) as those featuring an important contribution of the germylium and stannylium ion forms R_3E^+•HO–ER_3.[59]

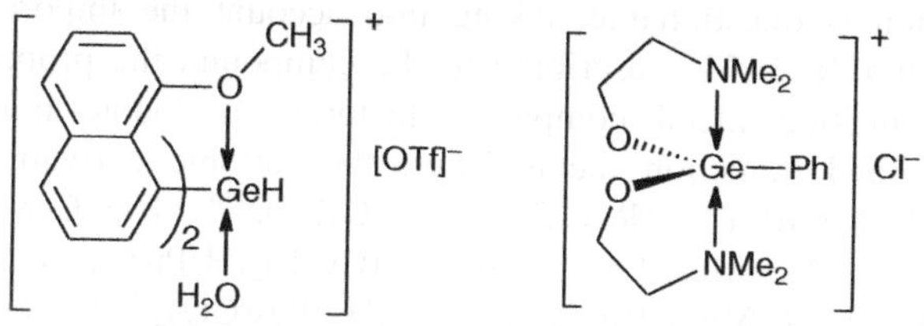

25⁺•[OTf]⁻ **26⁺•Cl⁻**

Scheme 1.22

Such a classification was based on the crystal structure peculiarities of these ionic species, namely, lack of cation–anion interactions, considerably stretched Ge–O [1.90 Å (av.)] and Sn–O [2.12 Å (av.)] bonds and observable flattening about the Ge [348° (av.)] and Sn (352.8°) centers.

An interesting Ge-containing dication **27**$^{2+}$•2I$^-$ was prepared by the reaction of the N-heterocyclic carbene (NHC)–GeI_2 complex with an excess of carbene (Scheme 1.23).[81] The Ge center in **27**$^{2+}$ revealed no bonding contacts to the iodide ions (closest Ge–I distance of 5.96 Å), however, the iodides were weakly interacting with the methyl H atoms (3.11 Å). The geometry about the Ge atom was pyramidal with Ge–C bond distances of 2.070(6) Å (slightly longer than the average Ge–C bond lengths of 1.90–2.05 Å).[81] Given the strong Lewis basicity of NHC ligands and the presence of nucleophilic THF solvent and iodide counteranions, it is hard to imagine both positive charges being situated exclusively on the Ge center, which corresponds to [Ge^{2+}←$(NHC)_3$] contribution **A**. Another resonance form **B** [Ge^-–$(NHC^+)_3$], in which a negatively charged Ge atom is bound to the three positively charged NHC ligands as the result of intramolecular charge transfer, may also significantly contribute to the overall Ge–NHC ligand bonding situation. Apparently, the real nature of this dication **27**$^{2+}$•2I$^-$ is somewhere between these two resonance extremes **A** and **B**, in which the Ge center possesses some, but by no means full, germyl dication character.

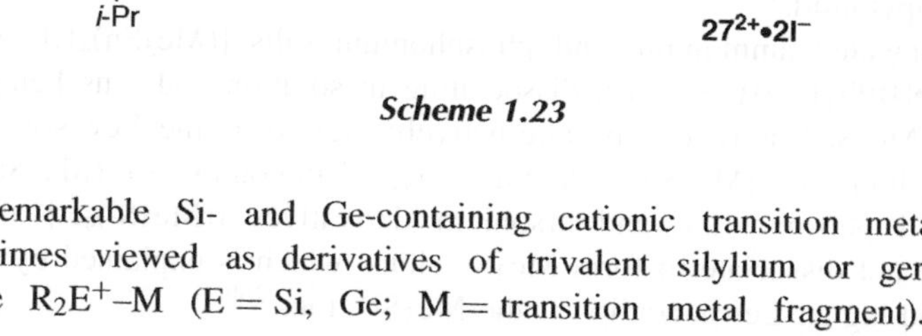

27²⁺•2I⁻

Scheme 1.23

Several remarkable Si- and Ge-containing cationic transition metal complexes were sometimes viewed as derivatives of trivalent silylium or germylium ions of the type R_2E^+–M (E = Si, Ge; M = transition metal fragment).[1h] However, such a classification is questionable, taking into account the importance of another resonance contributor $R_2E{=}M^+$, ascribing to the compound the properties of silylene (or germylene) transition metal complexes featuring a double bond between the group 14 element and transition metal. Thus, the cationic derivatives $[(\eta^5\text{-}C_5Me_5)(PMe_3)_2Ru{=}Si(SR)_2]^+{\bullet}BPh_4^-$ (R = Et, *p*-MeC_6H_4),[82a] $[(\eta^5\text{-}C_5Me_5)(PMe_3)_2Ru{=}SiR_2]^+{\bullet}[B(C_6F_5)_4]^-$ (R = Me, Ph),[82b] [*trans*-$(Cy_3P)_2(H)Pt{=}Si(SEt)_2]^+{\bullet}[BPh_4]^-$,[82c] and neutral $(\eta^5\text{-}C_5Me_5)(PMe_3)_2RuSi[S(p\text{-}MeC_6H_4)]Os(CO)_4$,[82d] were described by Tilley *et al.* as the base-free silylene complexes, in which sp^2-hybridized Si centers manifested short bonds to transition metals, planar geometry and very low-field ^{29}Si NMR resonances (259.4–311.0 ppm).

Similarly, the cationic Ge complex **28⁺**, which had no bonding contacts to its triflate counteranion, revealed a planar geometry about the cationic Ge1 center (360°) and short Ge1–W1 bond distance [2.487(2) Å] (Scheme 1.24).[83] Based on such structural peculiarities, the most important resonance structure of **28⁺** was described as Me(W2)Ge1=W1⁺, with the significant Ge1=W1 double bond character due to the π back-donation from tungsten to electron-deficient germanium, rather than the Me(W2)Ge1⁺–W1 contributor, featuring a full cationic charge situated on the germanium center.[83] The marked shortening of the Si–Fe and Ge–Fe bonds in the related base-stabilized μ-silylyne and μ-germylyne diiron complexes was realized in terms of the partial double bond character of Si=Fe and Ge=Fe.[84]

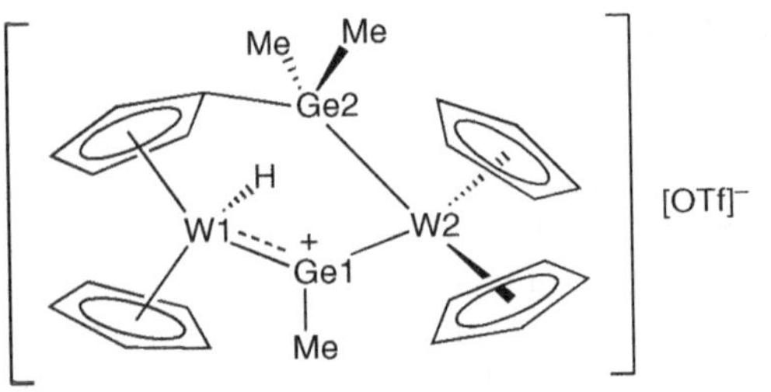

28⁺•[OTf]⁻

Scheme 1.24

Many intramolecularly stabilized stannyl cations have been synthesized, and some of them will be described below. In the early stages, several cationic derivatives with no bonding interaction to the low-coordinating anions were reported as featuring diagnostically pentacoordinate cationic Sn centers stabilized by the intra- or intermolecular coordination of *n*-donating ligands (:NR_2, :N≡C–R, :OH_2, etc.) at the apical positions of a trigonal bipyramid.[85]

The perstannylated ammonium and phosphonium salts $[(Me_3Sn)_4E]^+{\bullet}[OTf]^-$ and $[(Me_3Sn)_4E]^+{\bullet}[BPh_4]^-$ (E = N, P), dissociating in solution and thus being a masked source of the Me_3Sn^+ ions, can be alternatively viewed as the Lewis-base-stabilized trimethylstannylium ions $[Me_3Sn^+{\leftarrow}{:}E(SnMe_3)_3]$.[26a] In contrast to $[(Me_3Sn)_4E]^+$ ions, the persilylated phosphonium and arsonium derivatives $[(Me_3Si)_4E]^+{\bullet}[B(C_6F_5)_4]^-$ (E = P, As) do not dissociate to form Me_3Si^+ ions, which is explained by the stronger Si–E bonds and higher Lewis acidity of the Me_3Si^+ ions.[26b]

A family of stannyl cations **29–31**, stabilized by intramolecular side-on π-coordination to the C≡C bond of an alkynylborate fragment, was synthesized by Wrackmeyer *et al.* (Scheme 1.25).[86] The X-ray diffraction analysis of **29** revealed a somewhat pyramidal geometry about the cationic Sn atom (351.1°), which was oriented towards the C≡C bond, thus allowing their effective through-space interaction.[86a] The π-coordination of the cationic fragment to a carbon–carbon triple bond was also clearly manifested in the appreciable shielding of the Sn atoms observed in the ^{119}Sn NMR spectra of these derivatives.

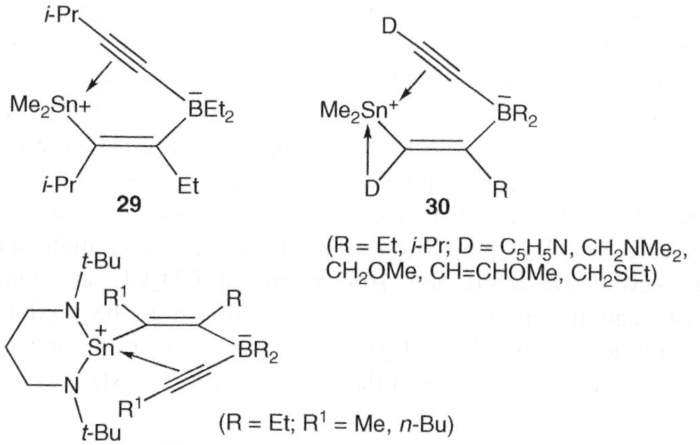

29 **30** **31**

Scheme 1.25

Likewise, the isostructural plumbyl cations **32**, intramolecularly π-coordinated to the C≡C bond of an alkynylborate moiety, were prepared by the same research group (Scheme 1.26).[87]

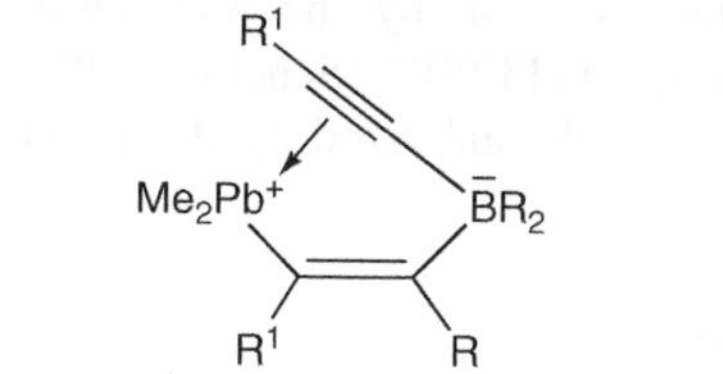

32

Scheme 1.26

The intramolecular stabilization of the cationic Pb center occurs at the expense of its departure from the idealized trigonal-planar geometry, leading to a slight pyramidalization about the Pb atom [355.7° for **32** (R = ^{I}Pr, R^1 = Me)]. The low-coordination of the metallic center in **32** was manifested in its ^{207}Pb NMR resonances observed in the range 630–750 ppm, shifted low-field by 600–800 ppm with respect to tetraorganoplumbanes.

Michl *et al.* reported the crystal structure of the cationic derivative n-Bu_3Sn^+•$[CB_{11}Me_{12}]^-$, prepared according to Scheme 1.4, B (Section 1.2.3).[12a] Lacking bonding interaction to the solvent, n-Bu_3Sn^+ was weakly coordinated to the Me groups of the carborane anion $[CB_{11}Me_{12}]^-$ featuring a long Sn–C(Me) bond distance (av. 2.81 Å) and not quite perfectly planar geometry around the Sn atom (353.1°). The resonance of the Sn atom in n-Bu_3Sn^+•$[CB_{11}Me_{12}]^-$ was observed at 454.3 ppm,[12a] far upfield from the 1500–2000 ppm range estimated for the trimethylstannylium ion,[41] indicating that the same cation aggregation through Me coordination exists also in solution. Employing the same synthetic approach, Michl *et al.* prepared a series of novel germylium, stannylium and plumbylium ion derivatives Me_3Sn^+•$[CB_{11}Me_{12}]^-$ (E = Ge, Sn, Pb) (Scheme 1.27).[12b] Similar to the above case of n-Bu_3Sn^+•$[CB_{11}Me_{12}]^-$, all Me_3Sn^+ cations revealed a remarkable interaction with the Me groups of the $CB_{11}Me_{12}^-$ counterion featuring long E–C(Me) bond distances of 2.5–3.0 Å (EXAFS). This was manifested in the appearance of NMR chemical shifts of the central nuclei of Me_3Sn^+•$[CB_{11}Me_{12}]^-$ in the relatively high field: 335.9 ppm in CD_2Cl_2 at −60 °C (E = Sn) and 1007.4 ppm in CD_2Cl_2 at room temperature (E = Pb).[12b] The cation–anion interaction was estimated to be predominantly ionic with some contribution (from 1/4 to 1/3) from covalent bonding, and the strength of this interaction was found to increase in the order: $Me_3Pb^+ < Me_3Sn^+ \ll Me_3Ge^+$.

$$\underset{[E = Ge, Sn]}{Me_3E\text{–}EMe_3\ (\text{or } Me_4Pb)} + CB_{11}Me_{12}\bullet \xrightarrow{\text{pentane}} \underset{[E = Ge, Sn, Pb]}{2\ Me_3E^+\bullet CB_{11}Me_{12}^-}$$

Scheme 1.27

An interesting example of the n-donor-supported silacyclopropenylium ion is the derivative **33+•TPFPB−**, in which the cationic spiro-Si center is shared by the six-membered ring N-heterocyclic unit and three-membered ring cyclopropene moiety. This was recently prepared by Driess *et al.* by the protonation of N-heterocyclic silacyclopropene **34** with $[H(OEt_2)_2]^+$•$TPFPB^-$ (Scheme 1.28).[88] The highly electrophilic silylium center in **33+** is effectively stabilized by the intramolecular electron donation from the adjacent N atom.

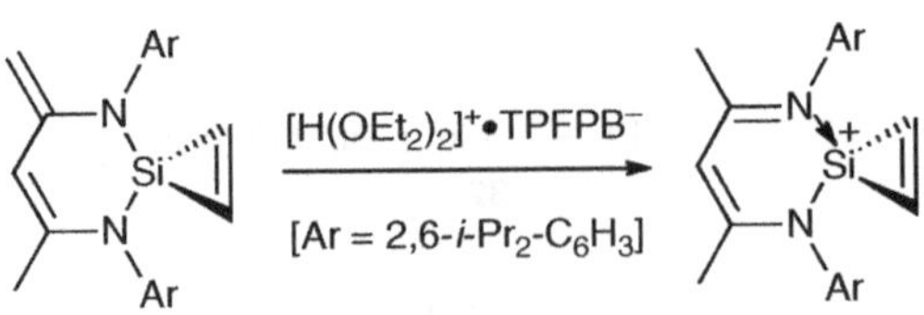

34 **33+•TPFPB−**

Scheme 1.28

1.6.2 Free (Noncoordinated) Cations

1.6.2.1 Cyclic π-Conjugated Cations

The heavy group 14 element centered cations stabilized by cyclic π-conjugation were the first isolable representatives of the family of cationic species free from observable interactions with either external (counteranions, nucleophilic solvents) or internal (intramolecular n- or π-donors) coordination. The most important of these, stable 2π-electron aromatic cyclotrigermenylium, cyclotrisilenylium and disilacyclopropenylium derivatives, 6π-electron aromatic silatropylium derivative, homoaromatic cyclotetrasilenylium, bishomoaromatic germacyclopropenylium, and trishomoaromatic cationic Ge-cluster derivatives, as well as transient aromatic 2-silaimidazolium and bishomoaromatic 7-silanorbornadienylium salts, will be discussed in detail in the Chapter 6, devoted to the issue of aromaticity among the group 14 organometallics.

Among other important examples of the cyclic π-conjugated cations, one should mention the very remarkably stable $(\eta^5\text{-}Me_5C_5)Si^+$ cation derivative **35⁺•TPFPB⁻**, which was prepared by the oxidation of decamethylsilicocene **36** with $(Me_5C_5H_2)Si^+$•TPFPB⁻ in CH_2Cl_2 (Scheme 1.29).[89]

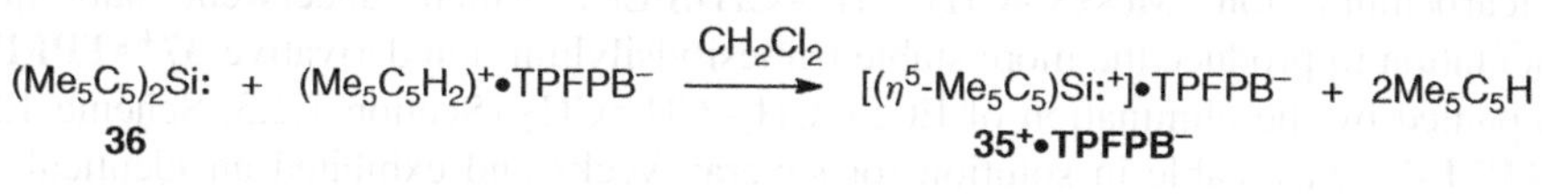

Scheme 1.29

In the crystalline form, the cation **33⁺** exhibited only a weak interaction with the TPFPB⁻ anion, featuring a nearly ideal pentagonal-pyramidal geometry (Scheme 1.30).

Such a pentagonal-pyramidal configuration of **35⁺•TPFPB⁻** was maintained also in solution, as shown by the observation of only a single resonance at 2.23 ppm for all five Me groups of the Me_5C_5 unit even at low temperature. Moreover, a resonance of the Si atom of **35⁺•TPFPB⁻** was observed at very high field (−400.2 ppm), diagnostic of the π-complexes of a divalent Si atom. Overall, both X-ray diffraction and NMR spectroscopy studies confirmed the structure of **33⁺** as a cationic π-complex featuring the η^5-Me_5C_5 ligand bound to a 'naked' Si center. Alternatively, the $(\eta^5\text{-}Me_5C_5)Si^+$ cation **35⁺** can be viewed as a *pentacoordinated* analog of the highly challenging *monohaptocoordinated* silyliumylidene ion $(\eta^1\text{-}Me_5C_5)Si^+$. Reactivity of **35⁺•TPFPB⁻** has proved to be very attractive. Thus, it reacts with $(Me_3Si)_2NLi$ to produce transient

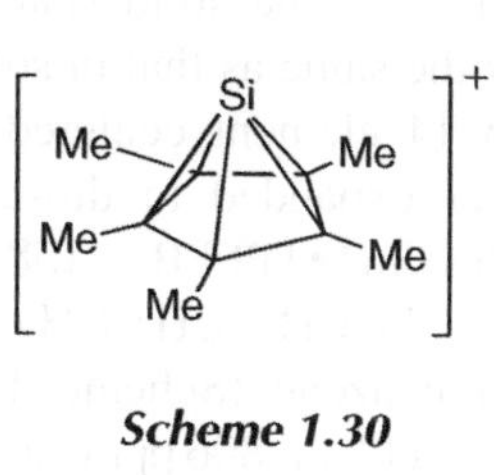

Scheme 1.30

silylene $[(Me_3Si)_2N](Me_5C_5)Si$:, undergoing subsequent dimerization finally to form the rather unusual *trans*-1,2-diaminodisilene $(\eta^1\text{-}Me_5C_5)[(Me_3Si)_2N]Si{=}Si[N(SiMe_3)_2](\eta^1\text{-}Me_5C_5)$, in which both Me_5C_5 groups are σ-bonded to the Si atoms.[89]

1.6.2.2 Acyclic Cations Lacking π-Conjugation[90]

The most challenging tricoordinate cations R_3E^+ (R = alkyl, aryl, silyl; E = Si–Pb) featuring an ideal trigonal-planar geometry and 'freedom' from both counteranions and solvents, thus being the least electronically perturbed by coordination to either external or internal nucleophiles, are very attractive targets for long-standing experimental and theoretical pursuits.

The first experimental breakthrough was achieved by Lambert and his group, who reported preparation of the free trimesitylsilylium ion Mes_3Si^+•TPFPB$^-$ (**37**$^+$•**TPFPB**$^-$) in 1997.[91] Because the classical hydride transfer reaction between Mes_3SiH and Ph_3C^+•TPFPB$^-$ was unsuccessful due to the large steric bulk of the Mes groups, a novel synthetic approach involving allyl substituents as leaving groups (the so-called 'allyl leaving group approach')[8] was developed. Thus, allyltrimesitylsilane $Mes_3Si–CH_2–CH{=}CH_2$ was reacted with $[Et_3Si(C_6H_6)]^+$•TPFPB$^-$ to form a transient β-silylcarbenium ion $Mes_3Si–CH_2–CH^+–CH_2SiEt_3$, which underwent subsequent fragmentation to produce the more stable trimesitylsilylium ion derivative **37**$^+$•**TPFPB**$^-$ accompanied by the elimination of $Et_3Si–CH_2–CH{=}CH_2$ (Section 1.2.3, Scheme 1.3).[8] **37**$^+$•**TPFPB**$^-$ was stable in solution for several weeks and exhibited an identical ^{29}Si NMR resonance at 225.5 ppm in different aromatic solvents[8,91] and with different counteranions,[1n,8] thus reliably suggesting the 'free' state of its **37**$^+$ cation in solution. However, in the presence of nucleophilic reagents (CD_3CN, Et_3N) this resonance was greatly shifted to the high-field region, pointing to the highly electrophilic nature of silylium ion **37**$^+$. Subsequent computations confirmed the status of Mes_3Si^+ as a free cation lacking observable coordination to solvent, in which the calculated ^{29}Si NMR chemical shift value was very close to the experimental value: 230.1 (GIAO/HF) and 243.9 (GIAO/DFT) ppm vs 225.5 ppm.[92]

However, because TPFPB$^-$ derivatives often form oils or clathrates precluding their crystallization, **37**$^+$•**TPFPB**$^-$ has failed to crystallize. To obtain crystals suitable for X-ray crystallography, the TPFPB$^-$ counteranion was replaced with Reed's carborane $CB_{11}HMe_5Br_6^-$ anion. Accordingly, **37**$^+$•**[CB$_{11}$HMe$_5$Br$_6$]**$^-$ was synthesized by the reaction of $Mes_3Si–CH_2–CH{=}CH_2$ with Et_3Si^+•$[CB_{11}HMe_5Br_6]^-$ in benzene.[93,94] The X-ray diffraction analysis of **37**$^+$•**[CB$_{11}$HMe$_5$Br$_6$]**$^-$ revealed that this is indeed a trigonal-planar sp^2-silylium ion truly free from covalent interaction with either counteranion or solvent (Figure 1.1).[93] The solid state ^{29}Si NMR resonance of **37**$^+$• **[CB$_{11}$HMe$_5$Br$_6$]**$^-$ was practically the same as that in solution: 226.7 ppm vs 225.5 ppm.

The range of the heavy group 14 element centered cations prepared by the 'allyl leaving group approach' was later expanded to those of germanium and tin. Thus, the trimesitylstannylium ion Mes_3Sn^+•TPFPB$^-$ (**38**$^+$•**TPFPB**$^-$) was successfully formed by the treatment of $Mes_3Sn–CH_2–CH{=}CH_2$ with $[Et_3Si(C_6H_6)]^+$•TPFPB$^-$ or $[Et_3SiCH_2CPh_2{}^+]$•TPFPB$^-$ in benzene (Scheme 1.3).[8] The cationic Sn atom in **38**$^+$•**TPFPB**$^-$ resonated at a low field of 806 ppm, a signal that was independent of the solvent used, thus testifying to the noncoordinated nature of the stannylium ion

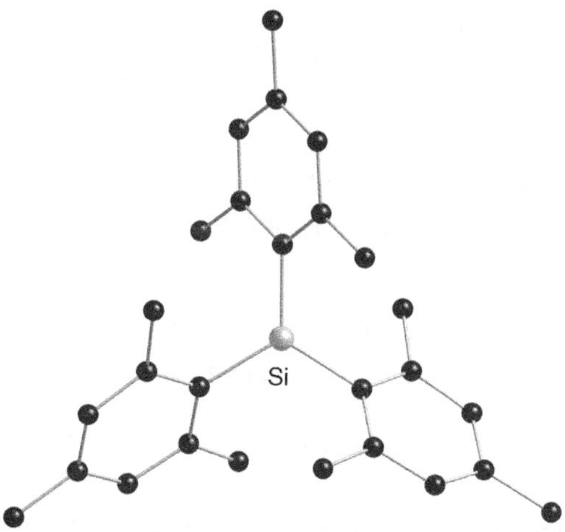

Figure 1.1 Crystal structure of the silylium ion derivative Mes_3Si^+•TPFPB$^-$ **37^+•TPFPB$^-$** (hydrogen atoms and TPFPB$^-$ counterion are not shown)

38^+. The trimesitylgermylium ion derivative Mes_3Ge^+•TPFPB$^-$ (**39^+•TPFPB$^-$**) was also synthesized by the 'allyl leaving group approach', and the degree of its cationic character was estimated to be comparable with those of the analogous trimesitylsilylium **37^+** and trimesitylstannylium **38^+** ions.[8] The tridurylsilylium ion Dur_3Si^+ prepared by the same experimental procedure featured a ^{29}Si NMR resonance at 226.8 ppm, a value that was very close to that of the Mes_3Si^+ cation of 225.5 ppm.[95] This was reasonably interpreted as the manifestation of the 'free' status of the Dur_3Si^+ cation. However, the heavier tridurylstannylium ion Dur_3Sn^+ exhibited a ^{119}Sn NMR chemical shift at 720 ppm, which was shifted to lower frequency compared with that of the Mes_3Sn^+ cation (806 ppm), on which basis the degree of Dur_3Sn^+ true stannylium ion character was estimated as ca. 70%.[95]

Utilizing the same synthetic strategy, Lambert *et al.* prepared the remarkable Tip_3Sn^+•TPFPB$^-$, **40^+•TPFPB$^-$**, by the reaction of $Tip_3Sn–CH_2–CH{=}CH_2$ with Ph_3C^+•TPFPB$^-$.[96] In the crystalline form, **40^+** revealed no bonding interaction with either TPFPB$^-$ counteranion or solvent, or with the methine hydrogens of the *i*-Pr groups, implying that **40^+** is indeed a 'free' trigonal-planar stannylium ion. The resonance of the cationic Sn atom of **40^+** was observed at 714 ppm, which was taken as evidence for its stannylium ion nature in solution (GIAO calculation at the MPW1PW91 level provided a similar value of 763 ppm).

As was mentioned in Section 1.2.5, the one-electron oxidation of the isolable free radicals represents one of the most attractive and straightforward methods for the synthesis of stable cations of the heavy group 14 elements. Thus, oxidation of the stable silyl radical $(t\text{-}Bu_2MeSi)_3Si$• with Ph_3C^+•TPFPB$^-$ in toluene resulted in the formation of a transient silylium ion salt $(t\text{-}Bu_2MeSi)_3Si^+$•TPFPB$^-$ (**41^+•TPFPB$^-$**), stabilized in the form of its nitrilium complex [41←**:N≡C–CH_3]$^+$•TPFPB$^-$** (Scheme 1.31).[97] Without complexation, **41^+•TPFPB$^-$** underwent unavoidable isomerization through the fast 1,2-Me shift from the peripheral Si to the central cationic Si atom yielding a new silyl

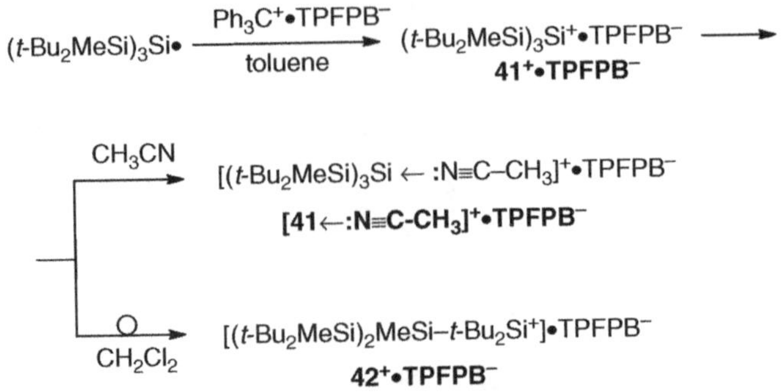

Scheme 1.31

cation **42$^+$•TPFPB$^-$** (Scheme 1.31). The driving force for this methyl migration seems to be an additional stabilization from which **42$^+$** can benefit through the hyperconjugative interaction of the silylium ion center with the adjacent Si–Si σ-bonds. Low-temperature NMR measurements at −50 °C allowed the direct observation of the cationic Si atom in **42$^+$•TPFPB$^-$**, whose low-field resonance of 303 ppm demonstrated its existence in solution as a 'noncoordinated' silylium ion.

In contrast to the unstable silylium ion **41$^+$**, the corresponding germylium ion $(t\text{-Bu}_2\text{MeSi})_3\text{Ge}^+$ and stannylium ion $(t\text{-Bu}_2\text{MeSi})_3\text{Sn}^+$ TPFPB$^-$ derivatives **43$^+$•TPFPB$^-$** and **44$^+$•TPFPB$^-$** were readily prepared by one-electron oxidation of the stable germyl $(t\text{-Bu}_2\text{MeSi})_3$Ge• and stannyl $(t\text{-Bu}_2\text{MeSi})_3$Sn• radicals in benzene (see Section 1.2.5, Scheme 1.6).[15] The crystal structure analysis of **43$^+$•TPFPB$^-$** confirmed that the cationic portion of the molecule **43$^+$** displayed no detectable interaction with either the TPFPB$^-$ counteranion or benzene molecules, thus proving that **43$^+$** is a genuine 'free' germylium ion in the solid state (Figure 1.2).[15a] **43$^+$** maintained its 'freedom' in solution as well, as was clearly seen in its solvent-independent ^{29}Si NMR resonance: 49.9 ppm in CD_2Cl_2, 49.9 ppm in $CDCl_3$ and 50.3 ppm in C_6D_6. However, because of its extremely high electrophilicity, **43$^+$** easily reacted with more nucleophilic solvents, such as acetonitrile and THF, either producing a nitrilium complex or causing a ring-opening polymerization of THF (see also Section 1.3). The electrophilicity of **43$^+$•TPFPB$^-$** greatly exceeds that of the 2π-electron aromatic cyclotrigermenylium derivative (see Chapter 6), because acyclic **43$^+$** lacks the highly stabilizing π-conjugation effects diagnostic of the cyclopropenylium-type skeleton of the cyclotrigermenylium ion. The Si–Ge bonds in germylium ion **43$^+$** were markedly stretched compared with those of the starting germyl radical $(t\text{-Bu}_2\text{MeSi})_3$Ge•: 2.5195(10) Å (av.) vs 2.4535(4) Å.[15a] This was attributed to the difference in the $4p_z(\text{Ge})$–$\sigma^*(\text{Si–C}_{t\text{-Bu}})$ hyperconjugation extent: more important for the $(t\text{-Bu}_2\text{MeSi})_3$Ge• radical, in which the $4p_z$-orbital is singly occupied, and less important for the $(t\text{-Bu}_2\text{MeSi})_3\text{Ge}^+$ cation **43$^+$**, in which the $4p_z$-orbital is vacant.

The stannylium ion **44$^+$•TPFPB$^-$** also features a perfect trigonal-planar geometry around the cationic sp^2-hybridized Sn atom, being free from any observable covalent interactions.[15b] $5p_z(\text{Sn})$–$\sigma^*(\text{Si–C}_{t\text{-Bu}})$ hyperconjugation also operated in this compound, resulting in the shortening of the Si–Sn bonds in the cation **44$^+$** compared with those

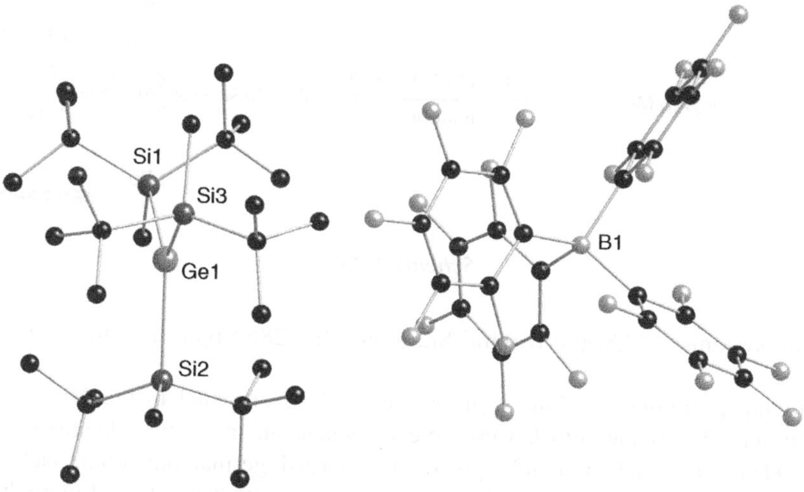

Figure 1.2 *Crystal structure of germylium ion derivative (t-Bu$_2$MeSi)$_3$Ge$^+$•TPFPB$^-$* ***43$^+$•TPFPB$^-$*** *(hydrogen atoms are not shown)*

in the radical (t-Bu$_2$MeSi)$_3$Sn•. Particularly noteworthy was the exceedingly low-field ^{119}Sn NMR resonance of **44$^+$**, undoubtedly pointing to its 'free' status in solution and representing the most deshielded Sn nucleus of all low-coordinate stannyl cations ever reported. This value of 2653 ppm by far exceeded the value expected on the basis of the ^{29}Si–^{119}Sn chemical shift empirical correlation (1500–2000 ppm) and the value (ca. 1000 ppm) calculated for a 'free' triorganostannylium ion,[42] being in reasonable agreement with the value of 2841 ppm calculated for the model (H$_3$Si)$_3$Sn$^+$ cation at the GIAO-B3LYP/6-311G(d) level.[15b]

Several other remarkable representatives of the stable cations based on the heavy group 14 elements were reported quite recently. Among them is the triarylgermylium derivative [2,6-(t-BuO)$_2$-C$_6$H$_3$]$_3$Ge$^+$•[Al{OC(CF$_3$)$_3$}$_4$]$^-$ **45$^+$•[Al{OC(CF$_3$)$_3$}$_4$]$^-$** prepared by the dehalogenation of bromogermane [2,6-(t-BuO)$_2$-C$_6$H$_3$]$_3$GeBr with the silver salt of the weakly coordinating anion Ag$^+$[Al{OC(CF$_3$)$_3$}$_4$]$^-$.[98] Although stable in its crystalline form, **45$^+$•[Al{OC(CF$_3$)$_3$}$_4$]$^-$** decomposed in solution at temperatures above −20 °C. Similar to the previously described **43$^+$•TPFPB$^-$**,[15a] triarylgermylium derivative **45$^+$•[Al{OC(CF$_3$)$_3$}$_4$]$^-$** manifested a cationic part **45$^+$** 'free' from any observable interactions with its counteranion **[Al{OC(CF$_3$)$_3$}$_4$]$^-$**, with trigonal-planar geometry about the tricoordinate Ge center.[98]

A cyclotetrasilenylium ion derivative **46$^+$•TPFPB$^-$**, prepared by the oxidation of the tetrasilyldisilene (t-Bu$_2$MeSi)$_2$Si=Si(SiMet-Bu$_2$)$_2$ with [Et$_3$Si(C$_6$H$_6$)]$^+$•TPFPB$^-$, was also 'free' in the solid state (Scheme 1.32).[99] The positive charge in the cationic portion **46$^+$** was delocalized over the three Si atoms (Si1, Si2 and Si3), which was seen in the essentially planar geometry around them and Si1–Si2/Si2–Si3 bond lengths intermediate between those of typical single and double bonds. The allylic nature of cation **46$^+$** was further supported by the observation of low-field ^{29}Si NMR resonances diagnostic of

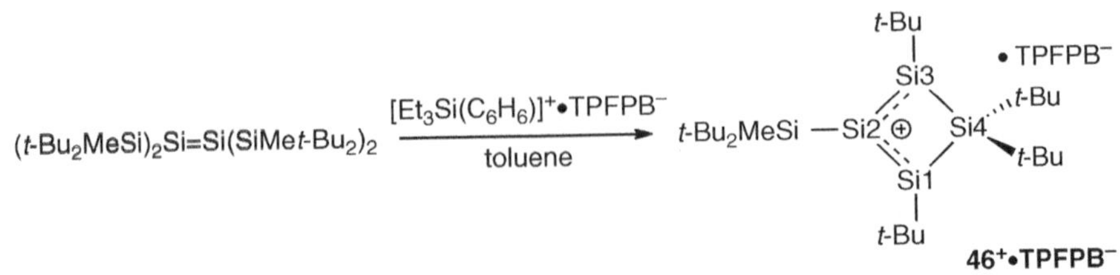

Scheme 1.32

the allylic systems: 183.8 ppm (central Si2 atom) and 286.8 ppm (terminal Si1 and Si3 atoms).

Employing a simple addition of macrocyclic [2.2.2]cryptand to the NHC complex of Cl(TfO)Ge: **47**, Baines *et al.* were able to isolate an interesting dicationic species **48^{2+}•2[OTf]$^-$** in which a doubly positively charged germanium atom, lacking any substituents, was encapsulated within a cryptand cavity (Scheme 1.33, Figure 1.3).[100a] The remote OTf^- counterions showed no any signs of bonding interaction with the Ge dication, which was at the first glance surprising giving the non-negligible nucleophilicity of the triflate anions. Such amazing chemical stability of the Ge^{2+} ion in **48^{2+}•2[OTf]$^-$** was attributed to an effective steric shielding of the cationic center by the three-dimensional network of [2.2.2]cryptand. Moreover, dicationic species **48^{2+}•2[OTf]$^-$** can even survive crystallization from the Lewis basic acetonitrile, which again pointed to an important steric protection role of the macrocycle. The Ge^{2+} of the cationic portion of the molecule **48^{2+}** was found to be only very weakly interacting with the oxygen and nitrogen *n*-donors of the [2.2.2]cryptand moiety, with the Ge–N [2.524(3) Å] and Ge–O [2.4856(16) Å] interatomic distances markedly exceeding those of the standard Ge–N and Ge–O single covalent bonds. The lack of observable

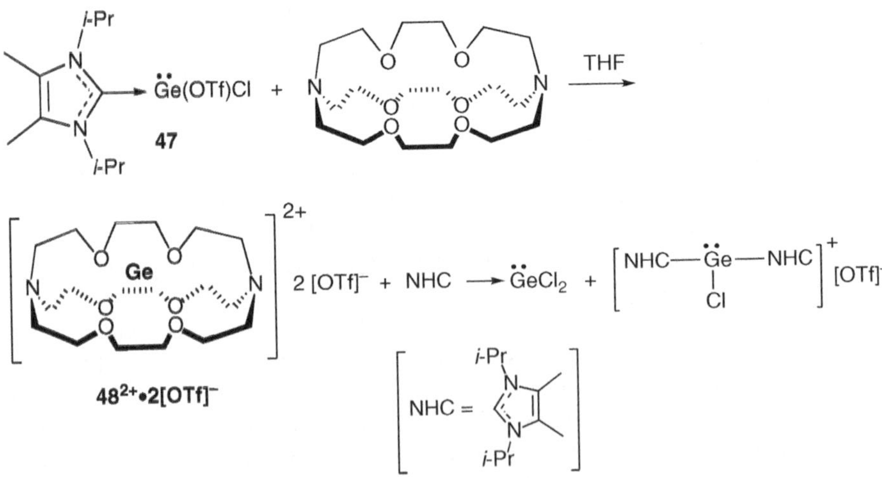

Scheme 1.33

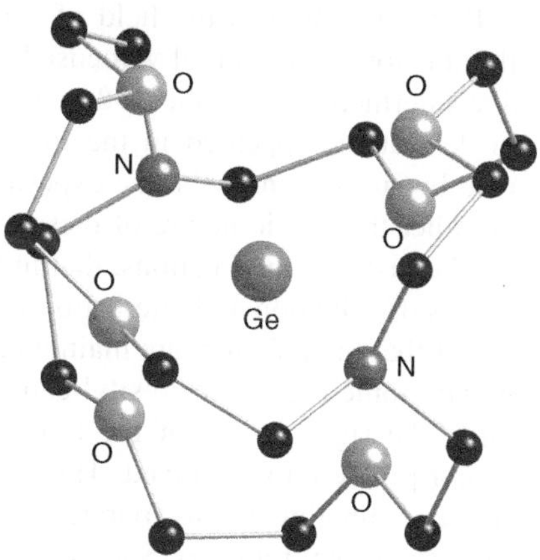

Figure 1.3 *Crystal structure of [2.2.2]cryptand-encapsulated Ge(II) dicationic derivative* ***48^{2+}•2[OTf]$^-$*** *(hydrogen atoms and TfO^- counteranions are not shown)*

bonding interactions between the germanium dicationic guest and [2.2.2]cryptand host was further supported by NBO computations at the PBE1PBE/6-311+G(2d,p) level, which revealed the pure 4s lone pair as the Ge valence orbital.[100a] Moreover, NPA analysis disclosed a high portion (+1.38) of the overall positive charge to be accommodated on the Ge center, even despite its donor–acceptor interactions with the cryptand n-donating heteroatoms. Indeed, the bond orders (WBI) for such Ge–N and Ge–O interactions were found to be exceedingly small: 0.11 and 0.10, respectively.

The high efficiency of macrocyclic structures in stabilizing Ge(II) dicationic species was further demonstrated by the complexation of Ge centers (with or without substituents) by crown ethers and azamacrocycles.[100b,c] In the resulting host–guest complexes, the cationic germanium atoms manifested highly variable coordination numbers depending on the size of the macrocyclic host and type of the counteranion used. Accordingly, the variety of isolated Ge(II)–macrocycle complexes is represented by a range of structural modifications: from complexes where a Ge(II) dication is sandwiched between two crown ether molecules to complexes in which a Ge center is encapsulated within the cavity of a crown ether/azamacrocycle.[100b,c]

1.7 Summary and Outlook

Carbenium ions constitute one of the most fundamental classes of organic chemistry species and have been studied for more than a hundred years. Their existence and important role as key reactive intermediates in a number of organic reactions (see Section 1.1) are now well established. The isolation and X-ray characterization of the stable carbocationic derivatives provided a great deal of invaluable information about their bonding and structure, largely supported by the conclusions drawn from state-of-the-art computational methods. The field of the heavy congeners of carbenium ions, in which the positive charge is centered on tricoordinate Si, Ge, Sn and Pb atoms, is not so mature.

Having started in the 1960–1970s, research in the field of cations of the heavy group 14 elements (first of all, silylium ions) was initially focused on the detection of such species as short-lived reactive intermediates. In the 1980s the first claims for the synthesis of stable silylium ion derivatives appeared in the literature (silyl perchlorates), which however were disproved by subsequent elaborate experimental and computational studies showing the covalent rather than ionic nature of these compounds. In contrast to the well-developed gas-phase chemistry of silyl cations, that in the condensed phase was greatly complicated by the effects of solvation and interaction with counteranions. At that time it became more or less clear that three factors are mainly responsible for the success or failure in the synthesis of truly ionic tricoordinate silyl cations: solvent, counteranion and substituents. Consequently, the right choice of the most appropriate solvents and counteranions was the first major problem to be solved. This was finally achieved by the utilization of extremely weakly coordinating anions (borates, carboranes) and solvents of particularly low nucleophilicity (benzene, toluene). Such experimental findings brought about the first milestone achievements in the synthesis of the stable silylium ion derivatives, $[Et_3Si(toluene)]^+$•$TPFPB^-$ by Lambert and i-Pr_3Si^+•$[CB_{11}H_6Br_6]^-$ by Reed, in the early 1990s. These silyl cations were not completely 'free', being coordinated to either solvent or counteranion, for which their true silylium ion nature and degree of their ionicity were heavily criticized. Despite this, one should undoubtedly recognize the crucial importance of their discoveries to the resolution of the 'free' silylium ion problem. This was finally achieved in the 2000s by a combination of the previous findings, weakly coordinating anions and low nucleophilic solvents, with the right choice of bulky susbtituents and culminated in the synthesis of the truly 'free' trigonal-planar silylium (Mes_3Si^+•$[CB_{11}HMe_5Br_6]^-$), germylium ((t-$Bu_2MeSi)_3Ge^+$•$TPFPB^-$) and stannylium [(t-$Bu_2MeSi)_3Sn^+$•$TPFPB^-$ and Tip_3Sn^+•$TPFPB^-$] ion derivatives. However, even after such discoveries the story of the heavy analogs of carbenium ions is not complete: there are still plenty of goals to achieve and problems to solve. Looking to the future, one should expect further developments of the general synthetic approaches to the stable silylium, germylium, stannylium and plumbylium ion derivatives. Systematic studies of the specific reactivity of such compounds, representing a novel class of powerful organometallic Lewis acids, are also highly desirable. As examples of such reactivity, one can mention the coupling of the 'heavy' cations R_3E^+ (E = Si–Pb) with a variety of nucleophiles forming E–C and E–E bonds or alkene polymerization initiated by the electrophilic addition of R_3E^+ to C=C bonds. Based on the effective interplay between sophisticated experimental techniques and modern computational methods, one can anticipate major breakthroughs in the above-mentioned fields in the near future.

1.8 References

1. Reviews on the chemistry of the heavier group 14 elements centered cations: (a) R. J. P. Corriu, M. Henner, *J. Organomet. Chem.* **1974**, *74*, 1; (b) H. Schwarz, in *The Chemistry of Organic Silicon Compounds* (Eds S. Patai, Z. Rappoport), John Wiley & Sons, Ltd, Chichester, **1989**, Chapter 7; (c) J. B. Lambert, W. J. Schulz, in *The Chemistry of Organic Silicon Compounds* (Eds S. Patai, Z. Rappoport), John Wiley & Sons, Ltd, Chichester, **1989**, Chapter 16; (d) P. D. Lickiss, *J. Chem. Soc., Dalton*

Trans. **1992**, 1333; (e) S. H. Strauss, *Chemtracts: Inorg. Chem*. **1993**, *5*, 119; (f) K. N. Houk, *Chemtracts: Org. Chem*. **1993**, *6*, 360; (g) P. Riviere, M. Riviere-Baudet, A. Castel, *Main Group Met. Chem*. **1994**, *17*, 679; (h) J. B. Lambert, L. Kania, S. Zhang, *Chem. Rev*. **1995**, *95*, 1191; (i) P. von R. Schleyer, *Science* **1997**, *275*, 39; (j) J. Belzner, *Angew. Chem., Int. Ed. Engl*. **1997**, *36*, 1277; (k) C. A. Reed, *Acc. Chem. Res*. **1998**, *31*, 325; (l) C. Maerker, P. von R. Schleyer, in *The Chemistry of Organic Silicon Compounds, Vol. 2* (Eds Z. Rappoport, Y. Apeloig), John Wiley & Sons, Ltd, Chichester, **1998**, Chapter 10; (m) P. Lickiss, in *The Chemistry of Organic Silicon Compounds, Vol. 2* (Eds Z. Rappoport, Y. Apeloig), John Wiley & Sons, Ltd, Chichester, **1998**, Chapter 11; (n) J. B. Lambert, Y. Zhao, S. M. Zhang, *J. Phys. Org. Chem*. **2001**, *14*, 370; (o) I. Zharov, J. Michl, in *The Chemistry of Organic Germanium, Tin and Lead Compounds, Vol. 2, Part 1* (Ed. Z. Rappoport), John Wiley & Sons, Ltd, Chichester, **2002**, Chapter 10; (p) T. Müller, *Adv. Organomet. Chem*. **2005**, *53*, 155; (q) V. Ya. Lee, A. Sekiguchi, *Acc. Chem. Res*. **2007**, *40*, 410; (r) V. Ya. Lee, A. Sekiguchi, in *Reviews of Reactive Intermediate Chemistry* (Eds M. S. Platz, R. A. Moss, M. Jones Jr), John Wiley & Sons, Ltd, Hoboken, **2007**, Chapter 2.

2. Those readers, who are interested in the vast area of hyper (penta- and hexa-)-coordinate cationic derivatives of the heavy group 14 elements, as well as their cation-radicals, are referred to the excellent reviews by Corriu (hypercoordinate cations) and Bock (cation-radicals): C. Chult, R. J. P. Corriu, C. Reye, J. C. Young, *Chem. Rev*. **1993**, *93*, 1371; H. Bock, W. Kaim, *Acc. Chem. Res*. **1982**, *15*, 9; H. Bock, B. Solouki, *Chem. Rev*. **1995**, *95*, 1161.
3. G. A. Olah, L. D. Field, *Organometallics* **1982**, *1*, 1485.
4. G. A. Olah, K. Laali, O. Farooq, *Organometallics* **1984**, *3*, 1337.
5. (a) J. B. Lambert, S. Zhang, *J. Chem. Soc., Chem. Commun*. **1993**, 383; (b) J. B. Lambert, S. Zhang, S. M. Ciro, *Organometallics* **1994**, *13*, 2430.
6. Z. Xie, J. Manning, R. W. Reed, R. Mathur, P. D. W. Boyd, A. Benesi, C. A. Reed, *J. Am. Chem. Soc*. **1996**, *118*, 2922.
7. M. Kira, T. Oyamada, H. Sakurai, *J. Organomet. Chem*. **1994**, *471*, C4.
8. J. B. Lambert, Y. Zhao, H. Wu, W. C. Tse, B. Kuhlmann, *J. Am. Chem. Soc*. **1999**, *121*, 5001.
9. (a) S. Fukuzumi, T. Kitano, K. Mochida, *J. Am. Chem. Soc*. **1990**, *112*, 3246; (b) S. Lochynski, B. Boduszek, H. J. Shine, *J. Org. Chem*. **1991**, *56*, 914; (c) A. Peloso, *J. Organomet. Chem*. **1974**, *67*, 423.
10. L. Doretti, S. Faleschini, *Gazz. Chim. Ital*. **1970**, *100*, 819.
11. M. Ichinohe, H. Fukui, A. Sekiguchi, *Chem. Lett*. **2000**, 600.
12. (a) I. Zharov, B. T. King, Z. Havlas, A. Pardi, J. Michl, *J. Am. Chem. Soc*. **2000**, *122*, 10253; (b) I. Zharov, T.-C. Weng, A. M. Orendt, D. H. Barich, J. Penner-Hahn, D. M. Grant, Z. Havlas, J. Michl, *J. Am. Chem. Soc*. **2004**, *126*, 12033.
13. (a) P. Jutzi, A. E. Bunte, *Angew. Chem., Int. Ed. Engl*. **1992**, *31*, 1605; (b) T. Müller, P. Jutzi, T. Kühler, *Organometallics* **2001**, *20*, 5619.
14. M. Ichinohe, Y. Hayata, A. Sekiguchi, *Chem. Lett*. **2002**, 1054.
15. (a) A. Sekiguchi, T. Fukawa, V. Ya. Lee, M. Nakamoto, M. Ichinohe, *Angew. Chem., Int. Ed*. **2003**, *42*, 1143; (b) A. Sekiguchi, T. Fukawa, V. Ya. Lee, M. Nakamoto, *J. Am. Chem. Soc*. **2003**, *125*, 9250.

16. In this section, only the reactions of $RR'R''E^+$ cations in the condensed phase will be discussed. For the numerous examples of gas-phase reactions of these ions, see the following leading references: *Reviews*: (a) H. Schwarz, in *The Chemistry of Organic Silicon Compounds* (Eds S. Patai, Z. Rappoport), John Wiley & Sons, Ltd, Chichester, **1989**, Chapter 7; (b) N. Goldberg, H. Schwarz, in *The Chemistry of Organic Silicon Compounds, Vol. 2, Part 2* (Eds Z. Rappoport, Y. Apeloig), John Wiley & Sons, Ltd, Chichester, **1998**, Chapter 18; (c) S. Fornarini, in *The Chemistry of Organic Silicon Compounds, Vol. 3* (Eds Z. Rappoport, Y. Apeloig), John Wiley & Sons, Ltd, Chichester, **2001**, Chapter 18; (d) J. M. Riveros and K. Takashima, in *The Chemistry of Organic Germanium, Tin and Lead Compounds, Vol. 2, Part 1* (Ed. Z. Rappoport), John Wiley & Sons, Ltd, Chichester, **2002**, Chapter 5; (e) T. A. Kochina, D. V. Vrazhnov, E. N. Sinotova, M. G. Voronkov, *Russ. Chem. Rev*. **2006**, *75*, 95; *Recent papers*: (f) T. A. Kochina, D. V. Vrazhnov, I. S. Ignatyev, *J. Organomet.Chem*. **1997**, *549*, 45; (g) T. A. Kochina, D. V. Vrazhnov, I. S. Ignatyev, V. D. Nefedov, E. N. Sinotova, *Russ. J. Gen. Chem*. **1999**, *69*, 905; (h) W. A. Tao, F. Wang, J. W. Denault, R. G. Cooks, *J. Chem. Soc., Perkin Trans. 2* **1999**, 2325; (i) C. E. C. A. Hop, D. A. Saulys, D. F. Gaines, J. E. Szulejko, T. B. McMahon, *J. Mass Spectrom*. **2000**, *35*, 1003; (j) N. Wade, C. Evans, F. Pepi, R. G. Cooks, *J. Phys. Chem. B* **2000**, *104*, 11230; (k) T. A. Kochina, E. V. Shchukin, V. D. Nefedov, E. N. Sinotova, *Russ. J. Gen. Chem*. **2000**, *70*, 1233; (l) Mustanir, M. Mishima, *J. Chem. Soc., Perkin Trans. 2* **2001**, 798; (m) E. V. Shchukin, T. A. Kochina, E. N. Sinotova, I. S. Ignat'ev, *Russ. J. Gen. Chem*. **2001**, *71*, 206; (n) T. A. Kochina, D. V. Vrazhnov, I. S. Ignatyev, *J. Organomet.Chem*. **2002**, *656*, 258; (o) L. A. Xavier, J. R. Pliego, J. M. Riveros, *Int. J. Mass Spectrom*. **2003**, *228*, 551; (p) T. A. Kochina, D. V. Vrazhnov, E. N. Sinotova, V. V. Avrorin, M. Yu. Katsap, Yu. V. Mykhov, *Russ. J. Gen. Chem*. **2002**, *72*, 1222; (q) T. A. Kochina, D. V. Vrazhnov, I. S. Ignat'ev, E. N. Sinotova, *Russ. J. Gen. Chem*. **2003**, *73*, 61; (r) E. A. Shishigin, V. V. Avrorin, T. A. Kochina, I. S. Ignat'ev, E. N. Sinotova, *Russ. J. Gen. Chem*. **2005**, *75*, 1395; (s) T. A. Kochina, D. V. Vrazhnov, E. N. Sinotova, I. S. Ignat'ev, *Russ. J. Gen. Chem*. **2005**, *75*, 69; (t) I. S. Ignat'ev, T. A. Kochina, *Russ. J. Gen. Chem*. **2005**, *75*, 1221; (u) T. A. Kochina, D. V. Vrazhnov, I. S. Ignat'ev, M. G. Voronkov, *Russ. J. Gen. Chem*. **2006**, *76*, 1774.
17. (a) J. B. Lambert, Y. Zhao, H. Wu, *J. Org. Chem*. **1999**, *64*, 2729; (b) J. B. Lambert, Y. Zhao, *J. Am. Chem. Soc*. **1996**, *118*, 7867.
18. (a) T. Müller, R. Meyer, D. Lennartz, H.-U. Siehl, *Angew. Chem., Int. Ed*. **2000**, *39*, 3074; (b) T. Müller, M. Juhasz, C. A. Reed, *Angew. Chem., Int. Ed*. **2004**, *43*, 1543.
19. G. A. Olah, X.-Y. Li, Q. Wang, G. Rasul, G. K. S. Prakash, *J. Am. Chem. Soc*. **1995**, *117*, 8962.
20. (a) C. A. Reed, N. L. P. Fackler, K.-C. Kim, D. Stasko, D. R. Evans, P. D. W. Boyd, C. E. F. Rickard, *J. Am. Chem. Soc*. **1999**, *121*, 6314; (b) C. A. Reed, K.-C. Kim, R. D. Bolskar, L. J. Mueller, *Science* **2000**, *289*, 101; (c) C. A. Reed, K.-C. Kim, E. S. Stoyanov, D. Stasko, F. S. Tham, L. J. Mueller, P. D. W. Boyd, *J. Am. Chem. Soc*. **2003**, *125*, 1796; (d) M. Juhasz, S. Hoffmann, E. Stoyanov, K.-C. Kim, C. A. Reed, *Angew. Chem., Int. Ed*. **2004**, *43*, 5352; (e) C. A. Reed, *Chem.*

Commun. **2005**, 1669; (f) E. S. Stoyanov, K.-C. Kim, C. A. Reed, *J. Am. Chem. Soc.* **2006**, *128*, 8500; (g) Y. Zhang, F. S. Tham, J. F. Nixon, C. Taylor, J. C. Green, C. A. Reed, *Angew. Chem., Int. Ed.* **2008**, *47*, 3801.
21. (a) D. Stasko, C. A. Reed, *J. Am. Chem. Soc.* **2002**, *124*, 1148; (b) T. Kato, C. A. Reed, *Angew. Chem., Int. Ed.* **2004**, *43*, 2908; (c) T. Kato, E. Stoyanov, J. Geier, H. Grützmacher, C. A. Reed, *J. Am. Chem. Soc.* **2004**, *126*, 12451.
22. C. Douvris, E. S. Stoyanov, F. S. Tham, C. A. Reed, *Chem. Commun.* **2007**, 1145.
23. C. Douvris, C. A. Reed, *Organometallics* **2008**, *27*, 807.
24. H.-U. Steinberger, C. Bauch, T. Müller, N. Auner, *Can. J. Chem.* **2003**, *81*, 1223.
25. (a) M. Johannsen, K. A. Jorgensen, G. Helmchen, *J. Am. Chem. Soc.* **1998**, *120*, 7637; (b) G. A. Olah, G. Rasul, G. K. Surya Prakash, *J. Am. Chem. Soc.* **1999**, *121*, 9615.
26. (a) M. Driess, C. Monsé, K. Merz, C. van Wüllen, *Angew. Chem., Int. Ed.* **2000**, *39*, 3684; (b) M. Driess, R. Barmeyer, C. Monsé, K. Merz, *Angew. Chem,. Int. Ed.* **2001**, *40*, 2308.
27. W. P. Neumann, H. Hillgärtner, K. M. Baines, R. Dicke, K. Vorspohl, U. Kobs, U. Nussbeutel, *Tetrahedron* **1989**, *45*, 951.
28. M. Schormann, S. Garratt, D. L. Hughes, J. C. Green, M. Bochmann, *J. Am. Chem. Soc.* **2002**, *124*, 11266.
29. J. M. Blackwell, W. E. Piers, R. McDonald, *J. Am. Chem. Soc.* **2002**, *124*, 1295.
30. W. A. Nugent, R. J. McKinney, R. L. Harlow, *Organometallics* **1984**, *3*, 1315.
31. (a) Y. Apeloig, in *The Chemistry of Organic Silicon Compounds* (Eds S. Patai, Z. Rappoport), John Wiley & Sons, Ltd, Chichester, **1989**, Chapter 2; (b) M. Karni, Y. Apeloig, J. Kapp, P. von R. Schleyer, in *The Chemistry of Organic Silicon Compounds, Vol. 3* (Eds Z. Rappoport, Y. Apeloig), John Wiley & Sons, Ltd, Chichester, **2001**, Chapter 1.
32. C. Maerker, J. Kapp, P. von R. Schleyer, in *Organosilicon Chemistry II* (Eds J. Weiss, N. Auner), Wiley-VCH, Weinheim, **1996**, p. 329.
33. (a) K. K. Das, K. Balasubramanian, *J. Chem. Phys.* **1990**, *93*, 5883; (b) J. Kapp, P. R. Schreiner, P. von R. Schleyer, *J. Am. Chem. Soc.* **1996**, *118*, 12154; (c) E. del Río, M. I. Menéndez, R. López, T. L. Sordo, *Chem. Commun.* **1997**, 1779.
34. G. Frenking, S. Fau, C. M. Marchand, H. Grützmacher, *J. Am. Chem. Soc.* **1997**, *119*, 6648.
35. Y. Apeloig, P. von R. Schleyer, *Tetrahedron Lett.* **1977**, 4647.
36. C.-H. Ottosson, K. J. Szabó, D. Cremer, *Organometallics* **1997**, *16*, 2377.
37. U. Pidun, M. Stahl, G. Frenking, *Chem. Eur. J.* **1996**, *2*, 869.
38. (a) D. Cremer, L. Olsson, H. Ottosson, *J. Mol. Struct. (Theochem)* **1994**, *313*, 91; (b) L. Olsson, C.-H. Ottosson, D. Cremer, *J. Am. Chem. Soc.* **1995**, *117*, 7460; (c) M. Arshadi, D. Johnels, U. Edlund, C.-H. Ottosson, D. Cremer, *J. Am. Chem. Soc.* **1996**, *118*, 5120.
39. C.-H. Ottosson, D. Cremer, *Organometallics* **1996**, *15*, 5495.
40. A. Bagno, G. Casella, G. Saielli, *J. Chem. Theory Comput.* **2006**, *2*, 37.
41. M. Arshadi, D. Johnels, U. Edlund, *Chem. Commun.* **1996**, 1279.
42. D. Cremer, L. Olsson, F. Reichel, E. Kraka, *Isr. J. Chem.* **1994**, *33*, 369.
43. J. Emsley, *The Elements*, Oxford University Press, Oxford, **1998**.

44. (a) S. K. Shin, J. L. Beauchamp, *J. Am. Chem. Soc*. **1989**, *111*, 900; (b) H. Tashiro, K. Kikukawa, K. Ikenaga, N. Shimizu, M. Mishima, *J. Chem. Soc. Perkin Trans. 2* **1998**, 2435.
45. S. Hollenstein, T. Laube, *J. Am. Chem. Soc*. **1993**, *115*, 7240.
46. G. A. Olah, G. K. S. Prakash, J. Sommer, *Superacids*, Wiley-Interscience, New York, **1985**.
47. C. Eaborn, *J. Organomet. Chem*. **1991**, *405*, 173.
48. (a) J. B. Lambert, W. J. Schulz Jr, *J. Am Chem. Soc*. **1983**, *105*, 1671; (b) J. B. Lambert, J. A. McConnell, W. J. Schulz Jr, *J. Am Chem. Soc*. **1986**, *108*, 2482; (c) J. B. Lambert, W. J. Schulz Jr, J. A. McConnell, W. Schilf, *J. Am Chem. Soc*. **1988**, *110*, 2201; (d) J. B. Lambert, W. Schilf, *J. Am Chem. Soc*. **1988**, *110*, 6364; (e) J. B. Lambert, L. Kania, W. Schilf, J. A. McConnell, *Organometallics* **1991**, *10*, 2578.
49. (a) G. A. Olah, L. D. Field, *Organometallics* **1982**, *1*, 1485; (b) G. K. S. Prakash, S. Keyaniyan, R. Aniszfeld, L. Heiliger, G. A. Olah, R. C. Stevens, H.-K. Choi, R. Bau, *J. Am Chem. Soc*. **1987**, *109*, 5123; (c) G. A. Olah, L. Heiliger, X.-Y. Li, G. K. S. Prakash, *J. Am. Chem. Soc*. **1990**, *112*, 5991; (e) G. A. Olah, G. Rasul, L. Heiliger, J. Bausch, G. K. S. Prakash, *J. Am. Chem. Soc*. **1992**, *114*, 7732.
50. J. B. Lambert, S. Zhang, C. L. Stern, J. C. Huffman, *Science* **1993**, *260*, 1917.
51. (a) C. A. Reed, Z. Xie, R. Bau, and A. Benesi, *Science* **1993**, *262*, 402; (b) S. P. Hoffmann, T. Kato, F. S. Tham, C. A. Reed, *Chem. Commun*. **2006**, 767.
52. L. Pauling, *Science* **1994**, *263*, 983.
53. (a) G. A. Olah, G. Rasul, X. Li, H. A. Buchholz, G. Sandford, G. K. Surya Prakash, *Science* **1994**, *263*, 983; (b) G. A. Olah, G. Rasul, G. K. Surya Prakash, *J. Organomet. Chem*. **1996**, *521*, 271.
54. F. Cacase, M. Attina, S. Fornarini, *Angew. Chem., Int. Ed. Engl*. **1995**, *34*, 654.
55. P. von R. Schleyer, P. Buzek, T. Müller, Y. Apeloig, H.-U. Siehl, *Angew. Chem., Int. Ed. Engl*. **1993**, *32*, 1471.
56. C. A. Reed, Z. Xie, *Science* **1994**, *263*, 985.
57. Z. Xie, D. J. Liston, T. Jelinek, V. Mitro, R. Bau, C. A. Reed, *J. Chem. Soc. Chem. Commun*. **1993**, 384.
58. Z. Xie, R. Bau, A. Benesi, C. A. Reed, *Organometallics* **1995**, *14*, 3933.
59. J. B. Lambert, S. M. Ciro, C. L. Stern, *J. Organomet. Chem*. **1995**, *499*, 49.
60. J. B. Lambert, B. Kuhlmann, *J. Chem. Soc., Chem. Commun*. **1992**, 931.
61. L. D. Henderson, W. E. Piers, G. J. Irvine, R. McDonald, *Organometallics* **2002**, *21*, 340.
62. (a) K. Hensen, T. Zengerly, P. Pickel, G. Klebe, *Angew. Chem. Int. Ed. Engl*. **1983**, *22*, 725; (b) K. Hensen, T. Zengerly, P. Pickel, G. Klebe, *Angew. Chem. Suppl*. **1983**, 973.
63. Z. Xie, R. Bau, C. A. Reed, *J. Chem. Soc., Chem. Commun*. **1994**, 2519.
64. (a) M. Kira, T. Hino, H. Sakurai, *Chem. Lett*. **1993**, 153; (b) M. Kira, T. Hino, H. Sakurai, *J. Am. Chem. Soc*. **1992**, *114*, 6697.
65. J. Belzner, D. Schär, B. O. Kneisel, R. Herbst-Irmer, *Organometallics* **1995**, *14*, 1840.

66. A. Bockholt, T. Braun, P. Jutzi, B. Neumann, A. Stammler, H.-G. Stammler, in *Organosilicon Chemistry V* (Eds N. Auner, J. Weis), Wiley-VCH, Weinheim, **2003**, 50.
67. (a) M. J. MacLachlan, S. C. Bourke, A. J. Lough, I. Manners, *J. Am. Chem. Soc*. **2000**, *122*, 2126; (b) S. C. Bourke, M. J. MacLachlan, A. J. Lough, I. Manners, *Chem. Eur. J*. **2005**, *11*, 1989.
68. (a) E. P. Kramarova, S. A. Pogozhikh, A. G. Shipov, V. V. Negrebetsky, S. N. Tandura, A. N. Shumskii, S. A. Artamkin, S. Yu. Bylikin, Yu. E. Ovchinnikov, Yu. I. Baukov, *Russ. Chem. Bull*. **2001**, *50*, 331; (b) E. P. Kramarova, A. A. Korlyukov, S. Yu. Bylikin, A. G. Shipov, Yu. I. Baukov, D. Kost, *Russ. Chem. Bull*. **2004**, *53*, 1135; (c) A. A. Korlyukov, S. A. Pogozhikh, Yu. E. Ovchinnikov, K. A. Lyssenko, M. Yu. Antipin, A. G. Shipov, O. A. Zamyshlyaeva, E. P. Kramarova, V. V. Negrebetsky, I. P. Yakovlev, Yu. I. Baukov, *J. Organomet. Chem*. **2006**, *691*, 3962.
69. A. Sekiguchi, Y. Murakami, N. Fukaya, Y. Kabe, *Chem. Lett*. **2004**, *33*, 530.
70. T. Küppers, E. Bernhardt, R. Eujen, H. Willner, C. W. Lehmann, *Angew. Chem., Int. Ed*. **2007**, *46*, 6346. A series of related bissilylated halonium ion derivatives $[Me_3Si–X–SiMe_3]^+$•$TPFPB^-$ (X = F, Cl, Br, I) was recently prepared by the reaction of Me_3SiX with $[Me_3Si]^+$•$TPFPB^-$: M. Lehmann, A. Schulz, A. Villinger, *Angew. Chem., Int. Ed*. **2009**, *48*, 7444.
71. S. Duttwyler, Q.-Q. Do, A. Linden, K. K. Baldridge, J. S. Siegel, *Angew. Chem., Int. Ed*. **2008**, *47*, 1719.
72. M. Driess, S. Yao, M. Brym, C. van Wüllen, *Angew. Chem., Int. Ed*. **2006**, *45*, 6730. A closely related example of five-membered ring germylium ion derivative $[CH_2CH\{(2,6\text{-}i\text{-}Pr_2\text{-}C_6H_3)N\}_2Ge{:}(\leftarrow OEt_2)]^+$•$TPFPB^-$ was also recently prepared by the treatment of N-heterocyclic germylene $[CH\{(2,6\text{-}i\text{-}Pr_2\text{-}C_6H_3)N\}]_2Ge{:}$ with $[H(OEt_2)_2]^+$•$TPFPB^-$: A. Schäfer, W. Saak, D. Haase, T. Müller, *Chem. Eur. J*. **2009**, *15*, 3945.
73. M. Stender, A. D. Phillips, P. P. Power, *Inorg. Chem*. **2001**, *40*, 5314.
74. H. V. R. Dias, Z. Wang, *J. Am. Chem. Soc*. **1997**, *119*, 4650.
75. D. L. Reger, P. S. Coan, *Inorg. Chem*. **1996**, *35*, 258.
76. A. Y. Khalimon, Z. H. Lin, R. Simionescu, S. F. Vyboishchikov, G. I. Nikonov, *Angew. Chem., Int. Ed*. **2007**, *46*, 4530.
77. (a) J. Schuppan, B. Herrschaft, T. Müller, *Organometallics* **2001**, *20*, 4584; (b) H.-U. Steinberger, T. Müller, N. Auner, C. Maerker, P. v. R. Schleyer, *Angew. Chem., Int. Ed. Engl*. **1997**, *36*, 626; (c) T. Müller, C. Bauch, M. Ostermeier, M. Bolte, N. Auner, *J. Am. Chem. Soc*. **2003**, *125*, 2158; (d) T. Müller, C. Bauch, M. Bolte, N. Auner, *Chem. Eur. J*. **2003**, *9*, 1746; (e) T. Müller, *Angew. Chem., Int. Ed*. **2001**, *40*, 3033; (f) T. Müller, in *Organosilicon Chemistry V* (Eds N. Auner, J. Weis), Wiley-VCH, Weinheim, **2003**, p. 34; (g) R. Meyer, K. Werner, T. Müller, *Chem. Eur. J*. **2002**, *8*, 1163; (h) N. Auner, G. Fearon, J. Weis, in *Organosilicon Chemistry III* (Eds N. Auner, J. Weis), Wiley-VCH, Weinheim, **1998**, p. 1; (i) N. Auner, T. Müller, M. Ostermeier, J. Schuppan, H.-U. Steinberger, in *Organosilicon Chemistry IV* (Eds N. Auner, J. Weis), Wiley-VCH, Weinheim, **2000**, p. 127;

(j) R. Panisch, M. Bolte, T. Müller, *J. Am. Chem. Soc*. **2006**, *128*, 9676; (k) R. Panisch, M. Bolte, T. Müller, *Organometallics* **2007**, *26*, 3524. See also the latest contributions: (l) A. Klaer, W. Saak, D. Haase, T. Müller, *J. Am. Chem. Soc*. **2008**, *130*, 14956; (m) Y. Yang, R. Panisch, M. Bolte, T. Müller, *Organometallics* **2008**, *27*, 4847; (n) A. Klaer, Y. Syha, H. R. Nasiri, T. Müller, *Chem. Eur. J*. **2009**, *15*, 8414.

78. (a) H. Schmidt, S. Keitemeyer, B. Neumann, H.-G. Stammler, W. W. Schoeller, P. Jutzi, *Organometallics* **1998**, *17*, 2149; (b) P. Jutzi, S. Keitemeyer, B. Neumann, H.-G. Stammler, *Organometallics* **1999**, *18*, 4778.
79. F. Cosledan, A. Castel, P. Riviere, J. Satgé, M. Veith, V. Huch, *Organometallics* **1998**, *17*, 2222.
80. V. N. Khrustalev, I. A. Portnyagin, I. V. Borisova, N. N. Zemlyansky, Yu. A. Ustynyuk, M. Yu. Antipin, M. S. Nechaev, *Organometallics* **2006**, *25*, 2501.
81. P. A. Rupar, V. N. Staroverov, P. J. Ragogna, K. M. Baines, *J. Am. Chem. Soc*. **2007**, *129*, 15138.
82. (a) D. A. Straus, S. D. Grumbine, T. D. Tilley, *J. Am. Chem. Soc*. **1990**, *112*, 7801; (b) S. K. Grumbine, T. D. Tilley, F. P. Arnold, A. L. Rheingold, *J. Am. Chem. Soc*. **1994**, *116*, 5495; (c) S. D. Grumbine, T. D. Tilley, F. P. Arnold, A. L. Rheingold, *J. Am. Chem. Soc*. **1993**, *115*, 7884; (d) S. D. Grumbine, T. D. Tilley, A. L. Rheingold, *J. Am. Chem. Soc*. **1993**, *115*, 358.
83. L. K. Figge, P. J. Carroll, D. H. Berry, *Angew. Chem., Int. Ed. Engl*. **1996**, *35*, 435.
84. (a) Y. Kawano, H. Tobita, H. Ogino, *Angew. Chem., Int. Ed. Engl*. **1991**, *30*, 843; (b) Y. Kawano, H. Tobita, M. Shimoi, H. Ogino, *J. Am. Chem. Soc*. **1994**, *116*, 8575; (c) J. Fujita, Y. Kawano, H. Tobita, H. Ogino, *Chem. Lett*. **1994**, 1353.
85. (a) G. van Koten, J. T. B. H. Jastrzebski, J. G. Noltes, A. L. Spek, J. C. Schoone, *J. Organomet. Chem*. **1978**, *148*, 233; (b) A. J. Crowe, P. J. Smith, P. G. Harrison, *J. Organomet. Chem*. **1981**, *204*, 327; (c) A. G. Davies, J. P. Goddard, M. B. Hursthouse, N. P. C. Walker, *J. Chem. Soc., Chem. Commun*. **1983**, 597; (d) W. A. Nugent, R. J. McKinney, R. L. Harlow, *Organometallics* **1984**, *3*, 1315.
86. (a) B. Wrackmeyer, S. Kundler, R. Boese, *Chem. Ber*. **1993**, *126*, 1361; (b) B. Wrackmeyer, S. Kundler, W. Milius, R. Boese, *Chem. Ber*. **1994**, *127*, 333; (c) B. Wrackmeyer, G. Kehr, S. Ali, *Inorg. Chim. Acta* **1994**, *216*, 51; (d) B. Wrackmeyer, K. H. von Locquenghien, S. Kundler, *J. Organomet. Chem*. **1995**, *503*, 289.
87. B. Wrackmeyer, K. Horchler, R. Boese, *Angew. Chem., Int. Ed. Engl*. **1989**, *28*, 1500.
88. S. Yao, Y. Xiong, C. van Wüllen, M. Driess, *Organometallics* **2009**, *28*, 1610.
89. P. Jutzi, A. Mix, B. Rummel, W. W. Schoeller, B. Neumann, H.-G. Stammler, *Science* **2004**, *305*, 84.
90. Lacking stabilizing but highly perturbing effects of π-conjugation, such cations still can (and do) benefit from the hyperconjugative interaction involving substituents' σ-bonds.
91. J. B. Lambert, Y. Zhao, *Angew. Chem., Int. Ed. Engl*. **1997**, *36*, 400.
92. T. Müller, Y. Zhao, J. B. Lambert, *Organometallics* **1998**, *17*, 278.

93. K.-C. Kim, C. A. Reed, D. W. Elliott, L. J. Mueller, F. Tham, L. Lin, J. B. Lambert, *Science* **2002**, *297*, 825.
94. P. P. Gaspar, *Science* **2002**, *297*, 785.
95. J. B. Lambert, L. Lin, *J. Org. Chem*. **2001**, *66*, 8537.
96. J. B. Lambert, L. Lin, S. Keinan, T. Müller, *J. Am. Chem. Soc*. **2003**, *125*, 6022.
97. M. Nakamoto, T. Fukawa, A. Sekiguchi, *Chem. Lett*. **2004**, *33*, 38.
98. C. Schenk, C. Drost, A. Schnepf, *Dalton Trans*. **2009**, 773.
99. S. Inoue, M. Ichinohe, T. Yamaguchi, A. Sekiguchi, *Organometallics* **2008**, *27*, 6056.
100. (a) P. A. Rupar, V. N. Staroverov, K. M. Baines, *Science* **2008**, *322*, 1360; (b) P. A. Rupar, R. Bandyopadhyay, B. F. T. Cooper, M. R. Stinchcombe, P. J. Ragogna, C. L. B. Macdonald, K. M. Baines, *Angew. Chem., Int. Ed*. **2009**, *48*, 5155; (c) F. Cheng, A. L. Hector, W. Levason, G. Reid, M. Webster, W. Zhang, *Angew. Chem., Int. Ed*. **2009**, *48*, 5152.

2

Heavy Analogs of Organic Free Radicals: Si-, Ge-, Sn- and Pb-Centered Radicals

2.1 Introduction

Organic free radicals are open shell species, that are typically defined as atoms (or a group of atoms) with an unpaired electron placed in the highest energy molecular orbital, typically referred to as a SOMO (singly occupied molecular orbital). Given the intrinsic electron deficiency of free radicals, one would expect them to feature high electronic demand and consequently extreme reactivity. Indeed, organic radicals are commonly considered as short-lived reactive intermediates, which in some particular cases can be stabilized with the appropriate choice of substituents to an extent that allows their direct EPR spectroscopic observation at room temperature conditions (*persistent* radicals) or even isolation and structural characterization (*stable* radicals).

Apart from their stability, the other fundamental issue in free radical chemistry is the configuration of the radical species, which is directly related to the nature of their SOMOs: whether they are of the p-type or the s/p-hybrid type. Accordingly, the former type of radicals is classified as π-radicals manifesting a trigonal-planar geometry, whereas the latter type of radicals represents σ-radicals featuring pyramidal configurations. Thus, the free radicals can adopt one of three possible structural variations: rigid pyramidal (**A**), rapidly inverting flexible pyramidal (**B**), and planar/flexible planar (**C**) (Scheme 2.1).

The first two types of radicals, **A** and **B**, belong to the class of σ-radicals, whereas **C**-type radicals should be identified as π-radicals. The interconversion between alkyl σ- and π-radicals is a low-energy process: for example, the pyramidal *tert*-butyl radical

Organometallic Compounds of Low-Coordinate Si, Ge, Sn and Pb: From Phantom Species to Stable Compounds
Vladimir Ya. Lee and Akira Sekiguchi

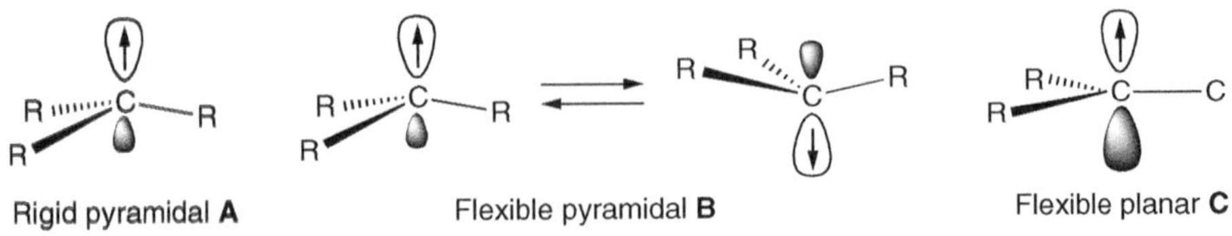

Scheme 2.1

(deviation of the radical carbon center from planarity is 19°) has a negligible barrier to inversion of ca. 0.5 kcal mol^{-1}.[1]

The preferred geometry of the free radicals is dictated by the effects of the substituents, both electronic and steric. The electronic effect of substituents can be simply visualized in terms of the frontier orbitals' interaction. Thus, it is well known that the electronegative π-donating and σ-accepting groups (halogens, amino groups, etc.) strongly favor pyramidalization at the radical center. This results from the attractive SOMO (unpaired electron in a p-orbital)–LUMO (antibonding σ^*-orbital of the C–X bond) interaction (second-order Jahn–Teller effect), which becomes progressively more favorable upon pyramidalization at the radical center (Scheme 2.2). Such an interaction lowers the SOMO energy level, which results in its stabilization. The other factor, contributing to the overall pyramidalization at the radical centers is the repulsive interaction between the unpaired electron in the SOMO and lone pair electrons in the heteroatom n-orbital. Conversely, the electropositive π-accepting (boryl) and σ-donating (silyl, germyl) groups have the totally opposite effect on the configuration of the radical centers, strongly favoring the flattening of the radical structures.

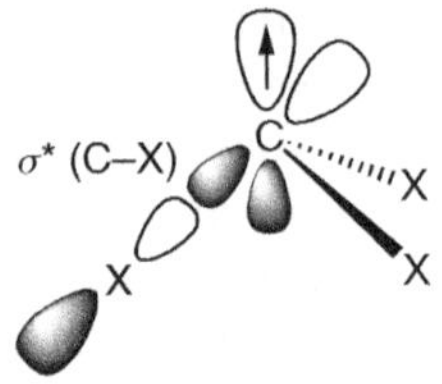

Scheme 2.2

On the other hand, steric bulkiness of the substituents also plays a crucial role in determining the radical configuration. Very large groups at the radical center prefer to be far away from each other in order to minimize their steric repulsion. Such a remote arrangement of the bulky substituents can be achieved by assuming a planar geometry at the radical center.

Direct detection of radical intermediates is possible by using one of the modern spectroscopic techniques: electronic paramagnetic resonance (EPR), electron nuclear double resonance (ENDOR), and chemically induced dynamic nuclear polarization (CIDNP), which will be discussed in more detail in Section 2.2.2.2. A large amount of structural information on the bonding situation and geometry of radicals in solution can be obtained from their EPR spectra, taking into account the general tendency for a marked increase in the values of the hyperfine coupling constants (hfccs), arising from the spin density of the ns-orbitals, with increasing s-character of the free radicals' SOMO. Thus, pyramidalization at the radical centers is manifested in larger values of the hfccs (σ-radicals), whereas smaller values of the hfccs are indicative of the more planar radicals (π-radicals).

As the most powerful tool for the structural elucidation of free radicals in the solid state, one should mention X-ray crystallography, which provides unique information about the real structure of the radical species in their crystalline form. Needless to say, the use of this comprehensive identification method is limited to the case of stable (isolable) radicals only (see Section 2.4).

The study of the free radicals centered on the heavy group 14 elements (Si, Ge, Sn and Pb) began several decades ago, and to date many transient species have been detected and identified using low-temperature spectroscopic techniques. The EPR spectra of some *persistent* radicals of the heavy group 14 elements were successfully measured at room temperature, and as a culmination of the progress in the field, several *stable* radicals with an unpaired electron centered on Si, Ge, Sn and Pb atoms were isolated and structurally characterized in their crystalline form. The experimental accomplishments in this field have been repeatedly reviewed, particularly those concerning silyl radical chemistry.[2] In the present chapter, we will briefly describe the progress in the field of heavy group 14 element radicals: from transient species (generation, identification and application) to *persistent* and *stable* radicals. Particular attention will be paid to the most important achievements and developments related to the synthesis and structural identification of the *stable* representatives of heavy group 14 element centered radicals.[3]

2.2 Early Studies: Transient Species RR′R″E•

2.2.1 Generation

Many synthetic methods have been employed for the generation of transient radicals of the heavy group 14 elements, and the most common of them will be briefly discussed below. In the vast majority of cases, the simple tricoordinate radicals of the type RR′R″E• [R, R′, R″ = H, alkyl (Me, Et, n-Pr, n-Bu, t-Bu, $Ph_nMe_{3-n}C$–CH_2, cyclo-C_3H_5, etc.), aryl (Ph, $Me_nC_6H_{5-n}$, 2,4,6-Me_3-C_6H_2, 2,4,6-Et_3-C_6H_2, 2,4,6-i-Pr_3-C_6H_2, etc.); E = Si, Ge, Sn, Pb] can be generated in one of the following three ways: (1) by the photolysis or thermolysis of the corresponding hydrides RR′R″EH in the presence of radical initiators $(t\text{-BuO})_2$[4–16] or AIBN[14] (Scheme 2.3, A); (2) by the γ-irradiation of R_4E[17–23] or photolysis of R_3E–Hg–ER_3[9,24] (Scheme 2.3, B and C); (3) by the thermolysis or photolysis of either R_3E–ER_3[9,11,13,25–30] or R_3E–ER_2–ER_3[31] (Scheme 2.3, D and E). Whereas the first two methods were widely utilized for the generation of a whole range of heavy group 14 element centered radicals, the last method has been employed mainly for the generation of germyl and stannyl radicals, being seldom used for the generation of silyl radicals because of the significant Si–Si bond strength. Among other experimental techniques reported to be suitable for the generation of radical species, one can mention the following methods: reduction of the chlorides R_3ECl with metallic sodium[32] (Scheme 2.3,

F), photolysis of anionic derivatives R_3ELi[33] and $(R_3E)_2Ca$[34] (Scheme 2.3, G and H), photolysis of heavy carbene analogs R_2E: [34] or $(R_2N)_2E$:[35,36] (Scheme 2.3, I), electrochemical generation of Ph_3Ge• radicals by the one-electron reduction of halides Ph_3GeX (X = Cl, Br, I)[37] (Scheme 2.3, J) and Cl_3Ge• radicals by the one-electron oxidation of $HGeCl_3$ (Scheme 2.3, K).[38]

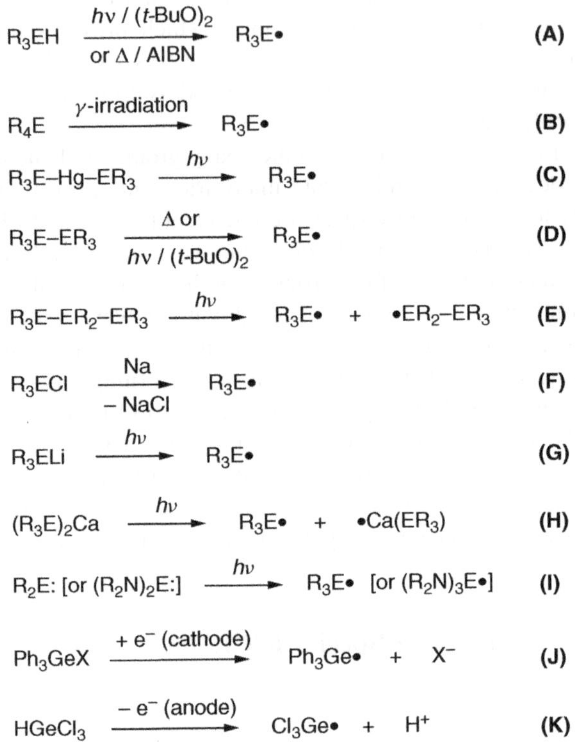

Scheme 2.3

2.2.2 Structure (Identification)

2.2.2.1 Electronic Spectroscopy

In the early stages of research in this field, the primary attention was focused on the simplest yet most fundamental parent species H_3E• (E = Si, Ge, Sn). Thus, the electronic spectrum of the H_3Si• radical, generated by the reaction of F• (or Cl•) upon the excimer laser photolysis of F_2 (or Cl_2) in the presence of H_4Si, revealed a strong vibrational progression between 365 nm and 410 nm, observed by resonance-enhanced multiphoton ionization spectroscopy (REMPI).[39] Likewise, the resonance of the H_3Ge• radical, generated by the reaction of either F• or Cl• with H_4Ge, was detected by REMPI spectroscopy in the range 370–430 nm.[40] The UV absorption band of the H_3Si• radical, generated by the flash photolysis of a mixture $CCl_4/H_4Si/N_2$, was observed in the region of 205–250 nm (absorption maximum at ca. 215 nm), indicative of its pyramidal structure.[41] The pyramidality of the F_3Si• radical, generated by the glow discharge of the disilane $F_3Si–SiF_3$, was deduced from the analysis of its rotational spectrum recorded during a recent microwave spectroscopy study.[42]

2.2.2.2 *EPR and CIDNP Spectroscopy*

2.2.2.2.1 EPR Spectroscopy

The first EPR spectra of the transient radicals H_3E• (E = Si, Ge, Sn) were recorded in inert gas matrices at low temperatures.[43] Accordingly, the following EPR parameters were reported at 4.2 K for the above-mentioned radical species [*g*-factor, hfcc for the central nuclei (^{29}Si, ^{73}Ge, ^{119}Sn) in mT, matrix inert gas]: H_3Si• (2.0030, 19, Xe); H_3Ge• (2.0073, 7.5, Xe); H_3Sn• (2.0170, 38, Kr).[43] The large values of the hfcc suggest the pronounced pyramidality of the H_3E• radicals, whose structural parameters were computationally estimated as follows (s-character of the bond orbital, bond angles in degrees): H_3Si• (0.285, 113.5); H_3Ge• (0.295, 115); and H_3Sn• (0.310, 117). The pyramidal geometry of the heavy analogs H_3E• is in sharp contrast to the planar environment of the prototypical methyl radical H_3C•, which revealed a resonance at $g = 2.0020$ in Xe matrix and for which the s-character of the bond orbital of 0.333 and bond angles of 120° were calculated.[43] The subsequent EPR reinvestigation of the H_3Si• radical, generated by either γ-irradiation of H_4Si adsorbed on a silica gel surface at 77 K[44] or by the reaction of H_4Si with H atoms in a Kr matrix,[45] disclosed very similar values for the ^{29}Si hfcc of 18.2 mT[44] and 19.0 mT.[45] The large hyperfine splitting, implying a large s-fraction of its SOMO, is consistent with the pyramidal configuration of the H_3Si• radical, in accordance with Pauling's hypothesis that the AX_3•-type radical is expected to be nonplanar if X is more electronegative than A.[46] The EPR spectrum of the H_3Ge• radical, generated in a matrix of nonmagnetic Xe isotopes, has also been reported.[47]

On further increasing the electronegativity of substituents, one should expect a progressive increase in the degree of pyramidalization. Thus, the EPR spectra of perhalogenated silyl radicals F_3Si•[48] and Cl_3Si•,[49] generated by either radiolysis of F_3SiH in an SF_6 matrix or irradiation of $SiCl_4$, revealed very large ^{29}Si hfcc values of 49.8 mT (F_3Si•),[48] 41.6 mT (Cl_3Si• in Cl_3SiCH_3 matrix)[49a] and 44.0 mT (Cl_3Si• in Cl_4Si matrix),[49b] thus pointing to a strong pyramidalization at the silicon radical centers because of the great difference in electronegativities of the Si and F atoms. The pyramidality of silyl radicals successively increased along the series $(CH_3)_3Si$• < $(CH_3)_2ClSi$• < $(CH_3)Cl_2Si$• < Cl_3Si•, as was manifested by the increasing values of their hfcc (129 < 215 < 308 < 440).[49b]

Simple trialkylsilyl radicals were proved to be pyramidal; however, the extent of their pyramidality depends on the steric bulk of their substituents. Thus, on going from Me_3Si• to Et_3Si• and *t*-Bu_3Si• radicals the geometry around the silicon radical centers becomes more and more planar (^{29}Si hfcc in mT): Me_3Si• (18.1, 18.3),[6,50] Et_3Si• (17.0),[51] *t*-Bu_3Si• (16.3).[52] An interesting example of the bicyclic bridgehead silyl radical $HC(CH_2)_3Si$• was reported as being pyramidal, a conclusion that was deduced from its small β-hydrogen hfcc of 0.15 mT.[53]

The introduction of electropositive silyl groups at the silicon radical centers results in a remarkable decrease in the ^{29}Si hfcc values upon increased silyl substitution, which implies remarkable planarity of the silyl-substituted silyl radicals (hfcc in mT): $Me_2(Me_3Si)Si\bullet$ (13.7), $Me(Me_3Si)_2Si\bullet$ (7.1), $(Me_3Si)_3Si\bullet$ (6.5).[54]

Aryl- and alkylgermyl radicals were also proved to be pyramidal. Thus, the geometries of phenyl-substituted Ge-centered radicals $Ph_3Ge\bullet$, $Ph_2MeGe\bullet$ and $PhMe_2Ge\bullet$ were reported as nonplanar, as deduced from the values of their α-CH_3 hfcc.[55] The EPR conformational analysis of the series of neopentyl $Me_{3-n}(t\text{-}BuCH_2)_nGe\bullet$ ($n = 1$–3) and (trimethylsilyl)methyl $Me_{3-n}(Me_3SiCH_2)_nGe\bullet$ ($n = 1$–3) germyl radicals revealed a nonequivalence of methylene protons due to the nonplanar geometry and slow inversion about the radical centers.[56] Likewise, the tris(2-methyl-2-phenylpropyl)germyl radical $(PhMe_2CCH_2)_3Ge\bullet$, generated by the reaction of the t-BuO• radical with R_3GeH, displayed nonequivalence of the methylene protons at $-120\,°C$ because of the hindered rotation around the Ge–C bonds.[57] This implies the existence of the $(PhMe_2CCH_2)_3Ge\bullet$ radical in the structural form of a rigid pyramid on the EPR timescale at low temperature.

The tris(2,4,6-triisopropylphenyl)stannyl radical $(2,4,6\text{-}i\text{-}Pr_3\text{-}C_6H_2)_3Sn\bullet$, spontaneously generated upon dissolving the distannane R_3Sn–SnR_3 in deoxygenated toluene, revealed a signal with a hyperfine ^{119}Sn splitting of 163 mT ($-140\,°C$, toluene matrix).[58] Judging from this value of the ^{119}Sn hfcc, the radical is more planar than the triphenylstannyl radical $Ph_3Sn\bullet$ (186.6 mT), but marginally more pyramidal than the trimethylstannyl radical $Me_3Sn\bullet$ (161.1 mT).

2.2.2.2.2 *CIDNP Spectroscopy*

Although chemically induced dynamic nuclear polarization (CIDNP) is an important alternative to the well-known EPR technique for direct spectroscopic studies of free radical reactions, it is generally much less familiar to organometallic chemists than its EPR counterpart. However, given the very high sensitivity of the CIDNP technique, which is well suited for investigations of short-lived (from nano- to microseconds) paramagnetic species, one should realize that CIDNP represents one of the most powerful spectroscopic methods for the study of radical reactions.[59] When such reactions are carried out directly in the probe of an NMR spectrometer, the CIDNP effects manifest themselves in the form of anomalous NMR resonances (enhanced absorption or emission signals) observed within the time of nuclear relaxation of the diamagnetic molecule (from several seconds to several minutes). The nonequilibrium population leading to an observation of unusual NMR lines is generated as a result of electron–nuclear interaction in the radical pair. Such radical pairs could be formed by the thermal/photolytic/radiolytic homolysis of the starting material, SET process between the donor and acceptor molecules, or occasional radical encounters in the bulk preceding the recombination.

The analysis of CIDNP effects observed in a radical pair provides invaluable information about the structure of short-lived paramagnetic species (either neutral or charged), molecular dynamics in the radical pair, and multiplicity of the reacting states. The observation of CIDNP effects gives clear evidence for the involvement of radical steps in the overall reaction pathway, although the question of whether the radical path is a major or a minor process is still ambiguous. The qualitative analysis of CIDNP effects can usually be done using Kaptein's rule.[60] According to this rule, the sign of the net CIDNP effect (Γ_N) observed in a high magnetic field is defined as the result of multiplication of the following parameters: $\Gamma_N = \mu \bullet \varepsilon \bullet \Delta g \bullet A$, where μ is the multiplicity of the radical pair precursor ('+' for a triplet and '−' for a singlet), ε is '+' for 'in-cage' and '−' for 'escape' recombination products, Δg is the sign of the difference in g-values of the radical with polarized nucleus and its radical partner in the radical pair, and A is the sign of the hfcc of the nucleus under study in the radical. The sign of Γ_N finally determines the phase of the NMR resonance of the nucleus under study: '+' for enhanced absorption (A) and '−' for emission (E). The CIDNP spectroscopy technique was successfully applied in the study of radical steps in a variety of group 14 organometallic transformations: reaction of benzyl chloride with Et_3GeM (M = Li, Na, K), photolysis of α-germyl– and α-stannyl ketones, homolytic addition of bromotrichloromethane to allyltriorganogermanes and -stannanes, homolytic substitution reactions, and reactions involving generation of transient germylenes and digermenes. All of these radical transformations were covered in a recent review by Taraban *et al.*, to which we direct interested readers.[59]

2.2.2.3 *Theoretical Calculations*

The parent species H_3E• (E = Si, Ge, Sn) are the most intensively computationally studied free radicals centered on the heavy group 14 elements. The halogen- and methyl-substituted derivatives of the type X_3E• and Me_3E• (X = F, Cl; E = Si, Ge, Sn) have also been the focus of recent theoretical investigations.

In complete accordance with the experimental results, a number of sophisticated calculations at different levels of theory have reliably established that the parent radicals H_3E• are pyramidal in their ground states, in sharp contrast to the methyl radical H_3C• which is well known to be planar. Thus, a systematic DFT study at the NL-SCF/TZ2P level disclosed the following trend in the changes of H–E–H bond angles along the series H_3E• (E = C, Si, Ge, Sn): 120.00° (H_3C•), 112.66° (H_3Si•), 112.44° (H_3Ge•) and 110.56° (H_3Sn•); that is, a steady increase in the degree of pyramidalization was observed descending group 14.[61] Accordingly, the inversion barriers of the pyramidal C_{3v} form of H_3E• species (through its planar D_{3h} form) also steadily increase going down group 14: 0.0 kcal/mol (H_3C•), 3.7 kcal/mol (H_3Si•), 3.8 kcal/mol (H_3Ge•), 7.0 kcal/mol (H_3Sn•).[61] This phenomenon has been typically explained in terms of the second-order Jahn–Teller effect, namely, by the mixing of energetically close ground (nonbonding SOMO) and excited (antibonding LUMO: E–H σ*-orbital) states, stabilizing the ground state by lowering its energy level and resulting in the pyramidalization at the E-radical center. The effect of pyramidalization is more pronounced for bigger and more electropositive heavy group 14 elements, because of a marked decrease in the SOMO–LUMO energy gap due to the raising of the SOMO and lowering of LUMO energy levels.

An alternative view, stressing the significance of steric effects on the overall geometry of H_3E• radical species, has been recently presented.[61] The authors concluded that H_3C• is planar mainly because of the great steric repulsion between the H atoms. In the particular case of H_3C•, such a steric effect outbalances the electronic effect, which always favors a pyramidal structure. For the heavy analogs of the methyl radical, H_3E• (E = Si, Ge, Sn), the longer E–H bonds decrease the steric repulsion between the hydrogen substituents and, accordingly, increase the degree of pyramidalization at the E radical center.

The influence of substituents R on the configuration of free radicals R_3E• was also studied computationally. Thus, the Me_3Si• radical was found to be pyramidal at the UHF/6-21G level with an inversion barrier of 13.3 kcal/mol,[62] whereas the carbon analog Me_3C• was essentially planar at the UHF/4-31G level with a negligible inversion barrier of 1.2 kcal/mol.[63] Two reasons are responsible for this striking difference: the marked decrease in the SOMO–LUMO energy gap for the Me_3Si• radical stabilizing its SOMO upon pyramidalization (electronic effect), and a decrease of the van der Waals repulsion of Me substituents on going from Me_3C• to Me_3Si• radicals because of the longer Si–C bonds (steric effect).[62] Replacement of hydrogen atoms in the H_3Si• radical successively with Me and F substituents results in the progressive increase in the degree of pyramidalization and inversion barriers calculated at the UHF/3-21G level: 5.1 kcal/mol (H_3Si•), 6.3 kcal/mol (H_2MeSi•), 13.4 kcal/mol (H_2FSi•).[62] Further increase in the degree of halogen substitution at the radical center of R_3E• causes even more severe pyramidalization and a sharp rise in the inversion barriers (see also Section 2.1). Thus, in the series of perhalosubstituted radicals of the type X_3E• (X = F, Cl; E = C, Si, Ge) both pyramidalization and inversion barriers markedly increase upon descending group 14 (X–E–X bond angles, inversion barriers, computational level): F_3C• (27.4 kcal/mol),[64] F_3Si• (106.44°, 68.1 kcal/mol, UHF/3-21G level),[62] Cl_3C• (116.1°, 2.97 kcal/mol, UHF/STO-3G level),[65] Cl_3Si• (108.0°, 39.40 kcal/mol, UHF/STO-3G level),[66] Cl_3Ge• (107.6°, 56.13 kcal/mol, UHF/STO-3G level).[66] However, subsequent calculations gave different estimates of the pyramidalization degree (UHF/3-21G* level) and inversion barriers [UMP2/3-21G*//UHF/6-31G*(3-21G*) levels] along the series of fluorides F_3E• and chlorides Cl_3E• (E = Si, Ge, Sn): F_3Si• (107.2°, 38.5 kcal/mol), F_3Ge• (106.1°, 28.7 kcal/mol), F_3Sn• (104.6°, 20.7 kcal/mol), Cl_3Si• (109.6°, 36.5 kcal/mol), Cl_3Ge• (108.3°, 25.6 kcal/mol), Cl_3Sn• (107.1°, 19.3 kcal/mol).[67] The general trend of an increase in pyramidalization at the heavy group element radical centers upon the increasing electronegativity of substituents was realized in terms of the above-discussed second-order Jahn–Teller effect; that is, stabilizing SOMO–LUMO orbital interactions. In sharp contrast, substitution with electropositive silyl groups causes an appreciable flattening of the silyl radical centers: the Si–Si–Si bond angle in the $(H_3Si)_3Si$• radical was estimated as 114.80° at the UMP2/DZP level. This value is intermediate between the two extremes, represented by the highly pyramidal F_3Si• (108.14°) and Cl_3Si• (109.51°) radicals and the ideal sp^2-type trigonal-planar radical configuration (120°).[68]

2.2.3 Synthetic Applications

Although many free radicals centered on the group 14 elements were claimed to be important reactive intermediates, whose participation in a number of organometallic reactions was more or less reliably proved, only two of them have received particular attention as synthetically useful species. These are the tri(*n*-butyl)stannyl n-Bu_3Sn• and tris(trimethylsilyl)silyl $(Me_3Si)_3Si$• radicals, promoting a vast majority of the radical chain reactions and are generated from the tri(*n*-butyl)stannyl hydride n-Bu_3SnH and tris(trimethylsilyl)silane $(Me_3Si)_3SiH$, respectively. Thus, the stannyl hydride n-Bu_3SnH is one of the most widely used mediators for the highly selective reduction of functional groups and formation of C–C bonds by either inter- or intramolecular cyclizations, typically proceeding in high yields.[69] There are, however, some drawbacks to using such a radical reagent, the most significant one being the incomplete removal of toxic tin by-products from the target compounds. Therefore, the silyl hydride $(Me_3Si)_3SiH$ was recently suggested as a valuable synthetic alternative to tri(*n*-butyl)stannyl hydride in many radical chain reactions.[2d,e,g,i] Both *n*-Bu_3SnH and $(Me_3Si)_3SiH$ were found to be very efficient in the reduction of a variety of functional groups: halides, chalcogen groups, thiono esters, isocyanides, etc. Aspects of the application of both *n*-Bu_3SnH and $(Me_3Si)_3SiH$ in organic synthesis with a number of practical examples have been thoroughly discussed in recent reviews,[2d,e,g,i,69] to which we direct interested readers. Below, we briefly mention some recent examples of the synthetic utilization of these radical reagents: *n*-Bu_3SnH, as a reagent for intramolecular cyclization of allylic propiolates,[70] intramolecular cyclization of enynes,[71] stannylformylation of 1,6-dienes accompanied by ring closure,[72] desulfonylation of heterocyclic α-sulfones,[73] diastereoselective 1,4-addition to α,β-unsaturated (−)-8-phenylmenthyl ester,[74] addition to ynals to produce 2-iodomethylene cyclopentanols and cyclohexanols,[75] and radical cyclization of oxime ethers;[76] $(Me_3Si)_3SiH$, as a reagent for reduction of carbonyl groups,[77] *Z–E* isomerization of some alkenes through the addition–elimination mecahnism,[78] and addition to β-alkenyloxyenones to form tetrahydrofuranyl systems.[79]

2.3 Persistent Radicals (Generation and Identification)

Free radicals are typically classified as *persistent*, if their lifetime is long enough[3a] to allow their preparation and direct EPR characterization at room temperature.[80] It should be noted that such lifetimes can vary broadly from several minutes to several months, depending on the substitution pattern at the radical center. In the present chapter we will use the above-described terminology for persistent radicals as radical species having relatively long lifetimes. However, those persistent radicals that can be isolated as individual compounds that are indefinitely stable at room temperature in the absence of air and moisture, and that can be even structurally characterized by X-ray crystallography, will be specifically named as *stable* radicals (the chemistry of *stable* radicals of heavy group 14 elements will be discussed in Section 2.4).

Many persistent radicals featuring an unpaired electron residing on a heavy group 14 element are currently known.[3a] In this section we will deal with only the most remarkable of them: the first representatives synthesized by Lappert *et al.*, and the most recent examples of persistent persilyl-substituted silyl radicals prepared by the groups of Matsumoto and Apeloig.

The first landmark syntheses of persistent radicals of heavy group 14 elements were published more than 30 years ago by Lappert *et al.*[35,81] The first member in this series, tris[bis(trimethylsilyl)methyl]stannyl radical $[(Me_3Si)_2CH]_3Sn\bullet$, was generated by the photolysis of the stable stannylene $[(Me_3Si)_2CH]_2Sn$: (see Chapter 4, Section 4.6.1.3) with visible light in benzene (Scheme 2.4).[35,81b]

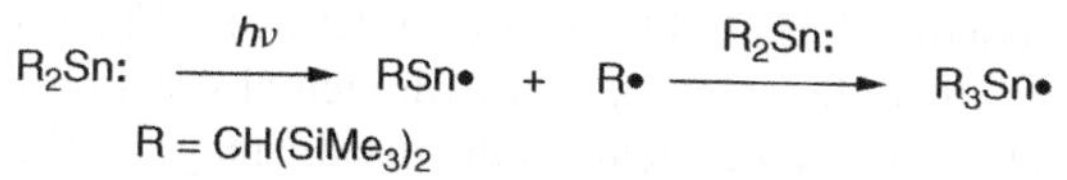

Scheme 2.4

This reaction pathway was explained assuming the initial photoinduced disproportionation of the stannylene $[(Me_3Si)_2CH]_2Sn$: into a pair of radicals $[(Me_3Si)_2CH]Sn$• and $(Me_3Si)_2CH$•, followed by the reaction of the starting stannylene $[(Me_3Si)_2CH]_2Sn$: with the stannyl radical $[(Me_3Si)_2CH]Sn$•, increasing the coordination number of tin from 2 to 3. The EPR spectrum of the $[(Me_3Si)_2CH]_3Sn$• radical revealed a central quartet ($g = 2.0094$) with a characteristic hyperfine coupling of the unpaired electron with the three equivalent protons of the $(Me_3Si)_2CH$ substituents. A pair of $^{119,117}Sn$ satellite signals with hfcc values of 169.8 and 177.6 mT was observed at higher gains. The magnitudes of these values, which were quite comparable with those of the transient Sn-centered radicals (see Section 2.2.2.2), suggest the appreciable pyramidality of the $[(Me_3Si)_2CH]_3Sn$• radical. This radical was proved to be remarkably persistent in benzene solution in the absence of air: no decrease in the intensity of its EPR signal was observed for more than a month at room temperature.

Similarly, a series of alkyl- or amino-substituted germyl and stannyl radicals R_3E• [R = $CH(SiMe_3)_2$, $N(SiMe_3)_2$; E = Si, Ge] was generated by the photolysis of the stable heavy carbene analogs R_2E: (see Chapter 4, Section 4.6) in benzene or hexane solutions at room temperature (Scheme 2.5, A).[81a,b] The silicon versions R_3Si• were prepared by the reaction of Si_2Cl_6 with six equivalents of RLi in benzene followed by photolysis of the reaction mixture (Scheme 2.5, B).[81a,b]

$$2\,R_2E{:} \xrightarrow{h\nu} R_3E\bullet + RE\bullet \qquad \textbf{(A)}$$

$$(R = CH(SiMe_3)_2,\ N(SiMe_3)_2;\ E = Si,\ Ge)$$

$$Si_2Cl_6 + 6\,RLi \xrightarrow{h\nu} R_3Si\bullet + (RSiCl_2)_2 + SiCl_4 + RSi\bullet \qquad \textbf{(B)}$$

Scheme 2.5

The EPR spectra of R_3E• radicals exhibited the anticipated splitting patterns: for example, the $[(Me_3Si)_2CH]_3Ge$• radical revealed a decet (coupling with ^{73}Ge nucleus, $I = 9/2$) of septets (coupling with the three equivalent ^{14}N nuclei, $I = 1$).[81a]

All of Lappert's radicals were rather stable in solution at room temperature (g-value, hfcc in mT, half-life time): $[(Me_3Si)_2CH]_3Si$• [2.0027, $a(^{29}Si) = 19.3$, ~10 min], $[(Me_3Si)_2CH]_3Ge$• [2.0078, $a(^{73}Ge) = 9.2$, at least 4 months], $[(Me_3Si)_2CH]_3Sn$• [2.0094, $a(^{117}Sn) = 169.8$ and $a(^{119}Sn) = 177.6$, ~1 year], $[(Me_3Si)_2N]_3Ge$• [1.9991, $a(^{73}Ge) = 17.1$, more than 5 months], $[(Me_3Si)_2N]_3Sn$• [1.9912, $a(^{117}Sn) = 317.6$ and $a(^{119}Sn) = 342.6$, ~3 months].[81a] Such remarkable persistence was attributed to the great steric bulkiness of the $(Me_3Si)_2CH$ or $(Me_3Si)_2N$ substituents, preventing the dimerization of radical species $[(Me_3Si)_2CH]_3E$• and $[(Me_3Si)_2N]_3E$•.[81a] The other factor contributing to the overall stabilization of Lappert's radicals is the low strength of the resulting E–H bonds (C–H: 104 kcal/mol; Si–H: 81 kcal/mol; Ge–H: 73 kcal/mol; Sn–H: 70 kcal/mol),[82] which disfavors radical hydrogen abstraction from the solvent.[81b] A third stabilizing factor for R_3E• radicals is the disadvantage of the disproportionation pathway typical for carbon-centered species, $R_3E\bullet \rightarrow R_2E{:} + R\bullet$, to form finally doubly bonded derivatives $R_2E{=}ER_2$.[35]

The large values of the hfccs are indicative of the pyramidal structures of all of the above-described persistent radicals. The increase in pyramidality of amino-substituted radicals $[(Me_3Si)_2N]_3E\bullet$ (E = Ge, Sn), compared with that of alkyl-substituted radicals $[(Me_3Si)_2CH]_3E\bullet$ (E = Si, Ge, Sn), is in accord with the greater electronegativity of N compared with that of C. However, the nonplanarity of amino-substituted radicals $[(Me_3Si)_2N]_3E\bullet$ (E = Ge, Sn) is less pronounced than that of $Cl_3Ge\bullet$ [$a(^{73}Ge)$ = 22.9 mT].[83] Thus, all persistent radicals of the type $R_3E\bullet$ [R = $CH(SiMe_3)_2$, $N(SiMe_3)_2$; E = Si, Ge, Sn] feature remarkable pyramidal geometry, being intermediate between the planar $CH_3\bullet$ and nearly tetrahedral $X_3E\bullet$ (X = F, Cl; E = Si, Ge, Sn) radicals, a tendency that is consistent with the increasing electronegativity difference between the central atom and the substituents.

A series of silyl radicals was recently synthesized by Matsumoto *et al.* Thus, the highly persistent tris(trialkylsilyl)silyl radicals $(Et_nMe_{3-n}Si)_3Si\bullet$ ($n = 1-3$) were generated by one of three methods: (1) hydrogen abstraction from hydrides $(Et_nMe_{3-n}Si)_3SiH$ by t-BuO• radical upon their photolysis in the presence of $(t\text{-BuO})_2$; (2) Si–Si bond breaking upon the photolysis of tetrakis(trialkylsilyl)silanes $(Et_nMe_{3-n}Si)_4Si$; and (3) central Si–Si bond breaking upon the photolysis of hexakis(trialkylsilyl)disilanes $(Et_nMe_{3-n}Si)_3Si–Si(SiMe_{3-n}Et_n)_3$ (Scheme 2.6).[84]

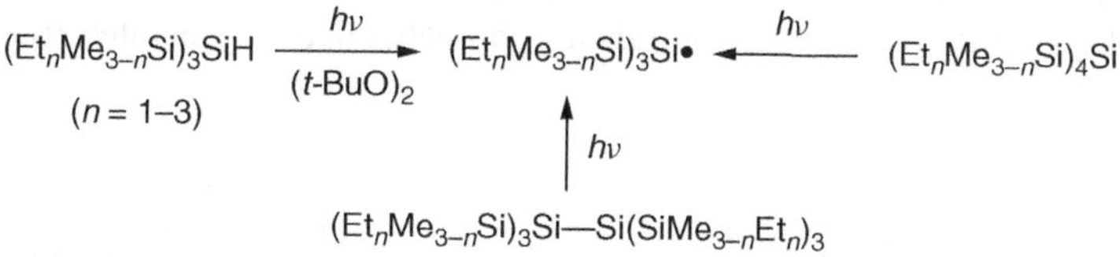

Scheme 2.6

All $(Et_nMe_{3-n}Si)_3Si\bullet$ radicals exhibited similar EPR spectra featuring two pairs of satellites from the coupling of the unpaired electron with the α-^{29}Si nucleus (larger hfcc) and β-^{29}Si nuclei (smaller hfcc). The EPR spectral parameters of these radicals, in comparison with the previously reported $(Me_3Si)_3Si\bullet$ radical, are given below [T in °C, g-value, $a(\alpha\text{-}^{29}Si)$ in mT, $a(\beta\text{-}^{29}Si)$ in mT]: $(Me_3Si)_3Si\bullet$ (−25, 2.0053, 6.38, 0.71);[85] $(EtMe_2Si)_3Si\bullet$ (15, 2.0060, 6.28, 0.71); $(Et_2MeSi)_3Si\bullet$ (15, 2.0060, 6.03, 0.73); $(Et_3Si)_3Si\bullet$ (15, 2.0063, 5.72, 0.79).[84] Clearly, upon increase in the steric bulk at the Si radical center with the increasing number of Et groups, silyl radicals become more and more planar [on going from $(Me_3Si)_3Si\bullet$ to $(Et_3Si)_3Si\bullet$], which is manifested in a progressive decrease in their $a(\alpha\text{-}^{29}Si)$ values. By the same token, delocalization of the radical spin density over the σ^*-orbitals of the Si–C bonds of the silyl substituents becomes more important in the same progression. In contrast to the less bulky $(Me_3Si)_3Si\bullet$ radical, whose EPR spectrum can be observed only at −25 °C,[85] all $(Et_nMe_{3-n}Si)_3Si\bullet$ radicals revealed a remarkable stability in solution at room temperature because of the significant steric shielding effect of the Et groups. The half-life time of these radicals depends on the generation method used. The most long-lived silyl radicals were formed from the hexakis(trialkylsilyl)disilanes: 3 h for $(EtMe_2Si)_3Si\bullet$, 1 day for $(Et_2MeSi)_3Si\bullet$, and 1.5 months for $(Et_3Si)_3Si\bullet$.[84]

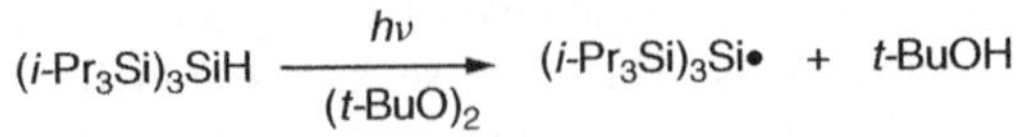

Scheme 2.7

Further increasing the steric bulkiness of the substituents, the same authors were able to prepare $(i\text{-Pr}_3\text{Si})_3\text{Si}\bullet$ by hydrogen abstraction from the corresponding hydride $(i\text{-Pr}_3\text{Si})_3\text{SiH}$ by the t-BuO• radical, photochemically generated from $(t\text{-BuO})_2$ (Scheme 2.7).[86]

The most interesting feature of the radical $(i\text{-Pr}_3\text{Si})_3\text{Si}\bullet$ is the remarkable planarity at the silyl radical center, which was deduced from its EPR parameters: $g = 2.0061$, $a(\alpha\text{-}^{29}\text{Si}) = 5.56\,\text{mT}$. The new radical $(i\text{-Pr}_3\text{Si})_3\text{Si}\bullet$ was even more persistent than the above-described $(\text{Et}_n\text{Me}_{3-n}\text{Si})_3\text{Si}\bullet$ radicals generated by the same method: its half-life at 15 °C was 5 days.[84]

A further member of the persistent persilyl silyl radicals family, namely the tris(*tert*-butyldimethylsilyl)silyl radical $(t\text{-BuMe}_2\text{Si})_3\text{Si}\bullet$, was independently generated by one of the following three methods: (1) one-electron oxidation of $(t\text{-BuMe}_2\text{Si})_3\text{SiNa}$ with either $\text{NO}^+\bullet\text{BF}_4^-$ or $\text{Ph}_3\text{C}^+\bullet\text{BPh}_4^-$; (2) one-electron reduction of $(t\text{-BuMe}_2\text{Si})_3\text{SiBr}$ with an equimolar amount of Na; and (3) hydrogen abstraction from the hydrosilane $(t\text{-BuMe}_2\text{Si})_3\text{SiH}$ by the t-BuO• radical, photochemically generated from $(t\text{-BuO})_2$ (Scheme 2.8).[87]

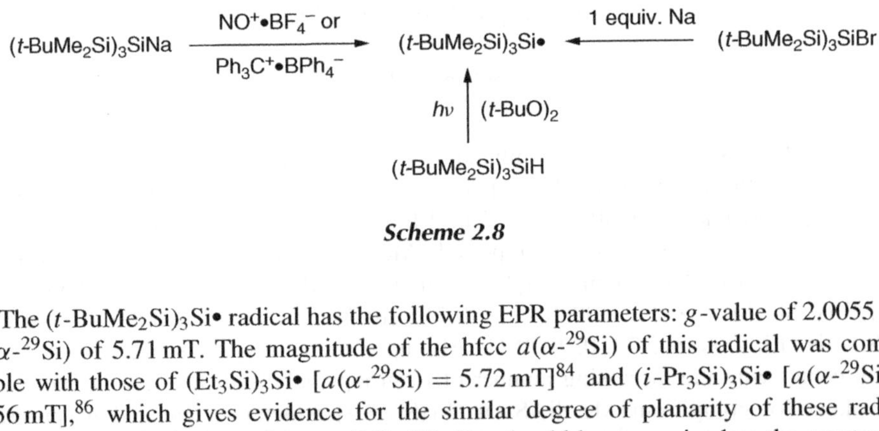

Scheme 2.8

The $(t\text{-BuMe}_2\text{Si})_3\text{Si}\bullet$ radical has the following EPR parameters: g-value of 2.0055 and $a(\alpha\text{-}^{29}\text{Si})$ of 5.71 mT. The magnitude of the hfcc $a(\alpha\text{-}^{29}\text{Si})$ of this radical was comparable with those of $(\text{Et}_3\text{Si})_3\text{Si}\bullet$ [$a(\alpha\text{-}^{29}\text{Si}) = 5.72\,\text{mT}$][84] and $(i\text{-Pr}_3\text{Si})_3\text{Si}\bullet$ [$a(\alpha\text{-}^{29}\text{Si}) = 5.56\,\text{mT}$],[86] which gives evidence for the similar degree of planarity of these radical species, although the last of these, $(i\text{-Pr}_3\text{Si})_3\text{Si}\bullet$, should be recognized as the most sterically crowded and, consequently, the most planar. The half-life of the $(t\text{-BuMe}_2\text{Si})_3\text{Si}\bullet$ radical depends on its generation conditions: ca. 1 day at 15 °C (generated by either oxidation or reduction reactions) and 10 h at 15 °C (generated by the hydrogen abstraction reaction).[87]

A family of persistent silyl-substituted silyl radicals was reported by Apeloig *et al*. The tris(pentamethyldisilyl)silyl radical $(\text{Me}_3\text{SiMe}_2\text{Si})_3\text{Si}\bullet$ was generated from various precursors: from the hydride $(\text{Me}_3\text{SiMe}_2\text{Si})_3\text{SiH}$ by hydrogen abstraction with $t\text{-Bu}_2\text{Hg}$, from the dimercury derivative $(\text{Me}_3\text{SiMe}_2\text{Si})_3\text{Si–Hg–Hg–Si}(\text{SiMe}_2\text{SiMe}_3)_3$

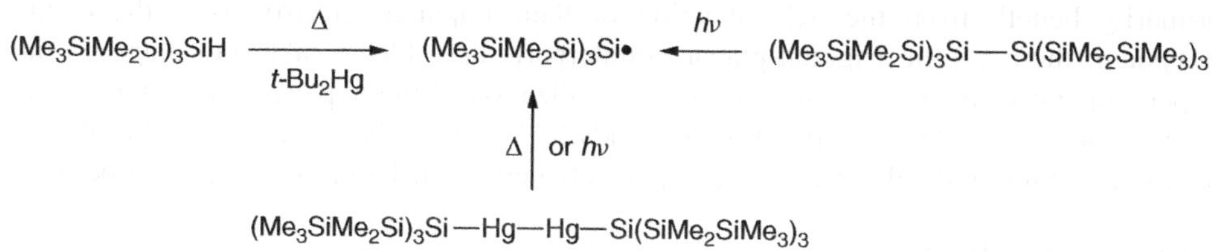

Scheme 2.9

by its thermolysis or photolysis, and from hexakis(pentamethyldisilyl)disilane $(Me_3SiMe_2Si)_3Si–Si(SiMe_2SiMe_3)_3$ by its photolysis (Scheme 2.9).[88]

The EPR spectrum of the silyl radical $(Me_3SiMe_2Si)_3Si•$ is represented by a central signal ($g = 2.0065$) accompanied by two pairs of satellites from α- and β-^{29}Si nuclei. The small value of $a(\alpha\text{-}^{29}Si)$ hfcc of 5.99 mT suggests an essentially planar configuration around the radical center, a conclusion that was further supported by computations. Thus, the Si–Si–Si bond angle in the optimized structure of the $(Me_3SiMe_2Si)_3Si•$ radical, calculated at the UB3LYP//6-31G*//UB3LYP/6-31G* level, was found to be 118.0° (cf. 120° for the ideal trigonal-planar sp^2-hybridized radical), whereas the same angle for the model $(H_3Si)_3Si•$ radical was estimated as 115.9°. Thus, comparing the magnitudes of the hfcc $a(\alpha\text{-}^{29}Si)$ values one should conclude that the $(Me_3SiMe_2Si)_3Si•$ radical (5.99 mT)[88] is more planar than the $(Me_3Si)_3Si•$ radical (6.38 mT)[85] but less planar than the $(i\text{-}Pr_3Si)_3Si•$ radical (5.56 mT).[86] Accordingly, the kinetic stability markedly increases in the order $(Me_3Si)_3Si• < (Me_3SiMe_2Si)_3Si• < (i\text{-}Pr_3Si)_3Si•$: the first radical species can be observed only at temperatures as low as −25 °C,[85] whereas the second has a half-life of 6 min at room temperature,[88] and the last has a half-life of 5 days at 15 °C.[86]

A series of rather unusual metal(Li, Hg)-substituted silyl radicals was also recently reported by Apeloig *et al.* [g-value, $a(\alpha\text{-}^{29}Si)$ in mT]: $Li(i\text{-}Pr_3Si)_2Si•$ (2.0073, 3.2);[89] $(thf)Li(t\text{-}Bu_2MeSi)_2Si•$ (2.0067, 3.33);[90] $t\text{-}BuHg–(t\text{-}Bu_2MeSi)_2Si–Hg–(t\text{-}Bu_2MeSi)_2Si•$ (1.984, 5.60);[91] $H(t\text{-}Bu_2MeSi)_2Si–Hg–(t\text{-}Bu_2MeSi)_2Si•$ (1.983, 5.60);[91] $(thf)Li–(t\text{-}Bu_2MeSi)_2Si–Hg–(t\text{-}Bu_2MeSi)_2Si•$ (1.984, 5.60).[91]

2.4 Stable Radicals

The groundbreaking developments of sophisticated experimental techniques during the last decade has enabled the isolation of stable representatives of free radical species of the heavy group 14 elements; that is, silyl, germyl, stannyl and plumbyl radicals. Such important progress in the isolation of stable derivatives opened unprecedented possibilities for their structural characterization in the crystalline form, which in turn allowed the direct comparison of the fundamental differences and similarities between the solution and solid-state structures of the free radical species.[3] All of the stable free radicals of the heavy group 14 elements can be classified into two large groups: neutral radicals and charged radicals (ion-radicals). Each of these two classes can be further subdivided into the two subclasses of cyclic and acyclic radicals with significantly distinctive physical and chemical properties. Thus, if the cyclic radicals can primarily benefit from the delocalization of their unpaired electron over the cyclic π-system, their acyclic trigonal-planar counterparts are able to take advantage of the hyperconjugative interaction through the delocalization of the unpaired electron over the antibonding σ^*-orbitals of neighboring bonds.[3b–e] All of these classes of the above-mentioned stable radicals of heavy group 14 elements will be briefly discussed below.

2.4.1 Neutral Radicals

2.4.1.1 Cyclic Radicals

The first radical derivative of this type, cyclotrigermenyl radical **1**, was reported by Power *et al*. more than a decade ago.[92] Cyclic radical **1**, having very large protective 2,6-Mes_2–C_6H_3 groups, was prepared by the reduction of chlorogermylene Cl(2,6-Mes_2–C_6H_3)Ge: with KC_8 in THF (Scheme 2.10).[92] The structure of **1**, featuring an unpaired electron placed on one of the Ge atoms and a double bond between the other two Ge atoms, represents an all-Ge version of the organic cyclopropenyl radical.

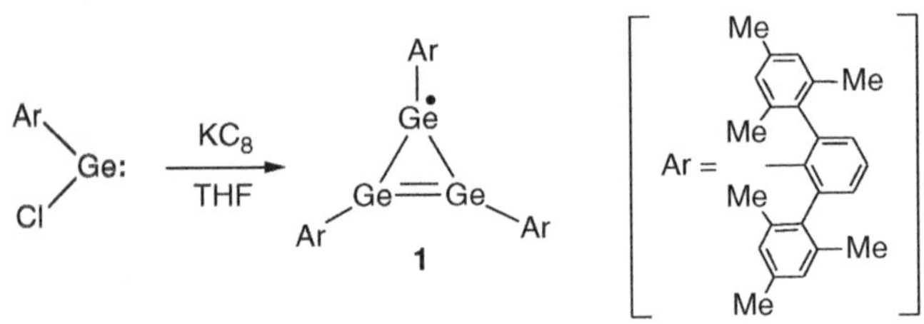

Scheme 2.10

The solution structure of **1** was deduced based on its EPR spectrum, which manifested a single resonance ($g = 2.0069$) with the hfcc $a(^{73}Ge) = 1.6\,mT$. Such a small value for the hfcc is indicative of the localization of an unpaired electron in an orbital with predominantly p-character, which implies the sp^2-hybridization at the Ge radical center and, consequently, the planar geometry around it.

The cyclotetrasilenyl radical **2**, a silicon variation of the cyclobutenyl radical, was recently reported by Sekiguchi *et al*. The synthesis of **2** was accomplished by the one-electron reduction of its cationic precursor, cyclotetrasilenylium ion **3$^+$•TPFPB$^-$**[93] (see Chapter 6, Section 6.3.3) with either t-Bu_3SiNa or KC_8 in diethyl ether (Scheme 2.11).[94]

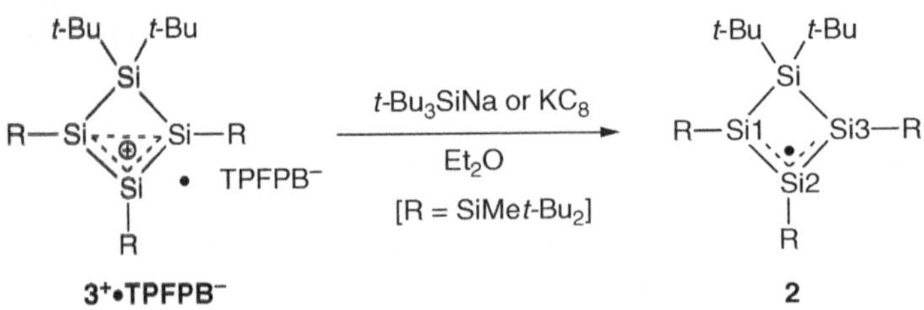

Scheme 2.11

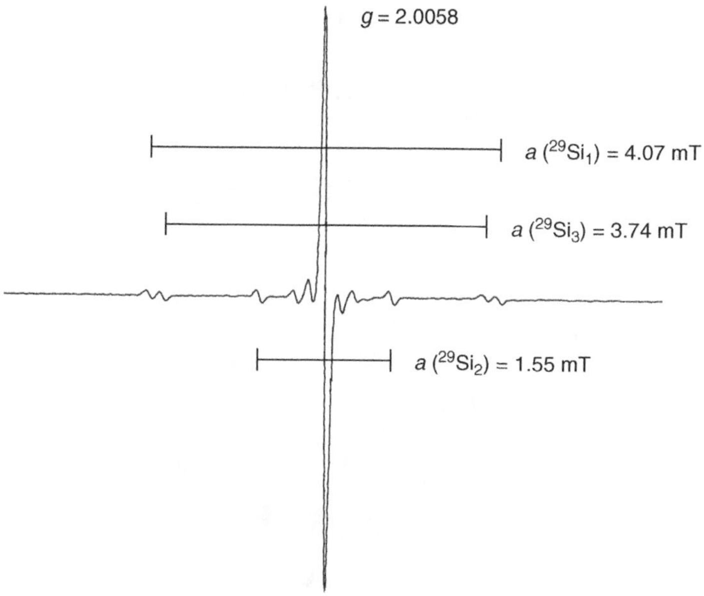

Figure 2.1 *EPR spectrum of cyclotetrasilenyl radical **2***

The four-membered Si_4-ring in **2** is nearly planar, with the Si1–Si2 and Si2–Si3 bonds being intermediate between the typical Si–Si and Si=Si bonds, thus suggesting an allylic-type structure for the cyclic radical **2**.

The EPR spectrum of **2** revealed a central signal at $g = 2.0058$ accompanied by three pairs of satellites [$a(^{29}Si) = 4.07$, 3.74 and 1.55 mT] resulting from the coupling of the unpaired electron with the two terminal (Si1 and Si3) and one central (Si2) nuclei of the delocalized allylic Si1–Si2–Si3 unit of radical **2** (Figure 2.1). The small magnitudes of these hfcc are consistent with the suggestion of the planarity for radical **2** in the liquid phase.

As the sole example of a stable bicyclic radical of heavy group 14 elements, 1,6,7-trigermabicyclo[4.1.0]hept-3-en-7-yl **4**, was recently prepared by Sekiguchi *et al.* by the one-electron oxidation of the corresponding bicyclic anion $\mathbf{5^-{\bullet}K^+}$ with $B(C_6F_5)_3$ in THF (Scheme 2.12).[95] The exclusive accommodation of the odd electron on the Ge3 atom can

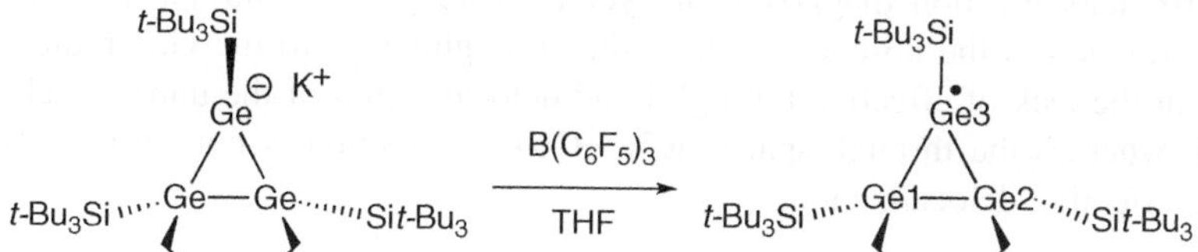

$\mathbf{5^-{\bullet}K^+}$ **4**

Scheme 2.12

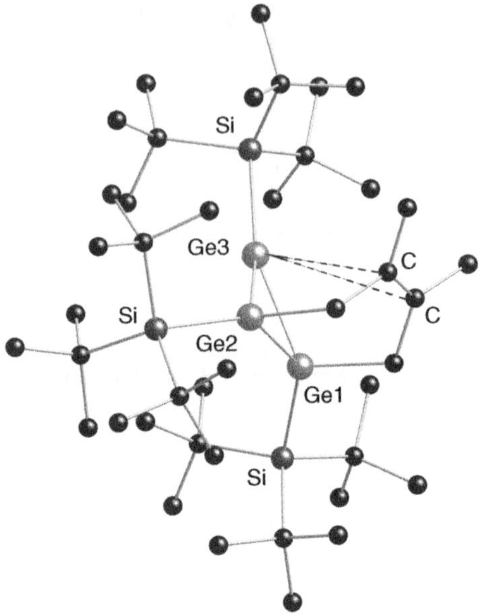

Figure 2.2 *Crystal structure of bicyclic radical* **4** *(hydrogen atoms are not shown)*

be clearly seen in the crystal structure of **4**, manifesting a diagnostically planar geometry at the tricoordinate Ge3 atom. The peculiar *endo*-conformation of **4** favors the stabilizing intramolecular through-space interaction between the SOMO(Ge3) and π(C=C) orbitals facilitated by their relative proximity: the Ge3–(C=C) distance of 3.63 Å is in the range of van der Waals interactions (Figure 2.2). Computations confirmed that the *endo*-form is marginally stabilized (in comparison with the *exo*-form) by the Ge3•---C=C interaction: a small preference of 0.59 kcal/mol in favor of the *endo*-conformer was found. However, in solution the *endo-exo* inversion at the Ge3-radical center can take place smoothly, resulting in the formation of a mixture of both conformers, which was clearly seen in the solution EPR spectrum of **4**, showing a pair of distinct signals ($g = 2.0210$ and 2.0223) of almost equal intensities.[95] The essential planarity of the Ge-radical centers in both *endo-* and *exo*-conformers was manifested in the characteristically small values of their hfcc (3.4 and 2.6 mT), testifying to the sp^2-hybridization of the Ge3 radical center.

Interestingly, bicyclic radical **4**[95] features a localized spin density, in contrast to the allylic-type delocalization diagnostic of cyclotetrasilenyl **2**[94]. The apparent reason for such a distinction is the absence of the π-bond neighboring to the Ge3 radical center, resulting in the lack of effective through-bond delocalization of the unpaired electron in radical **4**, whereas the through-space Ge3•---C=C interaction is not sufficiently strong to induce effective delocalization.

2.4.1.2 *Acyclic Tricoordinate Radicals*

The above-discussed cyclic silyl and germyl radicals have rather peculiar structural features caused by either through-bond or through-space intramolecular interaction of the unpaired electron with double bonds. The simple tricoordinate acyclic radicals of the type R_3E• (E = Si, Ge, Sn, Pb), lacking the stabilizing but highly electronically perturbing effects of cyclic π-delocalization, represent a more general and more challenging class of stable organometallic radicals.[3b–e] However, such highly desirable compounds became synthetically accessible only very recently, as a result of the right choice of most effectively stabilizing silyl substituents.

The first breakthrough in the preparation of tricoordinate R_3E• radical species lacking conjugation to π-bonds was achieved several years ago by Lee, Sekiguchi *et al.*, who reported the isolation and structural characterization of a homologous series of (t-$Bu_2MeSi)_3E$• (E = Si, Ge, Sn) radicals.[3b–e] All of these stable radicals, kinetically and thermodynamically stabilized by the bulky electropositive silyl substituents, were identically synthesized by a rather simple and straightforward method, clearly demonstrating the generality of this synthetic approach and allowing their preparation in gram-order scales.

The lightest member in the series, the silyl radical (t-$Bu_2MeSi)_3Si$• **6**, was prepared by the one-electron oxidation of the corresponding silylsodium derivative (t-$Bu_2MeSi)_3SiNa$ [generated by the reaction of (t-$Bu_2MeSi)_2SiBr_2$ with t-$Bu_2MeSiNa$) with a gentle oxidizing reagent, dichlorogermylene–dioxane complex $GeCl_2$•diox, in diethyl ether (Scheme 2.13, E = Si)].[96]

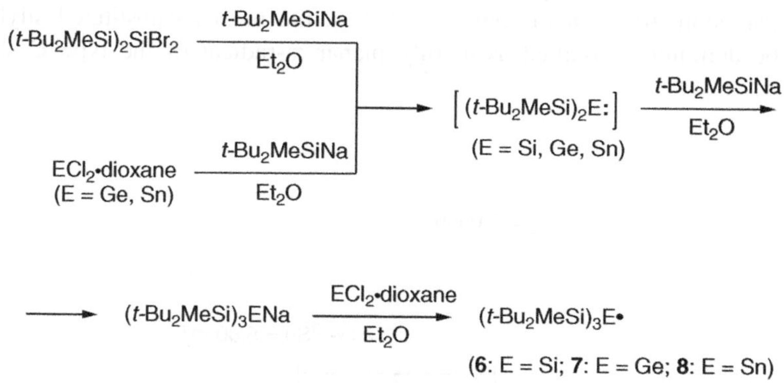

Scheme 2.13

The geometry around the radical center in **6** is perfectly trigonal-planar, implying a true sp^2-hybridization of the central Si atom and, hence, the localization of an unpaired electron in its $3p_z$-orbital (Figure 2.3).

The remarkable planarity of radical **6** was accounted for by the great bulkiness of the voluminous t-Bu_2MeSi groups, which prefer to be as far away from each other as possible to avoid their mutual steric repulsion. On the other hand, the important electron donation from the three electropositive silyl substituents results in a significant lowering of the inversion barrier at the silyl radical center, thus also promoting its planarization. The spatial arrangement of t-Bu_2MeSi substituents in radical **6** favors the hyperconjugative delocalization of the unpaired electron spin density over the antibonding σ^*-orbitals of the Si–$C_{t\text{-Bu}}$ bonds, which also contributes to the overall stabilization of the radical

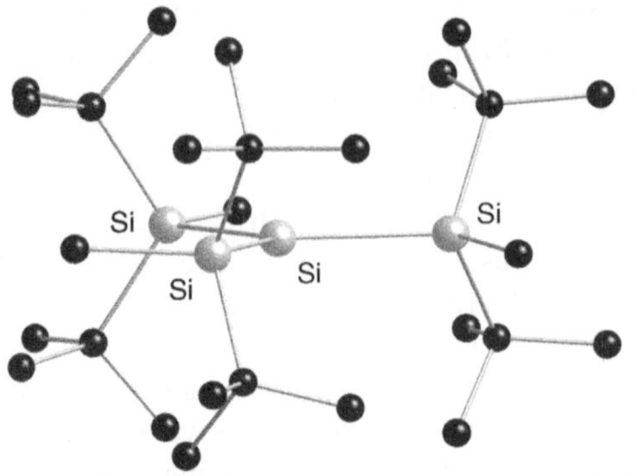

Figure 2.3 *Crystal structure of silyl radical (t-Bu_2MeSi)$_3$Si• **6** (hydrogen atoms are not shown)*

species. The solution structure of radical **6** was deduced from its EPR spectrum, which revealed a central resonance ($g = 2.0056$) along with a pair of satellites from the coupling of the odd electron with the paramagnetic ^{29}Si nuclei [hfcc: $a(\alpha\text{-}^{29}\text{Si}) = 5.80\,\text{mT}$ and $a(\beta\text{-}^{29}\text{Si}) = 0.79\,\text{mT}$] (Figure 2.4). The diagnostically small value of $a(\alpha\text{-}^{29}\text{Si})$ for radical **6** is indicative of the predominantly p-character of its SOMO, which testifies to the planar geometry of **6** in solution. That is, the persilyl-substituted silyl radical **6** should be definitely classified as a truly planar π-radical of the type **C** shown in Scheme 2.1.

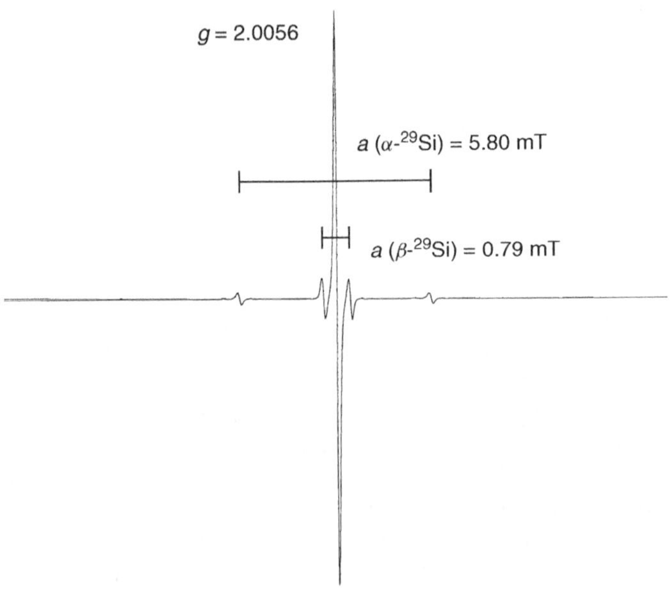

Figure 2.4 *EPR spectrum of silyl radical (t-Bu_2MeSi)$_3$Si• **6***

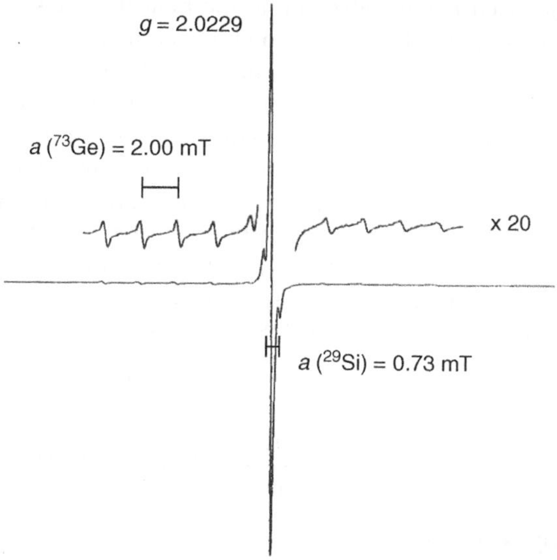

Figure 2.5 *EPR spectrum of germyl radical (t-Bu$_2$MeSi)$_3$Ge• 7*

Likewise, the germanium analog of **6**, persilyl-substituted germyl radical (t-Bu$_2$MeSi)$_3$Ge• **7**, was synthesized by a similar (even more straightforward) procedure (Scheme 2.13, E = Ge).[96] In this case, simple mixing of t-Bu$_2$MeSiNa and GeCl$_2$•diox complex in diethyl ether resulted in the formation of the intermediate germylsodium derivative (t-Bu$_2$MeSi)$_3$GeNa, which was subsequently oxidized with the second equivalent of GeCl$_2$•diox to produce the final radical **7**. The germyl radical **7** was isostructural with the above-described silyl radical **6**, featuring a trigonal-planar geometry around the radical center with the same spatial arrangement of the silyl substituents. The EPR spectrum of **7** displayed a central resonance ($g = 2.0229$) accompanied by a characteristic decet of the ^{73}Ge satellites with very small values of hfcc [$a(^{73}\text{Ge}) = 2.00\,\text{mT}$], thus pointing to a planar structure for germyl radical **7** in solution (Figure 2.5).

Utilization of the same synthetic procedure resulted in the successful synthesis of the first representative of the stable stannyl radicals, (t-Bu$_2$MeSi)$_3$Sn• **8**, prepared by the reaction of t-Bu$_2$MeSiNa with SnCl$_2$•diox in diethyl ether (Scheme 2.13, E = Sn).[97] Similar to the aforementioned silyl radical **6** and germyl radical **7**, stannyl radical **8** is perfectly trigonal-planar in the solid state. Its planarity is also retained in solution, as shown by the analysis of its EPR spectrum featuring a central signal ($g = 2.0482$) along with a pair of ^{119,117}Sn satellites with a diagnostically small hfcc [$a(^{119,117}\text{Sn}) = 32.9\,\text{mT}$]. Such a value is one order of magnitude smaller than those of all previously reported persistent stannyl radicals.

Undoubtedly, the amazing stabilbity and uniform structural and spectral behavior of the trigonal-planar π-radicals (t-Bu$_2$MeSi)E• (E = Si, Ge, Sn) **6–8** (both in the solid state and in solution) should be ascribed to the immediate impact of the bulky electropositive silyl substituents. In contrast, it is well known that simple alkyl and aryl substituents cause a highly pronounced pyramidalization at the radical centers, where the unpaired electron typically occupies an orbital with a high s-contribution (σ-radicals).[2e,f,i]

Given the intrinsic electron deficiency of the radicals ($(t\text{-Bu}_2\text{MeSi})\text{E}\bullet$ (E = Si, Ge, Sn) **6–8**, one would expect them to have high electronic demand and, consequently, enhanced reactivity toward one-electron reduction to form anionic derivatives $(t\text{-Bu}_2\text{MeSi})_3\text{E}^-\bullet\text{M}^+$[98] (see Chapter 3, Section 3.5.1) in which the octet rule will be satisfied (Scheme 2.14, **A**). Even more exciting, however, was the surprising ease of the one-electron oxidation of radicals $(t\text{-Bu}_2\text{MeSi})\text{E}\bullet$ (E = Si, Ge, Sn) **6–8** by strong Lewis acids such as $\text{Ph}_3\text{C}^+\bullet\text{B}(\text{C}_6\text{F}_5)_4^-$, thus providing straightforward access to the still more electron-deficient cationic species, namely, silylium,[99a] germylium,[99b] and stannylium[97] ion derivatives $(t\text{-Bu}_2\text{MeSi})_3\text{E}^+\bullet\text{B}(\text{C}_6\text{F}_5)_4^-$ (Scheme 2.14, **B**) (see Chapter 1, Section 1.6.2.2).

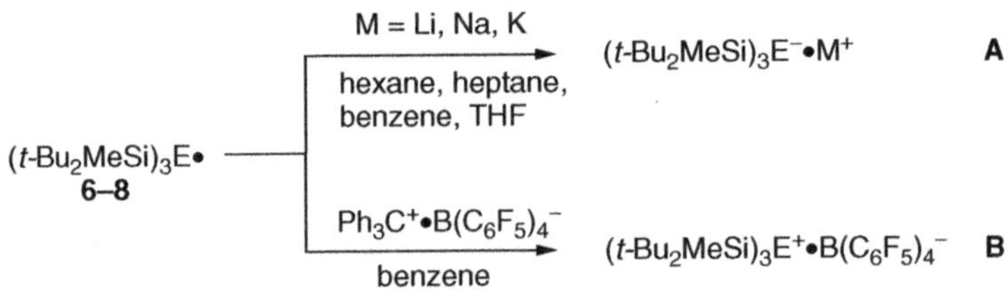

Scheme 2.14

Of particular interest is that the cations $(t\text{-Bu}_2\text{MeSi})_3\text{E}^+$ and anions $(t\text{-Bu}_2\text{MeSi})_3\text{E}^-$, generated by either oxidation or reduction of free radicals $(t\text{-Bu}_2\text{MeSi})_3\text{E}\bullet$ **6–8**, could be quantitatively reverted back to the starting **6–8** by reduction of cations or oxidation of anions, respectively, thus making a fully reversible redox system (Scheme 2.15, E = Si, Ge, Sn; R = SiMe*t*-Bu$_2$).

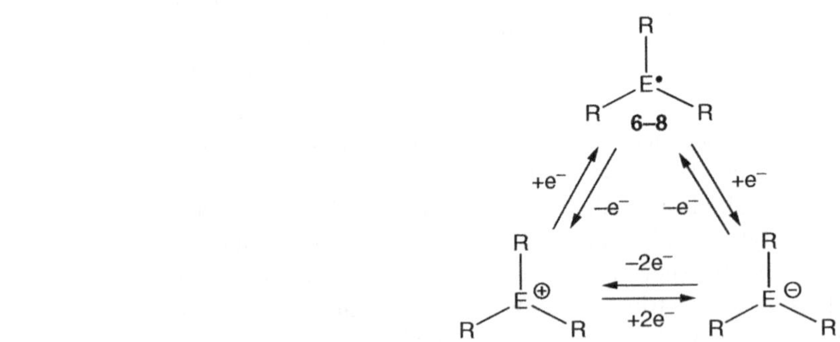

Scheme 2.15

The surprising readiness of radicals **6–8** to undergo one-electron oxidation is reasonably explained in terms of their rather low first ionization energy (IE) values. Thus, the first UV–PES bands were observed upon the gas-phase ionization of **6–8** at

*Table 2.1 Experimental PES band assignments for radicals **6**, **7** and **8** [calculated values ΔSCF (for the first IE) and TD–DFT (for the second IE) are given in parentheses (B3LYP/6-311G(d,p) (for Si and Ge radicals) and B3LYP/SDB-cc-pVTZ (for Sn radical) levels)] (all values are given in eV)*

MO	**6**	**7**	**8**
np_z(E·)(SOMO)	**6.15**	**6.0**	**5.8**
	(6.18)	(5.92)	(5.69)
σ (E–Si)	8.2	7.9	7.75
	(S 8.19; T 7.73)[a]	(S 7.99; T 7.54)[a]	(S 7.73; T 7.23)[a]
σ (Si–C)	8.75	8.8	8.8
σ (other bonds)	9.5	9.5	9.5

[a]TD–DFT calcuations of transitions to singlet (S) or triplet (T) states.

6.15 eV (for silyl radical **6**), 6.0 eV (for germyl radical **7**) and 5.8 eV (for stannyl radical **8**), corresponding to the removal of an unpaired electron from their easily accessible SOMOs/np_z(E•) (*n* – principal quantum number) (Table 2.1).[100]

The second IE values correspond to the removal of an electron from the E–Si$_{subst.}$ σ-bonds, and the trend in changes of these values clearly reflects the increase in the energy levels of the σ-orbitals of the bonds in the sequence Si–Si > Ge–Si > Sn–Si (Table 2.1). The higher energy parts of the UV–PES spectra of radicals **6–8** are nearly identical and correspond to the ionization of different σ-bonds of the molecules (Table 2.1). The first IE values for radicals **6–8** decrease on going from silyl to germyl to stannyl radicals, in complete accordance with the trend in the changes of the IE for the isolated atoms (Si 8.15 eV, Ge 7.90 eV, Sn 7.34 eV).[101] Consequently, stannyl radical **8** is the most easily oxidizable species, followed by germyl and silyl radicals **7** and **6**. Likewise, the first IEs of a series of permethylated derivatives $Me_3E–EMe_3$ (E = Si, Ge, Sn), corresponding to the removal of an electron from their HOMOs, which are largely localized on the E–E σ-bonds, decrease in the order Si–Si (8.68 eV) > Ge–Ge (8.60 eV) > Sn–Sn (8.20 eV), in accordance with the decrease in the E–E bond dissociation energy and IE of the isolated atoms.[102] Comparing the experimental first IE and calculated EA values for radicals **6–8**,[100] one can note that the steady decrease in the IE values (6.15, 6.0 and 5.8 eV) is manifested in an increase in the ease of oxidation and strengthening of reduction power in the order **6** < **7** < **8**, whereas the decrease in the EA values (1.120, 0.929 and 0.858 eV) gives evidence for a decrease in oxidation power: **6** > **7** > **8**. These tendencies are in line with those observed for the experimentally determined IE and EA values of isolated Si, Ge, and Sn atoms: 8.15, 7.90 and 7.34 eV (for IE) and 1.389, 1.233, and 1.112 eV (for EA).[101] The redox behavior of free radicals **6–8** in solution (CV) corroborates well the above-described trends in the gas phase (UV–PES). Thus, the steady decrease in the IE values on going from **6** to **8** (6.15, 6.0, 5.8 eV) is backed by the increase in the ease of oxidation in the same direction: **6** (0.40 V in *o*-dichlorobenzene, 0.80 V in THF) < **7** (0.28 V in *o*-dichlorobenzene, 0.60 V in THF) < **8** (–0.05 V in *o*-dichlorobenzene, 0.20 V in THF).[100b]

The first stable tricoordinate plumbyl radical $(Et(Me_3Si)_2Si)_3Pb$• **9** was very recently synthesized by Klinkhammer *et al.* employing the above-discussed synthetic methodology of Lee and Sekiguchi (see Scheme 2.13), namely, by the reaction of plumbylenes

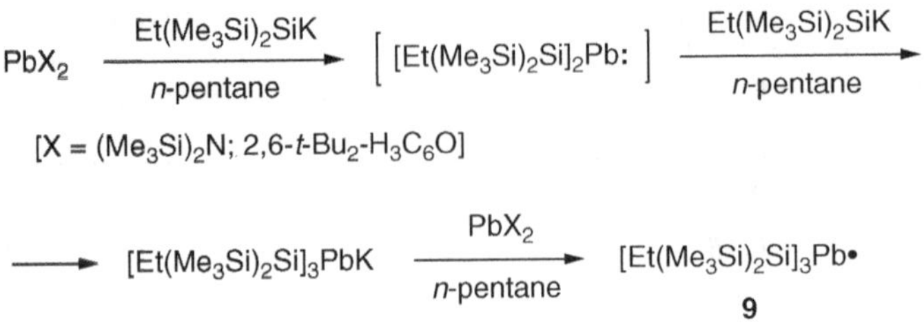

Scheme 2.16

$[(Me_3Si)_2N]_2Pb$: or $(2,6\text{-}t\text{-}Bu_2\text{-}H_3C_6O)_2Pb$: with $(Et(Me_3Si)_2Si)_3PbK$ in n-pentane (Scheme 2.16).[103] In contrast to other known transient R_3Pb• radicals, featuring markedly pyramidal geometries around their radical centers, sterically overloaded **9** was essentially planar (sum of the bond angles around the Pb atom = 355°), although marginally more pyramidal than the perfectly planar radicals **6–8** $(t\text{-}Bu_2MeSi)_3E$• (E = Si, Ge, Sn) (see above). Similarly, the unpaired electron of **9** was expected to reside in an orbital with high p-character and to be possibly hyperconjugatively delocalized (at least, in part) over the Si–Si σ^*-orbitals. Consequently, the Pb–Si bonds showed a significant contribution from the relativistically contracted Pb 6s-orbital (indeed, the mean Pb–Si bond length of 2.64 Å is markedly short). The room temperature EPR signal of **9** was broad $[g = 2.105, a(^{207}Pb) = 50.6\,mT]$, whereas the EPR spectrum in a frozen matrix revealed the axial symmetry expected for R_3Pb• radicals. The isotropic g value of 2.105 and the extent of anisotropy as a result of large spin-orbit coupling at the Pb atom were similar to those of related transient R_3Pb• radicals[20,32] but were much larger than those of analogous radicals of the lighter group 14 elements (Si, Ge, Sn).[96,97] Extending the synthetic approach of one-electron oxidation of the persilylstannanides and -plumbanides with the appropriate oxidizing reagents, the same authors prepared a series of persistent stannyl and plumbyl radicals, $[(Et(Me_3Si)_2Si]_3Sn$•, $[(Me_3Si)_3Si]_3Pb$•, $[(Me_3Si)_3Si]_3Sn$•, $[i\text{-}Pr(Me_3Si)_2Si]_3Sn$•, $[i\text{-}Pr(Me_3Si)_2Si]_3Pb$•, characterized by EPR spectroscopy and X-ray crystallography (for the last three radicals).[103b]

Plumbyl radical **9** was indefinitely stable below −20 °C in its crystalline form, whereas at room temperature, benzene solutions of **9** completely decomposed within one day (half-life at 25 °C is 8.5 h). Decomposition of **9** to produce metallic Pb and the tetrasilylplumbane $[Et(Me_3Si)_2Si]_4Pb$ followed first-order kinetics (with respect to **9**) with an activation energy of ca. 23.7 kcal/mol, a value that is consistent with Pb–Si bond cleavage being the rate-determining step.

A stable triarylgermyl radical Ar_3Ge• [Ar = 3,5-t-Bu_2-2,6-$(EtO)_2$-C_6H] **10** was very recently communicated by Drost *et al.*, utilizing the procedure originally applied for the synthesis of persilyl radicals **6–8**.[104] **10** manifested a planar geometry at the germanium radical center in the solid state (sum of the bond angles 359.5°) and pyramidal configuration in solution $[g = 2.0095, a(^{73}Ge) = 6.13\,mT]$.

2.4.2 Charged Radicals

The synthesis of the first stable charged radical species of the heavy group 14 elements became a reality in the past few years following the preparation of the appropriate precursors.[3b–e] Since then several representatives of charged radicals centered on the Si, Ge, and Sn atoms have been synthesized and structurally characterized. In this section, reviewing the recent accomplishments in the field of charged radicals as an independent class of highly reactive group 14 organometallics, we will separately deal with both anion-radicals and cation-radicals subdividing each category into the two subclasses of cyclic and acyclic representatives.

2.4.2.1 Anion-radicals

2.4.2.1.1 Cyclic Anion-radicals

The first representative of this type of stable charged radicals of heavy group 14 elements, the potassium salt of the cyclotetrasilane anion-radical $\mathbf{11}^{-\bullet}\bullet[\mathbf{K}^{+}(\mathbf{thf})_6]$, was recently synthesized by the reduction of dichloride **12** with potassium in THF (the doubly reduced cyclotetrasilane dianion was also isolated in the form of its dipotassium salt $\mathbf{13}^{2-}\bullet[\mathbf{K}^{+}(\mathbf{thf})_6]_2$) (Scheme 2.17).[105]

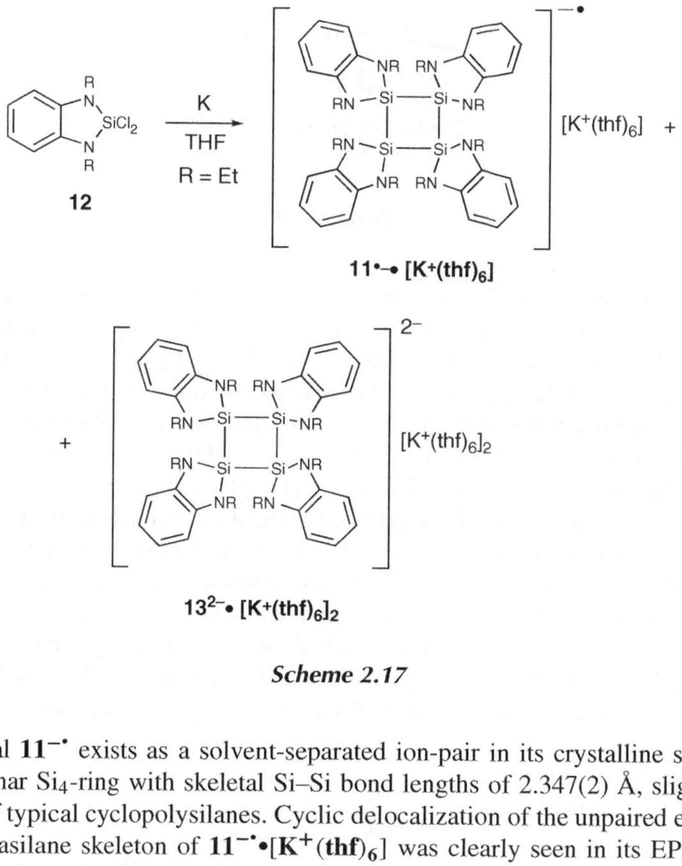

Scheme 2.17

Anion-radical $\mathbf{11}^{-\bullet}$ exists as a solvent-separated ion-pair in its crystalline state and has a square-planar Si_4-ring with skeletal Si–Si bond lengths of 2.347(2) Å, slightly shorter than those of typical cyclopolysilanes. Cyclic delocalization of the unpaired electron over the cyclotetrasilane skeleton of $\mathbf{11}^{-\bullet}\bullet[\mathbf{K}^{+}(\mathbf{thf})_6]$ was clearly seen in its EPR spectrum, which revealed a multiplet ($g = 2.0025$) of 15 lines (of total 17) with an hfcc of 0.35 mT because of the coupling of the unpaired electron with the eight equivalent ^{14}N nuclei ($I = 1$).

2.4.2.1.2 *Acyclic Anion-radicals*

2.4.2.1.2.1 Heavy Carbenes Anion-radicals

The heavy analogs of carbenes, that is silylenes, germylenes, stannylenes, and plumbylenes, typically have a singlet ground state with a σ^2-electronic configuration, implying the location of two paired electrons in the orbital of σ-symmetry corresponding to a $^1A_1$ electronic ground state (Scheme 2.18).[106] However, upon single electron transfer processes (oxidation or reduction), the electronic configuration of the heavy carbene analogs can be either changed or retained depending on the substitution pattern. Thus, for example, the silylene cation-radical was experimentally found to be of the σ-type, whereas the silylene anion-radical was calculated to be a π-radical.[107] The anion-radicals of dichlorogermylene, as well as its Lewis base complexes, were also calculated to be π-radicals, in contrast to their cation-radicals of σ-type.[108]

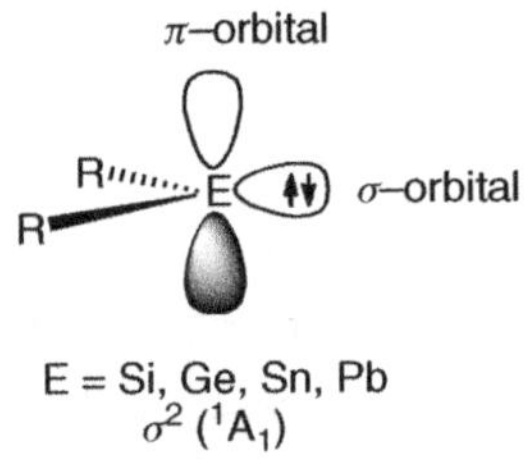

Scheme 2.18

The pioneering studies on the generation of anion-radicals of heavy carbene analogs with a lifetime sufficient for their direct EPR detection were reported at the beginning of the 1990s. The first report in this series was published by Sita *et al.*, who described the room temperature EPR spectrum of the reaction mixture obtained by the reduction of hexakis(2,6-diethylphenyl)cyclotristannane with potassium amalgam in THF.[109] The EPR resonance featuring a hyperfine splitting $a(^{119,117}Sn)$ of 15.2 mT was reasonably attributed to the anion-radical of bis(2,6-diethylphenyl)stannylene.

More recent developments in this field involve the direct reduction of the stable heavy carbene analogs (silylenes, germylenes and stannylenes) with alkali metals, providing further convincing evidence for the generation of anion-radicals of heavy carbenes. The generation of the first persistent germylene and stannylene anion-radicals was reported by Egorov *et al.* in 1995.[110] These anion-radicals were prepared by the reduction of the stable Lappert's germylene $[(Me_3Si)_2CH]_2Ge$: and stannylene $[(Me_3Si)_2CH]_2Sn$: with sodium mirror in THF (Scheme 2.19). Both the germylene anion-radical **$14^{-\bullet}$•Na^+**

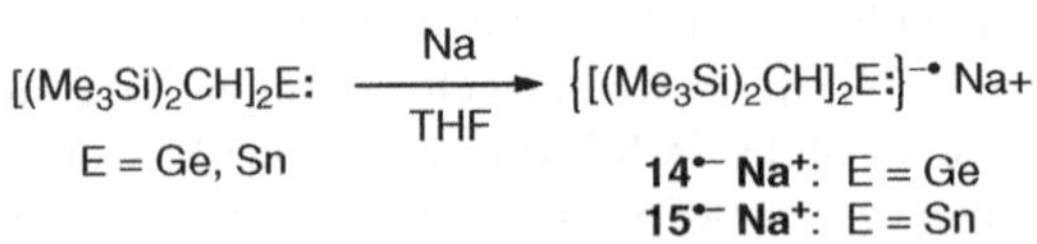

Scheme 2.19

and stannylene anion-radical **15$^{-\bullet}$•Na$^+$** were persistent in solution at room temperature, allowing their direct EPR characterization: their lifetimes were estimated as ca. 1.5 h (for **14$^{-\bullet}$•Na$^+$**) and several minutes (for **15$^{-\bullet}$•Na$^+$**). The germylene anion-radical **14$^{-\bullet}$•Na$^+$** displayed a strong resonance at $g = 2.0125$ as a triplet due to the coupling of an unpaired electron with the two C$\underline{H}$-protons of the alkyl substituents [$a(^1H) = 0.26$ mT]. The diagnostic coupling of an unpaired electron with the ^{73}Ge nucleus was manifested in the observation of eight (of total ten) weak satellites with the hfcc $a(^{73}Ge) = 1.25$ mT. The decay of the germylene anion-radical **14$^{-\bullet}$•Na$^+$** in solution obeys second-order kinetics, implying its apparent dimerization to form the diamagnetic dianion species.

The stannylene anion-radical **15$^{-\bullet}$•Na$^+$** was also characterized by its EPR spectrum, which displayed a broad singlet at $g = 2.0177$ accompanied by a pair of unresolved satellites arising from the coupling of the unpaired electron with both ^{119}Sn and ^{117}Sn nuclei [$a(^{119,117}Sn) = 11.6$ mT].[110] The characteristically small values of the hfcc $a(^{73}Ge)$ of 1.25 mT and $a(^{119,117}Sn)$ of 11.6 mT suggest that both germylene and stannylene anion-radicals **14$^{-\bullet}$•Na$^+$** and **15$^{-\bullet}$•Na$^+$** belong to the class of π-radicals with $^2B_1$ ground electronic state.

The preparation of the first persistent silylene anion-radical was reported recently by Kira *et al*. The sodium salt of the silylene anion-radical **16$^{-\bullet}$** with a lifetime of ca. 20 min at room temperature was generated by the reduction of stable silylene **17** with a potassium mirror in DME (Scheme 2.20).[111] The one-electron reduction of **17** with other alkali metals (Li, Na, K, Rb, Cs) resulted in the formation of the same silylene anion-radical **16$^{-\bullet}$** in the form of the corresponding alkali metal salt. All of these derivatives **16$^{-\bullet}$•M$^+$** (M = Li, Na, K, Rb, Cs) revealed the same EPR pattern, the main feature of which was a central multiplet ($g = 2.0077$) composed of more than 20 lines due to coupling with the 36 protons of the four Me_3Si-groups [$a(^1H) = 0.0189$ mT] and the 4 protons of the CH_2–CH_2 unit of the silacyclopentane skeleton [$a(^1H) = 0.0378$ mT]. The ^{29}Si satellites were observed as three sets of doublets, corresponding to coupling with the central Si [$a(^{29}Si) = 2.99$ mT] and the two different types of Si atoms of the Me_3Si groups [$a(^{29}Si) = 1.66$ and 1.30 mT]. These EPR parameters of **16$^{-\bullet}$** are similar to those of the transient lithio[bis(triisopropylsilyl)]silyl radical Li(i-$Pr_3Si)_2Si$•, generated along with other silyl radical species by the photolysis of $[(thf)_2Li(iPr_3Si)_2Si]_2Hg$: $g = 2.0073, a(^{29}Si_\alpha) = 3.20$ mT.[89]

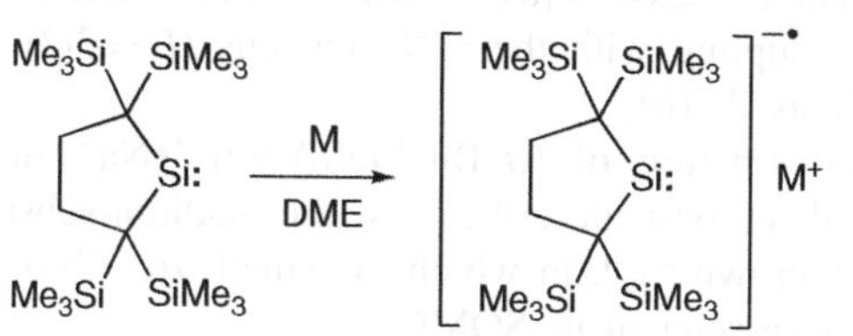

17 **16$^{\bullet-}$•M$^+$** (M = Li, Na, K, Rb, Cs)

Scheme 2.20

Among other important examples of the silylene anion-radicals (or alkali metal-substituted silyl radicals), one should mention those recently reported by Apeloig *et al.* (see also Section 2.3): Li(i-$Pr_3Si)_2Si$• generated by the photolysis of the silylmercury compounds [Li(i-Pr$_3$Si)$_2$Si–Hg–Si(Sii-Pr$_3$)$_2$Li],[89] and (thf)$_n$M(t-Bu$_2$MeSi)$_2$Si• (M = Li, Na) formed upon the photolysis of the hydridosilyllithium derivative [(t-Bu$_2$MeSi)$_2$SiHLi]$_2$ or by the reaction of isolable silylenoid (t-Bu$_2$MeSi)$_2$SiFLi(thf)$_3$ with t-BuLi or alkali metals (Li, Na).[90,112]

The isolation and structural characterization of the stable silylene anion-radical derivative [(t-Bu$_2$MeSi)$_2$Si:]$^{-•}$•M$^+$(crown ether)$_n$ (M = Li, Na) was achieved by Sekiguchi *et al.* quite recently. Both lithium and sodium derivatives were obtained by the reduction of the disilene (t-Bu$_2$MeSi)$_2$Si=Si(SiMet-Bu$_2$)$_2$ with 2.2 equiv. of MNp (lithium or sodium naphthalenide) in THF at −78 °C followed by the addition of 4.3 equiv. of crown ethers (Scheme 2.21).[113] The reaction presumably proceeds through the initial formation of 1,2-dianionic derivatives (t-Bu$_2$MeSi)$_2$(M)Si–Si(M)(SiMet-Bu$_2$)$_2$ upon the two-electron reduction of the starting disilene, which then undergo breaking of the central Si–Si bond upon the complexation of countercations with crown ethers. In the solid state, both salts [(t-Bu$_2$MeSi)$_2$Si:]$^{-•}$•M$^+$(crown ether)$_n$ (M = Li, Na) exist as solvent-separated ion pairs with no bonding contacts between the anionic and cationic portions of the molecule (Figure 2.6). The 'free' status of silylene anion-radical [(t-Bu$_2$MeSi)$_2$Si:]$^{-•}$ M$^+$ (solvent)$_n$ (M = Li, Na) is preserved in polar solvents, such as THF or DME, in which both Li and Na salts displayed a singlet resonance at $g = 2.0074$ [$a(\alpha$-^{29}Si$) = 2.91$ mT and $a(\beta$-^{29}Si$) = 1.02$ mT] with no hyperfine splitting due to either ^{7}Li or ^{23}Na nuclei being observed [Figure 2.7(a)].

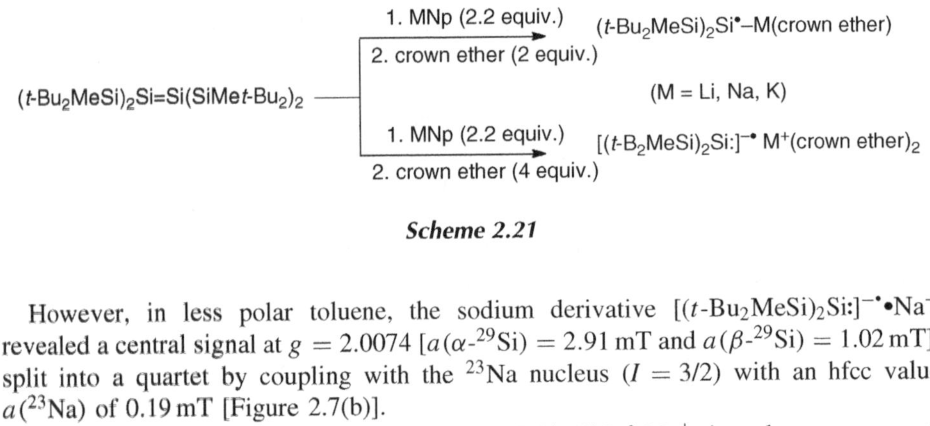

Scheme 2.21

However, in less polar toluene, the sodium derivative [(t-Bu$_2$MeSi)$_2$Si:]$^{-•}$•Na$^+$ revealed a central signal at $g = 2.0074$ [$a(\alpha$-^{29}Si$) = 2.91$ mT and $a(\beta$-^{29}Si$) = 1.02$ mT], split into a quartet by coupling with the ^{23}Na nucleus ($I = 3/2$) with an hfcc value $a(^{23}$Na$)$ of 0.19 mT [Figure 2.7(b)].

That is, the solution structure of [(t-Bu$_2$MeSi)$_2$Si:]$^{-•}$•Na$^+$ in toluene represents a contact ion pair and is best described as the sodium-substituted silyl radical [(t-Bu$_2$MeSi)$_2$Si$^•$–Na(15-crown-5)$_n$], in which the small $a(\alpha$-^{29}Si$)$ value of 2.91 mT is indicative of the high p-character of its SOMO.

Such alkali metal-substituted silyl radicals [(t-Bu$_2$MeSi)$_2$Si$^•$–M(crown ether)] were isolated when 2 (but not 4) equiv. of crown ether were added to the reaction mixture containing the dianionic derivative (t-Bu$_2$MeSi)$_2$(M)Si–Si(M)(SiMet-Bu$_2$)$_2$ (Scheme 2.21).[114]

The composition of the sodium salt [(t-Bu$_2$MeSi)$_2$Si$^•$–Na(15-crown ether-5)] as the alkali metal-substituted silyl radical was proved by the existence of a direct Si1–Na bond of 3.0744(8) Å in its crystalline state (Figure 2.8). The geometry around the

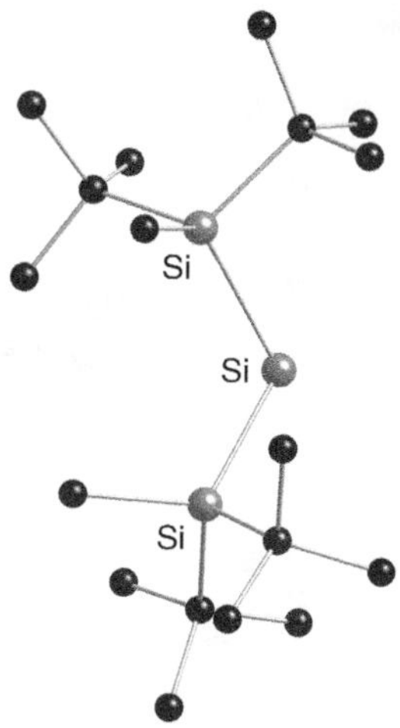

Figure 2.6 *Crystal structure of the silylene anion-radical* ***[(t-Bu₂MeSi)₂Si:]⁻•*** *(cationic portion of the molecule [Li⁺(12-crown-4)₂] and hydrogen atoms are not shown)*

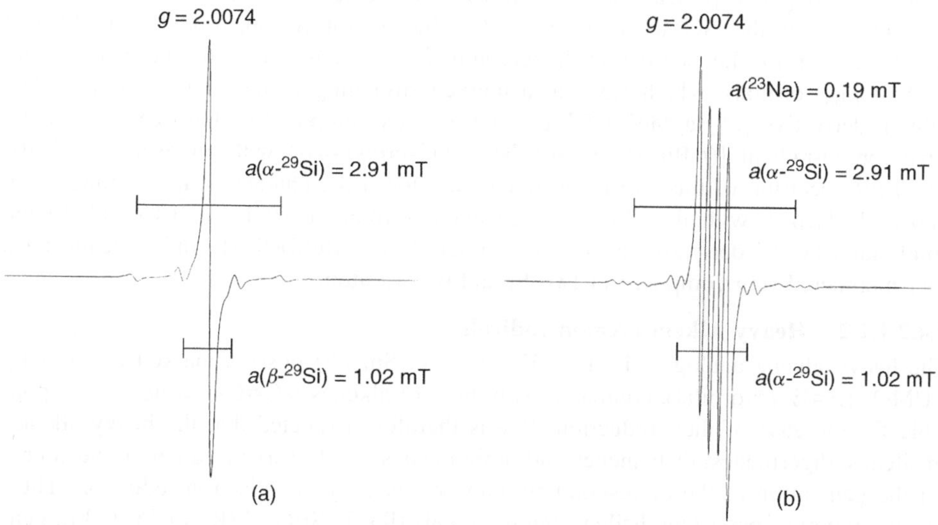

Figure 2.7 *EPR spectra of the silylene anion-radical* ***$[(t\text{-}Bu_2MeSi)_2Si{:}]^{-\bullet}\bullet Na^+$*** *at room temperature: (a) in DME; (b) in toluene*

central Si1 atom is nearly trigonal-planar (358°), and the lengths of the Si1–Si2 and Si1–Si3 bonds are quite normal: 2.3728(6) and 2.3685(7) Å, respectively. Such remarkable planarity of the alkali metal-substituted silyl radical definitely stems from its substitution pattern: all substituents (silyl groups and particularly alkali metals) are strongly σ-donating with respect to the anion-radical Si center, thus promoting its flattening. In effect, the silyl radical [(t-$Bu_2MeSi)_2Si^{\bullet}$–Na(15-crown ether-5)] (as well as its

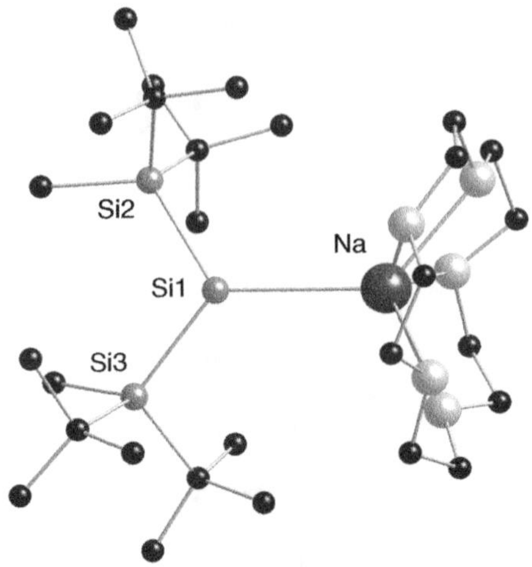

Figure 2.8 *Crystal structure of the Na-substituted silyl radical* ***[(t-Bu₂MeSi)₂Si–Na (15-crown-5)]*** *(hydrogen atoms are not shown)*

potassium congener $[(t\text{-Bu}_2\text{MeSi})_2\text{Si}^{\bullet}\text{–K(18-crown ether-6)]})$ are to be classified as π-radicals with the SOMO represented by the central Si $3p_z$-orbital. The SOMO can benefit from the stabilizing hyperconjugative interaction with the antibonding $\sigma^*(\text{Si–C}_{t\text{-Bu}})$ orbitals, which leads to a marked stretching of the $\text{Si–C}_{t\text{-Bu}}$ bonds. The lithium derivative $[(t\text{-Bu}_2\text{MeSi})_2\text{Si}^{\bullet}\text{–Li(thf)}]$ was also prepared by the treatment of the 'free' anion-radical $[(t\text{-Bu}_2\text{MeSi})_2\text{Si:}]^{-\bullet}\bullet[\text{Li}^+(\text{12-crown-4})_2]$ with an excess of LiBr. The EPR spectrum of this Li-salt in toluene disclosed the characteristic coupling of an unpaired electron with the ^{7}Li ($I = 3/2$) nucleus [quartet, $a(^7\text{Li}) = 0.16\,\text{mT}$]. In the solid state, the Li derivative exists as a dimer $[(t\text{-Bu}_2\text{MeSi})_2\text{Si}^{\bullet}\text{–Li(thf)}]_2$, featuring a four-membered ring composed of two Si and two Li atoms.

2.4.2.1.2.2 Heavy Alkenes Anion-radicals

The heavy alkene analogs >E=E< (E = Si, Ge, Sn, Pb) possess markedly low-lying LUMOs (E=E π^*-orbital), compared with those of alkenes >C=C<, which are responsible for the ease of their reduction. It was therefore expected that the heavy alkenes (disilenes, digermenes, distannenes and diplumbenes) could serve as an effective source for the generation of the corresponding anion-radicals by one-electron reduction. Thus, the first report of persistent disilene anion-radicals $[\text{R}_2\text{Si=SiR}_2]^{-\bullet}$ (R = i-Pr, t-Bu), generated by the reduction of 1,2-dihalides $\text{R}_2(\text{X})\text{Si–Si}(\text{X})\text{R}_2$ with alkali metals in THF, was published by Weidenbruch *et al.* in 1985.[115] More recently, Kira *et al.* mentioned the generation of the tetrasilyldisilene anion-radicals by the direct reduction of disilenes with potassium in DME.[116] However, the isolation and structural identification of the virtually stable anion-radical derivatives of disilenes was accomplished only a few years ago.

Thus, the first stable disilene anion-radical $\mathbf{18^{-\bullet}\bullet[Li(thf)_4]^+}$ was isolated as a solvent-separated ion pair by the reduction of the highly sterically congested disilene $(t\text{-Bu}_2\text{MeSi})_2\text{Si=Si}(\text{SiMe}t\text{-Bu}_2)_2$ with t-BuLi in THF (Scheme 2.22).[117] Upon

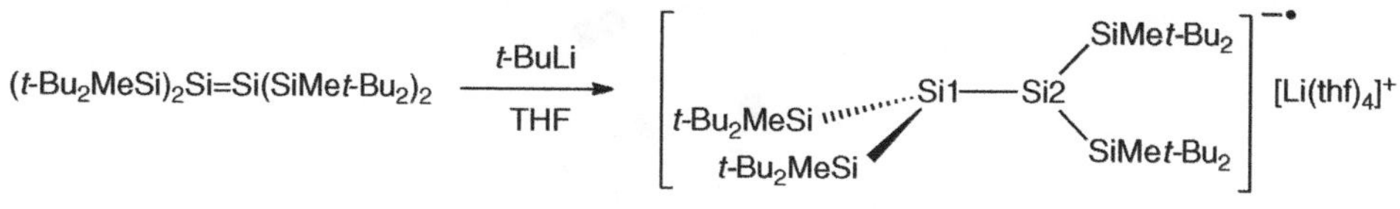

18$^{-\bullet}$**•[Li(thf)$_4$]**$^+$

Scheme 2.22

one-electron reduction, the central Si–Si bond of **18**$^{-\bullet}$ became highly twisted (88°) and stretched [2.341(5) Å vs 2.2598(18) Å in the starting disilene, ca. 3.6% elongation], which should be attributed to a decrease in the bond order on going from the starting disilene to a resulting disilene anion-radical. The most important structural peculiarity of anion-radical **18**$^{-\bullet}$ is the strikingly different environment around the two core Si atoms: nearly planar geometry around Si2 (359.9°) and more pyramidal around Si1 (352.7°). This phenomenon was rationalized in the framework of unpaired electron–negative charge separation: preferential accommodation of the single electron at the Si2 atom (radical center) and electron pair at the Si1 atom (anionic center). The EPR resonance of the anion-radical **18**$^{-\bullet}$ was observed at a field that is typical for silyl-substituted silyl radicals ($g = 2.0061$), accompanied by a pair of ^{29}Si satellites with an hfcc $a(^{29}\text{Si})$ of 2.45 mT, a value that was significantly smaller than that of the per(silyl)silyl radical (t-Bu$_2$MeSi)$_3$Si• **6** (5.80 mT).[96]

The tin homolog of **18**$^{-\bullet}$**•[Li(thf)$_4$]**$^+$, the distannene anion-radical **19**$^{-\bullet}$**•[K([2.2.2] cryptand)]**$^+$, was also successfully synthesized by a similar synthetic protocol; namely, by the direct reduction of the stable distannene (t-Bu$_2$MeSi)$_2$Sn=Sn(SiMet-Bu$_2$)$_2$ with a potassium mirror in THF in the presence of [2.2.2]cryptand (Scheme 2.23).[118] The crystal structure of **19**$^{-\bullet}$**•[K([2.2.2]cryptand)]**$^+$ is reminiscent of that of its silicon congener **18**$^{-\bullet}$**•[Li(thf)$_4$]**$^+$: a highly twisted (74°) and very long [2.8978(3) Å] Sn–Sn bond, which is almost 9% longer than that of its precursor distannene (Figure 2.9). The radical center–anionic center separation in the anion-radical **19**$^{-\bullet}$ was manifested in the distinctly different geometries around both Sn atoms: essentially planar around one of them (355.4° around Sn2, radical center) and markedly pyramidal around the other one (323.2° around Sn1, anionic center). Such unpaired electron–electron pair separation was preserved also in the solution of **19**$^{-\bullet}$**•[K([2.2.2]cryptand)]**$^+$, as was clearly seen

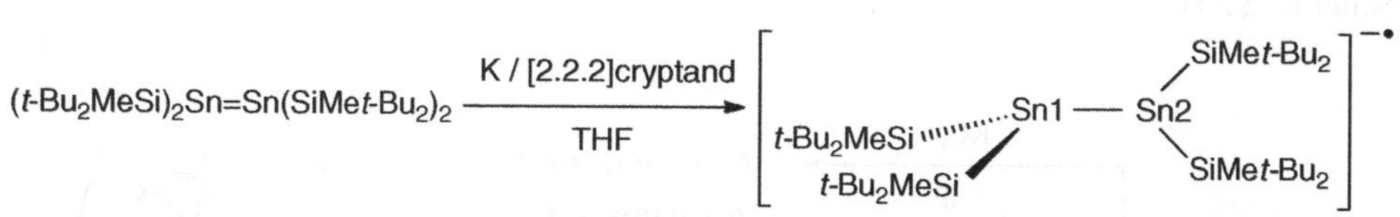

19$^{-\bullet}$**•[K([2.2.2]cryptand)]**$^+$

Scheme 2.23

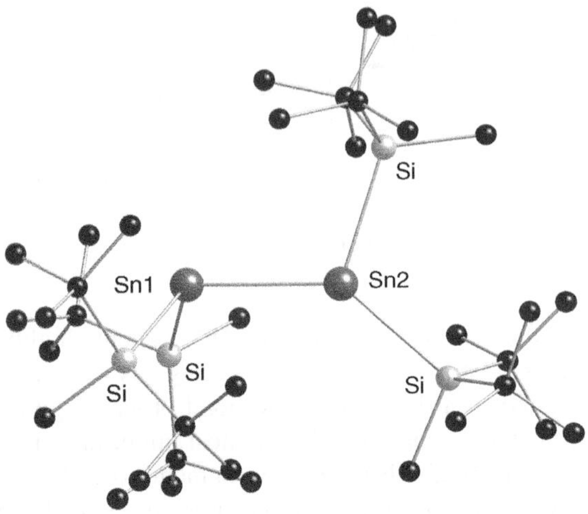

Figure 2.9 *Crystal structure of the distannene anion-radical* $\mathbf{19^{-\bullet}}$***•[K([2.2.2]cryptand)]***$^+$ *(cationic portion of the molecule [K([2.2.2]cryptand)]*$^+$ *and hydrogen atoms are not shown)*

in its EPR spectrum, which showed a central signal at $g = 2.0517$ along with the two pairs of $^{119,117}Sn$ satellites [$a(\alpha\text{-}^{119,117}Sn) = 34.0\,mT$ and $a(\beta\text{-}^{119,117}Sn) = 18.7\,mT$].

2.4.2.1.2.3 Heavy Alkynes Anion-radicals

The first representative of this type of compound, the potassium salt of the anion-radical of the valence isomer of distannyne $\mathbf{20^{-\bullet}}$**•[K(thf)**$_6$**]**$^+$, was synthesized by Power *et al.* by the reduction of chlorostannylene Ar(Cl)Sn: [Ar = 2,6-(2,4,6 – iPr$_3$-$C_6H_2)_2C_6H_3$] with potassium graphite in THF (Scheme 2.24).[119] Both Ar substituents at the Sn atoms are significantly *trans*-bent, whereas the Sn–Sn bond length of 2.8123(9) Å is quite normal for a Sn–Sn single bond. The EPR resonance of $\mathbf{20^{-\bullet}}$**•[K(thf)**$_6$**]**$^+$ was found at $g = 2.0691$ together with the $^{119,117}Sn$ satellites, whose hfcc values were simulated as 0.83 mT [$a(^{117}Sn)$] and 0.85 mT [$a(^{119}Sn)$]. Such small hfcc values are indicative of the localization of the unpaired electron in anion-radical $\mathbf{20^{-\bullet}}$**•[K(thf)**$_6$**]**$^+$ in an orbital of π-symmetry. All experimental (structural and spectral) data of $\mathbf{20^{-\bullet}}$**•[K(thf)**$_6$**]**$^+$ point to a formulation as a major contributor to the resonance structure in which each Sn atom bears a lone pair and the unpaired electron is accommodated over their 5p$_\pi$-orbitals (Scheme 2.25).

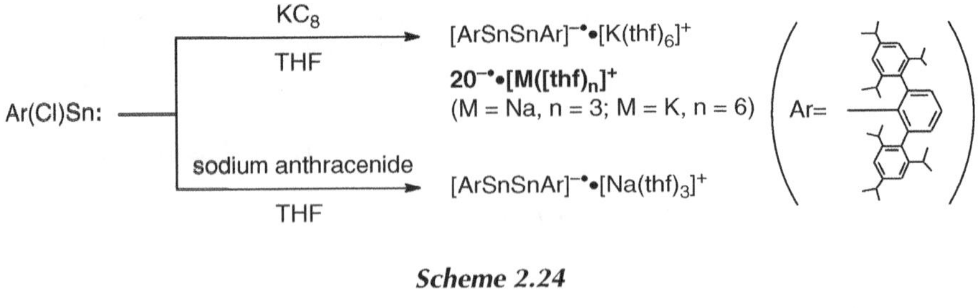

Scheme 2.24

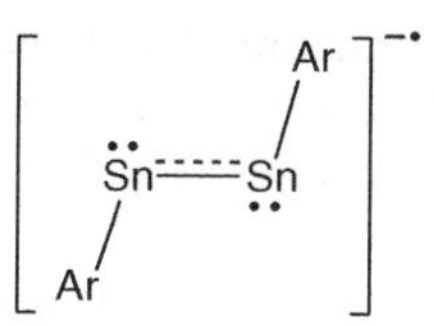

Scheme 2.25

Such a representation of the bonding between the two Sn atoms in anion-radical **20$^{-\bullet}$** with a formal Sn–Sn bond order of 1.5 and highly *trans*-bent Ar substituents (ca. 95°) is suggestive of the diminished Sn–Sn hybridization resulting from the poor $5p_{\pi}-5p_{\pi}$ orbital overlap to form weak π-bonds. The structural and spectral features of the sodium salt of the anion-radical **20$^{-\bullet}$•[Na(thf)$_6$]$^+$**, prepared similarly by the reduction of chlorostannylene Ar(Cl)Sn: with sodium anthracenide in THF, are almost identical to those of the above-described potassium derivative **20$^{-\bullet}$•[K(thf)$_6$]$^+$** (Scheme 2.24).[120]

In subsequent studies, the same authors extended the range of their stable charged radical species, preparing an anion-radical of the distannyne valence isomer with novel Ar* substituents [Ar*SnSnAr*]$^{-\bullet}$•[K(thf)$_6$]$^+$ [Ar* = 2, 6-(2,6 – iPr$_2$-C$_6$H$_3$)$_2$C$_6$H$_3$] (**21$^{-\bullet}$•[K(thf)$_6$]$^+$**), and the first anion-radicals of the digermyne valence isomers [ArGeGeAr]$^{-\bullet}$•Na$^+$ (**22$^{-\bullet}$•Na$^+$**) and [Ar*GeGeAr*]$^{-\bullet}$•K$^+$ (**23$^{-\bullet}$•K$^+$**).[121] All derivatives **21$^{-\bullet}$•[K(thf)$_6$]$^+$**, **22$^{-\bullet}$•Na$^+$** and **23$^{-\bullet}$•K$^+$** uniformly revealed the planarity of their core units and the *trans*-bent geometry of substituents. The Ge–Ge and Sn–Sn bonds were stretched upon reduction, being 2.32(1) Å on average for the germanium compounds and 2.81(1) Å for the tin derivatives. Such bond stretching amounts to ca. 1.3 and 5.2% elongation, respectively, compared with the starting neutral compounds. The other remarkable structural feature of all these anion-radicals was the great increase in *trans*-bending upon one-electron reduction. This was most pronounced in the case of tin derivatives, for which a dramatic increase (ca. 30°) in the *trans*-bending angle from the neutral precursor (Sn–Sn–C bond angle ca. 125°) to anion-radicals (Sn–Sn–C bond angles ca. 94–98°) was observed.[121] For the two germanium anion-radical salts **22$^{-\bullet}$•Na$^+$** and **23$^{-\bullet}$•K$^+$**, the increase in the *trans*-bending angle is less important (ca. 13–16°). Such great changes in the extent of *trans*-bending (especially in the case of tin compounds) cannot be simply explained by the effects of alkali metal countercations; instead, they should be associated with the changes in the internal electronic structure taking place upon reduction of the EE-unit (E = Ge, Sn). On the other hand, the increase in the *trans*-bending (the narrower E–E–C angles) of anion-radicals may give rise to an increase in p-character of the E–E σ-bonds, resulting in their elongation because of the greater average radius of the p-orbitals vs s-orbital. Moreover, since the LUMO of the neutral precursors ArEEAr and Ar*EEAr* seems to resemble a nonbonding lone pair combination rather than a π*-orbital (which is the case for classical alkynes), the addition of an electron to the LUMO is expected to have a small influence on the E–E bond distance. In other words, the lengthening of the E–E bonds upon their single-electron reduction is largely due to σ-bonding effects. In effect, the addition of a single electron to the LUMO of the neutral precursors generates a π-bond of order 0.5 yielding the overall E–E bond order of 1.5. Indeed, the Sn–Sn bonds in the anion-radicals of the distannyne valence isomers are close to the Sn–Sn single bonds in elemental tin, thus suggesting little multiple bond character. The EPR spectrum of the germanium anion-radical $\mathbf{23^{-\bullet}}$**•**$\mathbf{K^+}$ revealed a resonance at $g = 2.00382$ with a small $a(^{73}\mathrm{Ge})$ hfcc value of 0.75 mT, suggesting very low unpaired electron spin density at the Ge nucleus.

The reduction of the stable disilyne R–Si≡Si–R (R = Si *i*-Pr[CH($SiMe_3$)$_2$]$_2$)[122] with an equivalent amount of potassium graphite in THF produced the corresponding disilyne anion-radical $\mathbf{24^{-\bullet}}$**•[K(dme)$_4$]$^+$**, in which the cationic and anionic parts were well separated from each other (Scheme 2.26).[123] The central Si–Si bond distance of anion-radical $\mathbf{24^{-\bullet}}$ of 2.1728(14) Å is ca. 5% longer than that of the starting disilyne R–Si≡Si–R, within the normal range of disilenes >Si=Si< (Figure 2.10).

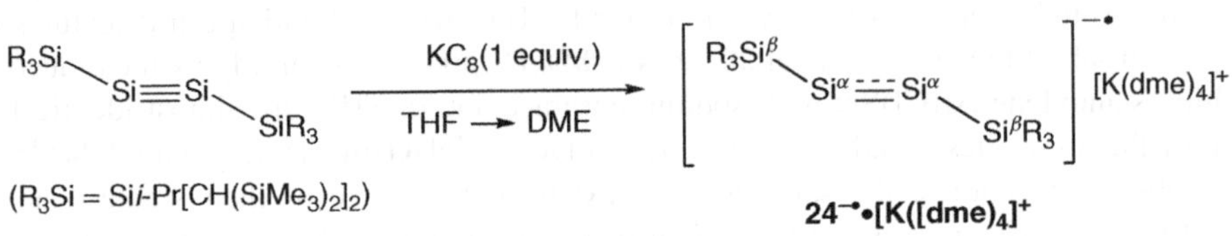

Scheme 2.26

The EPR spectrum of $\mathbf{24^{-\bullet}}$**•[K(dme)$_4$]$^+$** revealed a resonance at the diagnostically small value of $g = 1.99962$, observed as a triplet [$a(\delta\text{-}^1\mathrm{H}$ of *i*-Pr groups$) = 0.23\,\mathrm{mT}$] [hyperfine splitting with the δ-^{1}H of CH($SiMe_3$)$_2$ groups was not detected]. Two pairs of ^{29}Si satellites were observed with hfcc values $a(\alpha\text{-}^{29}\mathrm{Si}) = 3.92\,\mathrm{mT}$ and $a(\beta\text{-}^{29}\mathrm{Si}) = 2.24\,\mathrm{mT}$, of which the former is rather small and the latter is unusually large. The small $a(\alpha\text{-}^{29}\mathrm{Si})$ hfcc value of 3.92 mT was interpreted in terms of delocalization of an unpaired electron over the two central α-Si nuclei. This delocalization hypothesis was also supported by the structural peculiarities of anion-radical $\mathbf{24^{-\bullet}}$, whose (α-Si)–

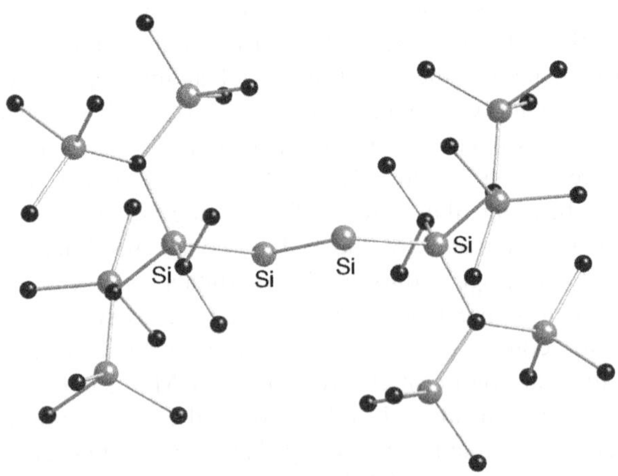

Figure 2.10 *Crystal structure of the disilyne anion-radical* $\mathbf{24^{-\bullet}}$***•[K(dme)$_4$]$^+$*** *(cationic portion of the molecule [K(dme)$_4$]$^+$ and hydrogen atoms are not shown)*

(α-Si)–(β-Si) bond angles were found to be essentially identical (112.8° and 114.0°), thus suggesting more or less equal distribution of the unpaired electron spin density between the two core α-Si atoms.

2.4.2.2 *Cation-radicals*

Only one example of an isolable cation-radical species based on the heavy group 14 elements has been reported to date, namely, the disilene cation-radical $[(t\text{-}Bu_2MeSi)_2Si{=}Si(SiMe\,t\text{-}Bu_2)_2]^{+\bullet}\bullet[B(C_6F_5)_4]^-$ **$25^{+\bullet}\bullet[B(C_6F_5)_4]^-$**.[124] This cationic derivative was prepared by the one-electron oxidation of the neutral disilene $(t\text{-}Bu_2MeSi)_2Si{=}Si(SiMe\,t\text{-}Bu_2)_2$ with an equivalent amount of $Ph_3C^+\bullet[B(C_6F_5)_4]^-$ in toluene (Scheme 2.27).

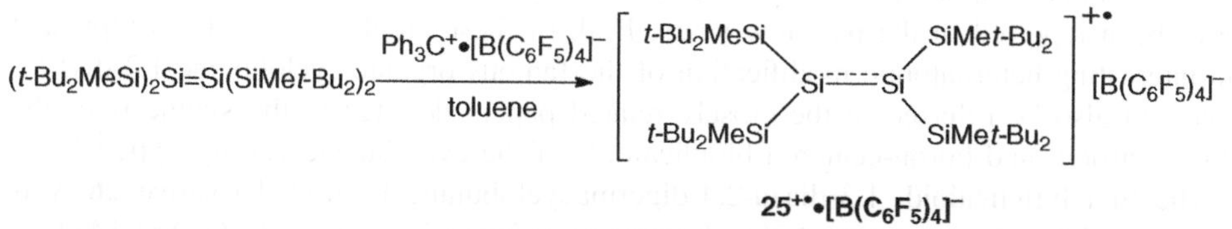

Scheme 2.27

In the solid state, cation-radical **$25^{+\bullet}$** is well separated from its anionic counterpart **$[B(C_6F_5)_4]^-$**. As expected, the Si–Si bond in **$25^{+\bullet}$** was stretched upon the single-electron oxidation of the starting disilene: 2.307(2) Å vs 2.2598(18) Å (ca. 2% elongation). Most importantly, the central Si–Si unit became highly twisted upon the oxidation: 64.9° in the cation-radical **$25^{+\bullet}$** vs 54.5° in the starting disilene. Both core Si atoms in **$25^{+\bullet}$** feature nearly planar geometry around them, which is in marked contrast to that of the above-discussed anion-radical **$18^{-\bullet}$** of the same disilene, manifesting the diagnostically planar environment around one of the central Si atoms and the distinctly pyramidal geometry around the other Si atom. The main reason responsible for the large twisting of the central Si–Si bond in cation-radical **$25^{+\bullet}$** is the stabilizing σ–π hyperconjugative interaction of the cation-radical center with the four σ-donating silyl groups. Such hyperconjugation was manifested in the stretching of the Si_{core}–$Si_{subst.}$ bonds: 2.463(2) and 2.466(2) Å (in **$25^{+\bullet}$**) vs 2.418 Å (av.) (in the starting disilene), which amounts to ca. 2% elongation.

The EPR spectrum of **$25^{+\bullet}\bullet[B(C_6F_5)_4]^-$** measured in the temperature range from 298 to 200 K in fluorobenzene revealed a single resonance at $g = 2.0049$ with a pair of ^{29}Si satellites [$a(^{29}Si) = 2.3$ mT]. This magnitude of the hfcc of 2.3 mT is markedly smaller than that of the per(silyl)silyl radical $(t\text{-}Bu_2MeSi)_3Si\bullet$ (5.8 mT), implying some extent of delocalization of the unpaired electron spin density over both core Si atoms in **$25^{+\bullet}$**. Such a delocalized solution structure of the disilene cation-radical **$25^{+\bullet}$** contrasts with that of the disilene anion-radical **$18^{-\bullet}$**, which features an unpaired electron at one central Si atom (sp^2-radical center) and the electron pair at the other central Si atom (sp^3-anionic center).

2.4.3 Stable Biradicals of the Heavy Group 14 Elements

Biradicals are typically defined as non-Kekulé structures that have an even number of valence electrons and lack at least one bond from the number required by the normal valency rules.[125] Two classes of biradicals can be distinguished, which differ from each other by the orientation of their spins: **triplet biradicals** featuring two electrons with parallel spins (↑↑), and **singlet biradicals** in which these two electrons are spin-paired (↑↓). A number of singlet and triplet biradicals have been directly observed as reactive intermediates, and some of them have even been isolated as stable compounds.[126] Stable biradicals of the heavy group 14 elements are not sufficiently represented to date, and there are still no examples of isolable triplet biradicals based on the Si, Ge, Sn and Pb atoms. We will describe below a couple of the most recent stable singlet biradicals (otherwise more appropriately named as biradicaloids, that is, 'closed shell species derived from a singlet biradical by a weak coupling between the radical centers'),[127] that have been reported in the past few years. Both of these singlet biradicaloids represent an interesting heteroatomic modification of the famous organic cyclobutane-1,3-diyl.[128] One can also be referred to the closely related papers devoted to the synthesis of the stable carbon- and boron-centered biradicaloids of the cyclobutane-1,3-diyl type.[129]

The first biradicaloid, 1,3-diaza-2,4-digermacyclobutane-1,3-diyl derivative **26**, was synthesized by the reaction of the digermyne valence isomer Ar*GeGeAr* [Ar* = 2,6-(2,6-i-Pr_2-C_6H_3)$_2$$C_6H_3$] with trimethylsilyl azide in hexane (Scheme 2.28).[130] The Ge_2N_2-core of **26** is planar with both N atoms featuring a trigonal-planar geometry around them (359.97°) and in-plane arrangement of the Me_3Si substituents (Figure 2.11).

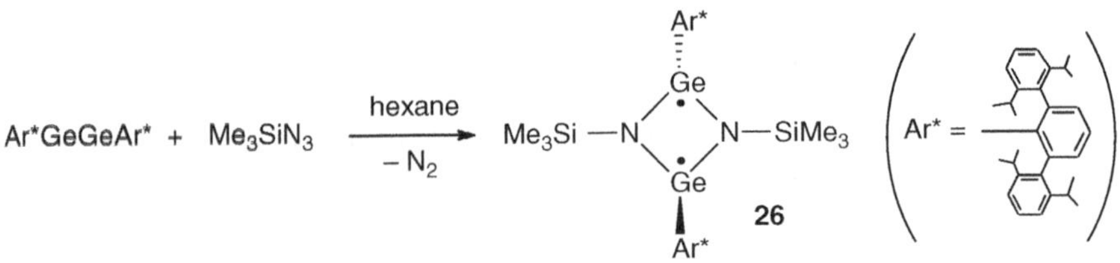

Scheme 2.28

In sharp contrast, both Ge atoms are distinctly pyramidal (322.10°) with a markedly out-of-plane *trans*-arrangement of the Ar* substituents. The formulation of **26** as a singlet biradical (biradicaloid) with odd electrons accommodated on the Ge centers was deduced from the trivalent state of both Ge atoms, which are separated from each other by 2.755 Å, clearly outside the distance of the covalent bonding interaction between them. Computations also revealed the absence of the Ge–Ge bonding, disclosing the nonbonding combination centered on the Ge atoms as the HOMO of the molecule. As was expected for the singlet biradicals, **26** was EPR-silent in the temperature range 77–300 K, whereas its normal NMR spectra were observed.

The other stable biradicaloid, 1,3-diaza-2,4-distannacyclobutane-1,3-diyl derivative **27**, was unexpectedly obtained by the reaction of the chloro(amino)stannylene dimer [Sn{N($SiMe_3$)$_2$}(μ-Cl)]$_2$ and AgOCN in diethyl ether (Scheme 2.29).[131]

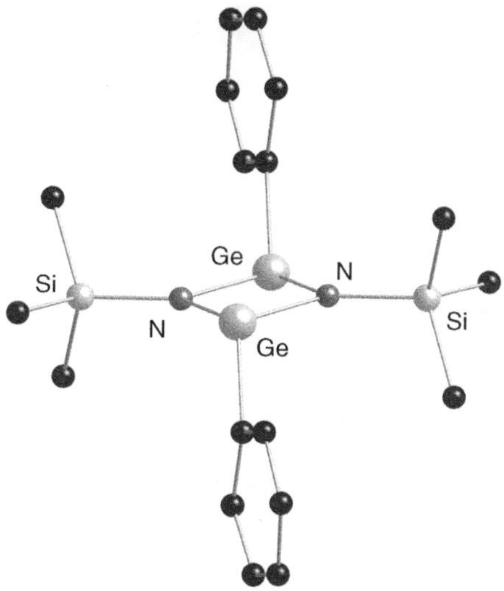

Figure 2.11 *Crystal structure of the Ge_2N_2-biradicaloid derivative **26** (2,6-i-$Pr_2C_6H_3$ substituents at the 2,6-positions of the aromatic rings, as well as hydrogen atoms, are not shown)*

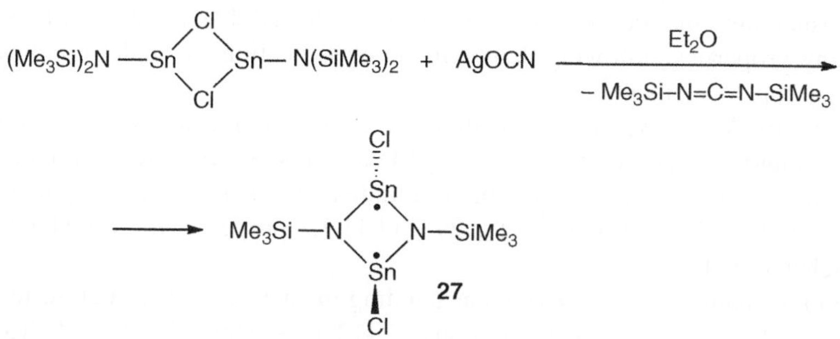

Scheme 2.29

Similar to the case of Power's biradicaloid, the Sn_2N_2-ring in **27** is planar with trigonal-planar geometry around both N atoms and pyramidal configuration around the Sn atoms (Figure 2.12).

The arrangement of substituents in **27** is also reminiscent of that in **26**; specifically, the nearly in-plane position of the Me_3Si groups and almost perpendicular situation of the Cl substituents at the Sn atoms. Such pyramidalization and tricoordination of both tins is indicative of the radical character of the Sn centers, which are separated by 3.398 Å, well beyond the range of Sn–Sn covalent interactions, indicating the absence of transannular 2,4-bonding. That is, **27** is to be considered as a singlet biradicaloid featuring two Sn radical centers, and, accordingly, **27** is diamagnetic exhibiting no EPR signals. Simple electron counting in **27** (two lone pairs from the N atoms and two odd electrons from the Sn atoms) gives rise to a total of six, implying the six π-electron four-center system

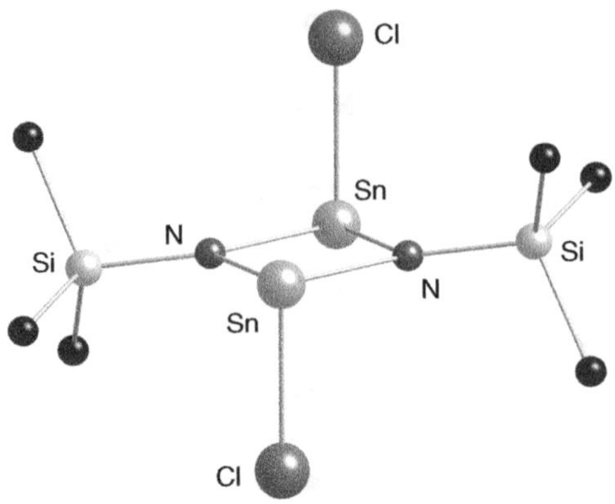

Figure 2.12 *Crystal structure of the Sn_2N_2-biradicaloid derivative* ***27*** *(hydrogen atoms are not shown)*

is the most appropriate bonding description of **27**. The triplet biradical is destabilized compared with the singlet biradical for **27** by 13.6 kcal/mol, which is ascribed to the σ-donating influence of the electropositive silyl groups at the N atoms, increasing the singlet-triplet energy gap Δ(S–T) to favor the singlet ground state.

Both germanium- and tin-containing biradicaloids **26** and **27** manifest similar structural and bonding properties, alllowing their reliable classification as singlet biradicaloids of the heavy group 14 elements.

As other remarkable examples, one should mention the recently prepared polycyclic (cage) compounds composed of heavy group 14 elements, which were reported to possess a more or less significant amount of singlet biradicaloid character. Among them are the derivatives of bicyclo[1.1.0]butane **28, 29** and [1.1.1]propellane/bicyclo[1.1.1]pentane **30**–**35** (Scheme 2.30).

The biradical nature of the transannular bonding in such cages was manifested in their diagnostically long E–E bridge separations: 2.412(1) Å (for **28**),[132] 2.5827(3) Å (for **29**),[133] 3.348(1) Å (for **30**),[109,134] 3.367(1) Å (for **31**),[109,134] 3.4226(6) Å (for **32**),[135] 3.363(1) Å (for **33**),[136] 2.869(2) Å (for **34**)[137] and 3.329(1) Å (for **35**)[138].

The heavy analogs of alkynes REER (E = Si, Ge, Sn, Pb), which have *trans*-bent structures that are distinctly different from the classical organic alkynes, were also reported to possess some degree of singlet biradical character. This implies the contribution of the biradical resonance form **D** along with the two degenerate zwitterionic forms **B** and **C** to the overall structure of the heavy alkyne analogs **A** (Scheme 2.31).[139] Such a biradical contribution is particularly important for the derivatives of disilynes RSiSiR and digermynes RGeGeR, whereas distannynes RSnSnR and diplumbynes RPbPbR analogs prefer the form in which both central atoms (Sn or Pb) feature a lone pair and are singly bonded (Scheme 2.31, **E**).[139c] Thus, recent computations assigned ca. 15% biradical character to both the 'out-of-plane' π-bond and nonbonding ('in-plane' π) electron pairs for the digermyne analogs, whereas the corresponding distannyne analogs have only ca. 4% biradical character.[139c] The particular bonding situation of the heavy analogs of alkynes and their valence isomers will be discussed in detail in Chapter 5.

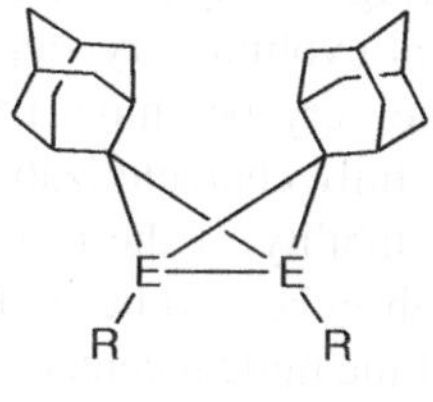

28 (E = Si, R = $SiMe_2t$-Bu)
29 (E = Ge, R = $SiMe_2t$-Bu)

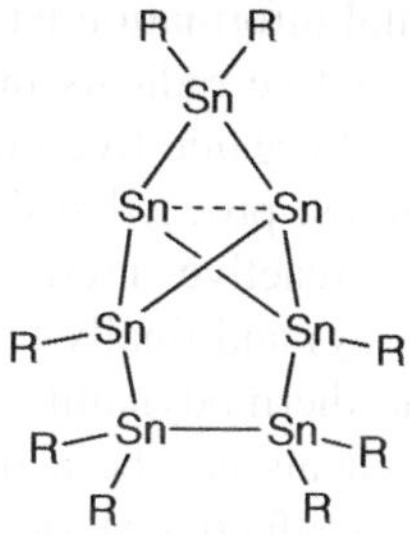

30 (R = 2,6-Et_2-C_6H_3)

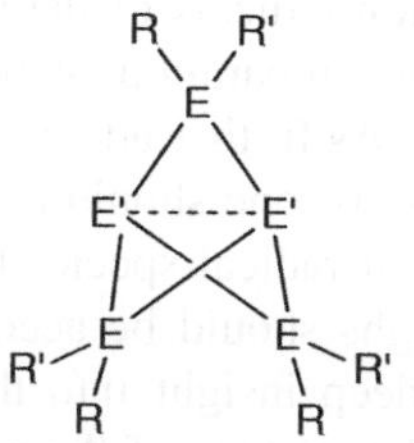

31 (E = E' = Sn; R =R' = 2,6-Et_2-C_6H_3)
32 (E = E' = Sn; R =R' = 2,6-(*i*-PrO)$_2$-C_6H_3)
33 (E = Sn, E' = Ge;
R = 2,6-[2,4,6-Me_3-C_6H_2]C_6H_3, R' = Cl)
34 (E = E' = Ge; R = R' = Mes)

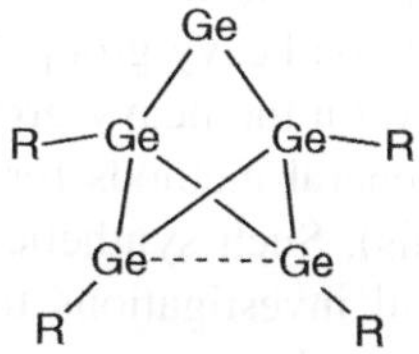

35 (R = $CH(SiMe_3)_2$)

Scheme 2.30

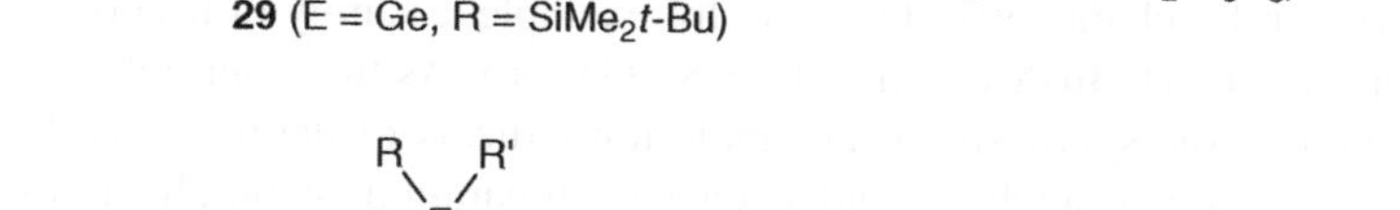

A **B** **C** **D** **E**

Scheme 2.31

2.5 Summary and Outlook

Organic free radicals constitute another class of important organic derivatives, which along with carbenium ions and carbanions are among the most fundamental organic species. Free radicals have been known for more than a century, since Gomberg's historical discovery of the first representative, triphenylmethyl radical $Ph_3C\bullet$, obtained by the treatment of triphenylmethyl chloride with metallic silver.[140] This was the first representative of a new class of organic derivatives, which were later called free radicals. The other milestone breakthrough at the beginning of the era of free radical chemistry was achieved by Paneth, who reported in 1929 the first evidence for the generation of the parent methyl radical by the thermal decomposition of tetramethyllead.[141] Since then, a great deal of experimental information has been accumulated supporting the real existence and key participation of free radicals in a number of chemical reactions. The chemistry of the heavy analogs of organic free radicals, featuring an unpaired electron centered on Si, Ge, Sn and Pb atoms, proceeded through several evolutionary steps: from the fleeting, often undetectable, reactive intermediates (at the very beginning) to the highly persistent species (midway) and finally to the isolable fully characterizable compounds (at the present time). The thermodynamic and kinetic stability of the heavy group 14 element analogs of free radicals can be achieved through either cyclic π-delocalization/acyclic hyperconjugation or effective steric protection of the radical center by bulky substituents. Taking advantage of such stabilizing factors, several stable representatives have been prepared and structurally characterized to date, for example, a family of tricoordinate planar π-radicals of the type $(t\text{-}Bu_2MeSi)_3E\bullet$ (E = Si, Ge, Sn). As the other culmination point, one should refer to the synthesis of the stable ion-radicals of the heavy carbene, heavy alkene and heavy alkyne analogs, and even the preparation of biradicaloids featuring two radical centers on heavy group 14 elements. As further advances in the chemistry of free radicals based on the heavy group 14 elements, one should expect the development of novel, more general methods for generation of radical species (first of all, persistent and stable radicals). Such synthetic breakthroughs should be accompanied by comprehensive theoretical investigations to obtain a deep insight into the bonding situations and, eventually, into the structure–reactivity relationships of the newly prepared radical species. The prospective synthetic applications of the free radicals of the heavy group 14 elements are among the most important tasks to be achieved. As the beginning of this process one can mention the utilization of the transient $n\text{-}Bu_3Sn\bullet$ and $(Me_3Si)_3Si\bullet$ radicals in organic synthesis (see Section 2.2.3). However, much more extensive investigations in the field of synthetic applications of the heavy group 14 element centered radicals (both transient and stable) as the initiators and reagents in radical substitution and addition reactions (especially in biological reactions) are required.

2.6 References

1. P. Pacansky, W. Koch, M. D. Miller, *J. Am. Chem. Soc.* **1991**, *113*, 317.
2. Reviews on the chemistry of the heavy group 14 element centered free radicals: (a) H. Sakurai, *J. Organomet. Chem. Library*, **1981**, *12* (Organomet. Chem. Rev.), 267; (b) H. Sakurai, in *Free Radicals, Vol. 2* (Ed. J. K. Kochi), John Wiley & Sons, Ltd, New York, **1973**, pp. 741–808; (c) M. F. Lappert and P. W. Lednor, *Adv. Organomet. Chem.* **1976**, *14*, 345; (d) C. Chatgilialoglu, *Acc. Chem. Res.* **1992**, *25*, 188; (e) C. Chatgilialoglu, *Chem. Rev.* **1995**, *95*, 1229; (f) J. Iley, in *The Chemistry of Organic Germanium, Tin and Lead Compounds* (Ed. S. Patai), John Wiley & Sons, Ltd, Chichester, **1995**, Chapter 5; (g) C. Chatgilialoglu, C. Ferreri, T. Gimisis, in *The Chemistry of Organic Silicon Compounds, Vol. 2, Part 2* (Eds Z. Rappoport, Y. Apeloig), John Wiley & Sons, Ltd, Chichester, **1998**, Chapter 25; (h) C. Chatgilialoglu, M. Newcomb, *Adv. Organomet. Chem.* **1999**, *44*, 67; (i) C. Chatgilialoglu, C. H. Schiesser, in *The Chemistry of Organic Silicon Compounds, Vol. 3* (Eds Z. Rappoport, Y. Apeloig), John Wiley & Sons, Ltd, Chichester, **2001**, Chapter 4; (j) M. W. Carland, C. H. Schiesser, in *The Chemistry of*

Organic Germanium, Tin and Lead Compounds, Vol. 2, Part 2 (Ed. Z. Rappoport), John Wiley & Sons, Ltd, Chichester, **2002**, Chapter 19.

3. Reviews on the chemistry of the persistent and stable heavy group 14 elements centered free radicals: (a) P. P. Power, *Chem. Rev*. **2003**, *103*, 789; (b) V. Ya. Lee, A. Sekiguchi, *Eur. J. Inorg. Chem*. **2005**, 1209; (c) V. Ya. Lee, A. Sekiguchi, *Acc. Chem. Res*. **2007**, *40*, 410; (d) V. Ya. Lee, A. Sekiguchi, in *Reviews of Reactive Intermediates Chemistry* (Eds R. A. Moss, M. S. Platz, M. Jones Jr), Wiley-Interscience, Hoboken, **2007**, Chapter 2; (e) V. Ya. Lee, M. Nakamoto, A. Sekiguchi, *Chem. Lett*. **2008**, *37*, 128.
4. M. A. Nay, G. N. C. Woodall, O. P. Strausz, H. E. Gunning, *J. Am. Chem. Soc*. **1965**, *87*, 179.
5. H. Sakurai, A. Hosomi, M. Kumada, *Bull. Chem. Soc. Jpn*. **1967**, *40*, 1551.
6. S. W. Bennett, C. Eaborn, A. Hudson, H. A. Hussain, R. A. Jackson, *J. Organomet. Chem*. **1969**, *16*, P36.
7. H. Sakurai, K. Mochida, M. Kira, *J. Am. Chem. Soc*. **1975**, *97*, 929.
8. M. Lehnig, *Tetrahedron Lett*. **1977**, 3663.
9. H. U. Buschhaus, M. Lehnig, W. P. Neumann, *J. Chem. Soc., Chem. Commun*. **1977**, 129.
10. S. K. Tokach, R. D. Koob, *J. Am. Chem. Soc*. **1980**, *102*, 376.
11. M. Lehnig, H. U. Buschhaus, W. P. Neumann, T. Apoussidis, *Bull. Soc. Chim. Belg*. **1980**, *89*, 907.
12. B. J. Cornett, K. Y. Choo, P. P. Gaspar, *J. Am. Chem. Soc*. **1980**, *102*, 377.
13. A. F. El-Farargy, M. Lehnig, W. P. Neumann, *Chem. Ber*. **1982**, *115*, 2783
14. A. F. El-Farargy, W. P. Neumann, *J. Organomet. Chem*. **1983**, *258*, 15.
15. K. Mochida, *Bull. Chem. Soc. Jpn*. **1984**, *57*, 796.
16. I. Safarik, A. Jodhan, O. P. Strausz, T. N. Bell, *Chem. Phys. Lett*. **1987**, *142*, 115.
17. G. S. Jackel, W. Gordy, *Phys. Rev*. **1968**, *176*, 443.
18. R. V. Lloyd, M. T. Rogers, *J. Am. Chem. Soc*. **1973**, *95*, 2459.
19. S. A. Fieldhouse, A. R. Lyons, H. C. Starkie, M. C. R. Symons, *J. Chem. Soc., Dalton Trans*. **1974**, 1966.
20. R. J. Booth, S. A. Fieldhouse, H. C. Starkie, M. C. R. Symons, *J. Chem. Soc., Dalton Trans*. **1976**, 1506.
21. S. K. Tokach, R. D. Koob, *J. Phys. Chem*. **1979**, *83*, 774.
22. E. Bastian, P. Potzinger, A. Ritter, H.-P. Schuchmann, C. von Sonntag, G. Weddle, *Ber. Bunsen-Ges. Phys. Chem*. **1980**, *84*, 56.
23. K. Matsumoto, M. Koshi, K. Okawa, H. Matsui, *J. Phys. Chem*. **1996**, *100*, 8796.
24. L. Gammie, I. Safarik, O. P. Strauz, R. Roberge, C. Sandorfy, *J. Am. Chem. Soc*. **1980**, *102*, 378.
25. J. A. Connor, R. N. Haszeldine, G. I. Leigh, R. D. Sedgwick, *J. Chem. Soc. A* **1967**, 768.
26. G. B. Watts, K. U. Ingold, *J. Am. Chem. Soc*. **1972**, *94*, 491.
27. H.-U. Buschhaus, W. P. Neumann, T. Apoussidis, *Liebigs Ann. Chem*. **1981**, 1190.
28. M. Lehnig, T. Apoussidis, W. P. Neumann, *Chem. Phys. Lett*. **1983**, *100*, 189.
29. K. Mochida, M. Wakasa, Y. Nakadaira, Y. Sakaguchi, H. Hayashi, *Organometallics* **1988**, *7*, 1869.
30. G. W. Slugget, W. L. Leigh, *Organometallics* **1992**, *11*, 3731.

31. C. Grugel, M. Lehnig, W. P. Neumann, J. Sauer, *Tetrahedron Lett.* **1980**, *21*, 273.
32. J. E. Bennett, J. A. Howard, *Chem. Phys. Lett.* **1972**, *15*, 322.
33. (a) K. Mochida, M. Wakasa, S. Ishizaka, M. Kotani, Y. Sakaguchi, H. Hayashi, *Chem. Lett.* **1985**, 1709; (b) K. Mochida, M. Wakasa, Y. Sakaguchi, H. Hayashi, *Chem. Lett.* **1986**, 773.
34. M. Westerhausen, T. Hildenbrand, *J. Organomet. Chem.* **1991**, *411*, 1.
35. P. J. Davidson, A. Hudson, M. F. Lappert, P. W. Lednor, *J. Chem. Soc., Chem. Commun.* **1973**, 829.
36. M. J. S. Gynane, D. H. Harris, M. F. Lappert, P. P. Power, P. Riviére, M. Riviére-Baudet, *J. Chem. Soc., Dalton Trans.* **1977**, 2004.
37. R. J. Boczkowski, R. S. Bottei, *J. Organomet. Chem.* **1973**, *49*, 389.
38. V. A. Petrosyan, M. E. Niyazymbetov, S. P. Kolesnikov, V. Ya. Lee, *Bull. Acad. Sci. USSR, Div. Chem. Sci.* **1988**, *37*, 1690.
39. R. D. Johnson III, J. W. Hudgens, *Chem. Phys. Lett.* **1987**, *141*, 163.
40. R. D. Johnson III, B. P. Tsai, J. W. Hudgens, *J. Chem. Phys.* **1988**, *89*, 4558.
41. P. D. Lightfoot, R. Becerra, A. A. Jemi-Alade, R. Lesclaux, *Chem. Phys. Lett.* **1991**, *180*, 441.
42. M. Tanimoto, S. Saito, *J. Chem. Phys.* **1999**, *111*, 9242.
43. (a) R. L. Morehouse, J. J. Christiansen, W. Gordy, *J. Chem. Phys.* **1966**, *45*, 1751; (b) G. S. Jackel, J. J. Christiansen, W. Gordy, *J. Chem. Phys.* **1967**, *47*, 4274; (c) G. S. Jackel, W. Gordy, *Phys. Rev.* **1968**, *176*, 443.
44. T. Katsu, Y. Yatsurugi, M. Sato, Y. Fujita, *Chem. Lett.* **1975**, 343.
45. (a) K. Nakamura, N. Masaki, S. Sato, K. Shimokoshi, *J. Chem. Phys.* **1985**, *83*, 4504; (b) K. Nakamura, N. Masaki, S. Sato, K. Shimokoshi, *J. Chem. Phys.* **1986**, *85*, 4204; (c) K. Nakamura, M. Okamoto, T. Takayanagi, T. Kawachi, K. Shimokoshi, S. Sato, *J. Chem. Phys.* **1989**, *90*, 2992.
46. L. Pauling, *J. Chem. Phys.* **1969**, *51*, 2767.
47. K. Nakamura, T. Takayanagi, M. Okamoto, K. Shimokoshi, S. Sato, *Chem. Phys. Lett.* **1989**, *164*, 593.
48. M. V. Merritt, R. W. Fessenden, *J. Chem. Phys.* **1972**, *56*, 2353.
49. (a) J. Roncin, *Mol. Cryst.* **1967**, *3*, 117; (b) C. Hesse, N. Leray, J. Roncin, *J. Chem. Phys.* **1972**, *57*, 749.
50. (a) S. W. Bennett, C. Eaborn, A. Hudson, R. A. Jackson, K. D. J. Root, *J. Chem. Soc. A* **1970**, 348; (b) P. J. Krusic, J. K. Kochi, *J. Am. Chem. Soc.* **1969**, *91*, 3938.
51. R. A. Jackson, *J. Chem. Soc., Perkin Trans. II* **1983**, 523.
52. R. A. Jackson, H. Weston, *J. Organomet. Chem.* **1984**, *277*, 13.
53. H. Sakurai, K. Ogi, A. Hosomi, M. Kira, *Chem. Lett.* **1974**, 891.
54. (a) J. Cooper, A. Hudson, R. A. Jackson, *Mol. Phys.* **1972**, *23*, 209; (b) R. A. Jackson, C. J. Rhodes, *J. Organomet. Chem.* **1987**, *336*, 45.
55. H. Sakurai, K. Mochida, M. Kira, *J. Organomet. Chem.* **1977**, *124*, 235.
56. K. Mochida, *Bull. Chem. Soc. Jpn.* **1984**, *57*, 796.
57. M. Lehnig, W. P. Neumann, E. Wallis, *J. Organomet. Chem.* **1987**, *333*, 17.
58. Generation and EPR studies of the $(2,4,6\text{-}i\text{-}Pr_3\text{-}C_6H_2)_3Sn^\bullet$ radical were repeatedly reported, however the most comprehensive discussion was given in references 27 and 28.

59. M. B. Taraban, O. S. Volkova, A. I. Kruppa, T. V. Leshina, in *The Chemistry of Organic Germanium, Tin and Lead Compounds, Vol. 2, Part 1* (Ed. Z. Rappoport), John Wiley & Sons, Ltd, Chichester, **2002**, Chapter 9.
60. R. Kaptein, *J. Chem. Soc., Chem. Commun*. **1971**, 732.
61. F. M. Bickelhaupt, T. Zigler, P. von R. Schleyer, *Organometallics* **1996**, *15*, 1477.
62. F. K. Cartledge, R. V. Piccione, *Organometallics* **1984**, *3*, 299.
63. K. N. Houk, M. N. Paddon-Row, *J. Am. Chem. Soc*. **1981**, *103*, 5046.
64. K. Morokuma, L. Pedersen, M. Karplus, *J. Chem. Phys*. **1968**, *48*, 4801.
65. J. Moc, Z. Latajka, H. Ratajczak, *Z. Phys. D: At., Mol. Clusters* **1986**, *4*, 185.
66. J. Moc, Z. Latajka, H. Ratajczak, *Chem. Phys. Lett*. **1987**, *136*, 122.
67. J. Moc, J. M. Rudzinski, H. Ratajczak, *Chem. Phys*. **1992**, *159*, 197.
68. M. Guerra, *J. Am. Chem. Soc*. **1993**, *115*, 11926.
69. (a) B. Giese, *Radicals in Organic Synthesis: Formation of Carbon-Carbon Bonds*, Pergamon Press, Oxford, 1986; (b) D. P. Curran, *Synthesis* **1988**, 417; (c) D. P. Curran, *Synthesis* **1988**, 489; (d) M. Malacria, *Chem. Rev*. **1996**, *96*, 289; (e) I. Ryu, N. Sonoda, D. P. Curran, *Chem. Rev*. **1996**, *96*, 177; (f) W. B. Motherwell, D. Crich, *Free Radical Chain Reactions in Organic Synthesis*, Academic Press, London, 1992; (g) D. P. Curran, N. A. Porter, B. Giese, *Stereochemistry of Radical Reactions*, VCH, Weinheim, 1995; (h) D. P. Curran, in *Comprehensive Organic Synthesis, Vol. 4* (Eds B. M. Trost, I. Fleming), Pergamon Press, Oxford, 1991, pp. 715–831; (i) C. P. Jasperse, D. P. Curran, T. L. Fevig, *Chem. Rev*. **1991**, *91*, 1237.
70. E. Lee, S. B. Ko, K. W. Jung, M. H. Chang, *Tetrahedron Lett*. **1989**, *30*, 827.
71. G. Stork, R. Mook Jr, *J. Am. Chem. Soc*. **1987**, *109*, 2829.
72. I. Ryu, A. Kurihara, H. Muraoka, S. Tsunoi, N. Kambe, N. Sonoda, *J. Org. Chem*. **1994**, *59*, 7570.
73. (a) S. F. Wnuk, M. J. Robins, *J. Am. Chem. Soc*. **1996**, *118*, 2519; (b) S. F. Wnuk, J. M. Rios, J. Khan, Y.-L. Hsu, *J. Org. Chem*. **2000**, *65*, 4169.
74. M. Nishida, A. Nishida, N. Kawahara, *J. Org. Chem*. **1996**, *61*, 3574.
75. M. Chareyron, P. Devin, L. Fensterbank, M. Malacria, *Synlett* **2000**, 83.
76. (a) T. Naito, D. Fukumoto, K. Takebayashi, T. Kiguchi, *Heterocycles* **1999**, *51*, 489; (b) T. Kiguchi, M. Okazaki, T. Naito, *Heterocycles* **1999**, *51*, 2711; (c) T. Naito, K. Nakagawa, T. Nakamura, A. Kasei, I. Ninomiya, T. Kiguchi, *J. Org. Chem*. **1999**, *64*, 2003; (d) T. Naito, J. S. Nair, A. Nishiki, K. Yamashita, T. Kiguchi, *Heterocycles* **2000**, *53*, 2611; (e) H. Miyabe, M. Ueda, N. Yoshioka, K. Yamakawa, T. Naito, *Tetrahedron* **2000**, *56*, 2413; (f) T. Kiguchi, K. Tajiri, I. Ninomiya, T. Naito, *Tetrahedron* **2000**, *56*, 5819; (g) H. Miyabe, H. Tanaka, T. Naito, *Chem. Pharm. Bull*. **2004**, *52*, 74.
77. A. Alberti, C. Chatgilialoglu, *Tetrahedron* **1990**, *46*, 3963.
78. (a) C. Ferreri, M. Ballestri, C. Chatgilialoglu, *Tetrahedron Lett*. **1993**, *34*, 5147; (b) C. Chatgilialoglu, M. Ballestri, C. Ferreri, D. Vecchi, *J. Org. Chem*. **1995**, *60*, 3826.
79. J. Cossy, L. Sallé, *Tetrahedron Lett*. **1995**, *36*, 7235.
80. The first solution EPR spectra of the transient radicals were observed in 1969 and 1972. Silyl radicals: see reerences 6 and 50b. Germyl radicals: see reference 6. Stannyl radicals: see reference 26.

81. (a) J. D. Cotton, C. S. Cundy, D. H. Harris, A. Hudson, M. F. Lappert, P. W. Lednor, *J. Chem. Soc., Chem. Commun*. **1974**, 651; (b) A. Hudson, M. F. Lappert, P. W. Lednor, *J. Chem. Soc., Dalton Trans*. **1976**, 2369.
82. R. A. Jackson, *Chem. Soc. Spec. Publ*. **1970**, 295.
83. J. Roncin, R. Debuyst, *J. Chem. Phys*. **1969**, *51*, 577.
84. S. Kyushin, H. Sakurai, T. Betsuyaku, H. Matsumoto, *Organometallics* **1997**, *16*, 5386.
85. C. Chatgilialoglu, S. Rossini, *Bull. Soc. Chim. Fr*. **1988**, 298.
86. S. Kyushin, H. Sakurai, H. Matsumoto, *Chem. Lett*. **1998**, 107.
87. M. Kira, T. Obata, I. Kon, H. Hashimoto, M. Ichinohe, H. Sakurai, S. Kyushin, H. Matsumoto, *Chem. Lett*. **1998**, 1097.
88. Y. Apeloig, D. Bravo-Zhivotovskii, M. Yuzefovich, M. Bendikov, A. I. Shames, *Appl. Magn. Reson*. **2000**, *18*, 425.
89. D. Bravo-Zhivotovskii, M. Yuzefovich, N. Sigal, G. Korogodsky, K. Klinkhammer, B. Tumanskii, A. Shames, Y. Apeloig, *Angew. Chem., Int. Ed*. **2002**, *41*, 649.
90. D. Bravo-Zhivotovskii, I. Ruderfer, S. Melamed, M. Botoshansky, B. Tumanskii, Y. Apeloig, *Angew. Chem., Int. Ed*. **2005**, *44*, 739.
91. D. Bravo-Zhivotovskii, I. Ruderfer, M. Yuzefovich, M. Kosa, M. Botoshansky, B. Tumanskii, Y. Apeloig, *Organometallics* **2005**, *24*, 2698.
92. M. M. Olmstead, L. Pu, R. S. Simons, P. P. Power, *Chem. Commun*. **1997**, 1595.
93. A. Sekiguchi, T. Matsuno, M. Ichinohe, *J. Am. Chem. Soc*. **2000**, *122*, 11250.
94. A. Sekiguchi, T. Matsuno, M. Ichinohe, *J. Am. Chem. Soc*. **2001**, *123*, 12436.
95. Y. Ishida, A. Sekiguchi, K. Kobayashi, S. Nagase, *Organometallics* **2004**, *23*, 4981.
96. A. Sekiguchi, T. Fukawa, M. Nakamoto, V. Ya. Lee, M. Ichinohe, *J. Am. Chem. Soc*. **2002**, *124*, 9865.
97. A. Sekiguchi, T. Fukawa, V. Ya. Lee, M. Nakamoto, *J. Am. Chem. Soc*. **2003**, *125*, 9250.
98. (a) M. Nakamoto, T. Fukawa, Y. Ya. Lee, A. Sekiguchi, *J. Am. Chem. Soc*. **2002**, *124*, 15160; (b) T. Fukawa, M. Nakamoto, V. Ya. Lee, A. Sekiguchi, *Organometallics*, **2004**, *23*, 2376.
99. (a) M. Nakamoto, T. Fukawa, A. Sekiguchi, *Chem. Lett*. **2004**, *33*, 38; (b) A. Sekiguchi, T. Fukawa, V. Ya. Lee, M. Nakamoto, M. Ichinohe, *Angew. Chem., Int. Ed*. **2003**, *42*, 1143.
100. (a) A. Chrostowska, A. Dargelos, A. Graciaa, P. Baylère, V. Ya. Lee, M. Nakamoto, A. Sekiguchi, *Organometallics*, **2008**, *27*, 2915; (b) J. Y. Becker, V. Ya. Lee, M. Nakamoto, A. Sekiguchi, A. Chrostowska, A. Dargelos, *Chem. Eur. J*. **2009**, *15*, 8480.
101. (a) M. Scheer, R. C. Bilodeau, C. A. Brodie, H. K. Haugen, *Phys. Rev. A* **1998**, *58*, 2844; (b) D. R. Lide (Ed.) *CRC Handbook of Chemistry and Physics*, CRC Press, Boca Raton, **2006–2007**, Chapter 10.
102. (a) L. Szepes, T. Korányi, G. Náray-Szabó, A. Modelli, G. Distefano, *J. Organomet. Chem*. **1981**, *217*, 35; (b) K. Mochida, S. D. Worley, J. Kochi, *Bull. Chem. Soc. Jpn*. **1985**, *58*, 3389; (c) K. Mochida, S. Masuda, Y. Harada, *Chem. Lett*. **1992**, 2281.

103. (a) C. Förster, K. W. Klinkhammer, B. Tumanskii, H.-J. Krüger, H. Kelm, *Angew. Chem., Int. Ed*. **2007**, *46*, 1156; (b) M. Becker, C. Förster, C. Franzen, J. Harthrath, E. Kirsten, J. Knuth, K. W. Klinkhammer, A. Sharma, D. Hinderberger, *Inorg. Chem*. **2008**, *47*, 9965.
104. C. Drost, J. Griebel, R. Kirmse, P. Lönnecke, J. Reinhold, *Angew. Chem., Int. Ed*. **2009**, *48*, 1962.
105. B. Gehrhus, P. B. Hitchcock, L. Zhang, *Angew. Chem., Int. Ed*. **2004**, *43*, 1124.
106. Reviews on the heavy carbene analogs R_2E: (E = Si, Ge, Sn, Pb): (a) W. P. Neumann, *Chem. Rev*. **1991**, *91*, 311; (b) P. P. Gaspar, R. West, in *The Chemistry of Organic Silicon Compounds, Vol. 2, Part 3* (Eds Z. Rappoport, Y. Apeloig), John Wiley & Sons, Ltd, Chichester, **1998**, Chapter 43; (c) N. Tokitoh, W. Ando, in *Reactive Intermediate Chemistry* (Eds R. A. Moss, M. S. Platz, M. Jones Jr), Wiley-Interscience, Hoboken, **2004**, Chapter 14.
107. L. B. Knight, M. Winiski, P. Kudelko, C. A. Arrington, *J. Chem. Phys*. **1989**, *91*, 3368.
108. V. Ya. Lee, A. A. Basova, I. A. Matchkarovskaya, V. I. Faustov, M. P. Egorov, O. M. Nefedov, R. D. Rakhimov, K. P. Butin, *J. Organomet. Chem*. **1995**, *499*, 27.
109. L. R. Sita, I. Kinoshita, *J. Am. Chem. Soc*. **1992**, *114*, 7024.
110. M. P. Egorov, O. M. Nefedov, T.-S. Lin, P. P. Gaspar, *Organometallics* **1995**, *14*, 1539.
111. S. Ishida, T. Iwamoto, M. Kira, *J. Am. Chem. Soc*. **2003**, *125*, 3212.
112. G. Molev, D. Bravo-Zhivotovskii, M. Karni, B. Tumanskii, M. Botoshansky, Y. Apeloig, *J. Am. Chem. Soc*. **2006**, *128*, 2784.
113. S. Inoue, M. Ichinohe, A. Sekiguchi, *J. Am. Chem. Soc*. **2007**, *129*, 6096.
114. S. Inoue, M. Ichinohe, A. Sekiguchi, *Organometallics* **2008**, *27*, 1358.
115. M. Weidenbruch, K. Kramer, A. Schäfer, J. K. Blum, *Chem. Ber*. **1985**, *118*, 107.
116. M. Kira, T. Iwamoto, *J. Organomet. Chem*. **2000**, *611*, 236.
117. A. Sekiguchi, S. Inoue, M. Ichinohe, Y. Arai, *J. Am. Chem. Soc*. **2004**, *126*, 9626.
118. (a) T. Fukawa, V. Ya. Lee, M. Nakamoto, A. Sekiguchi, *J. Am. Chem. Soc*. **2004**, *126*, 11758; (b) V. Ya. Lee, T. Fukawa, M. Nakamoto, A. Sekiguchi, B. L. Tumanskii, M. Karni, Y. Apeloig, *J. Am. Chem. Soc*. **2006**, *128*, 11643.
119. M. M. Olmstead, R. S. Simons, P. P. Power, *J. Am. Chem. Soc*. **1997**, *119*, 11705.
120. L. Pu, S. T. Haubrich, P. P. Power, *J. Organomet. Chem*. **1999**, *582*, 100.
121. L. Pu, A. D. Phillips, A. F. Richards, M. Stender, R. S. Simons, M. M. Olmstead, P. P. Power, *J. Am. Chem. Soc*. **2003**, *125*, 11626.
122. A. Sekiguchi, R. Kinjo, M. Ichinohe, *Science* **2004**, *305*, 1755.
123. R. Kinjo, M. Ichinohe, A. Sekiguchi, *J. Am. Chem. Soc*. **2007**, *129*, 26.
124. S. Inoue, M. Ichinohe, A. Sekiguchi, *J. Am. Chem. Soc*. **2008**, *130*, 6078.
125. J. A. Berson, *Acc. Chem. Res*. **1978**, *11*, 446.
126. (a) H. Grützmacher, F. Breher, *Angew. Chem., Int. Ed*. **2002**, *41*, 4006; (b) F. Breher, *Coord. Chem. Rev*. **2007**, *251*, 1007.
127. M. J. S. Dewar, E. F. Healy, *Chem. Phys. Lett*. **1987**, *141*, 521.
128. (a) R. Jain, G. J. Snyder, D. A. Dougherty, *J. Am. Chem. Soc*. **1984**, *106*, 7294; (b) R. Jain, M. B. Sponsler, F. D. Coms, D. A. Dougherty, *J. Am. Chem. Soc*. **1988**, *110*, 1356.

129. (a) E. Niecke, A. Fuchs, F. Baumeister, M. Nieger, W. W. Schoeller, *Angew. Chem., Int. Ed*. **1995**, *34*, 555; (b) T. Baumgartner, D. Gudat, M. Nieger, E. Niecke, and T. J. Schiffer, *J. Am. Chem. Soc*. **1999**, *121*, 5953. (c) E. Niecke, A. Fuchs, M. Nieger, *Angew. Chem., Int. Ed*. **1999**, *38*, 3028; (d) E. Niecke, A. Fuchs, M. Nieger, O. Schmidt, W. W. Schoeller, *Angew. Chem., Int. Ed*. **1999**, *38*, 3031; (e) E. Niecke, A. Fuchs, O. Schmidt, M. Nieger, *Phosphorus, Sulfur, Silicon Relat. Elem*. **1999**, *144–146*, 41; (f) H. Sugiyama, S. Ito, M. Yoshifuji, *Angew. Chem., Int. Ed*. **2003**, *42*, 3802; (g) D. Scheschkewitz, H. Amii, H. Gornitzka, W. W. Schoeller, D. Bourissou, G. Bertrand, *Science* **2002**, *295*, 1880; (h) W. W. Schoeller, A. Rozhenko, D. Bourissou, G. Bertrand, *Chem. Eur. J*. **2003**, *9*, 3611; (i) D. Scheschkewitz, H. Amii, H. Gornitzka, W. W. Schoeller, D. Bourissou, G. Bertrand, *Angew. Chem., Int. Ed*. **2004**, *43*, 585; (j) H. Amii, L. Vranicar, H. Gornitzka, D. Bourissou, G. Bertrand, *J. Am. Chem. Soc*. **2004**, *126*, 1344.
130. C. Cui, M. Brynda, M. M. Olmstead, P. P. Power, *J. Am. Chem. Soc*. **2004**, *126*, 6510.
131. H. Cox, P. B. Hitchcock, M. F. Lappert, L. J.-M. Pierssens, *Angew. Chem., Int. Ed*. **2004**, *43*, 4500.
132. T. Iwamoto, D. Yin, C. Kabuto, M. Kira, *J. Am. Chem. Soc*. **2001**, *123*, 12730.
133. T. Iwamoto, D. Yin, S. Boomgaarden, C. Kabuto, M. Kira, *Chem. Lett*. **2008**, *37*, 520.
134. (a) L. R. Sita, R. D. Bickerstaff, *J. Am. Chem. Soc*. **1989**, *111*, 6454; (b) L. R. Sita, I. Kinoshita, *J. Am. Chem. Soc*. **1990**, *112*, 8839; (c) L. R. Sita, I. Kinoshita, *J. Am. Chem. Soc*. **1991**, *113*, 5070.
135. C. Drost, M. Hildebrand, P. Lönnecke, *Main Group Metal Chem*., **2002**, *25*, 93.
136. A. F. Richards, M. Brynda, P. P. Power, *Organometallics* **2004**, *23*, 4009.
137. D. Nied, W. Klopper, F. Breher, *Angew. Chem., Int. Ed*. **2009**, *48*, 1411.
138. A. F. Richards, M. Brynda, M. M. Olmstead, P. P. Power, *Organometallics* **2004**, *23*, 2841.
139. (a) N. O. M. Malcolm, R. J. Gillespie, P. L. A. Popelier, *J. Chem. Soc., Dalton Trans*. **2002**, 3333; (b) C. Cui, M. M. Olmstead, J. C. Fettinger, G. H. Spikes, P. P. Power, *J. Am. Chem. Soc*. **2005**, *127*, 17530; (c) Y. Jung, M. Brynda, P. P. Power, M. Head-Gordon, *J. Am. Chem. Soc*. **2006**, *128*, 7185.
140. M. Gomberg, *J. Am. Chem. Soc*. **1900**, *22*, 757.
141. F. Paneth, W. Hofeditz, *Ber. Dtsch. Chem. Ges*. **1929**, *62*, 1335.

3

Heavy Analogs of Carbanions: Si-, Ge-, Sn- and Pb-Centered Anions

3.1 Introduction

Standard organic chemistry textbooks typically define a carbanion as a highly reactive species with a trivalent carbon that has a lone electron pair and bears a negative charge. In contrast to electron-deficient carbenium ions (see Chapter 1), carbanions, featuring eight valence electrons around the central carbon, are electron rich being strong nucleophiles and Lewis bases, typically more powerful than amines. The best-known examples of carbanionic species are definitely the Grignard reagents RMgX and organolithium reagents RLi, readily accessible and extremely useful organometallic derivatives extensively applied in synthetic organic chemistry. Known for more than a hundred years, metal salts of carbanions have been thoroughly studied from the viewpoint of their fundamental properties, such as basicity, stability, ion-pairing behavior and aggregation states.

Carbanions contain an sp^3-hybridized anionic carbon adopting a trigonal-pyramidal geometry with one of its tetrahedral valencies occupied by a lone electron pair. Such geometry is highly reminiscent of that of amines, for example NH_3, which is isoelectronic with the simplest methyl anion CH_3^-. This was indeed confirmed by PES measurements suggesting a barrier to inversion and pyramidal structure for CH_3^-,[1] that agreed well with the computational data, revealing an H–C–H bond angle for the methyl anion of 109.4°.[2] The preference for the pyramidal geometry of carbanions can be readily understood from simple orbital considerations. Thus, in a planar carbanion the lone pair would occupy a pure p-orbital, whereas in a pyramidalized carbanion the lone pair orbital has a substantial s-character (close to sp^3) gaining an additional stabilization through the

Organometallic Compounds of Low-Coordinate Si, Ge, Sn and Pb: From Phantom Species to Stable Compounds
Vladimir Ya. Lee and Akira Sekiguchi

lowering of its energy level. A PES study showed that increasing the amount of s-character of the anionic orbital on going from sp^3- to sp^2- and sp-hybridized carbanions results in an increase in the electron binding energy to 15.4 kcal/mol (for $H_2C{=}CH^-$ anion) and ca. 70 kcal/mol (for $HC{\equiv}C^-$ anion).[3]

Carbanions are intrinsically highly reactive because of the low electronegativity of carbon, appearing as reactive intermediates in a number of organic reactions. However, they can be stabilized to a marked extent by the effects of substituents in two ways. The first way provides stabilization of the carbanionic center by virtue of: (1) electronwithdrawing substituents through their inductive effect; (2) π-conjugative substituents through their resonance effect; and (3) highly polarizable substituents reducing the amount of negative charge on the carbon center. Among the substituents, that strongly stabilize carbanions, one should mention carbonyl, nitro, sulfonyl and cyano groups. In the second approach the carbanion is a part of a cyclic conjugated π-system in which the charge is delocalized over the ring carbon atoms. The most common situation is that where such delocalization leads to the formation of a particularly favorable aromatic anion; for example, the 6π-electron aromatic cyclopentadienide ion $C_5H_5^-$ (see Chapter 6). In contrast, $4n\pi$-electron antiaromatic systems, such as cyclopropenyl (4π) and cycloheptatrienyl (8π) anions, are highly destabilized.

The heavy analogs of carbanions, that is, silyl, germyl, stannyl and plumbyl anions, were prepared and synthetically utilized later than their organic counterparts. In particular, Gilman's group made many important pioneering contributions to silyllithium chemistry, as the most comprehensively studied field of the heavy group 14 element centered anions, in the 1950s and 1960s. Since then and to date, various aspects of the chemistry of alkali and alkaline earth metal salts of silanides, as well as their heavy homologs (germanides, stannanides and plumbanides), have been extensively investigated, including structure, stability, ion-pairing and aggregation states. Thus, for example, it was shown that the pyramidalization that is diagnostic of carbanions (see above) markedly increases descending group 14 from Si to Pb. The degree of such pyramidalization is totally governed by the substituent effects, which are quite parallel to those of free radicals (see Chapter 2), namely: electronic effect (increase in the inversion barrier because of the π-donating and σ-accepting substituents) and steric effect (flattening of the anionic structure upon the introduction of very bulky substituents). The two extremes are represented by the highly pyramidal X_3Si^- (small electronegative halogens X) and nearly planar $(R_3Si)_3Si^-$ (very bulky electropositive silyl groups) anions. Apart from the geometry, the nature of the interaction between the anionic and cationic portions of the molecule (covalent vs ionic), as well as the aggregation state of the anionic derivative (monomeric, dimeric or oligomeric), are of primary importance. The major instrumental tools for structural elucidation of such issues are NMR spectroscopy (in solution) and X-ray crystallography (in the solid state).

The chemistry of the heavy group 14 element centered anions, first of all that of silyllithiums, has been the subject of several preceding reviews, which have been published over the past two decades.[4] In the present chapter, we will briefly describe the progress that has been made in the whole story since its very beginning (synthesis, structure, reactions and synthetic applications of the anions based on the heavy group 14 elements), particularly dealing with the most recent advances in the field, such as the synthesis of sp^2-anions (silicon analogs of vinyllithium), cyclic and polycyclic anions and polyanions.

3.2 Synthesis

The methods for generation of silyl anions (both transient and stable) are the most extensively studied and developed.[4a,b,f,g,k] The choice of the particular synthetic approach depends on the type of substituents at the anionic Si center. Below, typical examples for preparation of silyl anions will be presented, followed by those for their heavy homologs; that is, for germyl, stannyl and plumbyl anions.

3.2.1 Alkyl-substituted Silyl Anions

1. Reduction of the Si–Si bond of disilanes with bases, such as RM, ROM, MH (M = alkali metal) (Scheme 3.1, A–C).[5–7]
2. Transmetalation of the Si–Hg bond of silylmercury derivatives with Li or Mg (Scheme 3.1, D and E).[8,9]
3. Reduction of the Si–H bond of hydrosilanes with KH (Scheme 3.1, F).[7]
4. Reduction of the Si–X bond of very bulky halosilanes with alkali metals (Scheme 3.1, G).[10]

$$Me_3Si{-}SiMe_3 + MeLi \xrightarrow{Et_2O/HMPA} Me_3SiLi + Me_4Si \qquad \textbf{(A)}$$

$$Me_3Si{-}SiMe_3 + MeONa \xrightarrow{HMPA} Me_3SiNa + Me_3SiOMe \qquad \textbf{(B)}$$

$$Me_3Si{-}SiMe_3 + 2MH \xrightarrow{THF\ or\ DME} 2Me_3SiM + H_2\ (M = Na, K) \qquad \textbf{(C)}$$

$$(Me_3Si)_2Hg + 2Li \xrightarrow{THF} 2Me_3SiLi + Hg \qquad \textbf{(D)}$$

$$(Me_3Si)_2Hg + Mg \xrightarrow{DME} (Me_3Si)_2Mg + Hg \qquad \textbf{(E)}$$

$$Et_3SiH + KH \xrightarrow{DME} Et_3SiK + H_2 \qquad \textbf{(F)}$$

$$t\text{-}Bu_3SiX + 2M \xrightarrow{THF\ or\ heptane} t\text{-}Bu_3SiM + MX\ (M = Na, K;\ X = Br, I) \qquad \textbf{(G)}$$

Scheme 3.1

It should be noted, that the trialkylsilyl anions R_3Si^- could not be prepared by the direct reduction of the corresponding halosilanes R_3SiX (X = Cl, Br) with alkali metals [except for the very bulky t-Bu_3SiX (Scheme 3.1, G)].[10] The problem arises because after the initial formation of the disilane R_3Si–SiR_3, reductive cleavage of its Si–Si bond works well only for those compounds possessing at least one aryl group at the silicon center.

3.2.2 Aryl-substituted Silyl Anions

1. Reduction of the Si–Cl bond of chlorosilanes with Li or K (the synthetically most useful method) (Scheme 3.2, A–C).[11–13]
2. Reduction of the Si–Si bond of disilanes with alkali metals (Scheme 3.2, D and E).[14,15]
3. Reduction of the Si–H bond of hydrosilanes with KH (Scheme 3.2, F).[7]
4. Reduction of the Si–C bond of arylsilanes with alkali metals (Scheme 3.2, G).[16]

$$Ar_3SiCl + 2\ Li \xrightarrow{THF} Ar_3SiLi + LiCl \quad \textbf{(A)}$$

$$Ar_n(alkyl)_{3-n}SiCl + 2\ Li \xrightarrow{THF} Ar_n(alkyl)_{3-n}SiLi + LiCl \quad \textbf{(B)}$$

$$Ar_2SiHCl + 2\ Li\ (or\ LDMAN) \xrightarrow{THF} Ar_2SiHLi + LiCl \quad \textbf{(C)}$$

$$Ph_3Si–SiPh_3 + 2\ M \xrightarrow{DME} 2\ Ph_3SiM\ (M = Li, Na, K) \quad \textbf{(D)}$$

$$PhMe_2Si–SiMe_2Ph + 2\ Li \xrightarrow{THF} 2\ PhMe_2SiLi \quad \textbf{(E)}$$

$$Ph_3SiH + KH \xrightarrow{DME} Ph_3SiK + H_2 \quad \textbf{(F)}$$

$$Ph_4Si + 2\ Li \xrightarrow{THF} Ph_3SiLi + PhLi \quad \textbf{(G)}$$

Scheme 3.2

3.2.3 Silyl-substituted Silyl Anions

1. Reduction of the Si–Si bond of oligosilanes with organo- or silyllithium reagents (Scheme 3.3, A–D).[17–19]
2. Transmetalation of the Si–Hg bond of silylmercury derivatives with Li (Scheme 3.3, E).[20]
3. Reduction of the Si–Si bond of oligosilanes with Li (Scheme 3.3, F).[21]
4. Reduction of the Si–Cl bond of chlorooligosilanes with Li (Scheme 3.3, G).[22]

$$Me_3Si–SiAr_2–SiMe_3 + MeLi \xrightarrow{THF/Et_2O} Me_3Si–SiAr_2Li + Me_4Si\ (Ar = Ph, Mes) \quad \textbf{(A)}$$

$$(Me_3Si)_4Si + MeLi \xrightarrow{THF/Et_2O} (Me_3Si)_3SiLi + Me_4Si \quad \textbf{(B)}$$

$$(Me_3Si)_4Si + Ph_3SiLi \xrightarrow{THF} (Me_3Si)_3SiLi + Me_3Si–SiPh_3 \quad \textbf{(C)}$$

$$(Me_3Sn)_4Si + MeLi \xrightarrow{THF} (Me_3Sn)_3SiLi + Me_4Sn \quad \textbf{(D)}$$

$$[(PhMe_2Si)_2MeSi]_2Hg + 2\ Li \xrightarrow{toluene} 2\ (PhMe_2Si)_2MeSiLi + Hg \quad \textbf{(E)}$$

$$(Me_3Si)_3Si–Si(SiMe_3)_3 + 2\ Li \xrightarrow{THF} 2(Me_3Si)_3SiLi \quad \textbf{(F)}$$

$$Ph_3Si–Ph_2SiCl + 2\ Li \xrightarrow{THF} Ph_3Si–Ph_2SiLi + LiCl \quad \textbf{(G)}$$

Scheme 3.3

3.2.4 Silyl Anions with Functional Groups

3.2.4.1 (Hydrido)Silyllithium Derivatives

1. Reduction of the silane H_4Si with alkali metals (Scheme 3.4, A).[23]
2. Reduction of the Si–Cl bond of chlorosilanes with Li (Scheme 3.4, B).[24]
3. Reduction of the Si–Si bond of disilanes with Li (Scheme 3.4, C).[13a]

$$H_4Si + 2\,M \xrightarrow[\text{or diglyme}]{\text{DME, HMPA}} H_3SiM + MH \ (M = Na, K) \quad \textbf{(A)}$$

$$HMes_2SiCl + 2\,Li \xrightarrow{\text{THF}} HMes_2SiLi + LiCl \quad \textbf{(B)}$$

$$Ph_2HSi\text{–}SiHPh_2 + 2\,Li \xrightarrow{\text{THF}} 2\,Ph_2HSiLi \quad \textbf{(C)}$$

Scheme 3.4

3.2.4.2 *(Halo)Silyllithium Derivatives*

(Halo)silyllithium derivatives $R_2Si(Li)X$, otherwise known as silylenoids, are postulated as reactive intermediates in the reduction of 1,1-dihalosilanes R_2SiX_2 with Li (Scheme 3.5, **A**).[25] Silylenoids are typically thermally labile and readily undergo either intramolecular α-elimination of LiX to form silylenes R_2Si: or intermolecular dimerization to produce disilenes $R_2Si{=}SiR_2$, finally yielding linear and cyclic oligo- and polysilanes.

Perchlorosilyllithium derivatives can be generated by either halogen-lithium exchange between Cl_3SiBr and 2,4,6-tri-*t*-butylphenyllithium (Scheme 3.5, **B**)[26] or deprotonation of Cl_3SiH with amines (Scheme 3.5, **C**).[27] Salts of the trichlorogermyl anion can be generated similarly but in contrast to the Cl_3Si^- anion, its germanium analog Cl_3Ge^- readily tends to lose chloride ion and form dichlorogermylene Cl_2Ge:.[28]

$$R_2SiCl_2 + 2\,Li \xrightarrow{\text{THF}} [R_2Si(Li)Cl] + LiCl \quad \textbf{(A)}$$

$$Cl_3SiBr + 2,4,6\text{-}t\text{-}Bu_3\text{-}C_6H_2Li \longrightarrow Cl_3SiLi + 2,4,6\text{-}t\text{-}Bu_3\text{-}C_6H_2Br \quad \textbf{(B)}$$

$$Cl_3SiH + R_3N \xrightarrow{CH_3CN} Cl_3Si^-\ HNR_3^+ \ (R = i\text{-Pr}, n\text{-Bu}) \quad \textbf{(C)}$$

Scheme 3.5

3.2.4.3 *(Amino)Silyllithium Derivatives*

1. Reduction of the Si–Cl bond of (amino)phenylchlorosilanes with Li (Scheme 3.6, **A**).[29]
2. Transmetalation of the Si–Sn bond of (amino)silylstannanes with *n*-BuLi or *t*-BuLi (Scheme 3.6, **B**).[4a]

$$(Et_2N)_nPh_{3-n}SiCl + 2\,Li \xrightarrow{\text{THF}} (Et_2N)_nPh_{3-n}SiLi + LiCl \ (n = 1,2) \quad \textbf{(A)}$$

$$(Et_2N)Ph_2Si\text{–}SnMe_3 + n\text{-BuLi} \xrightarrow{\text{THF}} (Et_2N)Ph_2SiLi + n\text{-BuSnMe}_3 \quad \textbf{(B)}$$

Scheme 3.6

3.2.4.4 (Alkoxy)Silyllithium Derivatives

1. Reduction of the Si–Cl bond of (alkoxy)phenylchlorosilanes with the lithium 1-(dimethylamino)naphthalenide (LDMAN) (Scheme 3.7, A).[13c]
2. Transmetalation of the Si–Sn bond of (alkoxy)silylstannanes with *n*-BuLi (Scheme 3.7, B).[30] (t-BuO)Ph_2SiLi prepared by this method exhibits ambiphilic reactivity, indicative of its silylenoid character.

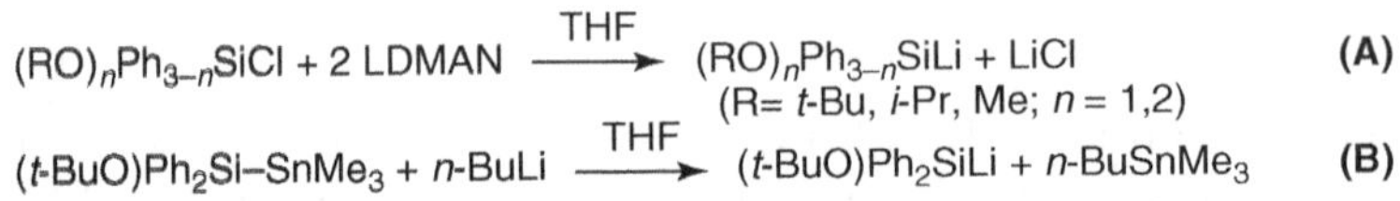

Scheme 3.7

3.2.4.5 (Acyl)Silyllithium Derivatives (Lithium Sila-enolates)

Reduction of acyltris(trimethylsilyl)silane R(O)C–$Si(SiMe_3)_3$ with $(Me_3Si)_3SiLi$ in THF generated a silicon version of lithium enolate (lithium sila-enolate), which can be trapped with Et_3SiCl to form quantitatively the corresponding siloxysilene (Scheme 3.8).[31]

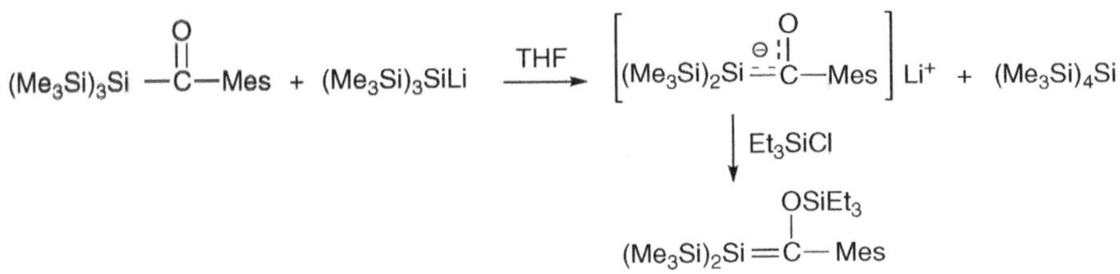

Scheme 3.8

3.2.5 Cyclic Anions

3.2.5.1 Silole, Germole and Stannole Anions and Dianions

Synthesis, structural features and aromaticity of these cyclic anions of the heavy group 14 elements will be discussed in detail in Chapter 6 (Section 6.3.2.2).

1. Silole monoanions can be most conveniently prepared by the reductive cleavage of the Si–Si bond of bis(siloles) with alkali metals (Scheme 3.9).[32]
2. Metallole dianions are readily available by the reduction of 1,1-dihalosiloles, 1,1-dihalogermoles and 1,1-diphenylstannoles with alkali metals (Scheme 3.10).[33]

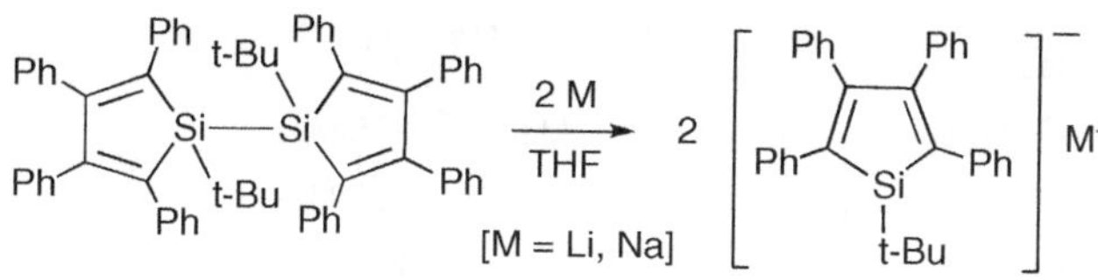

Scheme 3.9

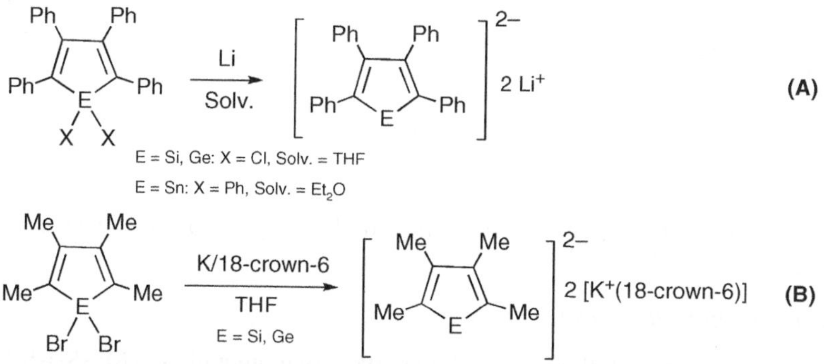

Scheme 3.10

3.2.5.2 Cyclic Oligosilyl Anions

1. As a particular case, dodecamethylcyclohexasilane *cyclo*-$(Me_2Si)_6$ can be reduced with Li to selectively cleave an exocyclic Si–CH_3 bond rather than endocyclic Si–Si bonds, producing undecamethylcyclohexasilanyllithium, which is stable at room temperature (Scheme 3.11, A).[34] Other reducing reagents, such as alkali metals (Na, K), organolithiums (MeLi, *n*-BuLi, PhLi), alkoxides ROM (R = Me, Et, *t*-Bu; M = Na, K) and Me_3SiK, also convert the cyclohexasilane *cyclo*-$(Me_2Si)_6$ in HMPA/Et_2O to its alkali metal derivatives, albeit in lower yields because of the unavoidable formation of ring-opening by-products (linear oligosilanes).[34]
2. The lower homolog, nonamethylcyclopentasilanylpotassium, cannot be prepared by the above-described method. Instead, it is available by the transmetalation of bis(nonamethylcyclopentasilanyl)mercury with Na/K alloy (Scheme 3.11, B).[4a] Alternatively, it can be prepared by the reaction of *cyclo*-$(Me_2Si)_6$ with *t*-BuOK in DME or diglyme.[35]

3.2.6 Polylithiosilanes

3.2.6.1 Tetralithiosilane

The preparation of $SiLi_4$ by the reaction of lithium vapor with $SiCl_4$ was first claimed three decades ago based on its subsequent trapping with MeCl to form $SiMe_4$, albeit in low yield (5–10%).[36] However, the real structure of the reduction product is yet to be unambiguously proved. Interestingly, tetrahedral $SiLi_4$ was found computationally to be only a saddle point on the PES whereas the minimum structure was represented by

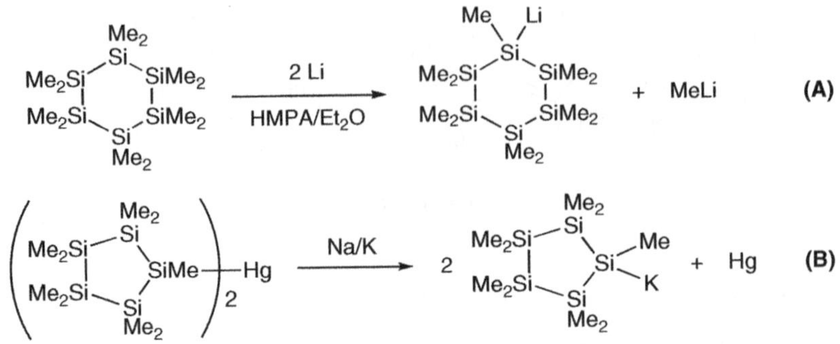

Scheme 3.11

its C_{2v} conformer featuring an inverted tetrahedral configuration of the Si atom (at the MP2/6-31G*//3-21G* level).[37]

3.2.6.2 1,1-Dilithiosilane Derivatives

The first representative of this class of reactive organometallics, $(Me_3Si)_2SiLi_2$, was prepared by the thermolysis of solid $(Me_3Si)_3SiLi$ under vacuum (Scheme 3.12).[38] Quenching with EtOD produced the expected $(Me_3Si)_2SiD_2$ in 35% yield, along with a number of other products. Formation of the target compound was also confirmed by the observation of the parent ion peaks of the monomer and dimer of dilithiosilane in the flash vaporization mass spectrum.

$$(Me_3Si)_3SiLi \xrightarrow[\text{2. EtOD}]{\text{1. 140–150°C}} (Me_3Si)_2SiD_2 + Me_3Si\text{–}SiMe_3 +$$
$$+ (Me_3Si)_3SiD + (Me_3Si)_4Si + (Me_3Si)_3Si\text{–}Si(SiMe_3)_3$$

Scheme 3.12

Stable derivatives of 1,1-dilithiosilane and 1,1-dilithiogermane were recently prepared and structurally characterized, and their chemistry will be discussed in Section 3.5.1.

3.2.6.3 1,2-Dilithiodisilane Derivatives

Several experimental procedures were reported to be effective for the generation of 1,2-dimetalladisilane derivatives.

1. Reaction of $(Me_3Si)_3SiLi$ or *t*-BuOK/18-crown-6 with hexakis(trimethylsilyl)disilane (Scheme 3.13, A).[21,39]
2. Reductive ring-opening of a cyclotrisilane with Li (Scheme 3.13, B).[40]
3. Reduction of tetrasilyldisilenes $(R_3Si)_2Si{=}Si(SiR_3)_2$ with Li (Scheme 3.13, C).[41]

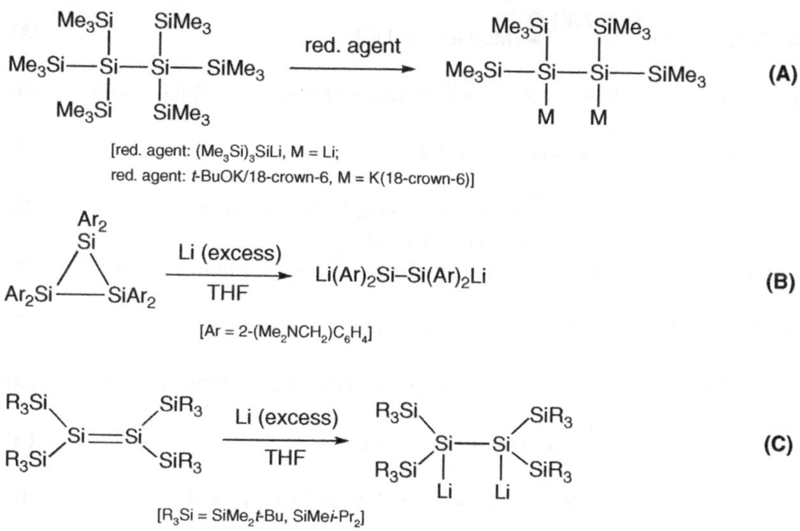

Scheme 3.13

3.2.6.4 1,3-Dilithiotrisilane and 1,4-Dilithiotetrasilane Derivatives

These dianionic species are readily available by the reductive ring-opening of some cyclic oligosilanes with Li (Scheme 3.14).[40,42]

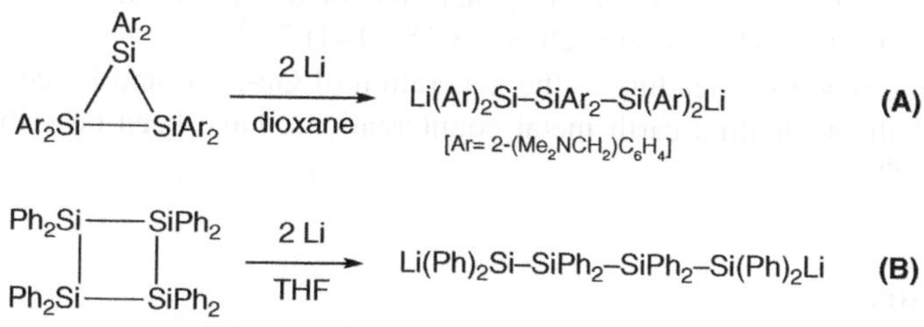

Scheme 3.14

3.2.7 Germyl, Stannyl and Plumbyl Anions

The most common methods for the synthesis of the heavier analogs of carbanions R_3EM (E = Ge, Sn, Pb; M = alkali metal), that is, germyl, stannyl and plumbyl anionic derivatives, are parallel to those described above for the silyl anions.[4h] They include: (1) reduction of the E–X bonds of halides R_3EX (R = alkyl, aryl; E = Ge, Sn, Pb; X = Cl, Br) with alkali metals (Scheme 3.15, A–C),[43–45]; and (2) reductive cleavage of the central E-E bond of digermanes, distannanes and diplumbanes R_3E-ER_3 (R = alkyl, aryl; E = Ge, Sn, Pb) with alkali metals or organolithium/organopotassium reagents (Scheme 3.15, D–G).[46–49]

$$Me_3GeCl + 2\,Li \xrightarrow{HMPA/Et_2O} Me_3GeLi + LiCl \quad \textbf{(A)}$$

$$Me_n(PhMe_2CCH_2)_{3-n}SnBr + 2\,Li \xrightarrow{THF} Me_n(PhMe_2CCH_2)_{3-n}SnLi + LiBr \quad \textbf{(B)}$$

$$Me_3PbBr + 2\,Li \xrightarrow{THF} Me_3PbLi + LiBr \quad \textbf{(C)}$$

$$Ph_3E\text{–}EPh_3 + t\text{-}BuOK \xrightarrow{DMPU} Ph_3EK + Ph_3EOt\text{-}Bu \quad [E = Ge, Sn, Pb] \quad \textbf{(D)}$$

[DMPU = *N,N'*-dimethylpropyleneurea]

$$(t\text{-}BuCH_2)_3Sn\text{–}Sn(CH_2t\text{-}Bu)_3 + K/naphthalene \xrightarrow{THF} 2\,(t\text{-}BuCH_2)_3SnK \quad \textbf{(E)}$$

$$n\text{-}Bu_3Pb\text{–}Pbn\text{-}Bu_3 + n\text{-}BuLi \xrightarrow{THF} n\text{-}Bu_3PbLi + n\text{-}Bu_4Pb \quad \textbf{(F)}$$

$$Mes_3Pb\text{–}PbMes_3 + 2\,Li \xrightarrow{THF} 2\,Mes_3PbLi \quad (Mes = 2,4,6\text{-trimethylphenyl}) \quad \textbf{(G)}$$

$$Mes_3GeH + t\text{-}BuLi \xrightarrow{THF} Mes_3GeLi + i\text{-}BuH \quad \textbf{(H)}$$

$$HPh_2Ge\text{–}GePh_2H + 2\,t\text{-}BuLi \xrightarrow{THF} LiPh_2Ge\text{–}GePh_2Li + BuH \quad \textbf{(I)}$$

$$R_3SnH + i\text{-}Pr_2NLi \xrightarrow{THF} R_3SnLi + i\text{-}Pr_2NH \quad [R = Me, n\text{-}Bu] \quad \textbf{(J)}$$

Scheme 3.15

In contrast to silylhydrides R_3SiH, direct metalation of germyl-, stannyl- and plumbylhydrides R_3EH (E = Ge, Sn, Pb) with a variety of reducing reagents [*n*-BuLi, *t*-BuLi, *i*-Pr_2NLi, NaH, $PhCH_2K$, $(Me_3Si)_2NK$, etc.] is readily available as one of the most synthetically attractive methods for the generation of anions because of the favorable polarization of the $E^{\delta-}$–$H^{\delta+}$ bond (Scheme 3.15, H–J).[50–52]

Some specific reactions, leading to the generation of Ge-, Sn- and Pb-centered anions with either alkali or alkaline earth metal countercations, have been described in a preceding review.[4h]

3.3 Structure

The most fundamental structural features of the heavy group 14 element centered anionic derivatives R_3EM (E = Si, Ge, Sn, Pb; M = alkali or alkaline earth metals) deals with the problems of geometry at the anionic center (pyramidal vs planar, inversion barrier), ion-pairing behavior reflecting the degree of anion–cation interaction (covalent vs ionic), and aggregation state both in the solid state and in solution (monomeric, dimeric or oligomeric).[4a,b,f–h,k] These issues are primarily addressed by NMR spectroscopy and X-ray crystallography studies, which are the most important experimental techniques for structural elucidation in solution and in the solid state, respectively.

3.3.1 NMR Spectroscopy (Condensed Phase Structure)

NMR spectroscopy is the most powerful experimental tool for probing the structures of carbanions in the condensed phase; the same is true for their heavy analogues, that is, silyl, germyl, stannyl and plumbyl anions. Providing a great deal of structural information on anionic derivatives in solution, NMR spectroscopy plays a key role analogous to that of EPR spectroscopy in the structural studies of radical species (see Chapter 2). Given the intrinsic electron-rich nature of the anionic centers, one should expect the appearance of their NMR signals in the diagnostic high-field region; however, the degree of shielding for the particular derivative depends on the substituents' pattern at the anionic center.

NMR spectroscopy was also successfully used for the study of geometry and configurational stability of silyllithium derivatives, which are generally markedly pyramidalized at silicon. Thus, the two methyl groups of the *i*-Pr substituents of the $Ph(i\text{-}Pr)_2SiLi$ derivative appeared to be nonequivalent, which was accounted for by their diastereotopicity because of configurational stability of the anionic silicon center on the NMR timescale up to 185 °C.[53] This allowed the determination of the lower limit for the inversion barrier at silicon as 24 kcal/mol, a value that was in fairly good agreement with the PES gas-phase estimate[54] and theoretical predictions[55] for the parent SiH_3^-.

NMR spectroscopy, particularly that of the ^{29}Si and ^{7}Li nuclei, also plays a decisive role in the investigation of ion-pairing phenomena. Thus, observation of scalar ^{29}Si–$^{7}Li(^{6}Li)$ coupling in both ^{29}Si and $^{7}Li(^{6}Li)$ NMR spectra is generally taken as direct evidence for a significant degree of covalency of the Si–Li bond [contact ion pair (CIP)], whereas absence of such coupling is usually interpreted in terms of an important ionic contribution to the bonding between Si and Li [solvent-separated ion pair (SSIP)]. For example, the ^{29}Si resonance of Ph_3Si^6Li [$I(^6Li) = 1$] was detected as a triplet with a coupling constant $^1J(^{29}Si\text{–}^6Li)$ of 17 Hz in 2-Me-THF at 173 K,[56] whereas a well-resolved quartet with $^1J(^{29}Si\text{–}^7Li) = 51$ Hz (THF, 173 K)[56] and 45 Hz (toluene-d_8, 193 K)[57] was observed in the ^{29}Si NMR spectrum for Ph_3Si^7Li [$I(^7Li) = 3/2$]. Such observations are indicative of the predominantly monomeric structure of Ph_3SiLi in solution and show at least a partially covalent nature for the Si–Li bonding (CIP), which was also supported by the essential solvent independence of the ^{29}Si NMR chemical shift of Ph_3Si^6Li: −9.2 ppm (2-Me-THF), −9.0 ppm (THF), −8.7 ppm (DME).[56] In contrast, for the carbon analog Ph_3CLi no coupling was observed down to 123 K,[58] which would be consistent with either ionic formulation of the C–Li bond or rapid intermolecular Li–Li exchange on the NMR timescale.

The other comprehensively studied silyllithium derivative, tris(trimethylsilyl)-silyllithium) $(Me_3Si)_3SiLi$, also revealed Si–Li bonding, manifested in the room temperature observation of a well-resolved quartet for the central anionic Si atom at −189.4 ppm with a $^1J(^{29}Si\text{–}^7Li)$ coupling constant of 38.6 Hz (in C_7H_8–C_6D_6).[59] This pointed to the monomeric state of $(Me_3Si)_3SiLi$ in solution and its existence as a CIP, in which the Si–Li bond is partially covalent. Likewise, other alkali metal derivatives $(Me_3Si)_3SiM$ (M = Li, Na, K, Rb, Cs), prepared by the transmetalation of $[(Me_3Si)_3Si]_2M'$ (M' = Zn, Cd, Hg) with the corresponding alkali metals in pentane or heptane, exhibited similar high-field resonances for their central Si atoms in benzene solution at −185.7 (M = K), −184.4 (M = Rb) and −179.4 (M = Cs) ppm, respectively.[60] One should note that the carbon analog of $(Me_3Si)_3SiLi$ is dimeric in the solid state, $\{[(Me_3Si)_3C]_2Li\}^-\bullet[Li^+(thf)_4]$, and undergoes a complicated exchange process in solution.[61]

The 1,2-dilithiodisilane $LiAr_2Si\text{–}SiAr_2Li$ [Ar = 2-$(Me_2NCH_2)C_6H_4$] (see Section 3.2.6, Scheme 3.13, B) is also monomeric at room temperature in THF solution, demonstrating the presence of Si–Li bonding by the observation of a resonance

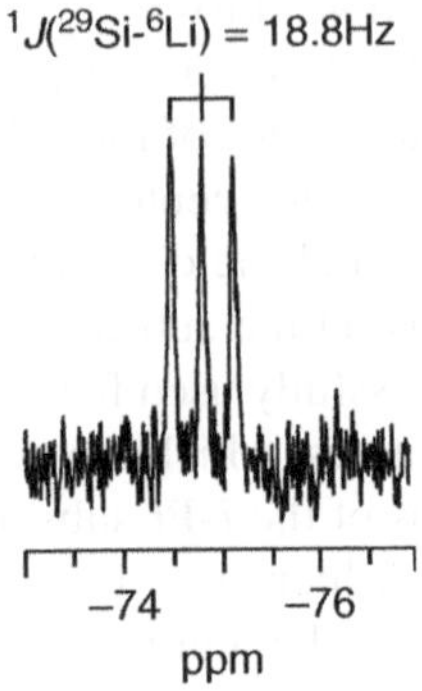

Figure 3.1 *The resonance of the anionic* $\underline{Si}$ *atom of* $Me_3Si–Me_2\underline{Si}^6Li$ *at −74.9 ppm (THF-*d_8*, 180 K)*

of the anionic silicon atoms as a quartet at −32.8 ppm with a $^1J(^{29}Si–^7Li)$ coupling constant of 36 Hz.[40]

The disilanyllithium $Me_3Si–Me_2Si^6Li$, prepared by the transmetalation of the silylmercury derivative $(Me_3SiMe_2Si)_2Hg$ with ^6Li in toluene according to the synthetic procedure described in Section 3.2.3 (Scheme 3.3, E), was also proved to be monomeric in THF-d_8 at 180 K.[62] The resonance of its anionic Si was observed at −74.9 ppm as a triplet because of coupling with one ^6Li nucleus [$^1J(^{29}Si–^6Li) = 18.8$ Hz] (Figure 3.1), which pointed to some observable extent of covalent bonding between the anionic and cationic portions of the molecule.[62]

Apart from geometrical considerations and ion-pairing behavior, NMR spectroscopy is very useful for revealing the aggregation states in silyllithium derivatives. The high charge densities on the anionic center and lithium countercation and the low dielectric constants of conventional NMR solvents (deuterated hydrocarbons, ethers) favor the formation of aggregates (dimers, tetramers, and higher oligomers) because of the strongly stabilizing electrostatic interactions, which outbalance the entropic disadvantage of oligomerization. As a representative example from organic chemistry, one should mention methyllithium which was reliably proved to exist as a tetramer $(MeLi)_4$ in solvents such as diethyl ether, thus forming a cubic structure.[63] The aggregation state of silyllithium derivatives (monomer vs dimer) could also be decided based on NMR spectroscopic data. Thus, the trisilanyllithium $(PhMe_2Si)_2MeSiLi$, conventionally synthesized by the transmetalation of $[(PhMe_2Si)_2MeSi]_2Hg$ with Li in toluene as depicted in Scheme 3.3, E (Section 3.2.3), was dimeric in toluene-d_8 featuring a high-field ^7Li NMR resonance at −0.54 ppm as a result of the shielding effect of the phenyl groups.[62b,64] However, this signal markedly moved down field (0.58 ppm, $\Delta\delta = 1.12$ ppm) when the NMR solvent was changed to the more polar THF-d_8, because of the dissociation of the dimeric structure $[(PhMe_2Si)_2MeSiLi]_2$ into the two monomers $(PhMe_2Si)_2MeSiLi$ in which the Li atoms were no longer shielded by phenyl groups.[62b,64]

For the anions centered on the group 14 elements heavier than Si, it was suggested that the degree of covalency of the E–Li bond (E = Ge, Sn, Pb) decreases descending group 14 (Ge > Sn > Pb) with a considerable amount of covalent character still attributable to the Ge–Li bond.[65] However, even stannyllithium derivatives sometimes manifest a significantly covalent Sn–Li interaction. Thus, n-Bu_3SnLi exhibited ^{119}Sn–^{7}Li coupling (402.5 Hz) in both ^{119}Sn and ^{7}Li NMR spectra measured in diethyl ether at 154 K, indicating its existence in solution as a CIP.[66] However, upon the addition of more than 2 equiv. of HMPA this CIP was transformed into the SSIP n-Bu_3Sn^{-}•$[Li(HMPA)_2]^{+}$ in ether or THF. The presence of the Sn–K bond in $[(t\text{-}BuCH_2)_3SnK]$•$3(\eta^6\text{-}C_7H_8)$ was manifested in the solid state ^{119}Sn CP-MAS NMR observation of a 1:1:1:1 quartet (I for $^{39}K = 3/2$) at $\delta_{iso} = -211$ ppm with a $^{1}J(^{119}Sn\text{–}^{39}K)$ coupling constant of 289 Hz.[67]

3.3.2 X-ray Crystallography (Crystal Structure)

Since the first crystallographic report on the silyllithium derivative Me_3SiLi in 1974, many other organometallic compounds featuring E–M bonds (E = heavy group 14 element, M = alkali or alkaline earth metal) have been structurally characterized.[4a,b,f–h] In this section we will deal with some representative examples of such compounds, of which silyllithium derivatives have been most extensively studied. The very first structurally characterized silyllithium was Me_3SiLi, which was proved to be hexameric in the crystalline form when crystallized from cyclohexane.[68] Its structure is characterized by the Li_6 six-membered ring adopting a markedly folded chair conformation, which otherwise may be better viewed as a greatly distorted octahedron with two opposite faces open (Figure 3.2). There are two types of lithium–lithium interaction in the hexamer: short (1,2-Li–Li distance) of 2.72 Å (av.) and long (1,3-Li–Li distance) of 3.25 Å (av.). Each Me_3Si group is η^3-coordinated to one of the Li_3 sides with Si–Li bond lengths of 2.68 Å (av.), which is close to the sum of the Si and Li covalent radii (2.69 Å).[68] Such silicon–lithium interaction may be considered as electron-deficient four-centered bonds; however, ionic contribution to the silicon–lithium bonding interaction cannot be ruled out.

Upon complexation with TMEDA, the hexameric structure $(Me_3SiLi)_6$ of the unsolvated silyllithium derivative is completely broken down on coordination by TMEDA to form a dimeric aggregate $(Me_3SiLi)_2$•3(TMEDA) [Si–Li = 2.70 Å (av.)], in which each Li atom is four-coordinate, being bonded to an Me_3Si substituent, one chelating TMEDA ligand and the Me_2N group of one other TMEDA molecule.[69]

The crystal structure of triphenylsilyllithium, one of the most readily available and widely used silyllithium derivatives, was determined only in 1993 as a THF-solvate Ph_3SiLi•3THF featuring a strongly pyramidalized Si center (the sum of the C–Si–C bond angles is 304°) and an Si–Li bond length of 2.672(9) Å (Figure 3.3).[70]

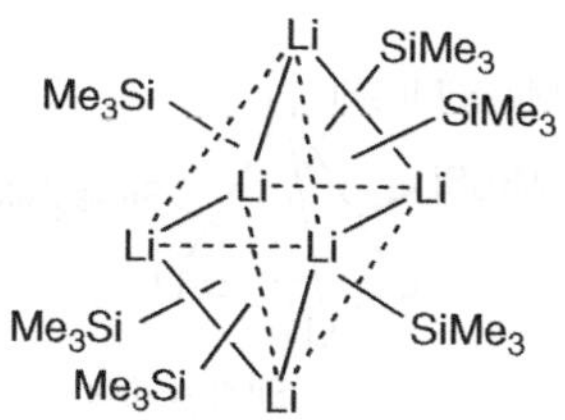

Figure 3.2 Schematic representation of the $(Me_3SiLi)_6$ hexamer

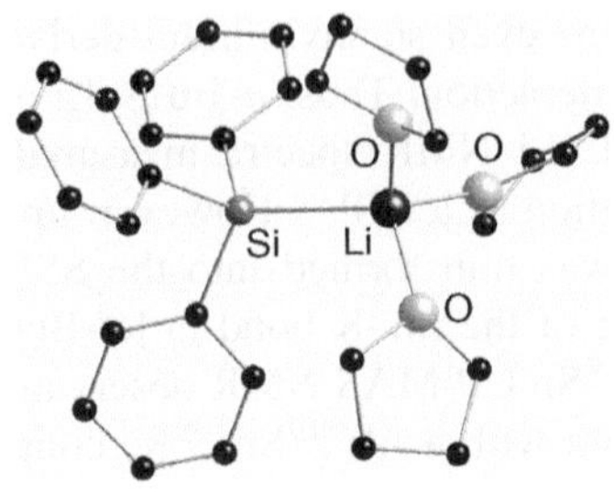

Figure 3.3 Crystal structure of $Ph_3SiLi{\bullet}3THF$ (hydrogen atoms are not shown)

This geometry of Ph_3SiLi is sharply different from that of $Ph_3CLi{\bullet}2Et_2O$, in which the central carbon adopts an essentially planar cofiguration.[71] Such a great structural difference is realized as a consequence of the much greater inversion barrier in the case of silicon.

Similarly, tris(trimethylsilyl)silyllithium was found to be a monomeric THF-solvate $(Me_3Si)_3SiLi{\bullet}3THF$, manifesting a distorted tetrahedral configuration of the anionic Si center (the sum of the Si–Si–Si bond angles is 308°) and typical Si–Li bond length of 2.669(13) Å.[59,70] In contrast, the structures of both tris(trimethylsilyl)silylrubidium and tris(trimethylsilyl)silylcesium (as toluene solvates) were dimeric with the alkali metals playing the bridging role: $[(Me_3Si)_3SiRb]_2{\bullet}C_7H_8$ and $[(Me_3Si)_3SiCs]_2{\bullet}3C_7H_8$.[60a]

The crystal structures of some other oligosilyllithium derivatives have also been reported. A THF-solvate of 1,4-dilithiooctaphenyltetrasilane $3THF{\bullet}[Li(Ph_2Si)_4Li]{\bullet}3THF$ (Section 3.2.6, Scheme 3.14, B) is monomeric, existing in a planar zigzag-shaped conformation with a Si–Li bond length of 2.714(10) Å.[72] Unsolvated pentamethyldisilanyllithium $(Me_3SiMe_2SiLi)_4$ has a tetrameric structure with the four lithium atoms at the vertices of the tetrahedron and an average Li–Li distance of 2.780 Å (Figure 3.4).[62a,73] Each Me_3SiMe_2Si unit is located above one of the faces of the Li_4 tetrahedron with the three nearly equal Si–Li distances of 2.683 Å (av.) testifying to the presence of four-center two-electron bonding. Short interatomic contacts between the lithium atoms and hydrogens of $SiMe_2$ units (2.08 Å) are indicative of Li···C–H agostic interactions. The bond angles around the anionic silicon are significantly narrowed, their sum amounting to only 307° (av.) because of the large contribution of s-character to the anionic orbital.

In accordance with the solution structure (Section 3.3.1), the trisilanyllithium $(PhMe_2Si)_2MeSiLi$ is dimeric in the crystalline state, forming a planar Si_2Li_2 four-membered ring with Si–Li distances of 2.664(5) and 2.778(7) Å.[73] The two lithium

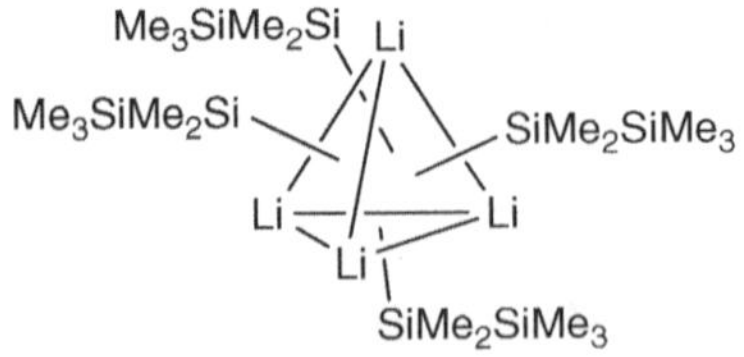

Figure 3.4 Schematic representation of the $(Me_3SiMe_2SiLi)_4$ tetramer

atoms of the dimer are surrounded by the four phenyl groups, and a weak Li–Ph π-interaction was deduced from their proximity.[73]

The first ever structurally characterized germyllithium derivative, $(Me_3Si)_3GeLi$•(n-donor) (n-donor = THF or PMDETA) exhibited Ge–Li bond lengths of 2.666(6) Å (THF donor) or 2.653(9) Å (PMDETA donor), which are longer than the sum of the covalent radii of the Ge and Li atoms (2.56 Å).[74a] The geometry around the anionic Ge center is rather far from the ideal tetrahedral, which is observed in an appreciable decrease of the Si–Ge–Si bond angles to 100.7° (av.). In the presence of 12-crown-4, tris(trimethylsilyl)germyllithium expectedly crystallized as a SSIP $(Me_3Si)_3Ge^-$•$[Li^+(12\text{-crown-}4)_2]$. The 'naked' $(Me_3Si)_3Ge^-$ ion also features very acute Si–Ge–Si angles of 102.7 and 100.5°, which were attributed to the large s-character of the anionic orbital on the Ge center and appreciable p-contribution to the Ge–Si σ-bonds.[74b]

The triphenylstannyl anion, prepared by the metalation of Ph_3SnH with n-BuLi in the presence of PMDETA, was isolated in the form of its lithium salt as a monomeric CIP, Ph_3SnLi•(PMDETA).[75a] The Sn–Li bond length of 2.871(7) Å (av.) was slightly longer than that expected for purely covalent bonding (2.74 Å). As in the above case of the germyl anion, the Sn atom in Ph_3SnLi•(PMDETA) is highly pyramidalized [average C–Sn–C bond angle = 96.1(2)°], a geometry that was previously observed for the 'naked' Ph_3Sn^- ion in the crystal structure of the SSIP Ph_3Sn^-•$[K^+(18\text{-crown-}6)]$ [average C–Sn–C bond angle = 96.9(2)°][75b]. Again, such a departure of the anionic Sn center from the ideal tetrahedrality was interpreted in terms of the high p-character of the Sn–C σ-bonds. Cryoscopic measurements in benzene proved the monomeric status of Ph_3SnLi in solution.[75a] Direct Sn–Li bonding contacts in Ph_3SnLi were also maintained in solution, which was manifested in the observation of unresolved $^{119,117}Sn$ satellites ($I = 1/2$) in the low-temperature ^7Li NMR spectrum (toluene-d_8, 183 K): $^1J(^{119,117}Sn\text{–}^7Li) = 412$ Hz.[75a] A Si–K bond length of 3.548(3) Å and highly pronounced pyramidal geometry about the anionic Sn center (275.1° around Sn) was found in the tris(*neo*-pentyl)stannylpotassium derivative, which crystallized from toluene as a solvate $[(t\text{-BuCH}_2)_3SnK]$•$3(\eta^6\text{-}C_7H_8)$.[67]

Triphenylplumbyllithium was synthesized by the cleavage of the Pb–Pb bond in $Ph_3Pb\text{–}PbPh_3$ with n-BuLi in the presence of PMDETA in toluene.[76] In the solid state, it exists as a monomeric CIP Ph_3PbLi•(PMDETA) nearly isostructural with the above-described tin analog Ph_3SnLi•(PMDETA).[75a] Similarly to the latter case of stannyllithium, the Pb–Li distance of 2.858(14) Å (average value of the two independent molecules in the unit cell) is somewhat long to be considered as a purely covalent bond (sum of the Pb and Li covalent radii = 2.81 Å). The Ph_3Pb^- ion is slightly more pyramidalized than its tin congener Ph_3Sn^- (282.9° around Pb vs 288.3° around Sn), which points to a dominance of the s-contribution to the Pb–Li bonding.[76]

Very few examples of the heavy group 14 element centered anions featuring direct bonding to alkaline earth metals have been reported. The first structurally characterized representatives were of the type $[(Me_3E)_2M]$•$(n\text{-donor})_n$ (E = Si, Ge, Sn; M = Mg, Ca; n-donor = DME, THF; n = 1, 2, 4). Thus, monomeric $[(Me_3Si)_2Mg]$•DME, prepared by the transmetalation of $(Me_3Si)_2Hg$ with Mg in DME, revealed a tetracoordinated Mg atom adopting a distorted tetrahedral configuration.[77a] Typically, the Si–Mg bond distance of 2.630(2) Å was longer than the sum of the Si and Mg covalent radii of

2.41 Å, whereas the Si–Mg–Si angle of 125.2(1)° was markedly greater than the normal tetrahedral value of 109.4° because of the steric repulsion between the bulky trimethylsilyl groups. $(Me_3Si)_2Mg$ can be also effectively stabilized by amino-ligands to form monomeric $[(Me_3Si)_2Mg]$•(TMEDA)[77b] and $[(Me_3Si)_2Mg]$•(TMDAP)[77c] derivatives, which exhibit structural characteristics very similar to those of $[(Me_3Si)_2Mg]$•DME.

Likewise, the germanium analog $[(Me_3Ge)_2Mg]$•2DME, prepared by the transmetalation of $(Me_3Ge)_2Hg$ with Mg in DME, had a Ge–Mg bond length of 2.719(6) Å and an acute Ge–Mg–Ge angle of 99.3(2)° (average values of the two independent molecules).[78]

The tin version, bis(trimethylstannyl) calcium, was isolated from the reduction of hexamethyldistannane with calcium in THF as a solvate $[(Me_3Sn)_2Ca]$•4THF (^{119}Sn NMR resonance −203 ppm).[79] The calcium atom is hexacoordinated by the two tin and four oxygen atoms in a distorted octahedral configuration with both Me_3Sn units situated at the apical positions. The Sn–Ca bond length was 3.2721(3) Å and the average value for the C–Sn–C angles was 96.5°, thus pointing to the predominantly s-character of the lone electron pair on the Sn center.

The 'silicon Grignard reagent', trimethylsilylmagnesium bromide, was prepared for the first time by the reduction of Me_3SiBr with activated Mg in toluene in the presence of TMEDA.[80] In crystalline form this compound exists as a centrosymmetric dimer $[Me_3SiMgBr$•$(TMEDA)]_2$ with two Br atoms bridging two Me_3SiMg•(TMEDA) units thus forming a planar Mg_2Br_2 ring. One of the Mg–Br distances of 2.534(4) Å is within the normal range for Grignard reagents, whereas the other one is much longer [3.220(4) Å], being 0.6 Å larger than the sum of the ionic radii of Mg^{2+} and Br^- (2.60 Å) thus indicating that the dimeric interaction is weak. The Si–Mg bond length was 2.630(4) Å and the Si–Mg–Br bond angle was 128.7(2)°. When bidentate TMEDA was replaced with tridentate PMDTA, the dimeric structure was broken down into a monomeric structure $Me_3SiMgBr$•(PMDTA), in which three coordination sites of the pentacoordinate Mg are occupied by the PMDTA ligand.[80] The Mg–Br and Si–Mg distances of 2.599(4) and 2.651(6) Å, respectively, were somewhat longer than those in the dimer $[Me_3SiMgBr$•$(TMEDA)]_2$ [2.534(4) and 2.630(4) Å], whereas the Si–Mg–Br bond angle of 105.0(2)° was markedly smaller than that in the dimer [128.7(2)°]. This was a result of the greater steric requirements of the triamine ligand PMDTA.

The magnesium salts of the tris(cyclopentadienyl)stannyl and -plumbyl anions were prepared by the reaction of stannocene and plumbocene Cp_2E (E = Sn, Pb) with Cp_2Mg and isolated as the SSIP $[(\eta^3\text{-}Cp)_3E^-]_2$•$[Mg^{2+}(thf)_6]$ (E = Sn, Pb). In both $(\eta^3\text{-}Cp)_3E^-$ anions the metallic Sn and Pb centers were insignificantly pyramidalized with the sum of the bond angles around them being 356.4°.[81] The $(\eta^3\text{-}Cp)_3E^-$ anion preserved its 'freedom' in solution, which was demonstrated by the observation of a single Cp resonance in the ^1H NMR spectra measured at low temperature: 5.77 ppm at −85 °C (E = Sn) and 5.56 ppm at −90 °C (E = Pb).

3.4 Reactions and Synthetic Applications

Carbanions are crucial intermediates in a wide variety of classical organic reaction pathways forming C–C bonds: E1 elimination (preparation of alkenes), nucleophilic addition to alkenes (Michael reaction), nucleophilic aromatic substitution (formation of Meisenheimer complexes), and rearrangements (Wittig rearrangement, Favorskii rearrangement, 1,2-phenyl group migration). On the other hand, as soft nucleophiles, carbanions are very good candidates for S_N2 reactions, which can be exemplified by the reaction of organolithium reagents (alkyl-, aryl- and alkenyllithiums) and Grignard reagents with primary alkyl halides, alkyl tosylates, and alkyl sulfates and sulfonates. The reactions of the alkali and alkaline earth metal derivatives of the heavy group 14 elements are to a large extent similar to those of the classical organolithium and organomagnesium reagents. Thus, perhaps the most important synthetic applications of silyllithium derivatives R_3SiLi is their use as silylating reagents in substitution reactions with R′X involving replacing of halogens (or other good leaving groups X, such as triflates) with silyl groups R_3Si to form new Si–C, Si–Si bonds, and other bonds to silicon:[4a,b]

$$R_3SiLi + R'X \rightarrow R_3SiR' + LiX$$

The most readily available and commonly used silyl anion reagents in organosilicon chemistry are Ph_3SiLi and $(Me_3Si)_3SiLi$, whereas synthetic organic chemistry requires application of much less sterically demanding silyllithium reagents, such as Me_3SiLi and $PhMe_2SiLi$.[4b] These and many other silyllithiums have found widespread applications in a variety of synthetic transformations: (1) S_N2 reactions with organic halides to form the corresponding substitution products; (2) 1,2-addition to C=C bonds of alkenes to make silyl-substituted alkanes; (3) 1,2-addition to C≡C bonds of alkynes to produce silyl-substituted alkenes: and (4) 1,2-addition to the carbonyl groups >C=O to form, after an aqueous work-up, silyl-substituted alcohols. The reactivity of silylmagnesium derivatives has been far less studied; however, generally it is the same as that of silyllithiums, resulting in the formation of identical products.[4b]

The reactivity of alkali metal derivatives of the group 14 elements heavier than silicon (germyl, stannyl and plumbyl anions) is parallel to that of silyllithium reagents. The four most important types of reactions are: (1) nucleophlic substitution at carbon with organic halides; (2) nucleophilic substitution at the metal with main group or transition metal halides; (3) nucleophilic addition to π-bonds; and (4) single electron transfer (SET) reactions.[4h]

The first group of reactions, nucleophilic substitution at carbon, typically proceeding by the classical S_N2 pathway with inversion of configuration at the carbon center, represents one of the most effective methods for forming E–C bonds (E = Ge, Sn, Pb):

$$R_3ELi + R'X \rightarrow R_3ER' + LiX$$

However, if there is great steric hindrance around the reaction center E, alkali metal–halogen exchange may take place instead of the desired substitution, resulting in the formation of halides R_3EX instead of coupling products R_3ER'. As a representative example, one can mention the reaction of R_3E^- derivatives with aromatic acyl chlorides ArCOCl providing straightforward access to synthetically useful acylgermanes, -stannanes and -plumbanes R_3ECOAr.

The second class of reactions, nucleophilic substitution at a metal, is closely related to the first class. Such transmetalation reactions provide access to a variety of organometallic derivatives featuring bonds between the heavy group 14 elements and main group or transition metals. In addition to halogens, triflates have been widely used as good leaving groups.

The third group of reactions of the heavy group 14 element centered anions R_3E^- (E = Ge, Sn, Pb) involves their nucleophilic 1,2-addition to π-bonds; first of all, to carbonyl groups >C=O (aldehydes and ketones) yielding the corresponding α-metallated alcohols >C(ER_3)–OH. Anionic derivatives R_3E^- can also undergo nucleophilic addition to C≡C triple bonds (1,2-addition) and to C=C–C=C 1,3-diene systems (1,4-addition).

The last class of reactions involves the SET processes, that is, exchange of a single electron between the two closed-shell diamagnetic molecules through the intermediate formation of radical species. The anionic derivatives R_3EM (E = heavy group 14 element, M = alkali metal), possessing readily accessible high-lying HOMOs and playing the role of electron donor, can react with various functional compounds A, having low-lying LUMOs (for example, strongly conjugated systems) and acting as electron acceptors, by a SET mechanism (Scheme 3.16). Elementary steps involving SET processes were proposed for different types of reactions of the heavy group 14 element centered anions: substitution reactions at carbon, substitution reactions at the metal, and addition reactions to conjugated compounds.

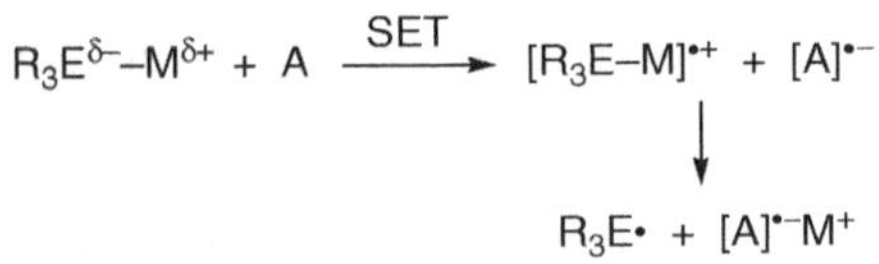

Scheme 3.16

As one of the most important practical uses of silyllithium derivatives, particularly in the field of material science, one should mention their application to the synthesis of polysilane dendrimers.[4g,m] Dendrimers are highly branched, regular, three-dimensional, tree-like monodisperse macromolecules, later generations of which have a globular shape.[82] They are highly attractive as very promising precursors for new types of advanced materials, especially dendrimers with functional groups on their surface featuring specific properties, such as solubility, viscosity and thermal behavior. Dendritic molecules have been known since the pioneering work of Vögtle *et al.* in 1978,[83] and at present two synthetic strategies for their synthesis are commonly employed: the divergent approach developed by Tomalia *et al.*[84] and Newcome *et al.*[85] in 1985 and the convergent approach introduced by Hawker and Fréchet[86] in 1990. Both methods involve repetition of the reaction sequence; thus allowing stepwise growth of the dendrimers, generation-by-generation.

Si-containing dendrimers possess three main types of bonding at the branching points: Si–O (siloxane dendrimers), Si–C (carbosilane dendrimers) and Si–Si (polysilane dendrimers). Siloxane dendrimers, as the first heteroatom-containing representatives, have been known since 1989,[87] whereas carbosilane dendrimers, chemically and thermally stable derivatives prepared up to high generations, were synthesized for the first time in 1992.[88] Polysilane dendrimers have been known since 1995, when the groups of Lambert, Suzuki and Sekiguchi independently reported the synthesis of such dendrimers with homocatenated chains of silicon atoms.[89–91] Lambert *et al.*[89] and Suzuki *et al.*[90] synthesized the same first generation dendrimer $MeSi[SiMe_2Si(SiMe_3)_3]_3$, containing 15 silicon atoms, by the coupling of $MeSi(SiMe_2Cl)_3$ (prepared by two different methods) and $(Me_3Si)_3SiLi$ (Scheme 3.17). In due course, Lambert *et al.* prepared and characterized a series of other polysilane dendrimers (up to the second generation):[92] $MeSi[SiMe(SiMe_3)_2]_3$,[92a] $MeSi[SiMe_2Si(SiMe_3)_3]_3$,[92a] $HSi[SiMe_2SiMe(SiMe_3)_2]_3$,[92a] $MeSi[SiMe_2SiMe(SiMe_3)_2]_3$,[92b] $MeSi[SiMe_2Si(SiMe_3)_2SiMe_2SiMe_2Si(SiMe_3)_3]_3$.[92c]

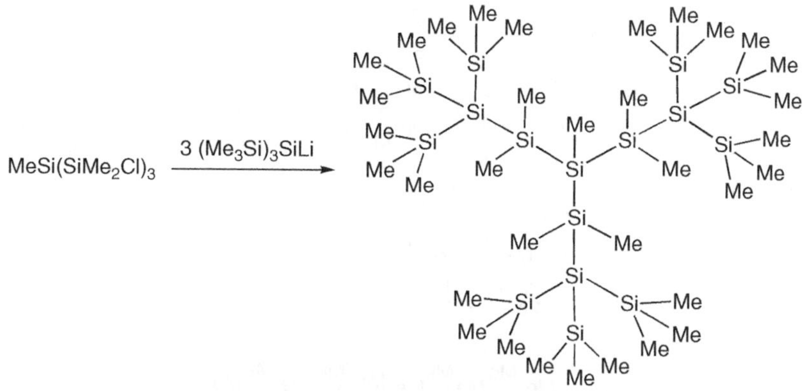

Scheme 3.17

Sekiguchi *et al.* synthesized the first representative of the second-generation polysilane dendrimers by the divergent method using silyllithium as a versatile building block (Scheme 3.18).[91] The silyllithium $(PhMe_2Si)_2MeSiLi$, the key reagent in this process, was prepared by the transmetalation of the silylmercury derivative with lithium as depicted in Scheme 3.3, E (Section 3.2.3). The synthesis started with the reaction of $PhMe_2SiCl$ with $(PhMe_2Si)_2MeSiLi$ in toluene to form $(PhMe_2Si)_3SiMe$ **1** possessing a core and three branching points. **1** was subsequently treated with TfOH in dichloromethane followed by the addition of either $(PhMe_2Si)_2MeSiLi$ or $(Me_3Si)_2MeSiLi$ to yield the first-generation dendrimers $MeSi[SiMe_2SiMe(SiMe_2Ph)_2]_3$ **2a** or $MeSi[SiMe_2SiMe(SiMe_3)_2]_3$ **2b**, respectively. Finally, treatment of **2a** with TfOH in dichloromethane followed by coupling with $(Me_3Si)_2MeSiLi$ resulted in the formation of the second-generation permethylated dendrimer **3** (Scheme 3.18). Polysilane dendrimer **3** contains in total 31 silicon atoms and 30 Si–Si bonds; its longest chain is composed of 11 silicon atoms. This highly branched structure of **3** clearly affected its electronic properties: the extinction coefficient $\varepsilon = 9.64 \times 10^4$ at $\lambda = 279$ nm was nearly two times as large as that of the first-generation dendrimer **2b** ($\varepsilon = 4.91 \times 10^4$ at $\lambda = 269$ nm).

Applying the above-described divergent technique, Sekiguchi *et al.* synthesized a first-generation hybrid dendrimer with alternating Si and Ge atoms in its chains.[93] A branching reagent, the silyllithium $(PhMe_2Ge)_2MeSiLi$, was prepared by the transmetalation of the silylmercury derivative $[(PhMe_2Si)_2MeSi]_2Hg$ with lithium in diethyl ether. The initiator core $(PhMe_2Ge)_3SiMe$ **4** was obtained in the first step of divergent growth by the coupling of $(PhMe_2Ge)_2MeSiLi$ with $PhMe_2GeCl$ (Scheme 3.19). The

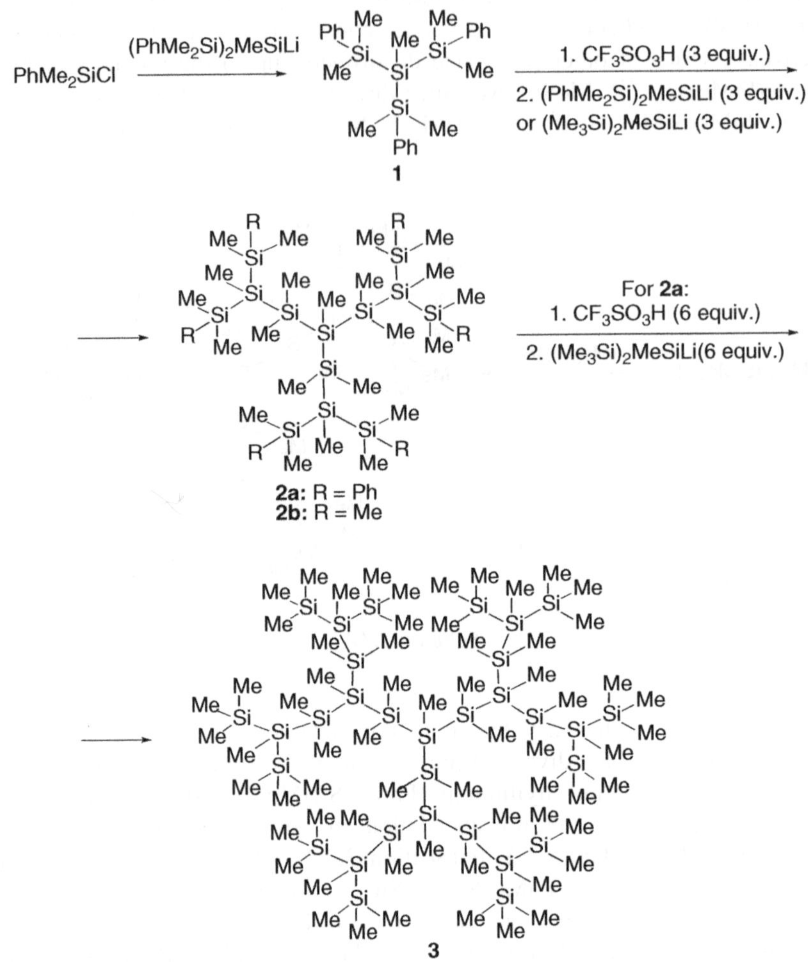

Scheme 3.18

first-generation phenyl-substituted dendrimer **5a** was formed upon the treatment of **4** with TfOH in dichloromethane, followed by reaction with $(PhMe_2Ge)_2MeSiLi$. The hybrid permethyl-substituted dendrimer **5b** was prepared from **5a** by the standard procedure: reaction of **5a** with TfOH in dichloromethane, with subsequent treatment with an excess ammonium chloride, followed by methylation with MeMgI in THF (Scheme 3.19).

In the crystal structure of **5b**, the *anti* (143.9–165.2°) and *orthogonal* (73.6–95.5°) conformations of the alternating –Si–Ge–Si–Ge– chains, which are defined by the E–E′–E–E′ (E, E′ = Si, Ge) torsional angles, clearly dominate, whereas *gauche* conformers were not found in the range of 49–61°.

A family of single- and double-core oligosilane dendrimers was prepared by Krempner *et al.*, who employed oligosilylpotassium derivatives as building blocks for the construction of dendritic systems.[94] Such dendrimers benefited from the extensive

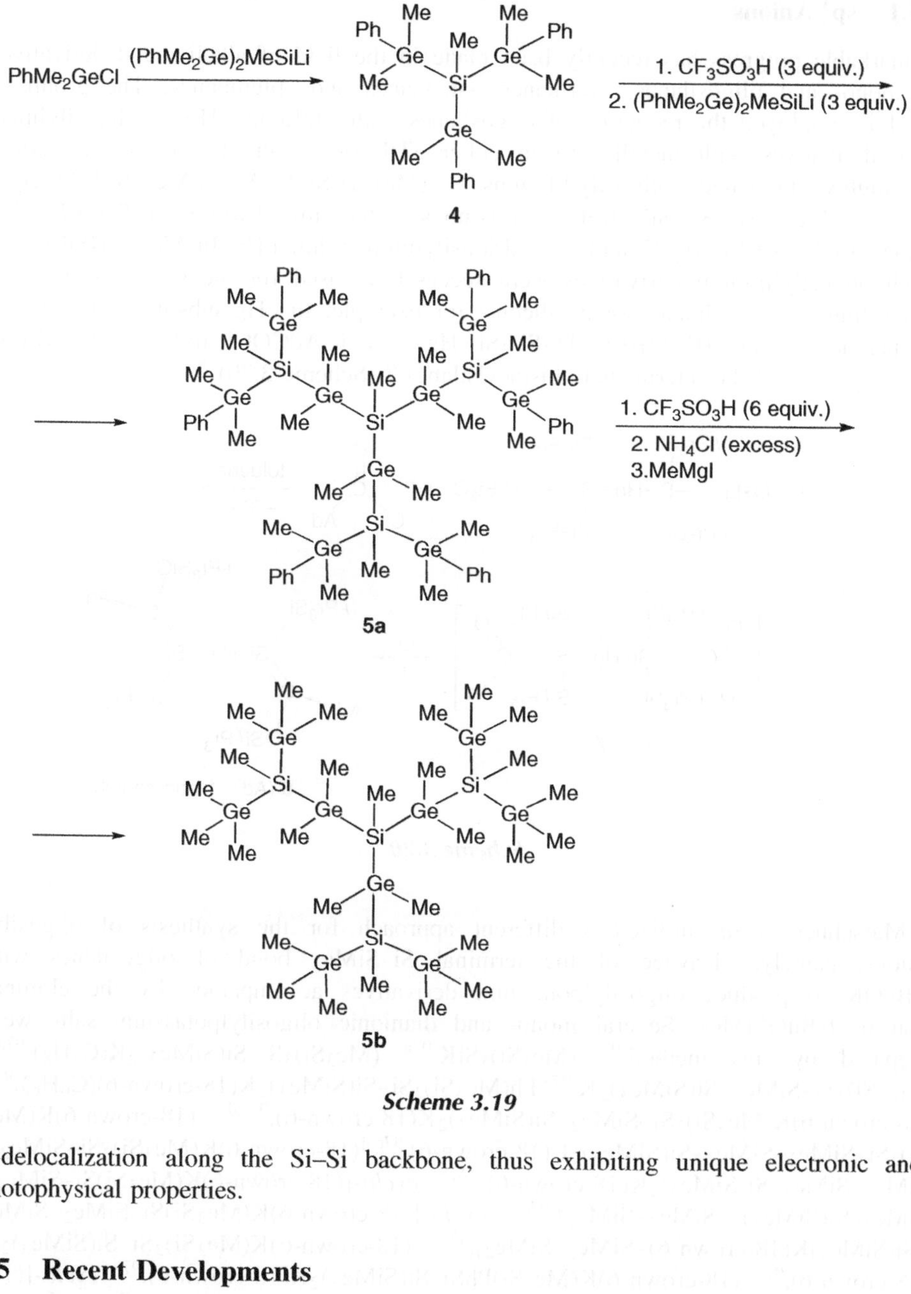

Scheme 3.19

σ-delocalization along the Si–Si backbone, thus exhibiting unique electronic and photophysical properties.

3.5 Recent Developments

Many anionic derivatives of the heavy group 14 elements have been prepared over the last two decades; and these achievements were comprehensively covered in preceding reviews.[4a,b,f–h] Therefore, in this section we will briefly deal only with the most recent advances, mainly made since 2000.

3.5.1 sp^3 Anions

Remarkable progress has recently been made in the field of alkali metal derivatives of mono- and oligosilanes, -germanes, -stannanes and -plumbanes. The group of Apeloig employed the reduction of oligosilanes with MeLi in THF or oligosilylmercury derivatives with metallic Li in either THF or hexane to produce a variety of highly branched oligosilyllithiums:[95] $(Me_3Si)_3Si–SiMe_2–SiMe_2–Si(SiMe_3)_2Li(thf)_3$,[95a] $(Me_3SiMe_2Si)_3SiLi(thf)_3$ and its nonsolvated dimer $[(Me_3SiMe_2Si)_3SiLi]_2$,[95b] $[(thf)_2Li(i\text{-}Pr_3Si)_2Si]_2Hg$,[95c] and the hydridosilyllithium dimer $[(t\text{-}Bu_2MeSi)_2HSiLi]_2$.[95d] Such oligosilyllithium derivatives were successfully used for the synthesis of other interesting low-coordinate organosilicons; for example, the Hg-substituted bis(silene) **6**, prepared from $[(Et_2O)_2Li(i\text{-}Pr_3Si)_2Si]_2Hg$[95c] and AdCOCl followed by Brook rearrangement of the intermediate bis(acylsilane) **7** (Scheme 3.20).[96]

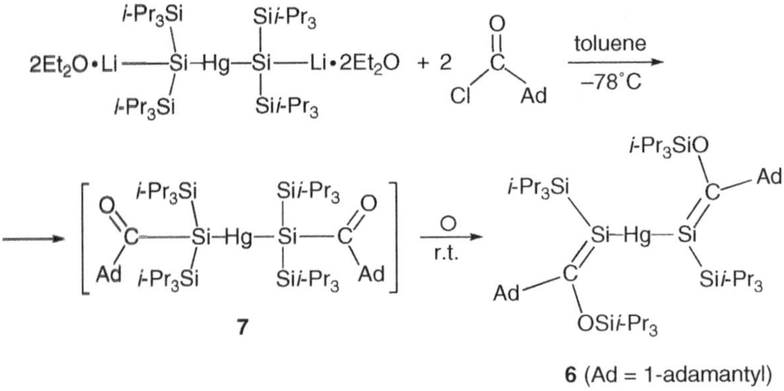

Scheme 3.20

Marschner *et al.* utilized a different approach for the synthesis of oligosilyl anions, namely, cleavage of the terminal $Si–SiMe_3$ bond of oligosilanes with t-BuOK to produce oligosilylpotassium derivatives accompanied by the elimination of $t\text{-}BuOSiMe_3$. Several mono- and dianionic oligosilylpotassium salts were prepared by this method:[97] $(Me_3Si)_3SiK$,[97a] $(Me_3Si)_3Si–Si(SiMe_3)_2K(C_7H_8)$,[97a,b] $(Me_3Si)_3Si–SiMe_2–Si(SiMe_3)_2K$,[97a] $Ph(Me_3Si)_2Si–Si(SiMe_3)_2K(18\text{-crown-}6)(C_6H_6)$,[97b] $(18\text{-crown-}6)K(Me_3Si)_2Si–SiMe_2–Si(SiMe_3)_2K(18\text{-crown-}6)$,[97c,d] $(18\text{-crown-}6)K(Me_3Si)_2Si–SiMe_2–SiMe_2–Si(SiMe_3)_2K(18\text{-crown-}6)$,[97c,d] $(18\text{-crown-}6)K(Me_3Si)_2Si–SiMe_2–SiMe_2–SiMe_2–Si(SiMe_3)_2K(18\text{-crown-}6)$,[97d] *cyclo*-$[(18\text{-crown-}6)K(Me_3Si)Si–SiMe_2–SiMe_2–Si(SiMe_3)_2–SiMe_2–SiMe_2]$,[97e] *cyclo*-$[(18\text{-crown-}6)K(Me_3Si)Si–SiMe_2–SiMe_2–Si(SiMe_3)K(18\text{-crown-}6)–SiMe_2–SiMe_2]$,[97e] $(18\text{-crown-}6)K(Me_3Si)_2Si–Si(SiMe_3)_2K(18\text{-crown-}6)$,[97f] $(18\text{-crown-}6)K(Me_3Si)PhSi–Si(SiMe_3)_2K(18\text{-crown-}6)$,[97f] *cyclo*-$[(18\text{-crown-}6)K(Me_3Si)Si–Si(SiMe_3)_2–Si(SiMe_3)K(18\text{-crown-}6)–Si(SiMe_3)_2]$,[97f] $\{\eta^5\text{-}C_5H_4[Si(SiMe_3)_2K(18\text{-crown-}6)]\}_2Fe$[97g] and $[Et_2N(Me_3Si)_2SiK(thf)_2]_2$.[97h] Alkali metal salts of per(silyl)stannyl anions are also available by this synthetic strategy: $(Me_3Si)_3SnM$(crown ether) (Na/15-crown-5, K/18-crown-6, Rb/18-crown-6) were synthesized by the reaction of $(Me_3Si)_4Sn$ with t-BuOK.[97i] Readily available and synthetically useful, silylpotassium derivatives have found an important application as building blocks in organosilicon chemistry, for example, for the preparation of transition metal complexes featuring silyl ligands (the 18-electron hafnocene disilene complex $Cp_2Hf(PMe_3)[\eta^2\text{-}(Me_3Si)_2SiSi(SiMe_3)_2]$)[97j] and disilenoid ($\beta$-fluorosilyl anion) $F(Me_3Si)_2Si–Si(SiMe_3)_2[K(18\text{-crown-6})]$.[97k,l]

A family of the homologous persilyl-substituted anions of the heavy group 14 elements $(t\text{-}Bu_2MeSi)_3E^-\bullet M^+$ **8a-c$^-$•M$^+$** (**a**: E = Si, M = Li; **b**: E = Ge, M = Li; **c**: E = Sn, M = Li, K) was prepared by Sekiguchi *et al.* by the straightforward one-electron reduction of the corresponding radicals $(t\text{-}Bu_2MeSi)_3E\bullet$[98] (see Chapter 2, Section 2.4.1.2) with alkali metals (Li, K) in different solvents.[99] Thus, silyl- and germyllithiums $(t\text{-}Bu_2MeSi)_3Si^-\bullet Li^+$ **8a$^-$•Li$^+$** and $(t\text{-}Bu_2MeSi)_3Ge^-\bullet Li^+$ **8b$^-$•Li$^+$** were readily available by the simple treatment of stable radicals $(t\text{-}Bu_2MeSi)_3Si\bullet$ and $(t\text{-}Bu_2MeSi)_3Ge\bullet$ with lithium in hexane (Scheme 3.21).[99a] Both **8a$^-$•Li$^+$** and **8b$^-$•Li$^+$** are monomeric and nonsolvated, featuring a remarkable flattening at the anionic E centers, with the sum of the angles around them totaling 359.1° and 352.4°, respectively (Figure 3.5). Given the general tendency of the heavy group 14 anions to adopt pyramidal geometry, it should be recognized that the diagnostic planarity of **8a,b$^-$•Li$^+$** has both steric and electronic origins. The steric factor is ascribed to the strong repulsion of the three very bulky $t\text{-}Bu_2MeSi$ groups, which prefer to occupy positions well separated from each other. Electronically, flattening around the anionic centers stems from the decrease in the inversion barrier because of the electropositive σ-donating silyl substituents [16.3 kcal/mol for $(H_3Si)_3Si^-$].[100] The other reason contributing to the overall flattening of the structures is the intramolecular Li···(C–H) agostic interaction, as shown by the Li···CH_3(t-Bu) bonding contacts of 2.518(5)–2.595(6) Å (for **8a$^-$•Li$^+$**) (Figure 3.5).[99a] Indeed, a SSIP $(t\text{-}Bu_2MeSi)_3Si^-\bullet[Li^+(thf)_4]$, lacking Si–Li bonding, is no longer planar, manifesting pronounced pyramidality around the central Si (349.5°) because of the absence of Li···(C–H) interaction. Such an agostic interaction is also not operative in the liquid phase at room temperature, and therefore rotation about the $Si_{centr.}–Si_{subst.}$ bonds in **8a$^-$•Li$^+$** is not restricted, giving rise to the observation of only one signal for the t-Bu groups in its ^1H NMR spectrum. However, at low temperatures (down to 220 K), a significant value of the activation energy for rotation about the Si–Si bonds, $E_a = 11.1 \pm 0.3$ kcal/mol, estimated on the basis of temperature-dependent ^1H NMR spectroscopy, testifies to the existence of an intramolecular Li···(C–H) agostic interactions in solution.

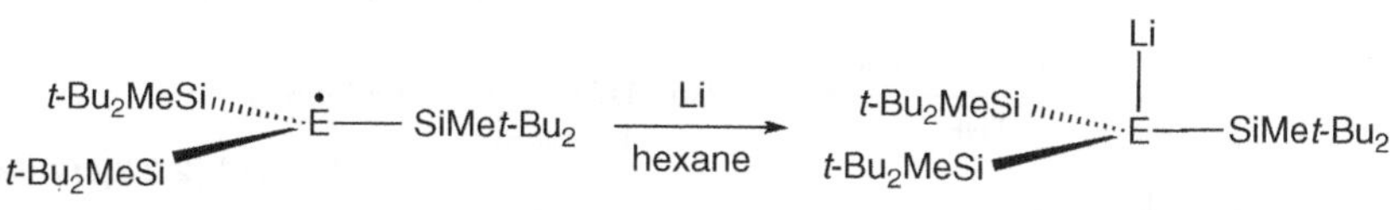

8a$^-$•Li$^+$(E = Si)
8b$^-$•Li$^+$(E = Ge)

Scheme 3.21

Being nonsolvated and therefore lacking Li^+-coordination to n-donor ligands, **8a$^-$•Li$^+$** exhibited a shorter Si–Li bond of 2.531(6) Å compared with those of typical

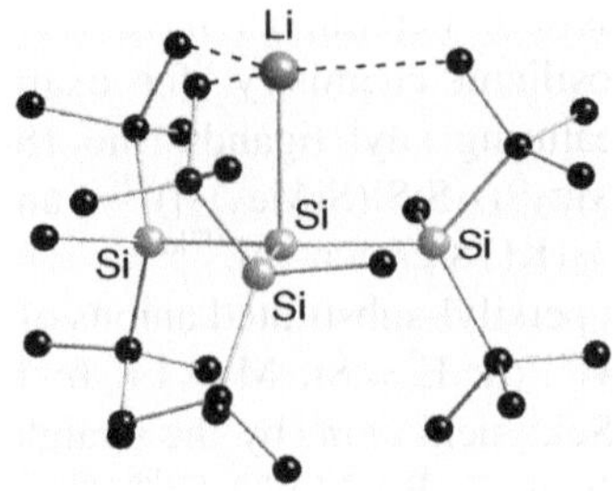

Figure 3.5 *Crystal structure of (t-Bu_2MeSi)$_3$$Si^-$•$Li^+$ (hydrogen atoms are not shown)*

solvated/coordinated silyllithiums (2.67–2.70 Å). On the other hand, the $Si_{centr.}$–$Si_{subst.}$ bonds in **8a**$^-$•**Li**$^+$ are markedly shortened to 2.3632(8) Å (av.) relative to those of the neutral radical precursor (*t*-Bu_2MeSi)$_3$Si• [2.4210(7) Å (av.)], because of the stabilizing hyperconjugation of the central Si anionic orbital and the adjacent σ*-orbitals of the Si–C(*t*-Bu) bonds.[99a]

Reduction of the stable stannyl radical (*t*-Bu_2MeSi)$_3$Sn• with alkali metals (Li, K) produced the corresponding salts of the stannyl anion (*t*-Bu_2MeSi)$_3$$Sn^-$•$M^+$**8c**$^-$•**M**$^+$ (M = Li, K) isolated in a variety of structural modifications depending on the reaction solvent used (in all structural forms, the geometry about the tin center was markedly pyramidal).[99b] Thus, in nonpolar heptane, a dimeric nonsolvated aggregate **[8c**$^-$•**Li**$^+$**]**$_2$ was formed, whereas in coordinating THF, this aggregation was broken down to produce a monomeric CIP as a THF-solvate **8c**$^-$•**[Li**$^+$**(thf)**$_2$**]** (Scheme 3.22).

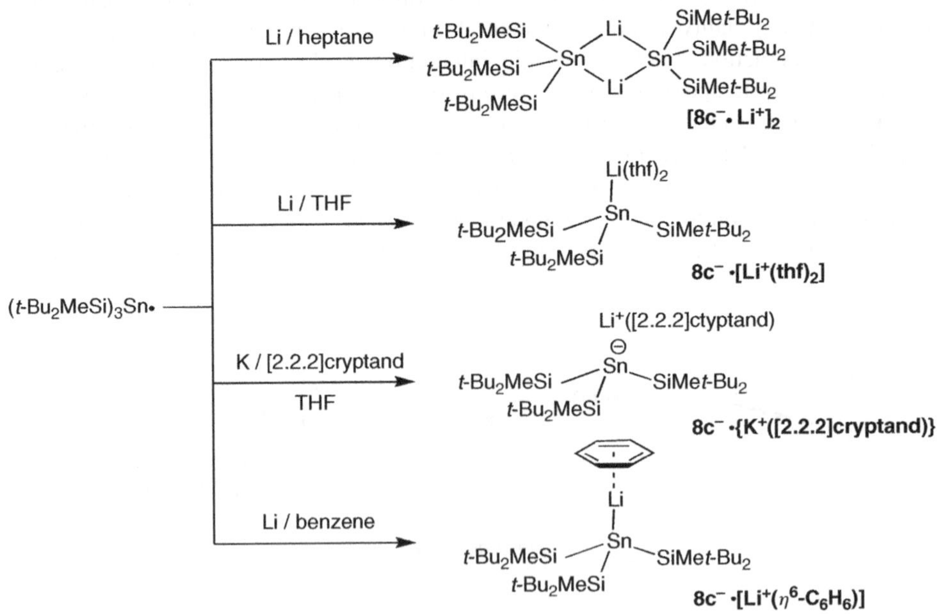

Scheme 3.22

Reduction of the stannyl radical with potassium in the presence of [2.2.2]cryptand resulted in the isolation of the SSIP **8c⁻•{K⁺([2.2.2]cryptand)}** featuring a 'naked' Sn anionic center. Upon reduction in benzene, the monomeric CIP **$8c^-$•[Li^+(η^6-C_6H_6)]** was produced in which the Li cation was π-complexed with the benzene molecule. The presence of the Li–benzene π-interaction in **$8c^-$•[Li^+(η^6-C_6H_6)]** in solution was deduced by the significant shielding of its Li nucleus, whose NMR resonance was observed in the characteristically high-field region at −3.4 ppm. Moreover, the existence of **$8c^-$•[Li^+(η^6-C_6H_6)]** as a CIP in C_6D_6 solution was proved by the observation of the ^{119}Sn NMR signal at −819.8 ppm as a quartet with a $^1J(^{119}Sn–^7Li)$ coupling constant of 572 Hz testifying to the integrity of the Sn–Li bonding (Figure 3.6).

A series of alkaline earth metal derivatives of tris(trimethylsilyl)silyl $[(Me_3Si)_3Si]_2M(thf)_n$[101a,b] and tris(trimethylsilyl)germyl $[(Me_3Si)_3Ge]_2M(thf)_n$[101c] (M = Ca and Sr, $n = 3$; M = Ba, $n = 4$) anions was prepared by the metathesis reaction between $(Me_3Si)_3SiK$[101b,d] and $(Me_3Si)_3GeK$[101c] with the alkaline earth metal iodides MI_2 in THF. In the solid state, all of these compounds crystallized as CIPs featuring Si–M and Ge–M bonds, which, however, can be broken down by the addition of strongly coordinating HMPA to form SSIPs (for example, $\{[(Me_3Si)_3Si]_2\}^{2-}$•$[Ba^{2+}(HMPA)_6]$). As expected, all of these alkaline earth metal salts, whether contacted or solvent-separated, revealed pronounced pyramidalization at the anionic Si and Ge centers.

Among other remarkable organosilicons possessing Si–alkaline earth metal bonds (CIPs), one should mention $[(Me_3Si)_3Si]_2Mg(thf)_2$,[102a] $[(Me_3Si)_3Si]_2Mg(TMEDA)$,[102b] $Ph[(Me_3Si)_3Si]Mg(thf)_2$,[102b] $[Me(Me_3Si)_2Si]_2Mg(thf)_2$,[102b] $[Me(Me_3Si)_2Si]_2Mg(1,4\text{-dioxane})_2$,[102b] $[Ph(Me_3Si)_2Si]_2Mg(thf)_2$,[102b] *cyclo*-$[(Me_3Si)_2Si–SiMe_2–SiMe_2–(Me_3Si)_2Si–Mg(thf)_2]$,[102b] $(t\text{-}Bu_3Si)_2Be$,[102c] and $(t\text{-}Bu_3Si)_2Mg(thf)_2$.[102c]

The heavy analogs of Grignard reagents, featuring the $R_3Si–Mg–Br$ linkage (CIPs), have also recently become available by the stoichiometric reaction of R_3SiM (M = Li, Na, K) derivatives with $MgBr_2$ in THF: $[(Me_3Si)_3Si]BrMg(thf)_2$,[102a]

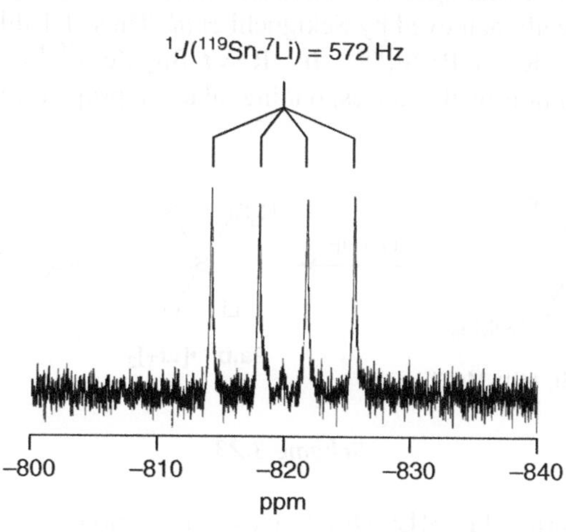

Figure 3.6 *^{119}Sn–^{7}Li scalar coupling in the ^{119}Sn NMR spectrum of (t-$Bu_2MeSi)_3$ Sn^-•[Li^+(η^6-C_6H_6)]*

$\{[Me(Me_3Si)_2Si]BrMg(thf)\}_2$,[102b] $\{(t\text{-}Bu_3Si)BrMg(thf)\}_2$.[102c] An interesting example of the SSIP $[(Ph_3Pb)_3Pb]^-$•$[MgBr(thf)_5]^+$, lacking Pb–Mg bonding, was unexpectedly formed upon the reaction of PhMgBr and $PbBr_2$ in THF.[103] In the $[(Ph_3Pb)_3Pb]^-$ anion, the four Pb atoms formed a trigonal pyramid with the central lead lone electron pair located at the apex. Such geometry caused a significant compression of the Pb–Pb–Pb angles to an average value of 93° (279° around the central anionic Pb) and stretching of the Pb–Pb bonds to 2.9759(5) Å (av.).

Significant progress has been made in the field of 1,1-dilithiosilanes of the heavy group 14 elements. As was mentioned in Section 3.2.6, the formation of the dilithiosilane derivative $(Me_3Si)_2SiLi_2$ was first claimed by Lagow *et al.* based on trapping reactions.[38] A decade later Tokitoh *et al.* reported the low-temperature generation and reactivity of the metastable diaryldilithiosilane Tbt(Dip)$SiLi_2$ (Tbt = 2,4,6-tris[bis(trimethylsilyl)methyl]phenyl and Dip = 2,6-diisopropylphenyl), prepared by the reduction of the corresponding dibromosilane Tbt(Dip)$SiBr_2$ with an excess of lithium naphthalenide in THF.[104] The dilithiosilane Tbt(Dip)$SiLi_2$ could be successfully trapped at −78 °C with a variety of electrophiles, such as D_2O, MeI, Me_2SiHCl and $Br(CH_2)_4Br$, to form the corresponding neutral products Tbt(Dip)SiD_2, Tbt(Dip)$SiMe_2$, Tbt(Dip)$Si(SiHMe_2)_2$ and *cyclo*-[Tbt(Dip)Si–CH_2–CH_2–CH_2–CH_2]. However, at temperatures above −50 °C, Tbt(Dip)$SiLi_2$ underwent unavoidable intramolecular migration of one of the lithium atoms from silicon (>$SiLi_2$ fragment) to replace hydrogen at the carbon center of the $(Me_3Si)_2C\underline{H}$ group (Tbt substituent), a process that was proved by trapping with D_2O. A germanium analog, the dilithiogermane Tbt(Dip)$GeLi_2$, was generated similarly by the reduction of Tbt(Dip)$GeBr_2$ with an excess of lithium naphthalenide in THF at −78 °C.[105] The formation of Tbt(Dip)$GeLi_2$ was confirmed by its successful trapping with D_2O and MeI. In contrast to the highly temperature-unstable dilithiosilane Tbt(Dip)$SiLi_2$ (see above), dilithiogermane Tbt(Dip)$GeLi_2$ is much more thermally robust, showing only a small sign of lithium migration even at room temperature.[105b]

The isolation and unambiguous characterization of stable representatives of dilithiosilanes were finally achieved by Sekiguchi *et al.* Thus, 1,1-dilithiosilanes R_2SiLi_2 **9a,b**$^{2-}$•$[Li^+]_2$[106] (**a**: R = i-Pr_3Si;[106a] **b**: R = t-Bu_2MeSi[106b]) were unexpectedly formed upon the reduction of the corresponding silacyclopropenes with lithium in THF (Scheme 3.23).

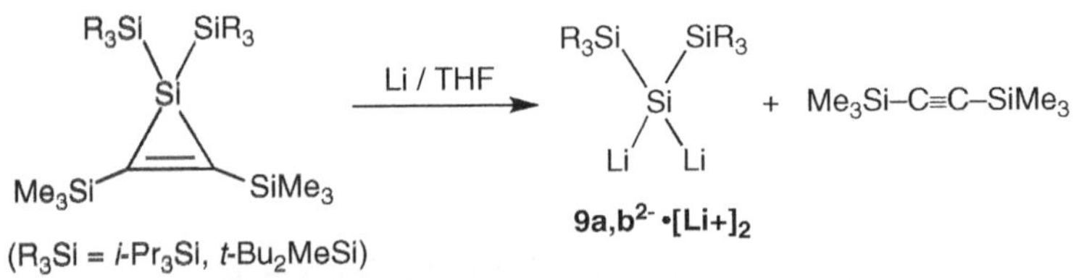

Scheme 3.23

In its crystalline form, **9a**$^{2-}$•$[Li^+(thf)]_2$ exists as a monomeric CIP with a nearly tetrahedral anionic silicon center and the two Si–Li bonding contacts of 2.549(7) and 2.55(1) Å, which are markedly shorter than those for the typical solvated monomeric

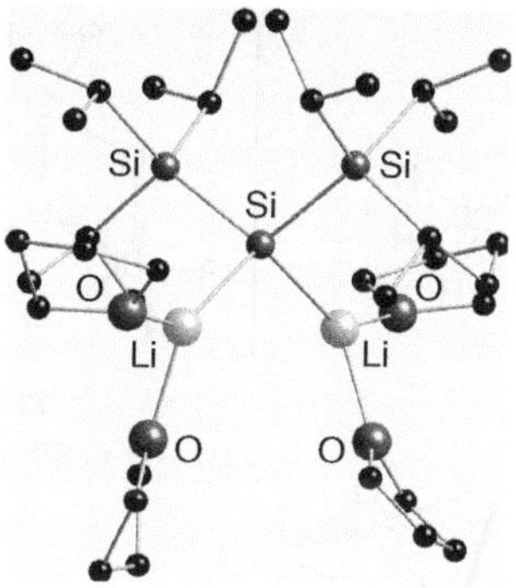

Figure 3.7 Crystal structure of $\mathbf{9a}^{2-}\bullet[Li^+(thf)]_2$ (hydrogen atoms are not shown)

silyllithiums (2.67–2.70 Å) (Figure 3.7).[106a] The monomeric composition of $\mathbf{9a}^{2-}\bullet[\mathbf{Li}^+(\mathbf{thf})]_2$ was retained in solution at 200 K, which was manifested by the observation of the anionic silicon resonance as a quintet with a $^1J(^{29}Si–^6Li)$ coupling constant of 15.0 Hz (I for $^6Li = 1$). Notably, the dianionic silicon center of $\mathbf{9a}^{2-}\bullet[\mathbf{Li}^+(\mathbf{thf})]_2$ was greatly shielded (−292.0 ppm) compared with the previously reported monoanionic $(Me_3Si)_3SiLi$ (−189.4 ppm).[59]

The above-described dilithiosilanes $\mathbf{9a,b}^{2-}\bullet[\mathbf{Li}^+]_2$[106] were also prepared in a variety of structural modifications by the reduction of silylmercury derivatives with lithium in hexane by Apeloig *et al.*[107] Thus, the unsolvated 1,1-dilithiosilane R_2SiLi_2 (R = SiMet-Bu_2) was synthesized in the form of its co-aggregate with R_2HSiLi of the type $[(R_2SiLi_2)(R_2HSiLi)_2]$ **10** (Scheme 3.24).[107a] **10** features a six-coordinate central anionic silicon ($\delta = -251$ ppm) bonded to four lithiums with Si–Li bond lengths of 2.613(8)–2.641(10) Å. Remarkably, upon dissolving in THF, **10** formed a solvated silyllithium aggregate **11** of formally trimeric composition $(R_2SiLi_2)_3(thf)_8$ (Scheme 3.24).[107b] The anionic part of **11** consists of the $[(R_2Si)_3Li_4]^{2-}$ cluster with a star-like tetralithium core, whereas the cationic part is represented by two lithium cations each solvated by four THF molecules. The central Si_3Li_3 hexagon in **11** is nearly planar and the central Li atom is 0.4 Å above this plane. The Si–Li bonds are in the normal range, whereas among the Li–Li bonding contacts two of them are regular [2.53(2) and 2.62(2) Å] and one is short [2.18(3) Å]. In the solid state NMR spectra of **11** the signals of the dianionic silicons were observed at −239 and −243 ppm (−254 ppm in THF solution), whereas the ^7Li resonances were found at 0.1 ppm (THF-solvated Li^+ cations) and 7 ppm (four core lithium atoms of the anionic part). $(i\text{-}Pr_3Si)_2SiLi_2$ was also obtained as a co-aggregate with t-BuLi of the type $\{[(i\text{-}Pr_3Si)_2SiLi_2]_2(t\text{-}BuLi)_2\}$**12**, the most interesting structural feature of which is the central Li_6 core composed of two distorted tetrahedra sharing a common Li–Li edge (Scheme 3.24).[107c]

To date, only a couple of stable 1,1-dilithiogermane derivatives have been isolated and structurally characterized, R_2GeLi_2 $\mathbf{13a,b}^{2-}\bullet[\mathbf{Li}^+]_2$ (**a**: R = i-Pr_3Si; **b**: R = t-Bu_2MeSi). Similarly to the above case of 1,1-dilithiosilane analogs $\mathbf{9a,b}^{2-}\bullet[\mathbf{Li}^+]_2$, the 1,1-dilithiogermanes $\mathbf{13a,b}^{2-}\bullet[\mathbf{Li}^+]_2$ were prepared by the reduction of germacyclopropenes with lithium in diethyl ether/THF (Scheme 3.25).[108]

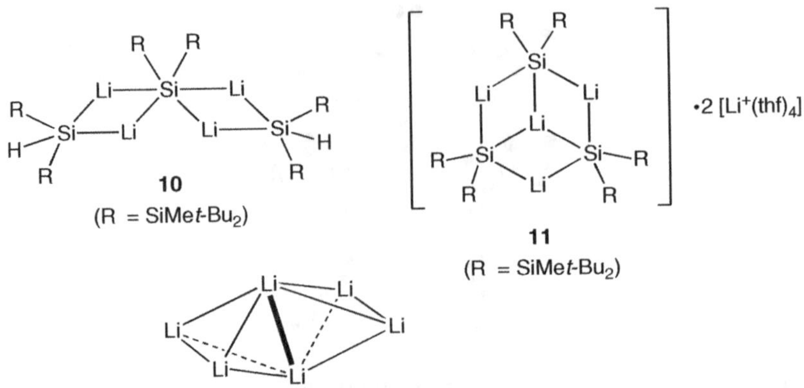

10 (R = SiMe*t*-Bu$_2$)

11 (R = SiMe*t*-Bu$_2$)

12 (Li$_6$-core)

Scheme 3.24

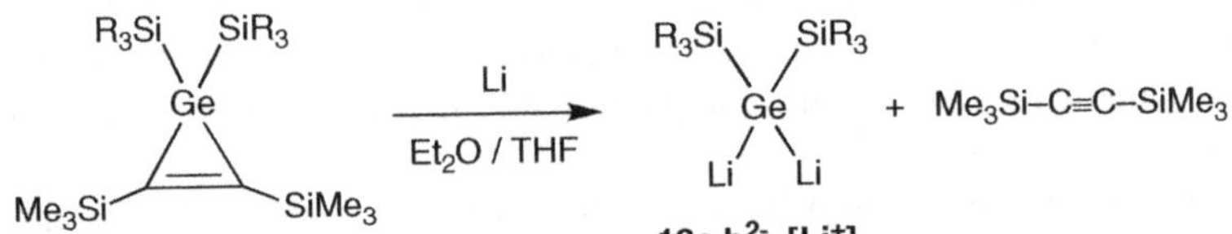

(R$_3$Si = *i*-Pr$_3$Si, *t*-Bu$_2$MeSi)

13a,b^{2-} •[Li$^+$]$_2$

Scheme 3.25

The crystal structure of **{13b^{2-}•[Li$^+$(thf)]$_2$}$_2$** is rather unusual, being represented as a dimeric CIP, in which the two lithium atoms are shared by the two germanium anionic centers, thus forming an almost regular rhombus, Ge_2Li_2 (Figure 3.8).[108] In the dimer, both Ge centers are pentacoordinate and bound to three Li atoms with Ge–Li bond distances of 2.649(6), 2.664(6) and 2.709(6) Å, which lie in the normal range of 2.61–2.76 Å.

The readily available 1,1-dilithiosilane **9b^{2-}•[Li$^+$]$_2$**[106b] and 1,1-dilithiogermane **13b^{2-}•[Li$^+$]$_2$**[108] found an important application as very useful synthons for the straightforward preparation of a wide range of doubly bonded and cyclic derivatives of the heavy group 14 elements, not accessible by any other synthetic routes. Thus, dilithio derivatives R_2ELi_2 (E = Si, Ge; R = SiMe*t*-Bu$_2$) have been used for the preparation of the *heavy analogs of alkenes* [disilenes and a silagermene $R_2Si{=}EAr_2$ (E = Si, Ge; Ar = Mes, Tip),[106b] silastannene $R_2Si{=}SnTip_2$,[109a] germastannene $R_2Ge{=}SnTip_2$,[109b] 1,3-disila-2-gallata- and -indataallenic anionic compounds $[R_2Si \cdots M \cdots SiR_2]^-$•$[Li^+(thf)_4]$ (M = Ga,In),[109c] 1,3-digerma-2-gallata- and -indataallenic anionic compounds $[R_2Ge \cdots M \cdots GeR_2]^-$•$[Li^+(thf)_4]$ (M = Ga, In),[109d] sila- and germaborenes $R_2Si{=}BMes$,[109e] $R_2Ge{=}BMes$[109e] and $R_2Si{=}BR'$ (R′ = 2,2,6,6-tetramethylpiperidino),[109f] tetrasilabuta-1,3-diene $R_2Si{=}Si(Mes){-}Si(Mes){=}SiR_2$,[109g] and the Schrock-type silylene complex $(\eta^5\text{-}C_5H_4Et)_2(PMe_3)Hf{=}SiR_2$,[109h]] and

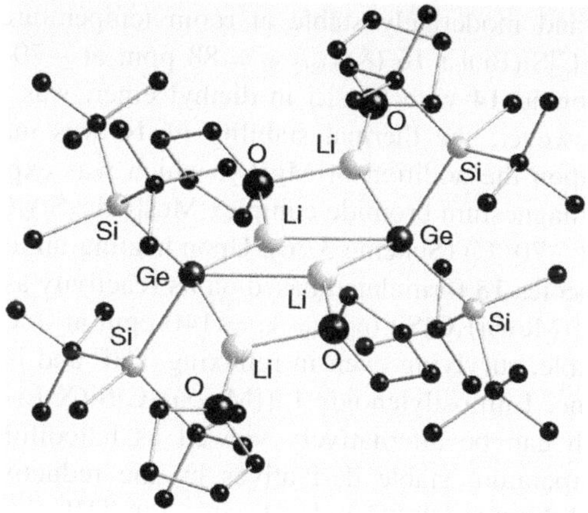

Figure 3.8 Crystal structure of $\{\mathbf{13b}^{2-}\bullet[Li^+(thf)]_2\}_2$ (hydrogen atoms are not shown)

heavy analogs of cyclopropenes [3,3-bis(di-*tert*-butylmethylsilyl)-1,2-bis(tri-*tert*-butylsilyl)cyclotrisilene,[109i] tetrakis(di-*tert*-butylmethylsilyl)cyclotrigermene,[109j] and tetrakis(di-*tert*-butylmethylsilyl)-1*H*-siladigermirene[109j])].

The 1,1-dilithiostannane derivatives are still unknown as stable/isolable species, although the ^{119}Sn and ^{13}C NMR chemical shifts of some dianions (Ph_2SnLi_2, Et_2SnLi_2), generated by the reduction of the corresponding dichlorostannanes R_2SnCl_2 (R = Ph, Et) with metallic lithium in THF have been recently measured.[110] The closest approach to the synthesis of the elusive 1,1-metallastannanes was recently demonstrated by Tokitoh *et al.*, who generated Tbt(Dip)$SnLi_2$/Tbt(Dip)SnK_2 by the reduction of dibromostannane Tbt(Dip)$SnBr_2$ in THF with an excess of lithium naphthalenide at −78 °C/potassium graphite at −110 °C.[111] The formation of both dilithiostannane Tbt(Dip)$SnLi_2$ (^{119}Sn NMR: $\delta = -362.2$ ppm) and dipotassiostannane Tbt(Dip)SnK_2 was confirmed by their low-temperature trapping reactions with DCl and MeI. Both Tbt(Dip)SnM_2 (M = Li, K) are metastable, surviving only at temperatures below −25 °C [for Tbt(Dip)$SnLi_2$] and −110 °C [for Tbt(Dip)SnK_2].

The chemistry of silylenoids, the silicon analogs of carbenoids of the general type $R_2Si(X)M$ (X = halogen, M = metal) that have the chemical properties of silylenes, has also advanced during the past decade. Usually encountered as reactive intermediates in organosilicon reactions, silylenoids have been stabilized to such an extent that they can be observed (and even isolated) at room temperature. Thus, the reduction of the tribromosilane $(Me_3Si)_3C$–$SiBr_3$ with lithium naphthalenide (2 equiv.) in THF produced a moderately stable species, which was originally claimed to be bromosilylene Br[$(Me_3Si)_3C$]Si:[112a] but later reformulated as a bromosilylenoid Br[$(Me_3Si)_3C$]Si(Br)Li **14** ($\delta_{\text{anionic Si}} = 106$ ppm).[112b,c] Typically for silylenoids, **14** exhibited ambiphilic reactivity: as a silyl anion nucleophile (trapping with *t*-BuOH and MeI) and as a silylene (trapping with MeOH, *i*-PrOH and 2,3-dimethylbuta-1,3-diene). On the other hand, **14** was inert towards Et_3SiH, which is known to be an effective trapping reagent for silylenes.[112b] Likewise, the chlorosilylenoid Cl[$(Me_3Si)_3C$]Si(Cl)Li **15** ($\delta_{\text{anionic Si}} =$ 87 ppm) was generated from $(Me_3Si)_3C$–$SiCl_3$ and lithium naphthalenide in THF, being stable below −30 °C and moderately stable at room temperature.[112b] The mesitylsilylenoid Mes[$(Me_3Si)_3C$]Si(Br)Li **16** ($\delta_{anionic\ Si}$ = 88 ppm at −70 °C), prepared by the reaction of bromosilylenoid **14** with MesLi in diethyl ether, was stable below −10 °C (Scheme 3.26).[112d] However, the thermal stability of **16** was markedly enhanced up to room temperature upon the addition of $MgBr_2$, which was explained by the formation of the silylenoid–magnesium bromide complex Mes[$(Me_3Si)_3C$]Si(Br)Li•$MgBr_2$ **17** ($\delta_{anionic\ Si}$ = 87 ppm at −70 °C) (Scheme 3.26). Upon heating up to 45 °C, **17** was transformed into the new species **18** formulated based on its reactivity as a weak $MgBr_2$ complex of a silylene Mes[$(Me_3Si)_3C$]Si: ($\delta_{anionic\ Si}$ = 140 ppm at 0 °C) (Scheme 3.26).[112d] **18** was remarkably stable, surviving even in refluxing THF and hexane but decomposing in refluxing toluene. Lithiosilylenoids Li[$(Me_3Si)_3C$]Si(X)Li **19** and **20** (**19:** X = Br; **20:** X = Cl), which can be alternatively viewed as halodilithiosilanes, were also prepared as room temperature stable derivatives by the reductive dehalogenation of $(Me_3Si)_3C$–SiX_3 with lithium naphthalenide (4 equiv.) in THF (Scheme 3.26).[112e,f] The composition of both **19** and **20**, which are believed to be formed via the intermediate halosilylenoids **14** and **15**, was established based on products of trapping reactions with MeOH and MeOD. The lithiosilylenoids **19** and **20** manifested the resonances of their central anionic Si atoms at −108 and −110 ppm at room temperature, which were markedly up-field shifted compared with those of halo- and mesitylsilylenoids **14**–**18**.

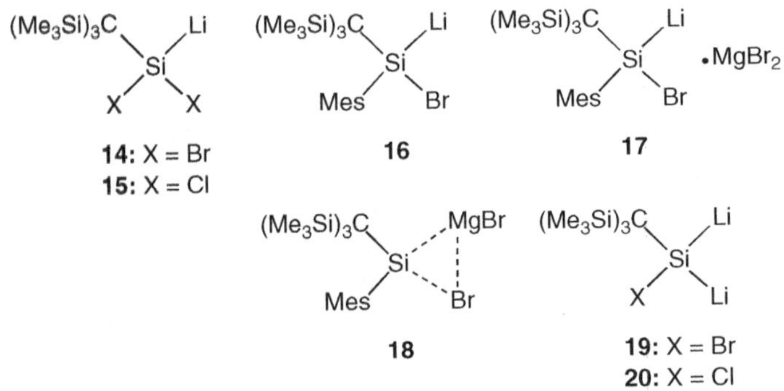

Scheme 3.26

The remarkable isolable monomeric silylenoid $R_2Si(F)[Li(thf)_3]$ (R = SiMe *t*-Bu_2) **21** was prepared by Apeloig *et al.* by the reaction of the fluorobromosilane $R_2Si(F)Br$ with the silyllithium R_3SiLi in THF.[113] In contrast to conventional silylenoids, **21** surprisingly exhibited not a tetracoordinate but a tricoordinate anionic silicon center, featuring Si···F and Li···F bonds and lacking direct Si···Li (interatomic distance 3.21 Å) bonding contact (Figure 3.9). The anionic Si is highly pyramidal (307.6°) and its bonding distance to F is long [1.698(3) Å], indicating a relatively weak Si–F covalent bond.

The resonance of the central silicon in **21** (107 ppm) was observed in the range of other stable silylenoids (see above). Computations revealed that the tetracoordinate form of silylenoid **21** with the Si–Li bond is destabilized by 4.8 kcal/mol compared with its tricoordinate isomer without such a bond.[113] This was also supported by NBO calculations,

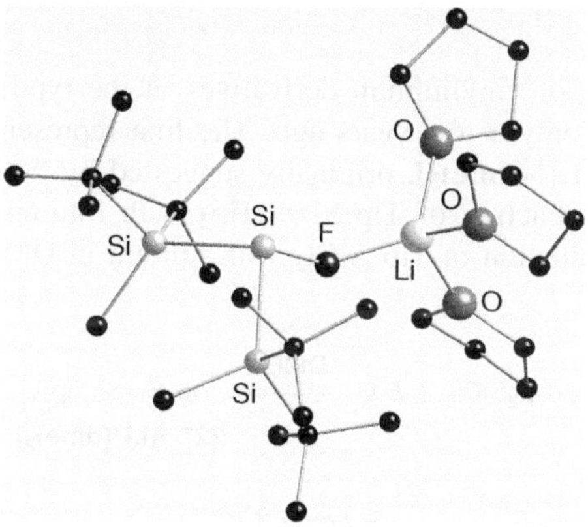

Figure 3.9 *Crystal structure of the isolable silylenoid **21** (hydrogen atoms are not shown)*

indicating a weakly covalent (highly ionic) Si–F bond and absence of a covalent interaction between Si–Li and Li–F (totally ionic Si–Li and Li–F bonds). Overall, silylenoid **21** is best viewed as the hybrid of the two resonance structures **21′a** and **21′b**, of which the former is the major contributor (ca. 75%) and the latter is the minor contributor (ca. 7%) (Scheme 3.27). The major extreme **21′a** is to be classified as a fluorine-substituted silyl anion $R_2(F)Si^-$ with a THF-solvated Li^+-countercation, whereas the minor extreme **21′b** consists of the R_2Si: silylene fragment and THF-solvated LiF molecule.

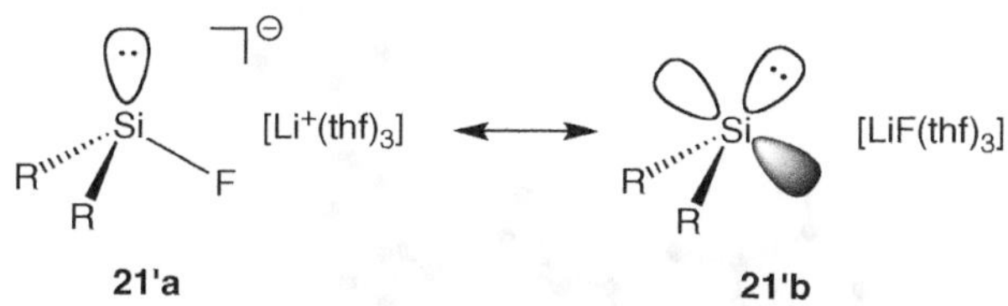

Scheme 3.27

As is typical for silylenoids, **21** exhibits versatile reactivity. Thus, it reacts as a nucleophile with MeCl, PhH_2SiCl, H_2O and MeOH; as an electrophile with MeLi; and as a silylene towards alkali metals (Li, Na). Upon heating (120 °C/30 min) **21** was transformed into the disilene $R_2Si{=}SiR_2$, by a pathway, that may involve the dimerization of transient silylenes R_2Si: formed through the α-elimination of LiF from **21**.[113]

Several other remarkable examples of silyl and germyl anionic derivatives have recently been reported. Among them are: a sulfur-substituted silylenoid $MesS(Mes)_2SiLi$,[114] ethynylsilyllithiums $RR^1(R^2C{\equiv}C)SiLi$ ($R = SiMe_2t$-Bu, $SiMe_3$; $R^1 = SiMe_2t$-Bu, $SiMe_3$, Ph, $C{\equiv}CSiMe_3$; $R^2 = SiMe_3$, Ph),[115] a borylsilyllithium (Mes)Tbt(Bpin)SiLi,[116] an alkali metal silanide with multidentate ligand $(MeOMe_2Si)_3SiM$ (M = Li, Na, K),[117] digermanyllithiums Me_3GeR_2GeLi (R = Me, Ph),[118] dialkali metal salts of digermene $[Ar(H)GeGe(H)Ar]^{2-}{\bullet}M_2^+$ (Ar = 2,6-(2,6-i-Pr_2-C_6H_3)$_2$-C_6H_3; M = Li, Na, K),[119] and alkali metal derivatives of the parent germyl anion H_3Ge^-.[120]

3.5.2 sp^2 Anions

Stable silicon analogues of vinyllithium derivatives of the type $R_2Si{=}Si(R)Li$ became synthetically accessible only a few years ago. The first representative, disilenyllithium $Tip_2Si{=}Si(Tip)Li$ **22$^-$•[Li$^+$(dme)$_2$]**, originally suggested by Weidenbruch *et al.* as the key intermediate in the reaction of $Tip_2Si{=}SiTip_2$ with lithium in DME,[121] was later isolated by the direct reduction of Tip_2SiCl_2 with lithium in DME (Scheme 3.28).[122]

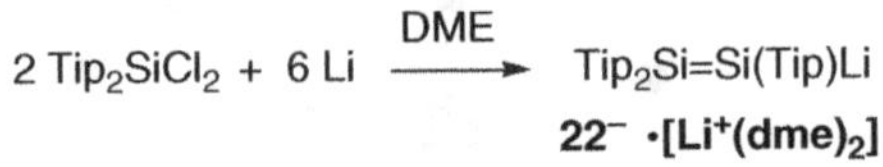

Scheme 3.28

The resonances of the doubly bonded silicons were observed in low-field at 94.5 and 100.5 ppm, of which the latter was assigned to an anionic Si center. In the solid state **22$^-$•[Li$^+$(dme)$_2$]** exists as a CIP with an Si–Li bond distance of 2.853(3) Å and Si=Si bond of 2.192(1) Å, which is markedly longer than that of the starting disilene $Tip_2Si{=}SiTip_2$ (2.144 Å) (Figure 3.10).

Another disilenyllithium derivative, $(t\text{-}Bu_2MeSi)_2Si{=}Si(Mes)Li$ **23$^-$•[Li$^+$(thf)$_3$]**, was prepared by the reductive cleavage of the central Si–Si bond of tetrasilabuta-1,3-diene $(t\text{-}Bu_2MeSi)_2Si{=}Si(Mes)–Si(Mes){=}Si(SiMe\,t\text{-}Bu_2)_2$ with t-BuLi (Scheme 3.29).[123]

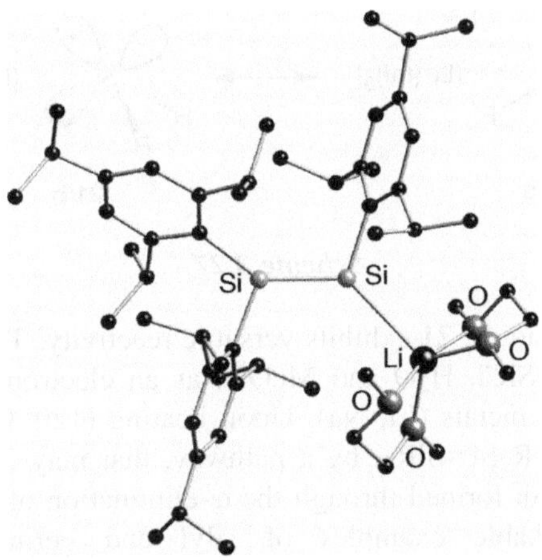

Figure 3.10 *Crystal structure of **22$^-$•[Li$^+$(dme)$_2$]** (hydrogen atoms are not shown)*

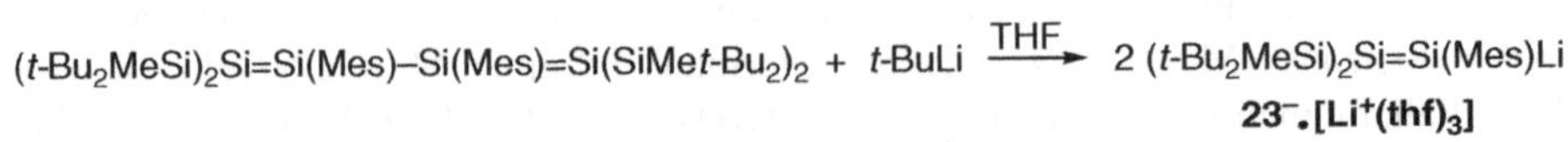

Scheme 3.29

The lithium-bound sp^2-Si atom was diagnostically deshielded and its resonance was observed at 277.6 ppm. This NMR assignment was supported by GIAO computations at the HF/6-311+G(2df,p)//B3LYP/6-31+G(d) level for the model $(H_3Si)_2Si{=}Si(Ph)Li$. 282.5 ppm for the anionic silicon. **23$^-$•[Li$^+$(thf)$_3$]** was isolated as a monomeric CIP with an unremarkably twisted Si=Si double bond (twist angle 15.2°) with a length of 2.2092(7) Å (cf. 2.192(1) Å for Si=Si in **22$^-$•[Li$^+$(dme)$_2$]**).

The reduction of the disilene $(t\text{-}Bu_2MeSi)_2Si{=}Si(SiMe\,t\text{-}Bu_2)_2$ with alkali metal naphthalenides MNp (M = Li, Na, K) in THF also resulted in the formation $(t\text{-}Bu_2MeSi)_2Si{=}Si(SiMe\,t\text{-}Bu_2)M$ **24a-c$^-$•M$^+$** (**a:** M = Li; **b:** M = Na; **c:** M = K) (Scheme 3.30).[124] The reaction proceeds through the intermediate formation of disilene 1,2-dianions $M(t\text{-}Bu_2MeSi)_2Si{-}Si(SiMe\,t\text{-}Bu_2)_2M$ followed by the β-elimination of $t\text{-}Bu_2MeSiM$ (Scheme 3.30).

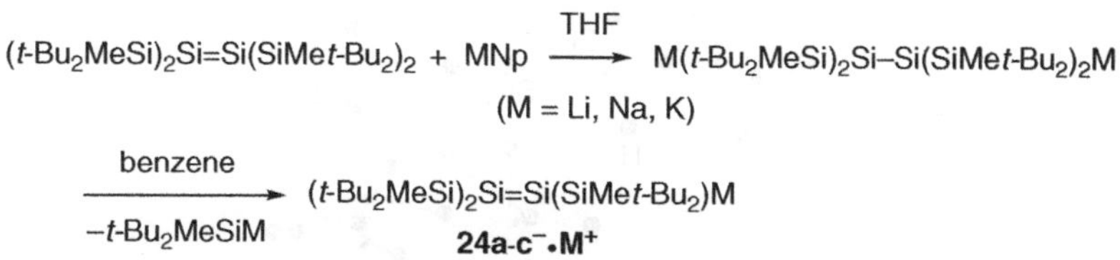

Scheme 3.30

Similarly to other disilenyllithium derivatives, **24a-c$^-$•M$^+$** manifested very low-field resonances of their anionic silicon centers observed at 328.4 (for **24a$^-$•Li$^+$**), 325.6 (for **24b$^-$•Na$^+$**) and 323.1 (for **24c$^-$•K$^+$**) ppm. The lithium salt was found to exist as a CIP **24a$^-$•[Li$^+$(thf)$_2$]** in its crystalline form, with planar geometry about both doubly bonded Si atoms and an Si–Li bond distance of 2.598(9) Å, markedly shorter than those of **22$^-$•[Li$^+$(dme)$_2$]** and **23$^-$•[Li$^+$(thf)$_3$]**. The Si=Si bond in **24a$^-$•[Li$^+$(thf)$_2$]** is not twisted (in contrast to the starting disilene) and its bond length of 2.1983(18) Å is quite similar to those of **22$^-$•[Li$^+$(dme)$_2$]** [2.192(1) Å] and **23$^-$•[Li$^+$(thf)$_3$]** [2.2092(7) Å].

A different approach for the preparation of disilenyllithium derivatives was demonstrated by the reaction of the stable disilyne $Dis_2i\text{-}PrSi{-}Si{\equiv}Si{-}Si\,i\text{-}PrDis_2$ with t-BuLi in THF.[125] The reaction mechanism was rationalized by assuming an initial SET process involving the intermediate formation of a disilyne anion-radical and t-Bu-radical as a key radical pair, followed by hydrogen abstraction by the disilyne anion-radical to yield finally disilenyllithium $H(Dis_2i\text{-}PrSi)Si{=}Si(Si\,i\text{-}PrDis_2)Li$ **25$^-$•Li$^+$** (Scheme 3.31). Overall, the formation of **25$^-$•Li$^+$** can be viewed as the result of the formal 1,2-addition of lithium hydride across the Si≡Si triple bond, a pathway that was originally established in the course of the reaction of stable disilenes with t-BuLi.[126]

The resonance of the anionic sp^2-Si atom in **25$^-$•Li$^+$** was observed at 165.0 ppm, whereas that of the H-substituted silicon center was found at 124.7 ppm [^1J (^{29}Si–^{1}H) = 155 Hz]. Disilenyllithium was structurally characterized as an SSIP **25$^-$•[Li$^+$(dme)$_3$]** after exchange of solvent from THF to DME (Scheme 3.31). The lithium cation is effectively solvated by the three DME molecules, thus leaving the disilenide $[H(Dis_2i\text{-}PrSi)Si{=}Si(Si\,i\text{-}PrDis_2)]^-$ as a free ion (Figure 3.11). Upon reduction, the Si=Si bond

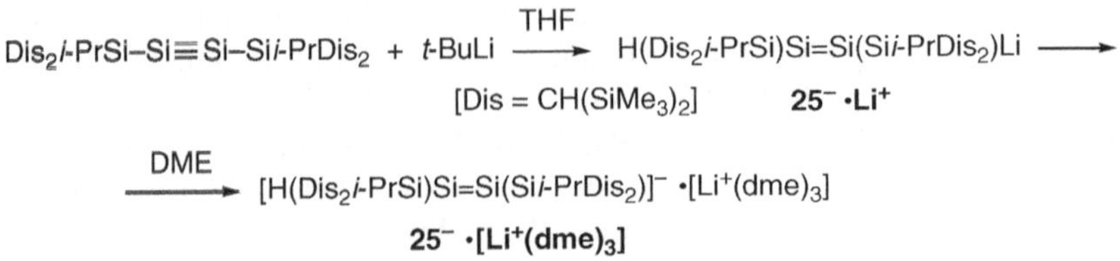

Scheme 3.31

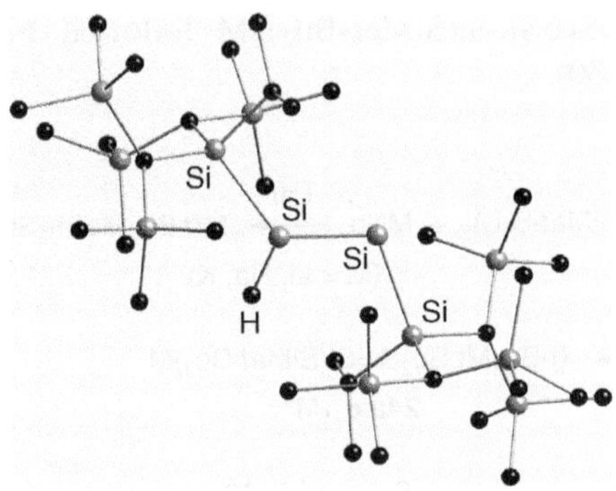

Figure 3.11 *Crystal structure of **25**⁻ (H atoms, except for Si–H, and cation [Li⁺(dme)₃] are not shown)*

in $\mathbf{25^{-}\bullet[Li^{+}(dme)_3]}$ of 2.2034(9) Å became ca. 7% stretched compared with that of the precursor disilyne [2.0622(9) Å]. Again, this value is very similar to those of other reported disilenyllithium derivatives (see above). The four silicons Si–Si≡Si–Si of $\mathbf{25^{-}}$ are nearly coplanar, and the very bulky Dis_2*i*-PrSi substituents are arranged in a *trans* fashion to minimize their steric repulsion.

Disilenyllithiums $R_2Si{=}Si(R)Li$ (R = Tip, SiMe*t*-Bu_2) found a successful utilization as nucleophilic reagents for the synthesis of novel disilenes (and other unusual molecules) inaccessible by other means.[127] Thus, they were used to prepare a silicon cluster featuring an unsubstituted vertex Si atom;[127a] a zirconium complex having disilene as an η^1-ligand;[127b] an unsaturated oligosilyl dianion;[127c] a *p*-phenylene-bridged bis(disilene);[127d] cyclic silenes;[127e,f] chlorocyclotrisilanes, cyclotrisilanides and a trisilaindane;[127g] an acyclic silyl anion-substituted silene;[127h] and boryl-substituted disilenes.[127i]

Power *et al.* reported the reduction of halides Cl(2,6-Tip_2-C_6H_3)E: (E = Ge, Sn) with alkali metals M (M = Na or K) in benzene forming dianionic derivatives $[(2,6\text{-}Tip_2\text{-}C_6H_3)EE(C_6H_3\text{-}2,6\text{-}Tip_2)]^{2-}\bullet 2M^+$ $\mathbf{26a,b^{2-}\bullet 2M^+}$ (**a:** E = Ge, M = Na; **b:** E = Sn, M = K) (Scheme 3.32), in which the two central heavy group 14 elements E can be viewed as being formally doubly bonded (Scheme 3.33).[128]

In the solid state, both $\mathbf{26a,b^{2-}\bullet 2M^+}$ derivatives are monomeric and nonsolvated with the countercations stabilized by sandwich-type π-coordination with two *ortho*-Tip groups. The Ge–Ge bond distance in $\mathbf{26a^{2-}\bullet 2Na^+}$ of 2.3943(13) Å is in the range for dimeric derivatives $(R_2Ge)_2$ [2.213(2)–2.443(2) Å], and the Sn–Sn separation in

$$\text{(2,6-Tip}_2\text{-C}_6\text{H}_3\text{)(Cl)E:} + \text{M} \xrightarrow{\text{benzene}} [\text{(2,6-Tip}_2\text{-C}_6\text{H}_3\text{)EE(C}_6\text{H}_3\text{-2,6-Tip}_2)]^{2-}\cdot 2\text{M}^+$$

$\mathbf{26a,b^{2-}\cdot 2M^+}$
(**a**: E = Ge, M = Na; **b**: E = Sn, M = K)

Scheme 3.32

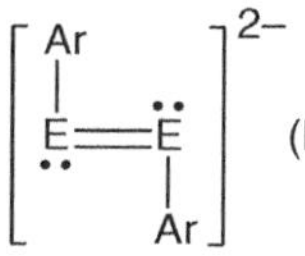

(E = Ge, Sn; Ar = 2,6-Tip$_2$-C$_6$H$_3$)

Scheme 3.33

$\mathbf{26b^{2-}\bullet 2K^+}$ of 2.7663(9) Å is also comparable with those found in the dimers $(R_2Sn)_2$ [2.768(1)–3.639(1) Å]. The unremarkable shortening of the formal E=E (E = Ge, Sn) double bonds in $\mathbf{26a,b^{2-}\bullet 2M^+}$ in comparison with single bonds might be attributed to the Coulombic repulsion between adjacent negative charges within the dianions $[E{=}E]^{2-}$. The E–M bonding contacts in both cases are longer than the sum of the covalent radii: for Ge---Na [3.121(3) and 3.074(3) Å in $\mathbf{26a^{2-}\bullet 2Na^+}$ vs 2.86 Å] and for Sn---K [3.579(2) and 3.591(2) Å in $\mathbf{26b^{2-}\bullet 2K^+}$ vs 3.36 Å].

The heavier analogs of vinyllithium of the type $R_2E{=}E(R)M$ (E = Ge, Sn, Pb; M = alkali metal) are still unknown as isolable derivatives, although the digermenyllithium $Ar_2Ge{=}Ge(Ar)Li$ (Ar = 2,6-diisopropylphenyl) was proposed to be formed in the reaction of digermene $Ar_2Ge{=}GeAr_2$ with lithium naphthalenide in DME, based on its quenching reaction with MeOH.[129]

3.5.3 Cyclic and Polycyclic Anions

Several remarkable organosilicon and organogermanium compounds, in which the anionic silicon or germanium is a part of a cyclic/polycyclic system, have been synthesized during the past decade. They will be briefly described below, except for those 6π-electron systems (heavy analogs of the cyclopentadienide ion and cyclobutadiene dianion), which will be discussed in detail in Chapter 6 (Section 6.3.2).

The smallest cyclic germyllithium, the 3-lithiodisilagermirane derivative $\mathbf{27^-\bullet Li^+}$, was synthesized by reduction of the Si=Ge bond of 1*H*-disilagermirene **28**[130] with *t*-BuLi in THF (Scheme 3.34).[126] The formation of $\mathbf{27^-\bullet Li^+}$, which is a product of the formal 1,2-addition of LiH across the Si=Ge bond, could be rationalized assuming an initial SET process involving intermediate formation of the anion-radical $\mathbf{28^{\bullet -}}$ and *tert*-butyl radical as a key radical pair, followed by hydrogen abstraction by $\mathbf{28^{\bullet -}}$ to form finally germyllithium $\mathbf{27^-\bullet Li^+}$ (Scheme 3.34).

The dipotassium salt of the cyclotetrasilane 1,3-dianion derivative $\mathbf{29^{2-}\bullet K_2^+}$ was prepared by the reduction of *trans*-1,3-dibromocyclotetrasilane **30** with an excess of potassium in benzene (Scheme 3.35).[131] The four-membered cyclotetrasilane ring is perfectly planar, whereas both anionic silicon centers are markedly pyramidalized (334.5°).

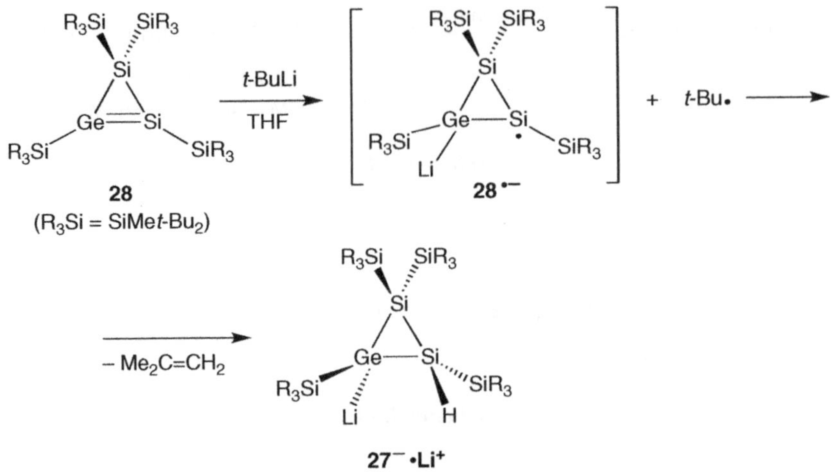

Scheme 3.34

In crystalline form, $\mathbf{29^{2-}{\bullet}K_2^+}$ exists as a CIP with its bonds from Si to K of 3.373(2) Å agreeing well with the sum of the covalent radii of the silicon and potassium atoms. The most striking feature of this salt is its supramolecular structure, composed of alternating $\mathbf{29^{2-}{\bullet}K_2^+}$ and C_6H_6 (crystallization solvent) units forming infinite columns. Within these columns, each potassium atom of $\mathbf{29^{2-}{\bullet}K_2^+}$ is η^2-coordinated by a benzene molecule [K–C bond lengths of 3.327(9) and 3.29(1) Å], and each benzene molecule coordinates two potassium atoms of different $\mathbf{29^{2-}{\bullet}K_2^+}$ molecules.

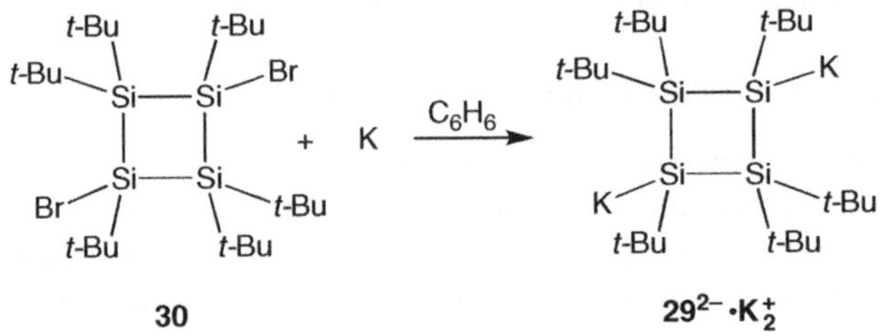

Scheme 3.35

Reduction of the 1,4-disila-Dewar benzene derivative **31** with lithium in THF produced an 8π-electron 1,4-dianionic cyclic system in the form of its dilithium salt $\mathbf{32^{2-}{\bullet}[Li^+(thf)]_2}$ which crystallized as a CIP (Scheme 3.36).[132] Interestingly, the six-membered Si_2C_4 ring of the $\mathbf{32^{2-}}$ dianion adopts a boat conformation with both [Li^+(thf)] units residing on the same side of the ring, despite the evident electrostatic repulsion produced by this spatial arrangement (Scheme 3.37). Apparently, this electrostatic disadvantage might be outbalanced by the extra stabilization gained through the significant interaction of Li^+ ions with the π-electron density of the C=C bonds. This was indeed observed in the crystal structure of $\mathbf{32^{2-}{\bullet}[Li^{+}(thf)]_2}$, in which C–Li bonding contacts of 2.271(4)–2.306(4) Å were found, resulting in stretching of the C=C bonds [1.381(3) and 1.385(3) Å vs. 1.356(6) Å in the starting **31**]. The resonance of the anionic silicons in $\mathbf{32^{2-}{\bullet}[Li^{+}(thf)]_2}$ was observed in toluene-d_8 at −47.6 ppm, whereas the doubly bonded carbon atoms resonated at 148.3 ppm. The latter signal appeared at a markedly higher field than that of the starting **31** (168.4 ppm), thus pointing to a partial rehybridization of the formally sp^2-carbons because of their interaction with Li^+ ions. Computationally, the boat form of the model compound of $\mathbf{32^{2-}{\bullet}[Li^{+}(thf)]_2}$ (Me_3Si instead of t-Bu_2MeSi, H instead of Et, THF molecules removed) was 6.7 kcal/mol more stable than the chair-type inverse sandwich form with the two lithium atoms situated above and below the six-membered ring.

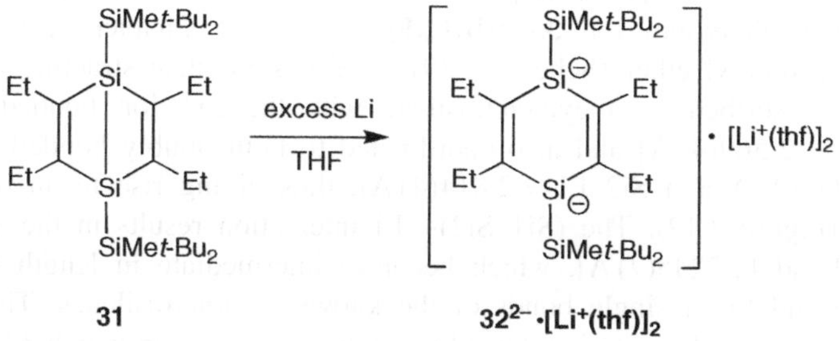

Scheme 3.36

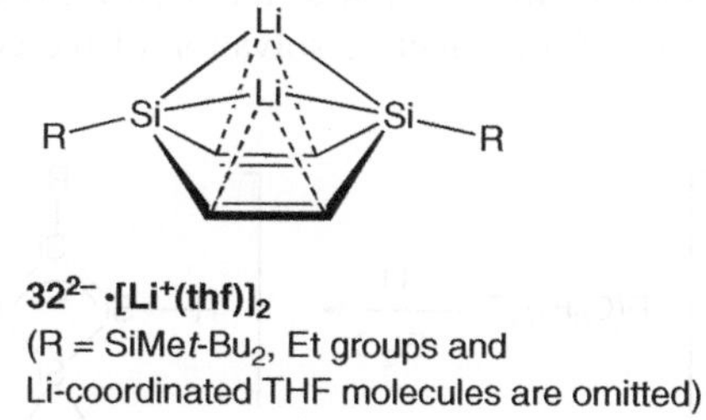

Scheme 3.37

The lithium salt of the cyclotetragermanide derivative $\mathbf{33^{-}{\bullet}[Li^{+}(dme)_3]}$ was prepared by Weidenbruch *et al.* by the reduction of the tetraaryldigermene $Tip_2Ge{=}GeTip_2$ (Tip = 2,4,6-triisopropylphenyl) with an excess of lithium in DME (Scheme 3.38).[133] The anionic part $\mathbf{33^{-}}$ of this ionic molecule featured a perfectly planar four-membered ring with two pairs of distinctly different skeletal Ge–Ge bonds of 2.5116(6) Å (within the range of Ge–Ge single bonds) and 2.3679(6) Å (within the range of Ge=Ge double bonds). Such crystallographic peculiarities led those authors to the suggestion of delocalization of the negative charge over the allylic-type Ge_3 fragment as part of a Ge_4 four-membered ring.

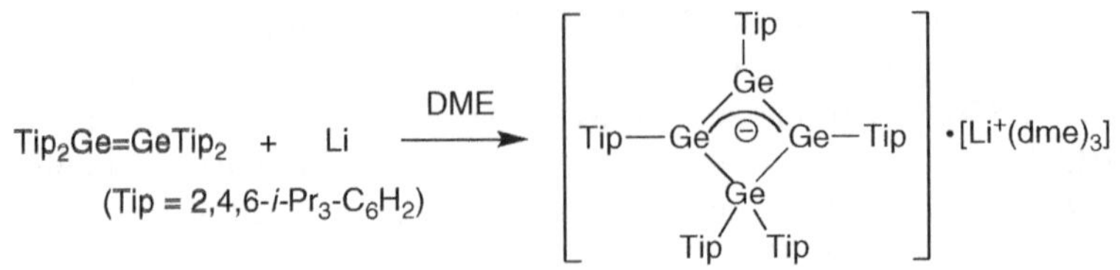

33⁻·[Li⁺(dme)₃]

Scheme 3.38

A silicon variation of the cyclotetragermanide salt **33^{-}•Li^{+}**, that is, the cyclotetrasilenide derivative **34^{-}•[Li^{+}(thf)]**, was synthesized by the two-electron reduction of the cyclotetrasilenylium ion **35^{+}•[$B(C_6F_5)_4$]$^{-}$**[134] (see Chapter 6, Section 6.3.3) with lithium in diethyl ether (Scheme 3.39).[135] The solid-state structure of **34^{-}•[Li^{+}(thf)]** is best described as a cyclotetrasilene with the Li^{+} ion bonded to the Si3 atom [Si3–Li = 2.569(4) Å] and also coordinated to both doubly bonded Si1 and Si2 [Si1–Li = 2.789(4) Å and Si2–Li = 2.814(4) Å], thus giving rise to an overall η^3-Li coordination (Figure 3.12). The (Si1–Si2)→Li interaction results in the stretching of the Si1–Si2 bond [2.2245(7) Å], which becomes intermediate in length between the Si=Si double and Si–Si single bonds of the known cyclotetrasilenes. The diagnostic trihaptocoordination of Li^{+} ion is retained in nonpolar toluene, in which **34^{-}•Li^{+}** exists as a CIP. In this CIP the negative charge is delocalized over the Si1–Si2–Si3 allylic-type anion, which is seen in the shielding of the terminal Si1 and Si3 atoms (−31.5 ppm) and very characteristic extreme deshielding of the central Si2 atom (273.0 ppm). However, in polar THF, the coordination of the Li^{+} ion to the Si_4-ring is broken, resulting in the formation of an SSIP because of the effective solvation of the countercation.

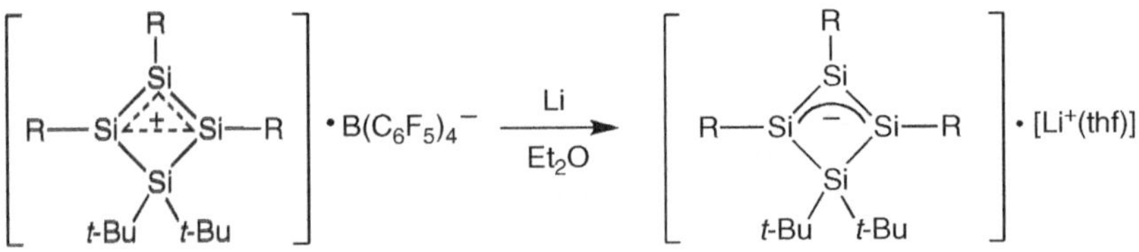

35⁺·B(C₆F₅)₄⁻ (R = SiMet-Bu₂) **34⁻·[Li⁺(thf)]**

Scheme 3.39

The bicyclic germyl anion derivative **36^{-}•[K^{+}(dme)$_2$]** is readily available by the reduction of the corresponding bicyclic germyl iodide **37** with potassium graphite in THF (Scheme 3.40).[136] All skeletal Ge–Ge bonds [2.4354(5)–2.5433(6) Å] are within the normal range for single bonds. The germylpotassium **36^{-}•[K^{+}(dme)$_2$]** crystallized as a CIP with a Ge–K bond distance of 3.4324(10) Å and the anionic Ge center being markedly pyramidalized (301.8°) However, in polar THF, the Ge–K bond dissociated

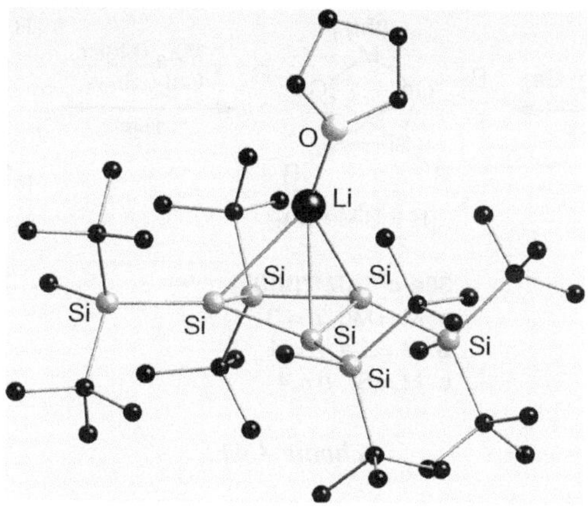

Figure 3.12 *Crystal structure of* ***34^-•[Li^+(thf)]*** *(H atoms are not shown)*

making the germyl anion **36^-** free from direct interaction with the countercation (SSIP). Interestingly, upon the one-electron oxidation of **36^-•[K^+(dme)$_2$]** with $B(C_6F_5)_3$, the corresponding bicyclic germyl radical was isolated as a stable derivative (see Chapter 2, Section 2.4.1.1).[136]

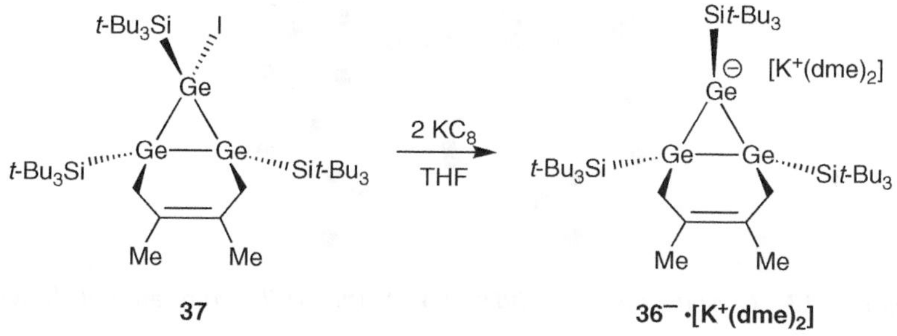

Scheme 3.40

The alkaline earth metal salts of the 1,3-disila-2,4-digermabicyclo[1.1.0]butane 2,4-dianion derivative **38a-c^{2-}•[M^{2+}(thf)$_n$]** (**a:** M = Mg, $n = 3$; **b:** M = Ca, $n = 4$; **c:** M = Sr, $n = 4$) were synthesized by the reduction of either *trans*-1,2-dichloro-$^3\Delta$-1,2,3,4-disiladigermetene **39** with alkaline earth metals M (Mg, Ca) in THF or 1,2-disila-3,4-digermacyclobutadiene dianion **40** with dihalides MX_2 ($MgBr_2$, CaI_2, SrI_2) in THF (Scheme 3.41).[137]

Bridgehead silicons of all **38a-c^{2-}•[M^{2+}(thf)$_n$]** compounds resonated in a very characteristic high-field region: −231.9 ppm (for the Mg salt), −220.2 ppm (for the Ca salt) and −225.2 ppm (for the Sr salt). The bridging Si–Si bonds of **38a-c^{2-}•[M^{2+}(thf)$_n$]** were short [2.3181(15) Å for the Mg salt], interplanar angles ϕ were acute [98.9° for the Mg salt] and bond angles θ were wide [133.65(6)° and 135.75(7)° for the Mg

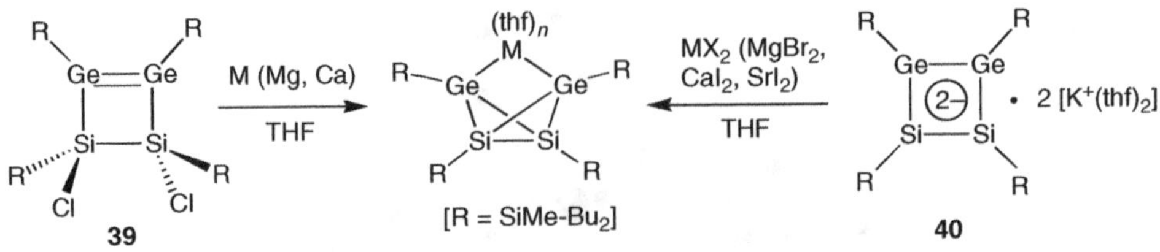

38a-c^{2-}•[M^{2+}(thf)$_n$]
a: M = Mg, $n = 3$
b: M = Ca, $n = 4$
c: M = Sr, $n = 4$

Scheme 3.41

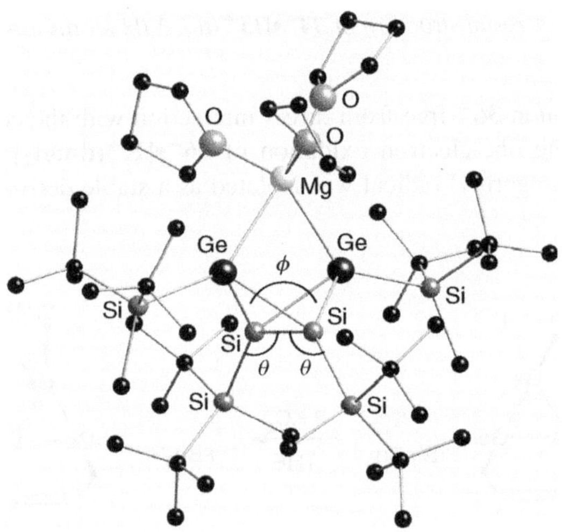

Figure 3.13 *Crystal structure of* ***38a^{2-}•[Mg^{2+}(thf)$_3$]*** *(H atoms are not shown)*

salt] (Figure 3.13).[137] These structural features of bicyclo[1.1.0]butane 2,4-dianions **38a-c^{2-}•[M^{2+}(thf)$_n$]**, with a capping alkaline earth metal playing a bridging role, match the properties of tricyclo[2.1.0.0^{2,5}]pentanes, in which a small interplanar angle ϕ forces the hybrid orbitals on the bridgehead carbons to form short bent C–C bridging bonds.

A highly strained tricyclic silylpotassium derivative, potassium 3-phenyl-2,3,4,5-tetrakis(di-*tert*-butylmethylsilyl)pentasilatricyclo[2.1.0.0^{2,5}]pentan-1-ide **39$^-$•K$^+$**, was very recently synthesized by the reduction of 3-chloro-3-phenyl-1,2,4,5-tetrakis(di-*tert*-butylmethylsilyl)pentasilatricyclo[2.1.0.0^{2,5}]pentane precursor **40** with KC_8 (Scheme 3.42).[138]

As is typical for the bridgehead silicons, the resonances of both Si1 and Si2 in **39$^-$•K$^+$** were observed in a diagnostic high-field region with the anionic Si1 center being far more shielded than the neutral Si2: −253.9 ppm vs −142.5 ppm. Interestingly, the Si1-bound potassium cation, lacking coordination to nucleophilic solvent, was intramolecularly

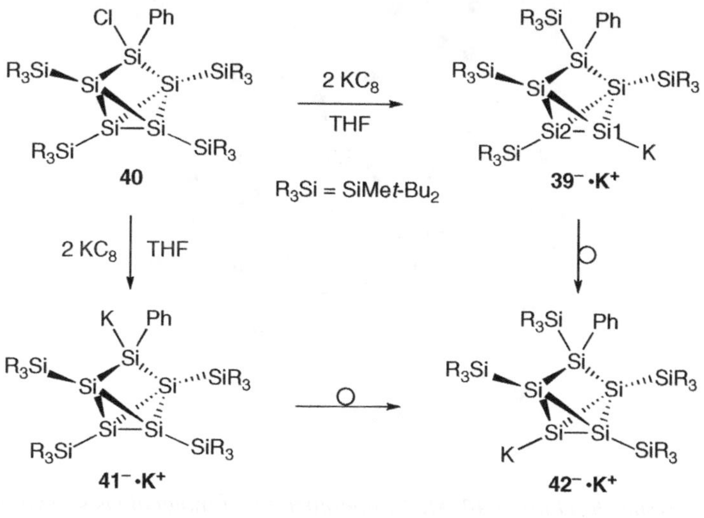

Scheme 3.42

stabilized by the π-interaction with the phenyl ring in **39⁻•K⁺** (Figure 3.14). Such $K^+\cdots C_6H_5$ π-coordination caused a remarkable bending of the Si1–K bond towards the benzene ring, resulting in the structurally rather unfavorable inverted tetrahedral configuration at the anionic Si1 center ('umbrella' configuration). DFT computations on the H_3Si-substituted model well reproduced the experimental structure, thus supporting the importance of such π-interaction (stabilization energy ca. 5.9 kcal/mol), which outbalanced the obvious disadvantage of the geometrical constraints. Accompanying NBO calculations revealed the large fraction of s-character in the exocyclic Si1 orbital holding a negative charge.

The mechanism of the formation of **39⁻•K⁺** may involve the initial reduction of **40** giving anticipated silylpotassium **41⁻•K⁺** with the anionic charge at the bridge Si, followed by 1,3-migration resulting in the formation of the silylpotassium **42⁻•K⁺** featuring a negative charge at the bridgehead Si (Scheme 3.42).[138] Such isomerization might be caused by a general trend for accommodation of the anionic charge in the orbital of greater s-character. The final bridgehead-to-bridgehead 1,2-migration, driven by the strongly favorable $K^+\cdots C_6H_5$ π-coordination, results in the formation of **39⁻•K⁺**.

The class of highly reactive Zintl anions/Zintl phases, that is, the ligand-free polyhedral compounds of the heavy group 14 elements formed by the reduction of elements from Si to Pb with alkali or alkaline earth metals and involving both ionic and covalent interactions, will not be considered in this chapter.[139]

3.6 Summary and Outlook

Carbanions, together with carbenium ions and free radicals, are among the most fundamental reactive intermediates, participating in a vast number of organic reactions. Carbanions are very useful in forming carbon–carbon bonds, and acting as nucleophiles both in substitution and addition reactions (Section 3.4). They are particularly well suited

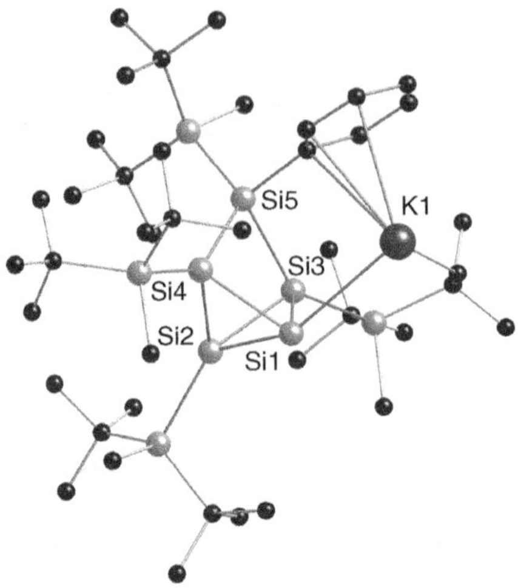

Figure 3.14 *Crystal structure of* ***40⁻•K⁺*** *(monomeric half-molecule is shown, H atoms are not shown)*

for S_N2 reactions, one of the most general textbook mechanisms in organic chemistry. The analogs of carbanions, based on Si, Ge, Sn and Pb atoms, also constitute one of the most important classes of reactive group 14 organometallics. The structural interest in the systematic studies of such species lies in the following areas: (1) geometry about the anionic centers (from tetrahedral to planar); (2) ion-pairing behavior (from covalent to ionic); and (3) aggregation state (from monomeric to oligomeric). Despite the great progress that has been made in the field of heavy analogs of carbanions during the past decades, a number of problems are still awaiting solutions. These concern, for example, clarification of the nature of the anion–cation interaction, which is generally represented by the two bonding extremes: CIP (cation directly bonded to anion) and SSIP (cation is efficiently solvated and the resulting cation-solvent complex interacts with the anion). As was demonstrated by the preceding studies (Section 3.3), the contribution of both CIP and SSIP, that is, the degree of covalency and ionicity of the particular anion–cation bonding situation, is strongly influenced by many factors, including solvent, countercation, and presence or absence of complexing reagent (crown ethers). Thus, polar solvents typically favor full charge separation and formation of the SSIP, whereas in nonpolar media, anionic derivatives often exist in the form of the CIP. The situation is even more complicated by the nature of the countercation; for example, cesium usually prefers CIP, whereas lithium in many cases forms a SSIP in which it can take advantage of the strong electrostatic interaction with *n*-donating solvents. The other issue, on which much still remains to be done, is related to the problem of the aggregation states of anionic derivatives. A wide variety of aggregates with different reactivities may simultaneously be present in solution, resulting in the observation of distinctly different reaction pathways and affecting the range of synthetic utility. Future experimental and computational work is expected to provide deep insight into this problem. Synthetic applications of the heavy group 14 element anionic derivatives are still very limited. As representative examples, one can mention Me_3SiLi and n-Bu_3SnLi, which are by far the most widely employed organic chemistry reagents. The field of the practical utilization of other anionic derivatives in synthetic chemistry is open for further investigations.

3.7 References

1. G. B. Ellison, P. C. Engelking, W. C. Lineberger, *J. Am. Chem. Soc.* **1978**, *100*, 2556.
2. D. A. Dixon, D. Feller, K. A. Peterson, *J. Phys. Chem. A* **1997**, *101*, 9405.
3. K. M. Ervin, S. Gronert, S. E. Barlow, M. K. Gilles, A. G. Harrison, V. M. Bierbaum, C. H. DePuy, W. C. Lineberger, G. B. Ellison, *J. Am. Chem. Soc.* **1990**, *112*, 5750.
4. Recent reviews on the heavy group 14 element centered anions: (a) K. Tamao, A. Kawachi, *Adv. Organomet. Chem.* **1995**, *38*, 1; (b) P. D. Lickiss, C. M. Smith, *Coord. Chem. Rev.* **1995**, *145*, 75; (c) P. Riviere, M. Riviere-Baudet, J. Satgé, in *Comprehensive Organometallic Chemistry II, Vol. 2* (Ed. G. Wilkinson), Pergamon Press, Oxford, **1995**, Chapter 5; (d) A. G. Davies, P. J. Smith, in *Comprehensive Organometallic Chemistry II, Vol. 2* (Ed. G. Wilkinson), Pergamon Press, Oxford, **1995**, Chapter 6; (e) P. G. Harrison, in *Comprehensive Organometallic Chemistry II, Vol. 2* (Ed. G. Wilkinson), Pergamon Press, Oxford, **1995**, Chapter 7; (f) J. Belzner, U. Dehnert, in *The Chemistry of Organic Silicon Compounds, Vol. 2* (Eds Z. Rappoport, Y. Apeloig), John Wiley & Sons, Ltd, Chichester, **1998**, Chapter 14; (g) A. Sekiguchi, V. Ya. Lee, M. Nanjo, *Coord. Chem. Rev.* **2000**, *210*, 11; (h) P. Riviere, A. Castel, M. Riviere-Baudet, in *The Chemistry of Organic Germanium, Tin and Lead Compounds, Vol. 2, Part 1* (Ed. Z. Rappoport), John Wiley & Sons, Ltd, Chichester, **2002**, Chapter 11; (i) H.-W. Lerner, *Coord. Chem. Rev.* **2005**, *249*, 781; (j) M. Saito, M. Yoshioka, *Coord. Chem. Rev.* **2005**, *249*, 765; (k) V. Ya. Lee, A. Sekiguchi, in *Reviews of Reactive Intermediate Chemistry* (Eds M. S. Platz, R. A. Moss, M. Jones Jr), Wiley-Interscience, Hoboken, **2007**, Chapter 2; (l) V. Ya. Lee, A. Sekiguchi, *Acc. Chem. Res.* **2007**, *40*, 410; (m) For the application of silyllithium derivatives in the synthesis of polysilane dendrimers, see the recent review: M. Nanjo, A. Sekiguchi, in *Silicon-Containing Dendritic Polymers* (Eds P. R. Dvornic, M. J. Owen), Springer, **2009**, Chapter 4; (n) For the synthesis and reactivity of disilenyllithium derivatives, see the recent review: D. Scheschkewitz, *Chem. Eur. J.* **2009**, *15*, 2476.
5. W. C. Still, *J. Org. Chem.* **1976**, *41*, 3063.
6. H. Sakurai, A. Okada, M. Kira, K. Yonezawa, *Tetrahedron Lett.* **1971**, 1511.
7. R. J. P. Corriu, C. Guérin, *J. Chem. Soc., Chem. Commun.* **1980**, 168.
8. (a) T. F. Schaaf, J. P. Oliver, *J. Am. Chem. Soc.* **1969**, *91*, 4327; (b) W. H. Ilsley, M. J. Albright, T. J. Anderson, M. D. Glick, J. P. Oliver, *Inorg. Chem.* **1980**, *19*, 3577; (c) N. S. Vyazankin, G. A. Razuvaev, E. N. Gladyshev, S. P. Korneva, *J. Organomet. Chem.* **1967**, *7*, 353; (d) E. N. Gladyshev, E. A. Fedorova, L. O. Yuntila, G. A. Razuvaev, N. S. Vyazankin, *J. Organomet. Chem.* **1975**, *96*, 169; (e) E. Hengge, N. Holtschmidt, *J. Organomet. Chem.* **1968**, *12*, P5.
9. A. R. Claggett, W. H. Ilsley, T. J. Anderson, M. D. Glick, J. P. Oliver, *J. Am. Chem. Soc.* **1977**, *99*, 1797.

10. (a) N. Wiberg, in *Frontiers in Organosilicon Chemistry* (Eds A. R. Bassindale, P. P. Gaspar), Royal Society of Chemistry, London, **1991**, p. 263; (b) N. Wiberg, G. Fisher, P. Karampatses, *Angew. Chem., Int. Ed. Engl.* **1984**, *23*, 59; (c) N. Wiberg, H. Schuster, A. Simon, K. Peters, *Angew. Chem., Int. Ed. Engl.* **1986**, *25*, 79; (d) N. Wiberg, K. Schurz, *J. Organomet. Chem.* **1988**, *341*, 145.
11. (a) M. V. George, D. J. Peterson, H. Gilman, *J. Am. Chem. Soc.* **1960**, *82*, 403; (b) H. Wagner, U. Schubert, *Chem. Ber.* **1990**, *123*, 2101; (c) J. Meyer, J. Willnecker, U. Schubert, *Chem. Ber.* **1989**, *122*, 223.
12. E. W. Colvin, *Silicon in Organic Synthesis*, Butterworths, London, **1981**, pp. 134–140.
13. (a) H. Gilman, W. Steudel, *Chem. Ind. (London)* **1959**, 1094; (b) M. Weidenbruch, K. Kramer, K. Peters, H. G. von Schnering, *Z. Naturforsch., Teil B*, **1985**, *40B*, 601; (c) K. Tamao, A. Kawachi, *Organometallics* **1995**, *14*, 3108.
14. (a) H. Gilman, T. C. Wu, *J. Am. Chem. Soc.* **1951**, *73*, 4031; (b) A. G. Brook, H. Gilman, *J. Am. Chem. Soc.* **1954**, *76*, 278.
15. H. Gilman, G. D. Lichtenwalter, *J. Am. Chem. Soc.* **1958**, *80*, 608.
16. M. Porchia, N. Brianese, U. Casellato, F. Ossola, G. Rossetto, P. Zanella, R. Graziani, *J. Chem. Soc., Dalton Trans.* **1989**, 677.
17. A. G. Brook, A. Baumegger, A. J. Lough, *Organometallics* **1992**, *11*, 310.
18. (a) H. Gilman, C. L. Smith, *J. Organomet. Chem.* **1968**, *14*, 91; (b) G. Gutekunst, A. G. Brook, *J. Organomet. Chem.* **1982**, *225*, 1.
19. (a) W. Biffar, T. Gasparis-Ebeling, H. Nöth, W. Storch, B. Wrackmeyer, *J. Magn. Res.* **1981**, *44*, 54; (b) R. H. Heyn, T. D. Tilley, *Inorg. Chem.* **1990**, *29*, 4051.
20. A. Sekiguchi, M. Nanjo, C. Kabuto, H. Sakurai, *J. Am. Chem. Soc.* **1995**, *117*, 4195.
21. H. Gilman, R. L. Harrell Jr, *J. Organomet. Chem.* **1967**, *9*, 67.
22. O. W. Steward, G. L. Heider, J. S. Johnson, *J. Organomet. Chem.* **1979**, *168*, 33.
23. (a) M. A. Ring, D. M. Ritter, *J. Am. Chem. Soc.* **1961**, *83*, 802; (b) M. A. Ring, D. M. Ritter, *J. Phys. Chem.* **1961**, *65*, 182; (c) E. Amberger, E. Mühlhofer, *J. Organomet. Chem.* **1968**, *12*, 55; (d) H. Bürger, R. Eujen, H. C. Marsmann, *Z. Naturforsch., B: Chem. Sci.* **1974**, 29B, 149; (e) O. Mundt, G. Becker, H. M. Hartmann, W. Schwarz, *Z. Anorg. Allg. Chem.* **1989**, *572*, 75; (f) F. Fehér, M. Krancher, M. Fehér, *Z. Anorg. Allg. Chem.* **1991**, *606*, 7.
24. D. M. Roddick, R. H. Heyn, T. D. Tilley, *Organometallics* **1989**, *8*, 324.
25. (a) P. Boudjouk, U. Samaraweera, R. Sooriyakumaran, J. Chrusciel, K. R. Anderson, *Angew. Chem., Int. Ed. Engl.* **1988**, *27*, 1355; (b) T. Tsumuraya, S. A. Batcheller, S. Masamune, *Angew. Chem., Int. Ed. Engl.* **1991**, *30*, 902; (c) R. Corriu, G. Lanneau, C. Priou, F. Soulairol, N. Auner, R. Probst, R. Conlin, C. J. Tan, *J. Organomet. Chem.* **1994**, *466*, 55.
26. H. Oehme, H. Weiss, *J. Organomet. Chem.* **1987**, *319*, C16.
27. (a) S. Nozakura, S. Konotsune, *Bull. Chem. Soc. Jpn.* **1956**, *29*, 322; (b) K. Naumann, G. Zon, K. Mislow, *J. Am. Chem. Soc.* **1969**, *91*, 7012; (c) S. C. Bernstein, *J. Am. Chem. Soc.* **1970**, *92*, 699; (d) R. A. Benkeser, K. M. Foley, J. B. Grutzner, W. E. Smith, *J. Am. Chem. Soc.* **1970**, *92*, 697; (e) R. A. Benkeser, K. M. Foley, J. M. Gaul, G. S. Li, *J. Am. Chem. Soc.* **1970**, *92*, 3232; (f) R. A. Benkeser, *Acc.*

Chem. Res. **1971**, *4*, 94; (g) P. Boudjouk, S. Kloos, A. B. Rajkumar, *J. Organomet. Chem.* **1993**, *443*, C41.

28. (a) V. A. Petrosyan, M. E. Niyazymbetov, S. P. Kolesnikov, V. Ya. Lee, *Bull. Acad. Sci. USSR, Div. Chem. Sci.* **1987**, 422 (translated from *Izv. Akad. Nauk SSSR, Ser. Khim.* **1987**, 461); (b) V. A. Petrosyan, M. E. Niyazymbetov, S. P. Kolesnikov, V. Ya. Lee, *Bull. Acad. Sci. USSR, Div. Chem. Sci.* **1989**, 1690 (translated from *Izv. Akad. Nauk SSSR, Ser. Khim.* **1988**, 1889); (c) Review: S. P. Kolesnikov, S. N. Tandura, O. M. Nefedov, in *The Chemistry of Organic Germanium, Tin and Lead Compounds, Vol. 2, Part 2* (Ed. Z. Rappoport), John Wiley & Sons, Ltd, Chichester, **2002**, Chapter 20.
29. (a) K. Tamao, A. Kawachi, Y. Ito, *J. Am. Chem. Soc.* **1992**, *114*, 3989; (b) K. Tamao, A. Kawachi, Y. Ito, *Organometallics* **1993**, *12*, 580.
30. K. Tamao, A. Kawachi, *Angew. Chem., Int. Ed. Engl.* **1995**, *34*, 818.
31. (a) J. Ohshita, Y. Masaoka, S. Masaoka, M. Ishikawa, A. Tachibana, T. Yano, T. Yamabe, *J. Organomet. Chem.* **1994**, *473*, 15; (b) J. Ohshita, S. Masaoka, M. Ishikawa, *Organometallics* **1996**, *15*, 2198; (c) J. Ohshita, S. Masaoka, Y. Masaoka, H. Hasebe, M. Ishikawa, A. Tachibana, T. Yano, T. Yamabe, *Organometallics* **1996**, *15*, 3136.
32. J.-H. Hong, P. Boudjouk, *J. Am. Chem. Soc.* **1993**, *115*, 5883.
33. (a) J.-H. Hong, P. Boudjouk, S. Castellino, *Organometallics* **1994**, *13*, 3387; (b) R. West, H. Sohn, U. Bankwitz, J. Calabrese, Y. Apeloig, T. Mueller, *J. Am. Chem. Soc.* **1995**, *117*, 11608; (c) R. West, H. Sohn, D. R. Powell, T. Müller, Y. Apeloig, *Angew. Chem., Int. Ed. Engl.* **1996**, *35*, 1002; (d) W. P. Freeman, T. D. Tilley, G. P. A. Yap, A. L. Rheingold, *Angew. Chem., Int. Ed. Engl.* **1996**, *35*, 882; (e) W. P. Freeman, T. D. Tilley, L. M. Liable-Sands, A. L. Rheingold, *J. Am. Chem. Soc.* **1996**, *118*, 10457; (f) M. Saito, R. Haga, M. Yoshioka, *Chem. Lett.* **2003**, *32*, 912; (g) M. Saito, R. Haga, M. Yoshioka, K. Ishimura, S. Nagase, *Angew. Chem., Int. Ed.* **2005**, *44*, 6553.
34. A. L. Allred, R. T. Smart, D. A. Van Beek Jr, *Organometallics* **1992**, *11*, 4225.
35. F. Uhlig, P. Gspaltl, M. Trabi, E. Hengge, *J. Organomet. Chem.* **1995**, *493*, 33.
36. J. A. Morrison, R. L. Lagow, *Inorg. Chem.* **1977**, *16*, 2972.
37. P. von R. Schleyer, A. E. Reed, *J. Am. Chem. Soc.* **1988**, *110*, 4453.
38. S. K. Mehrotra, H. Kawa, J. R. Baran Jr, M. M. Ludvig, R. J. Lagow, *J. Am. Chem. Soc.* **1990**, *112*, 9003.
39. R. Fischer, T. Konopa, J. Baumgartner, C. Marschner, *Organometallics* **2004**, *23*, 1899.
40. J. Belzner, U. Dehnert, D. Stalke, *Angew. Chem., Int. Ed. Engl.* **1994**, *33*, 2450.
41. M. Kira, T. Iwamoto, D. Yin, T. Maruyama, H. Sakurai, *Chem. Lett.* **2001**, 910.
42. H. Gilman, D. J. Peterson, A. W. Jarvie, H. J. S. Winkler, *J. Am. Chem. Soc.* **1960**, *82*, 2076.
43. K. M. Baines, K. A. Mueller, T. K. Sham, *Can. J. Chem.* **1992**, *70*, 2884.
44. J. C. Podesta, A. B. Chopa, N. N. Giagante, A. E. Zuniga, *J. Organomet. Chem.* **1995**, *494*, 5.
45. B. Wrackmeyer, K. Horchler, *Z. Naturforsch., B: Chem. Sci.* **1989**, 44b, 1195.
46. E. Buncel, T. K. Venkatachalam, *Heteroatom Chem.* **1994**, *5*, 201.

47. P. B. Hitchcock, M. F. Lappert, G. A. Lawless, B. Royo, *J. Chem. Soc., Chem. Commun.* **1993**, 554.
48. I. Suzuki, T. Furuta, Y. Yamamoto, *J. Organomet. Chem.* **1993**, *443*, C6.
49. R. Villazana, H. Sharma, F. Cervantes-Lee, K. H. Pannell, *Organometallics* **1993**, *12*, 4278.
50. A. Castel, P. Rivière, J. Satgé, Y. H. Ko, D. Desor, *J. Organomet. Chem.* **1990**, *397*, 7.
51. A. Castel, P. Rivière, J. Satgé, D. Desor, M. Ahbala, C. Abdenadher, *Inorg. Chim. Acta* **1993**, *212*, 51.
52. (a) W. Reimann, H. G. Kuivila, D. Farah, T. Apoussidis, *Organometallics* **1987**, *6*, 557; (b) T. S. Kaufman, *Synlett* **1997**, 1377.
53. J. B. Lambert, M. Urdaneta-Pérez, *J. Am. Chem. Soc.* **1978**, *100*, 157.
54. M. R. Nimlos, G. B. Ellison, *J. Am. Chem. Soc.* **1986**, *108*, 6522.
55. Y. Apeloig, in *The Chemistry of Organic Silicon Compounds* (Eds S. Patai, Z. Rappoport), John Wiley & Sons, Ltd, Chichester, **1989**, Chapter 2, p. 201.
56. U. Edlund, T. Lejon, T. K. Venkatachalam, E. Buncel, *J. Am. Chem. Soc.* **1985**, *107*, 6408.
57. H. V. R. Dias, M. M. Olmstead, K. Ruhlandt-Senge, P. P. Power, *J. Organomet. Chem.* **1993**, *462*, 1.
58. D. Seebach, J. Gabriel, R. Hässig, *Helv. Chim. Acta* **1984**, *67*, 1083.
59. A. Heine, R. Herbst-Irmer, G. M. Sheldrick, D. Stalke, *Inorg. Chem.* **1993**, *32*, 2694.
60. (a) K. W. Klinkhammer, W. Schwarz, *Z. Anorg. Allg. Chem.* **1993**, *619*, 1777; (b) K. W. Klinkhammer, G. Becker, W. Schwarz, in *Organosilicon Chemistry II* (Eds N. Auner, J. Weis), VCH, Weinheim, **1996**, p. 493.
61. A. G. Avent, C. Eaborn, P. B. Hitchcock, G. A. Lawless, P. D. Lickiss, M. Mallien, J. D. Smith, A. D. Webb, B. Wrackmeyer, *J. Chem. Soc., Dalton Trans.* **1993**, 3259.
62. (a) A. Sekiguchi, M. Nanjo, C. Kabuto, H. Sakurai, *Organometallics* **1995**, *14*, 2630; (b) M. Nanjo, A. Sekiguchi, H. Sakurai, *Bull. Chem. Soc. Jpn.* **1998**, *71*, 741.
63. (a) L. M. Seitz, T. L. Brown, *J. Am. Chem. Soc.* **1966**, *88*, 2174; (b) P. West, R. Waack, *J. Am. Chem. Soc.* **1967**, *89*, 4395.
64. A. Sekiguchi, M. Nanjo, C. Kabuto, H. Sakurai, *Angew. Chem., Int. Ed. Engl.* **1997**, *36*, 113.
65. R. H. Cox, E. G. Janzen, W. B. Harrison, *J. Magn. Reson.* **1971**, *4*, 274.
66. H. J. Reich, J. P. Borst, R. R. Dykstra, *Organometallics* **1994**, *13*, 1.
67. P. B. Hitchcock, M. F. Lappert, G. A. Lawless, B. Royo, *J. Chem. Soc., Chem. Commun.* **1993**, 554.
68. (a) T. F. Schaaf, W. Butler, M. D. Glick, J. P. Oliver, *J. Am. Chem. Soc.* **1974**, *96*, 7593; (b) W. H. Ilsley, T. F. Schaaf, M. D. Glick, J. P. Oliver, *J. Am. Chem. Soc.* **1980**, *102*, 3769.
69. B. Teclé, W. H. Ilsley, J. P. Oliver, *Organometallics* **1982**, *1*, 875.
70. H. V. R. Dias, M. M. Olmstead, K. Ruhlandt-Senge, P. P. Power, *J. Organomet. Chem.* **1993**, *462*, 1.
71. R. A. Bartlett, H. V. R. Dias, P. P. Power, *J. Organomet. Chem.* **1988**, *341*, 1.

72. G. Becker, H.-M. Hartmann, E. Hengge, F. Schrank, *Z. Anorg. Allg. Chem.* **1989**, *572*, 63.
73. M. Nanjo, A. Sekiguchi, H. Sakurai, *Bull. Chem. Soc. Jpn.* **1999**, *72*, 1387.
74. (a) S. Freitag, R. Herbst-Irmer, L. Lameyer, D. Stalke, *Organometallics* **1996**, *15*, 2839; (b) A. Heine, D. Stalke, *Angew. Chem., Int. Ed. Engl.* **1994**, *33*, 113.
75. (a) D. Reed, D. Stalke, D. S. Wright, *Angew. Chem. Int. Ed. Engl.* **1991**, *30*, 1459; (b) T. Birchall, J. A. Vetrone, *J. Chem. Soc., Chem. Commun.* **1988**, 877.
76. D. R. Armstrong, M. G. Davidson, D. Moncrieff, D. Stalke, D. S. Wright, *J. Chem. Soc., Chem. Commun.* **1992**, 1413.
77. (a) A. R. Claggett, W. H. Ilsley, T. J. Anderson, M. D. Glick, J. P. Oliver, *J. Am. Chem. Soc.* **1977**, *99*, 1797; (b) D. W. Goebel Jr, J. L. Hencher, J. P. Oliver, *Organometallics* **1983**, *2*, 746; (c) L. Rösch, J. Pickardt, S. Imme, U. Börner, *Z. Naturforsch., B: Chem. Sci.* **1986**, 41B, 1523.
78. L. Rösch, C. Krüger, A.-P. Chiang, *Z. Naturforsch., B: Chem. Sci.* **1984**, *39B*, 855.
79. M. Westerhausen, *Angew. Chem., Int. Ed. Engl.* **1994**, *33*, 1493.
80. R. Goddard, C. Krüger, N. A. Ramadan, A. Ritter, *Angew. Chem., Int. Ed. Engl.* **1995**, *34*, 1030.
81. (a) A. J. Edwards, M. A. Paver, P. R. Raithby, C. A. Russell, D. Stalke, A. Steiner, D. S. Wright, *J. Chem. Soc., Dalton Trans.* **1993**, 1465; (b) D. R. Armstrong, M. J. Duer, M. G. Davidson, D. Moncrieff, C. A. Russell, C. Stourton, A. Steiner, D. Stalke, D. S. Wright, *Organometallics* **1997**, *16*, 3340.
82. G. R. Newkome, C. N. Moorefield, F. Vögtle, *Dendritic Molecules. Concepts. Synthesis. Perspectives*, VCH, Weinheim, **1996**.
83. E. Buhleier, W. Wehner, F. Vögtle, *Synthesis* **1978**, 155.
84. (a) D. A. Tomalia, H. Baker, J. Dewald, M. Hall, G. Kallos, S. Martin, J. Roeck, J. Ryder, P. Smith, *Polym. J.* **1985**, *17*, 117; (b) D. A. Tomalia, H. Baker, J. Dewald, M. Hall, G. Kallos, S. Martin, J. Roeck, J. Ryder, P. Smith, *Macromolecules* **1986**, *19*, 2466.
85. G. R. Newkome, Z. Yao, G. R. Baker, V. K. Gupta, *J. Org. Chem.* **1985**, *50*, 2003.
86. C. J. Hawker, J. M. J. Fréchet, *J. Am. Chem. Soc.* **1990**, *112*, 7638.
87. E. A. Rebrov, A. M. Muzafarov, V. S. Papkov, A. A. Zhdanov, *Dokl. Akad. Nauk SSSR* **1989**, *309*, 376.
88. A. W. van der Made, P. W. N. M. van Leeuwen, *J. Chem. Soc., Chem. Commun.* **1992**, 1400.
89. J. B. Lambert, J. L. Pflug, C. L. Stern, *Angew. Chem., Int. Ed. Engl.* **1995**, *34*, 98.
90. H. Suzuki, Y. Kimata, S. Satoh, A. Kuriyama, *Chem. Lett.* **1995**, 293.
91. A. Sekiguchi, M. Nanjo, C. Kabuto, H. Sakurai, *J. Am. Chem. Soc.* **1995**, *117*, 4195.
92. (a) J. B. Lambert, J. L. Pflug, J. M. Denari, *Organometallics* **1996**, *15*, 615; (b) J. B. Lambert, E. Basso, N. Qing, S. H. Lim, J. L. Pflug, *J. Organomet. Chem.* **1998**, *554*, 113; (c) J. B. Lambert, H. Wu, *Organometallics* **1998**, *17*, 4904; (d) Review: J. B. Lambert, J. L. Pflug, H. Wu, X. Liu, *J. Organomet. Chem.* **2003**, *685*, 113.
93. M. Nanjo, A. Sekiguchi, *Organometallics* **1998**, *17*, 492.
94. (a) H. Reinke, C. Krempner, *J. Organomet. Chem.* **2003**, *685*, 134; (b) C. Krempner, S. Chtchian, H. Reinke, *Inorg. Chim. Acta* **2004**, *357*, 3733; (c) U. Jäger-Fiedler,

M. Köckerling, R. Ludwig, A. Wulf, C. Krempner, *Angew. Chem. Int. Ed.* **2006**, *45*, 6755; (d) C. Krempner, M. Köckerling, C. Mamat, *Chem. Commun.* **2006**, 720; (e) C. Krempner, H. Reinke, *Inorg. Chem. Commun.* **2006**, *9*, 259; (f) C. Krempner, H. Reinke, *Organometallics* **2007**, *26*, 2053; (g) C. Krempner, M. Köckerling, *Organometallics* **2008**, *27*, 346.

95. (a) Y. Apeloig, G. Korogodsky, D. Bravo-Zhivotovskii, D. Bläser, R. Boese, *Eur. J. Inorg. Chem.* **2000**, 1091; (b) Y. Apeloig, M. Yuzefovich, M. Bendikov, D. Bravo-Zhivotovskii, D. Bläser, R. Boese, *Angew. Chem. Int. Ed.* **2001**, *40*, 3016; (c) D. Bravo-Zhivotovskii, M. Yuzefovich, N. Sigal, G. Korogodsky, K. Klinkhammer, B. Tumanskii, A. Shames, Y. Apeloig, *Angew. Chem., Int. Ed.* **2002**, *41*, 649; (d) D. Bravo-Zhivotovskii, I. Ruderfer, S. Melamed, M. Botoshansky, B. Tumanskii, Y. Apeloig, *Angew. Chem., Int. Ed.* **2005**, *44*, 739.
96. D. Bravo-Zhivotovskii, R. Dobrovetsky, D. Nemirovsky, V. Molev, M. Bendikov, G. Molev, M. Botoshansky, Y. Apeloig, *Angew. Chem., Int. Ed.* **2008**, *47*, 4343.
97. (a) C. Marschner, *Eur. J. Inorg. Chem.* **1998**, 221; (b) C. Kayser, R. Fischer, J. Baumgartner, C. Marschner, *Organometallics* **2002**, *21*, 1023; (c) C. Kayser, G. Kickelbick, C. Marschner, *Angew. Chem., Int. Ed.* **2002**, *41*, 989; (d) R. Fischer, D. Frank, W. Gaderbauer, C. Kayser, C. Mechtler, J. Baumgartner, C. Marschner, *Organometallics* **2003**, *22*, 3723; (e) R. Fischer, T. Konopa, S. Ully, J. Baumgartner, C. Marschner, *J. Organomet. Chem.* **2003**, *685*, 79; (f) R. Fischer, T. Konopa, J. Baumgartner, C. Marschner, *Organometallics* **2004**, *23*, 1899; (g) H. Wagner, J. Baumgartner, C. Marschner, *Organometallics* **2007**, *26*, 1762; (h) M. Zirngast, J. Baumgartner, C. Marschner, *Eur. J. Inorg. Chem.* **2008**, 1078; (i) R. Fischer, J. Baumgartner, C. Marschner, F. Uhlig, *Inorg. Chim. Acta* **2005**, *358*, 3174; (j) R. Fischer, M. Zirngast, M. Flock, J. Baumgartner, C. Marschner, *J. Am. Chem. Soc.* **2005**, *127*, 70; (k) R. Fischer, J. Baumgartner, G. Kickelbick, C. Marschner, *J. Am. Chem. Soc.* **2003**, *125*, 3414; (l) M. Zirngast, M. Flock, J. Baumgartner, C. Marschner, *J. Am. Chem. Soc.* **2008**, *130*, 17460; Review: C. Marschner, *Organometallics* **2006**, *25*, 2110.
98. (a) A. Sekiguchi, T. Fukawa, M. Nakamoto, V. Ya. Lee, M. Ichinohe, *J. Am. Chem. Soc.* **2002**, *124*, 9865; (b) A. Sekiguchi, T. Fukawa, V. Ya. Lee, M. Nakamoto, *J. Am. Chem. Soc.* **2003**, *125*, 9250.
99. (a) M. Nakamoto, T. Fukawa, V. Ya. Lee, A. Sekiguchi, *J. Am. Chem. Soc.* **2002**, *124*, 15160; (b) T. Fukawa, M. Nakamoto, V. Ya. Lee, A. Sekiguchi, *Organometallics* **2004**, *23*, 2376.
100. M. Flock, C. Marschner, *Chem. Eur. J.* **2002**, *8*, 1024.
101. (a) W. Teng, U. Englich, K. Ruhlandt-Senge, *Angew. Chem., Int. Ed.* **2003**, *42*, 3661; (b) W. Teng, K. Ruhlandt-Senge, *Organometallics* **2004**, *23*, 2694; (c) W. Teng, K. Ruhlandt-Senge, *Organometallics* **2004**, *23*, 952; (d) D. M. Jenkins, W. Teng, U. Englich, D. Stone, K. Ruhlandt-Senge, *Organometallics* **2001**, *20*, 4600.
102. (a) J. D. Farwell, M. F. Lappert, C. Marschner, C. Strissel, T. Don Tilley, *J. Organomet. Chem.* **2000**, *603*, 185; (b) W. Gaderbauer, M. Zirngast, J. Baumgartner, C. Marschner, T. Don Tilley, *Organometallics* **2006**, *25*, 2599; (c) H.-W. Lerner, S. Scholz, M. Bolte, N. Wiberg, H. Nöth, I. Krossing, *Eur. J. Inorg. Chem.* **2003**, 666.
103. F. Stabenow, W. Saak, M. Weidenbruch, *Chem. Commun.* **2003**, 2342.

104. N. Tokitoh, K. Hatano, T. Sadahiro, R. Okazaki, *Chem. Lett.* **1999**, 931.
105. (a) N. Tokitoh, T. Sadahiro, K. Hatano, T. Sasaki, N. Takeda, R. Okazaki, *Chem. Lett.* **2002**, *31*, 34; (b) N. Tokitoh, K. Hatano, T. Sasaki, T. Sasamori, N. Takeda, N. Takagi, S. Nagase, *Organometallics* **2002**, *21*, 4309.
106. (a) A. Sekiguchi, M. Ichinohe, S. Yamaguchi, *J. Am. Chem. Soc.* **1999**, *121*, 10231; (b) M. Ichinohe, Y. Arai, A. Sekiguchi, N. Takagi, S. Nagase, *Organometallics* **2001**, *20*, 4141.
107. (a) D. Bravo-Zhivotovskii, I. Ruderfer, S. Melamed, M. Botoshansky, B. Tumanskii, Y. Apeloig, *Angew. Chem. Int. Ed.* **2005**, *44*, 739; (b) D. Bravo-Zhivotovskii, I. Ruderfer, S. Melamed, M. Botoshansky, A. Schmidt, Y. Apeloig, *Angew. Chem., Int. Ed.* **2006**, *45*, 4157; (c) D. Bravo-Zhivotovskii, G. Molev, V. Kravchenko, M. Botoshansky, A. Schmidt, Y. Apeloig, *Organometallics* **2006**, *25*, 4719.
108. A. Sekiguchi, R. Izumi, S. Ihara, M. Ichinohe, V. Ya. Lee, *Angew. Chem., Int. Ed.* **2002**, *41*, 1598.
109. (a) A. Sekiguchi, R. Izumi, V. Ya. Lee, M. Ichinohe, *J. Am. Chem. Soc.* **2002**, *124*, 14822; (b) A. Sekiguchi, R. Izumi, V. Ya. Lee, M. Ichinohe, *Organometallics* **2003**, *22*, 1483; (c) N. Nakata, R. Izumi, V. Ya. Lee, M. Ichinohe, A. Sekiguchi, *J. Am. Chem. Soc.* **2004**, *126*, 5058; (d) N. Nakata, R. Izumi, V. Ya. Lee, M. Ichinohe, A. Sekiguchi, *Chem. Lett.* **2008**, *37*, 1146; (e) N. Nakata, R. Izumi, V. Ya. Lee, M. Ichinohe, A. Sekiguchi, *Chem. Lett.* **2005**, *34*, 582; (f) N. Nakata, A. Sekiguchi, *J. Am. Chem. Soc.* **2006**, *128*, 422; (g) M. Ichinohe, K. Sanuki, S. Inoue, A. Sekiguchi, *Organometallics* **2004**, *23*, 3088; (h) N. Nakata, T. Fujita, A. Sekiguchi, *J. Am. Chem. Soc.* **2006**, *128*, 16024; (i) M. Ichinohe, M. Igarashi, K. Sanuki, A. Sekiguchi, *J. Am. Chem. Soc.* **2005**, *127*, 9978; (j) V. Ya. Lee, H. Yasuda, M. Ichinohe, A. Sekiguchi, *Angew. Chem., Int. Ed.* **2005**, *44*, 6378.
110. P. A. Chugunov, N. A. Troitskii, K. S. Nosov, M. P. Egorov, O. M. Nefedov, *Russ. Chem. Bull.* **2004**, *53*, 2327.
111. T. Tajima, N. Takeda, T. Sasamori, N. Tokitoh, *Eur. J. Inorg. Chem.* **2005**, 4291.
112. (a) M. E. Lee, H. M. Cho, M. S. Ryu, C. H. Kim, W. Ando, *J. Am. Chem. Soc.* **2001**, *123*, 7732; (b) M. E. Lee, H. M. Cho, Y. M. Lim, J. K. Choi, C. H. Park, S. E. Jeong, U. Lee, *Chem. Eur. J.* **2004**, *10*, 377; (c) Theoretical treatments: M. Flock, A. Dransfeld, *Chem. Eur. J.* **2003**, *9*, 3320; M. Flock, C. Marschner, *Chem. Eur. J.* **2005**, *11*, 4635; (d) Y. M. Lim, H. M. Cho, M. E. Lee, K. K. Baeck, *Organometallics* **2006**, *25*, 4960; (e) M. E. Lee, Y. M. Lim, J. Y. Son, W. G. Seo, *Chem. Lett.* **2008**, *37*, 680; (f) Y. M. Lim, M. E. Lee, J. Lee, Y. Do, *Organometallics* **2008**, *27*, 6375.
113. G. Molev, D. Bravo-Zhivotovskii, M. Karni, B. Tumanskii, M. Botoshansky, Y. Apeloig, *J. Am. Chem. Soc.* **2006**, *128*, 2784.
114. A. Kawachi, Y. Oishi, T. Kataoka, K. Tamao, *Organometallics* **2004**, *23*, 2949.
115. M. Kira, T. Kadowaki, D. Yin, K. Sakamoto, T. Iwamoto, C. Kabuto, *Organometallics* **2007**, *26*, 4890.
116. T. Kajiwara, N. Takeda, T. Sasamori, N. Tokitoh, *Organometallics* **2008**, *27*, 880.
117. C. Krempner, M. H. Chisholm, J. Gallucci, *Angew. Chem., Int. Ed.* **2008**, *47*, 410.
118. K. Mochida, S. Ogawa, N. Naito, A. El-Al Gaber, Y. Usui, M. Nanjo, *Chem. Lett.* **2007**, *36*, 414.
119. A. F. Richards, M. Brynda, P. P. Power, *J. Am. Chem. Soc.* **2004**, *126*, 10530.

120. W. Teng, D. G. Allis, K. Ruhlandt-Senge, *Chem. Eur. J.* **2007**, *13*, 1309.
121. M. Weidenbruch, S. Willms, W. Saak, G. Henkel, *Angew. Chem., Int. Ed. Engl.* **1997**, *36*, 2503.
122. D. Scheschkewitz, *Angew. Chem., Int. Ed.* **2004**, *43*, 2965.
123. M. Ichinohe, K. Sanuki, S. Inoue, A. Sekiguchi, *Organometallics* **2004**, *23*, 3088.
124. S. Inoue, M. Ichinohe, A. Sekiguchi, *Chem. Lett.* **2005**, *34*, 1564.
125. R. Kinjo, M. Ichinohe, A. Sekiguchi, *J. Am. Chem. Soc.* **2007**, *129*, 26.
126. V. Ya. Lee, A. Sekiguchi, *Chem. Lett.* **2004**, *33*, 84.
127. (a) D. Scheschkewitz, *Angew. Chem., Int. Ed.* **2005**, *44*, 2954; (b) T. Nguyen, D. Scheschkewitz, *J. Am. Chem. Soc.* **2005**, *127*, 10174; (c) K. Abersfelder, D. Güclü, D. Scheschkewitz, *Angew. Chem., Int. Ed.* **2006**, *45*, 1643; (d) I. Bejan, D. Scheschkewitz, *Angew. Chem., Int. Ed.* **2007**, *46*, 5783; (e) I. Bejan, D. Güclü, S. Inoue, M. Ichinohe, A. Sekiguchi, D. Scheschkewitz, *Angew. Chem., Int. Ed.* **2007**, *46*, 3349; (f) I. Bejan, S. Inoue, M. Ichinohe, A. Sekiguchi, D. Scheschkewitz, *Chem. Eur. J.* **2008**, *14*, 7119; (g) K. Abersfelder, D. Scheschkewitz, *J. Am. Chem. Soc.* **2008**, *130*, 4114; (h) S. Inoue, M. Ichinohe, A. Sekiguchi, *Angew. Chem., Int. Ed.* **2007**, *46*, 3346; (i) S. Inoue, M. Ichinohe, A. Sekiguchi, *Chem. Lett.* **2008**, *37*, 1044.
128. L. Pu, M. O. Senge, M. M. Olmstead, P. P. Power, *J. Am. Chem. Soc.* **1998**, *120*, 12682.
129. J. Park, S. A. Batcheller, S. Masamune, *J. Organomet. Chem.* **1989**, *367*, 39.
130. V. Ya. Lee, M. Ichinohe, A. Sekiguchi, N. Takagi, S. Nagase, *J. Am. Chem. Soc.* **2000**, *122*, 9034.
131. S. Kyushin, H. Kawai, H. Matsumoto, *Organometallics* **2004**, *23*, 311.
132. N. Nakata, T. Oikawa, T. Matsumoto, Y. Kabe, A. Sekiguchi, *Organometallics* **2006**, *25*, 5850.
133. H. Schäfer, W. Saak, M. Weidenbruch, *Angew. Chem., Int. Ed.* **2000**, *39*, 3703.
134. A. Sekiguchi, T. Matsuno, M. Ichinohe, *J. Am. Chem. Soc.* **2000**, *122*, 11250.
135. T. Matsuno, M. Ichinohe, A. Sekiguchi, *Angew. Chem., Int. Ed.* **2002**, *41*, 1575.
136. Y. Ishida, A. Sekiguchi, K. Kobayashi, S. Nagase, *Organometallics* **2004**, *23*, 4891.
137. V. Ya. Lee, K. Takanashi, M. Ichinohe, A. Sekiguchi, *Angew. Chem., Int. Ed.* **2004**, *43*, 6703.
138. V. Ya. Lee, T. Yokoyama, K. Takanashi, A. Sekiguchi, *Chem. Eur. J.* **2009**, *15*, 8401.
139. Reviews: (a) J. D. Corbett, *Chem. Rev.* **1985**, *85*, 383; (b) C. Belin, M. Tillard-Charbonnel, *Coord. Chem. Rev.* **1998**, *178–180 (Part 1)*, 529; (c) T. F. Fässler, *Coord. Chem. Rev.* **2001**, *215*, 347; (d) R. Nesper, in *Silicon Chemistry* (Eds P. Jutzi, U. Schubert), Wiley-VCH, Weinheim, **2003**, p. 171.

4

Heavy Analogs of Carbenes: Silylenes, Germylenes, Stannylenes and Plumbylenes

4.1 Introduction

Carbenes R_2C: are neutral organic compounds possessing a dicoordinate carbon center with two substituents and featuring only six electrons in its valence shell. Although formally isoelectronic with carbenium ions (also possessing six valence electrons), carbenes are not isostructural with them. In contrast to carbenium ions, whose central carbon forms three sp^2-hybrids and has a vacant p_z-orbital (Chapter 1), carbenes R_2C: can exist in two ground states distinguished from each other by the electronic filling of their orbitals. Lone pair electrons can be either paired and arranged in the n-orbital of σ-symmetry, thus leaving the p-orbital of π-symmetry vacant (*singlet* carbenes), or they can be unpaired and arranged in both the n-orbital (one electron) and the p-orbital (one electron) (*triplet* carbenes). The different electronic configurations of carbenium ions and carbenes results in their markedly distinctive reactivity: carbenium ions are highly electrophilic and behave as strong Lewis acids, whereas carbenes can be either ambiphilic manifesting dual Lewis acid/Lewis base character (singlet carbenes), or biradical (triplet carbenes). The question of electronic configuration, singlet vs triplet, is one of the most fundamental problems in carbene chemistry, the solution of which required numerous experimental and theoretical pursuits. It is now well established that the ground-state multiplicity of carbenes is directly related to the energy difference of their n- and p-orbitals: the singlet ground-state is favored by a large (at least 2 eV) n–p energy gap (for discussion, see Section 4.4.1). The overall multiplicity of carbenes is totally dictated by the influence of their substituents, both electronic and steric. Thus,

Organometallic Compounds of Low-Coordinate Si, Ge, Sn and Pb: From Phantom Species to Stable Compounds
Vladimir Ya. Lee and Akira Sekiguchi

the prototypical methylene H_2C: and bulky group substituted carbenes (t-Bu_2C:, Ad_2C:) are ground-state triplets, whereas Me_2C:, dihalocarbenes X_2C: and cyclopropenylidene are all singlets. Remarkable progress has been made recently in the chemistry of carbenes associated with the development of new experimental techniques and high-level computational methods, which allowed the direct observation of carbenes and elucidation of their structures and mechanisms of their reactions. Such advances finally culminated in the isolation of the first stable singlet carbenes and subsequent generation of the persistent triplet carbenes, breakthroughs pioneered by the groups of Bertrand, Arduengo and Tomioka nearly twenty years ago.

The heavy analogs of carbenes, in which the central carbon is replaced with silicon, germanium, tin or lead, have been extensively studied for the past several decades. At the initial stage, these species were generated as short-lived undetectable intermediates in the 1960s: one should mention the seminal work of Atwell and Weyenberg on the thermolysis of $MeO(Me_2Si)_2OMe$ generating transient Me_2Si:. In the 1970–1980s several heavy carbenes were successfully stabilized and studied under matrix isolation conditions in argon or hydrocarbon matrices at 77 K and below. Isolation and identification (including crystal structure analysis) of the stable heavy analogs of carbenes became possible after the introduction of appropriate substituents, stabilizing the carbene center by either thermodynamic or kinetic means. The first remarkably stable representatives were synthesized by the groups of Lappert, Veith and West.

Given the overwhelming number of experimental and computational reports in the field of heavy carbene analogs, we are unable (and we are not aiming) to cover the whole story. Instead, we briefly discuss the most fundamental aspects, such as the generation, identification, structure and reactivity of silylenes, germylenes, stannylenes and plumbylenes, particularly focusing on the chemistry of stable (or persistent) representatives in both singlet and triplet ground states.

4.2 Generation

Several experimental techniques have commonly been employed for the generation (and subsequent trapping) of transient heavy carbene analogs.[1] Most developed in the case of silylenes, these methods are also applicable for the generation of germylenes, stannylenes and plumbylenes.

4.2.1 Photolysis of Acyclic Oligo- and Polymetallanes

The Si–Si bonds of the oligo- or polysilanes chains can be cleaved upon photolysis in two ways, forming either silylenes or silyl radicals.[2] Thus, photolysis of both poly(di-n-butylsilane) and poly(di-n-hexylsilane) with shorter wavelength light [248 nm (pulsed) or 254 nm (CW)] resulted in both extrusion of silylenes R_2Si: (R = n-butyl or n-hexyl) trapped with Et_3SiH and homolytic Si–Si bond cleavage, whereas only polysilyl radicals were observed upon irradiation with longer wavelength light ($\lambda > 300$ nm).[3]

Among the most frequently used precursors for the photochemical generation of silylenes are trisilanes $RR'Si(SiMe_3)_2$, which readily eliminate hexamethyldisilane Me_3Si–$SiMe_3$ to form the corresponding silylenes $RR'Si$:. Thus, diphenylsilylene Ph_2Si: was produced by irradiation of $Ph_2Si(SiMe_3)_2$ along with the by-products formed from the isomeric silatriene (Scheme 4.1).[4]

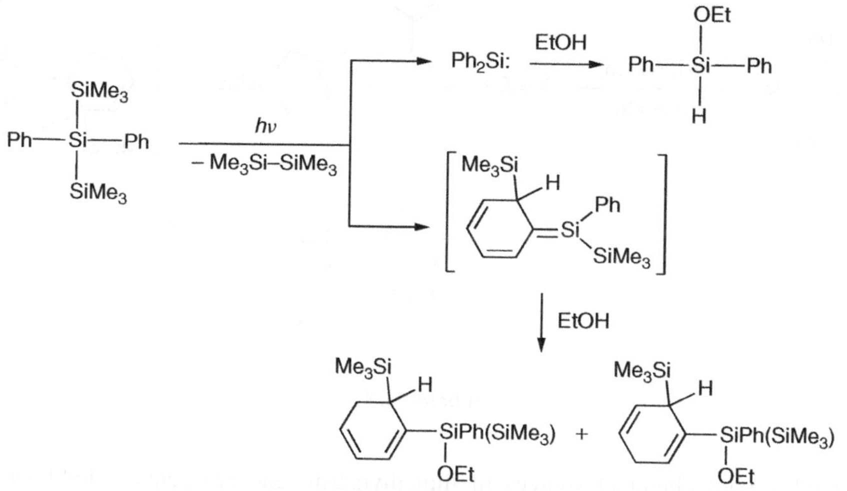

Scheme 4.1

Photolysis of $PhSi(SiMe_3)_3$ initially produced the silylene $Ph(Me_3Si)Si$:, which was trapped with EtOH to give an insertion product $Ph(Me_3Si)Si(H)OEt$, finally undergoing α-elimination of Me_3SiH to form another silylene $Ph(EtO)Si$: (Scheme 4.2).[5]

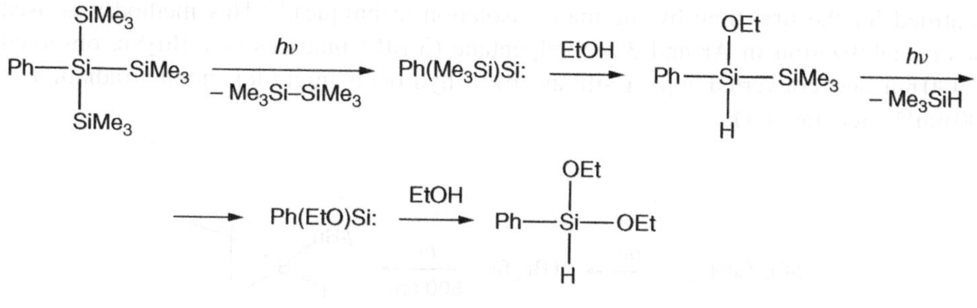

Scheme 4.2

Several other aryltris(trimethylsilyl)silanes $ArSi(SiMe_3)_3$ have also been used as convenient sources for transient silylenes $Ar(Me_3Si)Si$:.[6,7]

Likewise, diarylgermylenes Ar_2Ge: can be generated upon the photolysis of either bis(trimethylsilyl)diarylgermanes $Ar_2Ge(SiMe_3)_2$ or bis(trialkylgermyl)diarylgermanes $Ar_2Ge(GeR_3)_2$ by the α-elimination of $Me_3Si–SiMe_3$ or $R_3Ge–GeR_3$, respectively.[8–11] Diarylgermylenes R_2Ge: (R = Ph, Mes) and diethylgermylene Et_2Ge: were also generated from the digermanes $R_2(Cl)Ge–Ge(Cl)R_2$ and $Et_3Ge–Ge(Cl)Et_2$ by the α-elimination of germyl chlorides R_2GeCl_2 (R = Ph, Mes) and Et_3SiCl, respectively.[12] A surprising 1,2-Ph-migration from Ge to Si was observed upon the photolysis of $Ph–GeMe_2–SiMe_3$, resulting in the α-elimination of $PhSiMe_3$ and generation of Me_2Ge: along with the minor formation of transient germatriene (Scheme 4.3).[13]

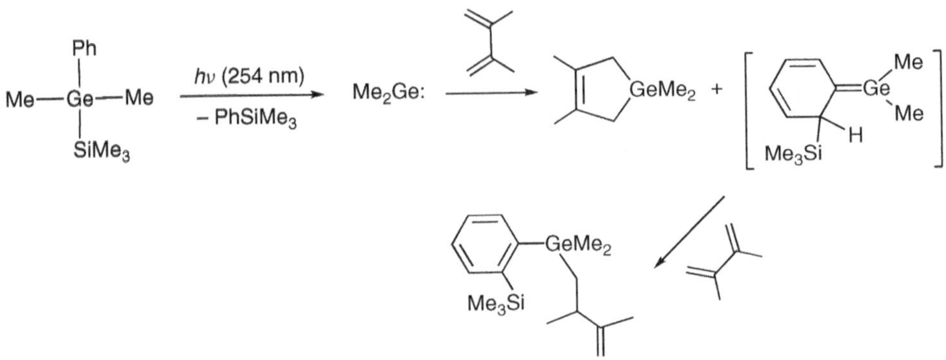

Scheme 4.3

The other photochemical sources of dimethylgermylene are pentamethyldigermane $Me_3Ge–Ge(H)Me_2$[14] and tetramethyldigermane $Me_2(H)Ge–Ge(H)Me_2$,[15] from which Me_2Ge: is generated upon the α-elimination of Me_3GeH and Me_2GeH_2, respectively. A convenient precursor for the parent, unsubstituted germylene is phenylgermane $PhGeH_3$, which cleanly produced H_2Ge: upon photolysis accompanied by the elimination of benzene PhH.[16]

Photochemical cleavage of diazidosilanes $RR'Si(N_3)_2$ provides an effective method for the generation of transient silylenes for low-temperature matrix studies (Me_2Si: was identified for the first time by the matrix isolation technique).[17] This method was used for the stabilization in Ar and 3-methylpentane (3-MP) matrices of t-Bu_2Si:, observed at 480 nm and converted into a silirane by a hydrogen migration upon irradiation at 500 nm[18] (Scheme 4.4).

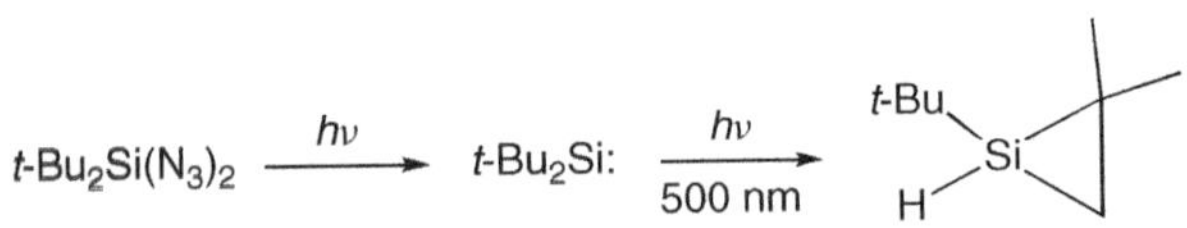

Scheme 4.4

Similarly, photolysis of dimethyldiazidogermane $Me_2Ge(N_3)_2$ opened a convenient route for the generation of one of the most extensively studied germylenes, Me_2Ge:.[19]

Several methylstannanes and methyldistannanes were tested as potential sources for the generation of Me_2Sn: in the gas phase for kinetic studies, the best results were obtained with Me_4Sn as the precursor (Scheme 4.5).[20] Dimethylstannylene Me_2Sn: can also be formed by one of the following routes: (a) photolysis of $PhCH_2–Sn(H)Me_2$ through homolytic cleavage of the $PhCH_2$–Sn bond forming a pair of stannyl $Me_2(H)Sn$• and benzyl $PhCH_2$• radicals followed by hydrogen abstraction giving free stannylene Me_2Sn: and toluene $PhCH_3$;[21] (b) microwave irradiation of Me_2SnH_2 in an Ar matrix at 10 K.[21]

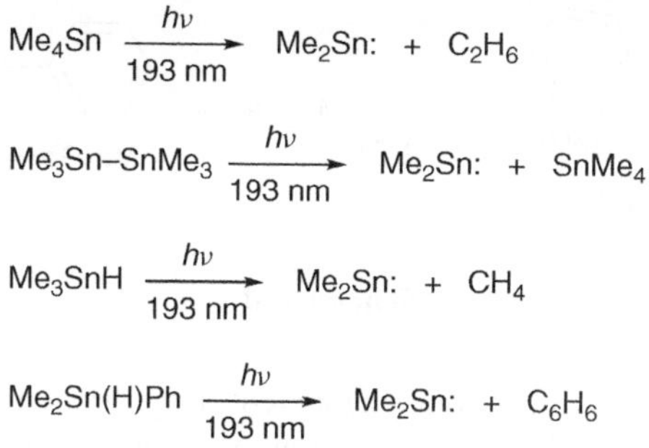

Scheme 4.5

4.2.2 Photolysis of Cyclic Metallanes

Photolytic ring contraction of the readily available medium-sized (five- and six-membered rings) cyclopolysilanes has been known as a convenient method for the generation of silylenes since the pioneering work of Ishikawa and Kumada in the 1970s.[22] The most famous example is represented by the 254 nm photolysis of dodecamethylcyclohexasilane *cyclo*-$(Me_2Si)_6$ resulting in the extrusion of Me_2Si: and formation of the ring contraction product, decamethylcyclopentasilane *cyclo*-$(Me_2Si)_5$ (Scheme 4.6).[17,22,23] This method was used in the first spectroscopical detection in 1979 of the dimethylsilylene Me_2Si: observed at 453 nm in a 3-MP matrix.[17] Cyclic silylenes can also be generated by such photochemical ring contraction conditions.[24]

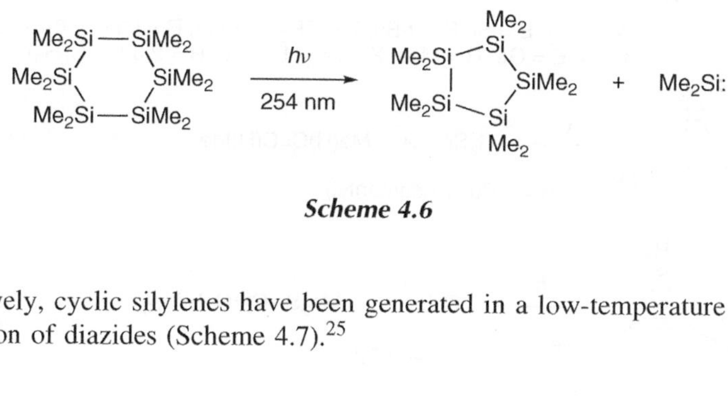

Scheme 4.6

Alternatively, cyclic silylenes have been generated in a low-temperature matrix upon the irradiation of diazides (Scheme 4.7).[25]

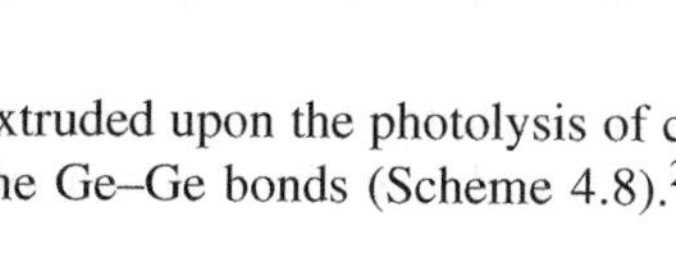

Scheme 4.7

Germylenes can also be extruded upon the photolysis of cyclic oligogermanes through the homolytic cleavage of the Ge–Ge bonds (Scheme 4.8).[26]

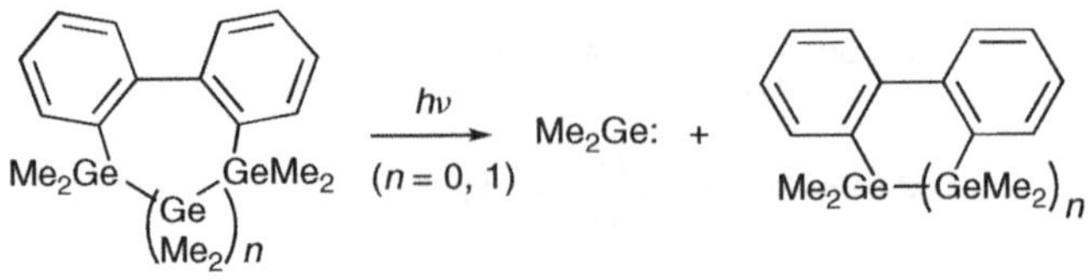

Scheme 4.8

Highly strained three- and four-membered ring systems have been widely employed as good sources for the generation of transient silylenes and germylenes (Scheme 4.9).[2a,8,27–37]

Cyclotristannanes *cyclo*-$(R_2Sn)_3$ with bulky aromatic substituents R (R = 9-phenanthryl, 2,6-Et_2-C_6H_3, 2,4,6-*i*-Pr_3-C_6H_2) can also serve as the photochemical and thermal precursors of stannylenes, which can be trapped with 2,3-dimethylbuta-1,3-diene or *n*-Bu_3SnH.[38]

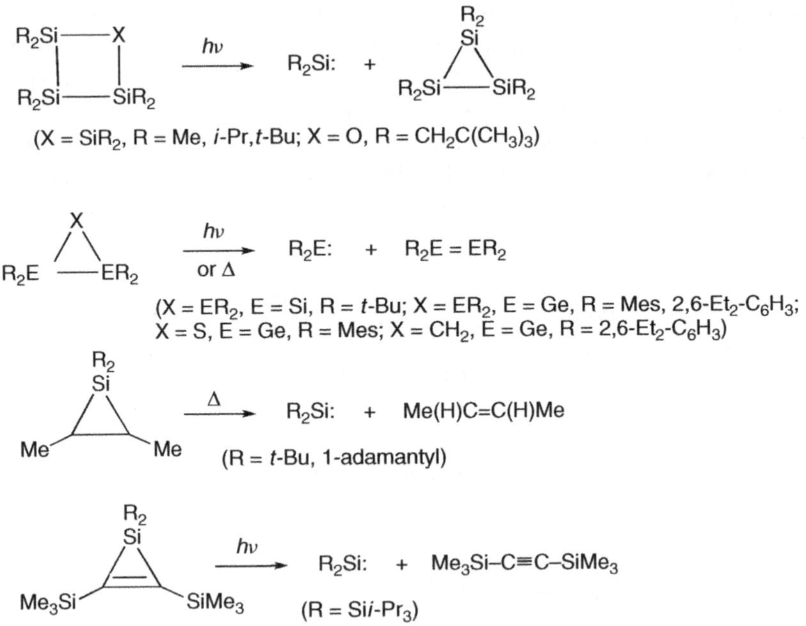

Scheme 4.9

Among the most frequently used precursors for the generation of transient silylenes and germylenes are 7-sila- and 7-germanorbornadienes, photochemical/thermal fragmentation of which provides a convenient route to substituted silylenes and germylenes (Scheme 4.10).[1a,c]

A ^1H CIDNP study on the photolytic fragmentation of the 7,7-dimethyl-7-silanorbornadiene derivative revealed that the reaction proceeds through the initial

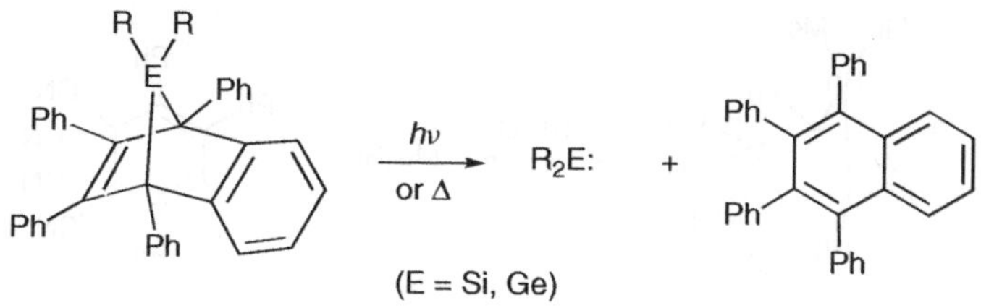

Scheme 4.10

reversible formation of a singlet biradical, followed by its S → T interconversion to the triplet biradical. The latter species finally underwent extrusion of Me_2Si: accompanied by the concomitant formation of tetraphenylnaphthalene as a side product (Scheme 4.11).[39]

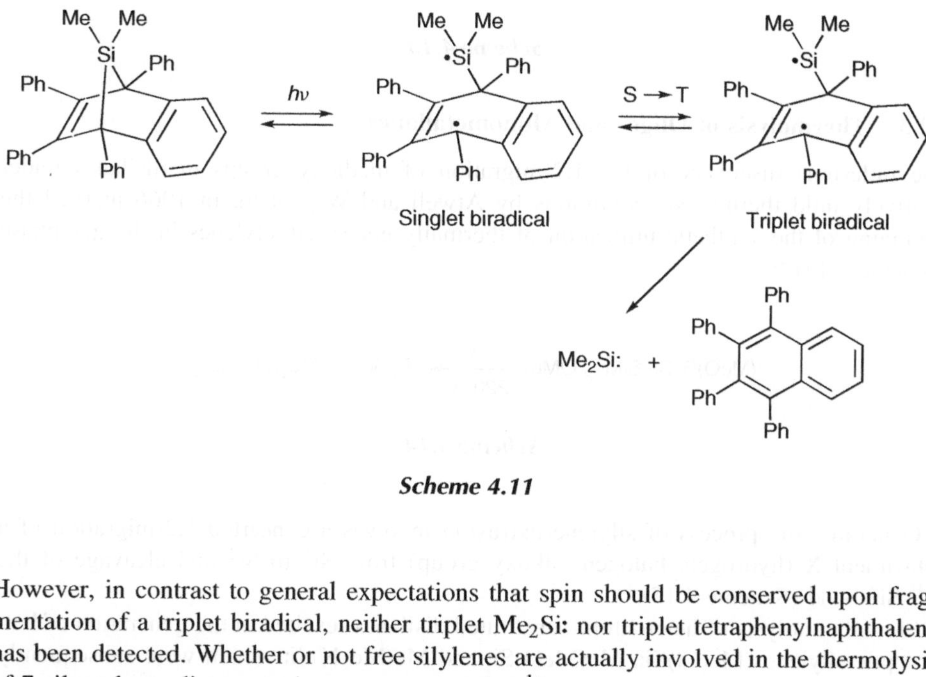

Scheme 4.11

However, in contrast to general expectations that spin should be conserved upon fragmentation of a triplet biradical, neither triplet Me_2Si: nor triplet tetraphenylnaphthalene has been detected. Whether or not free silylenes are actually involved in the thermolysis of 7-silanorbornadiene remains an open question.[1a]

Low-temperature stable 7,7-dimethyl-7-stannanorbornadiene was reported to undergo cycloreversion at temperatures above −20 °C with the quantitative formation of 5,5,6,6-tetracyano-1,2,3,4-tetraphenylcyclohexa-1,3-diene, thus implying generation of a transient dimethylstannylene Me_2Sn: (Scheme 4.12).[40]

1-Sila- and 1-germacyclopent-3-enes are common precursors for the photochemical generation of silylenes and germylenes accompanied by the elimination of buta-1,3-dienes (Scheme 4.13).[14,16d,41]

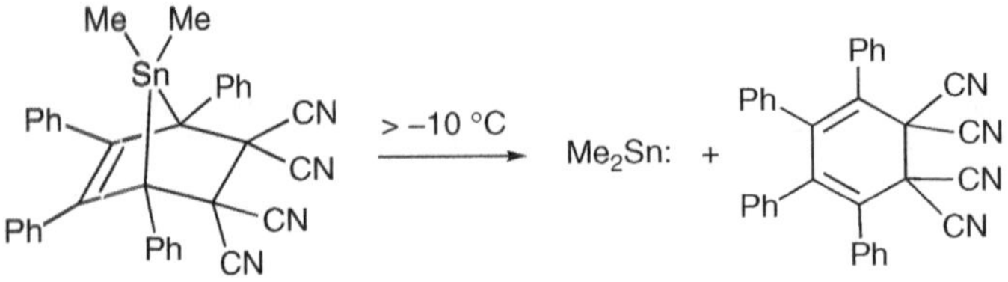

Scheme 4.12

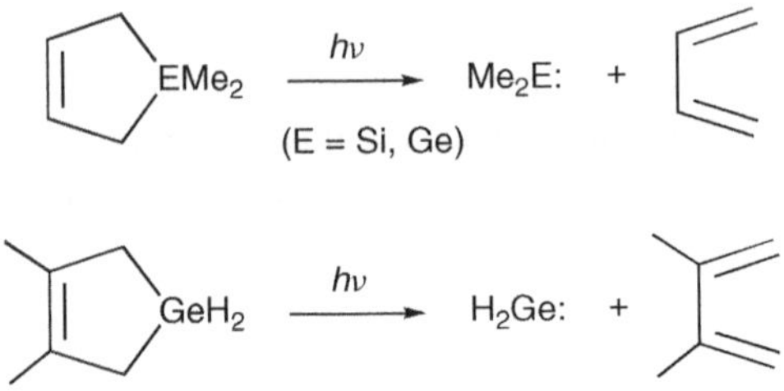

Scheme 4.13

4.2.3 Thermolysis of Oligo- and Monometallanes

The milestone discovery of the 1,2-migration of methoxy groups in disilanes under relatively mild thermolysis conditions by Atwell and Weyenberg in 1966 marked the beginning of the synthetic utilization of thermally generated silylenes in the gas phase (Scheme 4.14).[42]

$$\mathrm{(MeO)R_2Si{-}SiMe_2(OMe)} \xrightarrow[220\ ^\circ\mathrm{C}]{\Delta} \mathrm{R_2Si\!:} + \mathrm{Me_2Si(OMe)_2}$$

Scheme 4.14

Generally, the process of silylene extrusion involves a concerted 1,2-migration of a substituent X (hydrogen, halogen, alkoxy group) from Si^1 to Si^2 and cleavage of the Si^1–Si^2 bond (Scheme 4.15).[1a]

Flow thermolysis of heteroatom-substituted disilanes resulted in the generation of heteroleptic silylenes Me(Cl)Si:, Me(MeO)Si: and Me(Me_2N)Si:, which were subsequently trapped with 1,3-dienes, finally to give silacyclopentenes (Scheme 4.16).[43]

When generated in the presence of heterodienes (α,β-unsaturated aldehydes, ketones, diimines, diketones, iminoketones), the above-described silylenes produced a variety of five-membered ring heterocycles through the initial [1+2] cycloaddition of silylene to the C–heteroatom double bond to form a three-membered ring intermediate followed by its rearrangement.[44]

Germylenes can be generated by the thermal α-elimination from di- or oligogermanes (Scheme 4.17).[45]

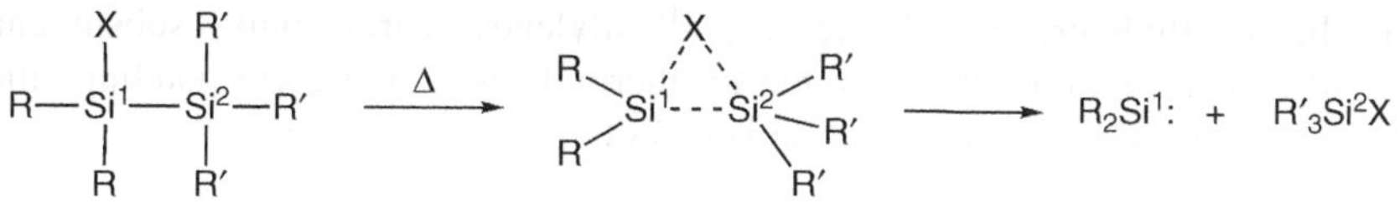

Scheme 4.15

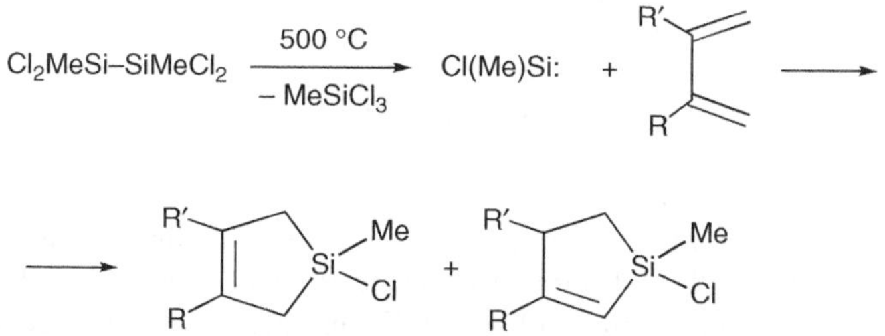

Scheme 4.16

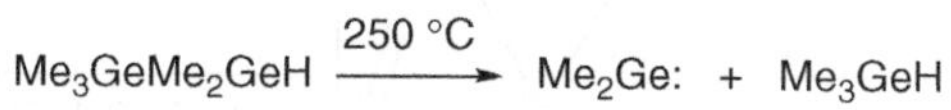

Scheme 4.17

Dimethylstannylene Me_2Sn: can be produced upon the thermolysis of dodecamethylcyclohexastannane *cyclo*-$(Me_2Sn)_6$ at 120 °C.[46] Temperature control is crucial, because below 120 °C the reaction is very slow, whereas above 130 °C $Me_2Sn{=}SnMe_2$ is also generated along with the target Me_2Sn:.

Apart from oligosilanes, several monosilanes have also been utilized as thermal sources for transient silylenes. Thus, the parent silylene H_2Si: was formed upon the pyrolysis of silane SiH_4 at temperatures above 380 °C and directly detected by intracavity laser spectroscopy.[47] However, this reaction, although technologically important, is of little use in the investigation of H_2Si: chemistry, because the latter species can be more easily generated from disilane Si_2H_6.

7-Sila- and 7-germanorbornadienes, already mentioned in Section 4.2.2 as very effective photochemical precursors for the generation of silylenes and germylenes, are also widely used as their thermal sources (Scheme 4.10).[1a–c] This method, known since the pioneering work of Gilman *et al.* on the thermal generation of Me_2Si:, [48] was then extended to the generation of silylenes with a variety of alkyl and aryl substituents.[1a,b] In contrast to the high-temperature generation of silylenes from 7-silanorbornadienes, their germanium analogs (7-germanorbornadienes) eliminated germylenes under relatively mild thermal conditions (70–150 °C).[49]

Other commonly used thermal precursors for silylenes are three-membered ring siliranes and five-membered ring silacyclopentenes. Mild thermolysis of siliranes leading to generation of silylenes has been known since the pioneering work of Seyferth *et al.* on hexamethylsilirane (Scheme 4.18).[50] Silylenes bearing bulky substituents (t-Bu_2Si:, Ad_2Si:) were similarly generated by thermolysis of the corresponding siliranes at 180–200 °C and were trapped with either $HSiEt_3$ or Et–C≡C–Et.[30,31]

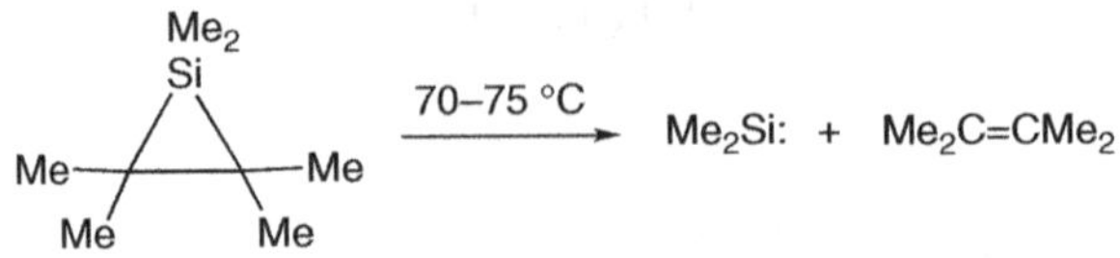

Scheme 4.18

Thermal cycloreversion of the silacyclopentenes resulted in the generation of silylenes trapped with 2,3-dimethylbuta-1,3-diene (silylene transfer reaction).[51] This gas phase process requires high temperatures (500–650 °C) and presumably proceeds through the extrusion of silylenes from an intermediate vinylsilirane (Scheme 4.19).

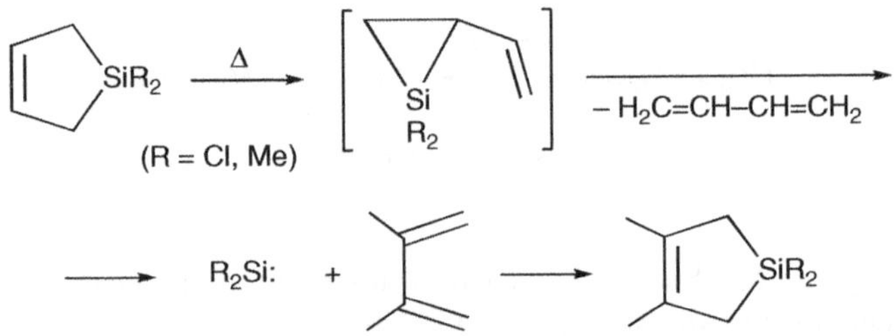

Scheme 4.19

In contrast to the above case, transfer of the germylene fragment from 1,1-dichloro-1-germacyclopent-3-ene to 2,3-dimethylbuta-1,3-diene proceeds under markedly milder conditions (Scheme 4.20).[52]

Scheme 4.20

4.2.4 α-Elimination of Silylenoids

Silylenoids, the silicon analogs of carbenoids, are α-heteroatom substituted tetracoordinate organosilicon compounds, which behave as masked dicoordinate silylenes (Scheme 4.21).

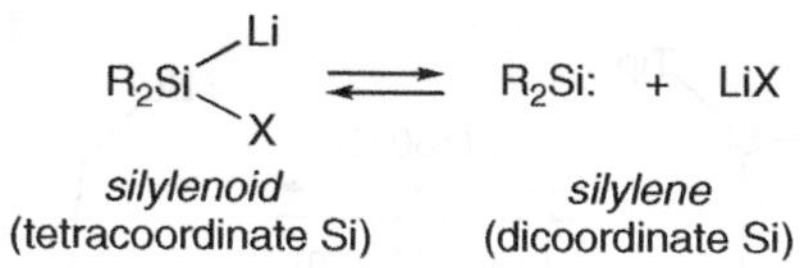

Scheme 4.21

They were originally proposed as reactive intermediates, for example, in the synthesis of polysilanes $(–R_2Si–)_n$ by the reduction of dichlorosilanes R_2SiCl_2 with alkali metals.[1a] The real participation of silylenoids in the reduction of functional group-substituted halosilanes was later proved in a series of publications from Tamao's group.[53] Thus, $(t\text{-BuO})Ph_2SiLi$ [prepared by the reduction of $(t\text{-BuO})Ph_2SiCl$ with LDMAN in THF at −78 °C, see Section 3.2.4, Scheme 3.7, A] showed electrophilic behavior diagnostic for silylenoids towards n-BuLi forming $n\text{-BuPh}_2SiLi$, whereas it lacked the reactivity typical for free silylenes (reactions with Et_2MeSiH, diphenylacetylene, dimethylbutadiene).[54] Interestingly, at 0 °C in THF, silylenoid $(t\text{-BuO})Ph_2SiLi$ undergoes bimolecular self-condensation to form the dimeric intermediate $(t\text{-BuO})Ph_2Si–SiPh_2Li$.

Because silylenoids and free silylenes can be reversibly generated from each other (Scheme 4.21), the intermediacy (and reactivity) of both species in α-dehalogenation processes remains an open, sometimes controversial, issue.[30,55–58]

Likewise, germylenes can be generated by α-dehalogenation of dihalogermanes. Thus, halogermylenes $[(Me_3Si)_3C](X)Ge$: (X = Cl, Br) were successfully formed (and trapped with isoprene) upon the reduction of trihalogermanes $[(Me_3Si)_3C]GeX_3$ with magnesium in THF.[59]

4.2.5 Heavy Alkene–Heavy Carbene Rearrangements

Although rearrangements of such types (for example, silene–silylene or disilene–silylsilylene) are not the common methods for generation of transient silylenes, several examples of them have been reported. Thus, 1-methylsilene $CH_3(H)Si{=}CH_2$ (generated in the gas phase from 1-methylsilacyclobutane) was reported to undergo 1,2-H-migration forming dimethylsilylene Me_2Si: (Scheme 4.22).[60]

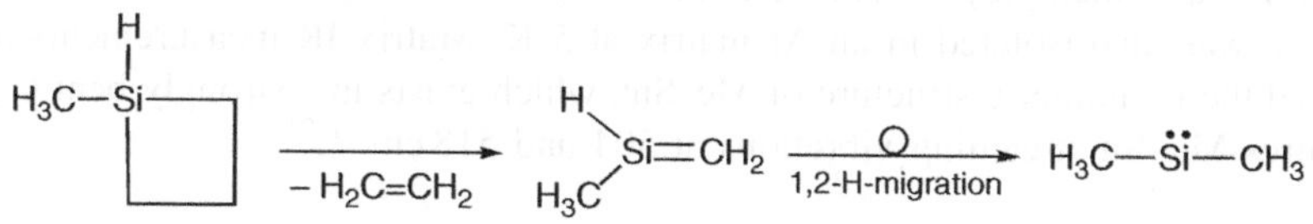

Scheme 4.22

Thermal isomerization of a silacyclobutadiene to a cyclopropenyl-substituted silylene is reversible in a low-temperature photochemical rearrangement (Scheme 4.23).[61]

Rearrangements of disilenes to silylsilylenes through 1,2-migration of a substituent from the doubly bonded silicon center to the adjacent silicon are also known (Scheme 4.24).[62]

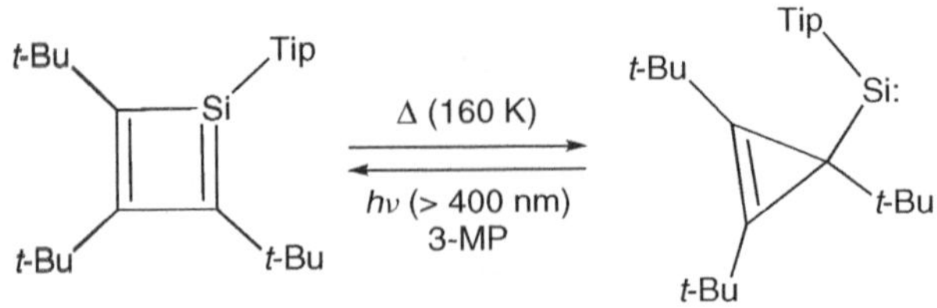

Scheme 4.23

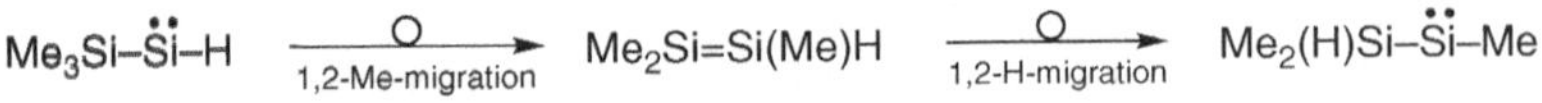

Scheme 4.24

4.3 Spectroscopic Identification

Much of the most important structural information on the transient analogs of carbenes of the heavy group 14 elements can be obtained from the electronic (UV) spectra of these species, photochemically generated and stabilized in inert matrices at low temperatures (matrix isolation technique): argon matrices at 5–30 K, or hydrocarbon (usually 3-MP) glasses at 77 K.[1a,b] Because the singlet ground state is well established for an overwhelming number of transient silylenes, the UV-absorption of the latter species is to be explained as the electronic transition between the $^1A_1$ ground and $^1B_1$ excited states (Figure 4.1), typically observed in the visible region.[1a]

For example, Me_2Si:, the best-studied silylene generated from a variety of precursors, absorbs at 450 nm in 3-MP at 77 K.[23] Likewise, transient germylenes and stannylenes have been observed under matrix isolation conditions.[1c] Thus, the UV absorption at 430 nm attributed to Me_2Ge:, generated from the cyclic trigermane and stabilized in a 3-MP matrix at 77 K, was reported as early as 1984.[26b] This was followed by a comprehensive study that unequivocally assigned UV and IR spectra of the species generated from $Me_2Ge(N_3)_2$ in an Ar matrix at 26 K to those of the free Me_2Ge:, a conclusion supported also by reactivity (insertion/addition) studies.[19a] Free Me_2Sn:, generated thermally from dodecamethylcyclohexastannane $(Me_2Sn)_6$[46] (or by microwave irradiation of Me_2SnH_2), was also isolated in an Ar matrix at 5 K. Matrix IR measurements reliably established the monomeric structure of Me_2Sn:, which exists in a strongly bent C_{2v} form and features Me–Sn stretching vibrations at 504 and 518 cm^{-1}.[21]

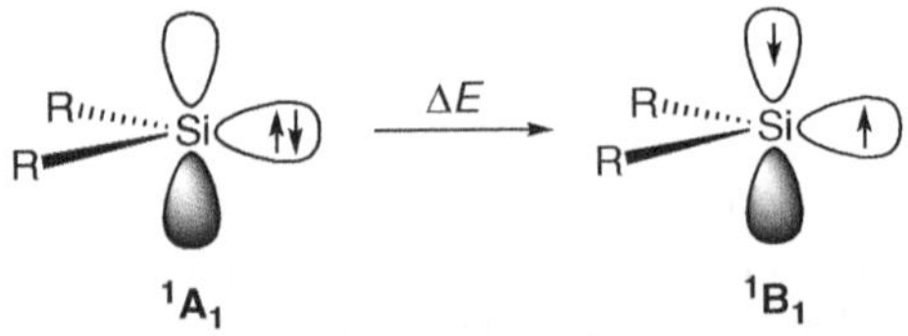

Figure 4.1 The lowest energy electronic transition between two singlet states: ground $^1A_1$ and excited $^1B_1$

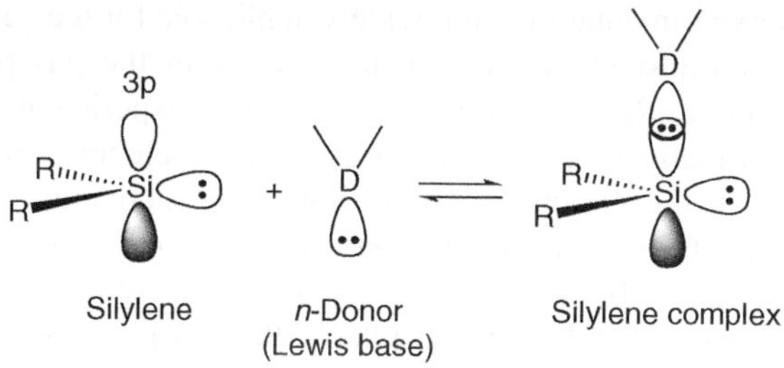

Figure 4.2 *Origin of the UV blue shift in the complexes of silylenes with n-donors (Lewis bases)*

Complexes of silylenes with Lewis bases revealed a diagnostic blue shift of the n–p electronic transition in their UV spectra compared with those of the free silylenes.[1a] This was ascribed to the effect of electron donation of the Lewis base lone pairs into the vacant 3p-orbital of the silylene center, elevating the energy level of the $^1B_1$ excited state (Figure 4.2). This was clearly demonstrated for the first time by the observation of a hypsochromic shift for mesitylsilylenes Mes(R)Si: (R = Mes, t-Bu, 1-adamantyl, 2,6-diisopropylphenoxy), photolytically generated from the corresponding trisilanes Mes(R)Si($SiMe_3$)$_2$ in either 3-MP or 2-Me–THF matrices.[63] The absorptions due to the free silylenes Mes(R)Si: observed at 425–577 nm in 3-MP were shifted to a shorter wavelength region of ca. 350 nm because of the formation of silylene complexes Mes(R)Si:(2-Me–THF).

The magnitude of the blue shift upon interaction of silylenes with Lewis bases serves as a qualitative measure of the strength of complexation, which depends on the nature of both base and silylene.[64] Thus, the stability of the dimethylsilylene complexes with Lewis bases increases in the following order (absorption band for silylene complexes): phosphines (390 nm for n-Bu_3P) < carbon monoxide (345 nm) < sulfides (322 nm for t-Bu_2S) < ethers (299 nm for Et_2O, 294 nm for 2-Me-THF) < amines (287 nm for Et_3N and n-Bu_3N).[1a] On the other hand, complexation of aryloxysilylene Mes(ArO)Si: (Ar = 2,6-i-Pr_2-C_6H_3) is markedly weaker than that of Mes_2Si:/Me_2Si:, because of π-donation from the substituent oxygen into the vacant 3p(Si:) orbital weakening the Lewis acidity of the former silylene: 346 nm [Mes(ArO)Si–n-Bu_3N complex] vs 325/287 nm (Mes_2Si–n-Bu_3N/Me_2Si–n-Bu_3N complexes).[1a,65]

Complexes of Mes_2Si: with alcohols (propan-2-ol, butan-2-ol) were also observed spectroscopically in a 3-MP matrix, the melting of which resulted in the fast formation of a silylene O–H insertion product Mes_2Si(Oi-Pr)H.[1a]

Lewis base complexes of diorganogermylenes, typically generated upon the photolysis of 7-germanorbornadienes or bis(trimethylsilyl)germanes in 3-MP at 77 K, also exhibited characteristic shifts to shorter wavelength regions (blue shifts) compared with the absorptions of free germylenes.[1b] Thus, complexes of Mes_2Ge: (absorption at 550 nm in the absence of Lewis bases) with n-donor ligands revealed the following long wavelength absorptions: 306 nm (for n-Bu_3P complex), 348 nm (for Me_2S complex), 352 nm (for thiophene complex), 360 nm (for 2-Me–THF complex), and 333 nm (for EtOH complex).[9]

Another powerful experimental method, widely applicable for the generation and spectroscopic detection of transient heavy carbene analogs in the gas phase, is the laser flash photolysis technique. Recent developments of such experimental procedures have made possible direct time-resolved kinetic studies for the elucidation of heavy carbene reaction mechanisms in the gas phase (transition states, activation barriers, rate constants, mechanistic aspects of insertion/addition reactions, etc.). Most of the experimental work in this field was done by Walsh *et al.*, who summarized their results in a series of comprehensive reviews.[1d–g] For the results of the gas-phase kinetic measurements of halosilylenes SiF_2, $SiCl_2$, $SiBr_2$ and SiHCl, we refer our readers to a review by Strausz *et al.*[1h]

Kinetic studies in the gas phase have the evident advantage of the absence of complications arising from nucleophilic solvents, which have a great tendency to be complexed with the Lewis acidic heavy carbene analogs. However, in some cases kinetic studies of silylene reactivity were carried out in the liquid phase: the rate constants for the reaction of Me_2Si:, generated by the photolysis of *cyclo*-$(Me_2Si)_6$ in methylcyclohexane solution, with MeOH and Et_3SiH were in a good agreement with those of the gas phase experiments.[66]

Complexation of the heavy carbene analogs with Lewis bases can also be observed under flash photolysis conditions: absorption of Me_2Si:, photolytically generated from *cyclo*-$(Me_2Si)_6$, was detected at 465 nm in cyclohexane, 340 nm in CH_3CN, 310 nm in THF and 270 nm in Et_3N.[67] Gas-phase kinetic studies concluded also that the formation of a pre-reaction complex between the heavy carbene analog (Lewis acid) and the reagent (Lewis base) can very often be considered as the initial step of the silylenes, germylenes, stannylenes and plumbylenes reactions.[1f] Because of the reversibility of this step, a negative activation energy for the overall reaction was observed in many cases.

The other physicochemical techniques used for the detection of short-lived heavy carbene analogs are vibrational spectroscopy (low-temperature matrix or gas phase IR and Raman spectroscopy), microwave spectroscopy, and gas phase UV–PES.[1i] One of the most comprehensively studied heavy carbenes is the prototypical silylene H_2Si:. Thus, H_2Si: generated from $PhSiH_3$ was detected by laser-induced fluorescence,[68] IR diode laser kinetic spectroscopy (high-resolution IR spectrum of SiH_2: in its ground state),[69] and the REMPI technique.[70] A complex of SiH_2: and HF was detected by IR spectroscopy under matrix isolation conditions among the products of the reaction of SiH_4 and F_2.[71] Among other fundamental silylenes detected by spectroscopic methods are: dihalosilylenes X_2Si: (X = halogen) studied using the He(I) PES technique[72] and dimethylsilylene Me_2Si: investigated under IR matrix isolation conditions.[73]

4.4 Structure

4.4.1 Multiplicity: Singlet vs Triplet

Ground-state multiplicity is a fundamental feature of carbenes, that dictates their overall reactivity. Thus, singlet carbenes ($^1\mathbf{A_1}$) with both a doubly occupied n-orbital of σ-symmetry and a vacant p-orbital of π-symmetry manifest an ambiphilic (Lewis acid/Lewis base) character, whereas triplet carbenes ($^3\mathbf{B_1}$) with two singly occupied

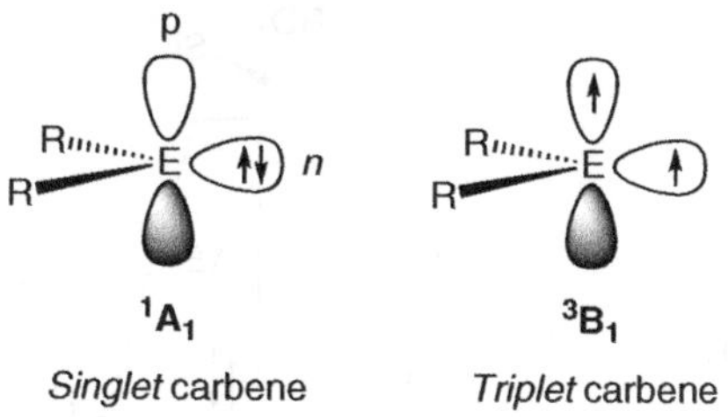

Figure 4.3 *Electronic configurations of carbenes and their heavy analogs (E = C–Pb)*

orbitals behave as biradicals (Figure 4.3, E = C).[74] A large n–p energy separation favors the singlet ground state: Hoffmann *et al.* determined the minimum value of 2 eV necessary for stabilization of the singlet ground state, whereas a value below 1.5 eV leads to a triplet ground state.[75] Thus, the parent carbene, methylene H_2C:, has a triplet ground state with the singlet lying 9.0 kcal/mol higher in energy.[76]

In sharp contrast to methylene, its heavy analogs H_2E: (E = Si, Ge, Sn, Pb) are well established to be ground-state singlets $^1A_1$ with largely destabilized triplet states $^3B_1$. The calculated singlet–triplet energy separation (ΔE_{ST}) steadily increases descending group 14: 21.0 kcal/mol for H_2Si: (experimental value 21.0 or 18.0 kcal/mol),[77] 23.1 kcal/mol for H_2Ge:, 23.8 kcal/mol for H_2Sn: and 41.0 kcal/mol for H_2Pb: (very large S–T separation because of the relativistic contraction of the 6s-orbital and significant contribution of spin–orbit coupling effects).[78] Likewise, in a series of halogen-substituted heavy carbenes X_2E: (X = F, Cl, Br, I) for a given halogen (X)ΔE_{ST} increases from Si to Pb, whereas for a given group 14 element (E) ΔE_{ST} decreases from F to I.[76a,79]

There are several reasons for the above-described preference for the singlet ground state for the heavy carbene analogs. One of them is related to the HOMO–LUMO energy gap, which is much greater in silylene H_2Si: (52 kcal/mol) than in carbene H_2C: (27 kcal/mol).[80] Accordingly, the energy gained in formation of a triplet state through the promotion of an electron from the n- to a p-orbital cannot outbalance this large HOMO–LUMO energy separation, thus favoring the singlet state for H_2Si:. The large HOMO–LUMO gap in heavy carbenes H_2E: (E = Si–Pb) could be rationalized by taking into account the high s-character of the atomic orbital accommodating the nonbonding electron pair (HOMO), much greater than that of the carbon analog H_2C:.[76b] This large share of the s-fraction suggests a lower energy level for this orbital and, consequently, greater HOMO–LUMO energy separation in heavy carbene analogs H_2E:.

The repulsion between two paired electrons in the n-orbital of carbene in its singlet ground state $^1A_1$ also contributes to the overall dominance of the singlet over the triplet. Because of the progressively increasing size of the lone pair orbital on going down group 14, the extent of such electron–electron repulsion is decreased for the heavier analogs of carbenes, thus stabilizing their singlet ground states.[76b]

4.4.2 Substituent Effects

Apeloig *et al.* comprehensively studied the influence of substituents (both electronic and steric) on the electronic transition of silylenes.[1j,81,82] Thus, they found that n-donating

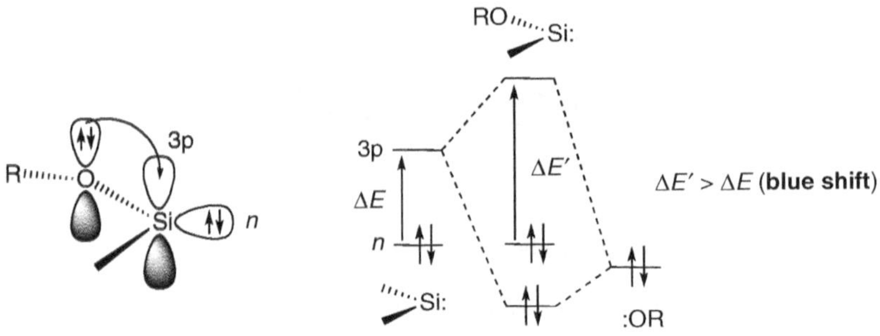

Figure 4.4 Electronic effect of the n-donating substituent (RO:): origin of the UV blue shift

substituents (NH_2, OH, SH) induce blue shifts in the UV spectra of silylenes because of the interaction of the substituent lone pair and the vacant 3p-orbital on the silylene center (Figure 4.4).[1j,81,82] Such an interaction raises the energy level of the unoccupied orbital, resulting in an overall increase in the n–3p (singlet–triplet) energy gap. The theoretically predicted blue shifts were observed experimentally: 496 nm for Mes(Me)Si:, 487 nm for Mes(Cl)Si:, 425 nm for Mes(MeO)Si: and 405 nm for Mes(Me_2N)Si:.[1a]

In contrast, π-donating groups (vinyl, phenyl) cause red shifts in the electronic spectra of silylenes, because of the effective three-orbital (π, π^* and 3p) interactions.[1j,81,82] The primary process involves interaction between the occupied π(substituent) and vacant 3p(Si)-orbitals, generating bonding and antibonding (π–3p) combinations (Figure 4.5, **A**). The secondary interaction between the antibonding (π–3p) combination and π^*-orbital of the substituents leads to a significant lowering of the 3p(Si)-orbital energy level and, consequently, to an overall red shift in the UV spectra of π-donor-substituted silylenes (Figure 4.5, **B**).

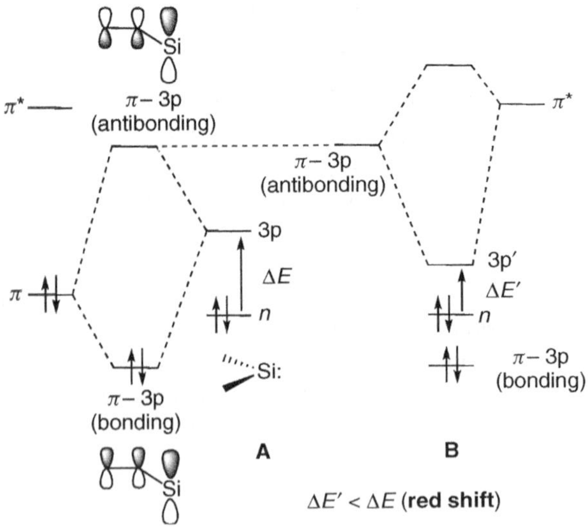

Figure 4.5 Electronic effect of the π-donating substituents (C=C): origin of the UV red shift

Inductive effects also play an important role in the stabilization/destabilization of frontier orbitals. Thus, electronegative σ-accepting substituents (F, CF_3) withdraw electron density from the doubly occupied n-orbital, stabilizing the $^1A_1$ ground state (Figure 4.1) and producing a blue shift.[1a,1j,81,82] In contrast, electropositive σ-donating substituents (SiH_3) stabilize the $^1B_1$ excited state (Figure 4.1) through their electron donation into the singly occupied s-type orbital, which results in a substantial red shift: 490 nm for Ph(Me)Si:, 625 nm for $Ph(Me_3Ge)Si$: and 660 nm for $Ph(Me_3Si)Si$:.[1a,1j,81,82]

Steric effects also contribute to the observation of either red or blue shifts in the UV spectra of silylenes R_2Si:. Thus, the bond angle R–Si–R around the silylene center strongly affects the first electronic transition. These angles are much larger in the excited $^1B_1$ state than in the ground $^1A_1$ state: 120.0° vs 93.2° for the parent H_2Si: (likewise, these angles are always wider in the $^3B_1$ triplet than in the $^1A_1$ singlet).[82] Upon widening of the R–Si–R bond angle beyond 93.2° and up to ca. 140° (intersection of the potential curves of the $^1A_1$ and $^1B_1$ states), the $^1A_1$ ground state tends to be destabilized and the $^1B_1$ excited state tends to be stabilized.[82] This results in an overall decrease in the ΔE energy gap and, consequently, in the UV red shift. This was indeed observed experimentally upon increasing of the steric congestion around the silylene centers by the use of bulky substituents in the following series of silylenes: 495 nm for Ph_2Si:, 530 nm for Ph(Mes)Si: and 577 nm for Mes_2Si:.[1j,82,83] In contrast, a blue shift was observed in the UV spectra of R_2Si: silylenes upon narrowing the R–Si–R bond angle: 469 nm for Et_2Si:, 449 nm for *cyclo*-[$(H_2C)_5Si$:] and 436 nm for *cyclo*-[$(H_2C)_4Si$:].[1j,83]

4.5 Reactions of Transient Species

As discussed above, heavy carbene analogs in their singlet states (Figure 4.3) feature two reactive sites, a filled n-orbital and a vacant p-orbital, manifesting both nucleophilic (Lewis base) and electrophilic (Lewis acid) behavior. Like carbenes, their heavy group 14 element analogs undergo the two most common types of reactions: insertion into σ-bonds (Figure 4.6, **A** and **B**) and cycloaddition to π-bonds (Figure 4.7). As was established both experimentally and theoretically, in the majority of cases the above processes proceed stepwise rather than simultaneously, with the *electrophilic* interaction preceding the *nucleophilic* interaction, leading to the intermediate formation of transient weakly bound ylide-type complexes.

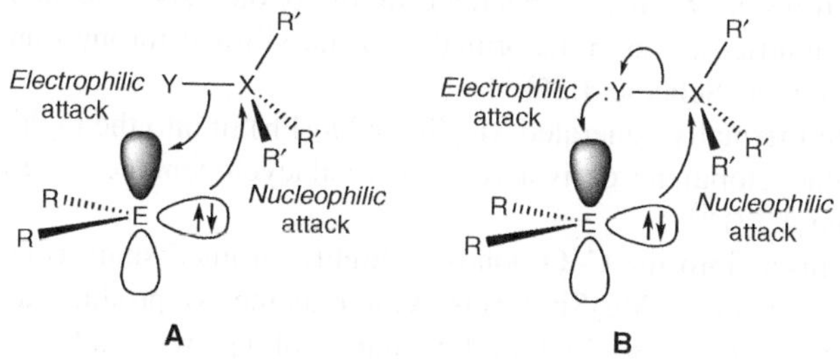

Figure 4.6 Orbital interactions involved in the insertion of heavy carbene analogs R_2E: ($E = Si–Pb$) into the X–Y σ-bond: (**A**) initial $\sigma_{X-Y} \rightarrow p_{(E:)}$ interaction; (**B**) initial $n_{Y:} \rightarrow p_{(E:)}$ interaction

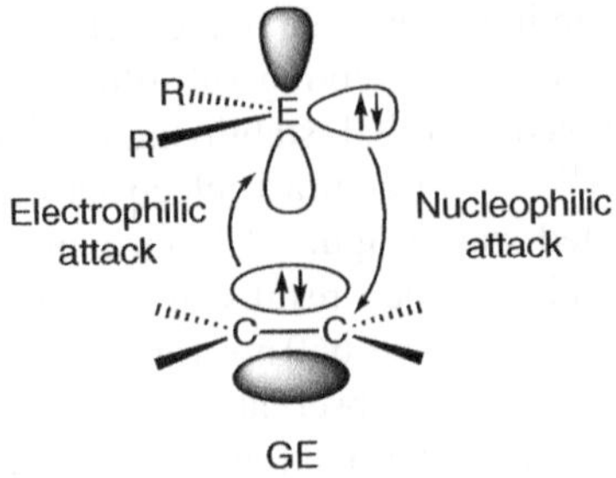

Figure 4.7 *Orbital interactions involved in the cycloaddition of heavy carbene analogs R_2E: ($E = Si$–Pb) to the C–C π-bond*

The most fundamental reactivity of the transient carbene analogs of the heavy group 14 elements, namely, insertion and cycloaddition processes, is briefly discussed below along with some other types of heavy carbene reactions (dimerization, rearrangements, complexation to transition metals).

4.5.1 Insertion into Single Bonds

As discussed in the previous section, insertion of silylenes R_2Si: into X–Y: σ-bonds commonly proceeds through the intermediate zwitterionic donor–acceptor complex $R_2Si^{-}\cdots Y^{+}$–X (silaylide) formed by the initial $n_{Y:} \rightarrow p_{(Si:)}$ interaction (Figure 4.6, **B**), leading finally to an insertion product.[1a,b] Silylenes, as well as their heavier homologs (germylenes, stannylenes and plumbylenes), insert into a variety of single bonds, including Si–H (first silylene reaction ever reported), Si–Si, Si–Halogen, Si–O, Si–N, C–H (intramolecular), C–Halogen, C–O, H–O and H–N bonds.[1a–c] Numerous adducts of silylenes and Lewis bases have been observed spectroscopically in frozen matrices: adducts with ethers, tertiary alcohols, tertiary amines and sulfides reverted to the starting silylenes upon warming but adducts with primary and secondary alcohols gave rise to the formation of the formal insertion products of silylenes into H–O bonds.[84]

The trapping of silylene (i-Pr_3Si)(t-Bu_3Si)Si:, generated by photolysis of the silirene precursor, with Me_3SiH resulted in the formation of the intermolecular Si–H insertion product (i-Pr_3Si)(t-Bu_3Si)Si(H)$SiMe_3$ along with the intramolecular C–H insertion by-product (Scheme 4.25).[57,58]

Likewise, the transient t-Bu_2Si:, generated by the photolysis of azide t-$Bu_2Si(N_3)_2$ in low-temperature matrices, was transformed into the silirane through an intramolecular C–H insertion process (Scheme 4.26).[63]

Insertion of the thermally generated Me_2Si: or Me(Ph)Si: into the highly strained Si–Si bond of 1,2-disilacyclobutene derivative gave trisilacyclopentenes as ring enlargement products (Scheme 4.27).[85]

Silylenes can insert into the C–O bonds of highly strained small ring systems. Thus, photochemically generated Me_2Si: reacts with oxetane to produce a five-membered ring insertion product, 1-oxa-2-silacyclopentane, along with allyloxydimethylsilane (Scheme 4.28).[86]

Silylenes are also active towards the Si–O bonds of alkoxysilanes (both cyclic and acyclic). Thus, photochemically generated Me_2Si: inserts into the Si–O bond

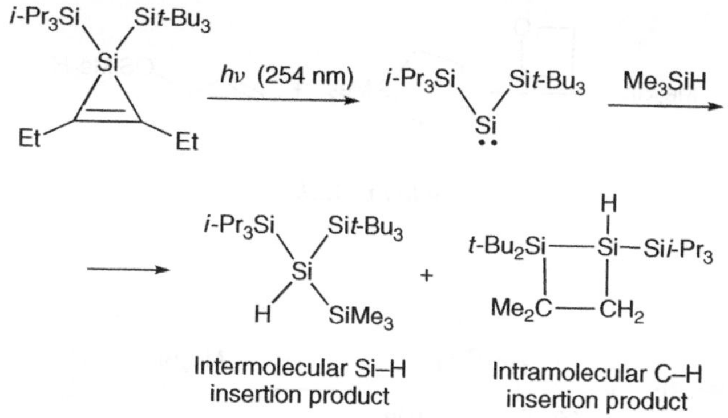

Scheme 4.25

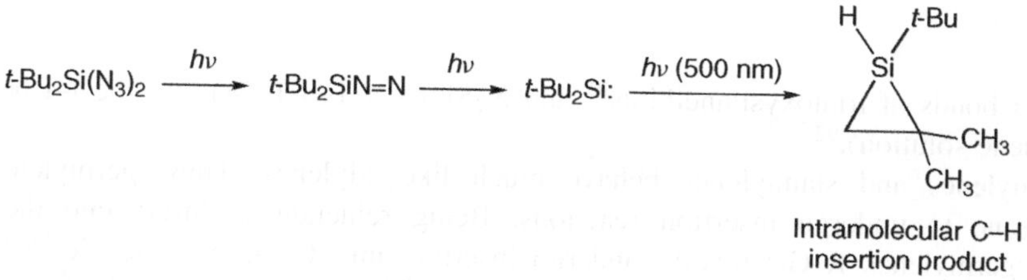

Scheme 4.26

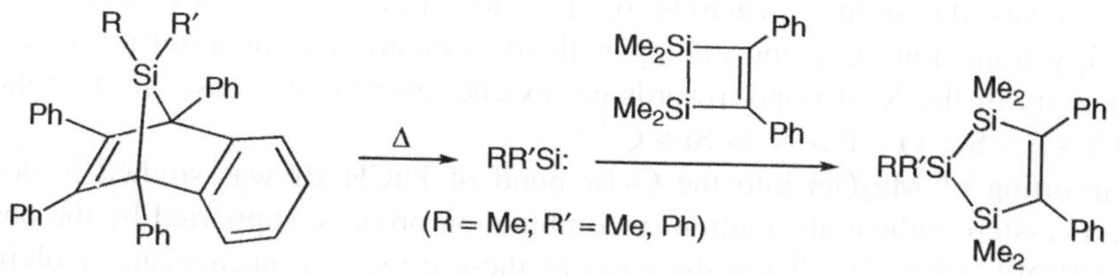

Scheme 4.27

of hexamethylcyclohexasiloxane forming 3,5,7-trioxa-1,2,4,6-tetrasilacycloheptane (Scheme 4.29).[87]

Reaction of the sterically hindered Mes_2Si: with *i*-PrOH resulted in the initial formation of the donor–acceptor complex (Mes_2Si:)←[:O*i*-Pr(H)], observed spectroscopically in a 3-MP matrix, which was then transformed into the formal O–H insertion product Mes_2Si(O*i*-Pr)H.[1a]

The N–H bond of primary and secondary amines RR′NH (R = R′ = –(CH_2–CMe_2)–; R = R′ = Et; R = H, R′ = *t*-Bu) is also a subject of Me_2Si: insertion to produce aminosilanes Me_2Si(H)NRR′ in rather good yields.[88]

Silylenes have been found to insert into other single bonds: H_2Si: into the Ge–H bond of GeH_4,[89] Me_2Si: into the Ta–H bonds of $Cp_2TaH(PMe_3)$ and Cp_2TaH_3,[90] Me_2Si: into

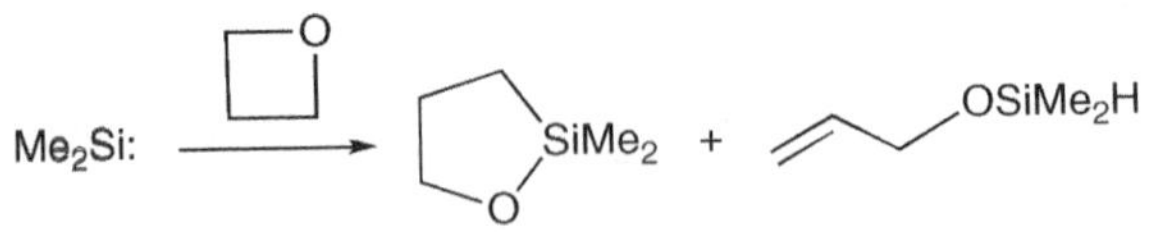

Scheme 4.28

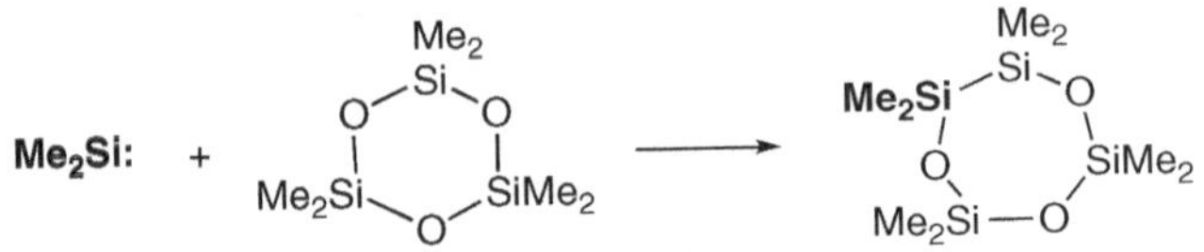

Scheme 4.29

the S–H bonds of trialoxysilanethiols[91] and X_2Si: (X = F, Cl, Br) into the I–I bond of I_2 (toluene solution).[92]

Germylenes and stannylenes behave much like silylenes. Thus, germylenes are also prone to undergo insertion reactions. Being reluctant to insert into the C–C bonds, germylenes R_2Ge: readily undergo insertion into Ge–C,[1c,93] Ge–Ge,[45,94] C–H (intramolecular),[95] C–O,[96] C–S,[9b,96] C–Sn,[97] O–H[96,98] and Sn–Halogen[49c,39,99] bonds. Recent calculations at the MP2 and B3LYP levels revealed that the interaction of germylenes with a variety of single bonds (C–H, Si–H, N–H, P–H, O–H, S–H, F–H, Cl–H) involves the initial formation of a donor–acceptor complex, followed by a high-energy transition state and ending with the formation of an insertion product.[100] The reactivity of the X–H bond towards germylene insertion decreases in the following order: Cl > F > S > O > P > N ≫ Si > C.[100]

The insertion of Me_2Ge: into the C–Br bond of $PhCH_2Br$ was studied in detail to reveal a two-step, radical abstraction–recombination process, supported by the observation of CIDNP effects.[94b,101] On the basis of these effects, a mechanism involving the formation of a caged singlet radical pair $[BrMe_2Ge^{\bullet}\ {}^{\bullet}CH_2Ph]^S$ as a key reactive intermediate starting from the singlet ground-state Me_2Ge: was proposed (Scheme 4.30).[94b] Similar radical pathways were suggested for the reactions of Me_2Ge: with CCl_4, CCl_3Br, $PhCH_2Cl$ and $PhCH_2I$.[94b]

In contrast, no indications of a radical pathway were found in the reaction of Me_2Ge: with alkynyl and propargyl halides, which pointed to a possible concerted mechanism of insertion (Scheme 4.31).[102]

In all of the above-discussed cases of germylene insertion into C–Halogen bonds, the formation of a donor–acceptor complex was proposed as the first step of the reaction. Thus, with benzyl halides a three-step procedure was suggested: (a) donor–acceptor complex formation; (b) radical abstraction; and (c) radical recombination.[1c]

Stannylenes R_2Sn:, albeit inert towards C–H and C–C bonds, can readily insert into weaker X–Y single bonds (X–Y = Sn–C,[103] Sn–Sn,[104] Sn–S,[104] Sn–Halogen,[1c,104] O–O[105] and S–S[105]), thus forming the expected insertion products X–R_2Sn–Y.

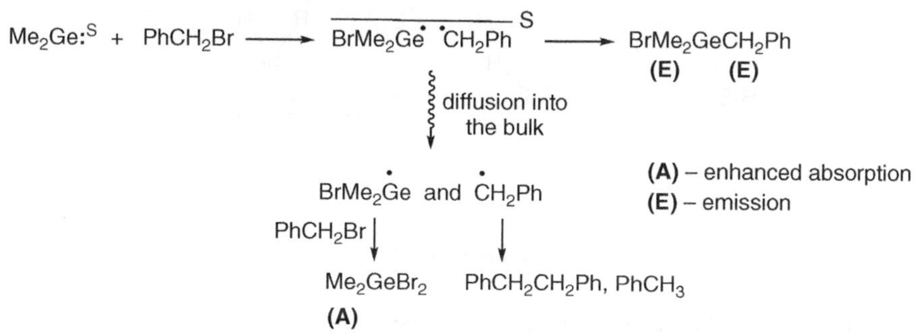

Scheme 4.30

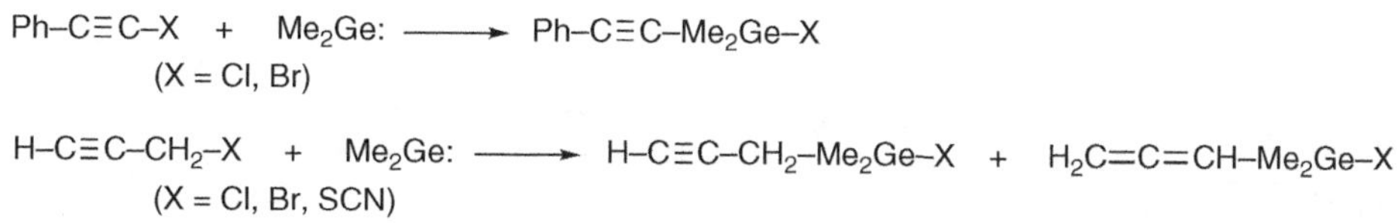

Scheme 4.31

4.5.2 Addition to Multiple Bonds

The addition reactions of silylenes, germylenes and stannylenes to alkenes, alkynes and 1,3-dienes are discussed below (for the related topics of addition of heavy carbene analogs to 1,3-heterodienes, cumulenes and heterocumulenes, ketones and thioketones, see the leading reviews by Gaspar and West,[1a] Tokitoh and Ando[1b] and Neumann[1c]).

4.5.2.1 Addition to Alkenes

Addition of carbenes to C=C double bonds is an important mechanistic probe for the establishment of carbene ground-state multiplicity. The famous Skell–Woodworth rule states: singlet carbenes add to alkenes in a concerted fashion forming cyclopropane derivatives stereospecifically, whereas triplet carbenes react stepwise and nonstereospecifically because of the intermediate formation of the triplet 1,3-biradical undergoing fast C–C bond rotation before its intersystem crossing (spin inversion) and ring closure.[106] Likewise, silylenes, which almost exclusively have a singlet ground state (for the exceptional cases of extremely rare triplet silylenes, see Section 4.6.2), are believed to add to alkenes in a concerted manner because these addition reactions were found to proceed stereospecifically.[1a] Thus, stereospecific addition of diadamantylsilylene Ad_2Si: to *cis*- and *trans*-but-2-ene led to the conclusion that it has a singlet ground state.[31] In general, addition of silylenes to alkenes results in the formation of three-membered ring silacyclopropanes (siliranes) (Scheme 4.32), the first stable representative of which was synthesized by Seyferth *et al.* in 1972.[107]

Di-*tert*-butylsilylene *t*-Bu_2Si:, generated by the photolysis of hexa-*tert*-butylcyclotrisilane,[29a] also reacts with alkenes (cyclopentene, $PhCH_2CH=CH_2$) to form stable siliranes.

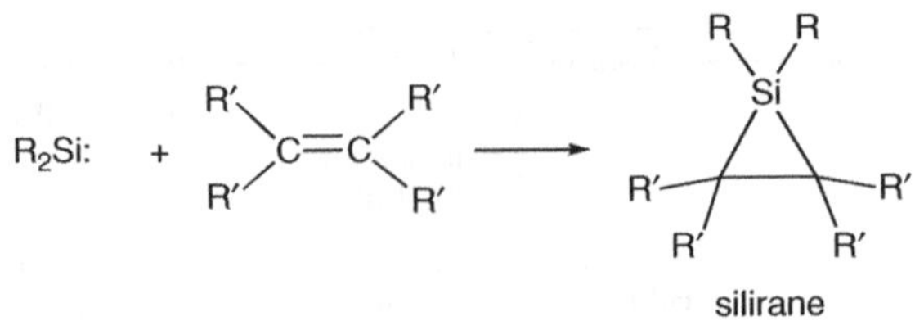

Scheme 4.32

Stable alkylidene siliranes were prepared by the addition of dimesitylsilylene Mes_2Si: to cumulated systems of 1-*tert*-butyl-3,3-dimethylallene and tetramethylbutatriene (Scheme 4.33).[108]

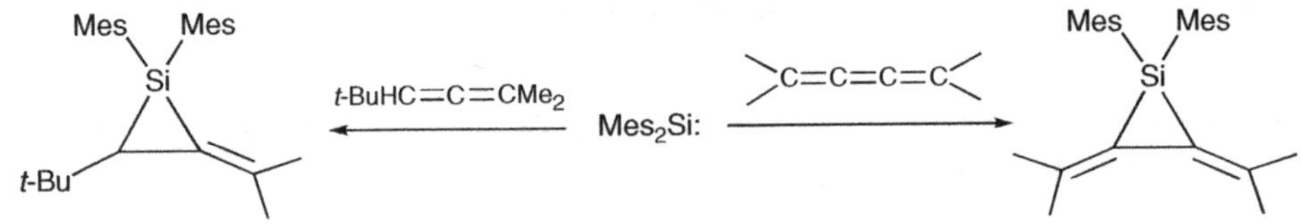

Scheme 4.33

Regiospecific addition of Dip_2Si:, generated by the photolysis of $Dip_2Si(SiMe_3)_2$ (Dip = 2,6-diisopropylphenyl), to a C=C bond localized at the six-membered ring–six-membered ring juncture of fullerene C_{60} produced a stable silirane derivative of C_{2v} symmetry.[109a] Reaction of the same silylene Dip_2Si: with the fullerene C_{70} resulted in the formation of two silirane adducts.[109b]

Upon the addition of t-Bu_2Si: to ethylene, the first silirane lacking substituents on its ring carbons was formed as a distillable liquid,[110] and addition of the same silylene to 2-methylstyrene produced another distillable silirane.[111]

Addition of transient germylenes to alkenes has been most comprehensively studied in the case of Me_2Ge:, readily available by the thermolysis/photolysis of the corresponding 7-dimethyl-7-germanorbornadiene.[1c] Because the polymerization of Me_2Ge: was found to proceed without a noticeable activation barrier, controlled only by diffusion, the addition reactions of Me_2Ge: were expected to be very fast to compete successfully with polymerization. Although simple alkenes (oct-1-ene, but-2-ene, cyclohexene, etc.) were inert towards free germylenes, activated alkenes, such as styrenes, readily reacted with Me_2Ge: forming 1,1-dimethyl-3,4-diphenyl-1-germacyclopentane derivatives as a 1:1 mixture of *syn*/*anti* isomers (Scheme 4.34).[112]

An initial formation of a transient germirane as a 1:1 adduct was assumed, which then underwent subsequent reaction with the second equivalent of styrene to form the final five-membered ring product.

The first stable germiranes were prepared in 1994 by Ando *et al.* by the reaction of the stable germylene Dis_2Ge: with either tetramethylbutatriene or *N*-phenylmaleimide (Scheme 4.35).[113]

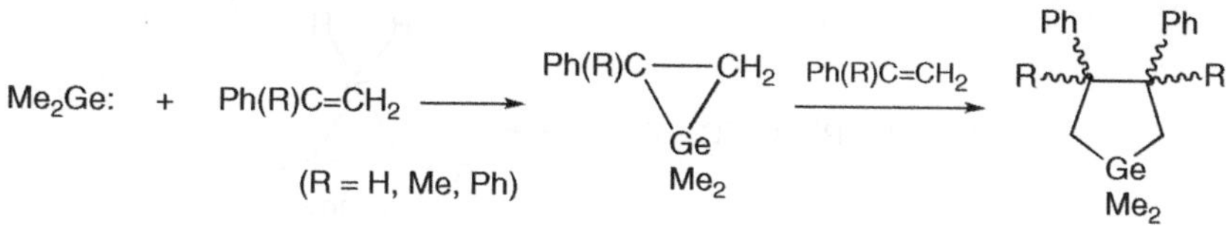

Scheme 4.34

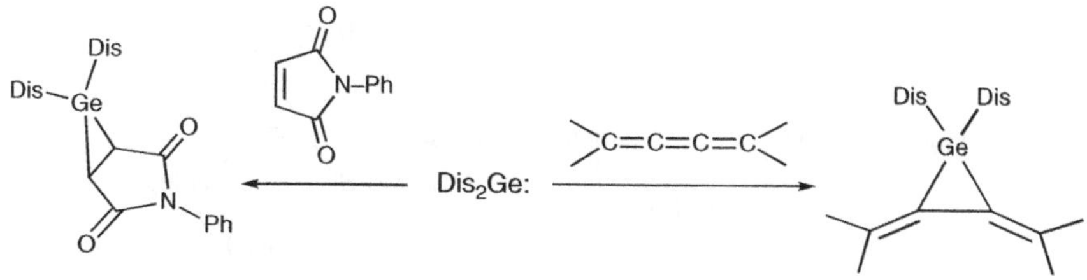

Scheme 4.35

In contrast to the above-described addition of t-Bu_2Si: to ethylene providing access to the stable silirane,[110] a similar reaction of the stable germylene Dis_2Ge: and ethylene afforded 1,2-digermacyclobutane via the germirane intermediate (Scheme 4.36).[114]

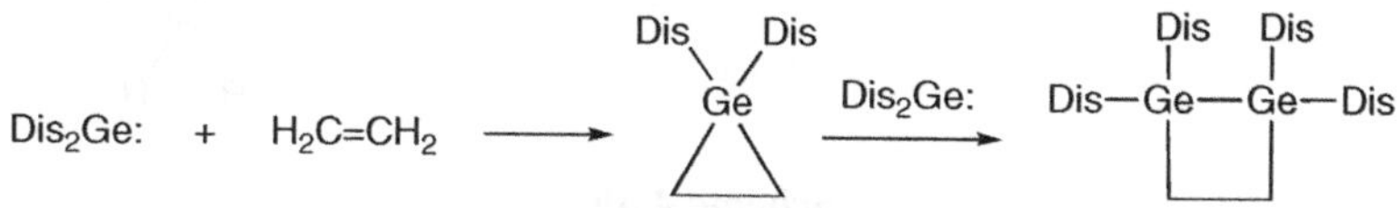

Scheme 4.36

4.5.2.2 Addition to Alkynes

Addition of silylenes to alkynes resulted in the formation of heavy cyclopropene analogs, silacyclopropenes (silirenes) (Scheme 4.37), the first stable representatives of which were synthesized in 1976 by the groups of Gaspar[115] and Seyferth[50a].

Among other remarkable stable silirene derivatives, one should mention polysilabicyclosilirenes available by the addition of Mes_2Si: [generated by the photolysis of $Mes_2Si(SiMe_3)_2$] with polysilacycloalkynes (Scheme 4.38),[116] and bis(silirenes) prepared by the reaction of 1,3-diynes R′–C≡C–C≡C–R′ with the photochemically generated silylenes R_2Si: (Scheme 4.39),[117] and bis(silirenes) from the thermolysis of bis(siliranes) in the presence of Me_3Si–C≡C–$SiMe_3$ (Scheme 4.40).[118]

Silylenes are also capable of undergoing addition to the C≡P triple bond of phosphaalkynes to form stable phosphasilirenes featuring endocyclic C=P double bonds (Scheme 4.41).[119] Phosphagermirenes can be similarly prepared by the cycloaddition of stable germylene Dis_2Ge: and phosphaalkyne t-Bu–C≡P.[120]

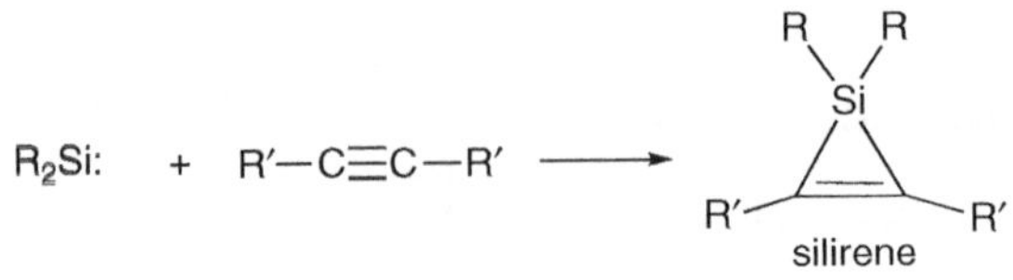

Scheme 4.37

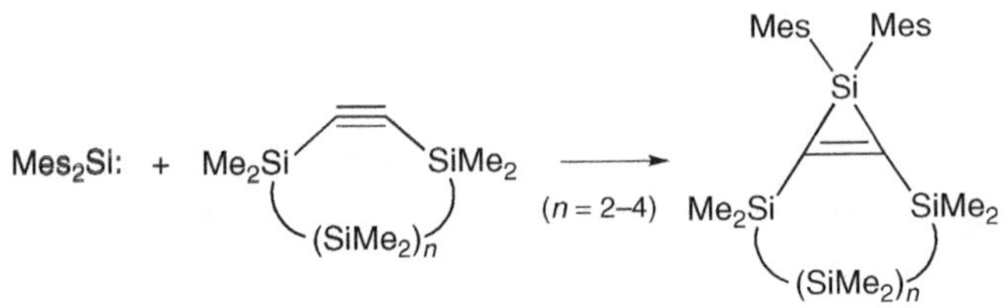

Scheme 4.38

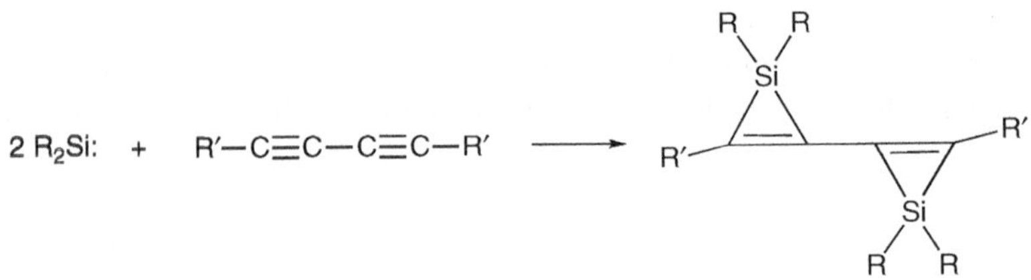

Scheme 4.39

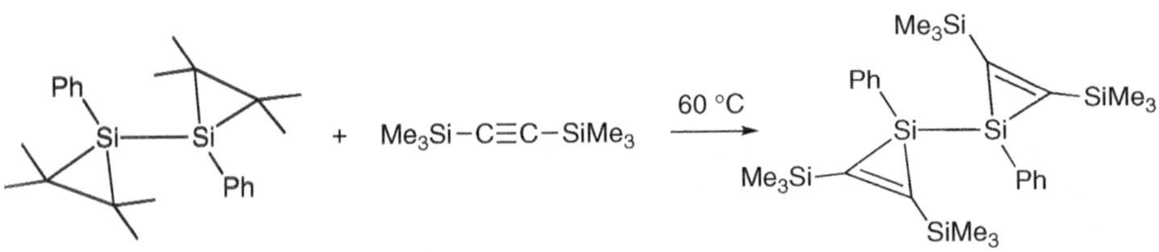

Scheme 4.40

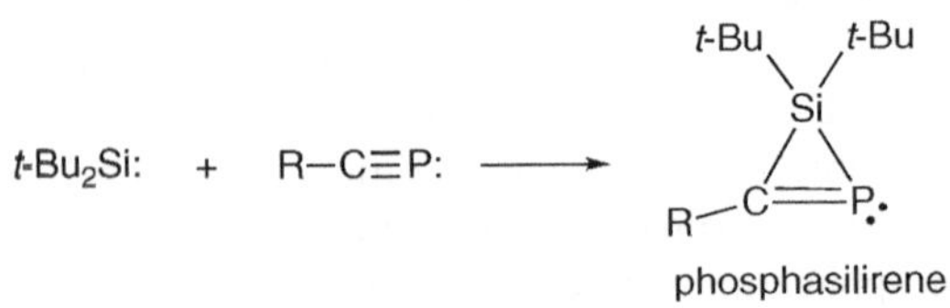

Scheme 4.41

The first stable germirene (and a new stable silirene) was synthesized by the reaction of the photochemically generated Me_2E: (E = Si, Ge) with a thiacycloheptyne derivative (Scheme 4.42).[121] The mechanism of this reaction was assumed to involve a stepwise addition of Me_2E: (E = Si, Ge) in its excited triplet state to the C≡C triple bond, an assumption supported by the observation of ^1H CIDNP effects.[121c]

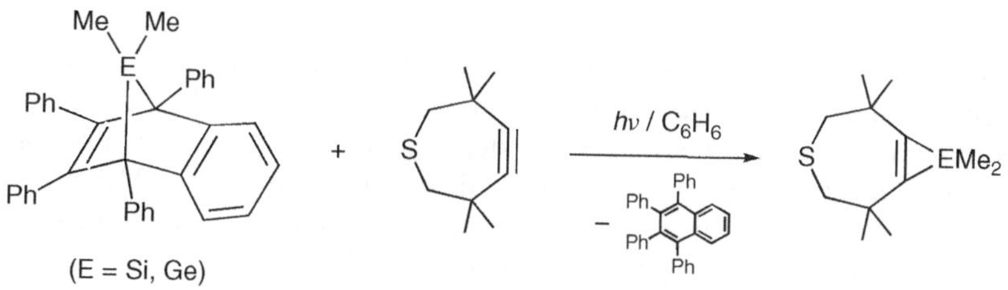

Scheme 4.42

Reaction of stannylene Dis_2Sn: with the same thiacycloheptyne results in the formation of the corresponding stannirene derivative, kinetically stabilized by the bulky R groups (Scheme 4.43, **A**).[122] In contrast, reaction of Dis_2Sn: with the less sterically hindered cyclooctyne produced a 1,2-distannacyclobut-3-ene derivative, formed by the insertion of the second stannylene unit into the strained Sn–C bond of the intermediate stannirene (Scheme 4.43, **B**).[103]

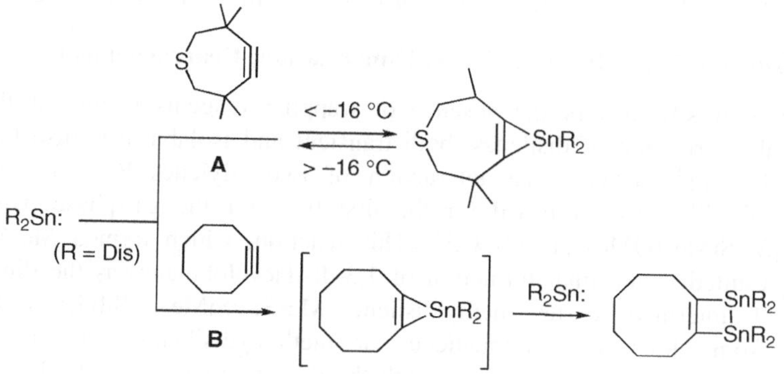

Scheme 4.43

4.5.2.3 *Addition to 1,3-Dienes*

Since the first reports of the formation of 1-silacyclopent-3-enes in the reactions of transient silylenes with conjugated 1,3-dienes, a stepwise mechanism involving a vinylsilirane intermediate has been commonly accepted (Scheme 4.44), but a concerted 1,4-addition pathway has not been completely ruled out.[1a]

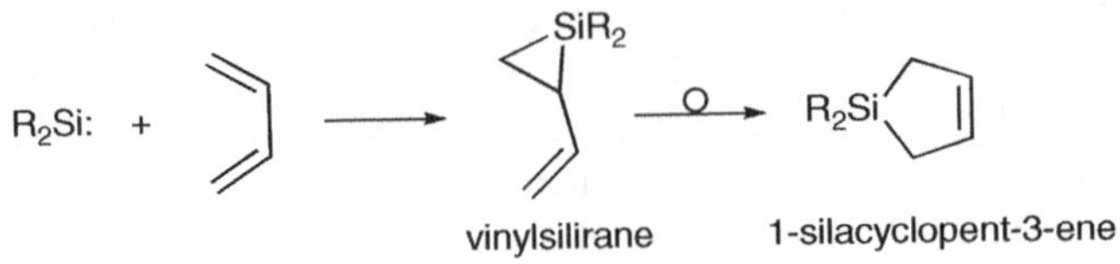

Scheme 4.44

Although the intermediate formation of vinylsiliranes was proved by trapping experiments,[1a] the mechanism (concerted or stepwise) of both the initial [1+2] cycloaddition and following rearrangement steps (Scheme 4.44) remains unclear.[123]

Examples of reactions of silylenes and 1,3-dienes are: addition of $(MeO)_2Si$: to 2,3-dimethylbuta-1,3-diene forming 1,1-dimethoxy-3,4-dimethyl-1-silacyclopent-3-ene;[124] and interaction of thermally generated Me(Cl)Si:, Me(MeO)Si: and $Me(Me_2N)Si$: with buta-1,3-diene, isoprene and 2,3-dimethylbuta-1,3-diene to yield mixtures of 1-silacyclopent-2- and 3-enes.[43]

The [1+4] cycloaddition pathway was also established for the reaction of free germylenes and conjugated 1,3-dienes yielding 1-germacyclopent-3-enes.[125] In contrast to the above-discussed silylene case, a concerted addition mechanism was suggested with the germylene playing the role of a nucleophile towards the electron-deficient dienes.[125,126]

Likewise, stable stannylene Dis_2Sn: was reactive towards 1,4-substituted 1,3-dienes X–CH=CH–CH=CH–X bearing electron-withdrawing substituents (X = COOMe, CN), whereas it was much less reactive towards dienes with electron-donating groups (X = Me, OMe).[127] Similarly to the reactivity of germylenes, stannylene acts as a donor, implying interaction of its HOMO (n-orbital) with the LUMO of the 1,3-diene.

4.5.3 Disilenes/Silenes from Silylenes: Dimerization, Rearrangement

Dimerization of silylenes in the absence of trapping reagents is one of the major routes for the formation of disilenes, both transient and isolable (see also Chapter 5, Section 5.3.1.1.1).[83] Bimolecular interaction of two silylenes R_2Si: to produce a disilene $R_2Si{=}SiR_2$ was proposed for the first time for the gas-phase pyrolysis of $(MeO)Me_2Si{-}SiMe_2(OMe)$ at 600 °C.[128] This reaction, which formed Me_2Si: at the first step, resulted in the final formation of 1,3-disilacyclobutanes as the dimerization products of intermediate tetramethyldisilene $Me_2Si{=}SiMe_2$. Silylenes thermally generated from either silanorbornadienes or methoxydisilanes were reacted with anthracene yielding bicyclic derivatives with the disilane units as the bridge, implying initial dimerization of silylenes to disilenes followed by the trapping of the latter species with anthracene (Scheme 4.45).[129]

Disilenes formed by the dimerization of silylenes bearing small substituents are kinetically unstable and readily undergo subsequent dimerization to form four-membered rings (see above) or polymerization. However, disilenes with bulky substituents are thermally stable and can be isolated (see Chapter 5, Section 5.3.1.1.1). Thus, the milestone tetramesityldisilene $Mes_2Si{=}SiMes_2$ was synthesized in this way by the dimerization of the transient dimesitylsilylene Mes_2Si: generated by the photolysis of the trisilane precursor (Scheme 4.46).[130]

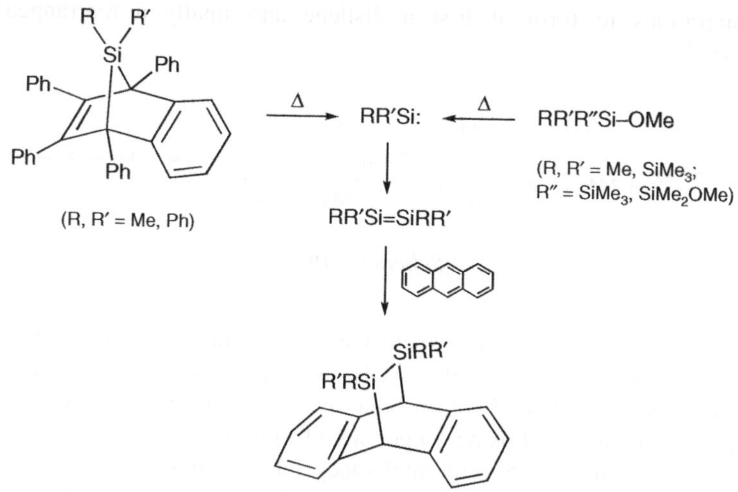

Scheme 4.45

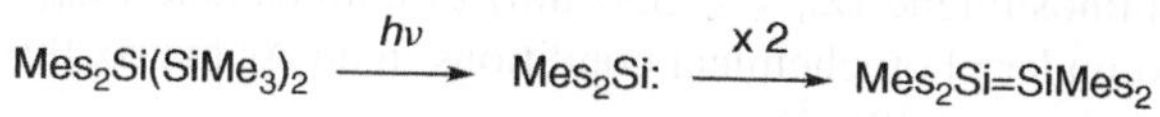

Scheme 4.46

Photolysis of cyclotrisilanes with bulky aryl substituents provided another route to stable disilenes formed both directly and by silylene dimerization (Scheme 4.47).[36a,c]

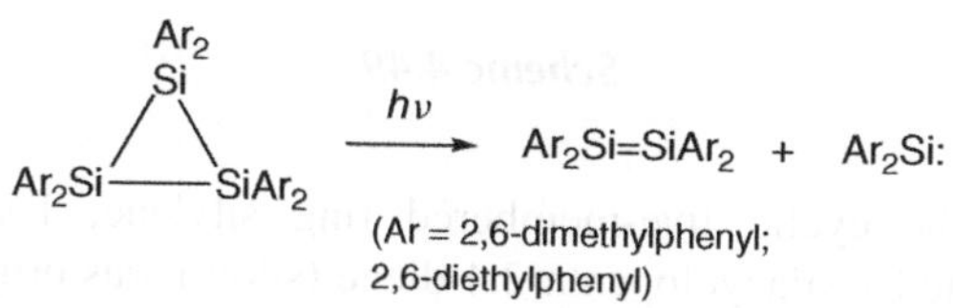

Scheme 4.47

Dimerization of the transient silylenes to disilenes was also detected in the gas phase. Thus, Me_2Si:, generated by the flash photolysis of *cyclo*-$(Me_2Si)_6$ in cyclohexane, dimerized to form $Me_2Si{=}SiMe_2$ following second-order kinetics at the diffusion-controlled rate.[131] Likewise, dimerization of Mes_2Si: to $Mes_2Si{=}SiMes_2$ also occurred at the diffusion-controlled limit.[132]

Intramolecular silylene-to-disilene rearrangements represent another route to the transient disilenes. Interconversions of silylenes and disilenes are reversible (see Section 4.2.5), and 1,2-migration of substituents may proceed in both directions: from silylene to form disilene and from disilene to produce silylene. This was deduced using D-labeled silylene $Me_3Si(CD_3)Si$:, which underwent two consecutive

CH_3/CD_3-migrations to form at first a disilene and finally a rearranged silylene (Scheme 4.48).[133]

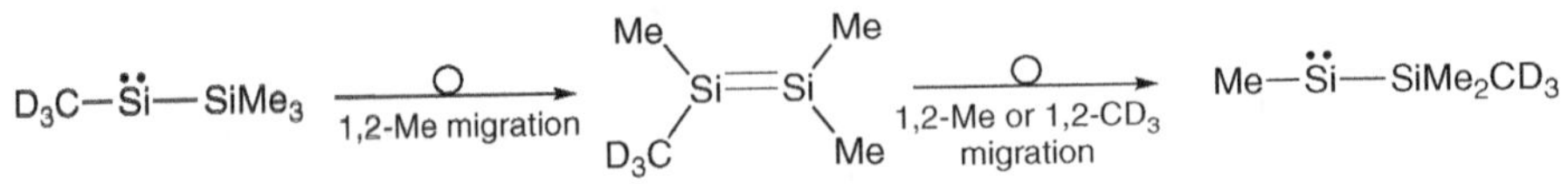

Scheme 4.48

Silylenes can be photochemically transformed into silenes through 1,2-migration of a substituent. Thus, the parent silene $H_2Si{=}CH_2$ isomerized to methylsilylene $H(H_3C)Si$: via a 1,2-hydrogen shift upon 254 nm irradiation, while the reverse reaction (methylsilylene to silene) was observed using longer wavelength irradiation (320 nm).[134]

As discussed in Section 4.2.5, a thermal silacyclobutadiene-to-cyclopropenyl silylene isomerization is reversible (see Scheme 4.23), and silylene can revert to the starting silacyclobutadiene under photochemical conditions (>400 nm, 3-MP, 77 K).[61]

An interesting sequence of isomerizations, silaimine (no UV absorptions observed at $\lambda > 300$ nm) to aminosilylene ($\lambda_{max} = 330$ nm) to aminosilene ($\lambda_{max} = 256$ nm), was reported to proceed under photochemical conditions in an Ar matrix through consecutive 1,2-H-migrations (Scheme 4.49).[135]

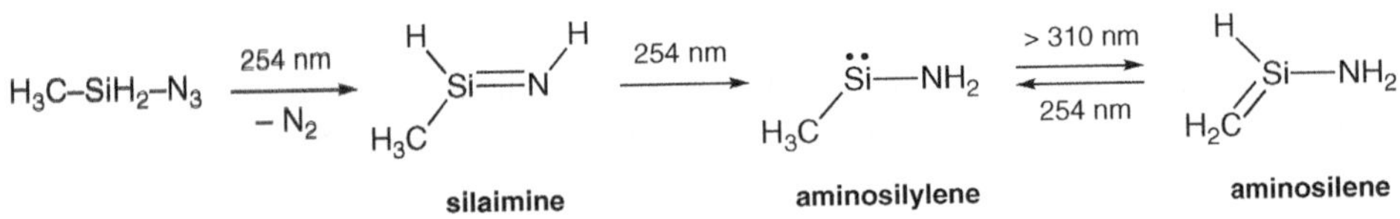

Scheme 4.49

Isomerization of the cyclic five-membered ring silylene, 1-silacyclopent-3-en-1-ylidene, to unsubstituted 1-silacyclopenta-2,4-diene (silole) was originally confirmed by the characterization of the silole dimer[136] and was later supported by IR and UV measurements in matrix isolation experiments, in which the isomeric silacyclopentadienes underwent photochemical interconversion (Scheme 4.50).[25,137]

4.5.4 Complexation to Transition Metals

Transition metal carbene complexes are among the most fundamental organometallic derivatives, well known since the groundbreaking discovery of the first representative $(OC)_5W{=}C(OMe)R$ (R = Me, Ph) by Fischer and Maasboel in 1964.[138] Such compounds can be classified into two major classes with distinctly different electronic properties: Fischer carbene complexes and Schrock carbene complexes. Fischer complexes feature a transition metal–carbene bond formed by the donor–acceptor interaction of two singlet (closed-shell) fragments: carbene-to-metal σ-donation and metal-to-carbene π-back donation (Figure 4.8). This metal–carbon bond has partial double bond character

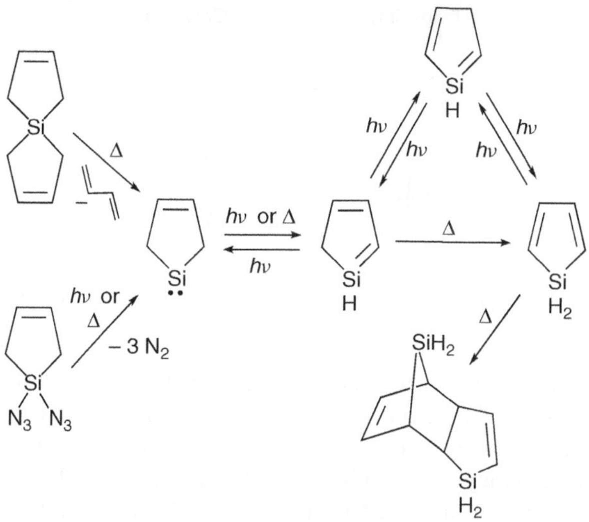

Scheme 4.50

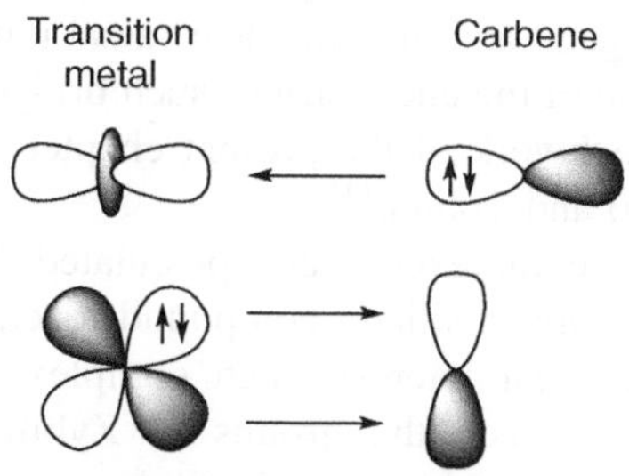

Figure 4.8 *Transition metal–carbene bonding in Fischer-type complexes*

and is polarized towards the metal as $M^{\delta-} = C^{\delta+}$, being prone to nucleophilic attack at the carbene center. Fischer-type complexes are typically formed by the interaction of heteroatom-substituted singlet carbenes (aminocarbenes, alkoxycarbenes) with large energy separations between their singlet and triplet ground states and transition metals in low oxidation states.

In contrast, Schrock carbene complexes manifest an essentially covalent metal–carbene bond formed by the coupling of two triplet fragments from both carbene and transition metal (Figure 4.9). This metal–carbon bond is considered to be a true double bond, which is polarized in the opposite direction to Fischer complexes, $M^{\delta+} = C^{\delta-}$, being susceptible to electrophilic attack at the carbene center. Schrock-type complexes can be formed by the interaction of carbenes with a small singlet–triplet energy gap (dialkylcarbenes, alkylidenes) with early transition metals in their highest oxidation states.

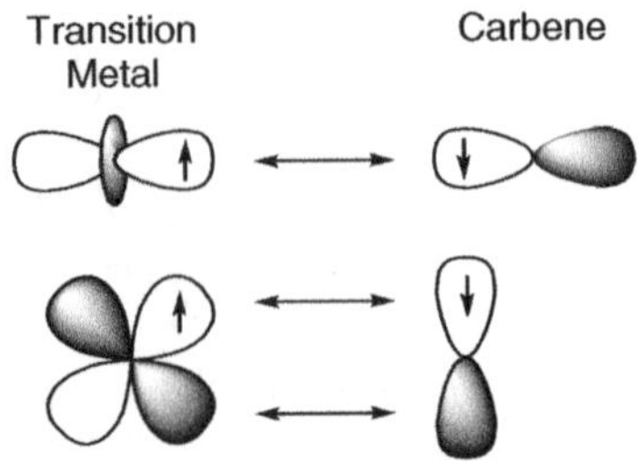

Figure 4.9 *Transition metal–carbene bonding in Schrock-type complexes*

The coordination chemistry of the heavy carbene analogs (first of all, germylenes and stannylenes, and to a smaller extent plumbylenes) has been extensively developed since the pioneering contribution by Marks in 1971.[139] Major advances in the field of transition metal germylene/stannylene/plumbylene complexes, comprehensively summarized in Petz's review,[140] are not discussed in this section. Instead, we focus on the coordination chemistry of silylene complexes, which were synthetically elusive until recently. A large family of bridged silylene complexes, featuring two transition metal fragments connected by a silylene bridge, can be subdivided into two classes: complexes possessing a metal–metal bond and manifesting both σ- and π-type interactions between the sp^2-hybridized silylene ligand and the dinuclear metal unit, and complexes lacking a metal–metal bond. Discussion of the chemistry of such bridged silylene (and germylene) complexes is outside the framework of the present chapter; instead we refer interested readers to a review by Ogino and Tobita.[141]

Transition metal silylene complexes were postulated long ago as the reactive intermediates in a number of organosilicon compound reactions catalysed by transition metals.[142] However, despite great interest, such complexes were not synthesized as stable derivatives until 1987, when the groups of Zybill and Tilley independently reported the base-stabilized transition metal silylene complexes $(OC)_4Fe{=}Si(Ot\text{-}Bu)_2[OP(NMe_2)_3]$[143] (Scheme 4.51) and $[Cp^*(Me_3P)_2Ru{=}SiPh_2(NCMe)]^+{\bullet}BPh_4^-$[144] (Scheme 4.52), respectively [a base-stabilized bis(silylene)iron complex was prepared in the following year by Ogino *et al.*[145]]. Because of the presence of a Lewis basic ligand (LB), a resonance form $M^-{-}Si{-}LB^+$, involving substantial metal–silicon single bond character with the positive charge localized largely on the Lewis base, considerably contributed to the overall ground-state structure of these silylene complexes.

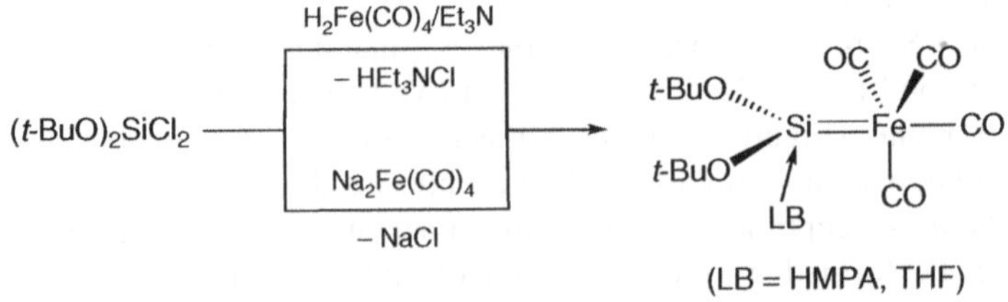

Scheme 4.51

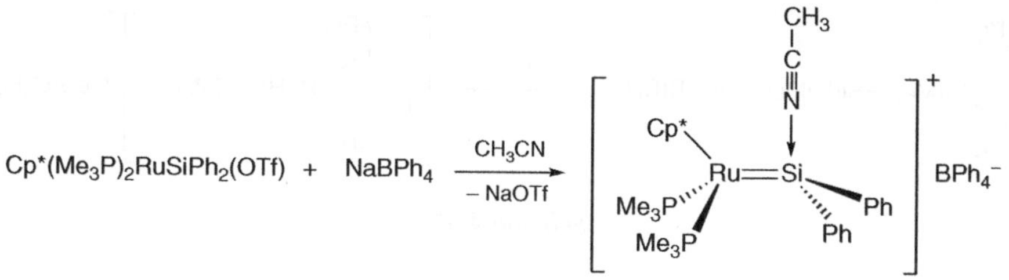

Scheme 4.52

Base-free transition metal silylene complexes, lacking the electronically perturbing effects of Lewis base coordination, can be prepared via one of three major synthetic routes. The first one involves *abstraction of an anionic group* (typically, Cl^- or OTf^-) from the silyl ligand, producing cationic silylene complexes of the group 8–10 metals. This approach was successfully employed for the synthesis of the first isolable base-free silylene complexes $[Cp^*(Me_3P)_2Ru{=}Si(SR)_2]^+{\bullet}BPh_4^-$ ($R = p$-tol, Et) (Scheme 4.53).[146]

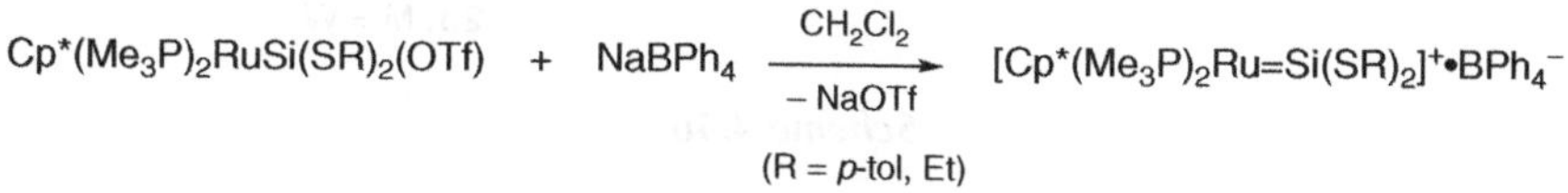

Scheme 4.53

The second synthetic route is based on the *direct coordination of free silylenes* (both stable and transient) to transition metal fragments. This method has been widely utilized for the preparation of transition metal complexes of N-heterocyclic silylenes.[147] Transient photochemically generated Mes_2Si: was trapped with low-valent platinum derivatives $(R_3P)_2Pt$ ($R = i$-Pr, Cy) to form platinum silylene complexes $(R_3P)_2Pt{=}SiMes_2$ (Scheme 4.54).[148]

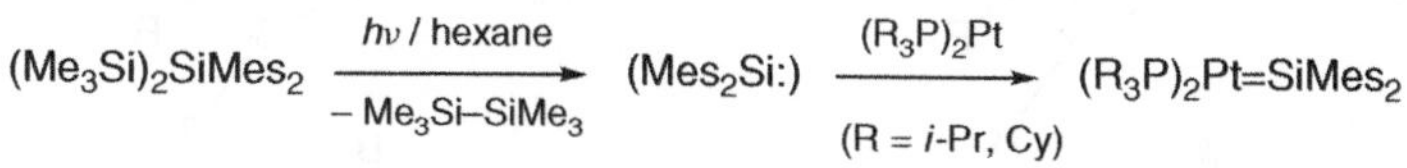

Scheme 4.54

The third approach to silylene complexes takes advantage of the *high migratory aptitude of the α substituent of a silyl ligand* (first of all, hydrogen), which is prone to move from silicon to a vacant coordination site of a transition metal (the so-called 1,2-shift). This synthetic methodology was employed for the first time in the preparation of the stable cationic platinum silylene complex (Scheme 4.55).[149]

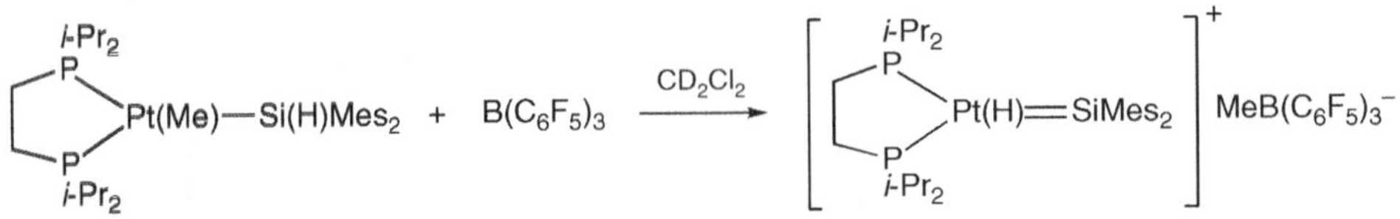

Scheme 4.55

A new synthetic pathway was recently discovered based on the reaction of the dianionic derivative tetrasilabicyclo[1.1.0]butane-2,4-diide $\mathbf{1}^{2-}\bullet\mathbf{Ca}^{2+}$ with the group 6 metallocene dichlorides Cp_2MCl_2 (M = Mo, W) producing unprecedented bicyclic silylene complexes $Cp_2M[\eta^1\text{-}Si_4(SiMe\textit{t}\text{-}Bu_2)_4]$ **2** (**2a:** M = Mo; **2b:** M = W) (Scheme 4.56).[150]

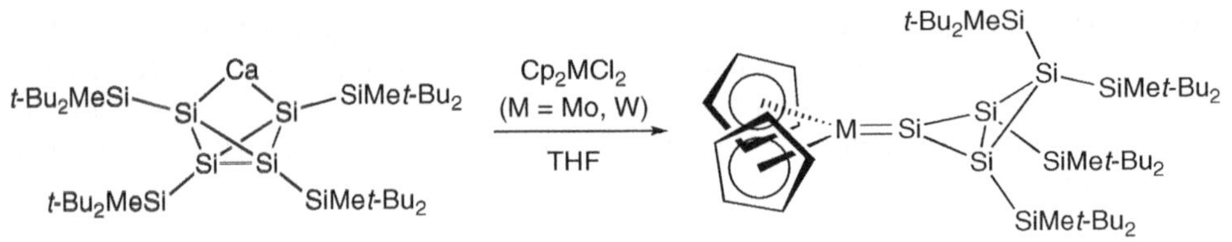

$\mathbf{1}^{2-}\bullet\mathbf{Ca}^{2+}$

2a: M = Mo
2b: M = W

Scheme 4.56

The diagnostic extreme deshielding of the transition metal-bound silicon atoms (323.6 ppm for **2a** and 260.9 ppm for **2b**) along with the large $^1J_{Si\text{–}W}$ coupling constant value of 278.2 Hz for **2b** testify to the presence of the base-free silylene ligand and M=Si double bond. Interestingly, such neutral bicyclic silylene complexes **A** are stabilized by the contribution of zwitterionic resonance form **B** featuring a homoaromatic cyclotetrasilenylium ion (Scheme 4.57).

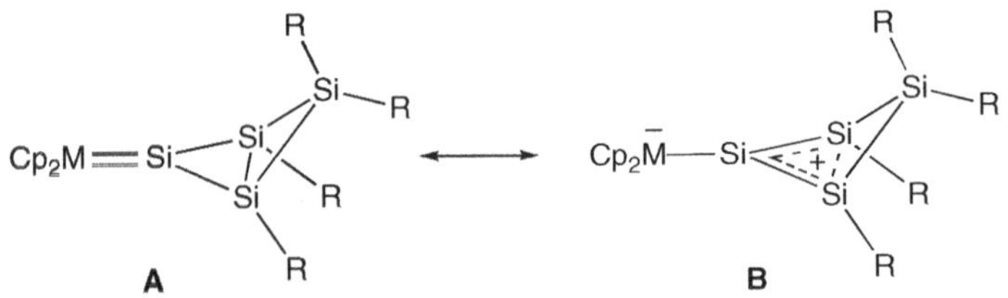

M = Mo, W; R = SiMe*t*-Bu$_2$

Scheme 4.57

The structural manifestation of the resonance extreme **B** was seen in the characteristic stretching of the W1–Si1 bond [2.4202(14) Å], shortening of the skeletal Si1–Si2 and Si1–Si4 bonds [2.3290(18) Å and 2.3231(19) Å] and extraordinary elongation of the Si2–Si4 bridge [2.4170(16) Å] (Figure 4.10).

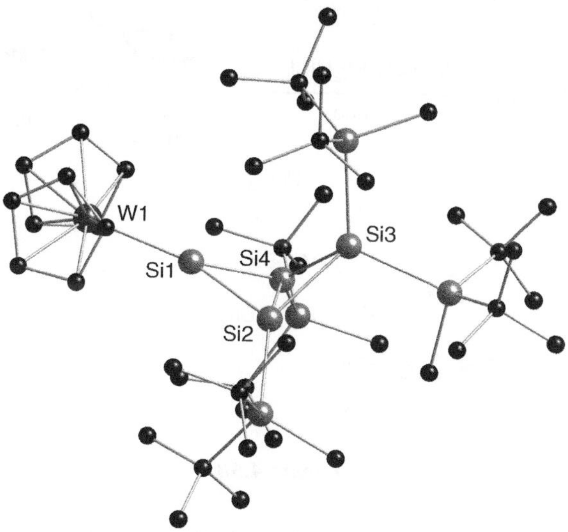

Figure 4.10 *Crystal structure of tungsten–silylene complex* **2b** *(hydrogen atoms are not shown)*

Inspection of the frontier MOs of **2b** model $Cp_2W[\eta^1\text{-}Si_4(SiH_3)_4]$ supported the presence of a homoaromatic contribution manifested in the bonding (HOMO–2) and antibonding (HOMO–1) π(W=Si)–σ(Si–Si bridge) interactions. Negative values of the nucleus-independent chemical shift (NICS) are also indicative of the Si_3-ring homoaromaticity: NICS(1) = −11.9 and NICS(−1) = −13.

The availability of isolable silylene complexes with a variety of transition metals has allowed systematic studies on the role of silylene fragments in catalytic transformations. Thus, preliminary investigations revealed the dramatic influence of the degree of silylium ion character of silylene complexes on their reactivity towards alkenes.[142]

Base-free silylene complexes manifest characteristic low-field resonances (typically observed at 200–370 ppm) of their electrophilic silylene centers, which feature a pronounced Lewis acidity. Thus, in terms of general classification of carbene complexes, the silylene derivatives described above can be better recognized as Fischer (or Fischer-like) complexes.

The only silylene complex reliably assigned to the Schrock-type class has recently been synthesized by the coupling of the 1,1-dilithiosilane derivative $(t\text{-}Bu_2MeSi)_2SiLi_2$ with the hafnocene dichloride $(\eta^5\text{-}C_5H_4Et)_2HfCl_2$ (Scheme 4.58).[151]

The initially formed 16-electron hafnium–silylene complex **3** could not be isolated because of its thermal lability; however, it was stabilized by complexation with PMe_3 forming the 18-electron complex $(\eta^5\text{-}C_5H_4Et)_2(Me_3P)Hf{=}Si(SiMe\,t\text{-}Bu_2)_2$ **4**. The Hf=Si bond in **4** of 2.6515(9) Å was ca. 5% shorter than the usual Hf–Si single bonds, indicating considerable double bond character. As is typical for base-free silylene complexes, **4** exhibited a remarkably deshielded resonance of its silylene Si atom observed at 295.4 ppm. Formulation of both **3** and **4** as Schrock-type silylene complexes was based on their reactivity. Thus, **3** reacts with MeOH and D_2O quantitatively

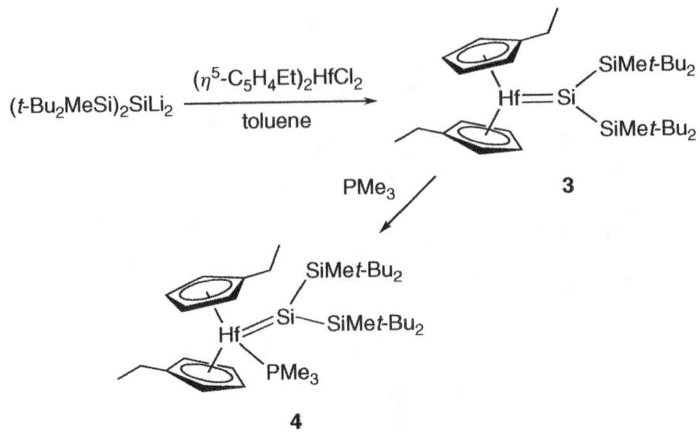

Scheme 4.58

yielding $(t\text{-}Bu_2MeSi)_2SiH_2$ and $(t\text{-}Bu_2MeSi)_2SiD_2$ with the regioselectivity corresponding to a bond polarization $M^{\delta+} = Si^{\delta-}$ expected for the Schrock-type complexes. Accordingly, NPA calculations revealed a substantial charge separation in the model **4** $(\eta^5\text{-}C_5H_4Et)_2(Me_3P)Hf{=}Si(SiMe_3)_2$ as follows: Si (−0.34) and Hf (+0.78).

4.6 Stable/Persistent Silylenes, Germylenes, Stannylenes and Plumbylenes

4.6.1 Singlet Species

There are several fundamental classes of thermally stable organometallic compounds possessing divalent group 14 element centers. Dihalogermylenes, -stannylenes and -plumbylenes X_2E: (X = halogen, E = Ge–Pb), as well as their complexes with *n*-donor ligands (Lewis bases), were the first known and thoroughly studied stable divalent species of the heavy group 14 elements.[152] The remarkable decamethylsilicocene $(\eta^5\text{-}C_5Me_5)_2Si$:, prepared by Jutzi *et al.*, features a higher coordinate Si atom and is therefore better described as a monomeric silicon(II) compound, rather than a true silylene possessing a dicoordinate silylene center.[153] The large family of stable carbene analogs of the heavy group 14 elements R_2E: (E = Si–Pb) includes very important and practically useful heterocyclic silylenes, germylenes, stannylenes and plumbylenes, in which the carbene center E is substituted by *n*-donating R ligands. These compounds have been known since the pioneering work of Veith *et al.* on the synthesis of the first N-heterocyclic germylenes, stannylenes and plumbylenes, many new representatives of which were prepared in due course.[154] The first N-heterocyclic silylenes, as the silicon version of Arduengo's carbenes, were synthesized by West *et al.*, and this chemistry was subsequently developed by the groups of Lappert, Heinicke and others.[147,155] The chemistry of all of the above-mentioned classes of stable carbene analogs has been frequently and comprehensively reviewed,[147,152–155] therefore, it is not covered in this chapter (aspects of the 6π-electron aromaticity of N-heterocyclic heavy carbene analogs are briefly discussed in the final chapter, Section 6.3.2.2). Numerous heavy carbene analogs, in which the carbene center is stabilized by the intramolecular/intermolecular coordination of *n*-donating ligands (typically, N-, P-, O-, S- and halogen groups), being essentially tri- or even higher coordinate,[154a] are also outside the framework of this book.

4.6.1.1 Silylenes

The above-described heterocyclic heavy carbene analogs are sterically protected by bulky groups at their heteroatoms and electronically stabilized by both inductive (decrease of the σ-electron density at >E: by electronegative heteroatoms) and resonance (π-donation from the heteroatom lone pair into the vacant p-orbital at >E:) effects. Such important thermodynamic stabilization is achieved at the expense of significant electronic perturbation at the heavy carbene center, which causes a remarkable modification of its behavior, which is manifested in the observation of physical and chemical properties not typical for classical carbenes. From this viewpoint, of particular interest is the cyclic dialkylsilylene **5**, which is the least electronically perturbed among all known isolable silylenes being stabilized only by kinetic means with bulky Me_3Si-groups.

The synthesis of **5** was achieved by the reductive dehalogenation of dibromosilacyclopentane **6** with potassium graphite performed in THF (Scheme 4.59).[156]

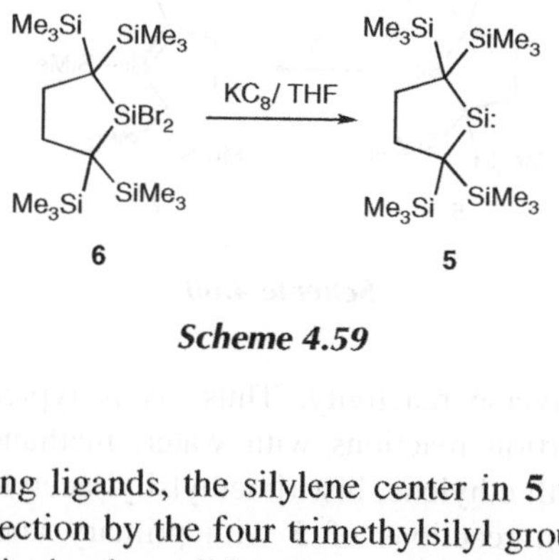

6 **5**

Scheme 4.59

Lacking adjacent *n*-donating ligands, the silylene center in **5** was kinetically stabilized via its effective steric protection by the four trimethylsilyl groups (Figure 4.11).

Silylene **5** is monomeric in the solid state, as demonstrated by its crystal structure analysis: the shortest distance between the silylene centers is 7.210(1) Å. The longest UV absorption of **5** in hexane, corresponding to the $n \rightarrow 3p$ electronic transition, was observed at 440 nm ($\varepsilon = 500$), a value that is close to that of dimethylsilylene found at 453 nm in the 3-MP matrix at 77 K.[17] The resonance of the divalent silicon atom was observed in the extreme low-field region at 567.4 ppm (calculated value for H_3Si-substituted model = 602 ppm), whereas those of the cyclic diaminosilylenes were measured at much higher fields (from 78 to 117 ppm). Such spectroscopic features (low-energy $n \rightarrow 3p$ transition and remarkable deshielding of the silicon atom) definitely point to a very small extent of electronic perturbation of the silylene center in **5**, although σ(C–Si)$\rightarrow$ 3p(Si) electron donation may also play a role in the overall stabilization of the silylene.

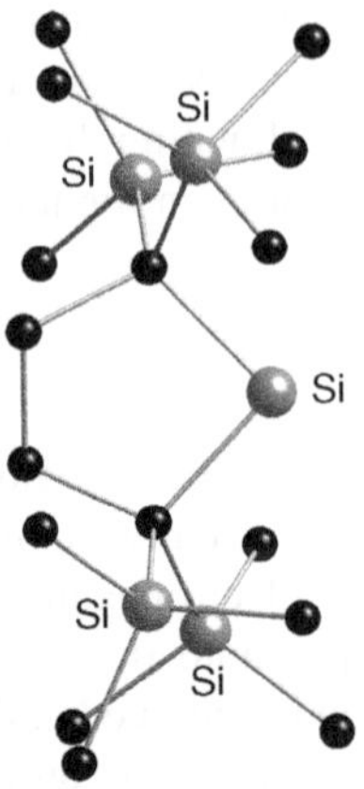

Figure 4.11 Crystal structure of the cyclic dialkylsilylene **5** (hydrogen atoms are not shown)

Silylene **5** underwent thermal isomerization at room temperature in hexane forming an isomeric silene **7** through the facile 1,2-migration of a Me_3Si-group from carbon to the silylene center (Scheme 4.60).[156]

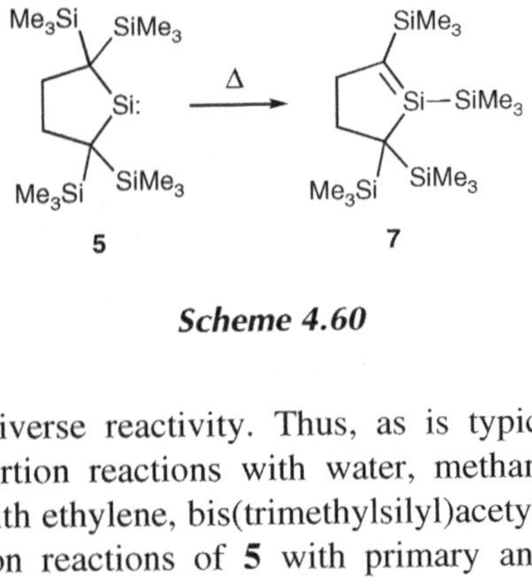

Scheme 4.60

Silylene **5** exhibited diverse reactivity. Thus, as is typical for transient silylenes, **5** readily underwent insertion reactions with water, methanol and triethylsilane, and cycloaddition reactions with ethylene, bis(trimethylsilyl)acetylene and 2,3-dimethylbuta-1,3-diene.[157] The insertion reactions of **5** with primary and secondary alkyl halides R–Cl were believed to proceed through the initial formation of a Lewis acid–Lewis base complex $[R_2Si:]^{\delta-}\cdots[Cl–R]^{\delta+}$, which was expected to undergo either homolytic or heterolytic R–Cl bond cleavage, or concerted R–Cl bond insertion, depending on the alkyl halide and reaction conditions used.[158] Likewise, silylene **5** smoothly inserted into the Si–Cl bonds of chlorosilanes.[157] Among other remarkable examples of the reactivity of **5**, one should mention its reactions with organic radicals giving the corresponding adducts,[159] the one-electron reduction of **5** generating silylene anion-radical $[\mathbf{5}]^{-\bullet}$ (see Chapter 2, Section 2.4.2.1.2.1),[160] and the interaction of **5** with arenes and alkenes believed to reflect the reactivity of the biradical silylene in its $^1B_1$ photoexcited state.[161] Silylene **5** was also utilized for the design of novel unsaturated systems: silicon–chalcogen doubly bonded derivatives by reactions with chalcogens,[162] silaketenimines by reactions with isocyanides,[163] and trisilaallene and related trimetallaallenic compounds by reactions with metal halides and subsequent reduction[164] (the chemistry of such unsaturated organosilicon systems is discussed in Chapter 5). Silylene **5** was also employed as a novel ligand in coordination chemistry and its palladium complexes have been recently reported.[165]

A remarkable diarylsilylene Tbt(Mes)Si: **8** was generated upon the thermal dissociation of extraordinarily crowded disilenes, (*Z*)-Tbt(Mes)Si=Si(Mes)Tbt and (*E*)-Tbt(Mes)Si=Si(Mes)Tbt, at 70 °C (Scheme 4.61).[166,167]

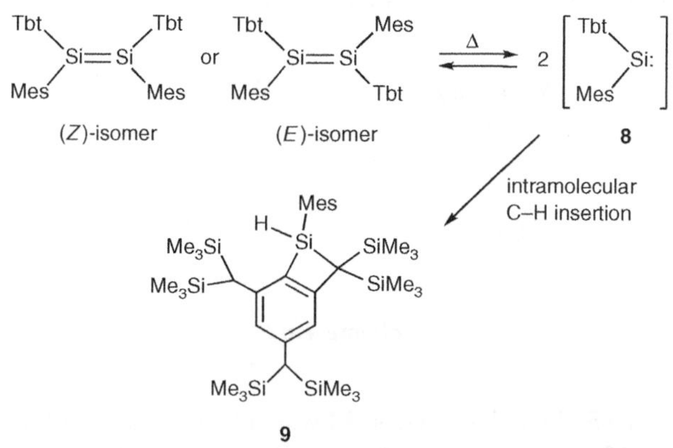

Scheme 4.61

Although not isolable in its individual form, silylene **8** was successfully trapped by a variety of reagents: methanol, triethylsilane, 2,3-dimethylbuta-1,3-diene, elemental chalcogens, alkenes and alkynes, even arenes such as benzene and naphthalene, producing the corresponding insertion and cycloaddition products.[166,167] **8** also readily reacted with heteroatom-containing multiply bonded compounds, such as carbon disulfide, nitriles, phosphaalkynes and isonitriles.[167,168] The last reaction with aromatic isonitriles is particularly important, resulting in the formation of the first stable silylene–Lewis base complexes of the type [Tbt(Mes)Si:]←C=N–Ar, which manifest reactivity typical of free silylenes, thus acting as masked silylenes[167,168b] (the chemistry of such silylene–Lewis base complexes is discussed in Chapter 5). In the absence of trapping reagents, silylene **8** underwent a thermal intramolecular C–H insertion of the silylene center into one of the [CH($SiMe_3$)$_2$] groups of a Tbt substituent to form benzosilacyclobutene derivative **9** (Scheme 4.61).[166a,167]

4.6.1.2 Germylenes

The very first stable dialkylgermylene [(Me_3Si)$_2$CH]$_2$Ge: **10** was synthesized in 1976 by Lappert *et al.* by the reaction of bis(amino)germylene [(Me_3Si)$_2$N]$_2$Ge: with Li[CH($SiMe_3$)$_2$] in diethyl ether.[169] While monomeric in solution, germylene **10** exists as a dimer, digermene [(Me_3Si)$_2$CH]$_2$Ge=Ge[CH($SiMe_3$)$_2$]$_2$, in its crystalline form. Increasing the steric bulk around the >Ge: center, Jutzi *et al.* prepared germylene [(Me_3Si)$_3$C][(Me_3Si)$_2$CH]Ge: **11** by the reaction of (η^5-Me_5C_5)[(Me_3Si)$_2$CH]Ge:

with $Li[C(SiMe_3)_3]$ in diethyl ether.[170] In contrast to Lappert's germylene **10**, this new germylene **11** was monomeric both in solution and in the solid state, which was shown by X-ray crystallography to exhibit a V-shaped configuration of the molecule (R–Ge–R bond angle of 111.3°) with the shortest nonbonding Ge···Ge separation of 5.704 Å. This dramatic difference in the aggregation states of crystalline **10** and **11** was attributed to the higher steric demands of $(Me_3Si)_3C$ vs $(Me_3Si)_2CH$ groups.

A stable cyclic dialkylgermylene **12** was prepared by Kira *et al.* similarly to the above-described cyclic dialkylsilylene **5**, namely, by the reduction of dichlorogermacyclopentane **13** with metallic sodium in toluene (Scheme 4.62).[171]

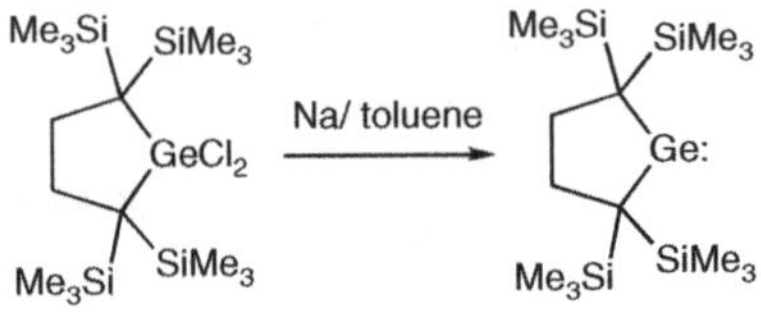

13 **12**

Scheme 4.62

Like dialkylsilylene **5**, dialkylgermylene **12** was shown to be monomeric. Thus, the UV absorption band of **12** observed at 450 nm (n–4p electronic transition) was essentially independent of the measurement temperature (77–293 K) or concentration of **12**, pointing to its monomeric state in solution. Moreover, the UV spectrum of **12** in THF was identical to that measured in hexane, which testified to the absence of germylene **12**–THF complexation, in contrast to the transient germylenes that readily form complexes with Lewis bases featuring characteristic blue shifts of their n–4p electronic transitions. The germylene, but not digermene, status of **12** in the solid state was seen in the nonbonding Ge···Ge separation of 7.243(6) Å, which is clearly beyond any bonding interactions.[171]

Several remarkable aryl-substituted germylenes, kinetically stabilized by bulky aryl groups, have been reported recently. The first isolable diarylgermylene Mes^*_2Ge: **13** was first prepared by the coupling of Mes*Li and $GeCl_2$–dioxane complex in THF by du Mont *et al.*, who reported the room temperature insertion of the germanium center of **13** into a C–H bond of one of the *o-tert*-butyl groups of a Mes* substituent, leading to germaindane derivative **14** (Scheme 4.63).[95] However, subsequent reinvestigation by Jutzi *et al.* revealed that the germylene **13** undergoes such rearrangement to **14** only in the presence of a Lewis acid (such as the starting $GeCl_2$–dioxane), and is readily isolable at room temperature in the absence of the latter.[172]

The shortest distance between the two germylene centers in the crystalline form of **13** is 8.64 Å, far exceeding the range of Ge–Ge bonding interactions. In contrast to other sterically overcrowded diarylgermylenes (λ_{max} = 550–580 nm), **13** exhibited a remarkable blue shift of the n(HOMO)→p(LUMO) transition in its UV-Vis spectrum observed at 405 nm in the solid matrix and at 430 nm in hexane solution.[172] This was explained by the specific configurational features of germylene **13** because of the extremely bulky Mes* substituents; the π-systems of the aryl groups are not coplanar with the p(Ge)-orbital leading to a destabilization of the latter, whereas the bond angle at the germylene center

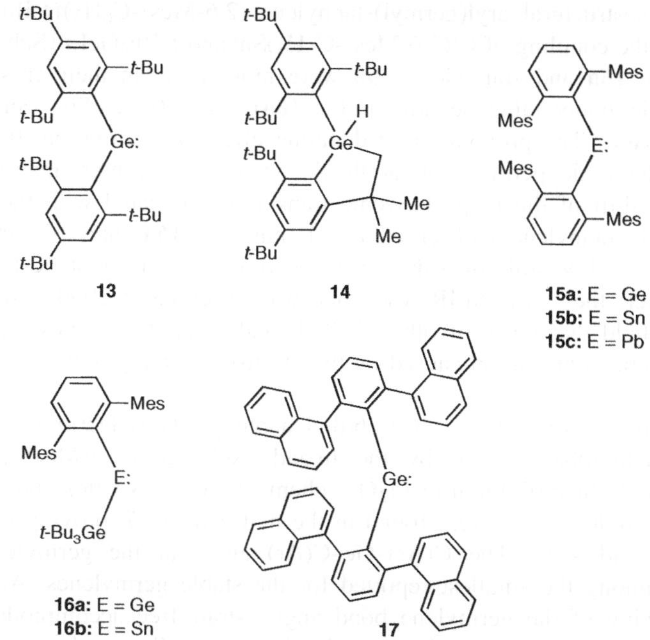

13 **14** **15a:** E = Ge **15b:** E = Sn **15c:** E = Pb

16a: E = Ge **16b:** E = Sn **17**

Scheme 4.63

is not as large as that of other bulky diarylgermylenes leading to a stabilization of the *n*-orbital. Overall, raising the energy level of the p-orbital and lowering the energy level of the *n*-orbital results in an increase of the HOMO–LUMO energy gap and hence the observation of a blue shift of the electronic transition of **13**.

Another stable diarylgermylene $(2,6\text{-Mes}_2\text{-C}_6H_3)_2$Ge: **15a** bearing two extremely bulky aryl substituents was readily available by the reaction of $(2,6\text{-Mes}_2\text{-C}_6H_3)$Li with $GeCl_2$–dioxane complex in Et_2O, as reported by Power *et al.*[173] The absorption maximum of **15a** was observed at 578 nm, that is, in a longer wavelength region than that of Mes*$_2$Ge: **13**. Germylene **15a** is monomeric in the solid state with the closest Ge···Ge separation of 8.409(1) Å. The C(Ar)–Ge–C(Ar) bond angle in **15a** of 114.4(2)° is the widest among those of germylenes with monodentate ligands. In a similar way, the heavier analogs of **15a**, namely, stannylene $(2,6\text{-Mes}_2\text{-C}_6H_3)_2$Sn: **15b** and plumbylene $(2,6\text{-Mes}_2\text{-C}_6H_3)_2$Pb: **15c**, were synthesized by the reaction of $(2,6\text{-Mes}_2\text{-C}_6H_3)$Li with $SnCl_2$ and $PbCl_2$, respectively (Scheme 4.63).[173] The structural features of both **15b** and **15c** are similar to those of the germanium version **15a:** UV-Vis absorptions of 553 nm (**15b**) and 526 nm (**15c**); NMR chemical shifts of 635 ppm (^{119}Sn for **15b**) and 3870 ppm (^{207}Pb for **15c**); closest distances between the heavy carbenes in the solid state of 8.392(3) Å (Sn···Sn for **15b**) and 8.388(1) Å (Pb···Pb for **15c**); C(Ar)–E–C(Ar) bond angles of 114.7(2)° (E = Sn for **15b**) and 114.5(6)° (E = Pb for **15c**).

The first stable germylene featuring an electropositive substituent, aryl(germyl) germylene $(2,6\text{-Mes}_2\text{-C}_6H_3)(t\text{-Bu}_3\text{Ge})$Ge: **16a**, was synthesized by Kira *et al.* by the reaction of $\text{Cl}(2,6\text{-Mes}_2\text{-C}_6H_3)$Ge: and t-Bu$_3$GeLi in toluene (Scheme 4.63).[174]

Likewise, the isostructural aryl(germyl)stannylene (2,6-Mes_2-C_6H_3)(t-Bu_3Ge)Sn: **16b** was formed by the coupling of Cl(2,6-Mes_2-C_6H_3)Sn: and t-Bu_3GeLi (Scheme 4.63).[174] Both germylene **16a** and stannylene **16b** were monomeric in their crystalline form: 9.14 Å (Ge···Ge nonbonding separation for **16a**) and 9.08 Å (Sn···Sn nonbonding separation for **16b**). The pronounced σ-donating effect of germyl substituents caused a significant electronic perturbation at the heavy carbene center, manifested in the remarkable red shift of the n–p electronic transition (compared with the other stable germylenes and stannylenes) observed at 719 nm (for **16a**) and 673 nm (for **16b**). This bathochromic UV shift of **16b** was in accord with the great deshielding of its stannylene center: the ^{119}Sn NMR resonance was observed at 2960 ppm, which is at a much lower field that those of other dialkyl- and diarylstannylenes, because of the large paramagnetic contribution caused by the electron-donating influence of its germyl substituents.

An interesting strain-free bis[2,6-di(1-naphthyl)phenyl]germylene **17** was prepared by Schmidbaur *et al.* by the ligand exchange of [($Me_3Si)_2N]_2$Ge: and [2,6-di(1-naphthyl)phenyl]lithium in Et_2O (Scheme 4.63).[175] No close bonding contacts between the germylene units were found in the unit cell of **17**, proving its monomeric status in the solid state. The C(Ar)–Ge–C(Ar) angle at the germylene center of 102.72(9)° is among the smallest reported for the stable germylenes. Apart from the extreme narrowing of the germylene bond angle, strain-free accommodation of both bulky aromatic substituents can be seen in the essentially nondistorted geometries of each of the aromatic rings. The calculated geometrical parameters of sterically unhindered diphenylgermylene Ph_2Ge: were very close to the experimental values found in **17**: Ge–C = 2.006 Å (in Ph_2Ge:) vs 2.036(2) and 2.030(2) Å in **17**; C–Ge–C = 101.6° (in Ph_2Ge:) vs 102.72(9)° in **17**. Thus, the two wing-like 2,6-di(1-naphthyl)phenyl substituents provide very effective kinetic stabilization of the germylene center in **17** without inducing destabilizing geometrical distortions.

An isolable diarylgermylene Tbt(Tip)Ge: **18** was generated by either reductive debromination of Tbt(Tip)$GeBr_2$ or cycloreversion of Tbt(Tip)-substituted germirene (Scheme 4.64).[176]

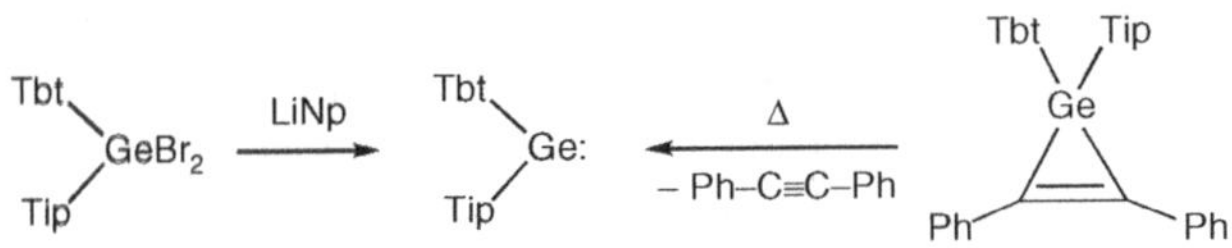

18

Scheme 4.64

It was suggested that germylene **18** is monomeric both in the solid state and in solution, although its crystal structure has not been determined. Replacement of a Tip group at the germylene center by a less bulky Mes group resulted in a dramatic decrease in the stability of the novel germylene Tbt(Mes)Ge:, which was found to be in equilibrium with its dimer, the digermene (E)-Tbt(Mes)Ge=Ge(Mes)Tbt.[177] In contrast, the above-described disilene (E)-Tbt(Mes)Si=Si(Mes)Tbt with the same substitution pattern underwent thermal dissociation into the silylene Tbt(Mes)Si: **8** only upon heating at 70 °C (see Section 4.6.1.1, Scheme 4.61).[166] Similarly to silylene **8**, germylene **18** readily formed the expected insertion and addition products with a variety of typical germylene scavengers, such as alkenes, dienes, alkynes, alcohols, hydrosilanes, and elemental chalcogens, and was also highly reactive towards heteroatomic compounds (carbon disulfide, nitrile oxide, geminal dihaloalkenes) and transition metal carbonyl complexes.[167]

4.6.1.3 Stannylenes

The first isolable dialkylstannylene $[(Me_3Si)_2CH]_2Sn$: **19** (and plumbylene $[(Me_3Si)_2CH]_2Pb$: **20**) were reported by Lappert *et al.* in the same very important paper published in 1976, that described the synthesis of the germanium analog $[(Me_3Si)_2CH]_2Ge$: **10** (Section 4.6.1.2).[169] Both $[(Me_3Si)_2CH]_2E$: (**19:** E = Sn; **20:** E = Pb) were conveniently prepared by the reaction of either Cl_2E: or $[(Me_3Si)_2N]_2E$: with $Li[CH(SiMe_3)_2]$ in diethyl ether. Similarly to the case of stable germylene **10**, dialkyl stannylene **19** was found to be monomeric in solution (cryoscopic measurements in benzene or cyclohexane), but dimeric in the solid state existing in the form of a distannene $[(Me_3Si)_2CH]_2Sn{=}Sn[CH(SiMe_3)_2]_2$ featuring a Sn=Sn double bond.

A heavier analog of the above-described silylene **5** (Section 4.6.1.1) and germylene **12** (Section 4.6.1.2), cyclic dialkylstannylene **21**, was synthesized by the reaction of 1,4-dilithio-1,1,4,4-tetrakis(trimethylsilyl)butane **22** with $SnCl_2$ in diethyl ether (Scheme 4.65).[178] The monomeric aggregation state of **21** in the solid state and in the condensed phase was deduced from its crystal and spectral (UV, NMR) characteristics, respectively. The longest wavelength UV absorption band of **21** was observed at 484 nm, and attributed to the *n*–5p electronic transition. The resonance of the stannylene center of **21** was found in the extremely low-field region at 2323 ppm in its ^{119}Sn NMR spectrum. Intermolecular Sn–Sn bonding contacts in the crystalline state of **21** were not found because the closest distances between the stannylene units were not shorter than 7.14 Å.

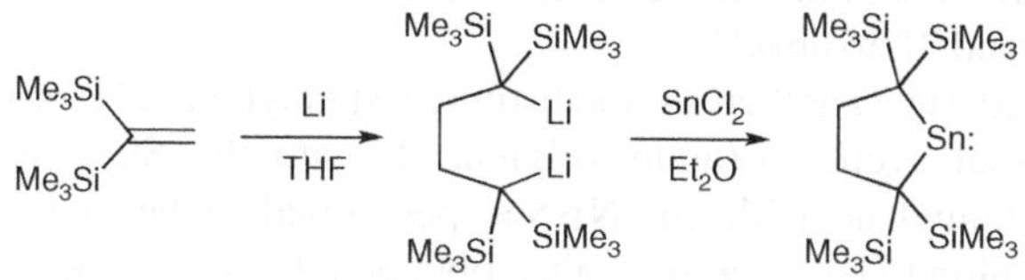

22 21

Scheme 4.65

Likewise, the seven-membered ring cyclic stannylene **23**, kinetically stabilized by steric shielding of the carbene center by the four adjacent Me_3Si-groups, was synthesized by the cyclization of dipotassium derivative $(thf)_2K(Me_3Si)_2C{-}SiMe_2{-}CH_2{-}CH_2{-}SiMe_2{-}C(SiMe_3)_2K(thf)_2$ **24** with $SnCl_2$ in diethyl ether (Scheme 4.66).[179] In its crystalline form, stannylene **23** exhibited no Sn–Sn bonding interaction (closest Sn···Sn contacts = 8.9 Å), which proved its monomeric status in the solid state. The C–Sn–C bond angle at the stannylene center of **23** (118°) was wide because of steric repulsion between the bulky trimethylsilyl groups. It was also markedly larger than the corresponding C–Sn–C angle of 86.7° in the five-membered ring stannylene **21** with a similar steric environment; such difference apparently stems from the geometrical changes on going from seven-membered ring **23** to five-membered ring **21**. As is typical for noncoordinated stannylenes, stabilized solely by kinetic means, dialkylstannylene **23** manifested a diagnostic low-field ^{119}Sn NMR resonance at 2299 ppm.

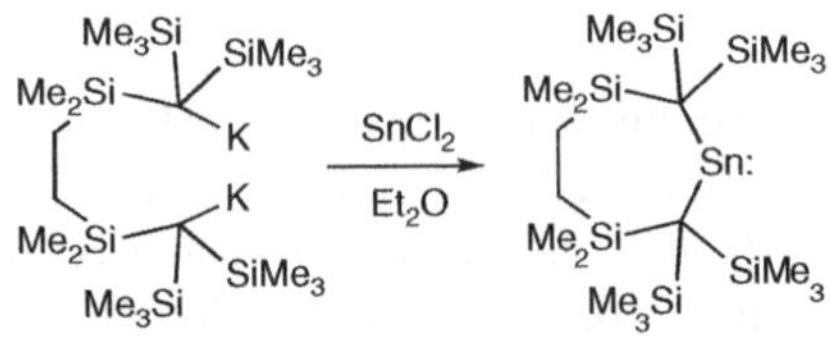

Scheme 4.66

Two stannylenes featuring electropositive silyl substituents have been reported. The first one in this series, bis[tris(trimethylsilyl)silyl]stannylene $[(Me_3Si)_3Si]_2Sn$: **25**, was prepared by Klinkhammer and Schwarz by the treatment of $[(Me_3Si)_2N]_2Sn$: with $(Me_3Si)_3SiK$ in pentane.[180] Although the most informative ^{119}Sn NMR resonance of **25** was not observed, the independence of its UV-Vis absorptions (559 and 838 nm) on the sample concentrations was indicative of the predominantly monomeric form of stannylene **25** in solution, which was also confirmed by cryoscopic measurements. In the solid state, however, **25** is dimeric, forming distannene $[(Me_3Si)_3Si]_2Sn{=}Sn[Si(SiMe_3)_2]_2$ with a Sn=Sn double bond. In contrast, isostructural plumbylene $[(Me_3Si)_3Si]_2Pb$: **26**, similarly prepared by the reaction of $[(Me_3Si)_2N]_2Pb$: with $(Me_3Si)_3SiK$ in pentane, was found to be monomeric both in crystalline form (wide Si–Pb–Si bond angles of 113.6–115.7° for the four independent molecules of **26**) and in solution (UV-Vis absorption at 578 and 1056 nm).[180]

Lappert *et al.* reported another silyl-substituted stannylene **27**, formed by the insertion of their original cyclic bis(amino)silylene **28** into the Sn–N bond of Lappert's stable bis(amino)stannylene $[(Me_3Si)_2N]_2Sn$:, performed in hexane (Scheme 4.67).[181] The isostructural plumbylene **29** was also obtained by the same synthetic protocol (Scheme 4.67).[181] Because the ^{119}Sn and ^{207}Pb NMR resonances of stannylene **27** and plumbylene **29** were not reported, their status in solution remains an open question. In the solid state, however, both **27** and **29** were reliably proved to be monomeric: wide Si–E–Si bond angles of 106.8° (for **27:** E = Sn) and 105.8° (for **29:** E = Pb), very long Sn···Sn nonbonding separations for **27** of more than 10 Å.

The first diarylstannylene kinetically stabilized by extremely bulky Mes*-groups, Mes*$_2$Sn: **30**, was synthesized by the treatment of $[(Me_3Si)_2N]_2Sn$: with Mes*Li in toluene (Scheme 4.68).[182] Stannylene **30** was not associated, being monomeric both in solution (cryoscopic measurements) and in the solid state (shortest Sn···Sn nonbonding distance of 8.84 Å). In the ^{119}Sn NMR spectrum of **30** two signals (instead of the expected one) were observed at 961 and 1105 ppm, which is a surprisingly high-field region compared with other base-free stannylenes. These two resonances coalesced at

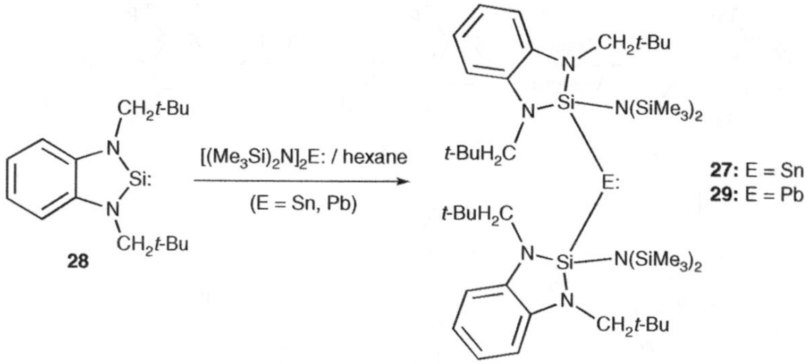

Scheme 4.67

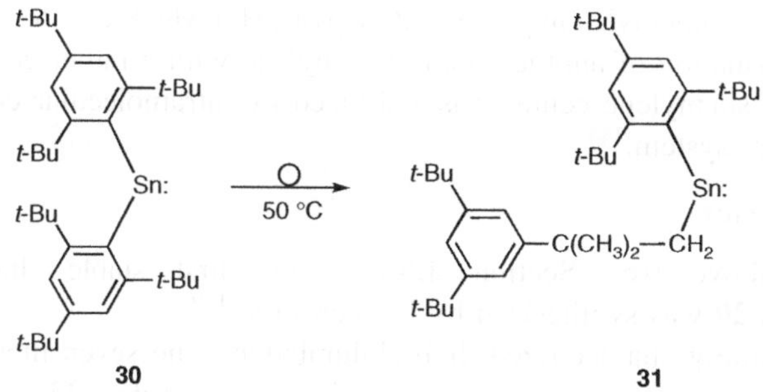

Scheme 4.68

50 °C into a singlet at 980 ppm, which was attributed to a new, less sterically encumbered stannylene **31**, formed by the thermal isomerization of **30** in solution (Scheme 4.68).[182]

Several heteroleptic stannylenes have also recently been synthesized. Thus, Eichler and Power prepared stannylstannylene **32**, as a valence isomer of distannene, by the reaction of the chlorostannylene **33** with MeLi (or MeMgBr).[183] The process was proposed to proceed through the initial methylation of **33** to give methylstannylene **34**, followed by its insertion into another molecule of MeLi forming stannyl anion derivative **35**, which then reacts with the starting chlorostannylene **33** to finally produce stannylstannylene **32** (Scheme 4.69). In contrast, when using the bulky *t*-BuLi the reaction stopped at the initial step to form heteroleptic alkyl(aryl)stannylene **36** (Scheme 4.69).[183]

Stannylenes **32** and **36** are monomeric and revealed nearly identical bond angles at the stannylene centers: 101.2° for the C–Sn–Sn angle in **32** and 101.6° for the C–Sn–C angle in **36**. Both **32** and **36** featured very low-field shifted ^{119}Sn NMR resonances at 2856.9 ppm for **32** and 1904 ppm for **36**, of which the extreme deshielding of the stannylene center in stannylstannylene **32** was explained by the strong paramagnetic contribution because of the more important σ-donating effect of the stannyl substituent in **32** vs the alkyl group in **36**.

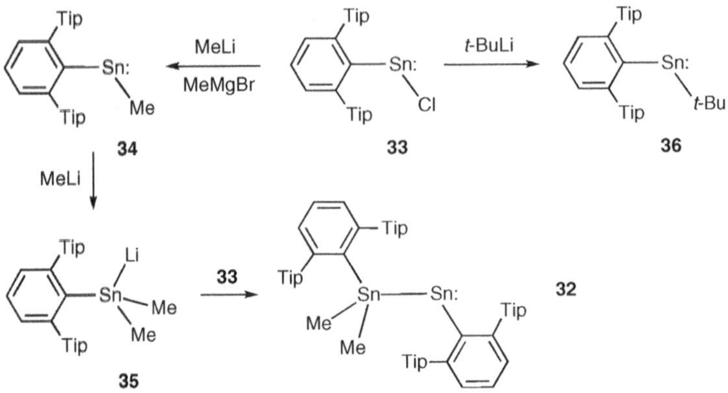

Scheme 4.69

Photolysis of diazomethylstannylene (2,6-Tip_2-C_6H_3)[$Me_3SiC(=N_2)$]Sn: resulted in the unexpected formation of another stannylstannylene with a rather complicated structure, in which the stannylene center was stabilized by intramolecular coordination to a cycloheptatrienyl π-system.[184]

4.6.1.4 Plumbylenes

As mentioned above (see Section 4.6.1.3), the first stable dialkylplumbylene [$(Me_3Si)_2CH$]$_2$Pb: **20** was synthesized by Lappert *et al.*[169]

The first structurally characterized dialkylplumbylene, the seven-membered ring **37**, was prepared by the same procedure as that for its tin analog **23** (Scheme 4.70).[185] Similarly to stannylene **23**, plumbylene **37** featured long intermolecular nonbonding Pb···Pb contacts of 8.91 Å and a wide C–Pb–C bond angle of 117.1°, indicative of its monomeric form in the crystalline state. As a consequence, the ^{207}Pb NMR resonance of **37** was observed in the extremely low-field region of 10050 ppm.

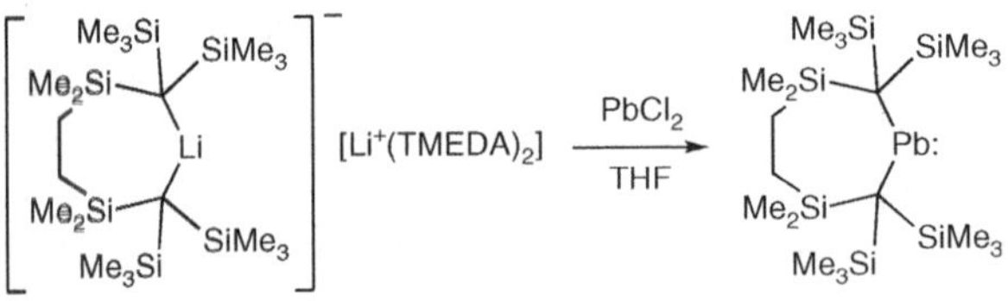

Scheme 4.70

Apart from the above-described plumbylene (2,6-Mes_2-C_6H_3)$_2$Pb: **15c** (Section 4.6.1.2),[173] a few other diarylplumbylenes have been synthesized. Thus, reaction of $PbCl_2$ with the Grignard reagent (2-*t*-Bu-4,5,6-Me_3C_6H)MgBr in THF produced the expected monomeric plumbylene (2-*t*-Bu-4,5,6-Me_3C_6H)$_2$Pb: **38** (Scheme 4.71).[186] The spectral and structural data of **38** were typical for diarylplumbylenes: 6927 ppm

(^{207}Pb NMR), 490 nm (UV-Vis), 103.0° (C–Pb–C bond angle). On the other hand, reaction of $PbCl_2$ with Mes*MgBr resulted in the unexpected formation of heteroleptic aryl(alkyl)plumbylene **39**, believed to be produced through the isomerization of the initially formed homoleptic diarylplumbylene Mes*$_2$Pb: **40**.[186] Heteroleptic plumbylene **39** revealed a more NMR-shielded and less crowded situation around the carbene center (compared with that of diarylplumbylene **38**): 5067 ppm (^{207}Pb NMR) and 94.8° (C–Pb–C bond angle). Attempts to prepare heteroleptic plumbylene by the disproportionation of diarylplumbylene (2-*t*-Bu-4,5,6-$Me_3C_6H)_2$Pb: **38** and disilylplumbylene [($Me_3Si)_3Si]_2$Pb: **26** resulted in the formation of the desired aryl(silyl)plumbylene (2-*t*-Bu-4,5,6-Me_3C_6H)[($Me_3Si)_3Si$]Pb: **41** [7545 ppm (^{207}Pb NMR), 610 nm (UV-Vis)]. In the solid state, however, **41** exists as a plumbylene dimer {(2-*t*-Bu-4,5,6-Me_3C_6H)[($Me_3Si)_3Si$]Pb}$_2$ **42** with a short lead–lead separation of 3.3695(11) Å (other related plumbylene dimers were reported by the end of the 1990s[187]).

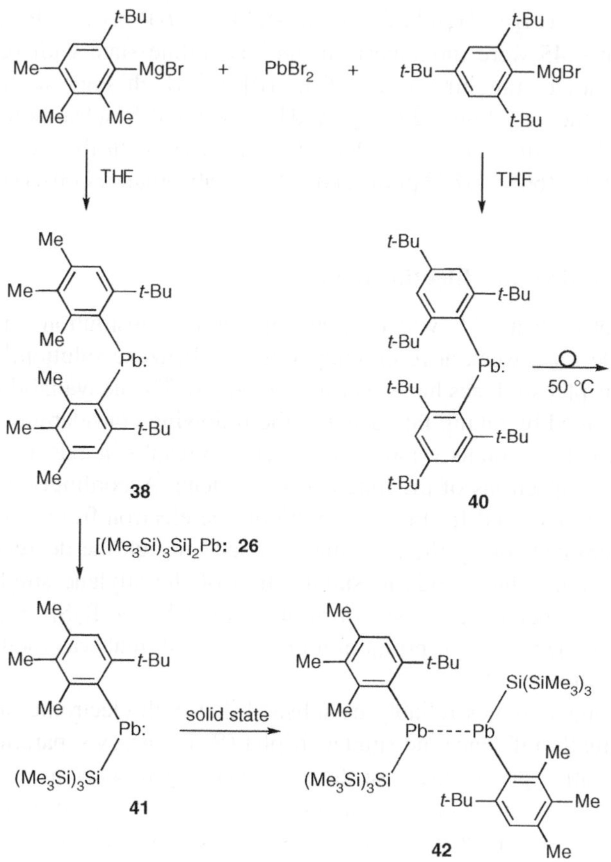

Scheme 4.71

Tokitoh *et al.* reported the isolation of the plumbylene Tbt_2Pb: **43**, in which the lead carbene center was effectively protected by extremely bulky Tbt groups.[188] Diarylplumbylene **43** was prepared by two consecutive additions of equivalent amounts of TbtLi to the stable diaminoplumbylene $[(Me_3Si)_2N]_2Pb$: in Et_2O. Similarly to other kinetically stabilized monomeric plumbylenes, **43** exhibited an extraordinarily low-field resonance of its plumbylene center (9751 ppm in the ^{207}Pb NMR spectrum) and long wavelength absorption (610 nm in the UV-Vis spectrum). This was corroborated by the crystallographic data of **43**: the shortest Pb···Pb nonbonding intermolecular distance of 11.60 Å suggested its monomeric status in the crystal, and its very wide C–Pb–C bond angle of 116.3° indicated a huge steric repulsion between the bulky Tbt groups.

Heteroleptic aryl(silyl)plumbylene $(2,6\text{-}Mes_2\text{-}C_6H_3)[(Me_3Si)_3Si]Pb$: **44** was available by the ligand exchange reaction of bis(silyl)plumbylene $[(Me_3Si)_3Si]_2Pb$: **26**[180] and arylcopper derivative $(2,6\text{-}Mes_2\text{-}C_6H_3)Cu$ performed in benzene.[189] Being sufficiently kinetically stabilized, plumbylene **44** was monomeric both in the solid state (C–Pb–C bond angle = 109.2°) and in solution (10510 ppm in the ^{207}Pb NMR spectrum).

A series of mixed alkyl(aryl)plumbylenes $R(2,6\text{-}Tip_2\text{-}C_6H_3)Pb$: (R = Me, *t*-Bu, Ph) **45** was prepared by Power *et al.* by the reaction of the readily available pyridine complex of bromo(aryl)plumbylene $Br(2,6\text{-}Tip_2\text{-}C_6H_3)Pb$:(Py) with MeMgBr, *t*-BuLi and PhLi, respectively.[190] All plumbylenes **45** were monomeric in the crystalline state with relatively narrow C–Pb–C bond angles ranging from 95.6 to 101.4°, which suggests the different steric requirements of the very bulky $2,6\text{-}Tip_2\text{-}C_6H_3$ group and R substituents. The two-coordinate status of **45** in the condensed phase was manifested in their diagnostic low-field ^{207}Pb NMR shifts (6657–7853 ppm) and UV-Vis absorbances observed at 460–570 nm.

4.6.2 Triplet Species: Generation and Identification

In contrast to persistent triplet carbenes,[191] which upon appropriate substitution with bulky aryl groups can be stable for a week at room temperature in benzene solution,[192] neither isolable nor persistent triplet silylenes have ever been reported. The unavailability of triplet silylenes can be explained by taking into account the following considerations. The larger size of the silicon valence orbitals versus those of carbon results in a decrease in the repulsion between the two electrons of the lone pair of silylene. Accordingly, the singlet–triplet conversion energy required for the promotion of one electron from a 3*n*- to 3p-orbital cannot be compensated for by the advantage gained through the decrease in the electron–electron repulsion, which leads to stabilization of the silylene singlet ground state. Consequently, the general idea of stabilization of the silylene R_2Si: in its triplet ground state was to minimize the *n*–p promotion energy; a task that was totally governed by the effects of substituents R.

As the starting point, theoretical studies reliably established that both electronic and steric effects of substituents greatly influence the singlet–triplet (S–T) energy separation in silylenes.[1a] Thus, electronegative groups increase the S–T energy gap, whereas electropositive substituents reduce this gap; π-donors stabilize singlets while π-acceptors favor triplets. On the other hand, the singlet state is progressively destabilized relative to the triplet state upon widening the R–Si–R bond angle: the singlet and triplet potential energy curves finally cross and the stability order is reversed in favor of the triplet state.

In their remarkable contribution, Grev *et al.* suggested that the combination of both the electronic factors of electropositive substituents and the steric effects of bulky groups to widen the R–Si–R angle should be employed for the successful design of triplet ground-state silylenes.[193] In another landmark paper, Holthausen *et al.* elaborated on which bulky electropositive substituents are the best choices for making triplet silylenes synthetically accessible.[194] Employing DFT computations at the BLYP/DZVP-ECP level, they suggested the following candidates for triplet silylenes [ΔE_{S-T} in kcal/mol, R–Si–R bond angles (singlet/triplet) in degrees]: $(i\text{-}Pr_3Si)_2Si$: (1.4–1.7, 119.1/137.2), $(i\text{-}Pr(t\text{-}Bu)_2Si)_2Si$: (2.4, 121.0/142.2), $(t\text{-}Bu_3Si)_2Si$: (7.1, 130.9/147.5). Given the insignificant preference for the triplet state over the singlet state in $(i\text{-}Pr_3Si)_2Si$: and $(i\text{-}Pr(t\text{-}Bu)_2Si)_2Si$: silylenes, it was not possible to draw an unequivocal conclusion regarding their multiplicity, taking into account the observed uncertainty of computations of up to −3 kcal/mol. However, the $(t\text{-}Bu_3Si)_2Si$: silylene with the bulkiest $t\text{-}Bu_3Si$ substituents was reliably established to be a ground-state triplet, even considering the largest possible deviation of −3 kcal/mol.

The computational recommendations were verified by the subsequent experiments performed mainly in Gaspar's group. However, despite theoretical predictions, the addition of silylene $(i\text{-}Pr_3Si)_2Si$:, generated both thermally and photochemically from $(i\text{-}Pr_3Si)_3SiH$, to *cis*- and *trans*-but-2-enes was stereospecific, indicative of its singlet, rather than triplet, ground state.[195] Further increasing the steric bulk around the silylene center, Gaspar *et al.* generated $(t\text{-}Bu_3Si)(i\text{-}Pr_3Si)Si$: from the 3-phenyl-1-tri-*tert*-butylsilyl-1-triisopropylsilyl-1-silacyclopent-3-ene precursor **46** (Scheme 4.72).[196] In the absence of trapping reagents, photolysis of **46** resulted in the isolation of 1,2-disilacyclobutane **47** formally formed by the intramolecular insertion of the silylene center of $(t\text{-}Bu_3Si)(i\text{-}Pr_3Si)Si$: into the C–H bond of a *t*-Bu group. Room temperature formation of **47** was taken as indirect evidence that $(t\text{-}Bu_3Si)(i\text{-}Pr_3Si)Si$: reacts from its triplet ground state, because analogous intramolecular C–H insertion of singlet silylenes was predicted to require ca. 20 kcal/mol activation energy, which is only achievable at high temperatures.

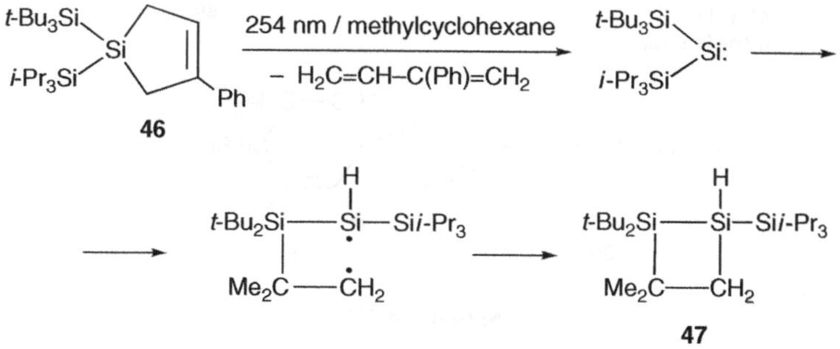

Scheme 4.72

However, the classification of $(t\text{-}Bu_3Si)(i\text{-}Pr_3Si)Si$: as a ground-state triplet was still ambiguous because no EPR signal of this silylene in a glass matrix can be observed. Moreover, recent elaborate calculations at the BLYP/DZVP and BLYP/TZVP levels showed that the previous calculations[194] systematically overestimated the stability of triplets, and accordingly only the $(t\text{-Bu}_3\text{Si})_2\text{Si}$: silylene was found to possess a triplet ground state with $\Delta E_{S-T} = 4.5$ kcal/mol, while all other sterically encumbered bis(silyl)silylenes [including silylene $(t\text{-Bu}_3\text{Si})(i\text{-Pr}_3\text{Si})\text{Si}$:] adopt a singlet ground state or feature nearly isoenergetic singlet/triplet ground states.[197] Given the closeness of the singlet and triplet energy levels for silylene $(t\text{-Bu}_3\text{Si})(i\text{-Pr}_3\text{Si})\text{Si}$:, one can imagine a facile switch between these two electronic states, which could be responsible for the observed reactivity of this silylene formally corresponding to the typical reactions from the triplet ground state.

The theoretical predictions on the triplet ground state of $(t\text{-Bu}_3\text{Si})_2\text{Si}$: silylene[194,197] were finally confirmed by the direct EPR observation of this silylene generated by the 254 nm photolysis of the silirene precursors in a methylcyclohexane matrix at 77 K.[198] A weak and broad resonance observed at 845 mT was attributed to $(t\text{-Bu}_3\text{Si})_2\text{Si}$: in its triplet ground state.

The Si–Si–Si bond angle in silylene $(t\text{-Bu}_3\text{Si})_2\text{Si}$: was estimated from the experimental EPR data to be ca. 125°, for which value a ground-state triplet is theoretically predicted.[194] Dependence of the EPR signal intensity as a function of temperature was linear over the whole temperature range studied (9–80 K), which was taken as further evidence of the triplet ground state of $(t\text{-Bu}_3\text{Si})_2\text{Si}$:. This was also corroborated by DFT calculations at the B3LYP/6-31G(d) level, which found a triplet state of $(t\text{-Bu}_3\text{Si})_2\text{Si}$: lying 5.6 kcal/mol lower than its singlet. Three products were isolated after annealing the methylcyclohexane glass matrix after the photolysis of **48a:** 1,2-disilacyclobutane **49** (35%), dihydrosilane $(t\text{-Bu}_3\text{Si})_2\text{SiH}_2$ **50** (4%) and 1,4-disilacyclohex-2-ene **51** (44%) (Scheme 4.73).

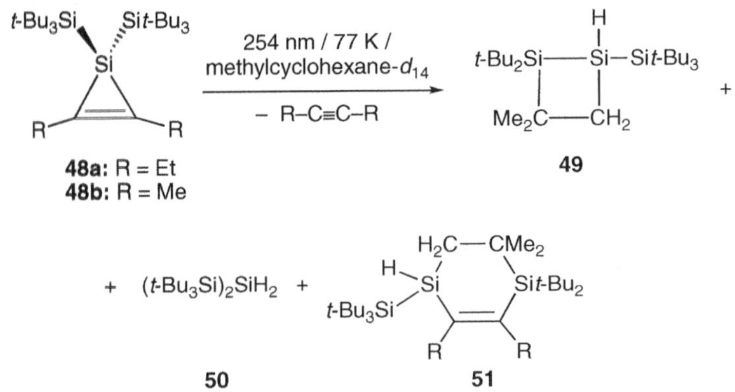

Scheme 4.73

Formation of the first two products **49** and **50** was attributed to the reactivity of $(t\text{-Bu}_3\text{Si})_2\text{Si}$: from its triplet ground state by the intramolecular insertion of the silylene center into the C–H bond of one of the t-Bu groups and by abstraction of two hydrogen atoms, respectively (Scheme 4.74). Six-membered ring product **51** might be formed starting with the photochemical ring-opening isomerization of silirene **48** to a transient

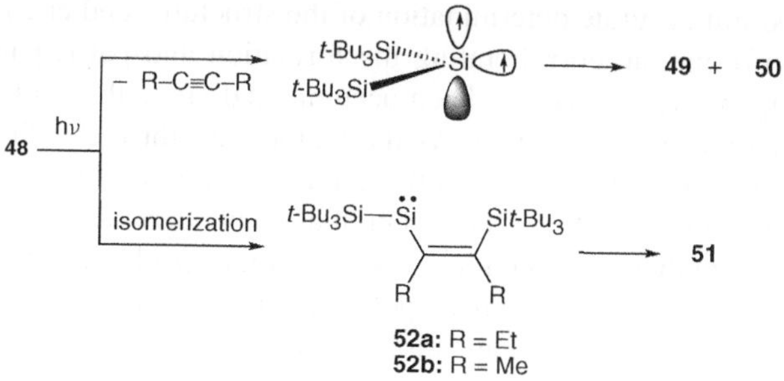

Scheme 4.74

silylene **52** through t-Bu_3Si group migration, followed by intramolecular insertion of the silylene center in **52** into the C–H bond of a t-Bu group.

Likewise, triplet ground-state alkali metal-substituted silylenes were generated very recently, confirming the theoretical prediction[199] that the presence of extremely electropositive substituents, such as Li, will lead to triplet silylenes with no need for the expansion of the R–Si–R bond angle at the divalent silicon because of bulky substituents R. Lithium- and potassium-substituted silylenes M(t-Bu_3Si)Si: (M = Li, K) were produced from the corresponding alkali metal-substituted silirenes **53** (**a:** M = Li; **b:** M = K) by their 254 nm photolysis in 2-Me-THF glass matrix at 14 K (Scheme 4.75).[200]

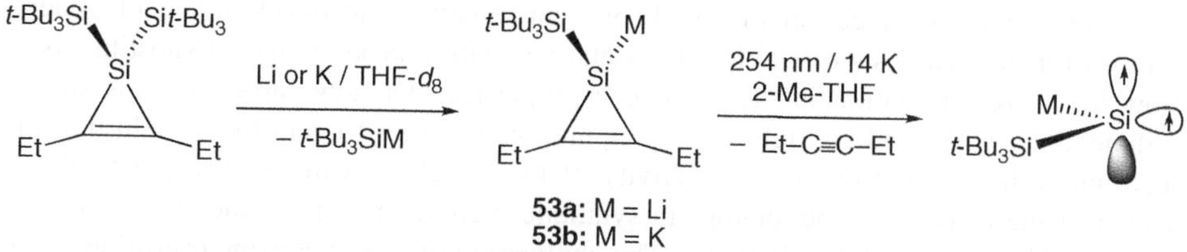

Scheme 4.75

The EPR resonances of the triplet silylenes M(t-Bu_3Si)Si: were observed at 790 mT (for M = Li) and 780 mT (for M = K). The triplet ground state of Li(t-Bu_3Si)Si: was confirmed by the temperature dependence of its 790 mT EPR signal intensity, which strictly followed Curie's law, IT = constant (I, EPR signal intensity; T, absolute temperature). This conclusion was supported by DFT calculations at the (U)B3LYP/6-31G(d) level, which revealed a preference for the triplet ground state of Li(t-Bu_3Si)Si: over the singlet state: $\Delta E_{S-T} = 9.4$ kcal/mol.

4.7 Summary and Outlook

Over the past several decades, the chemistry of carbenes as reactive intermediates has been greatly developed, mostly because of the advances in very fast spectroscopic techniques (nanosecond and picosecond laser flash photolysis) and modern computational methods, which permit accurate determination of the structures and energies of carbenes. A great deal of information, corroborating the formation and participation of carbenes in a number of organic reactions, has been accumulated since the start of active investigations more than half a century ago. As the major breakthrough achieved during the recent period, stabilization of carbenes to the extent of allowing their generation as persistent species (triplet carbenes) or even isolation as stable derivatives (singlet carbenes), is to be noted. The synthetic availability of stable singlet carbenes, which exhibit dual nucleophilic–electrophilic behavior, made possible intensive study of their characteristic reactions: dimerization to form alkenes, insertion, cycloadditions (of which cyclopropanation is the most fundamental), formation of ylides with Lewis bases or Lewis acids, and complexation with transition metals. The chemistry of the heavy analogs of carbenes, that is silylenes, germylenes, stannylenes and plumbylenes, also experienced a progressive evolution: from transient phantom species, neither detectable nor isolable, to fully characterizable indefinitely stable materials. As shown in this chapter, the electronic structure, multiplicity and philicity of the heavy carbene analogs are totally governed by the influence of substituents. Overall, these effects determine such very fundamental features of heavy carbenes as their stability (both thermodynamic and kinetic) and reactivity. Milestone discoveries of the first stable carbene analogs of the heavy group 14 elements opened up an unprecedented possibility for direct studies of their structural and chemical properties. One should particularly emphasize the practical importance of readily available and widely employed N-heterocyclic heavy carbenes, especially in the field of transition metal complexes. The story of the heavy analogs of carbenes is, however, very far from being complete; there are still many challenges to be overcome. As one of the major advances in the field, one can imagine the development of general methods for the synthesis of stable representatives with a broad range of substituents to allow for a great variation in the electronic properties of heavy carbenes. This should be done along with modern high-level computational studies to gain full insight into the electronic structure and chemical reactivity of the novel derivatives. The practical utilization of the heavy carbene analogs in synthetic organic and organometallic chemistry is still awaiting its broad development. The application of heavy carbene transition metal complexes in various catalytic transformations, involving organometallic compounds and featuring the participation of heavy carbene ligands, seems to be particularly attractive and promising.

4.8 References

1. Reviews on silylenes, germylenes, stannylenes and plumbylenes: (a) P. P. Gaspar, R. West, in *The Chemistry of Organic Silicon Compounds, Vol. 2* (Eds Z. Rappoport, Y. Apeloig), John Wiley & Sons, Ltd, Chichester, **1998**, Chapter 43; (b) N. Tokitoh, W. Ando, in *Reactive Intermediate Chemistry* (Eds R. A. Moss, M. S. Platz, M. Jones Jr), Wiley, Hoboken, **2004**, Chapter 14; (c) W. P. Neumann, *Chem. Rev.* **1991**, *91*, 311; (d) J. M. Jasinski, R. Becerra, R. Walsh, *Chem. Rev.* **1995**, *95*, 120613; (e) R. Becerra, R. Walsh, in *Research in Chemical Kinetics, Vol. 3* (Eds R. G. Compton, G. Hancock), Elsevier, Amsterdam, **1995**, p. 263; (f) S. E. Boganov, M. P. Egorov, V. I. Faustov, I. V. Krylova, O. M. Nefedov,

R. Becerra, R. Walsh, *Russ. Chem. Bull., Int. Ed.* **2005**, *54*, 483; (g) R. Becerra, R. Walsh, *PhysChemChemPhys* **2007**, *9*, 2817; (h) I. Safarik, V. Sandhu, E. M. Lown, O. P. Strausz, T. N. Bell, *Res. Chem. Intermed.* **1990**, *14*, 105; (i) S. E. Boganov, M. P. Egorov, V. I. Faustov, O. M. Nefedov, in *The Chemistry of Organic Germanium, Tin and Lead Compounds, Vol. 2* (Ed. Z. Rappoport), John Wiley & Sons, Ltd, Chichester, **2002**, Chapter 12: (j) Y. Apeloig, M. Karni, T. Müller, in *Organosilicon Chemistry II* (Eds N. Auner, J. Weiss), *VCH, Weinheim*, **1996**, p. 263; (k) Y. Mizuhata, T. Sasamori, N. Tokitoh, *Chem. Rev.* **2009**, *109*, 3479.

2. P. Trefonas III, R. West, R. D. Miller, *J. Am. Chem. Soc.* **1985**, *107*, 2737.
3. T. Karatsu, R. D. Miller, R. Sooriyakumaran, J. Michl, *J. Am. Chem. Soc.* **1989**, *111*, 1140.
4. M. Kira, T. Miyazawa, S. Koshihara, Y. Segawa, H. Sakurai, *Chem. Lett.* **1995**, 3.
5. K. Oka, R. Nakao, *Res. Chem. Intermed.* **1990**, *13*, 143.
6. M. Kira, T. Maruyama, H. Sakurai, *Chem. Lett.* **1993**, 1345.
7. S. G. Bott, P. Marshall, P. E. Wagenseller, Y. Wang, R. T. Conlin, *J. Organomet. Chem.* **1995**, *499*, 11.
8. S. Collins, S. Murakami, J. T. Snow, S. Masamune, *Tetrahedron Lett.* **1985**, *26*, 1281.
9. (a) W. Ando, T. Tsumuraya, A. Sekiguchi, *Chem. Lett.* **1987**, 317; (b) W. Ando, H. Itoh, T. Tsumuraya, *Organometallics* **1989**, *8*, 2759; (c) W. Ando, H. Itoh, T. Tsumuraya, H. Yoshida, *Organometallics* **1988**, *7*, 1880.
10. (a) M. Wakasa, I. Yoneda, K. Mochida, *J. Organomet. Chem.* **1989**, *366*, C1; (b) K. Mochida, I. Yoneda, M. Wakasa, *J. Organomet. Chem.* **1990**, *399*, 53.
11. S. Konieczny, S. J. Jacobs, J. K. Braddock-Wilking, P. P. Gaspar, *J. Organomet. Chem.* **1988**, *341*, C17.
12. P. Rivière, A. Castel, J. Satgé, D. Guyot, *J. Organomet. Chem.* **1986**, *315*, 157.
13. K. L. Bobbitt, V. M. Maloney, P. P. Gaspar, *Organometallics* **1991**, *10*, 2772.
14. R. Becerra, S. E. Boganov, M. P. Egorov, V. Ya. Lee, O. M. Nefedov, R. Walsh, *Chem. Phys. Lett.* **1996**, *250*, 111.
15. R. Becerra, M. P. Egorov, I. V. Krylova, O. M. Nefedov, R. Walsh, *Chem. Phys. Lett.* **2002**, *351*, 47.
16. (a) K. Saito, K. Obi, *Chem. Phys. Lett.* **1993**, *215*, 193; (b) K. Obi, M. Fukushima, K. Saito, *Appl. Surf. Sci.* **1994**, *79/80*, 465; (c) K. Saito, K. Obi, *Chem. Phys.* **1994**, *187*, 381; (d) R. Becerra, S. E. Boganov, M. P. Egorov, O. M. Nefedov, R. Walsh, *Chem. Phys. Lett.* **1996**, *260*, 433; (e) U. N. Alexander, N. A. Trout, K. D. King, W. D. Lawrance, *Chem. Phys. Lett.* **1999**, *299*, 291.
17. T. J. Drahnak, J. Michl, R. West, *J. Am. Chem. Soc.* **1979**, *101*, 5427.
18. K. M. Welsh, J. Michl, R. West, *J. Am. Chem. Soc.* **1988**, *110*, 6689.
19. (a) J. Barrau, D. L. Bean, K. M. Welsh, R. West, J. Michl, *Organometallics* **1989**, *8*, 2606; (b) J. Barrau, V. Balaji, J. Michl, *Organometallics* **1989**, *8*, 2034.
20. R. Becerra, S. E. Boganov, M. P. Egorov, V. I. Faustov, I. V. Krylova, O. M. Nefedov, R. Walsh, *J. Am. Chem. Soc.* **2002**, *124*, 7555.
21. P. Bleckmann, H. Maly, R. Minkwitz, W. P. Neumann, B. Watta, G. Olbrich, *Tetrahedron Lett.* **1982**, *23*, 4655.

22. (a) M. Ishikawa, M. Kumada, *J. Chem. Soc., Chem. Commun.* **1970**, 612; (b) M. Ishikawa, M. Kumada, *J. Organomet. Chem.* **1972**, *42*, 325; (c) M. Kumada, *J. Organomet. Chem.* **1975**, *100*, 127.
23. H. Vancik, G. Raabe, M. J. Michalczyk, R. West, J. Michl, *J. Am. Chem. Soc.* **1985**, *107*, 4097.
24. C. W. Carlson, K. J. Haller, X.-H. Zhang, R. West, *J. Am. Chem. Soc.* **1984**, *106*, 5521.
25. V. N. Khabashesku, V. Balaji, S. E. Boganov, O. M. Nefedov, J. Michl, *J. Am. Chem. Soc.* **1994**, *116*, 320.
26. (a) M. Kira, K. Sakamoto, H. Sakurai, *J. Am. Chem. Soc.* **1983**, *105*, 7469; (b) H. Sakurai, K. Sakamoto, M. Kira, *Chem. Lett.* **1984**, 1379.
27. H. Watanabe, E. Tabei, M. Goto, Y. Nagai, *J. Chem. Soc., Chem. Commun.* **1987**, 522.
28. H. Shizuka, K. Murata, Y. Arai, K. Tonokura, H. Tanaka, H. Matsumoto, Y. Nagai, G. Gillette, R. West, *J. Chem. Soc., Faraday Trans. 1* **1989**, *85*, 2369.
29. (a) A. Schäfer, M. Weidenbruch, S. Pohl, *J. Organomet. Chem.* **1985**, *282*, 305; (b) M. Weidenbruch, F. T. Grimm, M. Herrndorf, A. Schäfer, K. Peters, H. G. von Schnering, *J. Organomet. Chem.* **1988**, *341*, 335.
30. P. Boudjouk, U. Samaraweera, R. Sooriyakumaran, J. Chrusciel, K. R. Anderson, *Angew. Chem., Int. Ed. Engl.* **1988**, *27*, 1355.
31. D. H. Pae, M. Xiao, M. Y. Chiang, P. P. Gaspar, *J. Am. Chem. Soc.* **1991**, *113*, 1281.
32. (a) W. Ando, T. Tsumuraya, *Organometallics* **1988**, *7*, 1882; (b) T. Tsumuraya, S. Sato, W. Ando, *Organometallics* **1988**, *7*, 2015; (c) W. Ando, T. Tsumuraya, *Tetrahedron Lett.* **1986**, *27*, 3251; (d) W. Ando, T. Tsumuraya, *J. Chem. Soc., Chem. Commun.* **1987**, 1514; (e) T. Tsumuraya, S. Sato, W. Ando, *Organometallics* **1989**, *8*, 161; (f) T. Tsumuraya, S. Sato, W. Ando, *Organometallics* **1990**, *9*, 2061; (g) Y. Kabe, W. Ando, *J. Organomet. Chem.* **1994**, *482*, 131.
33. M. Ishikawa, T. Takaoka, M. Kumada, *J. Organomet. Chem.* **1972**, *42*, 333.
34. M. Ishikawa, T. Fuchikami, M. Kumada, *J. Am. Chem. Soc.* **1977**, *99*, 245.
35. H. Sakurai, Y. Kamiyama, Y. Nakadaira, *J. Am. Chem. Soc.* **1977**, *99*, 3879.
36. (a) S. Masamune, Y. Hanzawa, S. Murakami, T. Bally, J. F. Blount, *J. Am. Chem. Soc.* **1982**, *104*, 1150; (b) M. Masamune, Y. Hanzawa, D. J. Williams, *J. Am. Chem. Soc.* **1982**, *104*, 6136; (c) S. Masamune, S. Murakami, J. T. Snow, H. Tobita, D. J. Williams, *Organometallics* **1984**, *3*, 333; (d) J. T. Snow, S. Murakami, S. Masamune, D. J. Williams, *Tetrahedron Lett.* **1984**, *25*, 4191; (e) S. Collins, S. Murakami, J. T. Snow, S. Masamune, *Tetrahedron Lett.* **1985**, *26*, 1281.
37. P. P. Gaspar, A. M. Beatty, T. Chen, T. Haile, D. Lei, W. R. Winchester, J. Braddock-Wilking, N. R. Rath, W. T. Klooster, T. F. Koetzle, S. A. Mason, A. Albinati, *Organometallics* **1999**, *18*, 3921.
38. (a) J. Fu, W. P. Neumann, *J. Organomet. Chem.* **1984**, *272*, C5; (b) W. P. Neumann, J. Fu, *J. Organomet. Chem.* **1984**, *273*, 295; (c) S. Masamune, L. R. Sita, D. J. Williams, *J. Am. Chem. Soc.* **1983**, *105*, 630; (d) S. Masamune, L. R. Sita, *J. Am. Chem. Soc.* **1985**, *107*, 6390.

39. S. P. Kolesnikov, M. P. Egorov, A. M. Galminas, M. B. Ezhova, O. M. Nefedov, T. V. Leshina, M. B. Taraban, A. I. Kruppa, V. I. Maryasova, *J. Organomet. Chem.* **1990**, *391*, C1.
40. C. Grugel, W. P. Neumann, M. Schriewer, *Angew. Chem., Int. Ed. Engl.* **1979**, *18*, 543.
41. M. G. Steinmetz, C. Yu, *Organometallics* **1992**, *11*, 2686.
42. (a) W. H. Atwell, D. R. Weyenberg, *J. Organomet. Chem.* **1966**, *5*, 594; (b) W. H. Atwell, D. R. Weyenberg, *Angew. Chem., Int. Ed. Engl.* **1969**, *8*, 469.
43. J. Heinicke, B. Gehrhus, S. Meinel, *J. Organomet. Chem.* **1994**, *474*, 71.
44. J. Heinicke, B. Gehrhus, *J. Organomet. Chem.* **1992**, *423*, 13.
45. E. C. L. Ma, D. P. Paquin, P. P. Gaspar, *J. Chem. Soc.,Chem. Commun.* **1980**, 381.
46. B. Watta, W. P. Neumann, J. Sauer, *Organometallics* **1985**, *4*, 1954.
47. J. J. O'Brien, G. H. Atkinson, *J. Phys. Chem.* **1988**, *92*, 5782.
48. H. Gilman, S. G. Cottis, W. H. Atwell, *J. Am. Chem. Soc.* **1964**, *86*, 1596.
49. (a) W. P. Neumann, M. Schriewer, *Tetrahedron Lett.* **1980**, *21*, 3273; (b) M. Schriewer, W. P. Neumann, *Angew. Chem., Int. Ed. Engl.* **1981**, *20*, 1019; (c) M. Schriewer, W. P. Neumann, *J. Am. Chem. Soc.* **1983**, *105*, 897.
50. (a) D. Seyferth, D. C. Annarelli, S. C. Vick, *J. Am. Chem. Soc.* **1976**, *98*, 6382; (b) D. Seyferth, S. C. Vick, *J. Organomet. Chem.* **1977**, *125*, C11.
51. (a) E. A. Chernyshev, N. G. Komalenkova, S. A. Bashkirova, *J. Organomet. Chem.* **1984**, *271*, 129; (b) D. Lei, P. P. Gaspar, *Organometallics* **1985**, *4*, 1471; (c) P. P. Gaspar, D. Lei, *Organometallics* **1986**, *5*, 1276; (d) D. Lei, R.-J. Hwang, P. P. Gaspar, *J. Organomet. Chem.* **1984**, *271*, 1; (e) M. G. Steinmetz, C. Yu, *Organometallics* **1992**, *11*, 2686.
52. S. P. Kolesnikov, I. S. Rogozhin, M. G. Safarov, O. M. Nefedov, *Izv. Akad. Nauk SSSR, Ser. Khim.* **1981**, 1423.
53. K. Tamao, A. Kawachi, *Adv. Organomet. Chem.* **1995**, *38*, 1.
54. K. Tamao, A. Kawachi, *Organometallics* **1995**, *14*, 3108.
55. R. Corriu, G. Lanneau, C. Priou, F. Soulairol, N. Auner, R. Probst, R. Conlin, C. Tan, *J. Organomet. Chem.* **1994**, *466*, 55.
56. J. Grobe, T. Schierholt, in *Organosilicon Chemistry II* (Eds N. Auner, J. Weis), VCH, Weinheim, **1996**, pp. 317–319.
57. P. Jiang, P. P. Gaspar, *J. Am. Chem. Soc.* **2001**, *123*, 8622.
58. P. P. Gaspar, M. Xiao, D. H. Pae, D. J. Berger, T. Haile, T. Chen, D. Lei, W. R. Winchester, P. Jiang, *J. Organomet. Chem.* **2002**, *646*, 68.
59. H. Ohgaki, N. Fukaya, W. Ando, *Organometallics* **1997**, *16*, 4956.
60. (a) R. T. Conlin, Y.-W. Kwak, *J. Am. Chem. Soc.* **1986**, *108*, 834; (b) I. M. T. Davidson, S. Ijadi-Maghsoodi, T. J. Barton, N. Tillman, *J. Chem. Soc., Chem. Commun.* **1984**, 478.
61. D. B. Puranik, M. J. Fink, *J. Am. Chem. Soc.* **1989**, *111*, 5951.
62. B.-H. Boo, P. P. Gaspar, *Organometallics* **1986**, *5*, 698.
63. G. R. Gillette, G. H. Noren, R. West, *Organometallics* **1987**, *6*, 2617.
64. W. Ando, A. Sekiguchi, K. Hagiwara, A. Sakakibara, H. Yoshida, *Organometallics* **1988**, *7*, 558.

65. G. R. Gillette, G. Noren, R. West, *Organometallics* **1990**, *9*, 2925.
66. H. Shizuka, H. Tanaka, K. Tonokura, K. Murata, H. Hiratsuka, J. Ohshita, M. Ishikawa, *Chem. Phys. Lett.* **1988**, *143*, 225.
67. (a) G. Levin, P. K. Das, C. Bilgrien, C. L. Lee, *Organometallics* **1989**, *8*, 1206; (b) G. Levin, P. K. Das, C. L. Lee, *Organometallics* **1988**, *7*, 1231.
68. G. Inoue, M. Suzuki, *Chem. Phys. Lett.* **1984**, *105*, 641.
69. C. Yamada, H. Kanamori, E. Hirota, N. Nishiwaki, N. Itabashi, K. Kato, T. Goto, *J. Chem. Phys.* **1989**, *91*, 4582.
70. R. M. Robertson, M. J. Rossi, *J. Chem. Phys.* **1989**, *91*, 5037.
71. T. C. McInnis, L. Andrews, *J. Phys. Chem.* **1992**, *96*, 5276.
72. H. Bock, M. Kremer, M. Dolg, H.-W. Preuss, *Angew. Chem., Int. Ed. Engl.* **1991**, *30*, 1186.
73. G. Raabe, H. Vancik, R. West, J. Michl, *J. Am. Chem. Soc.* **1986**, *108*, 671.
74. D. Bourissou, O. Guerret, F. P. Gabbaï, G. Bertrand, *Chem. Rev.* **2000**, *100*, 39.
75. (a) R. Hoffmann, *J. Am. Chem. Soc.* **1968**, *90*, 1475; (b) R. Hoffmann, G. D. Zeiss, G. W. van Dine, *J. Am. Chem. Soc.* **1968**, *90*, 1485; (c) R. Gleiter, R. Hoffmann, *J. Am. Chem. Soc.* **1968**, *90*, 5457.
76. (a) M. Karni, Y. Apeloig, J. Kapp, P. von R. Schleyer, in *The Chemistry of Organic Silicon Compounds, Vol. 3* (Eds Z. Rappoport, Y. Apeloig), John Wiley & Sons, Ltd, Chichester, **2001**, Chapter 1 (and references therein); (b) Y. Apeloig, R. Pauncz, M. Karni, R. West, W. Steiner, D. Chapman, *Organometallics* **2003**, *22*, 3250 (and references therein).
77. J. Berkowitz, J. P. Greene, H. Cho, R. Ruscic, *J. Chem. Phys.* **1987**, *86*, 1235.
78. (a) K. Balasubramanian, A. D. McLean, *J. Chem. Phys.* **1986**, *85*, 5117; (b) K. Balasubramanian, *J. Chem. Phys.* **1988**, *89*, 5731.
79. (a) D. Dai, M. M. Al-Zahrani, K. Balasubramanian, *J. Phys. Chem.* **1994**, *98*, 9233; (b) M. Benavides-Garcia, K. Balasubramanian, *J. Chem. Phys.* **1992**, *97*, 7537; (c) M. Benavides-Garcia, K. Balasubramanian, *J. Chem. Phys.* **1994**, *100*, 2821.
80. B. T. Luke, J. A. Pople, M. B. Krogh-Jespersen, Y. Apeloig, J. Chandrasekhar, P. von R. Schleyer, *J. Am. Chem. Soc.* **1986**, *108*, 260.
81. Y. Apeloig, M. Karni, R. West, K. Welsh, *J. Am. Chem. Soc.* **1994**, *116*, 9719.
82. Y. Apeloig, M. Karni, *J. Chem. Soc., Chem. Commun.* **1985**, 1048.
83. M. J. Michalczyk, M. J. Fink, D. J. DeYoung, C. W. Carlson, K. M. Welsh, R. West, J. Michl, *Silicon, Germanium, Tin Lead Compd.* **1986**, *9*, 75.
84. G. R. Gillette, G. H. Noren, R. West, *Organometallics* **1989**, *8*, 487.
85. H. Sakurai, T. Kobayashi, Y. Nakadaira, *J. Organomet. Chem.* **1978**, *162*, C43.
86. T. Y. Gu, W. P. Weber, *J. Am. Chem. Soc.* **1980**, *102*, 1641.
87. H. S. D. Soysa, H. Okinoshima, W. P. Weber, *J. Organomet. Chem.* **1977**, *133*, C17.
88. T. Y. Gu, W. P. Weber, *J. Organomet. Chem.* **1980**, *184*, 7.
89. P. Zhu, M. Piserchio, F. W. Lampe, *J. Phys. Chem.* **1985**, *89*, 5344.
90. D. H. Berry, Q. Jiang, *J. Am Chem. Soc.* **1987**, *109*, 6210.
91. S. Konieczny, K. Wrzesien, W. Wojnowski, *J. Organomet. Chem.* **1993**, *446*, 73.
92. (a) P. L. Timms, *Chem. Soc. Rev.* **1996**, *25*, 93; (b) S. R. Church, C. G. Davies, R. Lümen, P. A. Mounier, G. Saint, P. L. Timms, *J. Chem. Soc., Dalton Trans.* **1996**, 227.

93. P. Bleckmann, R. Minkwitz, W. P. Neumann, M. Schriewer, M. Thibud, B. Watta, *Tetrahedron Lett.* **1984**, *25*, 2467.
94. (a) J. Barrau, M. El Amine, G. Rima, J. Satgé, *J. Organomet. Chem.* **1984**, *277*, 323; (b) J. Köcher, M. Lehnig, W. P. Neumann, *Organometallics* **1988**, *7*, 1201; (c) E. J. Bulten, J. G. Noltes, *J. Organomet. Chem.* **1969**, *16*, P8; (d) A. Castel, J. Escudié, P. Rivière, J. Satgé, M. N. Bochkarev, L. P. Maiorova, G. A. Razuvaev, *J. Organomet. Chem.* **1981**, *210*, 37.
95. L. Lange, B. Meyer, W.W. du Mont, *J. Organomet. Chem.* **1987**, *329*, C17.
96. J. Barrau, M. Bouchaut, H. Lavayssiere, G. Dousse, J. Satgé, *J. Organomet. Chem.* **1983**, *243*, 281.
97. J. Barrau, J. Satgé, *J. Organomet. Chem.* **1978**, *148*, C9.
98. M. F. Lappert, S. J. Miles, J. L. Atwood, M. J. Zaworotko, A. J. Carty, *J. Organomet. Chem.* **1981**, *212*, C4.
99. S. P. Kolesnikov, M. P. Egorov, A. S. Dvornikov, V. A. Kuz'min, O. M. Nefedov, *Metallorg. Khim.* **1989**, *2*, 799.
100. M.-D. Su, S.-Y. Chu, *J. Phys. Chem. A* **1999**, *103*, 11011.
101. J. Köcher, M. Lehnig, *Organometallics* **1984**, *3*, 937.
102. G. Billeb, H. Brauer, S. Maslov, W. P. Neumann, *J. Organomet. Chem.* **1989**, *373*, 11.
103. L. R. Sita, I. Kinoshita, S. P. Lee, *Organometallics* **1990**, *9*, 1644.
104. L. W. Gross, R. Moser, W. P. Neumann, K. H. Scherping, *Tetrahedron Lett.* **1982**, *23*, 635.
105. W. P. Neumann, A. Schwarz, *Angew. Chem., Int. Ed. Engl.* **1975**, *14*, 812.
106. (a) P. S. Skell, R. C. Woodworth, *J. Am. Chem. Soc.* **1956**, *78*, 4496; (b) P. S. Skell, A. Y. Garner, *J. Am. Chem. Soc.* **1956**, *78*, 5430.
107. (a) R. L. Lambert, D. Seyferth, *J. Am. Chem. Soc.* **1972**, *94*, 9246; (b) D. Seyferth, D. C. Annarelli, *J. Am. Chem. Soc.* **1975**, *97*, 2273; (c) D. Seyferth, *J. Organomet. Chem.* **1975**, *100*, 237.
108. (a) W. Ando, H. Saso, *Tetrahedron Lett.* **1986**, *27*, 5625; (b) H. Saso, W. Ando, *Chem. Lett.* **1988**, 1567; (c) T. Yamamoto, Y. Kabe, W. Ando, *Organometallics* **1993**, *12*, 1996; (d) W. Ando, T. Yamamoto, H. Saso, Y. Kabe, *J. Am. Chem. Soc.* **1991**, *113*, 2791.
109. (a) T. Akasaka, W. Ando, K. Kobayashi, S. Nagase, *J. Am. Chem. Soc.* **1993**, *115*, 1605; (b) T. Akasaka, E. Mitsuhida, W. Ando, K. Kobayashi, S. Nagase, *J. Chem. Soc., Chem. Commun.* **1995**, 1529.
110. P. Boudjouk, E. Black, R. Kumarathasan, *Organometallics* **1991**, *10*, 2095.
111. M. Weidenbruch, E. Kroke, H. Marsmann, S. Pohl, W. Saak, *J. Chem. Soc., Chem. Commun.* **1994**, 1233.
112. J. Köcher, W. P. Neumann, *Organometallics* **1985**, *4*, 400.
113. W. Ando, H. Ohgaki, Y. Kabe, *Angew. Chem., Int. Ed. Engl.* **1994**, *33*, 659.
114. H. Ohgaki, Y. Kabe, W. Ando, *Organometallics* **1995**, *14*, 2139.
115. R. T. Conlin, P. P. Gaspar, *J. Am. Chem. Soc.* **1976**, *98*, 3715.
116. F. Hojo, S. Sekigawa, N. Nakayama, T. Shimizu, W. Ando, *Organometallics* **1993**, *12*, 803.
117. (a) D. Ostendorf, L. Kirmaier, W. Saak, H. Marsmann, M. Weidenbruch, *Eur. J. Inorg. Chem.* **1999**, 2301; (b) L. Kirmaier, M. Weidenbruch, H. Marsmann, K.

Peters, H. G. von Schnering, *Organometallics* **1998**, *17*, 1237; (c) F. Meiners, W. Saak, M. Weidenbruch, *Organometallics* **2000**, *19*, 2835.
118. W. Ando, T. Shiba, T. Hidaka, K. Morihashi, O. Kikuchi, *J. Am. Chem. Soc.* **1997**, *119*, 3629.
119. A. Schäfer, M. Weidenbruch, W. Saak, S. Pohl, *Angew. Chem., Int. Ed. Engl.* **1987**, *26*, 776.
120. A. H. Cowley, S. W. Hall, C. M. Nunn, J. M. Power, *J. Chem. Soc., Chem. Commun.* **1988**, 753.
121. (a) A. Krebs, J. Berndt, *Tetrahedron Lett.* **1983**, *24*, 4083; (b) M. P. Egorov, S. P. Kolesnikov, Yu. T. Struchkov, M. Yu. Antipin, S. V. Sereda, O. M. Nefedov, *J. Organomet. Chem.* **1985**, *290*, C27; (c) M. P. Egorov, M. B. Ezhova, S. P. Kolesnikov, O. M. Nefedov, M. B. Taraban, A. I. Kruppa, T. V. Leshina, *Mendeleev Commun.* **1991**, *1*, 143.
122. L. R. Sita, R. D. Bickerstaff, *J. Am. Chem. Soc.* **1988**, *110*, 5208.
123. (a) T. J. Barton, S. A. Burns, P. P. Gaspar, Y. S. Chen, *Synth. React. Inorg. Met.-Org. Chem.* **1983**, *13*, 881; (b) D. Lei, P. P. Gaspar, *Res. Chem. Intermed.* **1989**, *12*, 103; (c) S. Zhang, R. T. Conlin, *J. Am. Chem. Soc.* **1991**, *113*, 4272; (d) K. L. Bobbitt, P. P. Gaspar, *J. Organomet. Chem.* **1995**, *499*, 17.
124. G. Mayer, H. P. Reisenauer, K. Schöttler, U. Wessolek-Kraus, *J. Organomet. Chem.* **1989**, *366*, 25.
125. W. P. Neumann, E. Michels, J. Köcher, *Tetrahedron Lett.* **1987**, *28*, 3783.
126. J. Köcher, W. P. Neumann, *J. Am. Chem. Soc.* **1984**, *106*, 3861.
127. R. Marx, W. P. Neumann, K. Hillner, *Tetrahedron Lett.* **1984**, *25*, 625.
128. R. T. Conlin, P. P. Gaspar, *J. Am. Chem. Soc.* **1976**, *98*, 868.
129. Y. Nakadaira, T. Kobayashi, T. Otsuka, H. Sakurai, *J. Am. Chem. Soc.* **1979**, *101*, 486.
130. R. West, M. J. Fink, J. Michl, *Science* **1981**, *214*, 1343.
131. M. Yamaji, K. Hamanishi, T. Takahashi, H. Shizuka, *J. Photochem. Photobiol. A* **1994**, *81*, 1.
132. R. T. Conlin, J. C. Netto-Ferreira, S. Zhang, J. C. Scaiano, *Organometallics* **1990**, *9*, 1332.
133. M. E. Lee, M. A. North, P. P. Gaspar, *Phosphorus SulfurSilicon* **1991**, *56*, 203.
134. G. Maier, G. Mihm, H. P. Reisenauer, D. L. Littman, *Chem. Ber.* **1984**, *117*, 2369.
135. (a) G. Maier, J. Glatthaar, in *Organosilicon Chemistry* (Eds N. Auner, J. Weis), VCH, Weinheim, **1994**, p. 131; (b) G. Maier, J. Glatthaar, *Angew. Chem., Int. Ed. Engl.* **1994**, *33*, 473.
136. D. Lei, Y.-S. Chen, B. H. Boo, J. Frueh, D. L. Svoboda, P. P. Gaspar, *Organometallics* **1992**, *11*, 559.
137. V. N. Khabashesku, V. Balaji, S. E. Boganov, S. A. Bashkirova, P. M. Matveichev, E. A. Chernyshev, O. M. Nefedov, J. Michl, *Mendeleev Commun.* **1992**, 38.
138. E. O. Fischer, A. Maasboel, *Angew. Chem., Int. Ed. Engl.* **1964**, *3*, 580.
139. T. J. Marks, *J. Am. Chem. Soc.* **1971**, *93*, 7090.
140. W. Petz, *Chem. Rev.* **1986**, *86*, 1019.
141. H. Ogino, H. Tobita, *Adv. Organomet. Chem.* **1998**, *42*, 223.
142. R. Waterman, P. G. Hayes, T. D. Tilley, *Acc. Chem. Res.* **2007**, *40*, 712.
143. C. Zybill, G. Müller, *Angew. Chem., Int. Ed. Engl.* **1987**, *26*, 669.

144. D. A. Straus, T. D. Tilley, A. L. Rheingold, S. J. Geib, *J. Am. Chem. Soc.* **1987**, *109*, 5872. Several other isolable transition metal complexes of the heavy group 14 element carbene analogs were prepared by Tilley and co-workers in due course. See, for example: P. G. Hayes, C. Beddie, M. B. Hall, R. Waterman, T. D. Tilley, *J. Am. Chem. Soc.* **2006**, *128*, 428 (osmium silylene complex Os=Si); and P. G. Hayes, C. W. Gribble, R. Waterman, T. D. Tilley, *J. Am. Chem. Soc.* **2009**, *131*, 4606 (osmium stannylene complex Os=Sn).
145. K. Ueno, H. Tobita, M. Shimoi, H. Ogino, *J. Am. Chem. Soc.* **1988**, *110*, 4092.
146. D. A. Straus, S. D. Grumbine, T. D. Tilley, *J. Am. Chem. Soc.* **1990**, *112*, 7801.
147. N. J. Hill, R. West, *J. Organomet. Chem.* **2004**, *689*, 4165.
148. (a) J. D. Feldman, G. P. Mitchell, J.-O. Nolte, T. D. Tilley, *J. Am. Chem. Soc.* **1998**, *120*, 11184; (b) J. D. Feldman, G. P. Mitchell, J.-O. Nolte, T. D. Tilley, *Can. J. Chem.* **2003**, *81*, 1127.
149. G. P. Mitchell, T. D. Tilley, *Angew. Chem., Int. Ed.* **1998**, *37*, 2524.
150. K. Takanashi, V. Ya. Lee, T. Yokoyama, A. Sekiguchi, *J. Am. Chem. Soc.* **2009**, *131*, 916.
151. N. Nakata, T. Fujita, A. Sekiguchi, *J. Am. Chem. Soc.* **2006**, *128*, 16024.
152. (a) G. Wilkinson, F. G. A. Stone, E. W. Abel (Eds), *Comprehensive Organometallic Chemistry, Vol. 2,* Pergamon Press, New York, 1982 [Chapter 10 (Germanium) by P. Riviere, M. Riviere-Baudet, J. Satgé; Chapter 11 (Tin) by A. G. Davies, P. J. Smith; Chapter 12 (Lead) by P. G. Harrison]; (b) V. Ya. Lee, A. A. Basova, I. A. Matchkarovskaya, V. I. Faustov, M. P. Egorov, O. M. Nefedov, R. D. Rakhimov, K. P. Butin, *J. Organomet. Chem.* **1995**, *499*, 27.
153. (a) P. Jutzi, in *The Chemistry of Organic Silicon Compounds, Vol. 2, Part 3* (Eds Z. Rappoport, Y. Apeloig), John Wiley & Sons, Ltd, Chichester, **1998**, Chapter 36; (b) T. Kühler, P. Jutzi, *Adv. Organomet. Chem.* **2003**, *49*, 1.
154. (a) J. Barrau, G. Rima, *Coord. Chem. Rev.* **1998**, 178–180, 593; (b) O. Kühl, *Coord. Chem. Rev.* **2004**, *248*, 411; (c) A. V. Zabula, F. E. Hahn, *Eur. J. Inorg. Chem.* **2008**, 5165; (d) S. Nagendran, H. W. Roesky, *Organometallics* **2008**, *27*, 457.
155. (a) R. West, M. Denk, *Pure Appl. Chem.* **1996**, *68*, 785; (b) M. Haaf, T. A. Schmedake, R. West, *Acc. Chem. Res.* **2000**, *33*, 704; (c) B. Gehrhus, M. F. Lappert, *J. Organomet. Chem.* **2001**, *617–618*, 209.
156. M. Kira, S. Ishida, T. Iwamoto, C. Kabuto, *J. Am. Chem. Soc.* **1999**, *121*, 9722.
157. M. Kira, T. Iwamoto, S. Ishida, *Bull. Chem. Soc. Jpn.* **2007**, *80*, 258.
158. S. Ishida, T. Iwamoto, C. Kabuto, M. Kira, *Chem. Lett.* **2001**, 1102.
159. T. Iwamoto, H. Masuda, S. Ishida, C. Kabuto, M. Kira, *J. Am. Chem. Soc.* **2003**, *125*, 9300.
160. S. Ishida, T. Iwamoto, M. Kira, *J. Am. Chem. Soc.* **2003**, *125*, 3212.
161. (a) M. Kira, S. Ishida, T. Iwamoto, C. Kabuto, *J. Am. Chem. Soc.* **2002**, *124*, 3830; (b) M. Kira, S. Ishida, T. Iwamoto, A. de Meijere, M. Fujitsuka, O. Ito, *Angew. Chem., Int. Ed.* **2004**, *43*, 4510.
162. T. Iwamoto, K. Sato, S. Ishida, C. Kabuto, M. Kira, *J. Am. Chem. Soc.* **2006**, *128*, 16914.
163. T. Abe, T. Iwamoto, C. Kabuto, M. Kira, *J. Am. Chem. Soc.* **2006**, *128*, 4228.

164. (a) S. Ishida, T. Iwamoto, C. Kabuto, M. Kira, *Nature* **2003**, *421*, 725; (b) T. Iwamoto, H. Masuda, C. Kabuto, M. Kira, *Organometallics* **2005**, *24*, 197; (c) T. Iwamoto, T. Abe, C. Kabuto, M. Kira, *Chem. Commun.* **2005**, 5190.
165. (a) C. Watanabe, T. Iwamoto, C. Kabuto, M. Kira, *Chem. Lett.* **2007**, *36*, 284; (b) C. Watanabe, T. Iwamoto, C. Kabuto, M. Kira, *Angew. Chem. Int. Ed.* **2008**, *47*, 5386.
166. (a) N. Tokitoh, H. Suzuki, R. Okazaki, K. Ogawa, *J. Am. Chem. Soc.* **1993**, *115*, 10428; (b) H. Suzuki, N. Tokitoh, R. Okazaki, J. Harada, K. Ogawa, S. Tomoda, M. Goto, *Organometallics* **1995**, *14*, 1016; (c) H. Suzuki, N. Tokitoh, R. Okazaki, *Bull. Chem. Soc. Jpn.* **1995**, *68*, 2471; (d) H. Suzuki, N. Tokitoh, R. Okazaki, *J. Am. Chem. Soc.* **1994**, *116*, 11572.
167. N. Tokitoh and R. Okazaki, *Coord. Chem. Rev.* **2000**, *210*, 251.
168. (a) N. Tokitoh, H. Suzuki, R. Okazaki, *Chem. Commun.* **1996**, 125; (b) N. Takeda, H. Suzuki, N. Tokitoh, R. Okazaki, S. Nagase, *J. Am. Chem. Soc.* **1997**, *119*, 1456.
169. (a) P. J. Davidson, D. H. Harris, M. F. Lappert, *J. Chem. Soc., Dalton Trans.* **1976**, 2268; (b) T. Fjeldberg, A. Haaland, B. E. R. Schilling, M. F. Lappert, A. J. Thorne, *J. Chem. Soc., Dalton Trans.* **1986**, 1551.
170. P. Jutzi, A. Becker, H. G. Stammler, B. Neumann, *Organometallics* **1991**, *10*, 1647.
171. M. Kira, S. Ishida, T. Iwamoto, M. Ichinohe, C. Kabuto, L. Ignatovich, H. Sakurai, *Chem. Lett.* **1999**, 263.
172. P. Jutzi, H. Schmidt, B. Neumann, H.-G. Stammler, *Organometallics* **1996**, *15*, 741.
173. R. S. Simons, L. Pu, M. M. Olmstead, P. P. Power, *Organometallics* **1997**, *16*, 1920.
174. W. Setaka, K. Sakamoto, M. Kira, P. P. Power, *Organometallics* **2001**, *20*, 4460.
175. G. L. Wegner, R. J. F. Berger, A. Schier, H. Schmidbaur, *Organometallics* **2001**, *20*, 418.
176. (a) N. Tokitoh, T. Matsumoto, K. Manmaru, R. Okazaki, *J. Am. Chem. Soc.* **1993**, *115*, 8855; (b) N. Tokitoh, K. Manmaru, R. Okazaki, *Organometallics* **1994**, *13*, 167; (c) N. Tokitoh, K. Kishikawa, T. Matsumoto, R. Okazaki, *Chem. Lett.* **1995**, 827.
177. K. Kishikawa, N. Tokitoh, R. Okazaki, *Chem. Lett.* **1998**, 239.
178. M. Kira, R. Yauchibara, R. Hirano, C. Kabuto, H. Sakurai, *J. Am. Chem. Soc.* **1991**, *113*, 7785.
179. C. Eaborn, M. S. Hill, P. B. Hitchcock, D. Patel, J. D. Smith, S. Zhang, *Organometallics* **2000**, *19*, 49. Closely related seven-membered ring dialkylstannylene stabilized by the B–H···Sn agostic interactions was very recently reported: K. Izod, C. Wills, W. Clegg, R. W. Harrington, *Organometallics* **2009**, *28*, 2211.
180. (a) K. W. Klinkhammer, W. Schwarz, *Angew. Chem., Int. Ed. Engl.* **1995**, *34*, 1334; (b) K. Klinkhammer, *Polyhedron* **2002**, *21*, 587.
181. B. Gehrhus, P. B. Hitchcock, M. F. Lappert, *Angew. Chem., Int. Ed. Engl.* **1997**, *36*, 2514.
182. M. Weidenbruch, J. Schlaefke, A. Schäfer, K. Peters, H. G. von Schnering, H. Marsmann, *Angew. Chem., Int. Ed. Engl.* **1994**, *33*, 1846.
183. B. E. Eichler, P. P. Power, *Inorg. Chem.* **2000**, *39*, 5444.

184. W. Setaka, K. Hirai, H. Tomioka, K. Sakamoto, M. Kira, *Chem. Commun.* **2008**, 6558.
185. C. Eaborn, T. Ganicz, P. B. Hitchcock, J. D. Smith, S. E. Sözerli, *Organometallics* **1997**, *16*, 5621. Closely related seven-membered ring dialkylplumbylene stabilized by the B–H···Pb agostic interactions was very recently reported: K. Izod, C. Wills, W. Clegg, R. W. Harrington, *Organometallics* **2009**, *28*, 2211.
186. M. Stürmann, M. Weidenbruch, K. W. Klinkhammer, F. Lissner, H. Marsmann, *Organometallics* **1998**, *17*, 4425.
187. (a) K. W. Klinkhammer, T. F. Fässler, H. Grützmacher, *Angew. Chem., Int. Ed.* **1998**, *37*, 124; (b) M. Stürmann, W. Saak, M. Weidenbruch, K. W. Klinkhammer, *Eur. J. Inorg. Chem.* **1999**, 579; (c) M. Stürmann, W. Saak, H. Marsmann, M. Weidenbruch, *Angew. Chem., Int. Ed.* **1999**, *38*, 187.
188. N. Kano, K. Shibata, N. Tokitoh, R. Okazaki, *Organometallics* **1999**, *18*, 2999.
189. J. Klett, K. W. Klinkhammer, M. Niemeyer, *Chem. Eur. J.* **1999**, *5*, 2531.
190. L. Pu, B. Twamley, P. P. Power, *Organometallics* **2000**, *19*, 2874.
191. (a) H. Tomioka, *Acc. Chem. Res.* **1997**, *30*, 315; (b) H. Tomioka, in *Reactive Intermediate Chemistry* (Eds R. A. Moss, M. S. Platz, M. Jones Jr), Wiley, Hoboken, **2004**, Chapter 9; (c) T. Itoh, K. Hirai, H. Tomioka, *Bull. Chem. Soc. Jpn.* **2007**, *80*, 138.
192. E. Iwamoto, K. Hirai, H. Tomioka, *J. Am. Chem. Soc.* **2003**, *125*, 14664.
193. R. S. Grev, H. F. Schaefer III, P. P. Gaspar, *J. Am. Chem. Soc.* **1991**, *113*, 5638.
194. M. C. Holthausen, W. Koch, Y. Apeloig, *J. Am. Chem. Soc.* **1999**, *121*, 2623.
195. P. P. Gaspar, A. M. Beatty, T. Chen, T. Haile, D. Lei, W. R. Winchester, J. Braddock-Wilking, N. P. Rath, W. T. Klooster, T. F. Koetzle, S. A. Mason, A. Albinati, *Organometallics* **1999**, *18*, 3921.
196. (a) P. Jiang, P. P. Gaspar, *J. Am. Chem. Soc.* **2001**, *123*, 8622; (b) P. P. Gaspar, M. Xiao, D. H. Pae, D. J. Berger, T. Haile, T. Chen, D. Lei, W. R. Winchester, P. Jiang, *J. Organomet. Chem.* **2002**, *646*, 68.
197. M. Yoshida, N. Tamaoki, *Organometallics* **2002**, *21*, 2587.
198. A. Sekiguchi, T. Tanaka, M. Ichinohe, K. Akiyama, S. Tero-Kubota, *J. Am. Chem. Soc.* **2003**, *125*, 4962.
199. (a) J. F. Harrison, R. C. Liedtke, J. F. Liebman, *J. Am. Chem. Soc.* **1979**, *101*, 7162; (b) M. E. Colvin, J. Breulet, H. F. Schaefer III, *Tetrahedron* **1985**, *41*, 1429; (c) K. Krogh-Jespersen, *J. Am. Chem. Soc.* **1985**, *107*, 537.
200. A. Sekiguchi, T. Tanaka, M. Ichinohe, K. Akiyama, P. P. Gaspar, *J. Am. Chem. Soc.* **2008**, *130*, 426.

5

Heavy Analogs of Alkenes, 1,3-Dienes, Allenes and Alkynes: Multiply Bonded Derivatives of Si, Ge, Sn and Pb

5.1 Introduction

Unsaturated hydrocarbons, alkenes, alkynes and dienes are among the most fundamental classes of organic compounds. The great variety and richness of organic chemistry are to a large extent caused by the disposition of carbon to form highly reactive carbon–carbon and carbon–heteroatom multiple bonds. Standard textbooks on organic chemistry describe the C=C double bond as one made of a σ-bond, formed by the linear end-on overlap of sp^2-hybrid orbitals of each carbon partner (sp^2–sp^2 interaction), and a π-bond, formed by the side-on overlap of unhybridized 2p-orbitals ($2p_\pi$–$2p_\pi$ interaction). Analogously, the carbon–carbon triple bond consists of a σ-bond, formed by the linear end-on overlap of the two sp-hybrids (sp–sp interaction), and two mutually orthogonal π-bonds, each formed by the side-on overlap of unhybridized 2p-orbitals ($2p_\pi$–$2p_\pi$ interaction). Because the s-character of the hybrid orbitals used for the formation of carbon–carbon bonds progressively increases from the single (sp^3–sp^3) to the double (sp^2–sp^2) and triple (sp–sp) bonds, C=C double bonds are shorter (and stronger) than C–C single bonds, and C≡C triple bonds are shorter (and stronger) than C=C double bonds. In accord with their hybridization types, the geometry of the carbon–carbon double bond in alkenes $R_2C{=}CR_2$ is planar with an idealized R–C=C bond angle of 120°, and the geometry of the carbon–carbon triple bond in alkynes R–C≡C–R is linear with an idealized R–C≡C bond angle of 180°.

Organometallic Compounds of Low-Coordinate Si, Ge, Sn and Pb: From Phantom Species to Stable Compounds
Vladimir Ya. Lee and Akira Sekiguchi

The analogs of alkenes $R_2E{=}ER_2$ and alkynes $RE{\equiv}ER$ of the heavy group 14 elements (E = Si–Pb) were originally thought to be synthetically inaccessible, because of the intrinsic incapability of the heavy main group elements to form multiple bonds, the so-called 'double bond rule': elements with the principal quantum number equal to or greater than three cannot form multiple bonds because of the significant Pauli repulsion of the electrons in the inner shells. Nevertheless, the milestone syntheses of the first isolable distannene by Lappert, disilene by West, and silene by Brook, as well as digermyne, distannyne and diplumbyne by Power, and disilyne by Sekiguchi overcame this prejudice, establishing a new fundamental class of highly reactive organometallics, stable alkene and alkyne analogs of the heavy group 14 elements. Despite the formal structural similarity with organic alkenes and alkynes, their heavy analogs revealed distinctly different properties: geometry, hybridization, and E–E bonding nature. The synthetic methods leading to the stable Si, Ge, Sn, and Pb analogs of alkenes, 1,3-dienes, allenes and alkynes, as well as their structural and bonding features, are discussed below.

5.2 Early Studies: Generation and Identification

For a long time, the heavy homonuclear analogs of alkenes of the type >E=E< (E = Si–Pb) have been speculated to participate in a number of organometallic transformations as reactive intermediates. Thus, the story of the transient disilenes >Si=Si< began with the famous claim of Kipping and Sands of the preparation of tetraphenyldisilene $Ph_2Si{=}SiPh_2$ by the simple reductive Wurtz coupling of Ph_2SiCl_2 with sodium.[1] However, subsequent studies showed that the product was not disilene but its dimer, octaphenylcyclotetrasilane *cyclo*-(Si_4Ph_8).[1]

The first convincing evidence for the intermediacy of disilenes was provided in 1972 in the classical experiment by Roark and Peddle, who generated the transient tetramethyldisilene $Me_2Si{=}SiMe_2$ by thermolysis of its 7,8-disilabicyclo[2.2.2]octa-2,5-diene precursor and successfully trapped it with anthracene forming the corresponding Diels–Alder adduct.[2]

Since then, the chemistry of transient disilenes has been remarkably developed by the efforts of many research groups, with the most important contributions made by Barton, Gaspar, Sakurai, Masamune *et al.* Such progress was repeatedly reviewed covering all aspects of the field: generation (both thermal and pho tochemical), identification (matrix-isolation technique in either argon matrix or rigid organic glass), and chemical trapping with a variety of trapping reagents.[3]

Accordingly, in the field of the heavier analogs of disilenes, one should recognize the pioneering work on digermenes >Ge=Ge< by Satgé *et al.*, who generated transient species by the photolysis of either poly(digermylmercury) compounds or cyclopolygermanes and trapped them with appropriate reagents, such as diacetyl, dimethyldisulfide and 2,3-dimethylbuta-1,3-diene.[3f,4]

Consideration of the transient alkene analogs of the heavy group 14 elements is beyond the scope of the present chapter, in which we specifically deal with their stable representatives. However, given the historical importance of the pioneering early studies, we refer interested readers to the comprehensive reviews on the chemistry of transient disilenes and digermenes.[3,4]

5.3 Stable Derivatives (Synthesis and Structure)

5.3.1 Heavy Analogs of Alkenes (including cyclic compounds)

5.3.1.1 Homonuclear Compounds

The structures and bonding of the alkene analogs of heavy group 14 elements, both in the crystalline form and in solution, are sharply distinctive from those of their organic counterparts (alkenes); therefore, before proceeding to the discussion on each class of heavy alkenes we will briefly comment on the specific structural features of the title compounds.

Structure and bonding

In contrast to organic alkenes, which are typically planar and feature diagnostically short double bonds (av. 1.34 Å for >C=C< vs av. 1.54 Å for >C–C<), their analogs of the heavy group 14 elements >E=E< (E = Si–Pb) revealed diverse types of structural modes sharply distinctive from those of their organic counterparts. Thus, apart from the stretching of the E=E bond descending group 14, there are two fundamental structural deformations characteristic of heavy alkenes >E=E< and leading to their departure from planarity: *trans*-bending at E centers and twisting about the E=E bond. The peculiar *trans*-bending geometry, resulting in the overall pyramidalization at the heavy group 14 elements, was rationalized in the framework of two different MO models providing alternative explanations for this phenomenon.[5]

The first qualitative model considered formation of the double bond as a result of the interaction of two monomeric carbene units, taking into account the different ground-state multiplicities of carbenes and their heavy analogs. Thus, interaction of the two monomeric carbenes, which typically have triplet ground states (or alternatively, singlet ground states with very low-lying triplets), results in a pair of covalent interactions between the singly occupied MO forming classical planar D_{2h} C=C double bonds (Figure 5.1, **A**). In contrast to carbenes, their heavy analogs (silylenes, germylenes, stannylenes and plumbylenes) are ground-state singlets (see Chapter 4); accordingly, their interaction in the manner described above for triplet carbenes would result in significant Pauli repulsion between the doubly occupied n-orbitals (Figure 5.1, **B**). Therefore, initial excitation to the triplet state (that is, the promotion of one electron from n- to p-level) is required to form a classical E=E bond, for which, however, the considerable energetic cost of the singlet–triplet energy gap (ΔE_{ST}) must be paid. Instead, the two heavy carbene fragments prefer to interact with each other in such a way that the electron pair of the n-orbital of one fragment donates its electron density into the vacant p-orbital of the other fragment (two equivalent donor–acceptor interactions), resulting in the formation of a nonclassical C_{2h} E=E bond manifesting *trans*-bent geometry (and pyramidalization) at the doubly bonded E centers (Figure 5.1, **C**).[5] One should recognize the pioneering contribution of Lappert's group to the field, who first proposed this donor–acceptor model to explain the bonding pattern in their remarkable *trans*-bent distannene $Dis_2Sn{=}SnDis_2$ (see Section 5.3.1.1.3) as early as 1976.[6]

Alternatively, the *trans*-bent geometry of heavy alkenes was rationalized as the result of a stabilizing mixing of E=E π- and σ^*-orbitals, which becomes possible upon bending at the E centers: the stronger the π–σ^* interaction, the larger the *trans*-bending (on

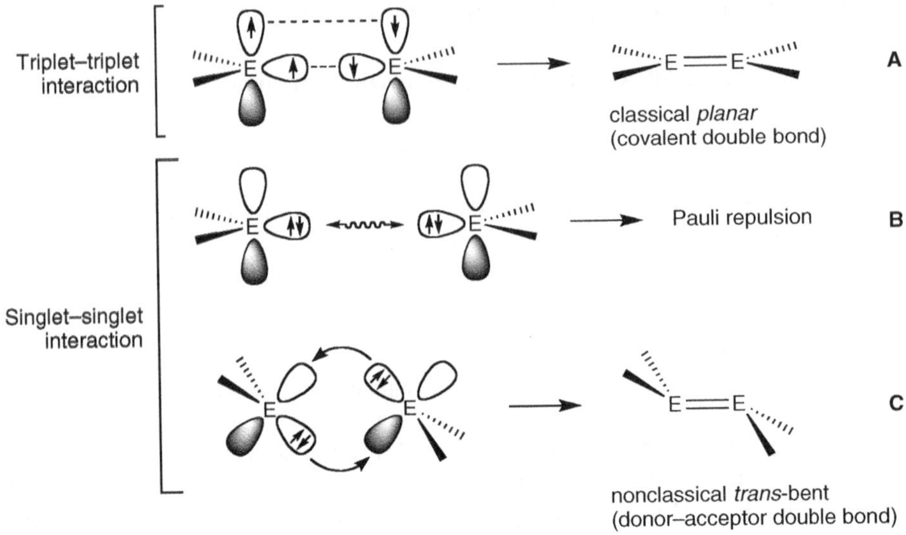

Figure 5.1 *Qualitative MO models representing the triplet–triplet interaction of the two monomeric carbenes forming a planar covalent C=C bond (**A**: E = C) and the singlet–singlet interaction of two monomeric carbene analogues of the heavy group 14 elements forming a trans-bent donor-acceptor E=E bond (**B** and **C**: E = Si–Pb)*

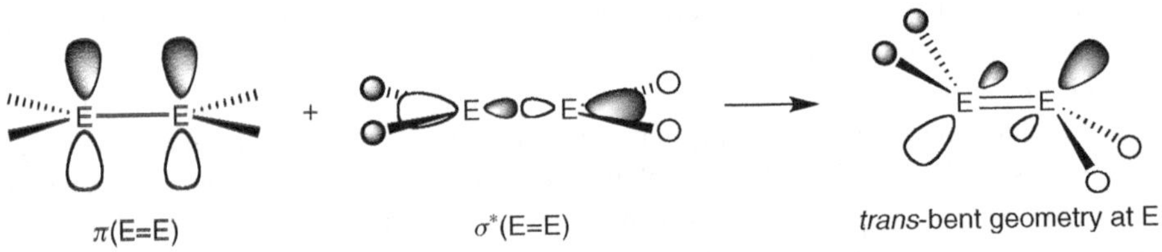

Figure 5.2 *Stabilizing π–σ^*(E=E) orbital mixing resulting in the trans-bending at the E centers*

the other hand, an alternative σ–π^* interaction was calculated to contribute much less to the overall *trans*-bending) (Figure 5.2). Such π–σ^* mixing, which appeared to be a manifestation of the second-order Jahn–Teller effect, depends on the π–σ^* energy separation and results in the lowering of the π-orbital energy level, and consequently, in its stabilization.[5]

The magnitude of the π–σ^* energy gap in the planar $H_2E{=}EH_2$ (E = C–Pb) significantly decreases on going down group 14, which leads to stronger π–σ^* mixing and larger bending at the heavier E. The electronegativity of the substituents in the heavy alkene analogs $R_2E{=}ER_2$ also has a profound effect on the degree of *trans*-bending. Thus, electron-accepting substituents R (F, OH, NH_2), which increase ΔE_{ST} and the degree of the π–σ^*-orbital interaction, cause remarkable *trans*-bending at E and weakening of the E=E bond. In contrast, electron-donating groups R (BH_2, SiH_3) reduce the ΔE_{ST} gap, producing much smaller bending deformations at E [in the limiting case of persilyl-substituted derivatives $(H_3Si)_2E{=}E(SiH_3)_2$ a planar geometry can be achieved] and strengthening of the E=E bond. In general, the manifestation of the

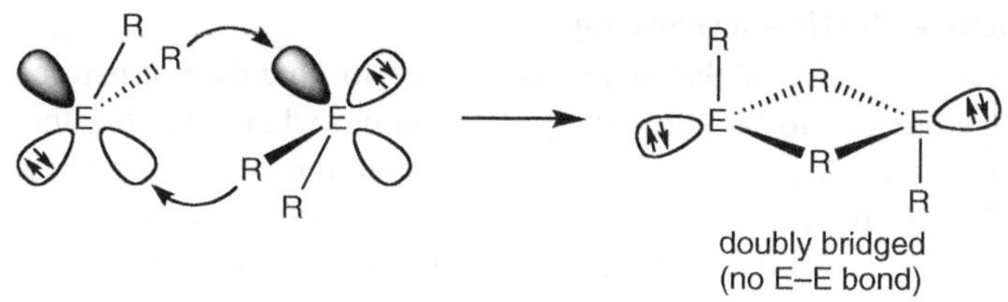

Figure 5.3 *Energetic preference for the doubly bridged isomer with no direct E–E bond over the trans-bent doubly bonded structure* ($\Delta E_{ST} > \frac{1}{2}E_{\sigma+\pi}$)

lone pair character at the formally doubly bonded element becomes more and more pronounced descending group 14. This can be realized taking into account the inherent reluctance of the heavy elements to hybridize because of the remarkably different radii of their ns- and np- ($n = 3$–6) orbitals, which leads to the gradual expulsion of s-electrons from bonding and a sharp decrease in the strength of the E=E double bond.[5]

The widely accepted so-called CGMT (Carter–Goddard–Malrieu–Trinquier) model[7] provided simple but clear quantitative estimates of the limits for stability of different E=E bonding modes: classical planar, nonclassical *trans*-bent, and nonclassical doubly bridged. Thus, a planar E=E bond is expected when its formation gains more energy (defined as overall bond energy $E_{\sigma+\pi}$) than that required for the singlet–triplet promotion of each heavy carbene partner: $\Delta E_{ST} < \frac{1}{4}E_{\sigma+\pi}$ (Figure 5.1, **A**, $H_2C{=}CH_2$). If the energy relation is reversed as $\Delta E_{ST} > \frac{1}{4}E_{\sigma+\pi}$, then a double donor–acceptor (dative) interaction is anticipated, resulting in the formation of a *trans*-bent structure (Figure 5.1, **C**, $H_2E{=}EH_2$ where E = Si, Ge, Sn). Overall, the general energetic requirement for the existence of the *trans*-bent form is: $\frac{1}{4}E_{\sigma+\pi} < \Delta E_{ST} < \frac{1}{2}E_{\sigma+\pi}$, whereas if $\Delta E_{ST} > \frac{1}{2}E_{\sigma+\pi}$ then direct E–E bonding is not favored and the doubly bridged isomer is predicted to be more stable than the *trans*-bent doubly bonded structure (Figure 5.3, $H_2Pb{=}PbH_2$).

In the case of stable derivatives, the above-discussed structural deformations, namely, *trans*-bending at E and twisting around the E=E bond, can be experimentally visualized in the crystal structures determined by X-ray diffraction analysis. Thus, the *trans*-bending angles α/α' are typically defined as the angles between the R–E–R/R′–E–R′ mean planes and the E–E bond axis (Figure 5.4, **A**), whereas the twisting angle β is defined as the dihedral angle between the R–E–R and R′–E–R′ mean planes (Figure 5.4, **B**).

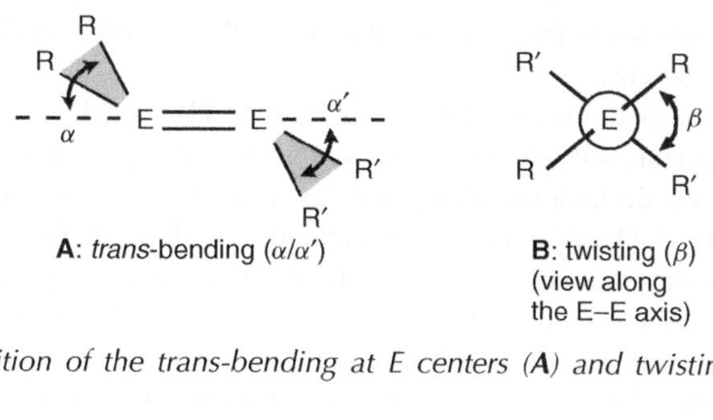

Figure 5.4 *Definition of the trans-bending at E centers* (***A***) *and twisting around the E=E bond* (***B***)

Solution structure: NMR spectroscopy

NMR spectroscopy is one of the major tools used to elucidate structures in solution. This method has proved to be particularly informative when studying the composition of the heavy alkene analogs $R_2E{=}ER_2$ (E = Si–Pb) in the condensed phase: ^{29}Si, ^{119}Sn and ^{207}Pb NMR spectroscopy were successfully applied for structural investigations of disilenes, distannenes and diplumbenes, respectively. The only exception is the ^{73}Ge nucleus, which, although magnetically active, has a rather large quadrupole moment and low sensitivity because of its low magnetogyric ratio and low natural abundance, which results in remarkable broadening of the ^{73}Ge resonance signals, greatly limiting the applicability of the method. The most fundamental NMR signature of heavy alkenes $R_2E{=}ER_2$ (E = Si–Pb) is the diagnostic low-field resonance of their formally sp^2-hybridized nuclei E. Such characteristic deshielding, which is parallel to that observed in the ^{13}C NMR spectra of unsaturated alkenes C=C compared with those of saturated alkanes, is commonly recognized as a reliable indication of the existence of doubly bonded species E=E in solution. Of all heteronuclear NMR methods mentioned above (^{29}Si, ^{119}Sn and ^{207}Pb), ^{29}Si NMR spectroscopy is the most advanced from the viewpoints of both experimental techniques and fundamental regularities. Below, we briefly deal with the remarkable recent progress made in the field of ^{29}Si NMR spectroscopy of disilenes, which helps in understanding the general trends in the chemical shifts of doubly bonded Si atoms and sheds light on the milestone issue of the chemical bonding in disilenes.

A comprehensive study, combining experimental [solid-state cross polarization magic angle spinning (CPMAS) ^{29}Si NMR] and computational [MP2/6-31G** and B3LYP/6-311+G(2df,p)] efforts, on a series of differently substituted disilenes RR′Si=SiRR′ (**a:** R = R′ = Mes; **b:** R = R′ = Tip; **c:** R = Mes, R′ = *t*-Bu; **d:** R = R′ = Dis; **e:** R = $SiMe_3$, R′ = Tip; **f:** R = R′ = SiMe*i*-Pr_2; **g:** R = R′ = Si*i*-Pr_3) was carried out by the research teams of West, Kira and Apeloig.[8] Thus, the ^{29}Si chemical shift tensors of disilenes were determined using the CPMAS technique. All studied disilenes revealed a remarkable anisotropy of their chemical shift tensor, indicative of a highly asymmetric electronic distribution around the silicon nuclei, as expected for a Si=Si π-bond. Typically, all disilenes exhibited significant deshielding along the axis lying in the molecular plane and perpendicular to both the π-bond and the Si–Si bond (δ_{11}) (Figure 5.5).

Moderate deshielding and strong shielding were observed along the in-plane axis δ_{22} disposed along the Si–Si bond and out-of-plane axis δ_{33} perpendicular to the molecular plane (Figure 5.5). Such observations closely match those in carbon chemistry, where doubly bonded carbons show large deshielding along the δ_{11} axis, resulting in the overall deshielding of the nucleus.

Accompanying computations on the parent disilene $H_2Si{=}SiH_2$ were in good agreement with the experimental observations, predicting strongest deshielding along the δ_{11} axis, intermediate deshielding along the δ_{22} axis and shielding along the δ_{33} axis (Figure 5.5, R = H).[8] Overall, the remarkable deshielding in the δ_{11} direction was attributed to a strong paramagnetic contribution to the δ_{11} component, which stems from the paramagnetic circular shielding current induced by the applied magnetic field. The extent of this paramagnetic contribution is determined by the Ramsey formula[9] and is inversely proportional to the energy difference between the interacting occupied and

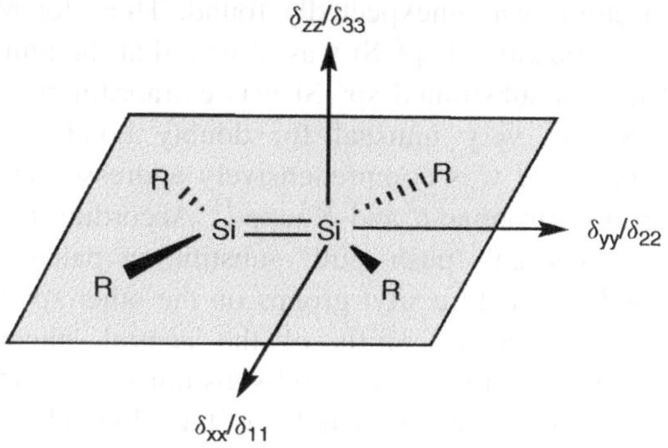

Figure 5.5 *Direction of the principal shielding tensor components in a planar disilene* $R_2Si{=}SiR_2$

vacant molecular orbitals that are coupled by the external magnetic field. In the case of disilenes, the occupied and vacant orbitals that mix upon the application of a magnetic field along the δ_{11} axis are σ(Si–Si) and π^*(Si=Si), respectively. Again, such computational results on disilenes nicely correspond with those on ethylene $H_2C{=}CH_2$, for which the strongest deshielding along the in-plane axis δ_{11} was also predicted.

The effects of disilenes' structural deformations (*trans*-bending at Si, twisting about the Si=Si bond, stretching of the Si=Si bond) on the chemical shielding tensor were also investigated in detail.[8] Remarkably, notable structural distortions, such as *trans*-bending or twisting by up to 25° or the presence of Si=Si–aryl conjugation, have little effect on the shielding tensors in aryl- and arylalkyldisilenes, leading to rather minor electronic perturbations of the Si=Si bond. In sharp contrast, the effect of stretching deformations on the chemical shielding tensors was calculated to be significant. Upon elongation of the Si=Si bond, all values of the chemical shielding tensor were shifted to low field; however, the strongest effect was noted for the δ_{11} component, which was deshielded to the greatest extent. Stretching of the Si=Si bond causes reduced overlap of both σ- and π-type orbitals, resulting in an appreciable decrease of the $\sigma_{Si\text{–}Si}$–$\pi^*_{Si=Si}$ energy separation and, consequently, a strong paramagnetic contribution to δ_{11} and deshielding of this component.

Bulky silyl substituents cause extreme deshielding in the δ_{11} direction, which was ascribed to the cooperative effect of the stretching of the Si=Si bond (see above) and decrease in the $\Delta E(\sigma_{Si\text{–}Si}\text{–}\pi^*_{Si=Si})$ because of the raising of the σ(Si–Si) energy level as a result of the electropositive substitution. Thus, for example, tetrasilyldisilenes (i-$Pr_2MeSi)_2Si{=}Si(SiMe$i$-$Pr_2)_2$ and (i-$Pr_3Si)_2Si{=}Si(Si$i$-$Pr_3)_2$ revealed record deshielding along the δ_{11} vector: 414 and 412 ppm, respectively, which results in the observation of their isotropic chemical shifts at the extremely low field of 143(solid)/145(solution) and 164(solid)/155(solution), respectively.[8]

The above-described hypothesis, based on the dominant role of the $\sigma_{Si\text{–}Si}$–$\pi^*_{Si=Si}$ interaction, was well suited to rationalize the low-field chemical shifts of the symmetrically substituted disilenes $R_2Si{=}SiR_2$. However, it failed to make sense of the ^{29}Si resonances of the unsymmetrically substituted disilenes of the type $R_2Si{=}Si(SiR_3)_2$ (R = alkyl or aryl group), for which a great difference between the isotropic chemical shifts of the doubly bonded silicon atoms was unexpectedly found. Thus, for $Mes_2Si{=}Si(Si\textit{i}\text{-}Pr_3)_2$ the resonance of the Mes-substituted sp^2-Si was observed at the remarkably low field of 152.3 ppm, whereas the silyl-substituted sp^2-Si was extraordinarily shielded, appearing in the region of −0.8 ppm, very unusual for doubly bonded silicon centers (see Section 5.3.1.1.1).[10] This issue was comprehensively addressed and interpreted in the recent theoretical study by Strohmann and Kaupp.[11] According to their computations, disilenes with an unsymmetrical 'push–pull' substitution pattern (σ-accepting alkyl groups on one sp^2-Si and σ-donating silyl groups on the other sp^2-Si) indeed exhibited greatly differing isotropic chemical shifts of the central silicon nuclei. Thus, in $Me_2Si{=}Si(SiH_3)_2$ the chemical shift of the silyl-substituted Si (−15.8 ppm) was calculated to be nearly 250 ppm more shielded than that of the alkyl-substituted Si (233.0 ppm) [at the BP86-GIAO/IGLO-III//B3LYP/6-31+G(d) level].[11] Supposing the preferential responsibility of the $\sigma_{Si\text{–}Si}\text{–}\pi^*_{Si=Si}$ mixing for the overall chemical shifts, one cannot explain the observed difference in the silicon resonances in $Me_2Si{=}Si(SiH_3)_2$, because the energy difference of the $\sigma\text{–}\pi^*$ excitation should be the same in the case of both sp^2-Si centers. Instead, the extreme deshielding of the Me-bound Si nucleus was attributed to a stronger interaction with the nuclear magnetic moment on a given nucleus, represented by the paramagnetic nuclear spin-electron orbit (PSO) term in the numerator of the Ramsey equation.[11] Accordingly, the more important the PSO term, the larger the paramagnetic contribution, and the stronger the deshielding at a given nucleus; this was the case of $Me_2Si{=}Si(SiH_3)_2$, in which the Me-substituted Si site featured a much greater PSO term than the H_3Si-substituted site, being by far more deshielded (233.0 vs −15.8 ppm). This can be visualized in the asymmetric charge distribution along the Si=Si bond, seen in both occupied and vacant MOs.

5.3.1.1.1 Disilenes >Si=Si<[3c–e,12]

Synthesis

The very first stable disilene, tetramesityldisilene $Mes_2Si{=}SiMes_2$ **1**, was reported by West *et al.* in their milestone paper published in 1981. It was prepared by the 254 nm photolysis of the trisilane precursor $Mes_2Si(SiMe_3)_2$ through the transient Mes_2Si:, which dimerized, giving **1** (Scheme 5.1).[13]

Scheme 5.1

Since then, several dozen other stable disilenes have been synthesized, isolated and, in majority of cases, structurally characterized. In general, all synthetic methods leading to stable disilenes can be classified into four main groups: (1) photolysis of acyclic trisilanes (**A**); (2) photolysis of cyclotrisilanes (**B**); (3) reductive dehalogenation of 1,1-dihalosilanes (**C**); and (4) reductive dehalogenation of 1,2-dihalodisilanes (**D**) (Scheme 5.2).

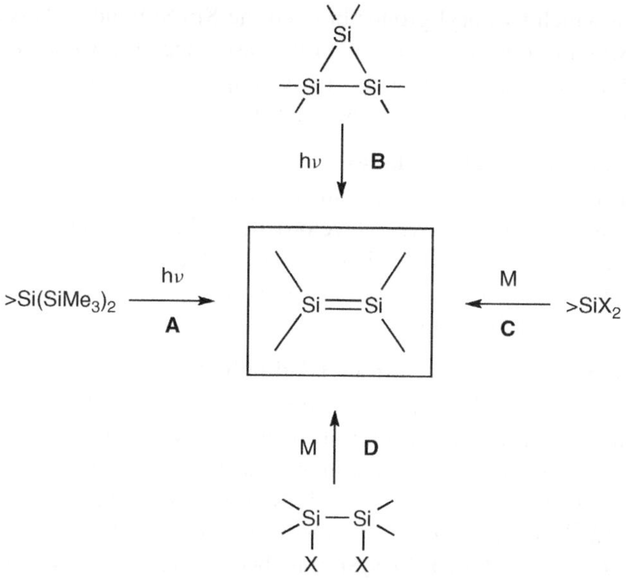

Scheme 5.2

Method A: photolysis of acyclic trisilanes

This method was mostly developed by West *et al.*, the authors of the first stable disilene **1**. Using this approach, they subsequently synthesized other stable disilenes: (*E*)- and (*Z*)-Mes(*t*-Bu)Si=Si(Mes)*t*-Bu **2**,[13c,d,14] (*E*)- and (*Z*)-[$(Me_3Si)_2N$](Mes)Si=Si[$N(SiMe_3)_2$]Mes **3**,[13c,14] (*E*)-Mes(Ad)Si=Si(Mes)Ad **4**,[15] (*E*)-Me_3Si(Tip)Si=Si($SiMe_3$)Tip **5**,[16] (*E*)-*t*-Bu(Tip)Si=Si(*t*-Bu)Tip **6**[16] and [(*E*)+(*Z*)]-Mes(Tip)Si=Si(Mes)Tip **7**[17] (Table 5.1).

The above-described tetraaryl- and dialkyldiaryldisilenes readily underwent either photochemical or thermal isomerizations, which were studied in detail by West's group.[14–16,18,19] Two distinctive types of disilene isomerizations were noted: *trans–cis* isomerization (*E*)-Ar(R)Si=Si(Ar)R ⇄ (*Z*)-Ar(R)Si=Si(Ar)R and 1,2-diaryl rearrangement $Ar_2Si{=}SiAr'_2$ ⇄ Ar(Ar′)Si=Si(Ar)Ar′. Mechanistic studies (kinetic measurements, trapping experiments) revealed that *cis–trans* isomerizations were most likely to proceed through rotation around the Si=Si bonds via a 90° twisted transition state.[14–16,18] Such rotation seems to be feasible because of the lowering of the rotation barriers in disilenes, which was attributed to both steric and electronic effects of the substituents. The activation energy for the thermal *cis–trans* isomerization of disilenes was estimated to be in the range 25.4–31.3 kcal/mol, and these values for the barrier to rotation about the silicon–silicon bond were taken as an experimental measure of the Si=Si π-bond strength.[15] The other type of disilene isomerization, namely 1,2-diaryl rearrangement in tetraaryldisilenes $Ar_2Si{=}SiAr'_2$ ⇄ Ar(Ar′)Si=Si(Ar)Ar′, was proposed to be dyotropic, involving a concerted intramolecular migration of the two aryl substituents to form a hypothetical 1,3-disilabicyclo[1.1.0]butane-type transition state in which two aryl groups bridged the Si=Si bond.[19] This highly ordered transition state was in good agreement with the large negative values of the activation entropy found for these rearrangements: for example, $\Delta S^{\ddagger} = -36 \pm 4$ cal/K·mol for the isomerization of $Mes_2Si{=}SiXyl_2$ to Mes(Xyl)Si=Si(Mes)Xyl.[19]

Method B: photolysis of cyclotrisilanes

This strategy for the synthesis of stable disilenes was originally employed by Masamune *et al.* They typically photolysed hexaarylcyclotrisilanes bearing bulky aryl groups, *cyclo*-(Ar_6Si_3) (Ar = 2,6-Me_2-C_6H_3 and 2,6-Et_2-C_6H_3), producing the corresponding tetraaryldisilenes $Ar_2Si{=}SiAr_2$ (**8**: Ar = 2,6-Me_2-C_6H_3; **9**: Ar = 2,6-Et_2-C_6H_3) as a result of the cycloreversion process (Table 5.1).[20]

Method C: reductive dehalogenation of 1,1-dihalosilanes

This is one of the most popular approaches for the preparation of stable disilenes: about 20 disilenes featuring different substitution patterns have been isolated (or generated) by this method. These are *alkyl-substituted* disilenes: $Dis_2Si{=}SiDis_2$ **10**,[21] fused tricyclic disilenes *cis*-**11** and *trans*-**11**;[22] *aryl-substituted* disilenes: $Mes_2Si{=}SiMes_2$ **1**,[23] $Tip_2Si{=}SiTip_2$ **12**,[24] Mes(Ar)Si=Si(Mes)Ar (**13**: Ar = 2,6-Et_2-C_6H_3; **14**: Ar = 2,6-*i*-Pr_2-C_6H_3),[25] (*E*)- and (*Z*)-Tbt(Mes)Si=Si(Tbt)Mes **15**,[26] (*E*)-Bbt(Br)Si=Si(Bbt)Br **16**,[27] (*E*)-Fc(Tip)Si=Si(Fc)Tip (Fc = ferrocenyl) **17**;[28] *silyl-substituted* disilenes: (*i*-Pr_2MeSi)$_2$Si=Si(SiMe*i*-Pr_2)$_2$ **18**,[29] (*t*-BuMe$_2$Si)$_2$Si=Si(SiMe$_2$*t*-Bu)$_2$ **19**,[29] (*i*-Pr_3Si)$_2$Si=Si(Si*i*-Pr_3)$_2$ **20**,[29] (*E*)-(*t*-BuMe$_2$Si)(*i*-Pr_2MeSi)Si=Si(SiMe$_2$*t*-Bu)(SiMe*i*-Pr_2) **21**,[30] (*E*)-Cl[(*t*-Bu_3Si)$_2$MeSi]Si=Si[SiMe(Si*t*-Bu_3)$_2$]Cl **22**,[31] (*E*)-(*t*-Bu_2MeSi)[(*t*-Bu_2MeSi)$_2$MeSi]Si=SiMe(SiMe*t*-Bu_2) **23**,[32] (*t*-Bu_2MeSi)$_2$Si=Si(SiMe*t*-Bu_2)$_2$ **24**,[33] fused bicyclic disilene **25**;[34] and *amino-substituted* disilenes: (*i*-Pr_2N)$_2$Si=Si(N*i*-Pr_2)$_2$ **26**[35] generated at low temperature in 3-MP by the dimerization of bis(amino)silylene (*i*-Pr_2N)$_2$Si:; and disilene **27**[36] with an exocyclic Si=Si bond formed as a tetramer of a stable N-heterocyclic silylene (Table 5.1).

Extraordinarily sterically overloaded disilenes Tbt(Mes)Si=Si(Tbt)Mes **15**[26] (both *E*- and *Z*-isomers) readily underwent thermal dissociation at 70 °C into diarylsilylene Tbt(Mes)Si: (see also Section 4.6.1.1).[26] The halogen-substituted disilene (*E*)-Bbt(Br)Si=Si(Bbt)Br **16** can be functionalized with either RMgBr or RLi, thus providing access to the substituted disilenes (*E*)-Bbt(Br)Si=Si(Bbt)R (R = Me, Et, *n*-Bu, Ph).[27]

Similarly to the above-mentioned tetraaryldisilenes, tetrasilyldisilenes are also prone to facile room temperature *E*–*Z* isomerization. Thus, dissolving pure disilene (*E*)-**21** in toluene results in the formation of a mixture of both *E*- and *Z*-isomers in a ratio of ca. 2:1.[30] The barrier for this (*E*)-**21** → (*Z*)-**21** isomerization, proceeding through rotation about the Si=Si bond, was estimated to be rather low: ca. 15 kcal/mol at 303 K (cf. 25.4–31.3 kcal/mol for tetraaryldisilenes). Such remarkable lowering of the rotation barrier was attributed to stabilization of the 90°-twisted transition state by $p\pi(Si_{sp^2})$–$\sigma/\sigma^*(Si_{sp^2}{-}Si_{sp^3})$ orbital mixing, facilitated by the high and low energy levels of the σ(Si–Si) and σ^*(Si–Si) orbitals, respectively.[30]

Method D: reductive dehalogenation of 1,2-dihalodisilanes

This is the other widely used method for the synthesis of heavy alkene derivatives. Several stable disilenes were readily accessible in this way, among them are *alkyl-substituted* disilene: *t*-Bu_2Si=Si*t*-Bu_2 **28**;[37] *aryl-substituted* disilene: $Mes_2Si{=}SiTip_2$

Table 5.1 *Crystallographic and spectral parameters of the structurally characterized disilenes*

Disilene (Synthetic method)	>Si=Si< (Å)	$\sum$(Si) (°)	>Si=Si< twist (°)	^{29}Si(Si=Si) (ppm)	Reference
$Mes_2Si{=}SiMes_2$ **1** (**A**)	2.143(2)	358.1 358.3	3.0		
1•C₇H₈ (**A**)	2.160(1)	NA	12.0	63.6	13
1•thf (**A**)	2.146	NA	13.0		
(*E*)-Mes(*t*-Bu)Si=Si(Mes)*t*-Bu **2** (**A**)	2.143(1)	359.8	NA	90.3	13c,d
(*E*)-Mes(Ad)Si=Si(Mes)Ad **4** (**A**)	2.138(2)	360.0	NA	87.1	15
(*E*)-Tip(Me_3Si)Si=Si(Tip)$SiMe_3$ **5** (**A**)	2.152(3)	NA	0	97.8	16
(*E*)-Tip(*t*-Bu)Si=Si(Tip)*t*-Bu **6** (**A**)	2.157(2)	NA	0	87.4	16
$Ar_2Si{=}SiAr_2$ (Ar = 2,6-Et_2-C_6H_3) **9** (**B**)	2.140(3)	360.0	10.0	NA	20b
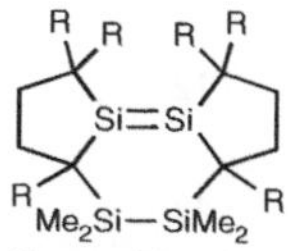 (R =$SiMe_3$) **11** (**C**)					
Trans-**11**	2.2687(7)	345.0 346.0	42.5	135.9	22
Cis-**11**	2.1767(6)	357.7 359.8	3.9	128.1	22
$Tip_2Si{=}SiTip_2$ **12** (**C**)	2.144	359.9	3.0	53.4	24
(*E*)-Tbt(Mes)Si=Si(Tbt)Mes **15** (**C**)	2.228(3)	357.2 358.9	8.7	66.5	26
(*Z*)-Tbt(Mes)Si=Si(Tbt)Mes **15** (**C**)	2.195(4)	358.8 359.3	14.0	NA	26
(*E*)-Bbt(Br)Si=Si(Bbt)Br **16** (**C**)	2.2264(8)	NA	NA	79.4	27
(*E*)-Fc(Tip)Si=Si(Fc)Tip **17** (**C**) (Fc = ferrocenyl)	2.1733(15)	352.3	30.7	72.6	28
(*i*-Pr_2MeSi)$_2$Si=Si(SiMe*i*-Pr_2)$_2$ **18** (**C**)	2.228(2)	NA	0	144.5	29
(*t*-BuMe_2Si)$_2$Si=Si(SiMe$_2$*t*-Bu)$_2$ **19** (**C**)	2.202(1)	NA	8.9	142.1	29
(*i*-Pr_3Si)$_2$Si=Si(Si*i*-Pr_3)$_2$ **20** (**C**)	2.251(1)	NA	0	154.5	29
(*E*)-(R_3Si)(R'_3Si)Si=Si(SiR_3)SiR'_3 **21** (**C**) (R_3Si = SiMe$_2$*t*-Bu, R'_3Si = SiMe*i*-Pr_2)	2.196(3)	360.0	0	141.8	30
(*E*)-Cl(R_3Si)Si=Si(Cl)SiR_3 **22** (**C**)	2.163(4)	359.6	3.8	NA	31
(*E*)-((R_3Si)$_2$MeSi)(R_3Si)Si=SiMe(SiR_3) **23** (**C**) (R_3Si=SiMe(SiT-Bu_3)$_2$)	2.1984(5)	359.5 359.8	29.0	103.8 (=*Si*Si$_2$) 158.9 (=*Si*Me)	32
(*t*-Bu_2MeSi)$_2$Si=Si(SiMe*t*-Bu_2)$_2$ **24** (**C**)	2.2598(18)	359.6 359.7	54.5	155.5	33
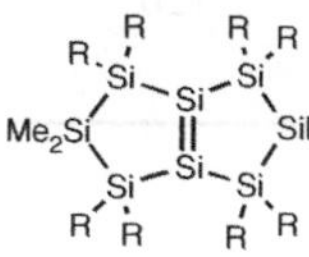 **25** (**C**) (R =$SiMe_3$)	2.180(3)	360.0	16.7	167.4	34

(continued overleaf)

Table 5.1 *Continued*

Disilene (Synthetic method)	>Si=Si< (Å)	∑(Si) (°)	>Si=Si< twist (°)	^{29}Si(Si=Si) (ppm)	Reference
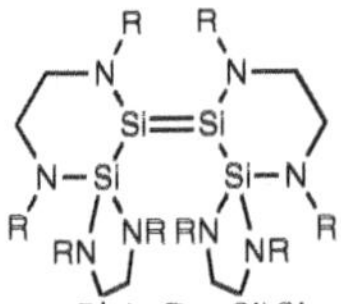 **27** (**C**) (R = *t*-Bu)	2.2890(14)	345.6 346.6	25.1	119.5	36
(*E*)-Ph(*t*-Bu$_3$Si)Si=Si(Ph)Si*t*-Bu$_3$ **30** (**D**)	2.182(2)	359.5	9.0	128.0	39a
(*t*-BuMe$_2$Si)$_2$Si=Si(Si*i*-Pr$_3$)$_2$ **31** (**D**)	2.2011(9)	360.0	28.0	142.0 152.7	40
(*t*-BuMe$_2$Si)$_2$Si=Si(SiMe*i*-Pr$_2$)$_2$ **32** (**D**) (2 independent molecules)	2.198(1) 2.1942(8)	360.0 360.0	9.0 11.7	132.4 156.6	40
(*E*)-Me$_3$Si(R$_3$Si)Si=Si(SiMe$_3$)SiR$_3$ **33** (**D**) (R$_3$Si = Si*i*-Pr$_3$)	2.1967(11)	358.4	NA	147.1	41
(Z)-Tip(Tip$_2$HSi)Si=Si(SiClTip$_2$)Tip **34** (**Other methods**)	2.2149(9)	NA	NA	96.0 102.7	42
(Z)-Tip(Tip$_2$HSi)Si=Si(SiBrTip$_2$)Tip **35** (**Other methods**)	2.2088(10)	NA	NA	95.6 104.1	42
(*E*)-R(η^1-Me$_5$C$_5$)Si=Si(R)(η^1-C$_5$Me$_5$) **36** (**Other methods**) [R = N(SiMe$_3$)$_2$]	2.1683(5)	359.1 359.5	4.9	NA	43
NHC:→S̈i=S̈i←:NHC **41** (**Other methods**) [NHC = 1,3-bis-(2,6-diisopropylphenyl)imidazol-2-ylidene]	2.2294(11)	NA	NA	224.5	44
(*t*-Bu$_2$MeSi)$_2$Si=Si(SiMe*t*-Bu$_2$)BR$_2$ (**Other methods**)					
R$_2$ = 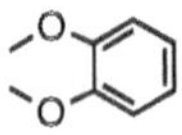 **43**	2.192(2)	360.0	7.1	111.7 (=*Si*Si) 127.3 (=*Si*B)	45
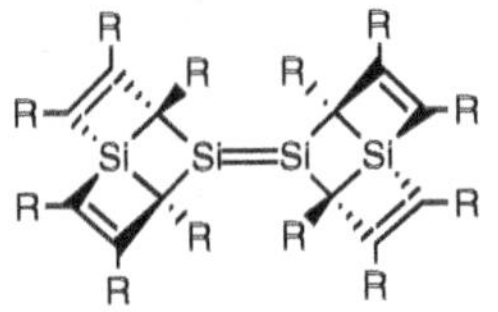 **44** (**Other methods**) [R = *t*-Bu]	2.2621(15)	360.0	12.1	102.5	46a
Tip[Tip$_2$(HO)Si]Si=Si[Si(OAr)Tip$_2$]Tip (**Other methods**)					
Ar = 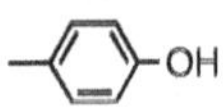 **47**	2.220(1)	NA	17.7	NA	12b
Ar = 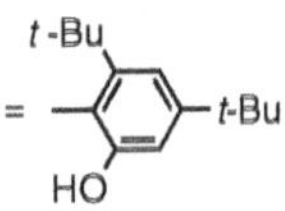 **48**	2.220(1)	NA	15.1	NA	12b

NA, Not available.

29;[38] and *silyl-substituted* disilenes: (E)-Ph(t-Bu_3Si)Si=Si(Ph)Sit-Bu_3 **30**,[39] (t-BuMe_2Si)$_2$Si=Si(Sii-Pr_3)$_2$ **31**,[40] (t-BuMe_2Si)$_2$Si=Si(SiMei-Pr_2)$_2$ **32**,[40] and (E)-Me_3Si(i-Pr_3Si)Si=Si(SiMe_3)Sii-Pr_3 **33**[41] (Table 5.1).

Other methods for the preparation of stable disilenes

There are several remarkable nontraditional synthetic routes to stable disilenes. Thus, Weidenbruch *et al.* prepared unsymmetrically substituted disilenes (Z)-Tip(Tip_2HSi)Si=Si(SiClTip$_2$)Tip **34** and (Z)-Tip(Tip_2HSi)Si=Si(SiBrTip$_2$)Tip **35** by the 1,4-addition of either HCl or HBr to the tetrasilabutadiene derivative Tip_2Si=Si(Tip)–Si(Tip)=SiTip_2 (Table 5.1).[42]

Jutzi *et al.* employed their highly electrophilic cationic derivative (η^5-Me_5C_5)Si^+•$B(C_6F_5)_4^-$, which upon reaction with lithium amide (Me_3Si)$_2$NLi produced at first silylene (Me_5C_5)[(Me_3Si)$_2$N]Si:, dimerizing to an isolable disilene (E)-(η^1-Me_5C_5)[(Me_3Si)$_2$N]Si=Si(η^1-C_5Me_5)[N(SiMe_3)$_2$] **36** (Table 5.1).[43a] Subsequent studies revealed that **36** underwent dissociation of its Si=Si bond in solution forming isolable silylene (Me_5C_5)[(Me_3Si)$_2$N]Si:, which was stable in solution.[43b]

By the coupling of 1,1-dilithiosilane derivatives (R_3Si)$_2$SiLi_2 (R_3Si = Sii-Pr_3, SiMet-Bu_2) with dichlorosilanes Ar_2SiCl_2 (Ar = Mes, Tip), Sekiguchi *et al.* synthesized novel disilenes featuring an unsymmetrical substitution pattern (R_3Si)$_2$Si=SiAr_2: (i-Pr_3Si)$_2$Si=SiMes_2 **37**, (i-Pr_3Si)$_2$Si=SiTip_2 **38**, (t-Bu_2MeSi)$_2$Si=SiMes_2 **39** and (t-Bu_2MeSi)$_2$Si=SiTip_2 **40**.[10]

A remarkable organosilicon derivative NHC:→$\ddot{\text{Si}}$=$\ddot{\text{Si}}$←:NHC (NHC = 1,3-bis(2,6-diisopropylphenyl)imidazol-2-ylidene) **41**, featuring a double bond between two silicon atoms in the zero formal oxidation state, was recently prepared by the reduction of NHC:→$SiCl_4$ complex with KC_8.[44] This unusual substituent-free and NHC ligand-supported disilene manifested extraordinarily deshielded ^{29}Si nuclei (δ = 224.5 ppm) and a normal Si=Si bond distance of 2.2294(11) Å (bond order WBI for the model compound = 1.73). Both Si atoms exhibited a highly pronounced *trans*-bent geometry with the two NHC-ligands coordinatively bound nearly perpendicular to the central Si=Si fragment [C–Si–Si bond angle = 93.37(5)°]. Such almost 90° *trans*-bending is consistent with the formulation of the Si–Si bonding orbitals as predominantly 3p in character with predominantly 3s-character of the Si lone pair orbitals.

Disilenyllithium derivatives of the type R_2Si=Si(R)Li have also been successfully utilized as convenient starting materials for the preparation of stable disilenes. Thus, interaction of the disilenide derivative (t-Bu_2MeSi)$_2$Si=Si(SiMet-Bu_2)Li with either B-chloropinacolborane or B-chlorocatecholborane resulted in the formation of the boryl-substituted disilenes **42** and **43** (Scheme 5.3, Table 5.1).[45]

The reduction of tri-*tert*-butylcyclopropenyltribromosilane **45** with potassium graphite unexpectedly produced a polycyclic lattice-framework disilene **44**, formed by multi-step reactions as a formal tetramer of tri-*tert*-butylcyclopropenylsilylyne (Scheme 5.4, Table 5.1).[46a] Featuring a rather long Si=Si bond of 2.2621(15) Å, disilene **44** exists in solution at room temperature in a thermal equilibrium with the corresponding silylene **46**, which was trapped with MeOH, CCl_4, CBr_4 and acetylenes (Scheme 5.4).[46b] The disilene **44** to silylene **46** dissociation energy was estimated to be as small as ca. 11.6 kcal/mol.

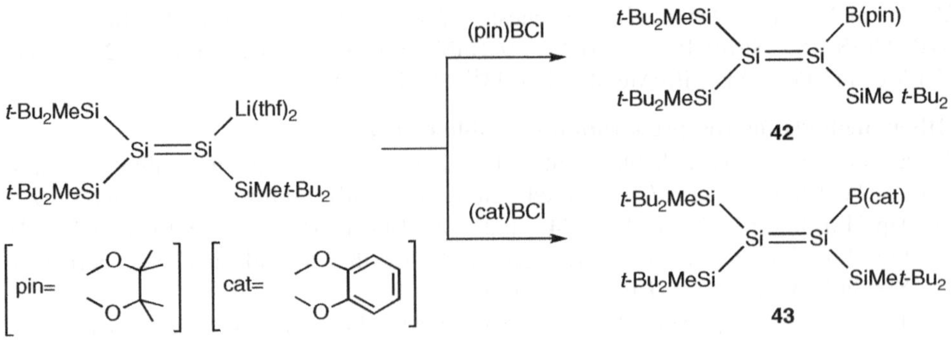

Scheme 5.3

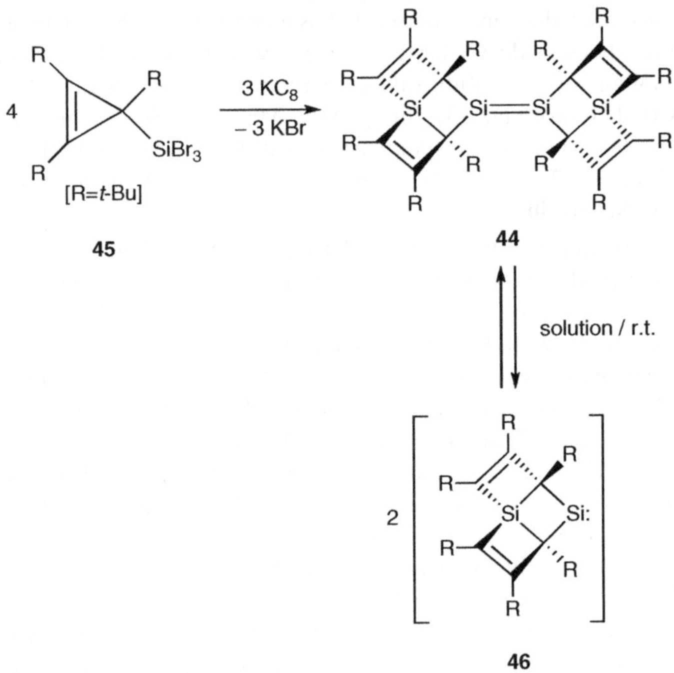

Scheme 5.4

^{29}Si NMR spectroscopy of disilenes

The resonances of the doubly bonded silicon atoms in disilenes are typically observed in the diagnostic low-field region, spanning a wide range of 53.4–167.4 ppm (for the origin of such deshielding, see Section 5.3.1.1, *Solution structure: NMR spectroscopy*). The magnitudes of the ^{29}Si NMR chemical shifts of the sp^2-silicons are totally governed by the influence of substituents, both electronic [decrease or increase in $\Delta E(\sigma_{\text{Si–Si}}-\pi^*_{\text{Si=Si}})$] and steric (stretching of the Si=Si bond). Accordingly, the Si resonances in tetraaryldisilenes $Ar_2Si{=}SiAr_2$ are the least deshielded (from 53.4 (for

12) to 66.5 [for (***E***)**-15**] ppm), whereas in tetraalkyldisilenes $Alkyl_2Si{=}SiAlkyl_2$ they are observed at a lower field [from 102.5 (for **44**) to 135.9 (for ***trans*-11**) ppm] (Table 5.1). Tetrasilyldisilenes, typically bearing very bulky σ-donating trialkylsilyl substituents, revealed a record deshielding of their doubly bonded Si atoms [from 132.4 (for **32**) to 167.4 (for **25**) ppm] (Table 5.1). As was discussed above, such extreme deshielding was ascribed to a remarkable decrease in $\Delta E(\sigma_{Si\text{–}Si}\text{–}\pi^*_{Si=Si})$ due to the raising of the $\sigma_{Si\text{–}Si}$ energy level produced by the electropositive silyl-substitution. Disilenes featuring mixed substitution patterns [aryl(alkyl)disilenes, aryl(silyl)disilenes, alkyl(silyl)disilenes], exhibited chemical shifts of their sp^2-Si in the region intermediate between those of disilenes with all-identical substituents (tetraaryldisilenes/tetraalkyldisilenes, tetraaryldisilenes/tetrasilyldisilenes, tetraalkyldisilenes/tetrasilyldisilenes).

In a peculiar case of unsymmetrically substituted disilenes $(R_3Si)_2Si{=}SiAr_2$, silyl-substituted sp^2-*Si* were extraordinarily shielded and observed in uncharacteristically high fields, whereas aryl-substituted sp^2-*Si* resonated at remarkably low fields: −0.8 and +153.2 ppm [for $(i\text{-}Pr_3Si)_2Si{=}SiMes_2$ **37**], +14.0 and +137.2 ppm [for $(i\text{-}Pr_3Si)_2Si{=}SiTip_2$ **38**], +8.2 and +148.6 ppm [for $(t\text{-}Bu_2MeSi)_2Si{=}SiMes_2$ **39**], +14.9 and +142.0 ppm [for $(t\text{-}Bu_2MeSi)_2Si{=}SiTip_2$ **40**].[10] Such unusual NMR spectral behavior was rationalized by considering the predominant importance of the PSO term, which represents the extent of the interaction with the nuclear magnetic moment, thus regulating the magnitude of paramagnetic contribution and degree of deshielding (see Section 5.3.1.1, *Solution structure: NMR spectroscopy*).[11]

Crystal structure of disilenes

Because the potential energy surface of disilenes is rather shallow and the Si=Si bonds are classified as 'soft' double bonds, the latter can be readily deformed in different ways (stretching, *trans*-bending, twisting), resulting in a great number of structural variations of disilenes (Figure 5.6). Given the 'softness' of the Si=Si bonds, one should also take into account the important influence of the crystal packing forces on the overall geometry of disilenes in the solid state.

The general tendencies of the influence of substituents on the configuration of the Si=Si bond are the following: electronegative substituents (F, OH, NH_2) increase ΔE_{ST} and the degree of π–σ^* orbital mixing to cause remarkable *trans*-bending at sp^2-Si, stretching and weakening of the Si=Si bond; electropositive groups (BH_2, SiH_3) reduce both ΔE_{ST} and the degree of π–σ^* interaction resulting in significant flattening at the sp^2-Si, shortening and strengthening the Si=Si bond (see Section 5.3.1.1, *Structure and bonding*). On the other hand, the effect of steric interaction between voluminous substituents, kinetically stabilizing the highly reactive Si=Si bond, also greatly contributes to the overall geometry of the double bond, resulting in its stretching and twisting.

Tetraalkyldisilenes, depending on the electronic nature and size of their alkyl substituents, are typically *trans*-bent, spanning the following range of geometrical parameters: from 2.1767(6) Å (for ***cis*-11**) to 2.2687(7) Å (for ***trans*-11**) (bond length); from 345.0° (for ***trans*-11**) to 360.0° (for **44**) (pyramidality); from 3.9° (for ***cis*-11**) to 42.5° (for ***trans*-11**) (twisting) (Table 5.1).

Tetraaryldisilenes are usually less deformed than their tetraalkyl-substituted counterparts, exhibiting the following range of structural characteristics: from 2.140(3) Å (for **9**)

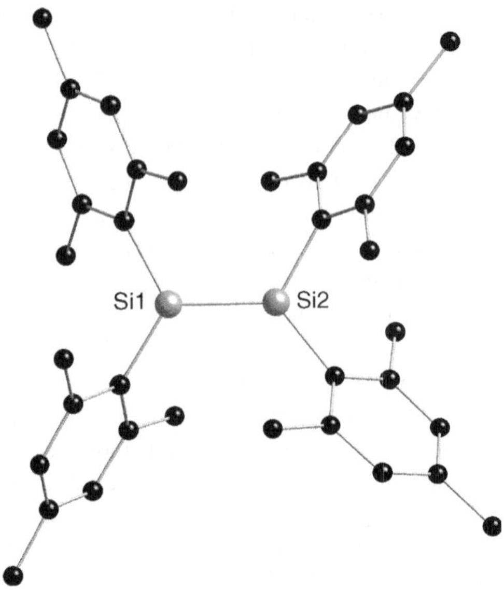

Figure 5.6 Crystal structure of tetramesityldisilene $Mes_2Si{=}SiMes_2$ **1** (hydrogen atoms are not shown)

to 2.228(2) Å (for **15**) (bond length); from 357.2° [for **(*E*)-15**] to 360.0° (for **9**) (pyramidality); from 3.0° (for **1** and **12**) to 14.0° [for **(*Z*)-15**] (twisting) (Table 5.1).

In agreement with theoretical predictions, tetrasilyldisilenes were expected to be the least deformed, and they indeed revealed the smallest degree of *trans*-bending: from 358.4° [for **(*E*)-33**] to 360.0° [for **19**, **(*E*)-21**, **25**, **31** and **32**] (pyramidality) (Table 5.1). However, in contrast to computational results, tetrasilyldisilenes showed rather long Si=Si bonds and extreme degrees of double bond twisting, which should be definitely attributed to the severe steric repulsion between the bulky trialkylsilyl substituents: from 2.180(3) Å (for **25**) to 2.2598(18) Å (for **24**) (bond length); from 0.0° [for **18, 19** and **(*E*)-21**] to 54.5° (for **24**) (twisting) (Table 5.1). Indeed, the most heavily substituted $(t\text{-}Bu_2MeSi)_2Si{=}Si(SiMe\,t\text{-}Bu_2)_2$ **24** exhibited the highest extent of structural deformations, manifesting the longest Si=Si bond among tetrasilyldisilenes of 2.2598(18) Å and greatest twisting of 54.5°, being nevertheless not *trans*-bent (sum of the bond angles around sp^2-Si atoms of 359.6° and 359.7°) in accord with its electropositive silyl-substitution.[33] Apparently, extreme twisting about the Si=Si bond in **24** causes partial breaking of the π-bond and, consequently, destabilization of the HOMO (Si=Si π-orbital), seen in the remarkable red shift in its electronic spectrum: λ_{max} = 612 nm (deep-blue disilene).

The longest Si=Si bond of 2.2890(14) Å was reported for the unusual bis(silyl) bis(amino)disilene **27**, which also exhibited strong bending (345.6° and 346.6°) and twisting (25.1°) deformations, existing in solution in thermal equilibrium with the monomeric silylene.[36] Both electronic and steric effects should account for such great structural distortions observed in **27**.

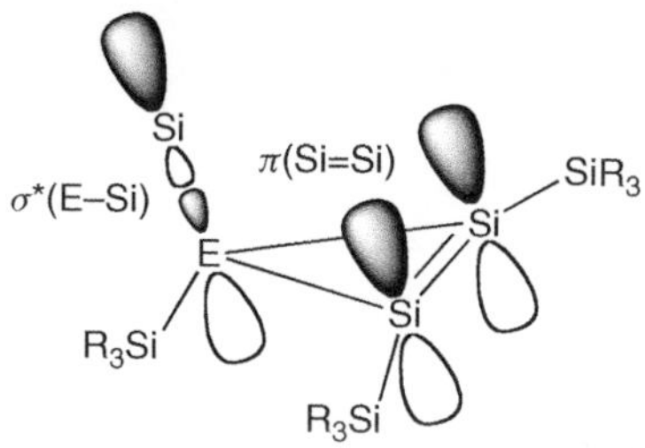

Figure 5.7 π(Si=Si)–σ*(E–Si) orbital mixing stabilizing the HOMO($\pi_{Si=Si}$) of the heavy cyclopropene

Cyclic disilenes with an endocyclic Si=Si bond

There are no universal methods for the design of cyclic disilenes; therefore, the particular approaches for each class of such derivatives are described below.

Three-membered ring compounds (heavy cyclopropenes)

The cyclopropene analogs of the heavy group 14 elements are especially challenging because of their unusual structures, which combine the properties of both three-membered rings and alkene fragments.[47] Given the intrinsically great ring strain of cyclopropene derivatives, one can expect their high reactivity and hence, instability. This unfavorable tendency can be counteracted by electropositive substitution with σ-donating silyl groups, which causes the effective π(Si=Si)–σ*(E–Si) orbital mixing that stabilizes the HOMO($\pi_{Si=Si}$) of the heavy cyclopropene molecule (Figure 5.7).

Computational studies revealed that, in contrast to carbon systems, the ring strain in cyclotrisilenes can be significantly relaxed by the cyclic π–σ* orbital interaction, which makes them even less strained than their saturated analogs, cyclotrisilanes.[48]

Indeed, all experimentally known stable cyclotrisilenes possess silyl substituents at the skeletal silicon atoms (both sp^2- and sp^3-Si) (Table 5.2). The isolable cyclotrisilene **49** was prepared by the reduction of $R_3Si–SiBr_2Cl$ (R = $SiMe_2t$-Bu) with potassium graphite in 1999 (Scheme 5.5).[49]

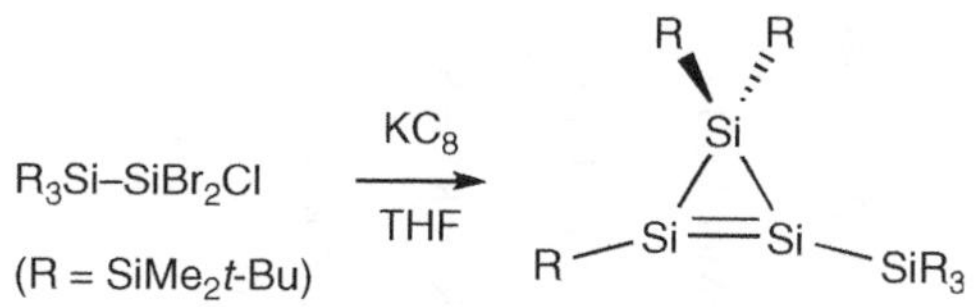

49

Scheme 5.5

The novel symmetrically substituted cyclotrisilene **50**[50] and the first hybrid cyclopropene of different heavy group 14 elements, 3*H*-disilagermirene **51** (Figure 5.8),[51] were synthesized by the Würtz-type reductive coupling of $RSiBr_3$ and R_2EX_2

Table 5.2 *Crystallographic and spectral parameters of the structurally characterized cyclic disilenes*

Cyclic disilene	>Si=Si< (Å)	Σ (Si) (°)	>Si=Si< twist (°)	Ring Folding (°)	^{29}Si NMR (Si=Si) (ppm)	Reference
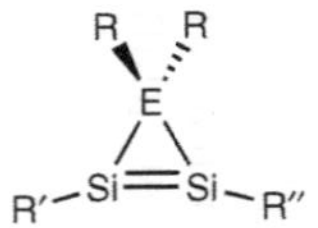						
49(E = Si; R = R′ = SiMe$_2$*t*-Bu; R″ = SiR$_3$)	2.132(2)	356.7 360.0	21.4		81.9 (=*Si*–R) 99.8 (=*Si*–SiR$_3$)	49
50 (E = Si; R = R′ = R″ = SiMe*t*-Bu$_2$)	2.138(2)	357.5 358.1	31.9		97.7	50
51 (E = Ge; R = R′ = R″ = SiMe*t*-Bu$_2$)	2.146(1)	356.9 357.4	37.0		107.8	51
55 (E = Si; R = SiMe*t*-Bu$_2$; R′ = R″ = Si*t*-Bu$_3$)	2.1612(8)	359.8 360.0	4.8		97.4	53
56 [E = Si; R′ = R″ = SiMe$_3$; R,R = –C(SiMe$_3$)$_2$CH$_2$CH$_2$(Me$_3$Si)$_2$C–]	2.1428(5)	359.1 360.0	8.6		142.9	54
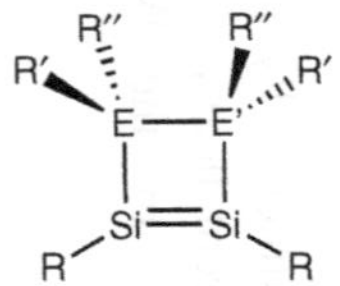						
59(E = E′ = Si; R = R′ = R″ = SiMe$_2$*t*-Bu)	2.174(4)	358.6	12.3	37.0	160.4	55
60(E = E′ = Si; R = R′ = Si*t*-Bu$_3$; R″ = I)	2.257(2)	359.2 359.6	30.8	28.0	164.4	56

61(E = E′ = Si; R = R′ = Si*t*-Bu$_3$; R″ = H)	2.360(2)	350.4	NA	23.0	141.3	57
62[E = E′ = C; R′ = R″ = H; R = SiMe(Si*t*-Bu$_3$)$_2$]	2.175(1)	359.8 360.0	NA	8.3	151.4	58
64(E = E′ = C; R = Si*i*-PrDis$_2$; R′ = Me; R″ = H)	2.1632(10)	358.1 358.4	32.2	1.0	152.1	59
65(E = E′ = Si; R = R′ = SiMe*t*-Bu$_2$; R″ = Br)	2.2014(9)	359.0 359.7	NA	33.7	141.4	60
66(E = E′ = Ge; R = R′ = SiMe*t*-Bu$_2$; R″ = Me)	2.1875(13)	355.3 357.6	NA	4.3	167.6	61
69(E = Si, E′ = Se; R = R′ = R″ = SiMe*t*-Bu$_2$; no substituents R′ and R″ on Se)	2.1706(12)	360.0	0.1	1.7	100.8 (=*Si*– Si) 192.1 (=*Si*– Se)	63
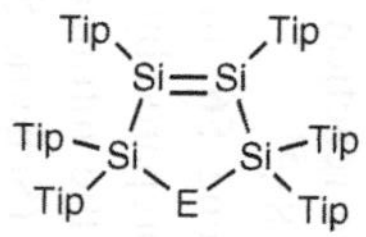 **72**(E = S)	2.170(1)	358.6 359.7	26.5	NA	86.0	64
73(E = Se)	2.181(1)	358.5 359.6	26.8	NA	90.3	64
74 (E = Te)	2.198(2)	358.6 359.3	28.2	NA	97.8	64

NA, Not available.

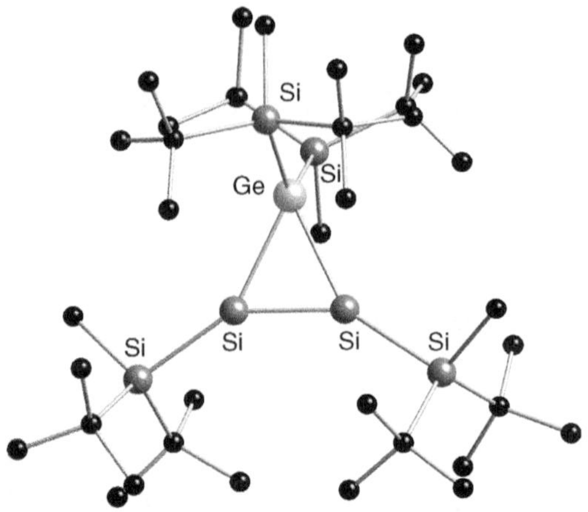

Figure 5.8 Crystal structure of 3H-disilagermirene **54** (hydrogen atoms are not shown)

(R = SiMe*t*-Bu_2, E = Si/Ge, X = Cl/Br) with sodium (Scheme 5.6). 3*H*-disilagermirene **51** underwent isomerization (both thermal and photochemical), involving formal 1,2-R-migration from Ge to Si skeletal atoms to form an isomeric 1*H*-disilagermirene featuring an endocyclic Si=Ge bond (see Section 5.3.1.2.1.5).[51]

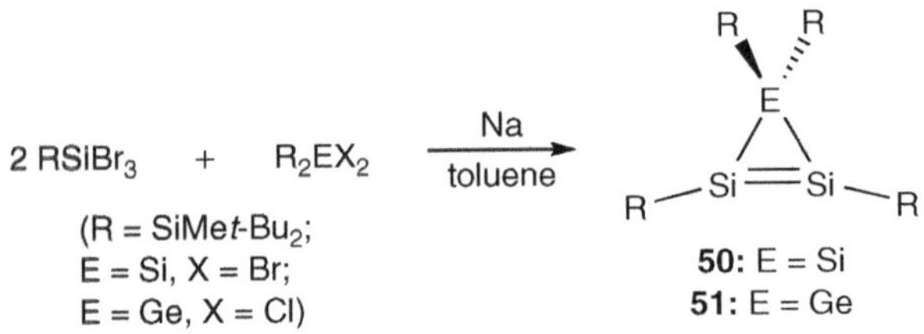

Scheme 5.6

Interestingly, the cyclotrisilene **52**, as the nearest homolog of **50**, distinguished from it by only a CH_2-unit, was synthesized from **50** in three steps (Scheme 5.7).[52] At first, the ring enlargement reaction with CH_2Cl_2 gave four-membered ring **53**, which was then reduced with KC_8 to produce trisilabicyclo[1.1.0]butane **54**, followed by its thermal isomerization to form cyclotrisilene **52**, representing an unprecedented alkyl-substituted heavy cyclopropene.[52]

The cyclotrisilene **55** with very bulky *t*-Bu_3Si substituents at the doubly bonded silicon atoms was available by the reaction of R_2SiLi_2 (R = SiMe*t*-Bu_2) and $R'SiBr_2SiBr_2R'$ (R′ = Si*t*-Bu_3) (Scheme 5.8).[53]

The unusual cyclotrisilene **56**, in which the Si_3-ring unit constituted a part of a spiroheptene system, was prepared by cleavage of one of the Si=Si bonds of tetrasilabuta-1,3-diene **57** followed by the intramolecular insertion of the resulting silylene **58** into the remaining Si=Si bond (Scheme 5.9).[54]

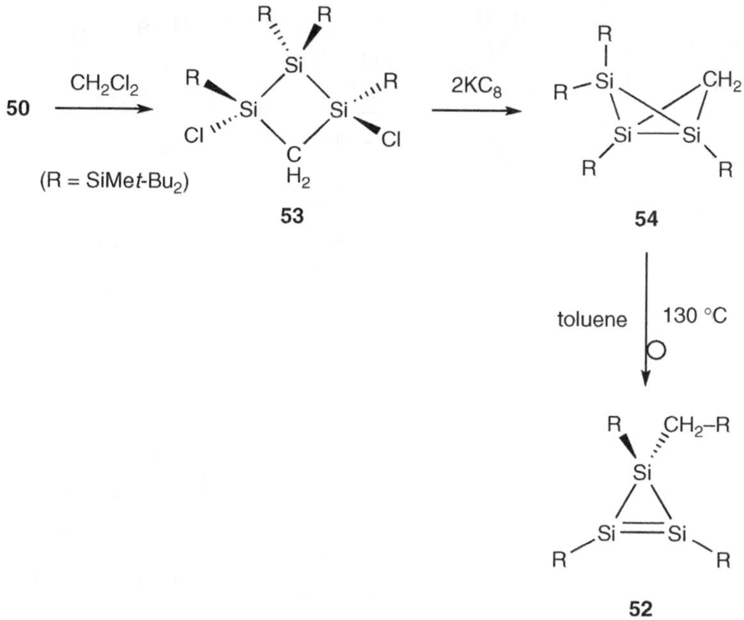

Scheme 5.7

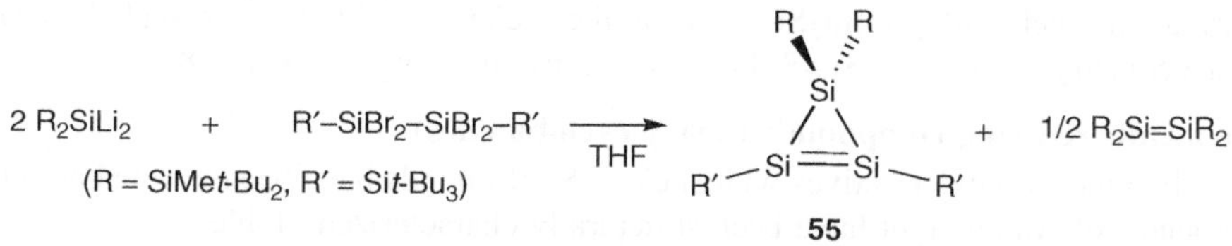

Scheme 5.8

The resonances of the doubly bonded Si atoms of cyclotrisilenes are typically observed at significantly higher fields than those of acyclic tetrasilyldisilenes (the low-field signal of 142.9 ppm in **56** is probably due to its peculiar spiro-structure): 81.9 (for **49**)–122.8 (for **52**) ppm vs 132.4 (for **32**)–167.4 (for **25**) ppm (Tables 5.1 and 5.2). This tendency is comparable with that in carbon chemistry, where the doubly bonded carbons in cyclopropene (108.7 ppm) resonate at higher fields than those in ethylene (123.5 ppm).

The Si=Si bonds in cyclotrisilenes are significantly shorter than those of acyclic tetrasilyldisilenes: 2.132(2) Å (for **49**)–2.1612(8) Å (for **55**) vs 2.180(3) Å (for **25**)–2.2598(18) Å (for **24**) (Tables 5.1 and 5.2). On the other hand, the double bonds in cyclotrisilenes are more twisted than those in tetrasilyldisilenes with the torsion angles about the Si=Si bond ranging from 21.4 (for **49**) to 37.0° (for **51**). Such remarkable twisting may result from the appreciable steric interaction between the bulky substituents at the doubly bonded Si atoms, which are eclipsed relative to each other. However, cyclotrisilene **55**,

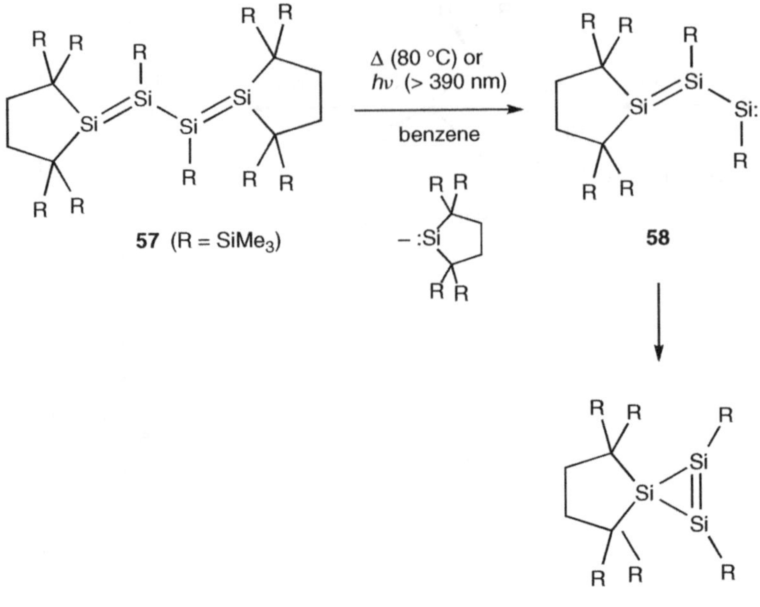

57 (R = $SiMe_3$)

58

56

Scheme 5.9

featuring extremely bulky *t*-Bu_3Si-groups at the skeletal sp^2-Si atoms, revealed a nearly planar geometry around the Si=Si bond with a twisting angle of just 4.8°.

Four-membered ring compounds (heavy cyclobutenes)

Currently, nine cyclic derivatives with skeletal Si=Si bonds have been isolated as stable compounds, of which eight have been structurally characterized (Table 5.2).

The first representative, cyclotetrasilene **59**, was prepared in 1996 by the co-reduction of a 1:2 mixture of $RSiBr_2SiBr_2R$ and R_2SiBr_2 (R = $SiMe_2$*t*-Bu) with lithium naphthalenide (Scheme 5.10).[55]

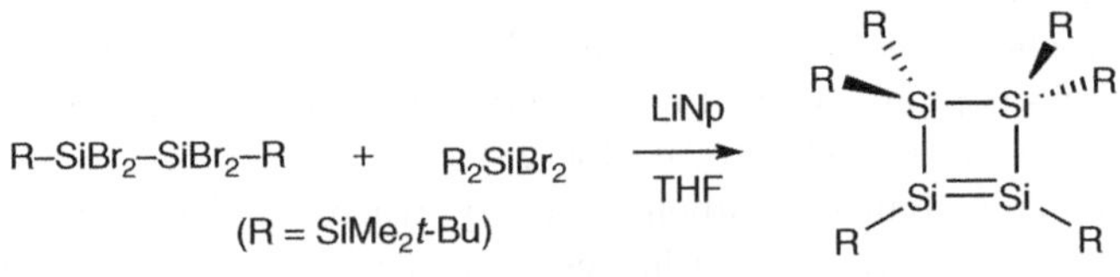

59

Scheme 5.10

Since then, other stable heavy cyclobutenes have been reported, prepared by different synthetic procedures: **60** by the reaction of tetrasilatetrahedrane with iodine (Scheme 5.11, A);[56] **61** by the reduction of R_2HSi–SiHBr–SiHBr–$SiHR_2$ with NaR

(Scheme 5.11, B);[57] **62** and **63/64** by the [2 + 2] cycloaddition of $CH_2=CH_2$ and *cis/trans*-MeCH=CHMe with isolable disilynes (Scheme 5.11, C and D);[58,59] **65** by the reduction of 1,2,3,4-tetrabromocyclotetrasilane with potassium graphite (Scheme 5.11, E);[60] **66** by the methylation of 1,2-disila-3,4-digermacyclobutadiene dianion derivative with dimethyl sulfate (Scheme 5.11, F);[61] **67** by the methylation of cyclotetrasilenylium ion derivative with methyllithium (Scheme 5.11, G).[62]

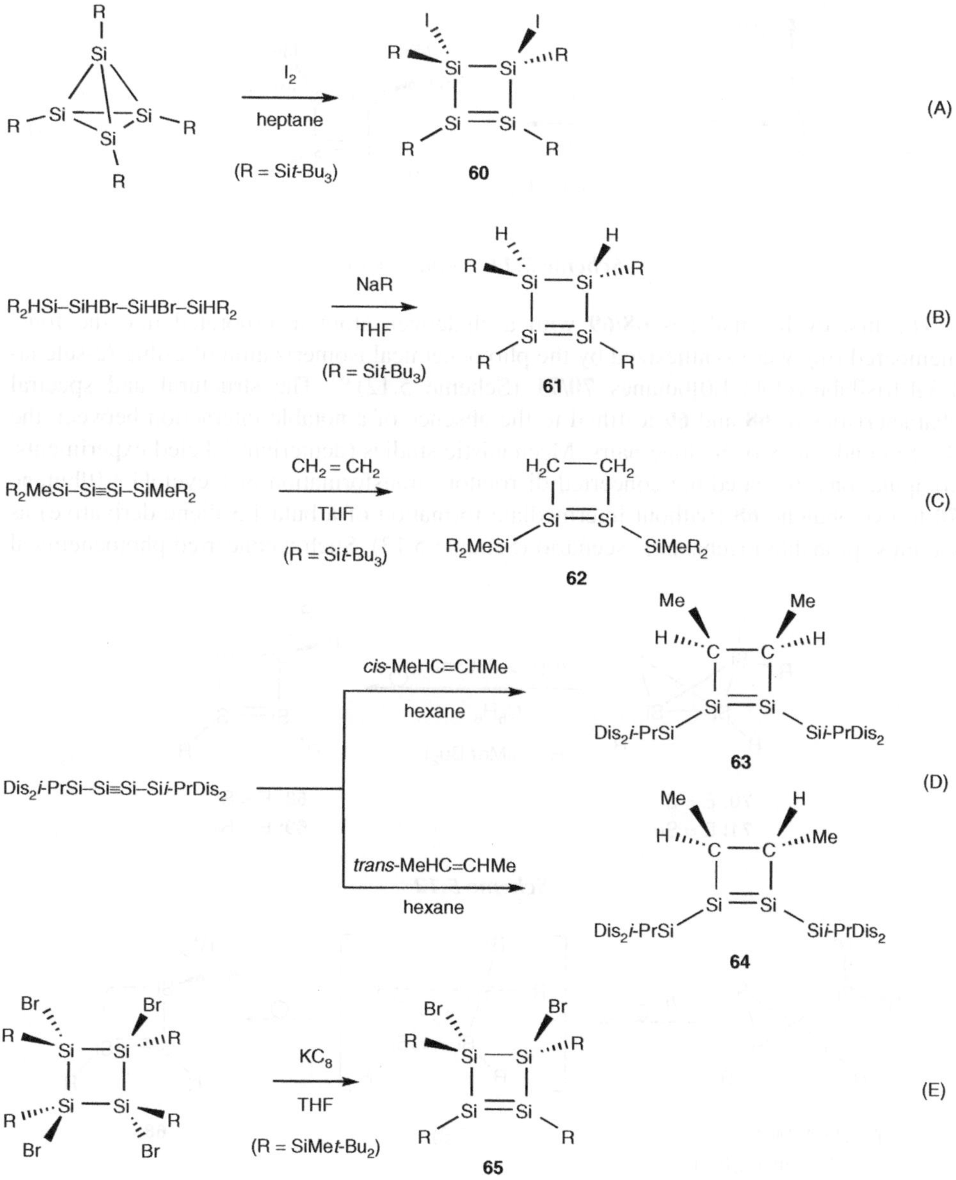

Scheme 5.11

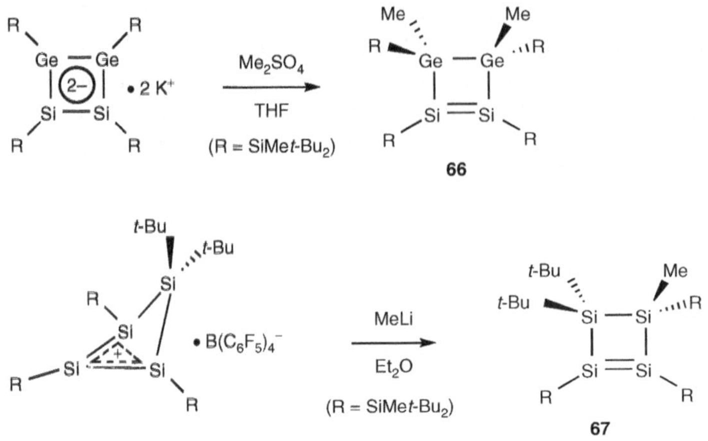

Scheme 5.11 *(continued)*

The first cyclic disilenes **68/69** with a chalcogen atom incorporated into the four-membered ring were synthesized by the photochemical isomerization of 2-thia-/2-selena-1,3,4-trisilabicyclo[1.1.0]butanes **70/71** (Scheme 5.12).[63] The structural and spectral characteristics of **68** and **69** testified to the absence of a notable interaction between the Si=Si bond and S or Se lone pairs. Mechanistic studies (deuterium-labeled experiments, computations) revealed the concerted disrotatory transformation of bicyclo[1.1.0]butane **70** to cyclobutene **68** (without intermediate formation of a buta-1,3-diene derivative) as the most probable mechanistic scenario (Scheme 5.13). Such a concerted photochemical

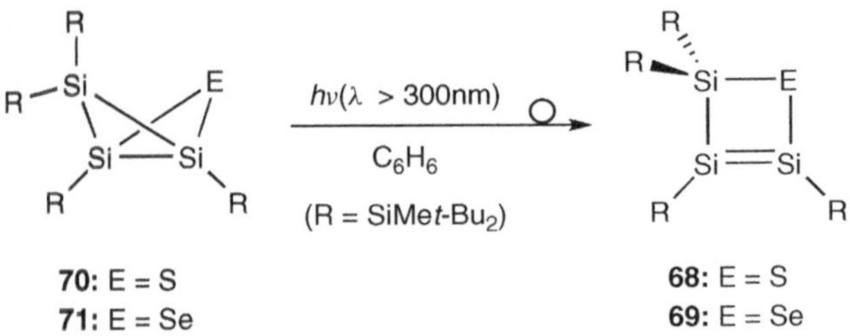

Scheme 5.12

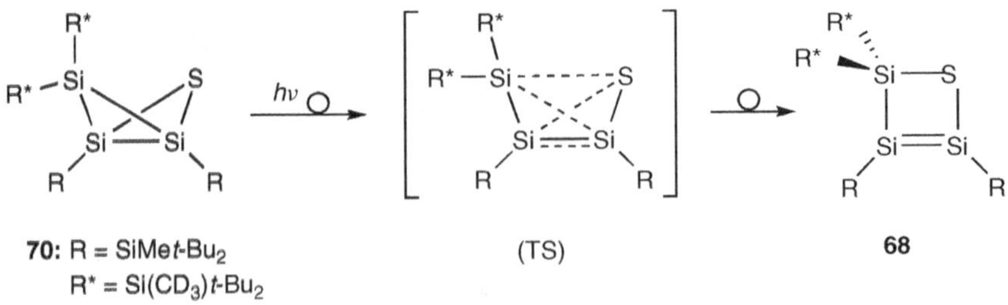

Scheme 5.13

transformation is to be classified as a symmetry-allowed [σ2s + σ2s] process obeying the Woodward–Hoffmann orbital symmetry rules.

The doubly bonded Si atoms of the heavy cyclobutenes resonated at much lower fields than those of the heavy cyclopropenes: 95.1 (for **68**)–195.0 (for **68**) ppm vs 81.9 (for **49**)–107.8 (for **51**) ppm (Table 5.2). This can be compared with the deshielding of sp^2-carbons in cyclobutene (137.2 ppm) relative to those in cyclopropene (108.7 ppm). The doubly bonded Si atoms of cyclotetrasilenes are even more deshielded than those of acyclic tetrasilyldisilenes: 95.1 (for **68**)–195.0 (for **68**) ppm vs 132.4 (for **32**)–167.4 (for **25**) ppm (Tables 5.1 and 5.2).

The Si=Si bonds in the heavy cyclobutenes are typically longer than those of the heavy cyclopropenes (again, parallel to the tendency in carbon chemistry where cyclopropenes are well known to have the shortest C=C bonds among cycloalkenes): 2.1632(10) Å (for **64**)–2.360(2) Å (for **61**) vs 2.132(2) Å (for **49**)–2.1612(8) Å (for **55**) (Table 5.2). The Si=Si bond of 2.360(2) Å in cyclotetrasilene **61** is extraordinarily stretched, being even longer than regular Si–Si single bonds (av. 2.34 Å).[57] The extent of other types of structural distortion (pyramidalization at the sp^2-Si, twisting of the Si=Si bond, folding of the four-membered ring) strongly depends on the substitution pattern: from 350.4 (for **61**) to 359.8/360.0° (for **62** and **69**) (pyramidalization); from 0.1 (for **69**) to 32.2° (for **64**) (twisting); from 1.0 (for **64**) to 37.0° (for **59**) (folding) (Table 5.2).

Five-membered ring compounds (heavy cyclopentenes)

Three five-membered ring compounds **72, 73** and **74** with an endocyclic Si=Si bond have been reported to date; all of them prepared by the [1 + 4] cycloaddition of chalcogens (S, Se, Te) to the isolable tetrasilabuta-1,3-diene derivative $Tip_2Si{=}Si(Tip){-}Si(Tip){=}SiTip_2$ (Scheme 5.14).[64] The doubly bonded silicons were observed as expected in a low-field region (86.0 ppm for **72**, 90.3 ppm for **73** and 97.8 ppm for **74**), although the extent of deshielding in heavy cyclopentenes was less important than that in the heavy cyclopropenes and cyclobutenes (Table 5.2). The Si=Si bonds in **72**–**74** were longer than those in heavy cyclopropenes, but comparable with those in heavy cyclobutenes: 2.170(1) Å for **72**, 2.181(1) Å for **73** and 2.198(2) Å for **74** (Table 5.2). One can note a steady stretching of the Si=Si bond on going from **72** to **74**, associated with the increase in the size of the chalcogen atom. The five-membered rings in heavy cyclopentenes **72**–**74** are nearly planar with insignificant pyramidalization at the sp^2-Si atoms (358.6/359.7° for **72**, 358.5/359.6° for **73**, 358.6/359.3° for **74**) and sizeable twisting about the Si=Si bond (26.5° for **72**, 26.8° for **73**, 28.2° for **74**) (Table 5.2).

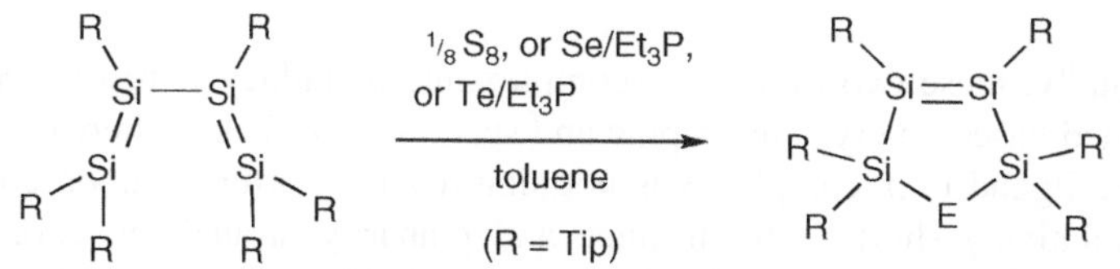

72: E = S
73: E = Se
74: E = Te

Scheme 5.14

Disilenes as ligands for transition metal complexes

Several stable organosilicon compounds, which can be considered as disilene–transition metal complexes (similar to alkene–transition metal complexes), have been reported during the past two decades: $L_nM[R_2Si{=}SiR_2]$ [$L = Ph_2PCH_2CH_2PPh_2$, $Cy_2PCH_2CH_2PCy_2$, Et_3P, Ph_3P, Cp, Me_3P, $Me_2PCH_2CH_2PMe_2$, CO, Cy_3P, $PhMe_2P$, (2,6-Me_2-C_6H_3)NC; M = Pt, Mo, W, Pd, Fe, Hf; R = *i*-Pr, Ph, Mes, Me, $SiMe_2t$-Bu, SiMe*t*-Bu_2, Cl, $SiMe_3$].[65–69] Most of them were prepared by indirect routes,[65–69] typically involving coupling of the 1,2-dilithiodisilane derivative with the transition metal dichloride (or alternatively, by the reaction of the 1,2-dichlorodisilane derivative with the alkali metal salt of the transition metal dianionic complex), although some were directly synthesized by the reaction of isolable disilenes and transition metal complexes.[65b, 68b] The bonding situation in these disilene–transition metal complexes was rationalized in the framework of the conventional Dewar–Chatt–Duncanson model,[70] similar to that of alkene–transition metal coordination compounds. According to this approximation, the bonding between disilene and the transition metal center is explained in terms of the interplay between the disilene→transition metal direct σ-donation and transition metal→disilene π-back-donation. Consequently, the structure of disilene complexes is totally determined by the balance between these two bonding types: if σ-donation prevails then the **π-complex** structure is formed, whereas when π-back-donation is more important, then a σ-complex, the so-called **metallacycle**, is expected (Scheme 5.15).

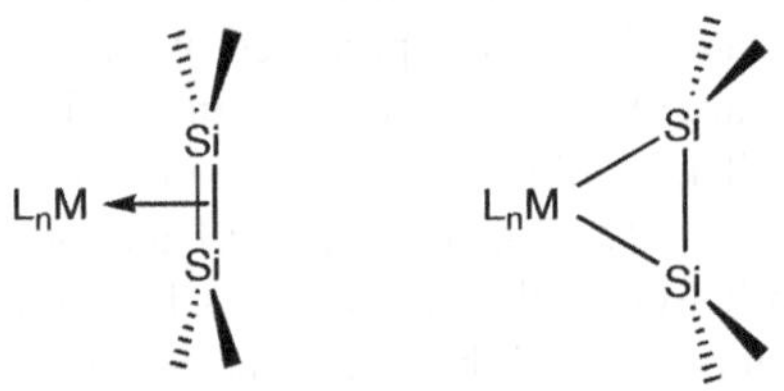

	A: π-complex	**B: metallacycle**
Si=Si bond:	*short*	*long*
Geometry at Si:	*planar*	*pyramidal*
^{29}Si NMR:	*low-field*	*high-field*

Scheme 5.15

Experimentally, these two limiting π-complex and metallacycle resonance forms can be distinguished based on crystallographic and spectral data. Thus, whereas the geometry of the disilene ligand in π-complexes is not significantly perturbed upon complexation, implying a relatively short Si–Si distance and planarity around the coordinated silicons (Scheme 5.15, **A**), metallacycles feature remarkable distortion of the original free disilene geometry taking place upon coordination: stretching of the Si–Si bond and pyramidalization at the coordinated silicons (Scheme 5.15, **B**). Moreover, the resonances of the coordinated Si atoms in π-complexes are expected to be observed at a lower field (approaching that of uncomplexed disilenes), whereas those in metallacycles should appear at a much higher field (as is typical for silacyclopropane systems).

Most of the known disilene complexes were characterized as metalladisilacyclopropanes of type **B** (Scheme 5.15) based on their structural and spectral features,[65–69] whereas some Pd-complexes of isolable tetrasilyldisilene were reported to possess a somewhat higher degree of π-complex character.[68e] The 14-electron disilene–palladium complex $(Cy_3P)Pd[(t\text{-}BuMe_2Si)_2Si{=}Si(SiMe_2t\text{-}Bu)_2]$ represented the highest contribution of the π-complex resonance extreme **A** (Scheme 5.15), attributed to the coordinative and electronic unsaturation at the transition metal resulting in the diminished extent of transition metal$\rightarrow$disilene π-back-donation.[68c] This was seen in the shortening of the silicon–silicon distance to 2.274(1) Å, insignificant pyramidalization at the coordinated Si atoms, and their remarkably low-field resonances of +65.3 ppm.[68c]

5.3.1.1.2 Digermenes >Ge=Ge<[5c,71]

Synthesis

Stable digermenes have been known since 1976, when Lappert's group communicated the first isolable derivative featuring a Ge=Ge double bond, $Dis_2Ge{=}GeDis_2$ **75**,[6a,b] whose crystal structure was reported several years later.[72] Tetraalkyldigermene **75** was prepared by the reaction of an isolable bis(amino)germylene $[(Me_3Si)_2N]_2Ge$: with DisLi [Dis = $CH(SiMe_3)_2$] (Scheme 5.16). Remarkably, although dimeric in its crystalline form (as was reliably established by X-ray crystallography), *digermene* **75** revealed chemical behavior matching that of the monomeric *germylene* Dis_2Ge:, implying the easy dissociation of the weak Ge=Ge donor–acceptor double bond in solution (see Section 5.3.1.1).

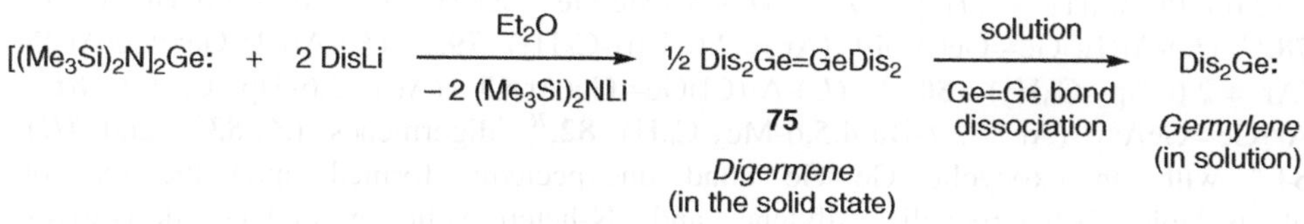

Scheme 5.16

The Ge=Ge double bond in **75** was not twisted and its length of 2.347(2) Å was determined for the first time, being notably shorter than that of the Ge–Ge single bonds (av. 2.44 Å) (Figure 5.9, Table 5.3). Given the intrinsically stronger bending at the heavier doubly bonded group 14 elements (see Section 5.3.1.1), it was not surprising that the digermene **75** manifested remarkable *trans*-bending of its Dis substituents with the sum of the bond angles at Ge centers amounting to 348.5°.

After Lappert's pioneering contribution, about forty digermenes (both acyclic and cyclic) were described as room temperature stable compounds, whose solid-state structures in the majority of cases were determined by X-ray diffraction analysis. There are

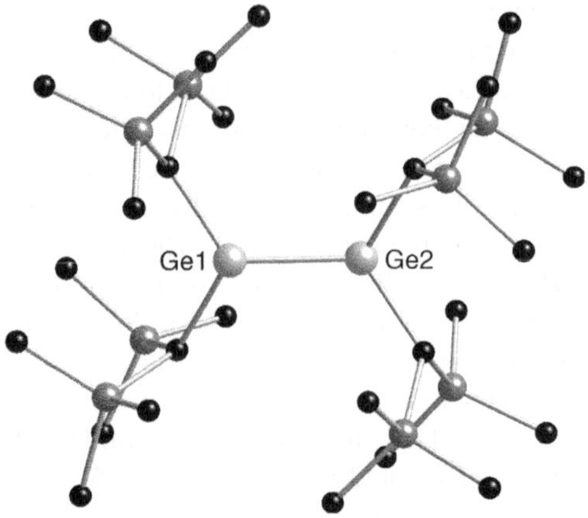

Figure 5.9 *Crystal structure of tetraalkyldigermene $Dis_2Ge{=}GeDis_2$ **75** (hydrogen atoms are not shown)*

four general synthetic procedures leading to isolation of stable acyclic digermenes: (1) dimerization of germylenes (A); (2) photolysis of cyclotrigermanes (B); (3) reductive dehalogenation of 1,1-dichlorogermanes (C); and (4) 1,2-addition or cycloaddition to the heavy analog of digermyne (D) (Scheme 5.17).

Method A: dimerization of germylenes

Apart from Lappert's compound **75**, several other stable digermenes have been synthesized by this approach by the coupling of RLi/RMgBr with :GeX_2-diox complex: (*E*)-Ar(Cl)Ge=Ge(Ar)Cl (Ar = 2,6-Mes_2-C_6H_3) **76**,[73] (*E*)-Ar(Cl)Ge=Ge(Ar)Cl [Ar = 2,6-(2,6-*i*-Pr_2-C_6H_3)$_2$-C_6H_3] **77**,[74] (*E*)-Ar(Me)Ge=Ge(Ar)Me (Ar = 2,6-Tip_2-C_6H_3) **78**,[75] (*E*)-Ar(Et)Ge=Ge(Ar)Et (Ar = 2,6-Tip_2-C_6H_3) **79**,[75] (*E*)-Ar(Ph)Ge=Ge(Ar)Ph (Ar = 2,6-Tip_2-C_6H_3) **80**,[75] (*E*)-Ar(Cl)Ge=Ge(Ar)Cl (Ar = 2,6-Tip_2-C_6H_3) **81**,[75] $Ar_2Ge{=}GeAr_2$ (Ar = 2-*t*-Bu-4,5,6-Me_3-C_6H) **82**,[76] digermenes (*Z*)-**83**[77] and (*E*)-**84**[78] with an exocyclic Ge=Ge bond unexpectedly formed upon the reaction of isolable N-heterocyclic silylene and N-heterocyclic germylene derivatives, $Tip_2Ge{=}GeTip_2$ **85** (crystallographic data reported for this digermene were not reliable because of disorder problems),[79,80] $Ar_2Ge{=}GeAr_2$ (Ar = 2,5-*t*-Bu_2-C_6H_3) **86**[81] and (*E*)-Bbt(Br)Ge=Ge(Bbt)Br **87**[82] (Table 5.3).

Similarly to Lappert's digermene **75**, some of the above-described digermenes were reported to be in equilibrium with the corresponding germylenes in solution $R_2Ge{=}GeR_2 \rightleftarrows 2\,R_2Ge$: (confirmed by UV-Vis studies and germylene trapping reactions): **82**,[76] **85**,[79] **87**.[82]

Method B: photolysis of cyclogermanes

As in the case of stable disilenes, this method was widely used by Masamune *et al.*: photolysis of hexaarylcyclotrigermanes bearing bulky aryl groups, *cyclo*-(Ar_6Ge_3) (Ar = 2,6-Me_2-C_6H_3 and 2,6-Et_2-C_6H_3), resulting in a cycloreversion to produce the corresponding tetraaryldigermenes $Ar_2Ge{=}GeAr_2$ (**88**: Ar = 2,6-Me_2-C_6H_3;[83] **89**: Ar = 2,6-Et_2-C_6H_3[84]) (Table 5.3). Digermene **88** with the smaller 2,6-Me_2-C_6H_3

Table 5.3 *Crystallographic parameters of the structurally characterized digermenes*

Digermene (Synthetic Method)	>Ge=Ge< (Å)	Σ(Ge) (°)	>Ge=Ge< twist (°)	Reference
$Dis_2Ge{=}GeDis_2$ **75** (**A**)	2.347(2)	348.5	0.0	72
(*E*)-Ar(Cl)Ge=Ge(Ar)Cl **76** (Ar = 2,6-Mes_2-C_6H_3) (**A**)	2.443(2)	333.8	NA	73
(*E*)-Ar(Cl)Ge=Ge(Ar)Cl **77** [Ar = 2,6-(2,6-*i*-Pr_2-C_6H_3)$_2$-C_6H_3] (**A**)	2.4624(4)	327.8	NA	74
(*E*)-Ar(Me)Ge=Ge(Ar)Me **78** (Ar = 2,6-Tip_2-C_6H_3) (**A**)	2.3173(3)	342.9	NA	75
(*E*)-Ar(Et)Ge=Ge(Ar)Et **79** (Ar = 2,6-Tip_2-C_6H_3] (**A**)	2.347(3)	343.0	NA	75
(*E*)-Ar(Ph)Ge=Ge(Ar)Ph **80** (Ar = 2,6-Tip_2-C_6H_3] (**A**)	2.3183(5)	348.4	NA	75
(*E*)-Ar(Cl)Ge=Ge(Ar)Cl **81** (Ar = 2,6-Tip_2-C_6H_3] (**A**)	2.363(2)	346.6	NA	75
$Ar_2Ge{=}GeAr_2$ **82** (Ar = 2-*t*-Bu-4,5,6-Me_3-C_6H] (**A**)	2.2521(8)	360.0	20.4	76
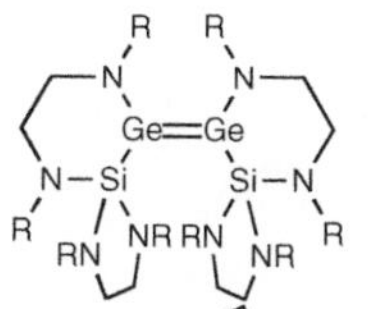 **(*Z*)-83** (R = *t*-Bu) (**A**)	2.454(2)	337.6 338.7	22.3	77
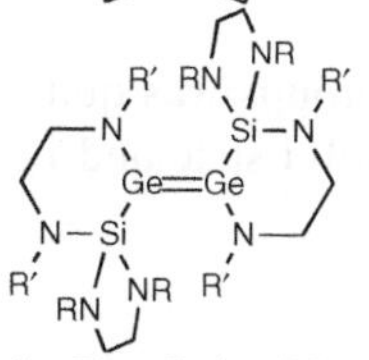 **(*E*)-84** (R = *t*-Bu, R' = *i*-Pr) (**A**)	2.460(1)	331.5	63.0	78
$Ar_2Ge{=}GeAr_2$ **86** (Ar = 2,5-*t*-Bu_2-C_6H_3] (**A**)	2.3643(4)	NA	NA	81
(*E*)-Bbt(Br)Ge=Ge(Bbt)Br **87** (**A**)	2.5087(7)	332.6	NA	82
$Ar_2Ge{=}GeAr_2$ **89** (Ar = 2,6-Et_2-C_6H_3] (**B**)	2.213(2)	358.4	10.0	84
(*Z*)-Mes(Ar)Ge=Ge(Mes)Ar **91** (Ar = 2,6-*i*-Pr_2-C_6H_3] (**C**)	2.301(1)	345.5	7.0	86
(*i*-Pr_2MeSi)$_2$Ge=Ge(SiMe*i*-Pr_2)$_2$ **92** (**C**)	2.267(1)	359.6	0.0	87
(*i*-Pr_3Si)$_2$Ge=Ge(Si*i*-Pr_3)$_2$ **94** (**C**)	2.298(1)	357.0	0.0	87
(*E*)-Tbt(Mes)Ge=Ge(Tbt)Mes **95** (**C**)	2.416(2)	355.3 356.6	NA	88b
(*E*)-Ar(H)Ge=Ge(Ar)H **96** [Ar = 2,6-(2,6-*i*-Pr_2-C_6H_3)$_2$-C_6H_3] (**D**)	2.3026(3)	NA	NA	89a
Ar$\ddot{G}$e=Ge(←:CN*t*-Bu)Ar **97** [Ar = 2,6-(2,6-*i*-Pr_2-C_6H_3)$_2$-C_6H_3] (**D**)	2.3432(9)	360.0 (three-coordinate Ge)	NA	91

NA, Not available.

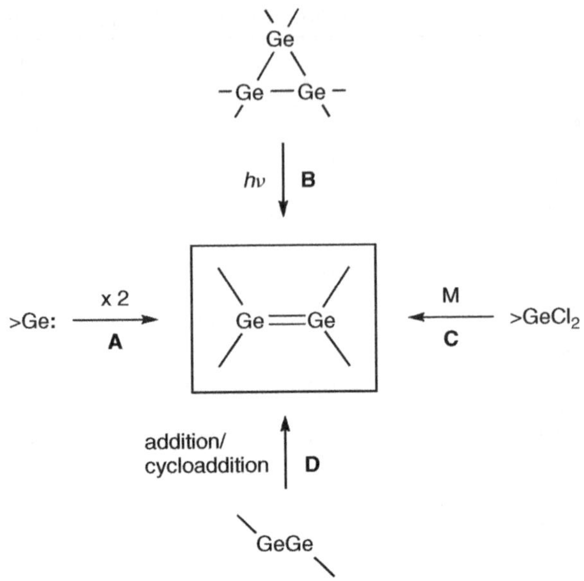

Scheme 5.17

substituents was metastable, whereas **89** featuring larger 2,6-Et_2-C_6H_3-groups was quite stable, being the first digermene to retain its Ge=Ge bond both in the solid state and in solution.

Method C: reductive dehalogenation of 1,1-dichlorogermanes

Several isolable digermenes have been prepared by this synthetic route, including several tetrasilyldigermenes: Ar_2Ge=$GeAr_2$ (Ar = 2,6-*i*-Pr_2-C_6H_3) **90**,[85] Tip_2Ge=$GeTip_2$ **85**,[85] (*Z*)-Mes(Ar)Ge=Ge(Mes)Ar (Ar = 2,6-*i*-Pr_2-C_6H_3) **91**,[86] (*i*-Pr_2MeSi)$_2$Ge=Ge(SiMe*i*-Pr_2)$_2$ **92**,[87] (*t*-BuMe_2Si)$_2$Ge=Ge(SiMe_2*t*-Bu)$_2$ **93**,[87] (*i*-Pr_3Si)$_2$Ge=Ge(Si*i*-Pr_3)$_2$ **94**[87] and (*E*)-Tbt(Mes)Ge=Ge(Tbt)Mes **95**[88] (Table 5.3).

Compound **95**, bearing extremely bulky Tbt and Mes substituents, underwent room temperature dissociation of the Ge=Ge bond, thus existing as a digermene Tbt(Mes)Ge=Ge(Tbt)Mes at low temperature (orange-yellow, 439 nm [$\pi \rightarrow \pi^*$]) and as a germylene Tbt(Mes)Ge: at room temperature (blue, 575 nm [$n \rightarrow$ p]).[88] The weakness of the Ge=Ge bond in **95**, and hence its ease of breaking, was manifested in its low bond dissociation energy of 14.7 kcal/mol. The germylene Tbt(Mes)Ge: formed upon the dissociation of digermene **95** was successfully trapped with MeOH, Et_3SiH and 2,3-dimethylbuta-1,3-diene.[88]

Tetraaryldigermene (*Z*)-Mes(Ar)Ge=Ge(Mes)Ar **91** underwent smooth *cis–trans* isomerization at room temperature in solution, forming an equilibrium mixture of (*Z*)- and (*E*)-isomers in a 2.1 ratio.[86] Because no germylenes could be trapped upon this rearrangement, the Ge=Ge bond dissociation–recombination pathway was ruled out. Instead, a simple rotation about the Ge=Ge bond was proposed as the most likely mechanistic scenario of this isomerization.

Method D: 1,2-addition or cycloaddition to the heavy analogs of digermyne

This is a very new method for the synthesis of stable digermenes, which became synthetically available only after the preparation of the germanium analog of alkynes ArGeGeAr [Ar = 2,6-(2,6-*i*-Pr_2-$C_6H_3)_2$-C_6H_3] (sometimes called a 'digermyne'). Thus, 1,2-addition of dihydrogen to ArGeGeAr resulted in the formation of hydrogen-substituted digermene (*E*)-Ar(H)Ge=Ge(Ar)H **96**,[89] whereas [2 + 2] cycloaddition of Ph–C≡C–Ph to ArGeGeAr produced a unique cyclobutadiene derivative featuring endocyclic Ge=Ge and C=C bonds[90] (details of this interesting compound are discussed in Section 5.3.2). Finally, reaction of ArGeGeAr with *t*-BuNC: yielded the 1.1 isonitrile complex Ar$\ddot{\text{G}}$e=Ge(←:CN*t*-Bu)Ar **97**, in which the isonitrile unit was coordinated to only one of the germanium atoms, thus supporting the importance of the dipolar resonance form Ar–$\ddot{\text{G}}$e$^-$ = Ge$^+$–Ar ↔ Ar–Ge$^+$ = $\ddot{\text{G}}$e$^-$–Ar contributing to the overall bonding of ArGeGeAr.[91]

Crystal structure of digermenes

In the absence of a magnetically useful nucleus (see Section 5.3.1.1), the major part of the structural information on digermenes can be extracted from their X-ray crystallographic studies. The general trends in the effect of substituents on the geometry of the Ge=Ge bond are quite parallel to those for disilenes, namely: electronegative substituents typically cause appreciable *trans*-bending at the sp^2-Ge, stretching and weakening of the Ge=Ge bond, whereas introduction of electropositive groups results in planarization at the sp^2-Ge, shortening and strengthening of the Ge=Ge bond. Steric repulsion between very bulky substituents may also contribute to the structural distortion of the Ge=Ge bond, causing its stretching and twisting. Compared with Si=Si bonds, Ge=Ge bonds are typically longer, weaker and more prone to structural distortions (*trans*-bending at Ge), which is quite expected given the well-documented increase in π–σ^* orbital mixing descending group 14.

Tetraalkyldigermenes are represented by only one example, Dis_2Ge=$GeDis_2$ **75**, for which a relatively long, *trans*-bent and not twisted Ge=Ge bond was reported: 2.347(2) Å (bond length); 348.5° (pyramidality); 0.0° (twisting) (Table 5.3).

Tetraaryldigermenes generally exhibit somewhat shorter and less *trans*-bent Ge=Ge bonds, which, however, may experience twisting deformation in the case of highly sterically demanding substituents: from 2.213(2) Å (for **89**) to 2.416(2) Å (for **95**) (bond length); from 345.5° (for **91**) to 360.0° (for **82**) (pyramidality); from 7.0° (for **91**) to 20.4° (for **82**) (twisting) (Table 5.3).

In accord with theoretical predictions, and similarly to tetrasilyldisilenes, tetrasilyldigermenes are deformed rather insignificantly and typically feature relatively short Ge=Ge bonds: from 2.267(1) Å (for **92**) to 2.298(1) Å (for **94**) (bond length); from 357.0° (for **94**) to 359.6° (for **92**) (pyramidality); 0.0° (for **92** and **94**) (twisting) (Table 5.3).

The geometry of digermenes with mixed substitution patterns depends on the nature of the substituents. Thus, aryl(alkyl)digermenes revealed moderate stretching and pyramidalization of the Ge=Ge bond: from 2.3173(3) Å (for **78**) to 2.347(3) Å (for **79**) (bond length); from 342.9° (for **78**) to 343.0° (for **79**) (pyramidality) (Table 5.3). An increase in the electron accepting power of substituents results in the stretching of the Ge=Ge bond and more pronounced *trans*-bending at Ge. Accordingly, aryl(halo)digermenes with electronegative halogen substituents exhibited the following structural characteristics: from

2.363(2) Å (for **81**) to 2.5087(7) Å (for **87**) (bond length); from 327.8° (for **77**) to 346.6° (for **81**) (pyramidality) (Table 5.3). In fact, the Ge=Ge distance of 2.5087(7) Å in **87** is the longest distance ever reported for a digermene (even exceeding the usual Ge–Ge single bond length of 2.44 Å), which pointed to the intrinsic weakness of this donor–acceptor-type bonding between the two germanium atoms.

The digermenes (*Z*)-**83** and (*E*)-**84**, featuring exocyclic Ge=Ge bonds substituted by both silicon and nitrogen atoms, represent another case of remarkable structural deformations (long Ge=Ge bond, highly pyramidal geometry at Ge, extreme twisting about the Ge=Ge bond): from 2.454(2) Å [for (*Z*)-**83**] to 2.460(1) Å [for (*E*)-**84**] (bond length); from 331.5° [for (*E*)-**84**] to 338.7° [for (*Z*)-**83**] (pyramidality); from 22.3° [for (*Z*)-**83**] to 63.0° [for (*E*)-**84**] (twisting) (Table 5.3). Digermene (*E*)-**84** manifests the most twisted Ge=Ge bond reported to date: 63.0°.

Cyclic digermenes with endocyclic Ge=Ge bonds

Seventeen organometallic derivatives possessing endocyclic Ge=Ge bonds have been reported in the literature: thirteen three-membered ring, two four-membered ring and two five-membered ring compounds. Their synthesis and particular structural features are described below.

Three-membered ring compounds (heavy cyclopropenes)

All stable cyclotrigermenes reported to date are stabilized by electropositive silyl substituents (Table 5.4). The first isolable cyclotrigermenes **98, 99** were readily available by the reaction of *t*-Bu_3EM (E = Si, M = Na; E = Ge, M = Li) with $GeCl_2$•diox (Scheme 5.18, Table 5.4).[92]

Several other stable cyclotrigermenes were synthesized by taking advantage of the readily available cyclotrigermenylium ion derivatives (see Chapter 6, Section 6.3.1). Thus, a series of unsymmetrically substituted cyclotrigermenes **100–103** was prepared by the coupling of tris(tri-*tert*-butylsilyl)cyclotrigermenylium tetrakis[3,5-bis(trifluoromethyl)phenyl]borate **104^+•B[3,5-$(CF_3)_2$-C_6H_3]$_4^-$** with the corresponding silyl-, germyl- and aryl-alkali metal salts RM (Scheme 5.19, Table 5.4).[93]

Likewise, treatment of tris(tri-*tert*-butylsilyl)cyclotrigermenylium tetrakis(2,3,5,6-tetrafluorophenyl)borate **104^+•B[2,3,5,6-F_4-C_6H]$_4^-$** with potassium halides KX (X = Cl, Br, I) resulted in the formation of halogen-substituted cyclotrigermenes **105–107** (Scheme 5.20, Table 5.4).[94]

Interaction of tetrachlorodigermane R–$GeCl_2$–$GeCl_2$–R (R = SiMe*t*-Bu_2) with either 1,1-dilithiosilane R_2SiLi_2 or dilithiogermane R_2GeLi_2 was proved to be another effective route for the design of heavy cyclopropene analogs with skeletal Ge=Ge bonds: 1*H*-siladigermirene **108** and cyclotrigermene **109** (Scheme 5.21, Table 5.4).[95] A three-membered ring dichloride **A** was proposed to be formed as the first intermediate, followed by its rapid Li↔Cl exchange process with the second equivalent of R_2ELi_2 (E = Si or Ge) to give digermenoid **B**, which finally underwent β-elimination of LiCl (Scheme 5.21). According to this mechanism, a half-equivalent of the corresponding disilene R_2Si=SiR_2 or digermene R_2G=GeR_2 should also be formed along with the cyclic products **108** or **109**, which was supported by experimental observations.[95]

Alkyl-substituted heavy cyclopropenes **110** and **111**, the nearest homologs of the above-described 1*H*-siladigermirene **108** and cyclotrigermene **109**, distinguished from

Table 5.4 *Crystallographic parameters of the structurally characterized cyclic digermenes*

Cyclic digermene	>Ge=Ge< (Å)	Σ(Ge) (°)	>Ge=Ge< twist (°)	Ring Folding (°)	Reference
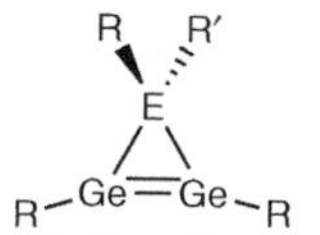					
98 (E = Ge; R = R′ = Si*t*-Bu_3)	2.239(4)	359.9	0.0		92
101 [E = Ge; R = Si*t*-Bu_3, R′ = $Si(SiMe_3)_3$]	2.264(2)	358.7 359.8	8.1		93
105 (E = Ge; R = Si*t*-Bu_3, R′ = Cl)	2.2723(8)	357.2 359.3	29.7		94
106 (E = Ge; R = Si*t*-Bu_3, R′ = Br)	2.2743(8)	343.7 359.4	34.9		94
107 (E = Ge; R = Si*t*-Bu_3, R′ = I)	2.2721(6)	344.8 359.6	35.3		94
108 (E = Si; R = R′ = SiMe*t*-Bu_2)	2.2429(6)	353.8 354.2	51.0		95a
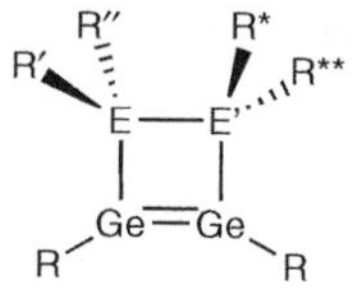					
116 (E = E′ = Si; R = R′ = R** = SiMe*t*-Bu_2; R″ = R* = Cl)	2.2911(4)	357.2 358.6	NA	28.3	97
118 [E = Ge, E′ = C; R = R′ = R″ = Tip; R* + R** : = N-(2-MeO-C_6H_4)]	2.2808(7)	NA	NA	NA	80
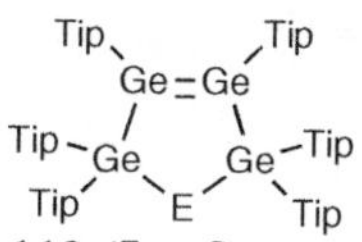					
119 (E = S)	2.2841(5)	348.4 351.4	NA	NA	98
120 (E = Se)	2.2975(5)	349.2 349.9	NA	NA	80

NA, Not available.

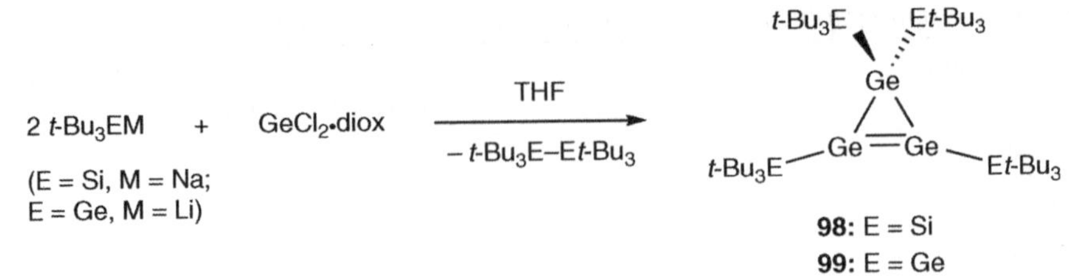

Scheme 5.18

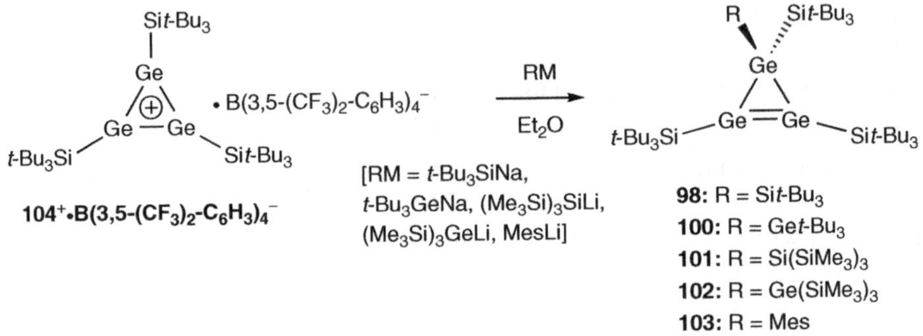

Scheme 5.19

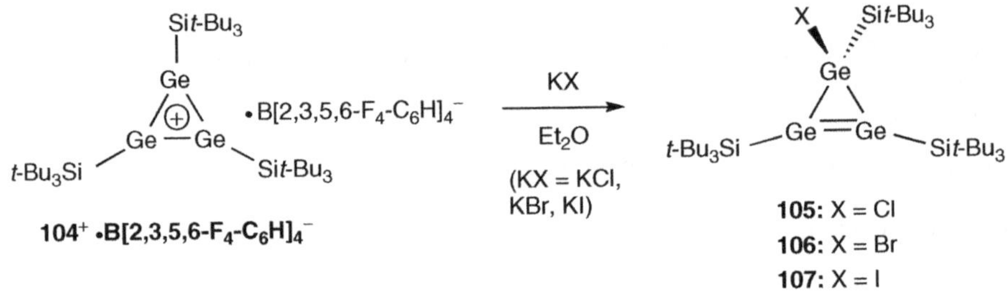

Scheme 5.20

them by only a CH_2-unit, were synthesized by the reductive dehalogenation of dichlorocyclobutane derivatives **112** and **113** with potassium graphite (Scheme 5.22).[96] In contrast to their silicon versions (see Section 5.3.1.1.1), the initially formed bicyclo[1.1.0]butanes **114** and **115** were not isolable and were isomerized at room temperature to more stable cyclopropene analogs **110** and **111** (Scheme 5.22).

As was the case for cyclotrisilenes, cyclotrigermenes revealed Ge=Ge bonds that are slightly shorter than those of acyclic tetrasilyldigermenes: 2.239(4) Å (for **98**) to 2.2743(8) Å (for **106**) vs 2.267(1) Å (for **92**) to 2.298(1) Å (for **94**) (Tables 5.3 and 5.4). The longest Ge=Ge bonds were found in halogen-substituted cyclotrigermenes **105**–**107**

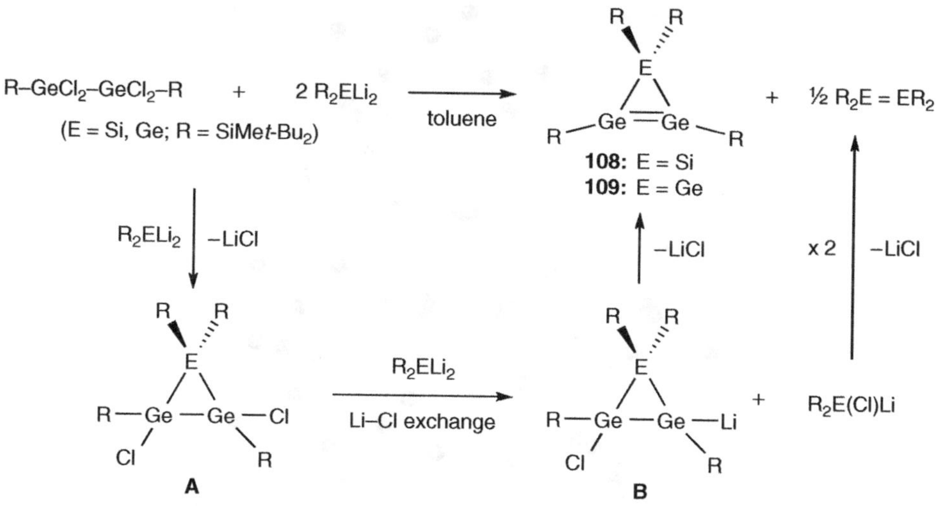

Scheme 5.21

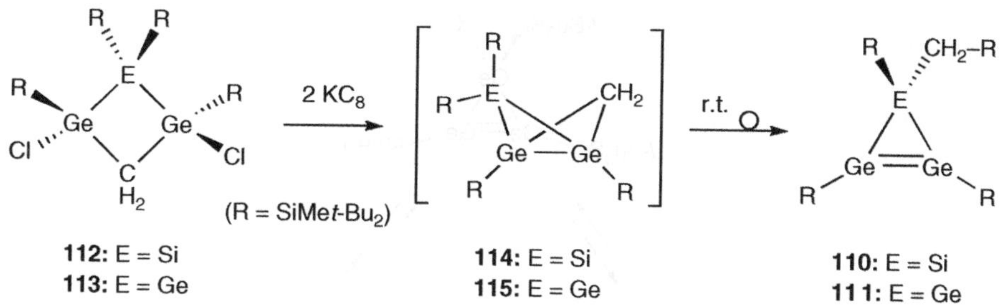

Scheme 5.22

[2.2721(6) –2.2743(8) Å], which was attributed to the effective π(Ge=Ge)–σ^*(Ge–X) orbital mixing facilitated by the electronegative halogens at the sp^3-Ge (Figure 5.10).

Interestingly, halogen-substituted cyclotrigermenes **105–107** disclosed dynamic behavior in solution, namely Cl/Br/I-walk over the cyclic Ge_3-skeleton, with the heavier halogens migrating more easily (Scheme 5.23).[94] Kinetic measurements indicated that such a rearrangement was most likely to proceed intramolecularly without intermediate formation of the cyclotrigermenylium ion, in contrast to the dissociation–recombination mechanism proposed for R_3C_3X rearrangements facilitated by the formation of the aromatic cyclopropenylium ion.

Cyclotrigermenes mostly manifest normal *trans*-bent geometry of the substituents at the doubly bonded Ge atoms, apart from cyclotrigermenes **101, 106** and **107**, which revealed unusual *cis*-bent arrangements as a result of electronic and steric interplay with the substituents at the sp^3-Ge atom.

With the exception of cyclotrigermene **98**, which exhibits perfectly planar geometry at the Ge=Ge bond, and in contrast to nontwisted tetrasilyldigermenes **92** and **94**,

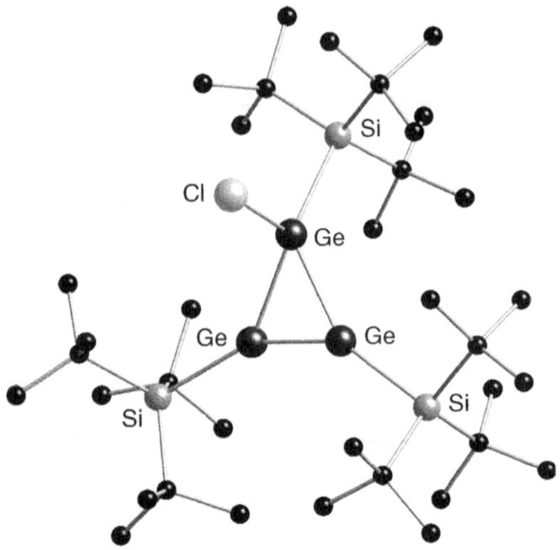

Figure 5.10 Crystal structure of Cl-substituted cyclotrigermene **105** (hydrogen atoms are not shown)

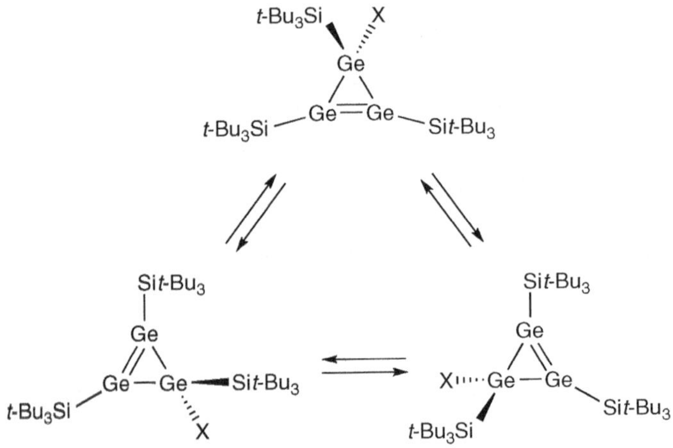

Scheme 5.23

cyclotrigermenes featured remarkable twisting about the Ge=Ge bond: from 8.1° (for **101**) to 60.5° (for **109**). Cyclotrigermene **109** revealed a record twisting of the endocyclic Ge=Ge bond in heavy cyclopropenes (60.5°), a value that closely approaches that of the most twisted digermene (*E*)-**84** with an exocyclic Ge=Ge bond (63.0°). Overall, the degree of twisting/bending at the endocyclic E=E bonds in cyclotrimetallenes depends on the nature of both the sp^2- and the sp^3-skeletal atoms. This can be clearly visualized by comparison of the twisting/bending angles for a series of heavy cyclopropenes of

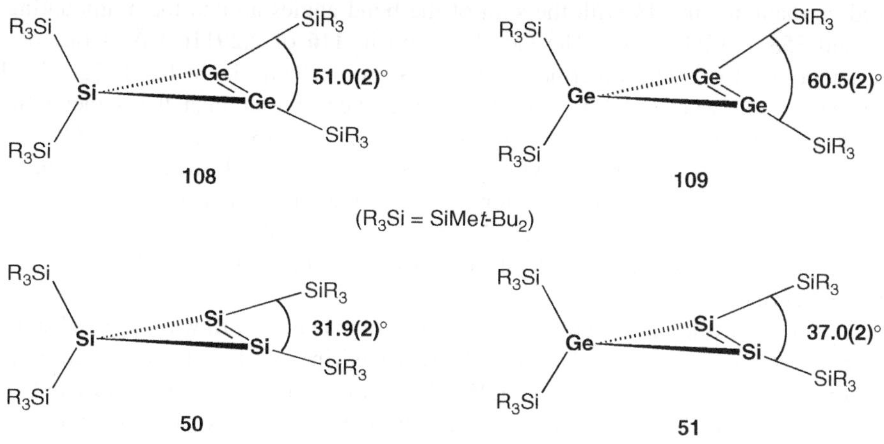

Figure 5.11

Si and Ge atoms with an identical substitution pattern (Figure 5.11).[95b] The following trends become obvious: (1) Ge=Ge bonds are more twisted than Si=Si bonds: 51.0° (in **108**) vs 31.9° (in **50**) and 60.5° (in **109**) vs 37.0° (in **51**); and (2) Ge=Ge (or Si=Si) bonds are more twisted when the sp^3-skeletal atom is Ge but not Si: 60.5° (in **109**) vs 51.0° (in **108**) and 37.0° (in **51**) vs 31.9° (in **50**). The former is due to the increase in *trans*-bending on descending group 14 (see Section 5.3.1.1), whereas the latter originates from the electronegativity difference between the Si and Ge atoms: more electronegative substituents R cause greater pyramidalization at the doubly bonded atoms E in heavy alkenes $R_2E{=}ER_2$.

Four-membered ring compounds (heavy cyclobutenes)

The first compound of this class, $^3\Delta$-1,2,3,4-disiladigermetene **116**, was unexpectedly formed upon the ring expansion reaction of either 3*H*-disilagermirene **51** or 1*H*-disilagermirene **117** with $GeCl_2$•diox (Scheme 5.24).[97]

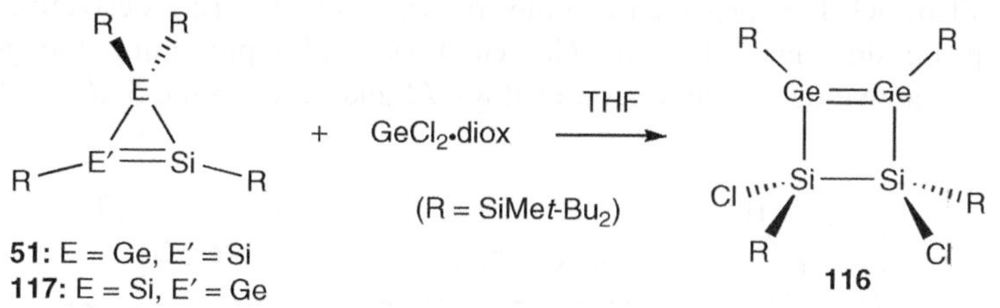

Scheme 5.24

The reaction of **51** with $GeCl_2$•diox forming **116** is, in effect, a rather curious transformation of a Si=Si bond into a Ge=Ge bond, representing a way of making a digermene from a disilene. There is an unremarkable pyramidalization at the doubly bonded germaniums in **116** with the sum of the bond angles around them amounting to 357.2 and 358.6° (Table 5.4). The Ge=Ge bond in **116** of 2.2911(4) Å is one of the longest reported for cyclic digermenes, whereas the skeletal Ge–Si bonds [2.3576(10) and 2.3585(9) Å] were notably squeezed, being even shorter than the skeletal Si–Si bond of 2.3722(10) Å. Accordingly, the Si–Cl bonds of 2.1325(10) and 2.1345(10) Å were evidently stretched compared with typical Si–Cl bonds. Such structural peculiarities (elongation of the Ge=Ge bond, shortening of the Ge–Si bonds and elongation of the Si–Cl bonds) were ascribed to π(Ge=Ge)–σ^*(Si–Cl) orbital mixing, facilitated by the electronegative chlorines at the skeletal Si atoms and folding of the Ge_2Si_2-ring (28.3°) favoring such hyperconjugation.

Another heavy cyclobutene with an endocyclic Ge=Ge bond **118** was prepared starting from tetragermabuta-1,3-diene Tip_2Ge=Ge(Tip)–Ge(Tip)=$GeTip_2$ by its reaction with 2-methoxyphenyl isonitrile (Scheme 5.25).[80] For this cyclic digermene **118** only the Ge=Ge bond distance of 2.2808(7) Å was reported, which was long compared with those in cyclotrigermenes (Table 5.4).

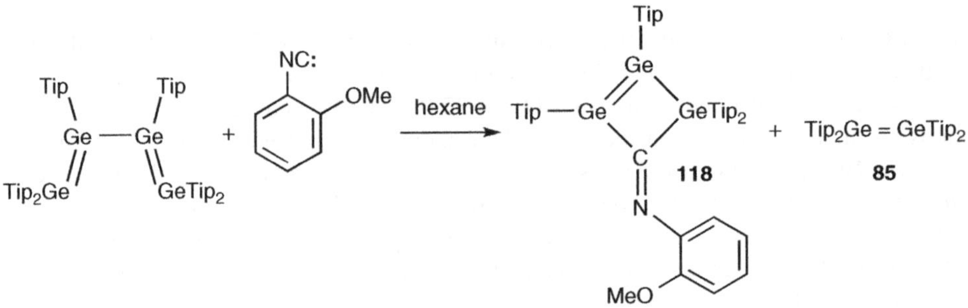

Scheme 5.25

Five-membered ring compounds (heavy cyclopentenes)

[1 + 4] Cycloaddition of sulfur or selenium to tetragermabuta-1,3-diene Tip_2Ge=Ge(Tip)–Ge(Tip)=$GeTip_2$ provided a route to five-membered ring derivatives **119** and **120**, each possessing an endocyclic Ge=Ge bond and representing the germanium versions of the above-described silicon analogs **72** and **73** (Scheme 5.26).[80,98]

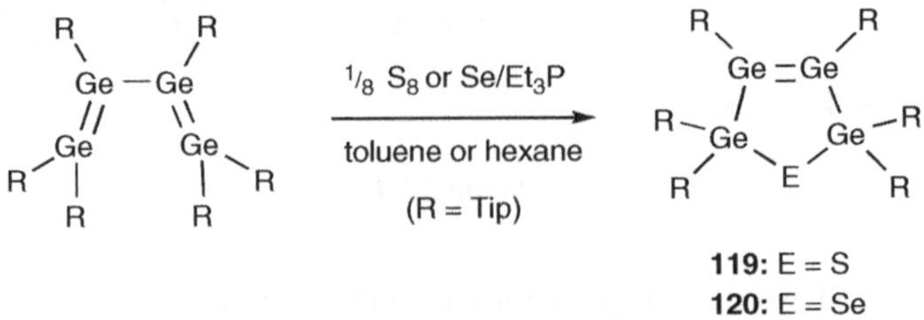

Scheme 5.26

As is typical for heavy cyclopentenes, the Ge=Ge bonds in both **119** and **120** were longer than those in heavy cyclopropenes but rather similar to those in heavy cyclobutenes: 2.2841(5) Å for **119** and 2.2975(5) Å for **120** (Table 5.4). The five-membered rings in both **119** and **120** are rather insignificantly folded, whereas the sp^2-Ge atoms are notably pyramidalized: 348.4/351.4° (in **119**) and 349.2/349.9° (in **120**) (Table 5.4).[80,98]

5.3.1.1.3 Distannenes >Sn=Sn<[5c,71b,d]

Synthesis

Tetraalkyldistannene $Dis_2Sn{=}SnDis_2$ **121** was the first isolable alkene analog of the heavy group 14 elements, reported by Lappert *et al.* as early as 1976,[6a,b,72b,99a,b] that is several years before the preparation of the most widely cited $Mes_2Si{=}SiMes_2$ **1** by West *et al.*[13a] The milestone synthesis of distannene **121** was accomplished by the coupling of an isolable bis(amino)stannylene $[(Me_3Si)_2N]_2Sn$: with alkyllithium derivative DisLi [Dis = $CH(SiMe_3)_2$] (Scheme 5.27, Table 5.5).[6a,b] Similarly to the analogous digermene **75**, distannene **121** was monomeric in solution (stannylene Dis_2Sn:) and dimeric in the solid state (distannene $Dis_2Sn{=}SnDis_2$). Again, similarly to the corresponding digermene **75**, distannene **121** exhibited pronounced pyramidal geometry at the tin atoms (sum of the bond angles 342°) and an undistorted Sn–Sn bond (twist angle 0°) (Table 5.5).[6a,b] The Sn=Sn bond length of 2.764(2) Å was determined for the first time, being remarkably long and comparable with those of typical Sn–Sn single bonds.

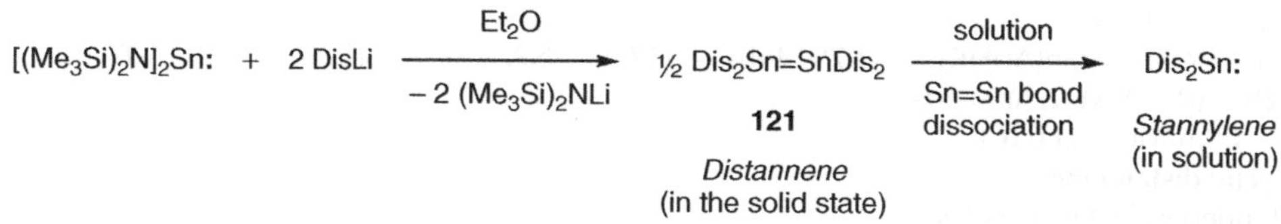

Scheme 5.27

Since then, eight other stable distannenes (including one three-membered ring cyclic derivative, cyclotristannene) have been isolated and characterized. They were synthesized by three general approaches: (1) dimerization of stannylenes (A); (2) photolysis of cyclotristannanes (B); and (3) ligand exchange between the homoleptic stannylenes R_2Sn: and R'_2Sn: to form unsymmetrically substituted distannenes (C) (Scheme 5.28).

Method A: dimerization of stannylenes

Several stable distannenes have been prepared by the dimerization of intermediate stannylenes: (*E*)-Ar(R)Sn=Sn(Ar)R [R = CH_2(4-*t*-Bu-C_6H_4), Ar = 2,6-(2,6-*i*-Pr_2-C_6H_3)$_2$-C_6H_3] **122**,[100] $Ar_2Sn{=}SnAr_2$ (Ar = 2-*t*-Bu-4,5,6-Me_3C_6H) **123**,[101] $[(Me_3Si)_3Si]_2Sn{=}Sn[Si(SiMe_3)_3]_2$ **124**,[102a,b] (*E*)-Mes[$(Me_3Si)_3Si$]Sn=Sn(Mes)[$Si(SiMe_3)_3$] **125**[102b,c] and (*t*-$Bu_2MeSi)_2Sn{=}Sn(SiMe$*t*-$Bu_2)_2$ **126**[103] (Table 5.5). All of these distannenes $R_2Sn{=}SnR_2$, except for the tetrasilyldistannene **126**, undergo dissociation of the Sn=Sn bond in solution into a pair of corresponding stannylenes R_2Sn:.

Table 5.5 *Crystallographic and spectral parameters of the structurally characterized distannenes*

Distannene (Synthetic Method)	>Sn=Sn< (Å)	Σ(Sn) (°)	>Sn=Sn< twist (°)	^{29}Sn NMR (ppm)	Reference
$Dis_2Sn{=}SnDis_2$ **121** (**A**)	2.764(2)	342.0	0.0	NA	6a,b
(*E*)-Ar(R)Sn=Sn(Ar)R **122** [R = CH_2(4-*t*-Bu-C_6H_4]; Ar = 2,6-(2,6-*i*-Pr_2-C_6H_3)$_2$-C_6H_3] (**A**)	2.7705(8)	329.7	NA	1205.7	100
		NA	13.0		
$Ar_2Sn{=}SnAr_2$ **123** (Ar = 2-*t*-Bu-4,5,6-Me_3C_6H) (**A**)	2.910(1)	320.9 354.9	44.0	1401.0 (sol.) 819.0 (CP-MAS)	101
$(R_3Si)_2Sn{=}Sn(SiR_3)_2$ **124** (R = $SiMe_3$) (**A**)	2.8247(6)	352.0	63.2	NA	102a,b
(*E*)-Ar(R_3Si)Sn=Sn(Ar)SiR_3 **125** (R = $SiMe_3$, Ar = Mes) (**A**)	2.7023(8)	NA	NA	NA	102b,c
(*t*-$Bu_2MeSi)_2Sn{=}Sn(SiMe$*t*-$Bu_2)_2$ **126** (**A**)	2.6683(10)	360.0	44.6	630.7	103
(*E*)-Ar(R_3Si)Sn=Sn(Ar)SiR_3 **128** [R = $SiMe_3$, Ar = 2,4,6-$(CF_3)_3$-C_6H_2] (**C**)	2.833(1)	337.9	NA	NA	106
(*E*)-Ar(R_3Si)Sn=Sn(Ar)SiR_3 **129** (R = $SiMe_3$, Ar = 2-*t*-Bu-4,5,6-Me_3-C_6H) (**C**)	2.7914(4)	337.4	NA	1506	107
Cyclic distannene (2 independent molecules): 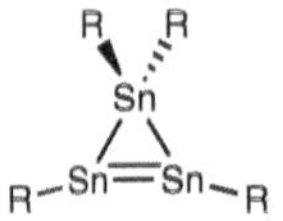 **130** (R = Si*t*-Bu_3)	2.601(3)/ 2.582(4)	350.8 355.8/ 359.6 360.0	NA	412.0	108

NA, Not available.

Method B: photolysis of cyclotristannanes

Similarly to the case of stable disilenes and digermenes, this method was developed by Masamune *et al.*: photolysis of a hexaarylcyclotristannane with very bulky Tip groups, *cyclo*-(Tip_6Sn_3), resulted in the formation of $Tip_2Sn{=}SnTip_2$ **127**.[104] Although not isolated in pure form, distannene **127** was proved to retain its Sn=Sn bond in the condensed phase as well, as demonstrated by the observation of its ^{119}Sn NMR

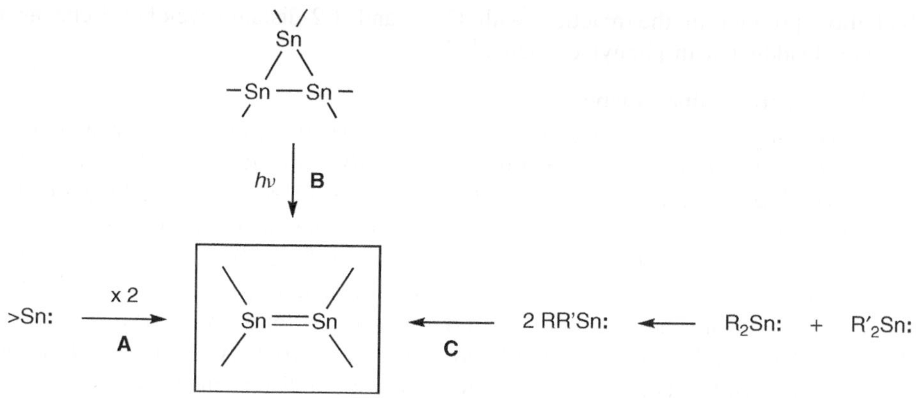

Scheme 5.28

resonance at 427.3 ppm.[104] This was later supported by the characteristic reactivity of **127**, corresponding to that expected for the Sn=Sn doubly bonded derivative.[105]

Method C: ligand exchange between the homoleptic stannylenes R_2Sn: and R'_2Sn: to form unsymmetrically substituted distannenes RR′Sn=SnRR′

This method was employed by the Weidenbruch and Klinkhammer groups, who prepared the following 'mixed' distannenes: (*E*)-Ar[$(Me_3Si)_3Si$]Sn=Sn(Ar)[$Si(SiMe_3)_3$] [Ar = 2,4,6-$(CF_3)_3$-C_6H_2] **128**[106] and (*E*)-Ar[$(Me_3Si)_3Si$]Sn=Sn(Ar)[$Si(SiMe_3)_3$] (Ar = 2-*t*-Bu-4,5,6-Me_3-C_6H) **129**.[107] Distannenes formed by this method also dissociate in solution into the corresponding stannylenes.

^{119}Sn NMR spectroscopy of distannenes

The ^{119}Sn NMR signals of the isolable acyclic distannenes in solution were observed in the low-field region well above +1 000 ppm, which is in the characteristic region of stannylene resonances (Table 5.5). This clearly pointed to a dissociation (complete or partial) of the Sn=Sn bond into the stannylenes >Sn: in solution. Accordingly, these distannenes react in the condensed phase as stannylenes but not as derivatives with real Sn=Sn double bonds. Although distannene Tip_2Sn=$SnTip_2$ **127** was reported to retain its structural integrity in solution [δ(^{119}Sn): 427.3 ppm], manifesting reactivity typical of the Sn=Sn bond, it could not be isolated as an individual compound.[104,105] To date, the only isolable acyclic distannene, for which the presence of its Sn=Sn bond both in the solid state and in solution has been reliably established, is tetrasilyldistannene (*t*-$Bu_2MeSi)_2Sn$=Sn(SiMe*t*-$Bu_2)_2$ **126** prepared by the reaction of *t*-$Bu_2MeSiNa$ with $SnCl_2$•diox in THF [surprisingly, when the same reaction was carried out in Et_2O a totally different product, the stable radical (*t*-$Bu_2MeSi)_3Sn$•, was formed; see Section 2.4.1.2].[103] The signal of the sp^2-Sn atoms in **126** was observed at 630.7 ppm in C_6D_6 solution, which is in the region of Sn=Sn bonds but clearly beyond the range of stannylene >Sn: resonances. The structural integrity of the Sn=Sn bond of **126** in solution was further corroborated by reactivity studies: **126** featured the chemical properties not of stannylene but of distannene exclusively, giving 1,2-dichlorodistannane as a

1,2-addition product in the reaction with CCl_4 and 1,2-distannacyclobut-3-ene as a [2 + 2] cycloadduct with phenylacetylene.[103b]

Crystal structure of distannenes

As expected, alkyl-, aryl- and (alkyl)aryl-substituted distannenes revealed pyramidalization at their tin centers far greater than that of their silyl-substituted counterparts: 320.9 to 354.9° (in **123**) vs 352.0° (in **124**) to 360.0° (in **126**) (Table 5.5). The length of the Sn=Sn bonds in distannenes is highly sensitive to both the electronic influence of substituents and their steric bulk, being mostly observed in the 2.70–2.83 Å range, values that are nearly equal to (or even longer than) those of regular Sn–Sn single bonds. The two extremes in distannenes are represented by the shortest (and strongest) Sn=Sn bond in **126** of 2.6683(10) Å and the longest bond in **123** of 2.910(1) Å (Table 5.5). Extraordinarily bulky substituents cause not only stretching of the distannene bond, as observed in tetraaryldistannene **123** and tetrasilyldistannene **124**, but also extreme twisting of their Sn=Sn bonds: 44.0° (in **123**) and 63.2° (in **124**).

Tetrasilyldistannene $(t\text{-}Bu_2MeSi)_2Sn{=}Sn(SiMe\textit{t}\text{-}Bu_2)_2$ **126**, which does not dissociate in solution, revealed very unusual structural features: its Sn=Sn bond of 2.6683(10) Å is very short and highly twisted (44.6°), and the geometry about the Sn atoms is ideally planar (sum of the bond angles 360°) (Table 5.5, Figure 5.12).[103] All of these structural peculiarities definitely result from silyl-substitution, which causes appreciable shortening and planarization of the Sn=Sn bond, producing at the same time its remarkable twisting because of the steric repulsion between the bulky substituents. Definitely, the novel structural motif in **126**, unprecedented in distannene chemistry (short, nonpyramidal and highly twisted Sn=Sn bond), does not adequately fit any of the existing >E=E< bonding models (E = heavy group 14 element): neither triplet–triplet interaction resulting in a pair of covalent bonds (Figure 5.1, **A**), nor singlet–singlet interaction producing two equivalent donor–acceptor bonds (Figure 5.1, **C**). Alternatively, the particular bonding situation in **126** was realized in the framework of a novel interaction mode: namely, out-of-plane (rather than in-plane) interaction of the two triplet stannylenes (Figure 5.13).

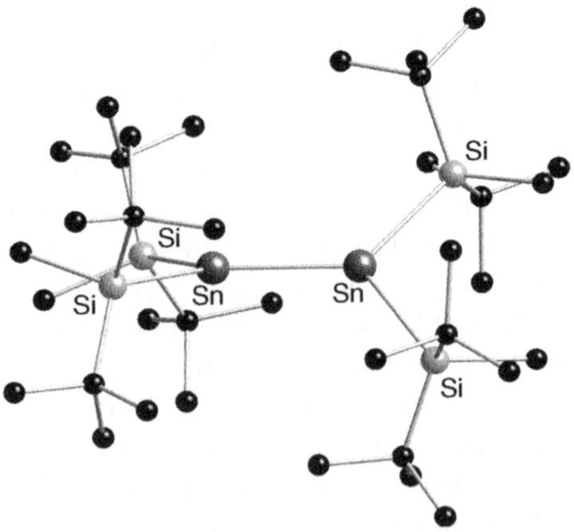

*Figure 5.12 Crystal structure of tetrasilyldistannene **126** (hydrogen atoms are not shown)*

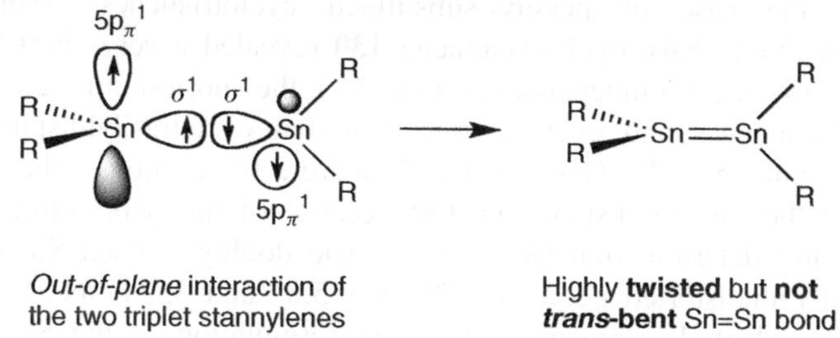

Figure 5.13 *Orbital interaction mode in tetrasilyldistannene* ***126****: out-of-plane interaction of the two triplet stannylenes forming a highly twisted but not trans-bent Sn=Sn bond*

The triplet state of stannylene ($(t\text{-Bu}_2\text{MeSi})_2\text{Sn}$: might be accessible because of its relatively small singlet–triplet energy separation $\Delta E_{S\text{-}T} = 8.5$ kcal/mol (B3LYP/SDD DFT level).[103b] This remarkable decrease in the $\Delta E_{S\text{-}T}$ in ($(t\text{-Bu}_2\text{MeSi})_2\text{Sn}$: is attributed to the electronic effects of the electropositive silyl substituents and their large steric bulk, widening the Si–Sn–Si bond angle to 109.6°. Computations predicted a progressive increase in $\Delta E_{S\text{-}T}$ values upon the decrease in the size and increase in the electronegativity of R groups in disubstituted stannylenes R_2Sn: ($\Delta E_{S\text{-}T}$ in kcal/mol): 8.5 (for R = SiMe*t*-Bu$_2$), 10.3 (for R = SiMe$_3$), 18.0 (for R = SiH$_3$), 26.5 (for R = H) and 29.2 (for R = Me) (B3LYP/SDD level).[103b]

Accordingly, deep-green distannene **126** revealed a remarkable red shift in its UV-Vis spectrum with the longest wavelength absorption at $\lambda = 670$ nm (compare with 494 nm observed for Tip$_2$Sn=SnTip$_2$ **127**).[103] This notable bathochromic shift in **126** was attributed to the high twisting of the Sn=Sn bond of 44.6°, which causes remarkable destabilization of its HOMO because of: (1) $5p_\pi$–$5p_\pi$(Sn=Sn)–σ(Sn–Si) orbitals mixing; and (2) reduced $5p_\pi$–$5p_\pi$ orbital overlap taking place upon twisting. Indeed, computations revealed that twisting about the Sn=Sn bond in **126** raises its HOMO energy level by 0.32 eV, resulting in the overall decrease in the HOMO–LUMO energy gap from 2.47 eV (*trans*-bent nontwisted distannene) to 2.18 eV (nonpyramidal twisted distannene).

Cyclic distannenes with endocyclic Sn=Sn bonds

Only one compound of this class is known to date: the three-membered ring cyclotristannene *cyclo*-(t-Bu$_3$Si)$_4$Sn$_3$ **130** prepared by Wiberg *et al.* by the reaction of silylsodium derivative t-Bu$_3$SiNa and stannylene (t-BuO)$_2$Sn: (Scheme 5.29, Table 5.5).[108]

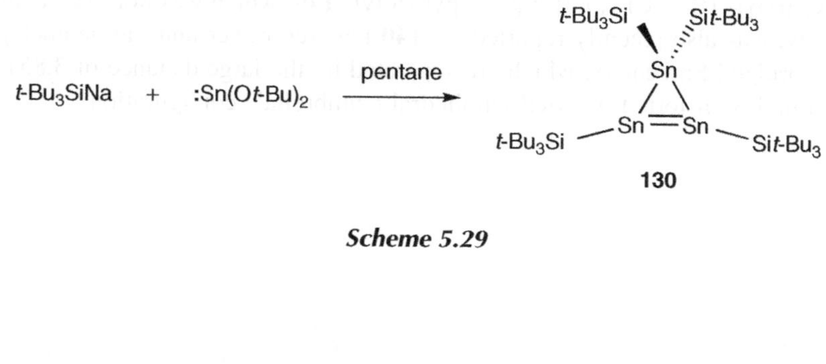

Scheme 5.29

Similarly to the case of persilyl-substituted cyclotrisilenes (Table 5.2) and cyclotrigermenes (Table 5.4), cyclotristannene **130** revealed a very short Sn=Sn bond of 2.601(3)/2.582(4) Å (two independent molecules), the shortest among all structurally characterized distannenes and even shorter than that of tetrasilyldistannene **126** of 2.6683(10) Å (Table 5.5).[108] One should, however, be cautious when considering these values for the tin–tin distance in **130** because of the poor refinement data of its crystal structure determination ($R_1 = 0.17$). The doubly bonded Sn atoms in **130** featured markedly undistorted geometry: 350.8/355.8° and 359.6/360.0° (for the two independent molecules). In agreement with the formulation of the Sn=Sn bond as short/strong and in contrast to most acyclic distannenes, cyclotristannene **130** did not dissociate in solution, as demonstrated by the observation of its sp^2-Sn resonances at 412.0 ppm, which is in the region diagnostic for doubly bonded tin centers.

An organotin derivative $Ar_2SnSnAr_2$ [Ar = 2,4,6-$(CF_3)_3$-C_6H_2] **131**, for which a formal Sn=Sn double bond could be drawn, showed an extraordinarily long tin–tin separation of 3.639(1) Å, being not a real distannene but a weakly bound stannylene dimer instead.[109]

Some organotin compounds revealed, instead of the two equivalent dative interactions expected for distannenes, zwitterionic structures $[R_2Sn:]^{\delta+} \rightarrow [\ddot{S}nR'_2]^{\delta-}$ with a single dative bond between the two largely different stannylene units: one of them playing the role of electron donor/Lewis base (deshielded and flattened Sn center) and the other one being an electron acceptor/Lewis acid (highly shielded and strongly pyramidal Sn center). Among such derivatives are: $(t\text{-}Bu_3Si)_2Sn \rightarrow Sn(Si\,t\text{-}Bu_3)_2$ **132**,[110] $Ar_2Sn \rightarrow Sn[1,8\text{-}(N\text{–}CH_2t\text{-}Bu)_2C_{10}H_6]$ [Ar = 2,6-$(Me_2N)_2C_6H_3$] **133** ([Sn–Sn = 3.087(2) Å],[111] $R_2Sn \rightarrow SnCl_2$ (R = $CH(SiMe_3)[8\text{-}C_9H_6N]$) **134** [Sn–Sn = 2.961(1) Å],[112] and {1-[N(*t*-Bu)=C(SiMe$_3$)CH]-2-[N(*t*-Bu)(SiMe$_3$)C=CH]C_6H_4}Sn→Sn{1,2-[N(*t*-Bu)(SiMe$_3$)C=CH]$_2C_6H_4$}**135** [Sn–Sn = 3.0087(3) Å].[113]

Given the rather shallow potential energy surface (PES) of the Sn_2R_4 derivatives, one can suggest great structural variations for valence isomers other than the distannenes $R_2Sn{=}SnR_2$, as was theoretically predicted for the Sn_2H_4 model by Trinquier in the early 1990s (Scheme 5.30, isomers **I–VI**).[7g,i]

Some of the above theoretical predictions were experimentally realized by Power *et al.*: *stannylstannylenes* **V** $Ar\ddot{S}n\text{–}SnMe_2Ar$ (Ar = 2,6-$Tip_2C_6H_3$) **136** with Sn–Sn = 2.8909(2) Å[114] and $Ar\ddot{S}n\text{–}SnPh_2Ar$ (Ar = 2,6-$Tip_2C_6H_3$) **137** with Sn–Sn = 2.9688(5) Å,[115] *trans doubly bridged* **III** $[ArSn(\mu\text{-}H)]_2$ (Ar = 2,6-$Tip_2C_6H_3$) **138** with a Sn–Sn separation of 3.1192(3) Å,[116] and *singly bridged* **VI** $ArSn(\mu\text{-}Br)Sn(Ar)[CH_2\text{-}4\text{-}i\text{-}Pr\text{-}C_6H_4]$ [Ar = 2,6-$(2,6\text{-}i\text{-}Pr_2\text{-}C_6H_3)_2$-$C_6H_3$] **139** with Sn–Sn = 2.9407(4) Å.[100] A novel *quadruply bridged distannene isomer* **VII** $[Sn(3,5\text{-}R_2\text{-}pz)_2]_2$ (R = CF_3, *t*-Bu; pz = pyrazolyl) **140**, which was not even considered theoretically, was also recently reported.[117] **140** featured rather unusual geometry at the quadruply bridged Sn centers, which are separated by the large distance of 3.851 Å and revealed a highly distorted inverted tetrahedral ('umbrella') configuration.

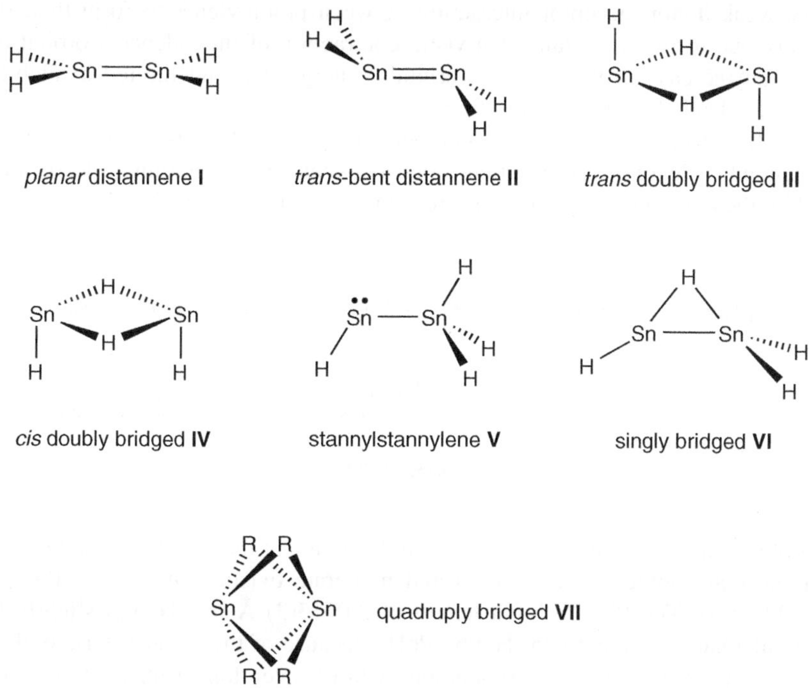

Scheme 5.30

5.3.1.1.4 *Diplumbenes >Pb=Pb<*[5c,71d]

Synthesis

As has already been discussed, for the heavier group 14 elements E with higher principal quantum numbers, the classification of the E=E double bond based solely on such subtle criteria (which work well only for carbon systems) as the length (short) and strength (strong) of the E=E bond becomes more and more problematic. Thus, for example, the Pb–Pb distances of 2.903 –3.947 Å observed between tricoordinate lead centers, which can formally be described as doubly bonded, are markedly longer than the standard Pb–Pb single bonds. Depending on the degree of separation between the two tricoordinate leads, such derivatives are sometimes classified as *diplumbenes* with a Pb=Pb bond and sometimes as *plumbylene dimers* with a weak association between two plumbylenes. However, distinguishing between these two classes is a difficult problem and the border between them is rather formal. In accord with the very long (and weak) double dative bonds between the two plumbylene units in either diplumbenes or plumbylene dimers, all of them readily undergo dissociation in solution to form monomeric plumbylenes.

This very weak donor–acceptor interaction between plumbylenes to form their dimers is definitely due to the important relativistic contraction of the valence s-orbital on Pb, leading to an increase in the s- and p-orbitals' energy gap and limited availability of s-electrons for lead–lead bond formation.

Tetrakis(2,4,6-triisopropylphenyl)diplumbene $Tip_2Pb{=}PbTip_2$ **141** was the first compound claimed to feature a true double bond between two tricoordinate leads and was prepared by the reaction of aryl Grignard reagent TipMgBr with $PbCl_2$ (Scheme 5.31).[118]

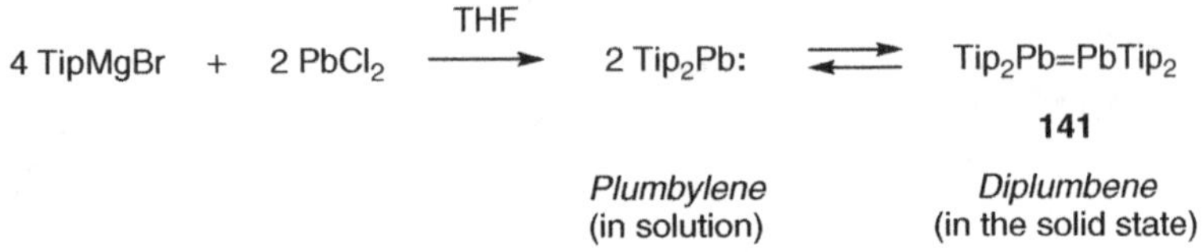

Scheme 5.31

The diplumbene **141** exhibited great pyramidalization at the doubly bonded lead atoms (sum of the bond angles 325.4/330.1°) and moderate twisting of the Pb=Pb bond of 21.7° (Table 5.6). The Pb=Pb bond length of 3.0515(3) Å in **141** was slightly longer than that calculated for the parent $H_2Pb{=}PbH_2$ because of the greater steric bulk of the voluminous Tip groups, and was comparable with (or even longer than) those of typical Pb–Pb single bonds.[118]

To date, four stable organolead derivatives have been described as diplumbenes (including **141**) and four as plumbylene dimers, being prepared by one of two general synthetic routes: (1) dimerization of homoleptic plumbylenes (A); and (2) dimerization of heteroleptic plumbylenes (B) (Scheme 5.32). No cyclic compounds with skeletal Pb=Pb double bonds have been isolated to date.

Method A: dimerization of homoleptic plumbylenes

Apart from diplumbene **141**, only $MgBr_2$-stabilized plumbylene dimer {$Mes_2Pb\cdots$[Br–$Mg(thf)_4$–Br]}$_2$ **147** has been prepared by this method, by the reaction of MesMgBr with $PbCl_2$.[119]

Interaction of ArMgCl (Ar = 2,4,6-Et_3-C_6H_2) with $PbCl_2$ resulted in the formation of a dimeric form $Ar_2Pb\cdots$[Br–$Mg(thf)_4$–Br]$\cdots PbAr_2$ **149**, featuring a novel structural variation: two plumbylene monomers linked by the bromine atoms of the linear Br–$Mg(thf)_4$–Br unit without direct Pb–Pb bonding.[122]

Method B: dimerization of heteroleptic plumbylenes

All other derivatives, featuring a greater or lesser degree of Pb=Pb double bond character, have been synthesized by the dimerization of heteroleptic plumbylenes. Among them are: *diplumbenes* (*E*)-Tip[$(Me_3Si)_3Si$]Pb=Pb(Tip)[$Si(SiMe_3)_3$] **142**,[119] (*E*)-Mes[$(Me_3Si)_3Si$]Pb=Pb(Mes)[$Si(SiMe_3)_3$] **143**,[102b,c] (*E*)-Ar(Me)Pb=Pb(Ar)Me [Ar = 2,6-(2,6-*i*-Pr_2-C_6H_3)$_2$-C_6H_3] **144**;[120] and *plumbylene dimers* (*E*)-{Ar[$(Me_3Si)_3Si$]Pb}$_2$ [Ar = 2,4,6-$(CF_3)_3$-C_6H_2] **145**,[106] (*E*)-{Ar[$(Me_3Si)_3Si$]Pb}$_2$ [Ar = 2-*t*-Bu-4,5,6-Me_3-C_6H] **146**[121] and [Ar(4-*t*-Bu-C_6H_4)Pb]$_2$ [Ar = 2,6-(2,6-*i*-Pr_2-C_6H_3)$_2$-C_6H_3] **148**.[120]

Table 5.6 *Crystallographic and spectral parameters of the structurally characterized diplumbenes*

Diplumbene or plumbylene dimer	>Pb=Pb< (Å)	Σ(Pb) (°)	>Pb=Pb< twist (°)	^{207}Pb NMR (ppm)	Reference
$Tip_2Pb{=}PbTip_2$ **141**	3.0515(3)	325.4 330.1	21.7	NA	118
(*E*)-Tip(R_3Si)Pb=Pb(Tip)SiR_3 **142** (R = SiMe_3)	2.9899(5)	339.0	NA	NA	119
(*E*)-Mes(R_3Si)Pb=Pb(Mes)SiR_3 **143** (R = SiMe_3)	2.9031(11)	NA	NA	NA	102b,c
(*E*)-Ar(Me)Pb=Pb(Ar)Me **144** [Ar = 2,6-(2,6-*i*-Pr_2-C_6H_3)$_2$-C_6H_3]	3.1601(6)	322.7	0.0	8738.0	120
(*E*)-[Ar(R_3Si)Pb]$_2$ **145** [Ar = 2,4,6-(CF_3)$_3$-C_6H_2; R = SiMe_3] (*plumbylene dimer*)	3.537(1)	336.8	NA	NA	106
(*E*)-[Ar(R_3Si)Pb]$_2$ **146** [Ar = 2-*t*-Bu-4,5,6-Me_3C_6H; R = SiMe_3] (*plumbylene dimer*)	3.3695(11)	NA	NA	7545.0	121
{Mes_2Pb[BrMg(thf$_4$)Br]}$_2$ **147** (*plumbylene dimer*)	3.3549(6)	NA	NA	NA	119
[Ar(Ar′)Pb]$_2$ **148** [Ar = 2,6-(2,6-*i*-Pr_2-C_6H_3)$_2$-C_6H_3; Ar′ = 4-*t*-Bu-C_6H_4] (*plumbylene dimer*)	3.947(1)	NA	74.2	7275.0	120

NA, Not available.

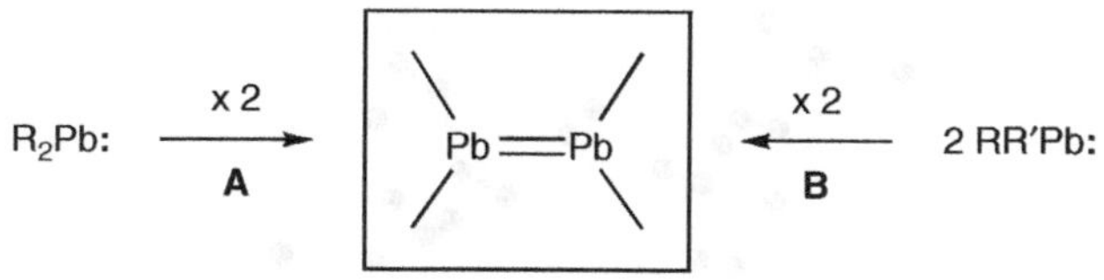

Scheme 5.32

^{207}Pb NMR spectroscopy of diplumbenes

Being dissociated in solution into plumbylenes because of the inherent weakness of the Pb–Pb bond (see above), diplumbenes and plumbylene dimers revealed either no signals, or very low-field ^{207}Pb NMR resonances above +7000 ppm, diagnostic of dicoordinate plumbylene monomers (Table 5.6).

Crystal structure of diplumbenes

Tetraaryl-, aryl(alkyl)- and aryl(silyl)diplumbenes and plumbylene dimers have been structurally characterized (Table 5.6, Figure 5.14). In accord with the general trend of progressively increasing pyramidalization at the heavier group 14 elements, diplumbenes/plumbylene dimers featured remarkable *trans*-bending at the lead centers: from 322.7° (in **144**) to 339.0° (in **142**). Depending on the extent of steric interaction between substituents, diplumbenes/plumbylene dimers manifest either undistorted (twist angle 0.0° in **144**) or extremely distorted (twist angle 74.2° in **148**) Pb–Pb bonds. Lead–lead separation in diplumbenes/plumbylene dimers spans a broad range of more than 1 Å, changing from 2.9031(11) Å (in **143**) to 3.947(1) Å (in **148**): 2.9031(11)–3.1601(6) Å (for *diplumbenes*) and 3.3549(6)–3.947(1) Å (for *plumbylene dimers*). Notably, all of these Pb–Pb distances are markedly longer than usual Pb–Pb single bonds.

Plumbylplumbylene **150**, the valence isomer of a diplumbene featuring a single Pb–Pb bond of 2.9928(3) Å between the divalent and tetravalent lead centers, was recently reported by Power *et al.* (Scheme 5.33).[123]

5.3.1.2 *Heteronuclear Compounds*

5.3.1.2.1 ***Group 14/Group 14 Combinations:*** **>E_{14} = E_{14}'<(E_{14}, E_{14}' = C, Si, Ge, Sn, Pb)**

5.3.1.2.1.1 Silenes >Si=C<[124]

Of all possible doubly bonded combinations between the different group 14 elements, silenes >Si=C< are the most extensively studied. The story of such compounds, initially as undetectable reactive intermediates, began with the seminal paper published in 1967 by Gusel'nikov and Flowers, who reported the generation of transient 1,1-dimethylsilene $Me_2Si{=}CH_2$ upon the pyrolysis of 1,1-dimethylsilacyclobutane (Scheme 5.34).[125]

The following advances in transient silenes chemistry were associated with their direct detection in an argon matrix at 8–20 K.[126] However, the major breakthrough in the field

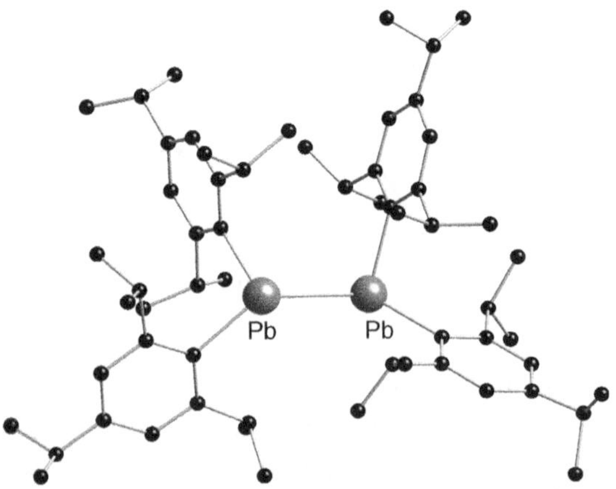

Figure 5.14 Crystal structure of tetraaryldiplumbene ***141*** *(hydrogen atoms are not shown)*

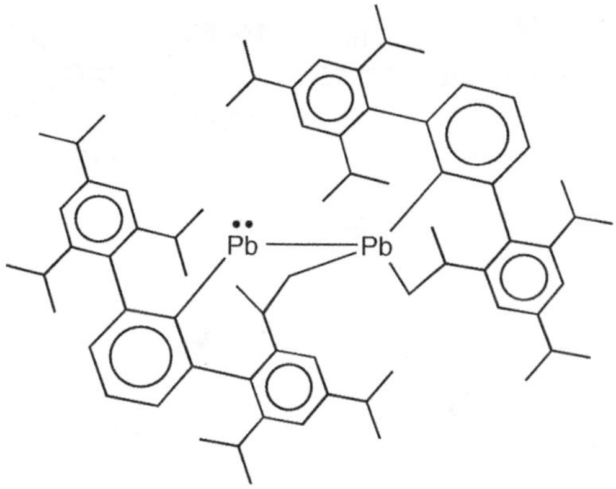

150

Scheme 5.33

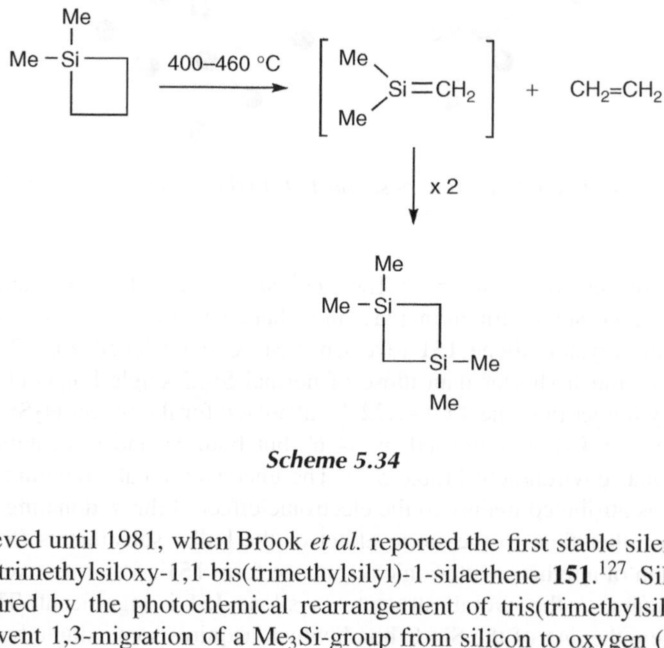

Scheme 5.34

was not achieved until 1981, when Brook *et al.* reported the first stable silene, namely 2-adamantyl-2-trimethylsiloxy-1,1-bis(trimethylsilyl)-1-silaethene **151**.[127] Silene **151** was cleanly prepared by the photochemical rearrangement of tris(trimethylsilyl)acylsilane, which underwent 1,3-migration of a Me_3Si-group from silicon to oxygen (Scheme 5.35, Figure 5.15, Table 5.7).[127a,b] Given the very effective steric protection of the reactive Si=C bond by its bulky substituents, silene **151** was indefinitely stable under an inert atmosphere. As the most diagnostic signature for the presence of the Si=C double bond in **151**, both Si and C atoms were expected to be markedly deshielded, similar to the case of organic alkenes featuring strongly deshielded C=C atoms.

Indeed, the carbon resonance of the Si=C bond in **151** was found at a very low field at 214.2 ppm; however, the silicon atom resonated at 41.8 ppm, that is, at higher fields than one would expect based on its formal sp^2-hybridization state (the reason for this

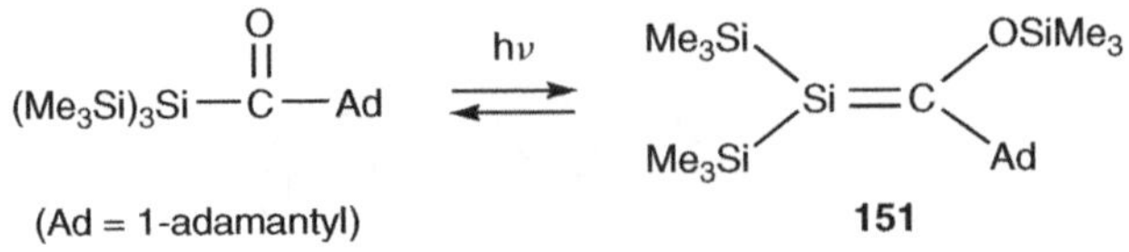

Scheme 5.35

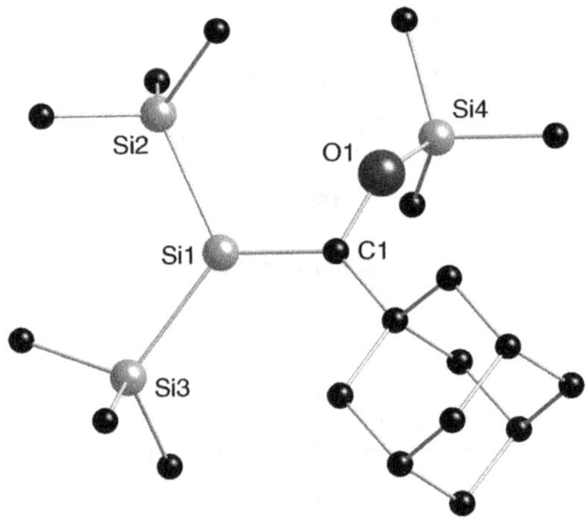

Figure 5.15 *Crystal structure of silene* ***151*** *(hydrogen atoms are not shown)*

discrepancy is discussed below).[127a,b] The $^{1}J(^{29}Si–^{13}C)$ coupling constant in **151** was large (84.3 Hz), consistent with an increase in s-character of the hybrids used for making this bond. In the crystal, silene **151** revealed a Si=C bond length of 1.764(3) Å,[127b,c] a value that was much shorter than those of normal Si–C single bonds (1.87–1.91 Å), albeit markedly longer than the 1.69–1.72 Å calculated for the parent $H_2Si{=}CH_2$.[124a,128] The Si=C bond in **151** was twisted by 14.6° but both Si and C centers featured an undistorted planar environment (Table 5.7). The unconventional stretching of the Si=C bond in **151** was attributed mainly to the electronic effect of the π-donating siloxy-group at the C atom rather than to the steric effect of the bulky substituents (Scheme 5.36). The contribution of another zwitterionic resonance form **151′**, featuring oxonium ion and silyl anion centers, to the overall structure of silene **151** is significant. This results in the observed lengthening of the Si=C bond, which acquires some single bond character, and accordingly to the deshielding of the sp^2-C and shielding of the sp^2-Si atoms.

The resonance contribution **151′** remarkably opposed and diminished the inherent $Si^{\delta+}{=}C^{\delta-}$ bond polarity, thus making it relatively unpolarized. This phenomenon of the so-called 'reversed polarity' of the $Si^{\delta-}{=}C^{\delta+}$ bond, comprehensively treated in a theoretical contribution by Apeloig and Karni,[128] is known to play a decisive role in controlling the overall reactivity of silenes. Many isolable silenes are stabilized by such a 'reversed polarization' electronic effect, particularly 2-silenolates, in which the silyl-anion resonance contribution is predominant (see below).

Table 5.7 *Crystallographic and spectral parameters of the structurally characterized silenes*

Silene	>Si=C< (Å)	Σ(Si) (°)	Σ(C) (°)	>Si=C< twist (°)	^{29}Si(Si=C)	^{13}C(Si=C)	Reference
151 (Scheme 5.35)	1.764(3)	360.0	359.7	14.6	41.8	214.2	127a-c
152 (Scheme 5.37)	1.702(5)	360.0	360.0	1.6	144.2	77.2	129a-c
152•THF (Scheme 5.37)	1.747(5)	348.7	358.8	NA	52.4	NA	129d,e
153	1.761(4)	341.8	359.2	NA	35.4	100.1	129f
154 (Scheme 5.39)	1.741(2)	359.8	359.9	4.6	51.7	196.8	130
155 (Scheme 5.40)	1.770(15)	359.6	359.8	NA	107.6	225.0	131
	1.758(15)	359.8	359.8				
158 (Scheme 5.41)	1.755(2)	359.6	359.8	NA	−71.9	159.9	132
159 (Scheme 5.43)	1.7431(17)	360.0	359.8	6.4	78.6	130.9	133
162 (Scheme 5.44)	1.775(3)	342.2	358.6	NA	21.9	214.6	134
165 (Scheme 5.45)	1.7459(15)	356.2	360.0	NA	75.0	155.2	135
166 (Scheme 5.46)	1.745(2)	360.0	359.9	4.5	79.2	188.4	136
168 (Scheme 5.48)	1.926(3)	317.8	359.7	NA	−78.7	274.1	139a

NA, Not available.

151 **151′**

Scheme 5.36

Several other silenes R(Me_3Si)Si=C($OSiMe_3$)R′ with different R and R′ groups were subsequently synthesized by Brook *et al.* using the same acylsilane photolysis synthetic route.[127b,d,e]

The electronic situation at the Si=C bond was very different in the second example of a stable silene, Me_2Si=C($SiMe_3$)SiMe*t*-Bu_2 **152**, reported by Wiberg *et al.* in 1983.[129] A salt elimination reaction resulted in the formation of a THF-solvate of the silene **152**, from which free crystalline **152** was obtained by the azeotropic distillation of THF with benzene (Scheme 5.37, Table 5.7).[129a–c] In the absence of π-donating groups, the remarkable thermal stability of silene **152** was primarily attributed not to electronic factors but to the high steric protection of its Si=C bond because of the very bulky groups at the C center (although the Me substituents at the Si center are surprisingly small). The substituents' pattern at the double bond in **152**, namely C_2Si=CSi_2, favored (and enhanced) its natural bond polarization $Si^{\delta+}=C^{\delta-}$, which according to theoretical predictions[128] should increase the degree of bond ionicity and result in its shortening. Indeed, the Si=C bond in **152** was determined to be planar (twist angle 1.6°, Table 5.7) with a length of only 1.702(5) Å, which is significantly shorter than

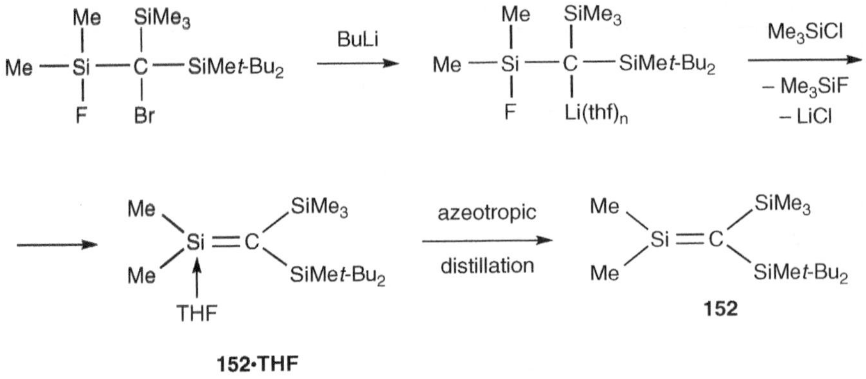

Scheme 5.37

that of Brook's silene **151** [1.764(3) Å][127b] and in good agreement with theoretical expectations.[124a,128] This Si=C bond was also ca. 9% shorter than those of typical Si–C bonds. This testifies to the lack of any notable contribution of a resonance form of the type **151′** (Scheme 5.36) in silene **152**, because of the absence of adjacent π-donor substituents. Accordingly, the carbon center of the Si=C bond in **152** was remarkably shielded compared with that of Brook's silene **151** (77.2 vs 214.2 ppm), whereas the silicon center in **152** was deshielded (144.2 vs. 41.8 ppm).[129b] It is interesting that silene **152** (and related silenes) readily underwent complexation with Lewis basic ligands (THF, Me_3N, Me_2EtN, pyridine, F^-), which coordinated at the electrophilic Si center to satisfy its high electronic demand.[129d–f] Such n-donor coordination resulted in the appreciable stretching of the Si=C bonds, and dramatic distortion of the Si center's geometry towards the tetrahedral configuration, accompanied by its NMR shielding [**152•THF** and $Me_2(Me_2EtN{:}\rightarrow)Si{=}C(SiMe_2Ph)_2$ **153**, Table 5.7]. These trends can be realized in the framework of the increasing contribution of the ylide-type resonance extreme **B**, as depicted in Scheme 5.38.

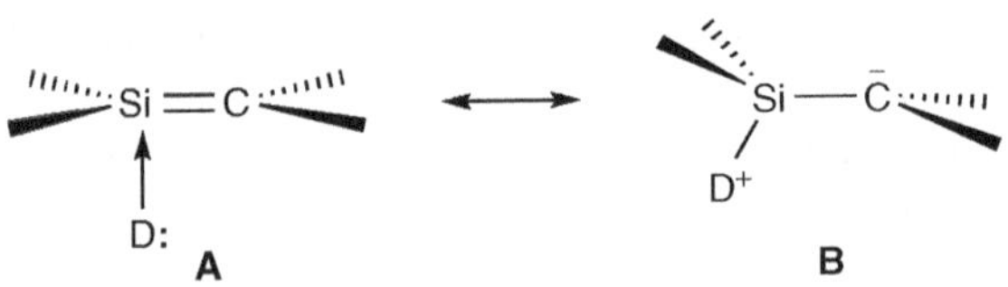

Scheme 5.38

Apeloig *et al.* used a different approach to synthesize stable silene **154**, a sila-Peterson reaction between the tris(silyl)silyllithium derivative and adamantanone (Scheme 5.39, Table 5.7).[130]

In contrast to the above-described Wiberg's silene **152**, **154** featured an opposite substitution pattern, namely $Si_2Si{=}CC_2$, which greatly affected its bonding situation and, consequently, spectral and structural characteristics. Given the remarkably relaxed intrinsic polarity of the $Si^{\delta+}{=}C^{\delta-}$ bond because of the inductive effects of its substituents, it

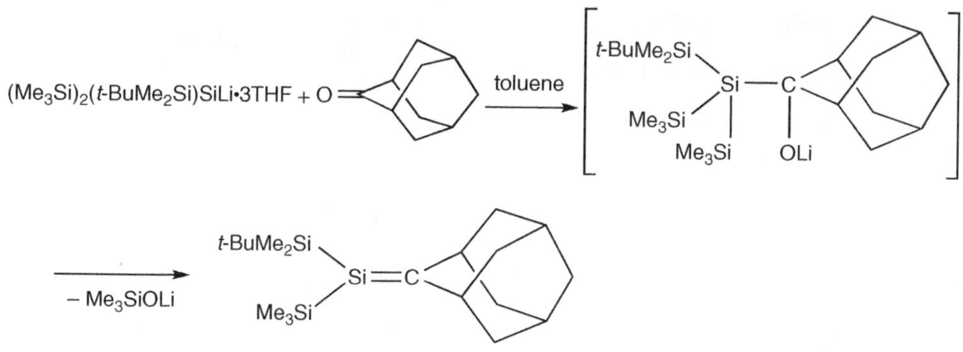

154

Scheme 5.39

was not surprising that the silicon center in **154** was shielded (51.7 vs 77.2 ppm), whereas the carbon center was deshielded (196.8 vs 144.2 ppm) compared with those of **152**. In accord with the above-mentioned discussion, this diminished Si=C bond polarity in **154** resulted in its lengthening to 1.741(2) Å, a value that was intermediate between those of Wiberg's silene **152** [1.702(5) Å] and Brook's silene **151** [1.764(3) Å]. Accordingly, the Si=C bond in **154** was essentially planar with a negligible twisting of 4.6° (Table 5.7). A decrease in Si=C bond polarity in silene **154** also significantly enhanced its kinetic stability.

The same group recently published the synthesis and structure of the first bis(silene) **155** containing two isolated noninteracting Si=C bonds linked by a mercury atom.[131] Bis(silene) **154** was synthesized by a Brook-type thermal rearrangement of a transient bis(acylsilane) **156** generated by the reaction of bis(lithiosilyl)mercury **157** and 1-adamantoyl chloride (Scheme 5.40, Table 5.7). Possessing siloxy groups at its unsaturated carbon centers, bis(silene) **155** featured properties resembling those of Brook's silene **151**: planar, relatively long Si=C bonds [1.770(15)/1.758(15) Å vs 1.764(3) Å] and significantly deshielded sp^2-C atoms (225.0 ppm vs 214.2 ppm), although the doubly bonded silicon atoms in **155** resonated at a lower field than those of **151** (107.6 ppm vs 41.8 ppm). Being effectively protected by the bulky i-Pr_3Si and adamantyl groups, bis(silene) **155** was amazingly inert towards water, methanol and acetone.

An interesting example of isolable 4-silatriafulvene derivative **158**, manifesting highly pronounced effects of the 'reversed polarity' of the $Si^{\delta-}=C^{\delta+}$ bond, was reported by Kira *et al.*[132] Silene **158** was prepared by the common route through the sila-Peterson reaction of $(t\text{-}BuMe_2Si)_3SiLi$ with di-*tert*-butylcyclopropenone (Scheme 5.41, Table 5.7).

The exocyclic Si=C bond in **158** was remarkably stabilized by the important contribution of the zwitterionic resonance form **158′** featuring a 2π-electron aromatic cyclopropenylium ion and silyl anion fragments (Scheme 5.42). Such a bonding contribution of **158′** to the overall structure of silene **158** was seen in the extreme shielding of its formally doubly bonded Si atom observed at −71.9 ppm (for sp^2-C: 159.9 ppm), as well as in the stretching of the Si=C bond to 1.755(2) Å (Table 5.7). The regioselectivity of methanol addition to **158** also confirmed the inverted $Si^{\delta-}=C^{\delta+}$ bond polarity.

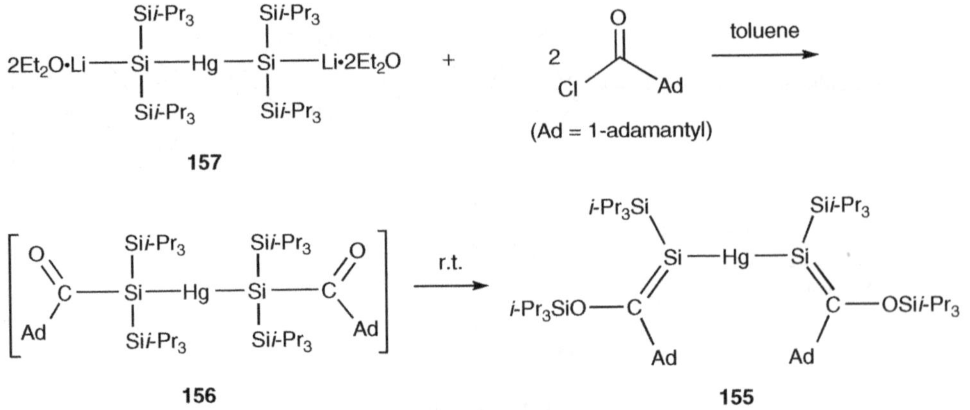

Scheme 5.40

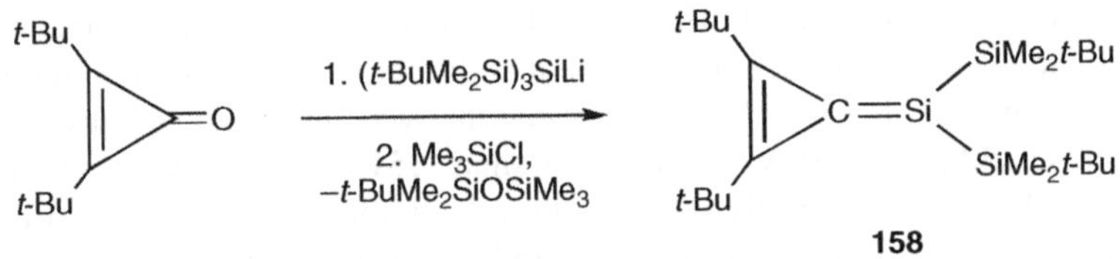

Scheme 5.41

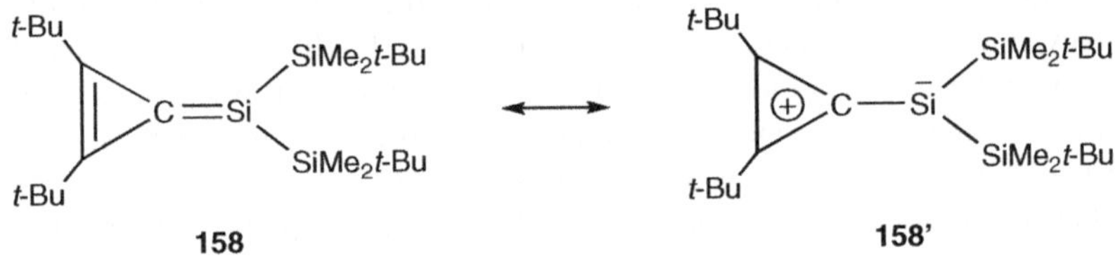

Scheme 5.42

Silene **159**, which had an unusual substitution pattern, where the carbon part of the Si=C bond was the same as that of Apeloig's silene **154** whereas the silicon part was substituted with π-donating ligands (a silyloxy-group and silyl-anion substituent), was also recently reported (Scheme 5.43, Table 5.7).[133] Silene **159** was formed by a sila-Peterson reaction of lithium disilenide derivative **160** and adamantanone through the proposed 1,2-disilaallene intermediate **161**. The above-described substitution pattern in **159** did not favor the 'reversed polarization' of the Si=C bond; therefore, its spectral and structural data were markedly different from those of Brook's silene **151** and Kira's silene **158**, namely: a normal (not stretched) Si=C bond of 1.7431(17) Å, and deshielded Si (78.6 ppm) and shielded C (130.9 ppm) sp^2-centers (Table 5.7).

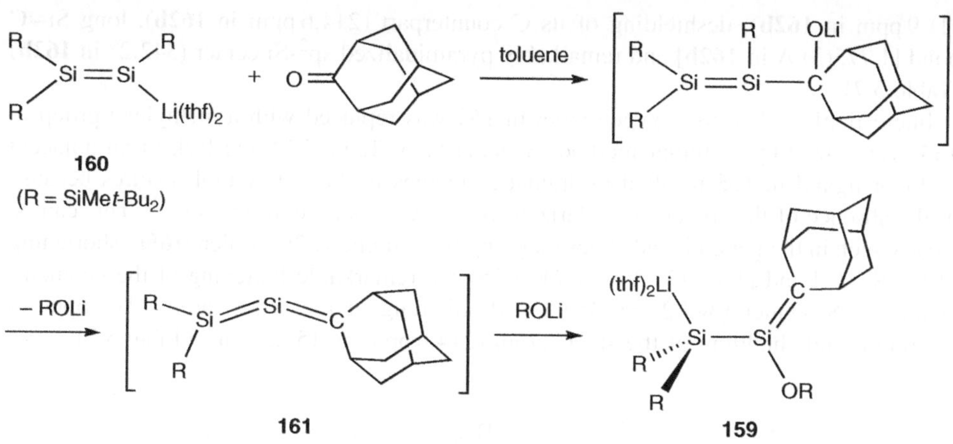

Scheme 5.43

Several cyclic compounds with endocyclic Si=C bonds became synthetically accessible in recent years. Thus, the four-membered ring cyclic silenes **162** were prepared by the reaction of lithium disilenide derivatives **163** with the acyl chlorides R′COCl (R′ = 1-adamantyl, *t*-Bu) (Scheme 5.44, Table 5.7).[134]

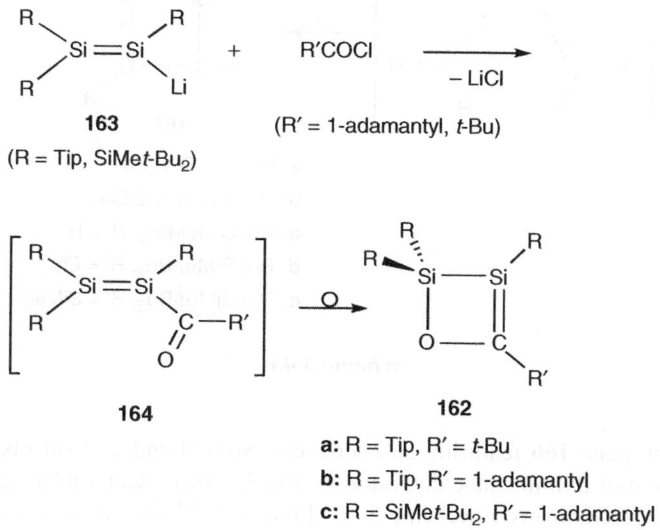

Scheme 5.44

The reaction was suggested to proceed through the intermediate acyl disilenes **164**, which, however, appeared to be undetectable down to 193 K. Given the presence of the skeletal oxygen (π-donor) next to the doubly bonded carbon atom in **162**, one can suggest the appreciable contribution of the 'reversely polarized' $Si^{\delta-}=C^{\delta+}$ bond. This was indeed the case, and was seen in the appreciable shielding of the sp^2-Si atom

(21.9 ppm in **162b**), deshielding of its C counterpart (214.6 ppm in **162b**), long Si=C bond [1.775(3) Å in **162b**] and remarkably pyramidalized sp^2-Si center (342.2° in **162b**) (Table 5.7).

Interestingly, when the oxygen atom in **162** was replaced with a methylene group in **165** synthesized by a similar method (Scheme 5.45, Table 5.7), the lack of an adjacent π-donor ligand in **165** resulted in dramatic changes in their structural features because of the absence of the 'reversely polarized' $Si^{\delta-}=C^{\delta+}$ bond contribution.[135] This can be clearly seen in the general tendencies on going from silene **162b** to silene**165a**: shortening of the Si=C bond [1.775(3) Å vs 1.7459(15) Å], remarkable flattening of the geometry at the sp^2-Si center (342.2° vs 356.2°), deshielding of the sp^2-Si atom (21.9 ppm vs 75.0 ppm) and shielding of the sp^2-C atom (214.6 ppm vs 155.2 ppm) (Table 5.7).

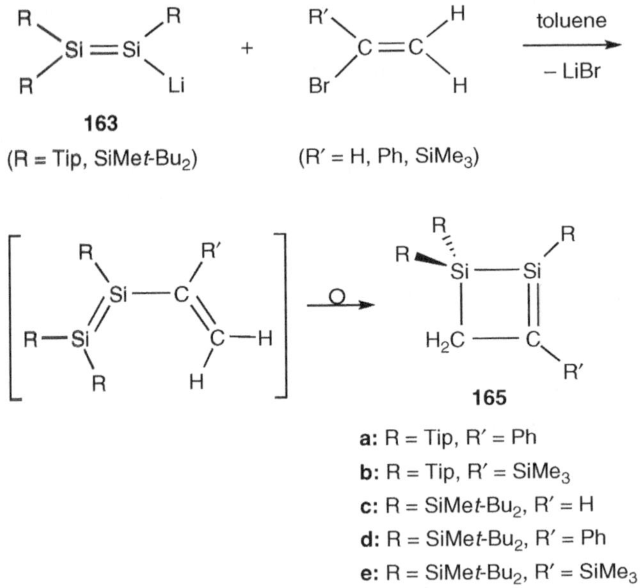

Scheme 5.45

Disilacyclopropene **166** featuring an endocyclic Si=C bond was surprisingly formed upon the reaction of dilithiosilane derivative $(t\text{-}Bu_3Si)_2SiLi_2$ with 1-adamantoyl chloride AdCOCl (Ad = 1-adamantyl) (Scheme 5.46, Table 5.7).[136] Given the traditional substitution pattern at the Si=C bond in **166**, one can expect a normal (not 'reversed') double bond polarization with physicochemical characteristics typical for such cases. Indeed, the Si=C bond in **166** was planar with a negligible twisting of 4.5° and 'normal' bond length of 1.745(2) Å (Table 5.7). Moreover, both doubly bonded Si and C centers resonated as expected at a low field, being observed at 79.2 and 188.4 ppm, respectively. The cyclic silene **166** exhibited high thermal stability and amazing chemical inertness, being unreactive towards air and methanol, which definitely stems from the effective steric protection of its Si=C bond by the extremely bulky $t\text{-}Bu_3Si$ and 1-adamantyl groups.

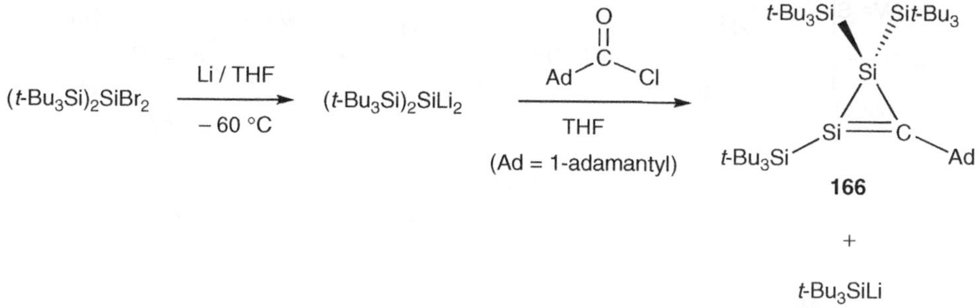

Scheme 5.46

Several other room temperature stable compounds were classified as silenes with Si=C bonds based on their low-field $^{29}Si/^{13}C$ NMR resonances and characteristic reactivity, although their crystal structure details were not reported.[137] Thus, a remarkable dimesityl-neopentylsilene $Mes_2Si{=}CH{-}CH_2t$-Bu **167** was prepared by Couret *et al.* by applying a novel synthetic approach: addition of t-BuLi to $Mes_2(F)Si{-}CH{=}CH_2$ followed by β-elimination of LiF from the resulting lithium derivative $Mes_2(F)Si{-}CH(Li){-}CH_2t$-Bu.[137a,b] The presence of the Si=C bond in **167** was manifested in the diagnostic deshielding of both the silicon (77.6 ppm) and the carbon (110.4 ppm) centers, and was further proved by characteristic addition and cycloaddition reactions with water, alcohols, azides, 1,3-dienes, aldehydes and ketones.

As species closely related to silenes, 2-silaenolates (2-silenolates) are of particular importance.[138,139] Given the full negative charge at the oxygen atom of 2-silenolates **A**, one can expect a significant contribution from the resonance form **B** featuring a negative charge at the silicon center adjacent to the carbonyl group (Scheme 5.47).

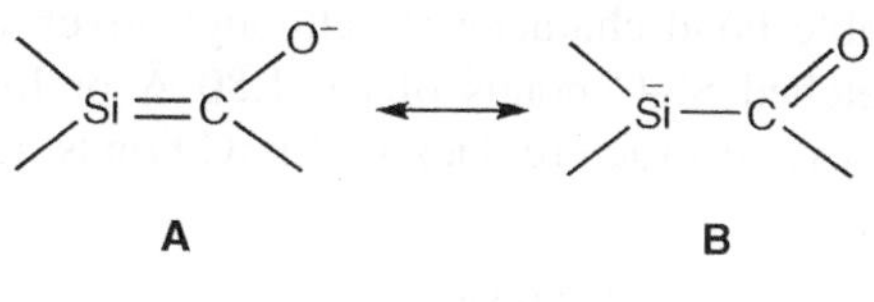

Scheme 5.47

Indeed, for the sole example of isolable 2-silenolates, prepared by the reduction of tris(trimethylsilyl)acylsilane with t-BuOK by Ottossou *et al.* (Scheme 5.48), the resonance extreme **168′** clearly dominated, being a major contributor to the overall electronic structure of **168** (Scheme 5.49).[139a]

This was seen in the extraordinary elongation of the silicon–carbon bond in **168** [1.926(3) Å] which was even longer than those of typical Si–C single bonds, shortening of the carbon–oxygen distance [1.245(3) Å] and remarkable pyramidalization at the central silicon (317.8°). In accord with the principal resonance contribution **168′**, the resonances of the formally doubly bonded Si and C atoms were extremely shifted to high and low fields, being observed at −78.7 ppm and 274.1 ppm, respectively.[139a] Overall,

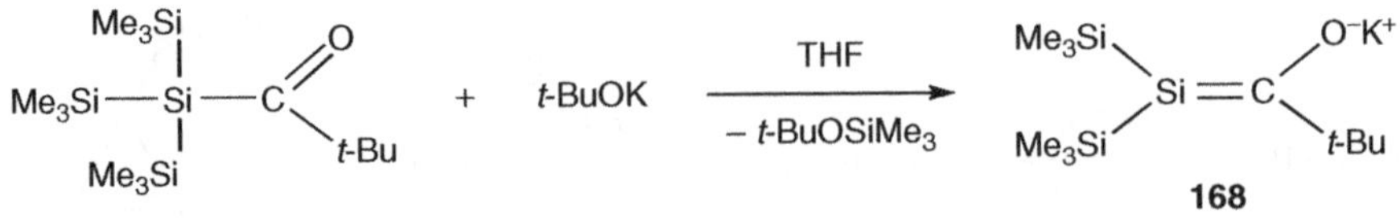

Scheme 5.48

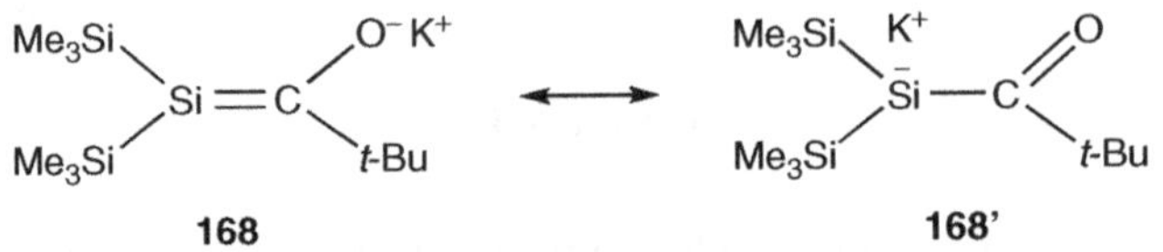

Scheme 5.49

2-silenolate **168** represents an extreme case of the above-discussed 'reversely polarized' $Si^{\delta-}=C^{\delta+}$ bond promoted by the strongly π-donating O^- substituent at the C atom. Compared with other 'reversely polarized' silenes (for example, Brook's silene **151**), 2-silenolate **168** manifests a maximum degree of such 'reverse polarization' greatly facilitated by the formation of a very strong and thermodynamically favorable C=O bond.

A series of silenes, stabilized by the intramolecular coordination of n-donor ligands (typically, amino group) to the highly electrophilic sp^2-silicon center, was recently reported by Oehme *et al.* (such donor-stabilized silenes are not discussed in the present chapter).[140]

Several 'transition metal η^2-silene complexes' were described as having some important degree of Si=C double bond character,[141] although given their peculiar structural features (first of all, stretched Si–C bonds of ca. 1.80 Å or longer), one can suggest their predominant metallacyclic structure, lacking Si=C bonds, rather than the transition metal–silene π-complex.

5.3.1.2.1.2 Germenes >Ge=C<[71a,c,124f,i]

The isolation of germenes was preceded by the identification of their unstable congeners as reactive intermediates [for example, $Me_2Ge{=}C(SiMe_3)_2$],[142a] diagnostically vanishing through [2 + 2] head-to-tail or head-to-head dimerization modes to form 1,3- or 1,2-digermacyclobutanes, respectively.[142]

As a major breakthrough, the first structurally characterized germenes were independently synthesized in 1987 by the groups of Berndt and Escudié. Thus, Berndt *et al.* prepared their isolable germene **169** by the coupling of electrophilic cryptocarbene and stable germylenes R_2Ge: (Scheme 5.50, Table 5.8).[143]

The presence of the Ge=C double bonds in **169** was manifested in the low-field shifts of their tricoordinate carbon centers (115 ppm in **169a** and 93 ppm in **169b**). Accordingly, the germanium–carbon distance in **169a** was diagnostically squeezed to 1.827(4) Å, being ca. 8% shorter than the Ge–C single bond in Me_4Ge. Both Ge and C centers in

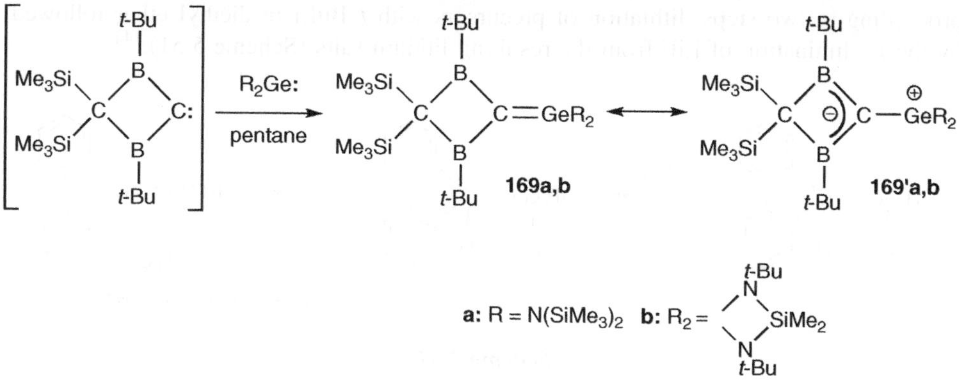

Scheme 5.50

Table 5.8 *Crystallographic and spectral parameters of the structurally characterized germenes*

Germene	>Ge=C< [Å]	Σ(Ge) [°]	Σ(C) [°]	>Ge=C< twist [°]	^{13}C(Ge=C)	Reference
169a (Scheme 5.50)	1.827(4)	359.9	359.7	36.0	115.0	143
170 (Scheme 5.51)	1.803(4)	360.0	360.0	6.0	NA	144b
171 (Scheme 5.52)	NA	360.0	360.0	34.0	NA	148a
172 (Scheme 5.52)	1.833(4)	360.0	360.0	12.7	NA	148b
173 (Scheme 5.52)	1.819(6)	360.0	360.0	0.5	116.9	148c
174 (Scheme 5.52)	1.819(2)	359.9	360.0	2.8	NA	148d
175 (Scheme 5.52)	1.840(4)	360.0	360.0	2.9	NA	148e
176 (Scheme 5.53)	1.771(16)	359.7	360.0	4.0	NA	149
177 (Scheme 5.54)	1.859(3)	359.2	358.9	6.4	196.6	150
179a (Scheme 5.55)	2.085(3)	303.0	357.4	48.8	145.5	151
180	2.102(12)	295.0	360.0	NA	158.8	152

NA, Not available.

169a featured a planar geometry similar to the case of alkenes; however, unlike the C=C bonds in alkenes, the Ge=C bond was remarkably twisted by 36° (Table 5.8).

Spectral and crystallographic data of **169** pointed to a notable contribution of ylide resonance form **169′** featuring '+' on germanium and '−' delocalized over the B–C–B fragment of the B_2C_2 four-membered ring (Scheme 5.50). This was seen in the appreciable shielding of the boron centers in **169** [$\delta(^{11}B)$: 66 ppm in **169a** and 65 ppm in **169b**] because of their partial negative charges, as well as in the relative shortening of the C(sp^2)–B bonds in **169a** [1.534(7) and 1.523(6) Å].

Escudié *et al.* employed a different approach to synthesize their stable germenes, namely, dehydrofluorination of the corresponding (fluorenyl)fluorogermane precursors proceeding in two steps: lithiation of precursors with *t*-BuLi in diethyl ether followed by the β-elimination of LiF from the resulting lithium salts (Scheme 5.51).[144]

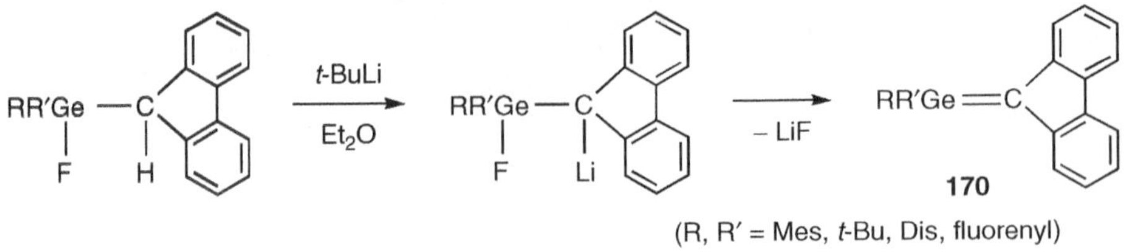

Scheme 5.51

The first isolable germene in this series, dimesityl(fluorenylidene)germane **170**, was prepared by this route in 1987 [Scheme 5.51 ($R = R' =$ Mes), Figure 5.16, Table 5.8).[144a,b] The remarkable thermal stability of germene **170** was primarily attributed to the effective steric protection of the Ge=C bond by the bulky Mes groups. Moreover, given the planar geometry around this double bond with a small twisting of 6° (Table 5.8), one can also suggest an additional stabilization of the Ge=C bond through its interaction with the fluorenylidene moiety, resulting in the substantial zwitterionic resonance contribution $Ge^{\delta+}=C^{\delta-}$ favored by the aromatic delocalization of the negative charge over the cyclopentadiene fragment of the fluorenylidene group.[144b] Such highly pronounced natural $Ge^{\delta+}=C^{\delta-}$ bond polarization, increasing the ionicity of this bond, was reasonably expected to cause its overall shortening.[128] Indeed, the Ge=C bond in **170** was found to be only 1.803(4) Å (mean value for the

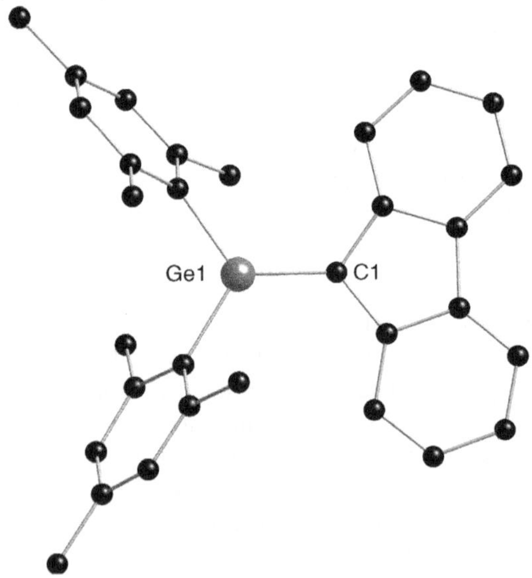

Figure 5.16 *Crystal structure of germene **170** (hydrogen atoms are not shown)*

two crystallographically independent molecules), being remarkably shorter than those of typical Ge–C single bonds and even shorter than that of Berndt's germene **169a** [1.827(4) Å].[144b]

In due course, Escudié *et al.* prepared other stable germenes employing the same synthetic approach (Scheme 5.51: R = R′ = Dis;[144c] R = Dis, R′ = Mes;[144c] R = *t*-Bu, R′ = fluorenyl;[144d] R = R′ = fluorenyl).[144d] In the absence of crystal structural data, the composition of these compounds as true germenes >Ge=C< was confirmed by trapping experiments with a variety of addition and cycloaddition reagents.[144c,d]

The same authors reported also the thermal generation of halogen-substituted germenes $Cp(X)Ge{=}CR_2$ (X = F, Cl; CR_2 = fluorenylidene), formed upon the β-elimination of Me_3SiX from their $Cp(X)_2Ge\text{–}(Me_3Si)CR_2$ precursors.[145] Because the room temperature isolation of such halogermenes was precluded as a result of their instability, they were trapped *in situ* with MeOH, $CHCl_3$ and H_2O to form the corresponding addition products.

Dimesitylneopentylgermene $Mes_2Ge{=}CH\text{–}CH_2$*t*-Bu, lacking resonance stabilization, was synthesized by Couret *et al.* similarly to the above-described isostructural dimesitylneopentylsilene $Mes_2Si{=}CH\text{–}CH_2$*t*-Bu **167**, that is, by the addition of *t*-BuLi to $Mes_2(F)Ge\text{–}CH{=}CH_2$ followed by β-elimination of LiF from the resulting lithium salt $Mes_2(F)Ge\text{–}CH(Li)\text{–}CH_2$*t*-Bu.[146] The presence of the Ge=C bond in germene $Mes_2Ge{=}CH\text{–}CH_2$*t*-Bu was proved by its chemical trappings with HCl, MeOH, $LiAlH_4$ and *t*-BuLi/H_2O.

Taking advantage of the readily available and highly reactive germylene Ar_2Ge: (Ar = 2-*t*-Bu-4,5,6-Me_3C_6H), generated by the thermal dissociation of digermene $Ar_2Ge{=}GeAr_2$ in solution,[147] Weidenbruch *et al.* synthesized a series of germenes **171**–**175**, in some of which the Ge=C bonds were in conjugation with the C≡C triple bond (Scheme 5.52, Table 5.8).[148] X-ray diffraction analysis revealed that the germene **171**, quite similar to Berndt's germene **169a**,[143] featured a planar geometry at both Ge and C sp^2-centers as well as a notably twisted Ge=C bond (Scheme 5.52, Table 5.8).[148a] The unexpected formation of **172** with an exocyclic Ge=C bond was realized in terms of the interaction of two molecules of the phosphaalkyne *t*-Bu–C≡P with two molecules of the germylene Ar_2Ge: (Scheme 5.52).[148b] A family of acetylene-linked bis(germenes) **173**–**175** was available through the initial double [1 + 2] cycloaddition of germylenes across the C≡C bonds followed by the ring-opening rearrangement of the intermediate bis(germirenes) (Scheme 5.52).[148c–e] The remarkable conjugation between the Ge=C and C≡C bonds in **173**–**175** was seen in the characteristic bathochromic shifts of their longest wavelength UV absorptions, observed at 500 nm (for **173**),[148c] 518 nm (for **174**)[148d] and 595 nm (for **175**).[148e] Moreover, the C–C bonds of the Ge=**C–C**≡C fragments were diagnostically shortened, suggesting the presence of a notable conjugation between the Ge=C and C≡C bonds.

The Ge=C bonds in **173**–**175** were in the range reported for other germenes: 1.819(6) Å in **73**,[148c] 1.819(2) Å in **174**[148d] and 1.840(4) Å in **175**[148e] (Table 5.8).

Among other isolable germenes, the two representing the shortest and longest extremes for the Ge=C bond length are of interest. The shortest Ge=C bond reported to date was prepared by Tokitoh *et al.* by the reaction of diarylgermylene Tbt(Tip)Ge: (see Chapter 4, Section 4.6.1.2) with carbon disulfide CS_2 (Scheme 5.53, Table 5.8).[149] The germene **176** featured a Ge=C bond length of only 1.771(16) Å, which is remarkably shorter than those of Berndt's germene **169** [1.827(4) Å][143] and Escudié's germene **170** [1.803(4)

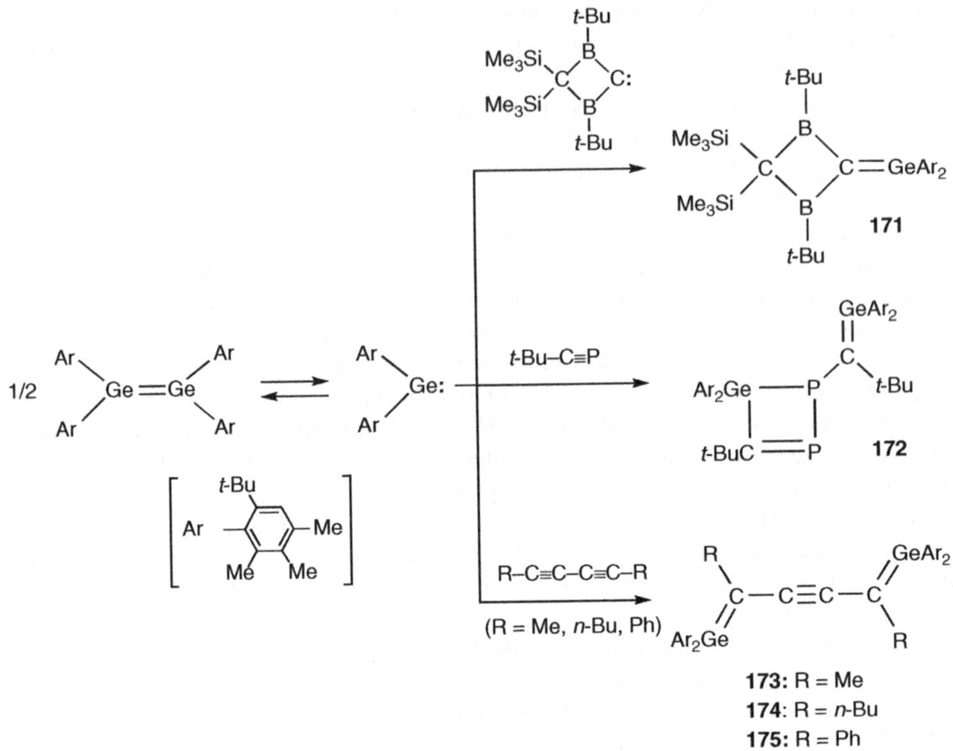

Scheme 5.52

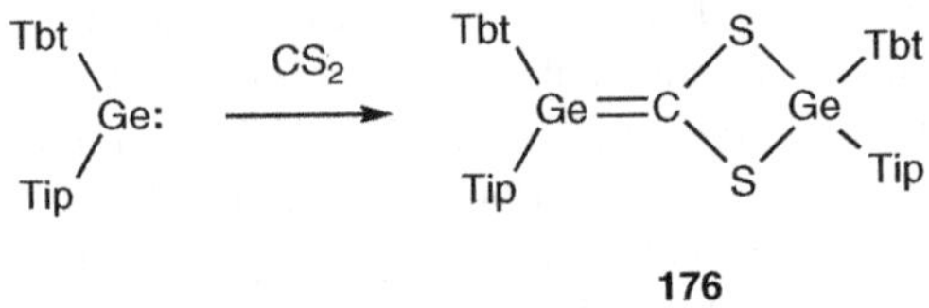

Scheme 5.53

Å].[144b] Such extraordinary Ge=C bond shortening in **176** is amazing, particularly given the presence of the two π-donating S substituents at the doubly bonded C atom. The Ge=C bond in **176** was almost ideally planar with a trigonal-planar geometry around both Ge and C centers and a very small twisting of only 4° (Table 5.8).

The longest Ge=C bond of 1.895(3) Å was reported by Lee *et al.* in 2002.[150] This bond was incorporated into the bicyclic skeleton of norbornene derivative **177**, which was unexpectedly obtained upon the reaction of benzaldehyde with the 2,4-disila-1-germatricyclo[2.1.0.0^{2,5}]pentane cage compound **178** (Scheme 5.54). Although electronic effects of the substituents at the Ge=C bond may also play a role in its lengthening, the major factor contributing to the overall stretching is probably steric repulsion between the bulky *t*-Bu_2MeSi substituents. Nevertheless, the Ge=C bond in **177** manifested nearly planar geometry with the sum of the bond angles around the doubly bonded Ge and C atoms of 359.2 and 358.9°, respectively, and an insignificant twisting of 6.4° (Table 5.8).

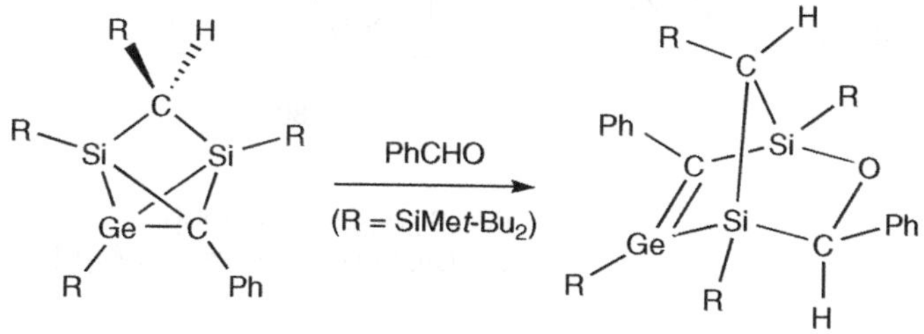

178 **177**

Scheme 5.54

An even longer Ge=C bond was reported by Schumann *et al.* for the cyclopropenylidene adduct of diaminogermylene. This compound **179a** was prepared by the reaction of *in situ* generated cyclopropenylidene derivative with $[(Me_3Si)_2N]_2Ge$: (Scheme 5.55, E = Ge).[151] The most prominent structural feature of **179a** was its extraordinarily long Ge=C bond of 2.085(3) Å, which significantly exceeded the range of typical Ge=C bonds, being even longer than standard Ge–C single bonds. Moreover, the Ge center was greatly pyramidalized with the sum of the bond angles 303.0°.

These structural peculiarities of **179a** indicate that its real structure is best described as the ylide resonance extreme **179′** rather than the germene form **179** (Scheme 5.56). This was also corroborated by the low-field resonance of the central carbon atom observed at 145.5 ppm, which is in the region characteristic for cyclopropenylium derivatives. Such a zwitterionic structure of **179a** is highly reminiscent of that of 4-silatriafulvene **158**,[132] which featured a very similar silyl anion-substituted cyclopropenylium composition. By the same method, Schumann *et al.* also synthesized the corresponding cyclopropenylidene adducts of divalent tin and lead, **179b** and **179c**, respectively (Scheme 5.55, E = Sn and E = Pb).[151] Both **179b** and **179c** revealed analogous ylide-type structures (Scheme 5.56, resonance form **179′**), which were seen in their diagnostic structural features: long E–C distances [2.303(9) Å in **179b** and 2.423(8) Å in **179c**], highly pyramidal E centers (299.9° in **179b** and 297.1° in **179c**), highly deshielded central C atoms (150.2 ppm in **179b** and 148.5 ppm in **179c**), and a notably shielded Sn center in **179b** (−44.7 ppm).

As a closely related example, an adduct of an NHC (1,3-dimesitylimidazolyl-2-ylidene) and GeI_2 **180** was described by Arduengo *et al.*[152] Quite similarly to the above-described case of pseudogermene **179a**, adduct **180** featured a very long Ge–C bond [2.102(12) Å] and remarkably pyramidal Ge center (295.0°), indicative of the predominant contribution of the charge-separated ylide form $>Ge^{-}–C^{+}<$ rather than the doubly bonded germene form >Ge=C<.

In a series of recent papers, Leung *et al.* reported the preparation of germavinylidenes and their derivatives, in which coordinatively unsaturated germanium centers were intramolecularly stabilized by nitrogen *n*-donor ligands.[153] Although these compounds

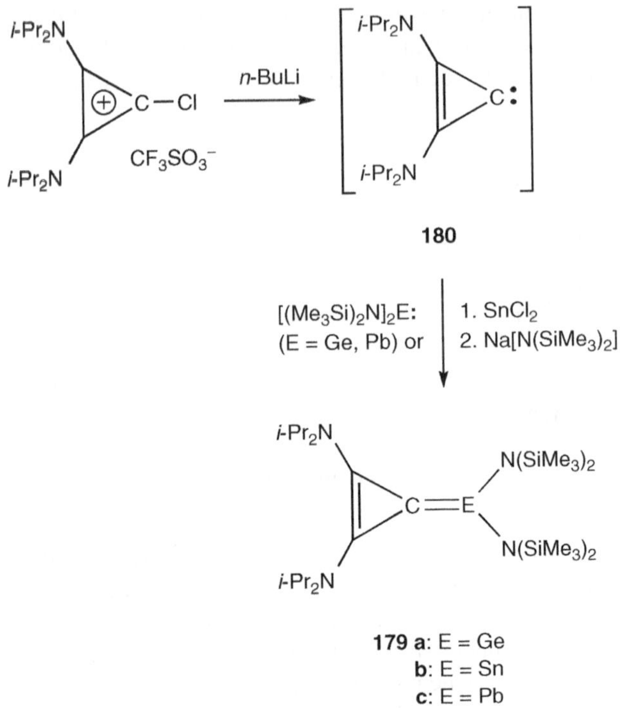

Scheme 5.55

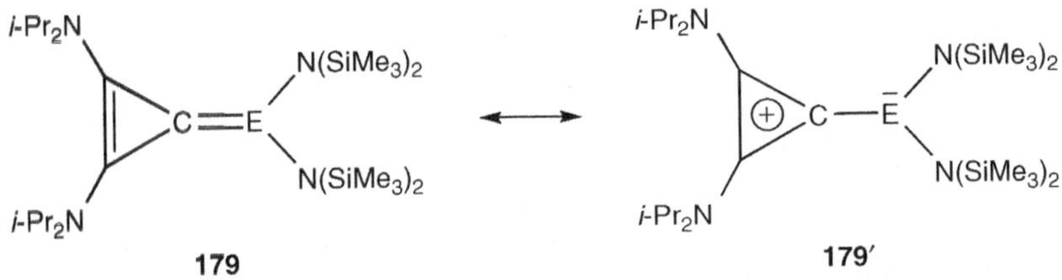

Scheme 5.56

were claimed to be germenes, given their specific structural features (rather long Ge–C distances well in the range of normal single bonds, pyramidality and tetracoordination of Ge centers), it seems problematic to classify them as true Ge=C doubly bonded species.

5.3.1.2.1.3 Stannenes >Sn=C<[71b,124f,i]

Several transient stannenes, such as $Me_2Sn{=}C(SiMe_3)_2$,[154] were generated as reactive intermediates prior to isolation of their stable counterparts.

The first isolable stannenes **181a,b** were synthesized by Berndt *et al.* in 1987 by the same method used for the preparation of stable germenes (Section 5.3.1.2.1.2, Scheme 5.50), namely, by the reaction of electrophilic cryptocarbene and stable stannylenes RR′Sn: (Scheme 5.57, Table 5.9).[155a,b]

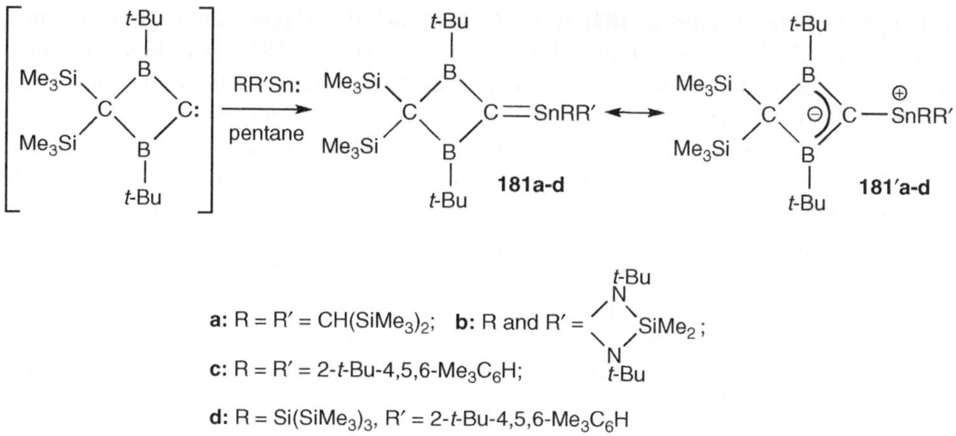

Scheme 5.57

Table 5.9 *Crystallographic and spectral parameters of the structurally characterized stannenes*

Stannene	>Sn=C< (Å)	Σ(Sn) (°)	Σ(C) (°)	>Sn=C< twist (°)	^{119}Sn(Sn=C)	^{13}C(Sn=C)	Reference
181a (Scheme 5.57)	2.025(4)	359.6	355.3	61.0	835.0	142.0	155a
181c (Scheme 5.57)	2.032(5)	359.7	359.8	36.0	374.2	137.0	157
181d (Scheme 5.57)	2.032(2)	357.7	358.1	12.0	NA	NA	148a
182c (Scheme 5.58)	2.003(5)	360.0	360.0	10.0	277.4	146.9	158d
182d (Scheme 5.58)	2.016(5)	359.9	359.9	28.5	270.0	144.9	159

NA, Not available.

In accord with its structure, doubly bonded Sn and C atoms in **181a,b** resonated in at a low field: 835 and 142 ppm (for **181a**),[155a] 647 and 91 ppm (for **181b**)[155b]. The Sn=C bond in **181a** of 2.025(4) Å was markedly shorter than the other Sn–C bonds in the same compound [2.152(5) and 2.172(4) Å], being ca. 6.4% shorter than typical Sn–C bonds (Table 5.9).[155a,b] This bond length of 2.025(4) Å was in relatively good agreement with the values calculated for the parent $H_2Sn{=}CH_2$ (1.982 Å,[156a] 2.063 Å,[156b] 1.945 Å).[156c] Compared with germene **169a**, stannene **181a** revealed stronger distortions of the Sn=C bond, manifested in the pyramidalization of its doubly bonded atoms (359.6° around Sn and 355.3° around C) and twisting of the Sn=C bond (61°). Similarly to the case of germenes **169**, stannenes **181** showed an important contribution from the ylide resonance form **181′** (Scheme 5.57), seen in the characteristic shielding of the boron atoms in **181** [$\delta(^{11}B)$: 64 ppm in **181a** and 50 ppm in **181b**] and shortening of the C(sp^2)–B bonds [1.510(6) and 1.494(7) Å].

Employing Berndt's cryptocarbene, Weidenbruch *et al.* prepared other stable stannenes **181c,d** by analogous procedures, as outlined in Scheme 5.57 (Table 5.9).[148a,157]

Although the boron centers in **181c** were shielded [$\delta(^{11}B)$: 63 ppm] and $C(sp^2)$–B bonds were shortened, the extent of the ylide resonance extreme **181′c** was less important than in the case of Berndt's stannenes **181a,b**, as demonstrated by the remarkable high field shift of the ^{119}Sn NMR resonance in **181c** compared with that of **181a** (374 vs 835 ppm).[157] The Sn=C bonds in both **181c** and **181d** were identical, 2.032(5) and 2.032(2) Å, and close to that in **181a** [2.025(4) Å]. However, in contrast to **181a**, both **181c** and **181d** featured significantly more planar Sn=C bonds (sum of the bond angles around Sn, sum of the bond angles around C, twisting angle): 359.6° vs 359.7° and 357.7°; 355.3° vs 359.8° and 358.1°; 61° vs 36° and 12° (Table 5.9).[148a,157]

A series of stable stannenes **182a-c** was prepared by Escudié *et al.* by the procedure previously used for the synthesis of isostructural germenes (see Section 5.3.1.2.1.2), namely, by the dehydrofluorination of (fluorenyl)fluorostannanes with *t*-BuLi in diethyl ether (Scheme 5.58, Table 5.9).[158]

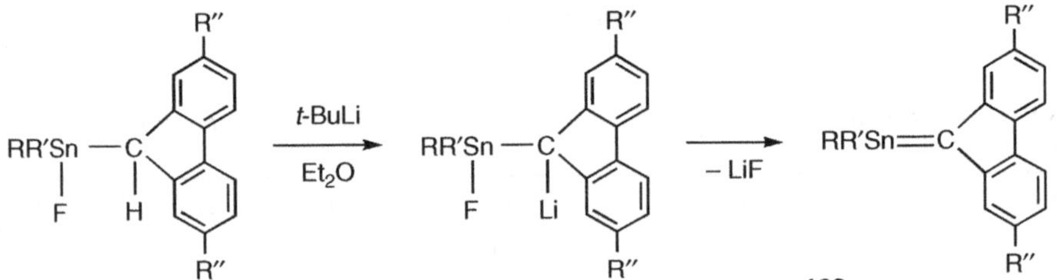

182

(R = Dis, Tip, Tbt; R′ = Dis, Mes, Tip; R″ = H, *t*-Bu)

a: R = R′ = Dis, R″ = H
b: R = R′ = Tip, R″ = H
c: R = R′ = Tip, R″ = *t*-Bu
d: R = Tbt, R′ = Mes, R″ = H

Scheme 5.58

With Dis and Tip groups on their tin atoms, both **182a** and **182b** proved to be room temperature stable, although not isolable as individual compounds.[158a,b] Consequently, their composition as stannenes >Sn=C< was established on the basis of NMR spectral characteristics: $\delta(^{119}Sn)$ 288.0 ppm (for **182b**) and $\delta(^{13}C)$ 133.9 ppm (for **182b**).[158b] Surprisingly, the tin center in **182b** was remarkably less deshielded (288.0 ppm) than one would expect for a true sp^2-Sn atom, which was explained by the complexation of diethyl ether solvent to the electron-deficient tin. Placing *t*-Bu groups onto the fluorenylidene moiety, Escudié *et al.* designed stannene **182c** as an isolable analog of **182b**, whose spectral characteristics were similar to those of **182b**: $\delta(^{119}Sn)$ 277.4 ppm and $\delta(^{13}C)$ 146.9 ppm (Scheme 5.58, Table 5.9).[158d] The stannene **182c** featured the shortest ever reported Sn=C bond of 2.003(5) Å, which was shorter than typical Sn–C bonds by ca. 7%. Both of the doubly bonded Sn and C centers in **182c** displayed undistorted trigonal-planar geometry, and the Sn=C bond was rather insignificantly twisted with a twisting angle of merely 10° (Table 5.9).

Utilizing the above-described method, Tokitoh *et al.* recently prepared another representative of stable stannenes **182d** bearing bulky Tbt and Mes groups at the tin center

(Scheme 5.58, Table 5.9).[159] The spectral and structural characteristics of the novel stannene **182d** were similar to those of other stable stannenes: low-field signals for both sp^2-Sn (270.0 ppm) and sp^2-C (144.9 ppm), a short Sn=C bond of 2.016(5) Å, trigonal-planar geometry around the doubly bonded Sn and C atoms and twisting of the Sn=C bond by 28.5° (Table 5.9). The predominant formulation of **182d** as a true stannene >Sn=C< rather than the ylide resonance form $>Sn^+–C^-<$ was deduced from the nonequivalence of all atoms of the fluorenylidene unit, implying restricted rotation about the Sn=C bond.

Several compounds with tricoordinate tin centers, which can formally be regarded as stannenes >Sn=C<, were actually classified as zwitterionic adducts of stannylenes with Lewis bases of the type $>Sn^-–C^+<$. They include the stannylene complexes with stable N-heterocyclic carbenes $Tip_2Sn=CR_2$ [R_2 = –N(*i*-Pr)–C(Me)=C(Me)–N(*i*-Pr)–] **183**,[160] $Cl_2Sn=CR_2$ [R_2 = –N(*i*-Pr)–C(Me)=C(Me)–N(*i*-Pr)–] **184**,[161] and the previously described adduct of diaminostannylene with cyclopropenylidene **179b** (Scheme 5.55, E = Sn).[151] The formulation of these compounds as Lewis acid–Lewis base adducts rather than true stannenes was based on their peculiar structural characteristics. Thus, their Sn–C bonds were extraordinarily stretched, 2.379(5) Å (in **183**), 2.290(5) Å (in **184**) and 2.303(9) Å (in **179b**), values that far exceeded those expected for Sn=C bonds and are well in the range of standard Sn–C single bonds. Another indication of the considerable contribution of the ylide resonance form $>Sn^-–C^+<$ was a remarkable pyramidalization at the tin centers amounting to 308.7° (in **183**), 282.0° (in **184**) and 299.9° (in **179b**).

5.3.1.2.1.4 Plumbenes >Pb=C<[5a]

No stable compounds featuring genuine >Pb=C< double bonds have been reported to date, reflecting increasing difficulties in synthesizing multiply bonded derivatives on descending the main groups of the periodic table. One can mention only Schumann's cyclopropenylidene adduct of diaminoplumbylene **179c** (see Scheme 5.55, E = Pb),[151] which, however, was described as the ylide-type Lewis acid–Lewis base complex $>Pb^-–C^+<$ **179c′** rather than the true plumbene >Pb=C< **179c**. This assignment was based on the extremely long Pb–C bond separation of 2.423(8) Å, substantial pyramidalization at the Pb center of 297.1° and remarkable deshielding of the central C atom observed at 148.5 ppm (which is in the region typical for cyclopropenylium derivatives).

5.3.1.2.1.5 Silagermenes >Si=Ge<[124i]

Silagermenes (otherwise known as germasilenes) became synthetically accessible as isolable derivatives only during the last decade. The first mention of the generation of silagermenes dates back to 1991, when Baines and Cooke reported the generation of the metastable $Mes_2Si=GeMes_2$ **185** by either thermolysis or low-temperature photolysis of hexamesitylsiladigermirane (Scheme 5.59).[162] Because silagermene **185** was stable only at temperatures below −70 °C, its formation was deduced from low temperature NMR and UV observations [80.6 ppm (sp^2-Si) and 414 nm (Si=Ge π–π^* transition)], as well as its trapping reaction with methanol. At room temperature, **185** underwent irreversible 1,2-migration of a Mes group from germanium to silicon to produce silylgermylene $Mes_3Si(Mes)Ge$:, which was subsequently trapped with Et_3SiH (Scheme 5.59). The possibility of such a silagermene–silyl germylene rearrangement was theoretically predicted by Grev *et al.*, who estimated that the singlet silylgermylene $H_3Si(H)Ge$: is 6.2 kcal/mol more stable than the isomeric *trans*-bent silagermene $H_2Si{=}GeH_2$.[163] One should, however, remain cautious about the generalization of this prediction, because the ease of 1,2-shifts of substituents is totally governed by their nature: whereas small groups can migrate smoothly, large substituents whose migratory aptitude is significantly suppressed are mostly reluctant to migrate.

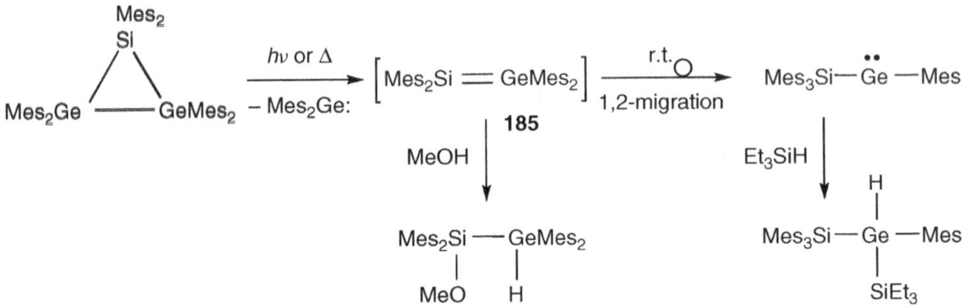

Scheme 5.59

The first isolable silagermene, 1*H*-disilagermirene **186**, featuring an endocyclic Si=Ge bond incorporated into a three-membered ring Si_2Ge-skeleton, was synthesized by Lee *et al.* in 2000.[164] This cyclic silagermene **186** was readily accessible by isomerization (either thermal or photochemical) of the 3*H*-disilagermirene **187** possessing a skeletal Si=Si bond (see Section 5.3.1.1.1) (Scheme 5.60). The composition of **186** as a compound with a Si=Ge bond was confirmed by the observation of the low-field resonance of the skeletal Si atom at 100.7 ppm. Computations at the DFT B3LYP/DZd level on the H_3Si-substituted models predicted 1*H*-disilagermirene **186** to be more stable than the isomeric 3*H*-disilagermirene **187** by ca. 2.3 kcal/mol, a value that was in fairly good agreement with the experimental estimate of ca. 3 kcal/mol.[164a] The thermodynamic preference of the silagermene **186** over the disilene **187** stems probably from the greater strength of the skeletal Si–Si bond in **186** compared with the skeletal Si–Ge bond in **187**. The Si=Ge bond length in the model H_3Si-substituted 1*H*-disilagermirene was calculated to be ca. 2.178 Å at the DFT B3LYP/DZd level,[164] which is in good agreement with the previous *ab initio* estimate on the parent $H_2Si{=}GeH_2$ of 2.184 Å.[163]

The nearest homolog of **186**, 1*H*-disilagermirene **188**, was prepared from 3*H*-disilagermirene **187** in three steps (Scheme 5.61).[52] At first, the ring expansion reaction of **187** with CH_2Cl_2 produced four-membered ring **189**, which was subsequently reduced with KC_8 yielding an isolable disilagermabicyclo[1.1.0]butane **190**. Finally, thermal isomerization of **190** resulted in the formation of 1*H*-disilagermirene **188** (Scheme 5.61).[52] The doubly bonded Si atom of **188** was more deshielded than that of the parent **186**: 126.6 ppm vs 100.7 ppm.

Reacting 1*H*-disilagermirene **186** with phenylacetylene, the same authors prepared the five-membered ring 1,2-disila-3-germacyclopenta-2,4-diene **191**, in which the Si=Ge double bond was structurally authenticated for the first time (Scheme 5.60,

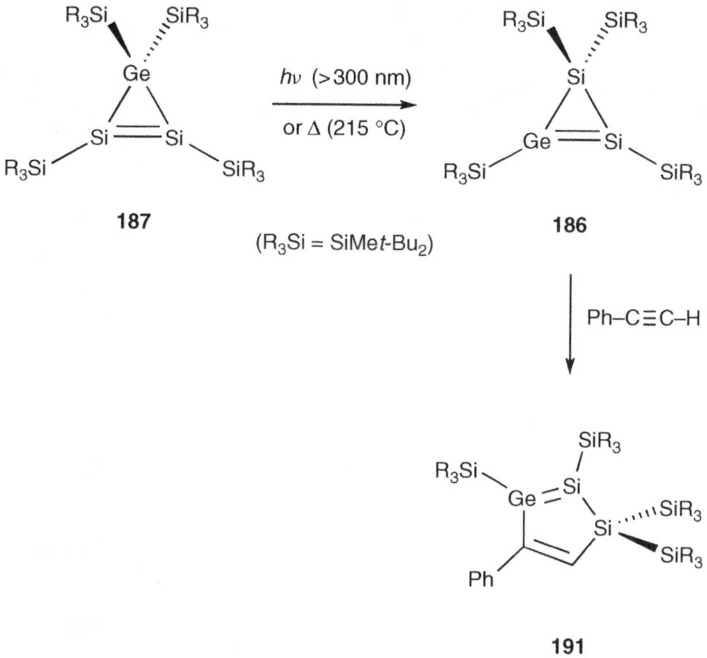

Scheme 5.60

Figure 5.17).[165] The doubly bonded silicon atom in **191** resonated as expected in the low field region at 124.2 ppm. The Si=Ge bond was twisted by 38.6° and its length of 2.250(1) Å was intermediate between those of typical Si=Si and Ge=Ge bonds, being notably longer than the computational estimates of 2.178 and 2.184 Å, evidently because of steric reasons resulting from its bulky substituents.[165a]

As was suggested, the mechanism for the unexpected formation of the heavy cyclopentadiene **191** from the heavy cyclopropene **186** may involve the initial [2 + 2] cycloaddition of phenylacetylene across the Si=Ge bond of **186**, yielding a highly strained bicyclo[2.1.0]pentene derivative **192** (Scheme 5.62).[165] Valence isomerization of the latter intermediate results in the formation of another transient species, cyclopentadiene derivative **193**, featuring Si=C and Ge=C bonds. Final migration of the silyl substituent in **193** would complete the reaction sequence, producing the most stable cyclopentadiene isomer **191** possessing Si=Ge and C=C bonds. The last step is apparently driven by the evident benefits gained upon the formation of the thermodynamically favorable carbon–carbon double bond. Computations at the DFT B3LYP/6-31G* level on the parent H-substituted derivatives revealed the following stability order of the cyclic $GeSi_2C_2$ isomers (in kcal/mol): +13.4 (**192′**), +14.3 (**193′**), 0.0 (**191′**).[165b]

Several other stable silagermenes, $(t\text{-}Bu_2MeSi)_2Si{=}GeMes_2$ **194**[166] and $(t\text{-}Bu_3Si)_2Si{=}GeMes_2$ **195**,[167] were synthesized by the coupling of readily available 1,1-dilithiosilane derivatives R_2SiLi_2 (R = SiMe*t*-Bu$_2$, Si*t*-Bu$_3$) with dichlorodimesitylgermane Mes_2GeCl_2. Both **194** and **195** exhibited unusually shielded sp^2-Si centers, whose resonances were observed at 22.4 and 18.7 ppm, respectively. This observation

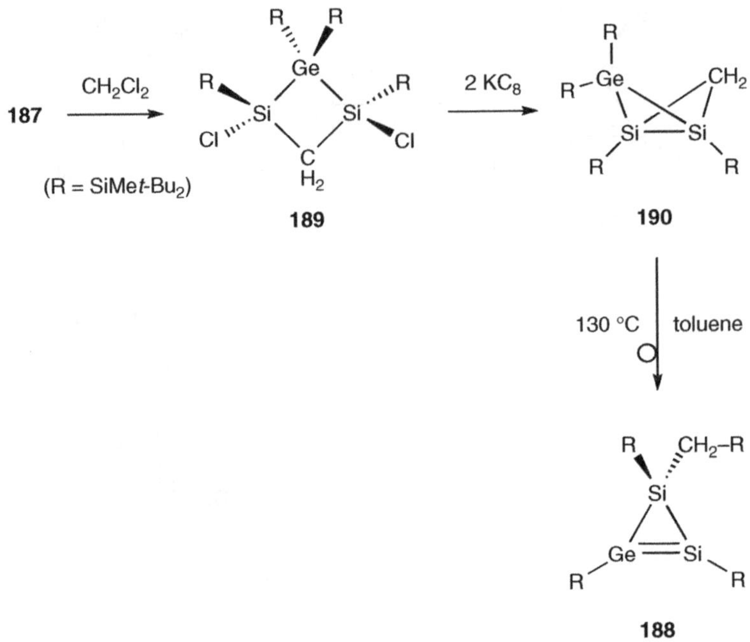

Scheme 5.61

was attributed to the effects of substitution patterns, namely, to the σ-electronic effects of the substituents (electron-donating silyl groups at Si and electron-withdrawing groups at Ge), which alter the Si=Ge bond polarity from the natural $Si^{\delta+}=Ge^{\delta-}$ to a reversed $Si^{\delta-}=Ge^{\delta+}$. This was corroborated by the regioselectivity of MeOH addition to **194** to form a single regioisomer (t-$Bu_2MeSi)_2$(H)Si–Ge(OMe)Mes_2.[166] The Si=Ge bond in **195** was longer than that of the cyclic silagermene **191**, 2.2769(8) Å vs 2.250(1) Å, apparently because of the stronger steric interactions in the former.[167] Although both doubly bonded Si and Ge atoms displayed planar geometry around them (360.0°), the Si=Ge bond was notably twisted by 24.7°. The silagermene **195** revealed a remarkable red shift in its UV spectrum, where the longest wavelength absorption was observed at 553 nm (π–π^* transition). This bathochromic shift most likely originates from the twisting of the Si=Ge bond, which results in the destabilization of the HOMO because of the reduced $3p_\pi$(Si)–$4p_\pi$(Ge) orbital overlap. The silagermene **195** was transformed under thermal conditions at 100 °C to a symmetrically substituted isomer (E)–[Mes(t-Bu_3Si)Si=Ge(Mes)Sit-Bu_3] **195a**.[167]

Silagermenes in which the Si=Ge bond is incorporated into an allenic fragment of the types >Si=Ge=Si< or >Ge=Si=Ge< are considered in Section 5.3.2.

5.3.1.2.1.6 Silastannenes >Si=Sn<[124i]

The formation of transient silastannenes was originally suggested upon the reaction of the stable N-heterocyclic silylene and isolable stannylenes; however, this assumption was not supported by either spectroscopic or trapping reaction evidence.[168]

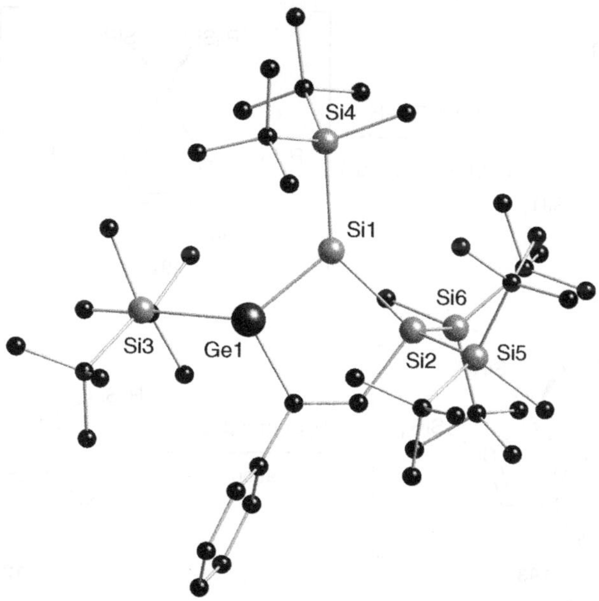

Figure 5.17 *Crystal structure of silagermene **191** (hydrogen atoms are not shown)*

The first (and still the sole) isolable silastannene $(t\text{-}Bu_2MeSi)_2Si{=}SnTip_2$ **196** was prepared in 2002 by the coupling of 1,1-dilithiosilane derivative $(t\text{-}Bu_2MeSi)_2SiLi_2$ with the diaryldichlorostannane Tip_2SnCl_2 (Scheme 5.63).[169]

Similarly to the above-described case of the silagermenes with reversed $Si^{\delta-}=Ge^{\delta+}$ bond polarization, silastannene **196** also featured a remarkable inversely polarized $Si^{\delta-}=Sn^{\delta+}$ double bond because of the electronic effects of its substituents. This was manifested in the appreciable deshielding of the sp^2-Sn nucleus [δ(^{119}Sn) 516.7 ppm] and significant shielding of the sp^2-Si nucleus [δ(^{29}Si) = 27.4 ppm]. The Si=Sn bond of 2.4188(14) Å in **196** was intermediate between the typical values of Si=Si (2.138–2.289 Å) and Sn=Sn (2.590–3.087 Å) bonds, being ca. 7% shorter than the standard Si–Sn bond length of 2.60 Å (Figure 5.18).[169]

The Si=Sn bond was remarkably twisted by 34.6°, thus reflecting a high degree of steric interaction between the bulky substituents. As expected, the Si=Sn bond in **196** was *trans*-bent, however, the bending pattern was quite unusual, being greater (26.2°) at Si and smaller (9.6°) at Sn, even though one might expect the opposite trend considering the well-known tendency for an increase in *trans*-bending descending group 14. These nontrivial structural features of the silastannene **196** were rationalized in terms of the unconventional unsymmetrical donor–acceptor interaction mode (Scheme 5.64), promoted by the reversed $>Si^{\delta-}=Sn^{\delta+}<$ double bond polarization discussed above. Then, the negatively polarized silicon moiety $>Si^{\delta-}$ would be bent more strongly than the positively polarized tin unit $>Sn^{\delta+}$, which was indeed observed experimentally. Accordingly, DFT B3LYP/DZd computations on the model silastannene $(H_3Si)_2Si{=}SnPh_2$ revealed a substantial Si=Sn bond polarization: −0.536 on Si and +1.400 on Sn (NPA charges).

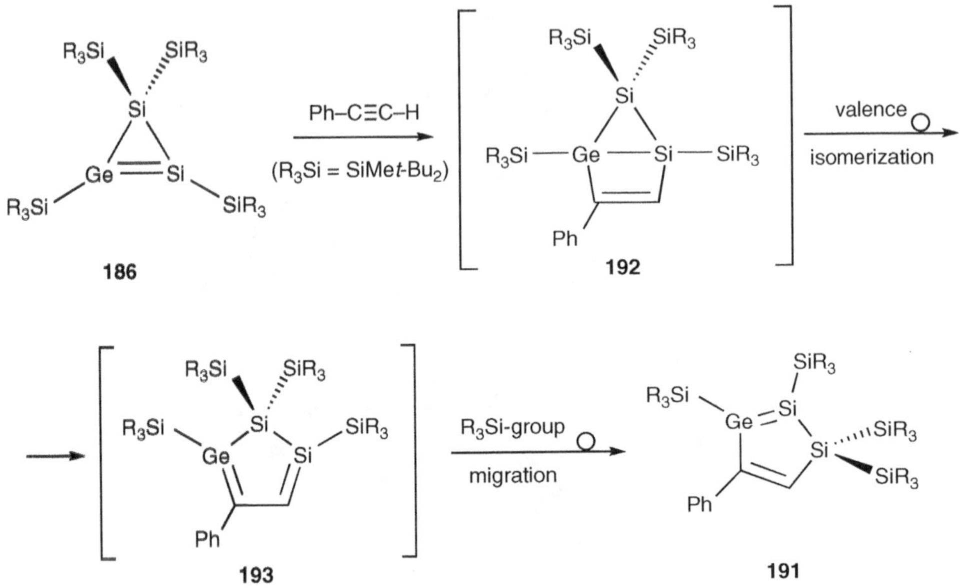

Scheme 5.62

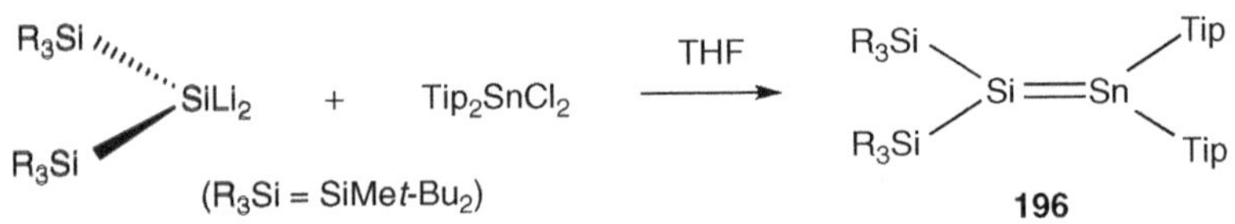

Scheme 5.63

5.3.1.2.1.7 Germastannenes >Ge=Sn<[124i]

The transient germastannene $Mes_2Ge{=}SnTip_2$ **197** was claimed to be formed upon the reductive dehydrofluorination of the (fluorostannyl)germane $Mes_2(H)Ge–Sn(F)Tip_2$ with *t*-BuLi by Escudié *et al.* in 1996.[170] The formulation of **197** as a true germastannene was based on the observation of a low-field ^{119}Sn resonance at 360 ppm and trapping reactions of the Ge=Sn bond with MeOH and PhCHO. However, at room temperature, germastannene **197** underwent unavoidable decomposition, forming a set of products including the three-membered ring germadistannirane *cyclo*-($Mes_2Ge–Tip_2Sn–Tip_2Sn$), which implied the initial dissociation of the Ge=Sn bond into germylene Mes_2Ge: and stannylene Tip_2Sn: fragments.

The first isolable structurally characterized germastannene $Tip_2Ge{=}SnTip_2$ **198** was prepared by Weidenbruch *et al.* by the straightforward one-pot reaction of TipMgBr, $GeCl_2$•diox complex and $SnCl_2$ (Scheme 5.65).[171] The presence of the sp^2-Sn in **198** was evidenced by its low-field resonance observed at 268.0 ppm. The Ge=Sn bond of 2.5065(5) Å was short, being ca. 4.6% shorter than typical Ge–Sn bonds (Figure 5.19). As expected for the heavy group 14 elements, the *trans*-bending at Sn in **198** was larger

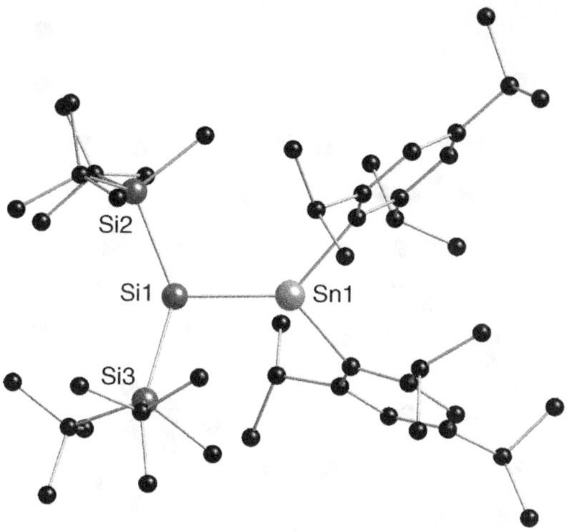

Figure 5.18 *Crystal structure of silastannene **196** (hydrogen atoms are not shown)*

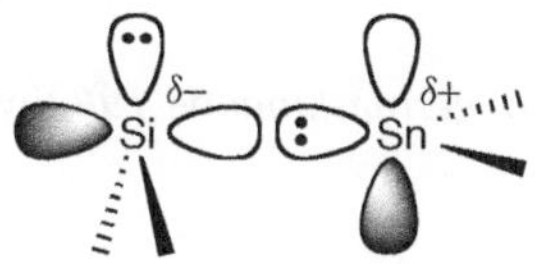

196

Scheme 5.64

(337.3°) than that at Ge (349.1°). Although stable in the solid state, germastannene **198** gradually decomposed in solution forming cyclotristannane *cyclo*-($Tip_2Sn)_3$ and digermene $Tip_2Ge{=}GeTip_2$.

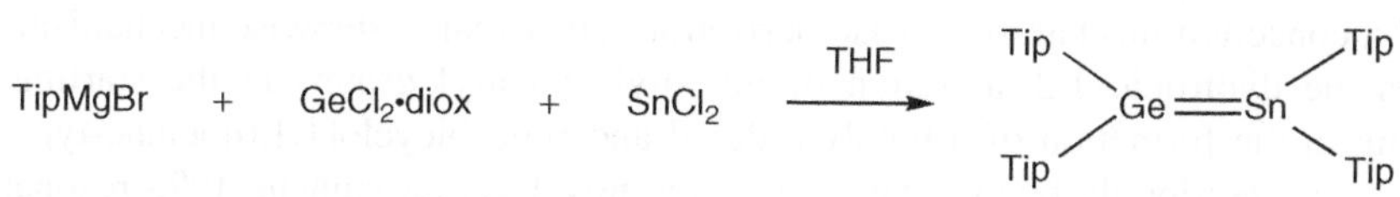

198

Scheme 5.65

Utilizing the same procedure as that used for silastannene **196**, namely, coupling of 1,1-dilithiogermane derivative (t-$Bu_2MeSi)_2GeLi_2$ and diaryldichlorostannane Tip_2SnCl_2, Sekiguchi *et al.* synthesized another stable germastannene (t-$Bu_2MeSi)_2Ge{=}SnTip_2$ **199** (Scheme 5.66).[172] In the absence of structural data, **199** was classified as a germastannene based on its characteristic low-field ^{119}Sn NMR resonance at 525.1 ppm, diagnostic of an sp^2-Sn atom. Similarly to the above-described silagermene **195**, germastannene **199** readily isomerized under very mild thermal conditions (50 °C) to form a symmetrically substituted isomer (t-$Bu_2MeSi)TipGe{=}Sn(SiMe$t$-$Bu_2)Tip$ **199a** as a single E-isomer

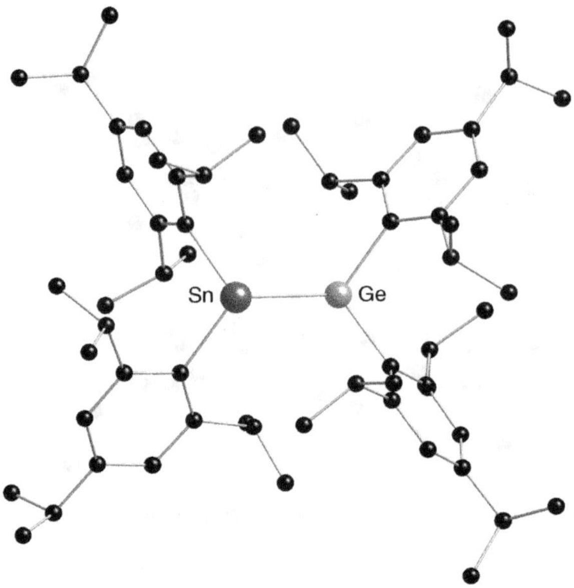

Figure 5.19 *Crystal structure of germastannene **198** (hydrogen atoms are not shown)*

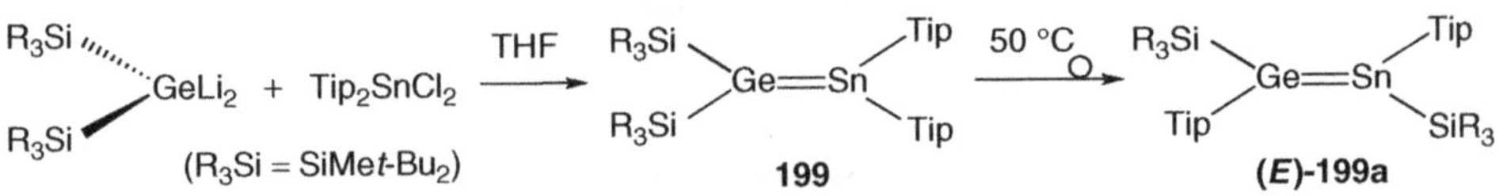

Scheme 5.66

(Scheme 5.66).[172] The results of the ^{1}H NMR kinetic measurements of the **199**→ **199a** isomerization (negative value for the activation entropy $\Delta S^{\ddagger}$ of −12.0 cal/K•mol) suggested a concerted mechanism for isomerization rather than a stepwise mechanism. This implies the dyotropic 1,2-migration of the silyl and aryl groups in the starting **199**, resulting in the formation of a highly ordered and rigid bicyclo[1.1.0]butane-type transition state. The doubly bonded Sn atom in the novel germastannene **199a** resonated at a higher field than that of starting **199** (373.4 ppm vs 525.1 ppm), evidently because of the different substitution pattern at the tin center.

The cyclic germastannene **200**, incorporating an endocyclic Ge=Sn bond into the four-membered ring skeleton of $^3\Delta - 1,2,3,4$-disilagermastannetene, was available by the unexpected ring expansion reaction of either $1H$- or $3H$-disilagermirene **186** or **187** with the $SnCl_2$•diox complex (Scheme 5.67).[173] The tin center of the Ge=Sn bond in **200** resonated in a diagnostic low field at 439.3 ppm. In contrast to germastannenes **197** and **198**, cyclic germastannene **200** was indefinitely stable both in the solid state and in solution, showing no signs of dissociation into the germylene and stannylene. The enormous thermal stability of the cyclic germastannene **200** was attributed to the stabilizing effect of the σ-donating silyl groups, as well as to possible π(Ge=Sn) →

σ^*(Si–Cl) orbital mixing resulting in the lowering of the π-energy level and overall stabilization of the HOMO.[173b]

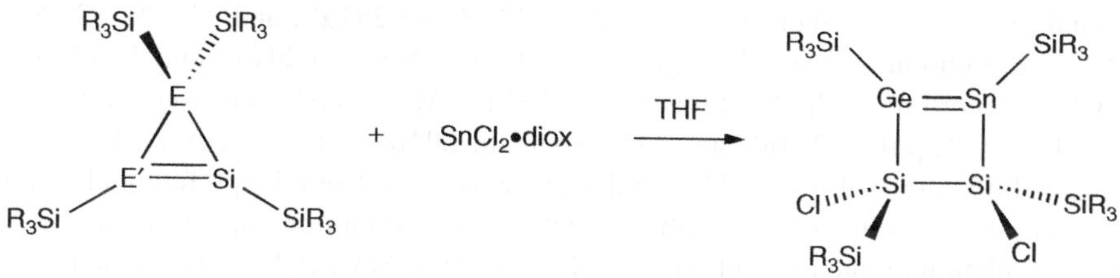

186 (E = Si, E′ = Ge; R$_3$Si = SiMe*t*-Bu$_2$)
187 (E = Ge, E′ = Si; R$_3$Si = SiMe*t*-Bu$_2$)

200

Scheme 5.67

5.3.1.2.2 Group 14/Group 13 Combinations: $>E_{14}=E_{13}-$ (E_{14} = Si, Ge; E_{13} = B, Ga, In)

Until 2004, no stable heteronuclear alkene analogs of group 13 and 14 elements $>E_{14}=E_{13}-$, apart from boraalkenes (methyleneboranes) $>C=B-$ featuring double bonds between the lightest elements of the groups, had been reported. This was documented in a comprehensive review by Power, who described several stable boraalkenes as well as their monoanions.[5a] As an early computational approach, Si–B and Si–Al π-bond energies in the parent $H_2Si{=}BH$ and $H_2Si{=}AlH$ were estimated as 27.1 and 14.1 kcal/mol, with the latter value indicative of rather weak bonding.[174]

The isolation of the first stable compounds in the $>E_{14}=E_{13}-$ series was achieved in 2004. The lithium salts of 1,3-disila-2-gallata- and -indataallenic anions $[R_2Si{\cdots}M{\cdots}SiR_2]^-$ **201a,b$^-$•[Li$^+$(thf)$_4$]** (**a**: M = Ga, R = SiMe*t*-Bu$_2$; **b**: M = In, R = SiMe*t*-Bu$_2$) were quantitatively available by the simple reaction of dilithiosilane derivative R_2SiLi_2 with 0.5 equiv. MCl_3 in THF (Scheme 5.68, E = Si).[175]

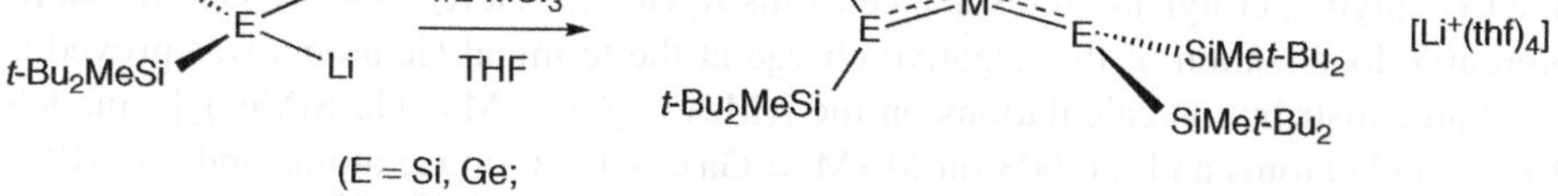

(E = Si, Ge;
M = Ga, In)

201a,b$^-$•[Li$^+$(thf)$_4$] (**a**: E = Si, M = Ga; **b**: E = Si, M = In)
202a,b$^-$•[Li$^+$(thf)$_4$] (**a**: E = Ge, M = Ga; **b**: E = Ge, M = In)

Scheme 5.68

Both **201a$^-$** and **201b$^-$** revealed no bonding interactions with the lithium countercation, representing free anions in their crystalline state. Both Si–Ga(In) bonds in **201a,b$^-$** are ca. 6–7% shorter than the sum of the silicon and gallium(indium) covalent radii: 2.2828(9), 2.2775(9) Å in **201a$^-$** and 2.4849(9), 2.4792(9) Å in **201b$^-$**.

In contrast to linear organic allenes, the anionic $[R_2Si{\cdots}M{\cdots}SiR_2]^-$ framework in **201a,b$^-$** is bent: 161.60(3) and 161.35(3)°, respectively. Moreover, the terminal silicons in the $[R_2Si{\cdots}M{\cdots}SiR_2]^-$ fragment are markedly pyramidalized, with the sum of the bond angles around them being 341.26, 342.96° in **201a$^-$**, and 324.39, 328.87° in **201b$^-$**, and pronounced *trans*-bent geometry: 41.63, 45.35° in **201a$^-$** and 53.17, 58.63° in **201b$^-$**. Accordingly, the Si atoms in the $[R_2Si{\cdots}M{\cdots}SiR_2]^-$ unit were significantly shielded: −79.9 ppm in **201a$^-$** and −77.6 ppm in **201b$^-$**. Most likely, such high-field chemical shifts, unusual for doubly bonded systems, stem from the preferential localization of the anionic charge at the silicon centers, an assumption supported by the NPA charge calculation on the model $[(Me_3Si)_2Si{\cdots}Ga{\cdots}Si(SiMe_3)_2]^-$: − 0.736 and −0.729 on Si and +0.627 on Ga. Overall, given the high polarity of Si–M bonds, the structure of allenic anions $[R_2Si{\cdots}M{\cdots}SiR_2]^-$ **201$^-$** is best described as the major contribution of the allylic silyl anion resonance extremes $R_2Si{=}M{-}SiR_2^- \leftrightarrow {}^-R_2Si{-}M{=}SiR_2$, rather than the contribution of the ate-type anion $R_2Si{=}M^-{=}SiR_2$.

Germanium versions of **201a,b$^-$**, namely, the lithium salts of 1,3-digerma-2-gallata- and -indataallenic anions $[R_2Ge{\cdots}M{\cdots}GeR_2]^-$ **202a,b$^-$•[Li$^+$(thf)$_4$]** (**a**: M = Ga, R = SiMe*t*-Bu$_2$; **b**: M = In, R = SiMe*t*-Bu$_2$) were prepared similarly by the reaction of dilithiogermane R_2GeLi_2 with 0.5 equiv. MCl_3 in THF (Scheme 5.68, E = Ge).[176] The structural characteristics of **202b$^-$** followed the same trends observed earlier for **201a,b$^-$**: short Ge–In bonds of 2.5453(4) and 2.5387(4) Å (ca. 7% shorter than the sum of the germanium and indium covalent radii), bent $R_2Ge{\cdots}M{\cdots}GeR_2$ framework [158.451(14)°], highly pyramidal germanium atoms (sum of the bond angles of 317.3 and 321.1°, *trans*-bent angles of 64.9 and 59.7°). Comparing geometries of **202b$^-$** and its silicon analog **201b$^-$**, one can note that the deviation from planarity (bending, twisting) becomes more pronounced in the case of heavier elements. That is not surprising given the well-known tendency for significant pyramidalization at the anionic centers descending group 14, which stems from the progressive increase in the s- and p-orbitals' energy separation, resulting in the inherent reluctance of the heavy group 14 elements to hybridize. This causes, in turn, the remarkable 'lone-pair character' and large share of s-fraction of the anionic orbital and, consequently, high p-character of the hybrid orbitals to substituents. Similarly to the case of the silicon versions **201a,b$^-$**, the constitution of 1,3-digerma-2-gallata- and -indataallenic anions **202a,b$^-$** is best described as the hybrid of the two allylic germyl anion resonance forms $R_2Ge{=}M{-}GeR_2^- \leftrightarrow {}^-R_2Ge{-}M{=}GeR_2$. Preferential localization of the negative charge at the terminal Ge atoms was proved by NPA charge distribution calculations on the $[(Me_3Si)_2Ge{\cdots}M{\cdots}Ge(SiMe_3)_2]^-$ models: −0.754 on Ge atoms and +0.603 on M (M = Ga), −0.806 on Ge atoms and +0.818 on M (M= In).

Synthesis of neutral heavy alkenes of $>E_{14}{=}E_{13}-$ type has been achieved only in the case of boron–silicon derivatives. However, in contrast to the well-established chemistry of boraalkenes >C=B–, synthesis of their silicon analogs >Si=B– (silaborenes) was hampered by the lack of suitable precursors and relative weakness of the bond (27.1 kcal/mol for $H_2Si{=}BH$ vs 53.7 kcal/mol for $H_2C{=}BH$)[174]. Initial attempts to prepare silaborenes employed the synthetically very useful dilithiosilane or -germane derivatives R_2ELi_2 (R = SiMe*t*-Bu$_2$; E = Si, Ge). However, their reaction with $MesBCl_2$ in THF afforded seven-membered ring compounds **203a,b** (**a**: E = Si; **b**: E = Ge) instead of the expected sila- or germaborenes, which were assumed to be initially formed as THF complexes **204a,b** (**a**: E = Si; **b**: E = Ge) followed by nucleophilic attack of the negatively polarized silicon or germanium centers on coordinated THF, which finally results in its ring opening and formation of heterocyclic compounds **203a,b** (Scheme 5.69).[177]

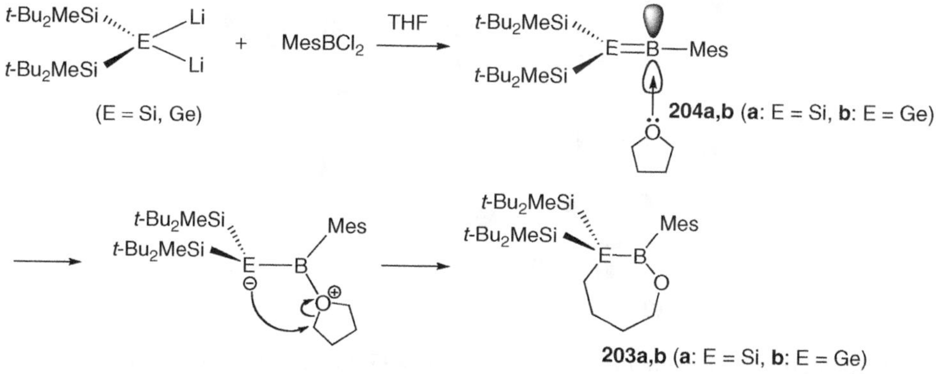

Scheme 5.69

The isolation of the first room temperature stable silaborene **205** was achieved by the reaction of the same dilithiosilane $(t\text{-}Bu_2MeSi)_2SiLi_2$ with dichloro(2,2,6,6-tetramethylpiperidino)borane using the less coordinating toluene instead of THF (Scheme 5.70).[178]

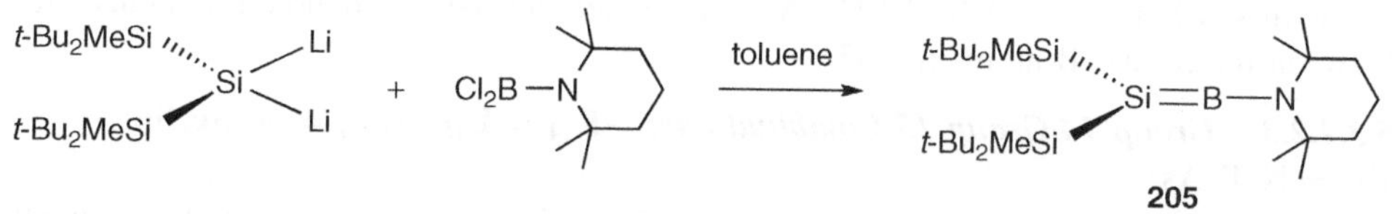

Scheme 5.70

The presence of an amino substituent at the doubly bonded boron also contributes to the overall thermodynamic stabilization of **205**, reducing the intrinsically high Lewis acidity of the boron center through π-donation from the nitrogen lone pair. This was manifested in the enormous shielding of the sp^2-Si atom in **205** observed in a high-field region very unusual for doubly bonded silicons (−128.1 ppm). This was explained by the σ-donation of the electropositive silyl substituents at Si and σ-acceptance of the amino substituent at B, leading to a reversed double bond polarization: $>Si^{\delta-}=B^{\delta+}$–. Such an explanation, however, does not take into account the paramagnetic contribution, which is well known to operate in the opposite direction and cause the remarkable low-field shift of silyl-substituted disilenes in their ^{29}Si NMR spectra. Apparently, the huge high-field shift of −128.1 ppm of silaborene **205** stems primarily from the important contribution of another resonance extreme, represented by the charge-separated formula $>Si^{-}–B=N^{+}<$

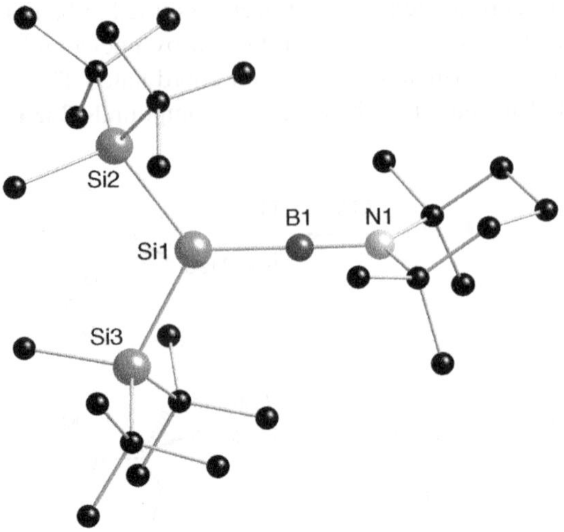

Figure 5.20 *Crystal structure of the silaborene **205** (hydrogen atoms are not shown)*

(GIAO ^{29}Si NMR chemical shift for the Me_3Si-substituted model: −94.3 ppm; NPA charge distribution: −0.268 on Si and +0.357 on B).[178] Such partial localization of the negative charge on the central Si1 can also be seen in the slight pyramidalization around it (sum of the bond angles 358.3°), which corresponds to a *trans*-bending of 13° (Figure 5.20). The Si1–B1–N1 framework is nearly linear [176.87(13)°] with a Si1–B1 bond distance of 1.8379(17) Å, which is ca. 10% shorter than the typical Si–B single bonds in silylboranes (2.038–2.1249 Å), thus indicating the true double bond character between the Si and B atoms in **205**.

5.3.1.2.3 Group 14/Group 15 Combinations: **>E_{14} = E_{15}– (E_{14} = Si, Ge, Sn; E_{15} = N, P, As)**

Stable derivatives of the several classes of >E_{14}=E_{15}– type (E_{14}, group 14 element; E_{15}, group 15 element) have been synthesized during the past two decades: silaimines >Si=N–, germaimines >Ge=N–, stannaimines >Sn=N–, phosphasilenes >Si=P–, phosphagermenes >Ge=P–, phosphastannenes >Sn=P– and arsasilenes >Si=As–. They are briefly described below.

5.3.1.2.3.1 Silaimines >Si=N–[179]

The first thermally stable silaimines were independently reported in 1986 by two German groups. Klingebiel *et al.* utilized a method based on the thermal β-elimination of LiCl from the immediate precursor, lithium (chlorosilyl)amide **206**, resulting in the formation of silaimine **207** (Scheme 5.71).[180] The presence of a doubly bonded Si atom in **207** was supported by the observation of a low-field ^{29}Si NMR resonance at 60.3 ppm.

Wiberg *et al.* applied a different approach: coupling of azido-di-*tert*-butylchlorosilane *t*-Bu_2SiClN_3 and tri-*tert*-butylsilyl sodium *t*-Bu_3SiNa in dibutyl ether forming silaimine *t*-Bu_2Si=N–Si*t*-Bu_3 **208** after elimination of NaCl and N_2 (Scheme 5.72).[181]

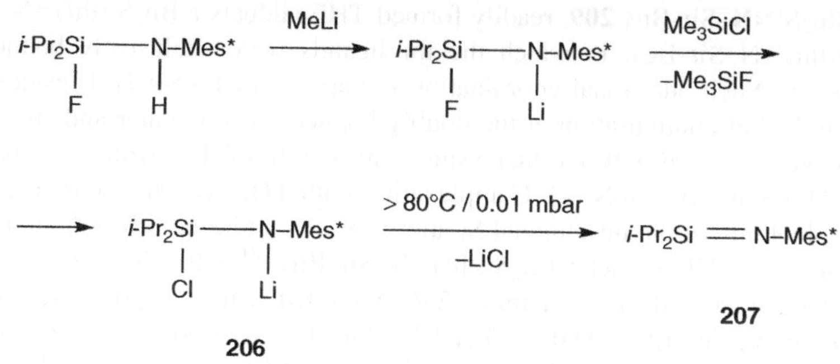

Scheme 5.71

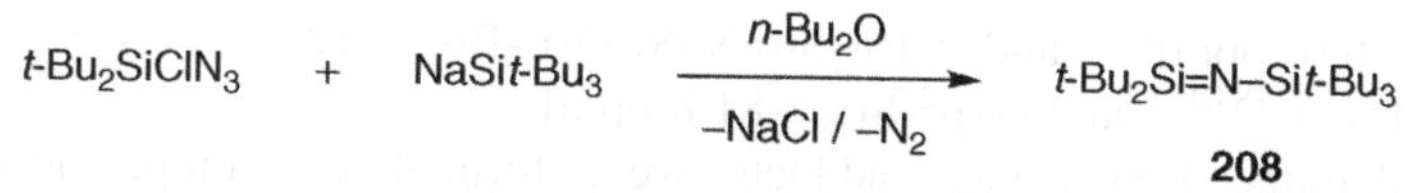

Scheme 5.72

The doubly bonded Si atom of **208** resonated at low field (78.29 ppm), and the geometry around it was found to be nearly planar (sum of the bond angles 359.9°). The Si=N double bond of 1.568(3) Å was markedly shorter than typical Si–N bonds (1.64–1.78 Å).[181] Unexpectedly, the Si=N–Si skeleton was essentially linear with the bond angle 177.8°, indicating that the nitrogen lone pair in **208** was stereochemically inert. Such surprising at first glance geometry at the N center (linear instead of anticipated trigonal) of the free silaimines was explained by taking into account the very low barrier to inversion at N because of the rather shallow bending potential energy surface of silaimines. Thus, although the $H_2Si{=}N{-}H$ model was calculated to possess a Si=N–H bond angle of 126.6°, which is substantially smaller than that of 177.8° in **208**, its barrier to linearization was exceedingly low: 3.3 kcal/mol (at the 6-31G* level) and 6.0 kcal/mol (at the MP4/6-31G* level).[182] Electropositive silyl groups at nitrogen further decrease the inversion barrier resulting in a nearly linear geometry of the $H_2Si{=}N{-}SiH_3$ model, manifesting a Si=N–Si bond angle of 175.6° and short Si=N bond of 1.549 Å (6–31G*).[182] The shortening and strengthening of the Si=N bond upon linearization is also remarkable: 1.572 Å in the bent form of $H_2Si{=}N{-}H$ vs 1.528 Å in its linear form. This stems from the increasing s-contribution to nitrogen hybridization in the linear form, which compensates for the angle distortion at nitrogen. Overall, the inversion barrier in silaimine $H_2Si{=}N{-}H$ is significantly lower than that of the isostructural imine $H_2C{=}N{-}H$ (32.7 kcal/mol). This huge difference is attributed mainly to the greater polarization of the $Si^{\delta++}{=}N^{\delta--}$ bond compared with the $C^{\delta+}{=}N^{\delta-}$ bond, because of the larger electronegativity difference between Si vs N (1.90 vs 3.04) compared with C vs N (2.55 vs 3.04).

Another important consequence of the remarkable $Si^{\delta+}{=}N^{\delta-}$ bond polarization was the highly pronounced tendency of free silaimines for complexation with Lewis bases to satisfy the great electronic demand at the electrophilic silicon center. This was verified by numerous experimental observations. Thus, silaimine **208**, as well as silaimine $Me_2Si{=}N{-}Si\textit{t}\text{-}Bu_3$ **209**, readily formed THF-adducts $\textit{t}\text{-}Bu_2Si(thf){=}N{-}Si\textit{t}\text{-}Bu_3$ and $Me_2Si(thf){=}N{-}Si\textit{t}\text{-}Bu_3$, in which the thf ligands were exclusively bound to the sp^2-Si atoms.[181] Such additional coordination [>O:(from thf)→Si=N–] resulted in the distorted tetrahedral configuration at the doubly bonded silicon center and stretching of the Si=N bond in accord with the increasing contribution of the zwitterionic resonance extreme [$>O^+$(from thf)–Si–N^-–]. Complexation with THF was also manifested in the remarkable shielding of the unsaturated Si atom: 78.29 ppm in the original silaimine **208** vs 1.08 ppm in its THF-adduct $\textit{t}\text{-}Bu_2Si(thf){=}N{-}Si\textit{t}\text{-}Bu_3$.[181] $\textit{t}\text{-}Bu_2Si(O{=}CPh_2){=}N{-}Si\textit{t}\text{-}Bu_3$, a benzophenone adduct of silaimine **208**, exhibited structural properties similar to those of the above-described THF-adduct $\textit{t}\text{-}Bu_2Si(thf){=}N{-}Si\textit{t}\text{-}Bu_3$: lengthening of the Si=N bond to 1.601(2) Å and a decrease in the Si=N–Si bond angle to 169.3°.[181b,c]

Klingebiel *et al.* similarly prepared several other stable free silaimines by the thermal elimination of LiCl from lithium (chlorosilyl)amides (see Scheme 5.71): $\textit{t}\text{-}Bu_2Si{=}N{-}Mes^*$ **210** [$\delta(sp^2\text{-}Si) = 63.1$ ppm],[183] $\textit{t}\text{-}Bu_2Si{=}N{-}Si(\textit{i}\text{-}Pr)_2\textit{t}\text{-}Bu$ **211** [$\delta(sp^2\text{-}Si) = 76.6$ ppm][184] and $\textit{t}\text{-}Bu_2Si{=}N{-}Si(Ph)\textit{t}\text{-}Bu_2$ **212** [Si=N = 1.573(2) Å, Si=N–Si = 168.3°[184,185] and $\delta(sp^2\text{-}Si) = 81.8$ ppm].[186]

Several silaimine–Lewis base adducts were formed by complexation of free silaimines. All of them exhibited spectral and structural properties that are characteristic for such adducts, namely, shielding of the unsaturated silicon center, stretching of the Si=N bond and pyramidalization at the sp^2-Si atom [$\delta(sp^2\text{-}Si)$ in ppm, Si=N bond length in Å, the sum of the bond angles around the sp^2-Si atom in degrees]: $\textit{t}\text{-}Bu_2Si(thf){=}N{-}Si(Me)\textit{t}\text{-}Bu_2$ **213** [3.60, 1.596(2), 347.8];[187] $\textit{t}\text{-}Bu_2Si(thf){=}N{-}Si(\textit{i}\text{-}Pr)_2\textit{t}\text{-}Bu$ **211•thf** [3.13, 1.599(2), 347.8];[184] $\textit{t}\text{-}Bu_2Si(thf){=}N{-}Si(Ph)\textit{t}\text{-}Bu_2$ **212•thf** (2.68, –, –); [184] $\textit{t}\text{-}Bu_2Si(pyridine){=}N{-}Si(\textit{i}\text{-}Pr)_2\textit{t}\text{-}Bu$ **211•pyridine** [−12.2, 1.606(2), 346.6];[184] $\textit{t}\text{-}Bu_2Si(pyridine){=}N{-}Si(Ph)\textit{t}\text{-}Bu_2$ **212•pyridine** [−12.6, 1.611(2), 347.0].[184]

Reaction of the stable N-heterocyclic silylene **214** with azide Ph_3CN_3 resulted in the formation of an isolable THF-stabilized silaimine **215** (Scheme 5.73).[188] Similarly to other silaimine complexes, **215** featured a short Si=N bond [1.599(4) Å] and insignificant departure of the doubly bonded Si atom from the ideal trigonal-planar geometry (sum of the bond angles 350°).

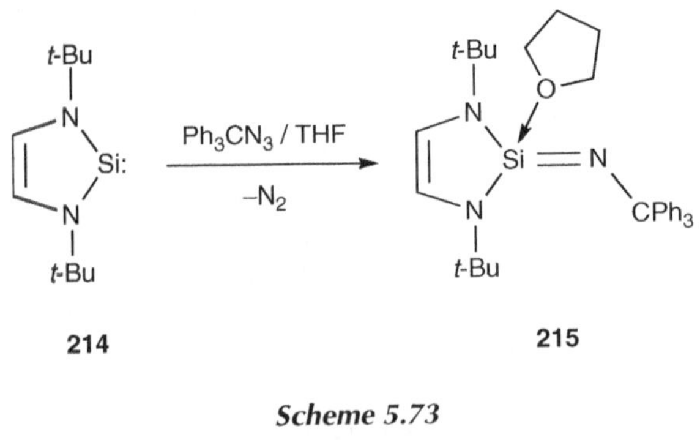

Scheme 5.73

In some cases, lithiation of (fluorosilyl)amine precursors $R_2(F)Si{-}NH{-}R'$ resulted in the formation of stable derivatives, considered to be formal LiF adducts of silaimines

216 (Scheme 5.74).[189] As is typical for free silaimines, LiF complexes **216** manifested relatively short Si=N bonds and almost linear geometry at their N centers.

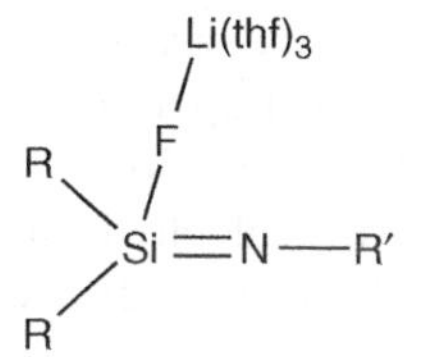

216

R = Me, *i*-Pr, *t*-Bu; R′ = Mes*

Scheme 5.74

Thermolysis of silatetrazoline **217** resulted in an unexpected [3 + 2] cycloreversion producing the novel silaimine *t*-Bu_2Si=N–Si(Cl)*t*-Bu_2 **218** after cycloelimination of *t*-Bu_3SiN_3 (Scheme 5.75).[190a] Upon treatment with NMe_2Et, silaimine **218** readily formed the Lewis base-supported complex *t*-Bu_2Si(NMe_2Et)=N–Si(Cl)*t*-Bu_2 **218•NMe₂Et** (Scheme 5.75).[190b] Surprisingly, the resonance of the doubly bonded Si atom of the silaimine was almost unaffected upon its complexation with the *n*-donor ligand: 3.0 ppm in **218** vs −3.4 ppm in **218•NMe_2Et**.

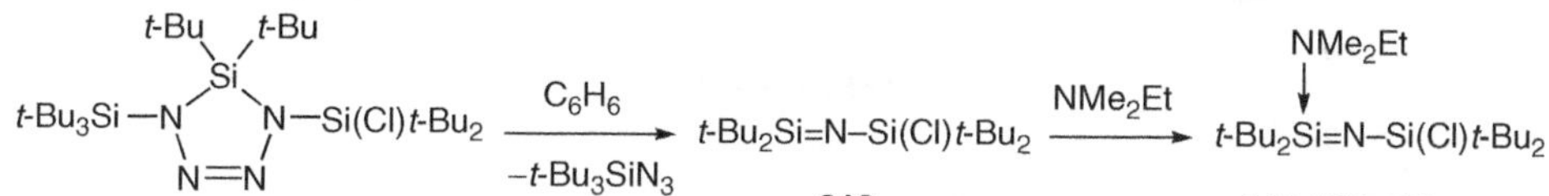

217 **218** **218•NMe_2Et**

Scheme 5.75

5.3.1.2.3.2 Germaimines >Ge=N–[71a,c,191a,b]

Isolation of the first stable germaimines was preceded by the direct observation of their transient congeners in low-temperature matrices. Thus, germaimine Mes_2Ge=N–Mes **219** was detected upon the photolysis of its Mes_3GeN_3 precursor in 3-MP matrix at 77 K accompanied by the elimination of molecular nitrogen.[192] Two absorption bands at 309 nm (π–π^* electronic transition) and 459 nm (n–π^* electronic transition) were assigned to **219**, which exhibited an intense yellow color.

The first room temperature stable germaimine ***trans*-220** was claimed by Glidewell *et al.* in 1987.[193] ***trans*-220** was reported to be prepared by the reaction of bis(amino)germylene $[(Me_3Si)_2N]_2$Ge: and dimethyl 2-diazomalonate $N_2C(COOMe)_2$, and its constitution was proved by trapping reactions with EtOH and water. Upon standing in solution at room temperature, ***trans*-220** was transformed into another compound, which was believed to be its isomer ***cis*-220** (Scheme 5.76). However, subsequent elaborate studies by Meller *et al.* showed that the originally claimed 'germaimine' ***trans*-220** was actually represented by its cyclic form **221**, which lacks a double bond between Ge and N (Scheme 5.76).[194a] The thermodynamically more favored compound ***iso*-221**, to which the structure of ***cis*-220** was mistakenly ascribed, appeared to be an isomer of **221** formed upon methyl migration from the oxygen of the MeO-group to the ring nitrogen (Scheme 5.76).[194b] Although **221** is not a germaimine, it reacts with *tert*-butyl isocyanate, benzaldehyde and acrolein as its synthetic equivalent to form products of formal [2 + 2] and [4 + 2] cycloadditions across the Ge=N bond. Apparently, initial ring opening and isomerization steps were involved in these unusual reactions of **221**.

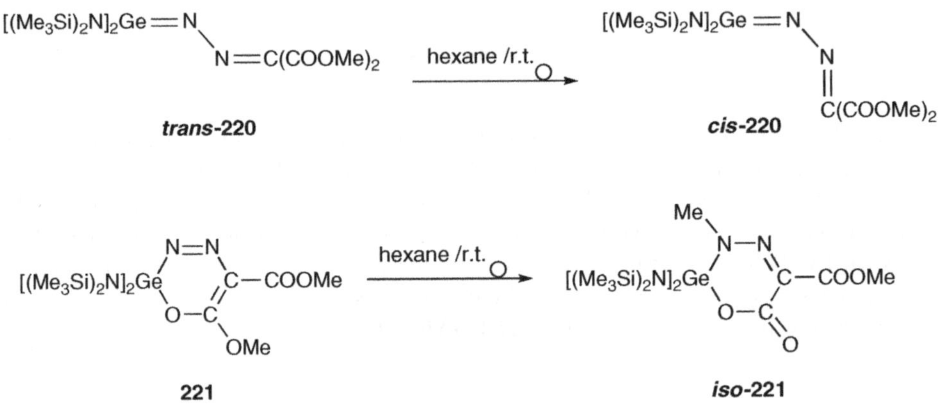

Scheme 5.76

In the following years, some other germaimines were reported to be stable and even isolable at ambient conditions: $[(Me_3Si)_2N]_2Ge{=}N{-}SiR_3$ available by the reaction of $[(Me_3Si)_2N]_2Ge$: with N_3SiR_3 (R = Et, O-*t*-Bu),[195] $Mes_2Ge{=}N{-}Mes$ **219** synthesized by the photolysis of Mes_3GeN_3 and characterized by its reactions with benzaldehyde and diphenylnitrone,[196] and $H_2Ge{=}N{-}CF_3$ prepared from GeH_4 and CF_3NO.[197] However, assignment of the last species $H_2Ge{=}N{-}CF_3$ to the class of germaimines is still doubtful, particularly given its somewhat ambiguous characterization based solely on IR/Raman spectroscopy data and some trapping reactions, its inexplicable resistance towards water and surprising thermal stability in the absence of sterically demanding groups on Ge.

The first really stable free germaimines **222** were isolated and structurally characterized by Meller *et al.* in 1991.[198] They were prepared by the reaction of sterically very crowded bis(amino)germylenes $[R(Me_3Si)N]_2Ge$: and 2,6-substituted phenylazides $R'N_3$ (Scheme 5.77).

The Ge=N bond lengths in **222a** and **222c** were found to be rather short, 1.703(2) and 1.691(3) Å, respectively, being in good agreement with the value of 1.681 Å calculated for the $H_2Ge{=}NH$ model (SCF).[199] Similarly to the case of silaimines, the geometry around the Ge=N bond was nearly planar with a N–Ge=N–C torsion angle of 178.4° (in **222a**) and 178.1° (in **222c**).

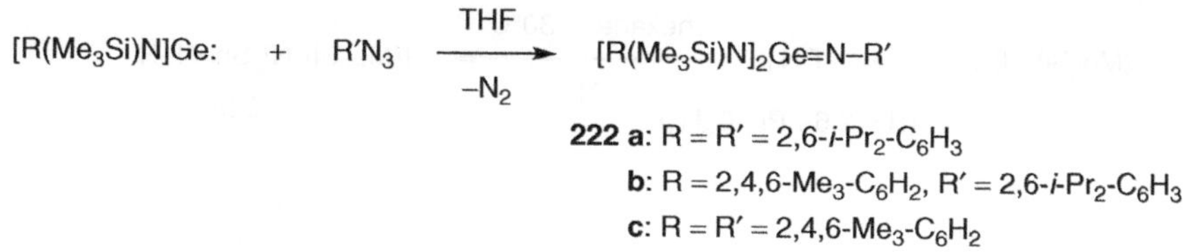

Scheme 5.77

A similar reaction of germylene Dis_2Ge: [Dis = $CH(SiMe_3)_2$] and diazidosilane $R_2Si(N_3)_2$ (R = *t*-Bu, Mes) resulted in the formation of *N*-(azidosilyl)germaimines $Dis_2Ge{=}N{-}Si(N_3)R_2$ **223** (Scheme 5.78).[200] Upon further treatment of **223** with a second equivalent of Dis_2Ge:, bifunctional bis(germaimines) **224** were isolated in high yields (Scheme 5.78). Both mono- and bis(germaimines) **223** and **224** manifested shortened Ge=N bonds of 1.704(5) Å (in **223a**) and 1.681(8) Å (in **224b**), respectively, as well as nearly planar environments around their double bonds, characterized by the C–Ge=N–Si torsion angles of 1.0° (in **223a**) and 3.2° (in **224b**).[200]

Several stabilized germaimines featuring Ge=N bonds supported by intramolecular coordination of *n*-donor ligands (N, O, Hal), in which the germanium center was to a large extent tetracoordinate, were reported in the 1990s.[201]

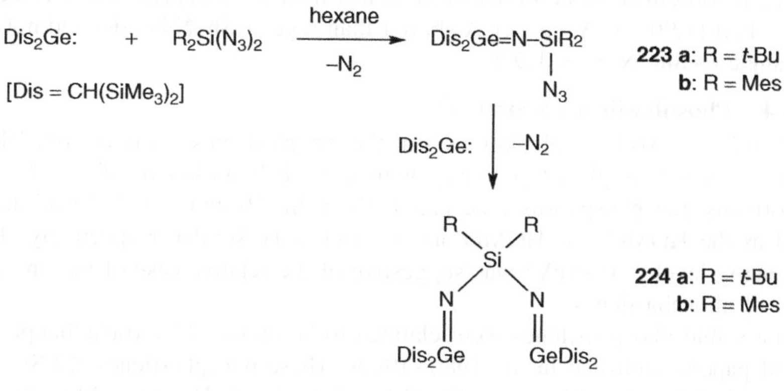

Scheme 5.78

5.3.1.2.3.3 Stannaimines >Sn=N–[71b,191a]

Only two stable stannaimines have been isolated and structurally characterized to date. The first of them, $[(Me_3Si)_2N]_2Sn{=}N{-}R$ (R = 2,6-*i*-Pr_2-C_6H_3) **225**, was prepared in 1993 by Meller's group by the reaction of bis(amino)stannylene $[(Me_3Si)_2N]_2Sn$: and (2,6-diisopropylphenyl)azide (2,6-*i*-Pr-C_6H_3)N_3 (Scheme 5.79).[202] The three N atoms are arranged in a trigonal-planar fashion around the Sn center (359.2°) in **225**, and the Sn=N bond of 1.921(2) Å is markedly shorter than normal Sn–N bonds.[202a]

Stannaimine **225** was stable below −30 °C, whereas above this temperature it rearranged into bicyclic **226** through the intramolecular addition of the C–H bond of one of the Me_2CH groups across the Sn=N bond (Scheme 5.79).[202a]

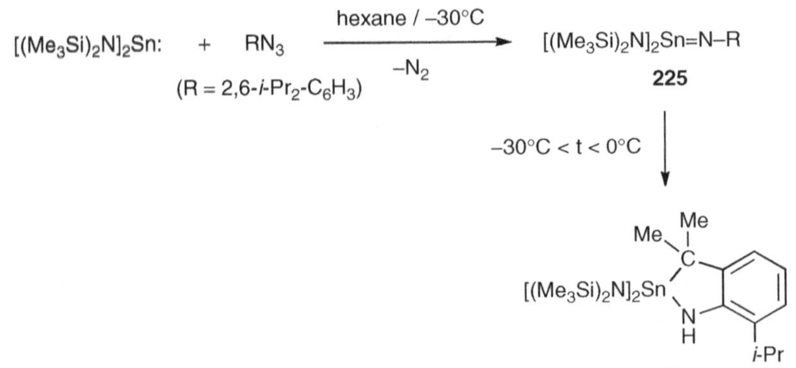

Scheme 5.79

A second stannaimine $Dis_2Sn{=}N{-}Si(N_3)t\text{-}Bu_2$ **227** was synthesized by Ando *et al.* by a procedure analogous to that used for preparation of the isostructural germaimine **223a** (see Scheme 5.78), namely, by the reaction of stable stannylene Dis_2Sn: [Dis = $CH(SiMe_3)_2$] and diazidosilane $t\text{-}Bu_2Si(N_3)_2$.[200b] As was expected for the doubly bonded system, the tricoordinate Sn atom resonated at low field at 340.4 ppm, whereas the Sn=N bond was short [1.905(5) Å, ca. 0.1 Å shorter than regular Sn–N bonds] and not twisted (torsion angle C–Sn=N–Si = 1.9°).

5.3.1.2.3.4 Phosphasilenes >Si=P–[203]

Early *ab initio* 6-31G** calculations on the simplest unsubstituted phosphasilene $H_2Si{=}PH$ predicted its planar geometry with a Si=P bond length of 2.061 Å.[204] As frontier orbitals, the phosphorus lone pair $n(P)$, $\pi(Si{=}P)$ and $\pi^*(Si{=}P)$ orbitals were identified as the HOMO – 1, HOMO and LUMO of $H_2Si{=}PH$, respectively. The low LUMO energy level (+0.85 eV) was suggestive of the relative ease of the one-electron reduction of phosphasilenes.

The first stable phosphasilenes were claimed to be prepared by Bickelhaupt *et al.* in a series of papers published in the 1980–1990s. These phosphasilenes RR′Si=PAr [R, R′ = *t*-Bu, Ph, Mes, Tip, Mes*, 2,6-Et_2-C_6H_3, 2,4,6-Et_3-C_6H_2; Ar = Mes, Tip, Mes*, 2,4,6-Et_3-C_6H_2, 2,4,6-Ph_3-C_6H_2, 2,6-Me_2-4-$(OC_8H_{17})C_6H_2$] **228** were prepared by the reductive dehydrochlorination of the corresponding precursors RR′Si(Cl)–P(H)Ar with *n*-BuLi.[205] Albeit not isolated in pure form, phosphasilenes **228** were reliably characterized by their diagnostically deshielded Si (148.7–199.0 ppm) and P (65.8–136.0 ppm) nuclei, pointing to the double bond character of the Si–P bond. Moreover, the $^1J(^{29}Si{-}^{31}P)$ coupling constants were very large (148.5–155.0 Hz), thus testifying to the high s-character of the Si–P bond, as expected for bonding between two elements in their sp^2-hybridization states.[205] Interestingly, phosphasilenes **228** exhibited no observable solvent effects on their ^{29}Si NMR chemical shifts or $^1J(^{29}Si{-}^{31}P)$ coupling constants, which was in sharp contrast to a remarkable shielding of the doubly bonded Si atoms in silaimines in polar solvents because of their coordination to the highly electrophilic silicon centers (see above). Apparently, this distinction in NMR behavior of silaimines and phosphasilenes stems from the markedly higher polarity of the Si=N bond compared with that of the Si=P bond.

The first phosphasilene **229**, whose crystal structure was reported, was prepared by the reaction of the lithium salt of the 1,3-diphospha-2-silaallyl anion derivative **230$^-$•[Li$^+$(Et$_2$O)$_2$]**[206a] with Ph_2PCl (Scheme 5.80).[206b] The Si=P double bond in **229** of 2.094(3) Å was remarkably shorter than the average Si–P single bond (2.25 Å), but was, however, longer than the value theoretically predicted for the parent $H_2Si{=}PH$ (2.061 Å),[204] apparently because of the steric hindrance caused by the bulky substituents.[206b] The Si=P bond in **229** was insignificantly twisted by 3.4°, whereas the doubly bonded Si atom was slightly pyramidalized with the sum of its bond angles being 356.7°. Such unexpected distortion at Si, contradicting the planar geometry of the parent $H_2Si{=}PH$,[204] can be best rationalized in terms of a second-order Jahn–Teller effect; that is, by mixing of the energetically close $HOMO_{\pi(Si=P)}$ and $(LUMO + 1)_{\sigma*(Si-P)}$ to lower the π-orbital energy and stabilize the HOMO. Both Si and P sp^2-centers were diagnostically deshielded, being observed at 180.2 and 128.7 ppm, respectively. The $^1J(^{29}Si{-}^{31}P)$ coupling constant in **229** of 203.0 Hz notably exceeded the range of previously reported values (148.5–155.0 Hz).

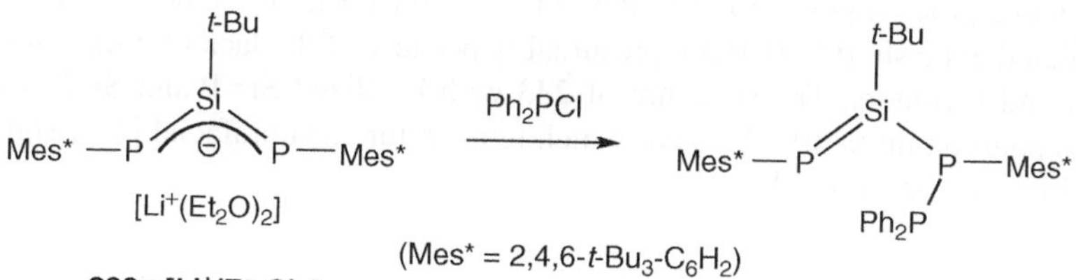

230$^-$•[Li$^+$(Et$_2$O)$_2$] **229**

Scheme 5.80

The chemistry of phosphasilenes $R_2Si{=}P{-}R'$ was further developed by Driess *et al.*, who employed the general approach of the thermal β-elimination of LiF from the lithium (fluorosilyl)phosphide derivatives $R_2Si(F){-}P(Li)R'$ (Scheme 5.81).[207] Thus, the isolable phosphasilenes $Tip_2Si{=}P{-}R$ (R = $SiMe_2t$-Bu, Sii-Pr_3) **231** were prepared by this route from $Tip_2Si(F){-}P(Li)R$.[207a] Interestingly, whereas the resonances of the doubly bonded silicon centers in **231** were observed as expected in the low field region (167.8, 178.4 ppm), their phosphorus partners were found to be shifted to a high field region (11.1, 17.7 ppm). Such unusually strong shielding of the P nuclei was attributed to the σ-donating influence of the trialkylsilyl groups on phosphorus. The characteristically large $^1J(^{29}Si{-}^{31}P)$ coupling constants in **231** (149, 160 Hz) were indicative of the true π-bonding between the Si and P atoms.[207a]

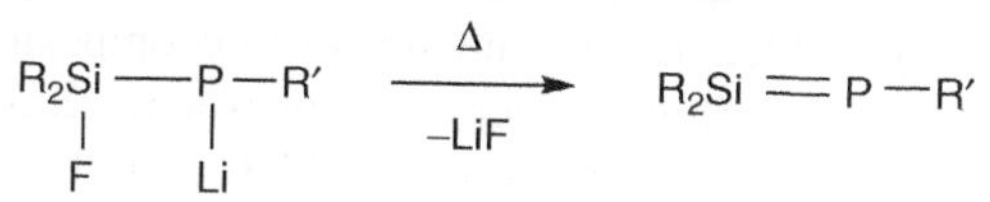

Scheme 5.81

In due course, a family of other phosphasilene derivatives RR′Si=P–R″ (R = *t*-Bu, Tip; R′ = Tip; R″ = $SiMe_3$, $SiMePh_2$, SiH*t*-Bu_2, Si*i*-Pr_3, $SiMe_2$*t*-Bu, $SiPh_3$, GeH*t*-Bu_2) **232** was prepared by the same synthetic procedure by Driess *et al.*[207b–e] Similarly to the previously reported **231**, the silicon parts of the Si=P bonds in phosphasilenes **232** were highly deshielded (doublets at 167.8–222.5 ppm with large silicon–phosphorus coupling constants of 149–161 Hz), whereas their phosphorus counterparts were remarkably shielded (from −31.8 to 28.1 ppm).[207c–e] The crystal structure of one of the above-discussed phosphasilenes **232**, namely Tip(*t*-Bu)Si=P–Si*i*-Pr_3, was also reported.[207c] In contrast to Niecke's phosphasilene **229**, the Si=P bond in Tip(*t*-Bu)Si=P–Si*i*-Pr_3 was not distorted, featuring a trigonal-planar coordinated doubly bonded silicon (sum of the bond angles = 359.9°) and a short Si=P bond of 2.062(1) Å, which was much shorter than the adjacent Si–P single bond [2.255(1) Å].

An interesting example of [1,3]-sigmatropic migration of a fluorine atom in 2-phospha-1,3-disilaallyl fluoride derivative **233** was also reported (Scheme 5.82).[208] Prepared by the traditional thermal *β*-elimination of LiF from the corresponding precursor **234**, phosphasilene **233** exhibited record shielding of its phosphorus center observed at −33.0 ppm. No crystal data for **233** were reported because of the positional disorder of both P and F atoms; however, the length of the Si=P bond in the unsubstituted model H_2Si=P–SiH_2F was calculated to be short (2.070 Å), presumably because of the lack of bulky substituents at both Si and P centers. The structure of **233** with localized Si=P and Si–F bonds was retained in solution up to 40 °C, above which temperature a suprafacial [1,3]-sigmatropic fluorine shift was observed.[208]

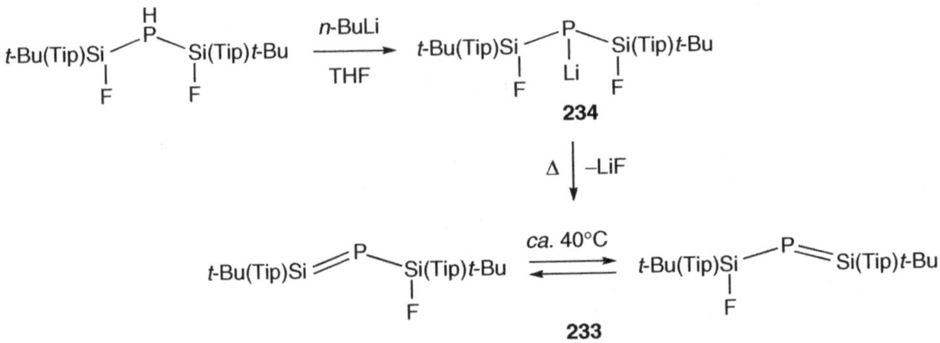

Scheme 5.82

Recently, a novel phosphasilene Tip(*t*-Bu_3Si)Si=PH **235** (as a mixture of *E*- and *Z*-isomers) was synthesized by the dehydrofluorination of (fluorosilyl)phosphine **236** with *n*-BuLi (Scheme 5.83).[209] Lacking bulky substituents at the phosphorus atom, **235** was surprisingly thermally stable, showing no signs of decomposition after refluxing in toluene solution for several days. Both isomeric phosphasilenes **235** manifested low-field resonances of the doubly bonded Si and P atoms and large Si=P coupling constants: *E*-isomer [249.8 ppm for ^{29}Si NMR, 123.1 ppm for ^{31}P NMR, $^{1}J(^{29}Si–^{31}P)$ = 157 Hz], *Z*-isomer [248.9 ppm for ^{29}Si NMR, 134.2 ppm for ^{31}P NMR, $^{1}J(^{29}Si–^{31}P)$ = 130 Hz].[209]

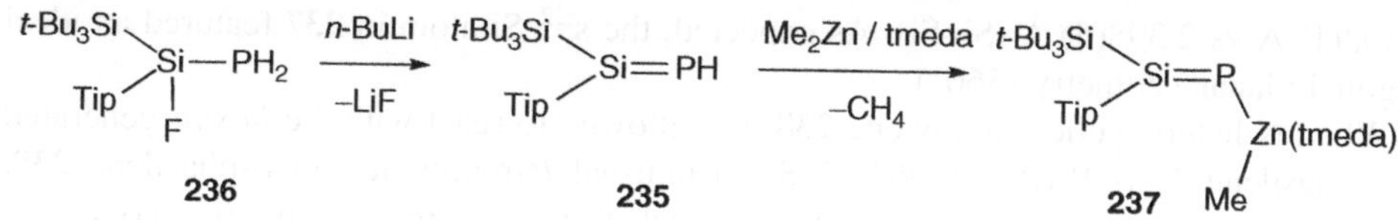

236 **235** **237**

Scheme 5.83

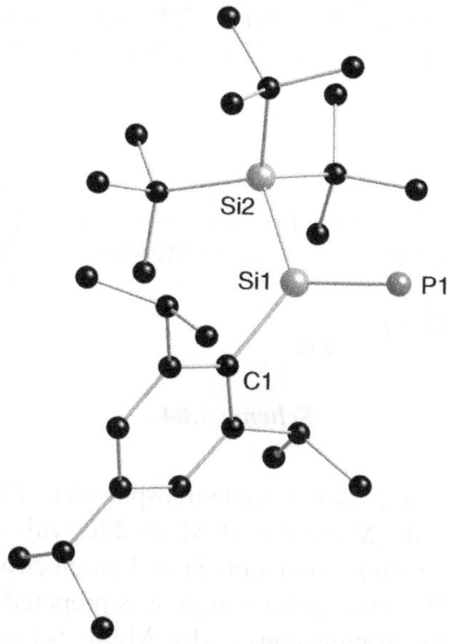

Figure 5.21 *Crystal structure of the P-unsubstituted phosphasilene **235** (hydrogen atoms are not shown)*

The tricoordinate silicon center in **235** adopts a trigonal-planar geometry, and the Si1=P1 bond of 2.094(5) Å is ca. 7% shorter than the typical Si–P single bond but is, however, slightly longer than those of the P-silyl-substituted phosphasilenes (Figure 5.21).

Phosphasilene **235** was further functionalized by metalation with Me_2Zn in the presence of TMEDA to furnish an unusual metal-substituted *P*-zinciophosphasilene **237** [only one example of ferriophosphasilene $Tip_2Si=P–Fe(CO)_2Cp$ was claimed to be generated, but was not isolated in pure form][210] (Scheme 5.83).[209] For obvious steric reasons, **237** was isolated solely as an *E*-isomer. The resonances of both doubly bonded atoms in **237** were observed at extremely low field: 203.7 ppm for ^{29}Si NMR [$^1J(^{29}Si–^{31}P) = 177$ Hz] and 227.8 ppm for ^{31}P NMR. The ^{29}Si resonance of 203.7 ppm in **237** was, however, remarkably shifted to higher fields compared with those of *E*/*Z*–**235** (249.8 and 248.9 ppm), a phenomenon that was attributed to the negative $n(P) \rightarrow \sigma^*$(Si–Si) hyperconjugation that is more efficient in **237** than in **235** because of the higher partial negative charge at the phosphorus atom in the former. The presence of such hyperconjugation was also supported by the shortening of the Si=P bond and slight stretching of the Si–Si bond in **237** compared with those of *E*/*Z*–**235**: 2.064(1) Å vs 2.094(5) Å (Si=P) and

2.402(1) Å vs 2.398(5) Å (Si–Si). As expected, the sp^2-Si atom in **237** featured an ideal trigonal-planar geometry (360°).

When N-heterocyclic plumbylene **238** was allowed to react with the *in situ* generated phosphasilene Tip(t-Bu_3Si)Si=PH_2 **235**, an unusual *P*-plumbyleniophosphasilene **239**, featuring a divalent group 14 element as a substituent on the phosphorus atom, was produced (Scheme 5.84).[211] Crystal structural analysis of **239** confirmed its composition as an *E*-isomer with a trigonal-planar geometry at the doubly bonded Si atom and a Si=P bond length of 2.0849(16) Å. Similarly to the *P*-zinciophosphasilene **237**, the resonances of both sp^2-Si and sp^2-P nuclei in **239** were extraordinarily shifted to low field: 227.0 ppm for ^{29}Si NMR [$^1J(^{29}Si-^{31}P) = 206$ Hz] and 292.5 ppm for ^{31}P NMR.

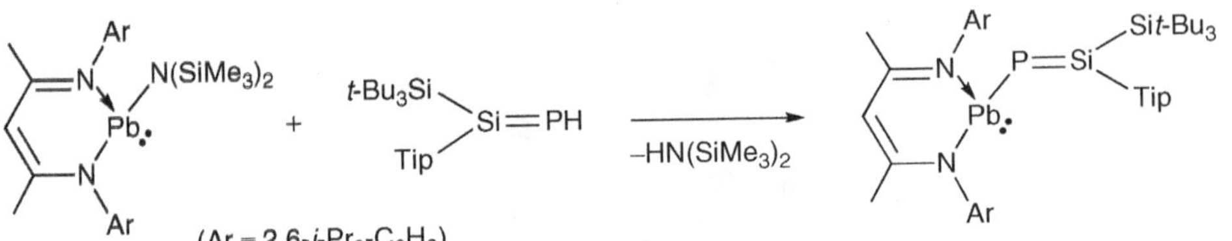

238 **235** **239**

Scheme 5.84

One of the latest representatives of the stable phosphasilene family was reported by Lee *et al.* This phosphasilene (t-$Bu_2MeSi)_2$Si=P–Mes* **240** with a novel 'push–pull' substitution pattern (electron-donating groups on Si and an electron-accepting group on P), relaxing the natural $Si^{\delta+}=P^{\delta-}$ bond polarization, was prepared by a convenient one-step coupling of readily available dilithiosilane (t-$Bu_2MeSi)_2SiLi_2$ (see Chapter 3, Section 3.5.1) and dichlorophosphine Mes*PCl_2 (Scheme 5.85).[212] As for other phosphasilenes, the resonance of the doubly bonded Si atom in **240** was observed in the low field region at 201.2 ppm as a doublet with a large $^1J(^{29}Si-^{31}P)$ coupling constant of 171.3 Hz. Accordingly, the phosphorus nucleus in **240** was extremely deshielded and found at 389.3 ppm, by far exceeding the range reported for other phosphasilenes (from −33.0 to 292.5 ppm). The huge deshielding of the doubly bonded atoms in **240** should definitely be ascribed to the σ-donating effect of the electropositive silyl substituents on the Si atom, which causes a remarkable decrease of the $n(P)_{HOMO}-\pi^*(Si{=}P)_{LUMO}$ energy gap. Such appreciable $n_P-\pi^*_{Si=P}$ orbital mixing in **240** results in a large paramagnetic contribution, which according to the Ramsey formula[9] is inversely proportional to the energy separation between the interacting orbitals, and is responsible for the overall low-field

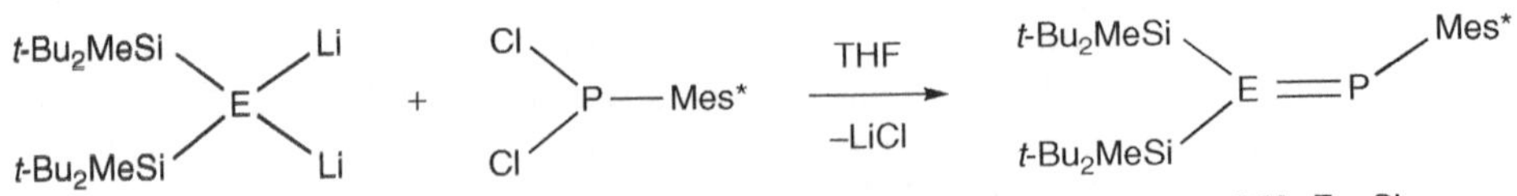

240: E = Si
243: E = Ge

Scheme 5.85

isotropic chemical shift. In full agreement with experimental observations, computations on a series of differently substituted phosphasilenes revealed a general tendency toward significant deshielding of the doubly bonded atoms taking place on decreasing the n–π^* energy gap upon the successive introduction of silyl substituents on the sp^2-Si center [$\Delta E(n-\pi^*)$ in eV, δ (^{29}Si) in ppm, δ (^{31}P) in ppm]: 4.33, 186.0, 85.4 [t-Bu_2Si=PMes]; 3.97, 205.2, 189.6 [(E)-t-Bu(Me_3Si)Si=PMes]; 3.73, 236.1, 322.6 [$(Me_3Si)_2Si$=PMes]. Accordingly, because of the decrease in the HOMO–LUMO energy gap, the longest UV absorption in **240** at 461 nm was substantially red shifted compared with those of phosphasilenes with aryl substituents on their Si atoms (331–343 nm).[212]

The length of the Si=P bond in **240** of 2.1114(7) Å, although the longest among other structurally characterized phosphasilenes because of its severe steric congestion, was still shorter than the sum of the Si and P covalent radii by ca. 7%. On the other hand, the Si=P bond exhibited no type of distortions, being neither pyramidalized at Si (360°) nor twisted [torsion angle C(Mes*)–P=Si–Si = 178.8°].

Cyclic voltammetry measurements of phosphasilene **240** (vs Ag/Ag^+, THF, 0.1 M [n-Bu_4N]ClO_4, r.t.) revealed a reversible one-electron reduction wave with the reduction potential $E_{1/2}$(red) = −1.78 V corresponding to the generation of an anion-radical of **240**, which was stable under electrochemical conditions.[212] This persistent anion-radical derivative of phosphasilene **240**, [(t-$Bu_2MeSi)_2Si$=P–Mes*]$^{-\bullet}$•K^+ **[240]**$^{-\bullet}$•K^+, was generated also by the chemical reduction of **240** with an equimolar amount of KC_8 in THF.[212] Anion-radical **[240]**$^{-\bullet}$•K^+ exhibited interesting solvent-dependent behavior, being more delocalized [Si=P]$^{-\bullet}$ in polar THF and more localized [Si$^{\bullet}$–P$^-$] in nonpolar benzene, which was manifested in its distinctly different EPR parameters measured in these two solvents [g-factor, a(^{31}P) in mT, a(^{29}Si) in mT]: 2.0083, 5.4, 5.0 (in THF); 2.0076, 1.5, 6.5 (in benzene).

A series of phosphasilenes with aryl substituents at the Si atoms, stabilized by bulky indacenyl groups, was synthesized by dehydrobromination of the corresponding phospha(bromosilanes) with 1,8-diazabicyclo[5.4.0]undec-7-ene (DBU).[213] Bis(phosphasilene), in which the two Si=P units are connected by a phenylene linker, was also available by this method. Spectral characteristics of these phosphasilenes were indicative of the presence of π-conjugation between the Si=P bonds and aromatic π-systems.

5.3.1.2.3.5 Phosphagermenes >Ge=P–[71a–c,191a]

The three stable phosphagermenes **241a-c** were reported by Escudié *et al.*, and were synthesized by the traditional procedure of reductive dehydrofluorination of (fluorogermyl)phosphine precursors **242** (Scheme 5.86).[214]

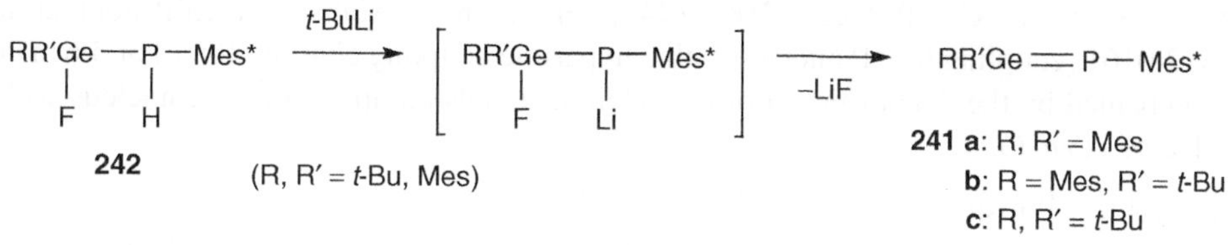

Scheme 5.86

The very first isolable phosphagermene, $Mes_2Ge{=}P{-}Mes^*$ **241a**, was prepared by this route in 1985[214a] and its crystal structure was determined three years later.[214b] Whereas the presence of the Ge=P bond of **241a** in solution was demonstrated by its low-field ^{31}P NMR resonance (175.4 ppm),[214a] the double bond character of the germanium–phosphorus covalent interaction in the solid state was manifested in its remarkably short bond of 2.138(3) Å, which was ca. 8.5% shorter than standard Ge–P single bonds.[214b] As was typically observed for phosphasilenes (see above), the Ge=P bond in **241a** was nearly planar and the geometry about the germanium center was also not distorted from planarity (360°).

Likewise, other stable phosphagermenes *t*-Bu(Mes)Ge=P–Mes* **241b** and *t*-$Bu_2Ge{=}P{-}Mes^*$ **241c** revealed spectral and structural characteristics similar to those of their predecessor **241a**, namely: low-field ^{31}P NMR resonances [157.4 and 169.2 ppm (for *E*- and *Z*-isomers of **241b**), 156.6 ppm for **241c**], and planar and short Ge=P bonds [2.144(3) Å for (*E*)–**241b**].[214c] Interestingly, during the synthesis of phosphagermene **241b**, only the (*Z*)-isomer was initially formed as a kinetic product, which quantitatively isomerized into the thermodynamically more stable (*E*)-isomer. Such isomerization followed first-order reaction kinetics with the isomerization barrier estimated to be 22.3 kcal/mol.

A novel phosphagermene $Mes_2Ge{=}P{-}Tip$ was also prepared (however, not isolated in pure form) by the one-step reaction of Mes_2GeF_2 and *in situ* generated $TipPLi_2$.[215] Its composition as a phosphagermene was demonstrated by the characteristic low-field resonance of its phosphorus nucleus (145.3 ppm) and cycloaddition reactions with some aldehydes, nitrones and α,β-unsaturated carbonyl compounds.

The most recent stable phosphagermene **243** was synthesized by Lee *et al.* similarly to the isostructural phosphasilene **240** as outlined in Scheme 5.85: by a one-step reaction of dilithiogermane $(t\text{-}Bu_2MeSi)_2GeLi_2$ (see Chapter 3, Section 3.5.1) and dichlorophosphine Mes^*PCl_2.[212] The phosphorus nucleus in phosphagermene **243** was even more deshielded than that in phosphasilene **240**, 416.3 vs 389.3 ppm, because of the higher electronegativity of germanium. The Ge=P bond in **243** of 2.1748(14) Å was longer than those reported before [2.138(3)–2.144(3) Å], however, it was shorter than the sum of the Ge and P covalent radii by ca. 7%. Again, similarly to phosphasilene **240**, phosphagermene **243** featured a planar geometry around the Ge=P bond: neither pyramidalization at Ge (360°) nor twisting [torsion angle C(Mes*)–P=Ge–Si = 178.9°] was observed. Electrochemical reduction of **243** (vs Ag/Ag^+, THF, 0.1 M $[n\text{-}Bu_4N]ClO_4$, r.t.) generated a stable phosphagermene anion-radical: reversible one-electron reduction wave with the reduction potential $E_{1/2}(\text{red}) = -1.75$ V.[212] Reduction of phosphagermene **243** with 1 equiv. of KC_8 in THF produced the same persistent anion-radical as its potassium salt $[(t\text{-}Bu_2MeSi)_2Ge{=}P{-}Mes^*]^{-\bullet}\bullet K^+$ $\mathbf{[243]^{-\bullet}\bullet K^+}$, which exhibited an EPR doublet at $g = 2.0161$ [coupling to ^{31}P nucleus with a hyperfine coupling constant (hfcc) of 5.6 mT] accompanied by the diagnostic set of ten satellite signals (coupling to ^{73}Ge nucleus with a hfcc of 2.5 mT).[212]

5.3.1.2.3.6 Phosphastannenes >Sn=P–[71b,191a]

Although structural information on the Sn=P double bond is still not available, a couple of phosphastannenes **244** were reported as isolable derivatives prepared by Escudié *et al.* by the reductive dehydrofluorination of (fluorostannyl)phosphine precursors **245** (Scheme 5.87).[215b,216]

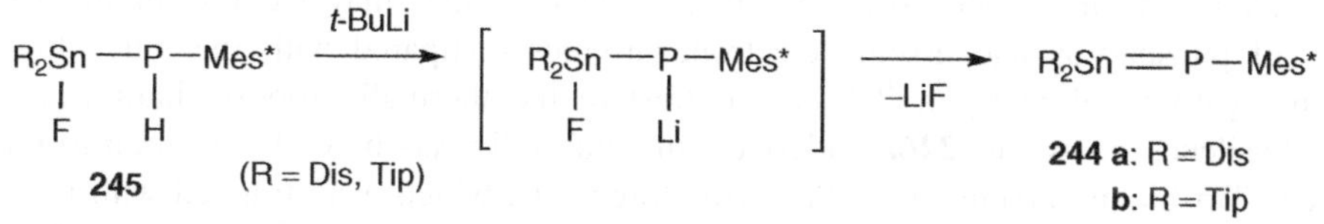

Scheme 5.87

The structure of the first phosphastannene $Dis_2Sn{=}P{-}Mes^*$ **244a** was proved by its NMR spectral data: low-field resonances for both tin and phosphorus nuclei (658.3 and 204.7 ppm) and large $^1J(^{117/119}Sn{-}^{31}P)$ coupling constants of 2191 and 2295 Hz were indicative of the Sn=P double bond.[216a] The longest wavelength absorptions of **244a** were observed at 350 and 460 nm, which were attributed to $\pi \rightarrow \pi^*$ and $n \rightarrow \pi^*$ electronic transitions, respectively.

Likewise, another phosphastannene $Tip_2Sn{=}P{-}Mes^*$ **244b** exhibited deshielded tin and phosphorus nuclei that resonated at 499.5 and 170.7 ppm, respectively, as well as large $^1J(^{117/119}Sn{-}^{31}P)$ coupling constants of 2110 and 2208 Hz.[215b,216b] The latter values were significantly greater than those for typical Sn–P single bonds (ca. 1000 Hz).

5.3.1.2.3.7 Arsasilenes >Si=As–[217]

The doubly bonded derivatives of arsenic and heavy group 14 elements are known only as >Si=As–combinations, the so-called arsasilenes. To date, only five stable arsasilenes **246** have been prepared, all of them by Driess *et al.* by the thermal β-elimination of LiF from the isolable lithium salts of (fluorosilyl)arsides **247** (Scheme 5.88).[207b–d,218]

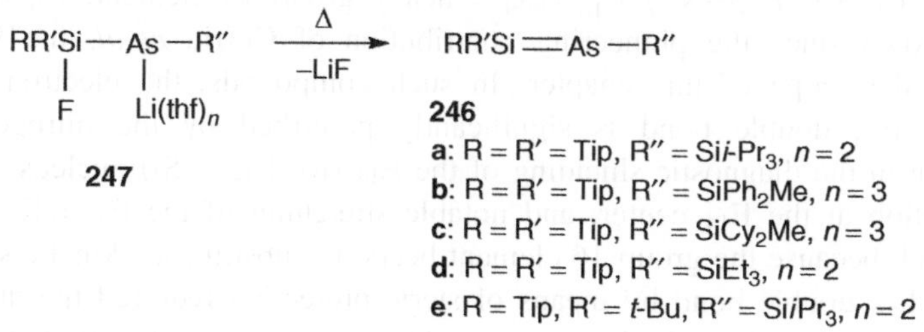

Scheme 5.88

The first room temperature stable arsasilene $Tip_2Si{=}As{-}Si\textit{i}\text{-}Pr_3$ **246a** was synthesized by this method in 1992.[218] The presence of the Si=As bond in **246a** was supported by observation of the low-field resonance in its ^{29}Si NMR spectrum at 179.1 ppm, diagnostic of an sp^2-Si center.[218a]

Since then, several other arsasilenes **246b–d** with Tip groups on Si and different silyl substituents on As have been prepared employing the same synthetic strategy.[207b,d]

In these arsasilenes, the doubly bonded Si nuclei resonated as expected in the low field region, being observed at 183.6 ppm (for **246b**) and 187.0 ppm (for **246c**).[207d] Similarly to phosphasilenes and in contrast to silaimines, arsasilenes do not tend to form adducts with *n*-donor solvents such as THF, which may stem from the similarity of phosphorus and arsenic electronegativities compared with the much higher electronegativity of nitrogen.[207b,d] In contrast to the thermally robust silaimines and phosphasilenes, arsasilene **246a** underwent thermal >Si=As– bond dissociation forming silylene Tip_2Si: and arsinidene $(i\text{-}Pr_3Si)As$: fragments, which were trapped with Et_3SiH to produce $Tip_2Si(SiEt_3)H$ and $(i\text{-}Pr_3Si)As(SiEt_3)H$, respectively.[207b] Thus, the Si=As bond is less stable than its Si=N and Si=P homologs. Further distinction between the Si=P and Si=As bonds was observed in their different electrochemical behavior: while phosphasilenes underwent a reversible one-electron reduction at −1.74 to −1.81 V, their arsenic counterparts (arsasilenes) exclusively exhibited irreversible reduction waves at −2.01 to −2.04 V.[207b]

The first (and still the sole) structurally characterized arsasilene $Tip(t\text{-}Bu)Si{=}As{-}Si\,i\text{-}Pr_3$ **246e** was reported in 1995.[207c] The Si=As bond distance of 2.164(1) Å was in good agreement with the theoretical prediction of 2.168 Å calculated for the parent arsasilene $H_2Si{=}AsH$.[207b] Accordingly, the doubly bonded Si atom in **246e** featured a trigonal-planar geometry (sum of the bond angles 359.9°), and was largely deshielded (228.8 ppm).[207c]

5.3.1.2.4 Group 14/Group 16 Combinations $>E_{14} = E_{16}$ **(E_{14} = Si, Ge, Sn; E_{16} = O, S, Se, Te)**

Because the chemistry of ketone analogs in which the carbonyl oxygen is replaced by heavy group 14 elements (so-called 'heavy ketones') has been repeatedly and comprehensively reviewed in recent years by Tokitoh *et al.*,[219] here we only briefly discuss the most important advances in the field, referring for details to the original reviews. Numerous examples of intramolecularly supported derivatives of the type $>E_{14}(\leftarrow:N){=}E_{16} \leftrightarrow >E_{14}(-N^+){-}E_{16}^-$ (E_{14} = heavy group 14 element; E_{16} = group 16 element), known since the pioneering contribution of Corriu *et al.* in 1989,[220] are also outside the scope of this chapter. In such compounds, the electronic situation of the $E_{14}{=}E_{16}$ double bond is significantly perturbed by the nitrogen electron donation seen in the diagnostic shielding of the E_{14} (for E_{14} = Si) nucleus, remarkable pyramidalization at the E_{14} center, and notable stretching of the $E_{14}{=}E_{16}$ bond. On the other hand, because the group 16 element bears no substituents, kinetic stabilization of the $E_{14}{=}E_{16}$ double bond by means of steric protection required the utilization of particularly bulky ligands at the heavy group 14 element. Examples of such kinetically stabilized heavy ketones are described below.

5.3.1.2.4.1 $>E_{14}{=}O$ Combinations

To date, no examples of stable derivatives of this type have been isolated. The problems in their synthesis are primarily associated with the great polarity of the $[>E_{14}^{\delta+} = O^{\delta-} \leftrightarrow >E_{14}^{+}{-}O^{-}]$ bond caused by the vastly different electronegativities of the heavy group 14 element and oxygen atoms. Such bond polarization results in the extraordinarily high tendency of such compounds to undergo oligomerization proceeding (in contrast to carbon analogs) without barriers.

Transient representatives, such as dimethylsilanone $Me_2Si{=}O$ and dimethylgermanone $Me_2Ge{=}O$, were generated in the gas phase by vacuum pyrolysis of the corresponding precursors, stabilized in low-temperature argon matrices and identified by IR spectroscopy supported by frequency calculations.[221]

To relieve the high $Si^{\delta+}{=}O^{\delta-}$ natural bond polarity, Driess *et al.* employed the 'push–pull' approach to the silanone problem, employing an n-donor for stabilization of Lewis acidic silicon and a π-acceptor for stabilization of Lewis basic oxygen centers. This methodology allowed isolation of the stable silaformamide–borane complex **248**, prepared by the addition of $H_2O{\bullet}B(C_6F_5)_3$ to the N-heterocyclic silylene **249** (Scheme 5.89).[222a]

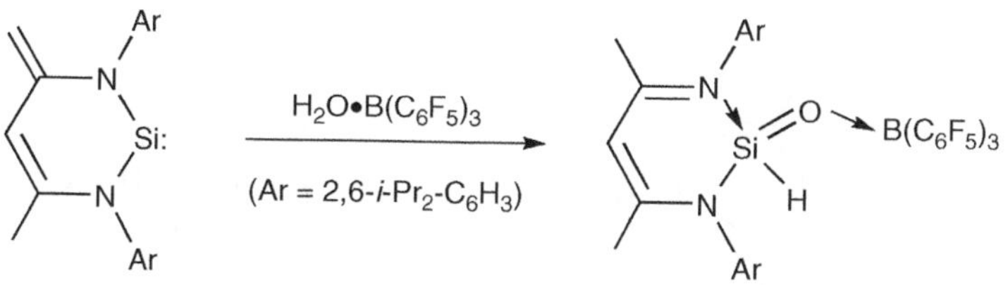

249 **248**

Scheme 5.89

The silicon–oxygen distance in **248** of 1.552(2) Å was remarkably shortened compared with the typical Si–O bonds in Si–O–C or Si–O–Si (ca. 7% shortening), and was comparable with the value of 1.537 Å calculated for the parent silaformamide $HSi(O)NH_2$. The high Si–O bond order was further supported by the observation of absorption at 1 112 cm^{-1} in the IR spectrum of the ^{18}O-labeled **248**, assigned to the $Si{=}^{18}O$ stretching vibration (extrapolation for the $Si{=}^{16}O$ stretching frequency: 1 165 cm^{-1}).[222a] From the large value of the corresponding force constants, the Si–O bond order in **248** was estimated as 1.83, which was corroborated by NBO computations showing a significant π-bonding interaction between the σ^*(Si–N)–n(O lone pair) orbitals.

As the other candidate for the elusive Si=O bond, an isolable silanoic silylester **250** was synthesized by the low-temperature oxidation of siloxysilylene **251** with N_2O (or CO_2) (Scheme 5.90).[222b] Silylester **251**, for which substantial Si=O double bond character was also established, featured a rather short silicon–oxygen bond distance of 1.579(3) Å.

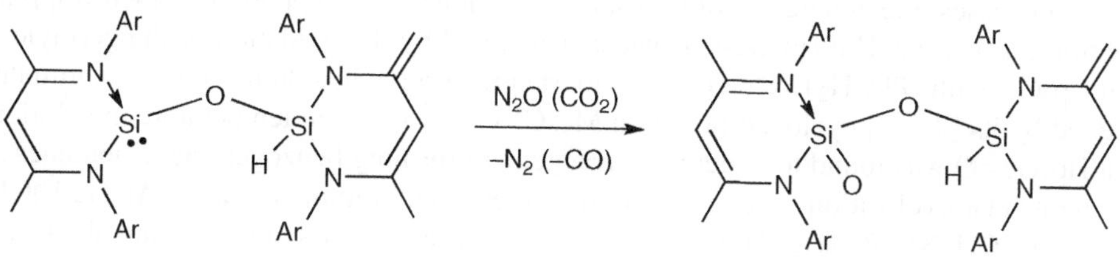

251 **250**

Scheme 5.90

The problem of thermodynamic stabilization of the Si=O double bond was even more closely approached by the synthesis of N-heterocyclic carbene(NHC)-supported silanone **252**, available by the low-temperature treatment of NHC–silylene complex **253** with N_2O (Scheme 5.91).[222c] The Si=O bond of 1.541(2) Å in **252** was even shorter than those in the above-mentioned silaformamide–borane complex **248** [1.552(2) Å] and silanoic silylester **250** [1.579(3) Å], indicative of a strong silicon–oxygen bonding interaction in **252**. However, the stabilization of the (NHC:→)Si=O bond in **252** by NHC lone-pair electron donation occurs at the expense of electronic perturbation of the double bond, manifested in the important contribution of the ylide-like $(NHC)^+$–Si–O^- resonance extreme seen in the shortening of the Si–N and Si–C bond distances and notably tetrahedral geometry at the Si center.

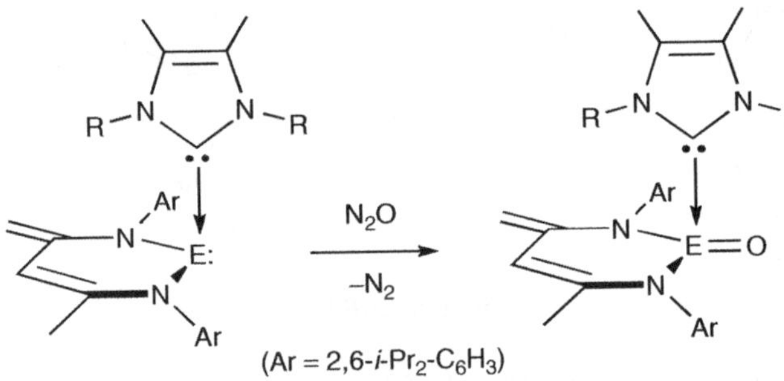

253: E = Si, R = Me
255: E = Ge, R = Me
257: E = Ge, R = *i*-Pr

252: E = Si, R = Me
254: E = Ge, R = Me
256: E = Ge, R = *i*-Pr

Scheme 5.91

The germanium versions of **252**, NHC-supported germanones **254** and **256**, were prepared by an analogous procedure: exposure of NHC–germylene complexes **255** and **257** to N_2O (Scheme 5.91).[222d] Similarly to their silicon analogs, germanones **254** and **256** revealed remarkably shortened Ge=O bonds of 1.672(3) and 1.664(2)/1.670(2) Å, respectively, that is, ca. 7% shorter than the typical Ge–O single bonds in germoxanes. Again, structural features of both **254** and **256** were indicative of the important ylide-like resonance contribution $(NHC)^+$–Ge–O^-.

In some cases, the generation of transient germanones was supported by their trapping reactions. Thus, the Tbt(Tip)Ge=O species generated by the oxidation of the germylene Tbt(Tip)Ge: with $(PhCH_2)_3N^+$–O^- was reported to survive in solution for a few minutes (proved by its [2 + 3] cycloaddition with MesCNO).[223] At room temperature in solution, Tbt(Tip)Ge=O was found to undergo rearrangement forming benzogermacyclobutenes as intramolecular cyclization products.[223] In the other case, lithium derivative $Ar_2Ge(F)OLi$ (Ar = 2,6-$[CH_2N(i\text{-}Pr)_2]_2C_6H_3$) was reported to behave as a synthetic equivalent of a germanone Ar_2Ge=O complexed with LiF.[224] The characteristic germanone reactivity of $Ar_2Ge(F)OLi$ was evidenced by its trapping reactions with water, methanol and *t*-BuLi/H_2O.

5.3.1.2.4.2 >E_{14}=S Combinations

Computationally, the Si=S bond was predicted to be more stable (and hence more accessible) than the Si=O bond, because of the remarkably smaller polarization of the former: $H_2Si^{+0.67}=S^{-0.42}$ vs $H_2Si^{+1.00}=O^{-0.68}$.[225] Experimentally, all isolable heavy ketones of the >E_{14}=S type, that is, silane-/germane-/stannane-/plumbanethiones, have been prepared by one of two general routes: desulfurization of the corresponding diaryltetrathiasilolanes/-germolanes/-stannolanes/-plumbolanes with Ph_3P (method **A**, E = Si–Pb) or by the direct reaction of stable silylenes/germylenes/stannylenes with styrene episulfide or Me_3P=S as the sulfur source (method **B**, E = Si–Sn) (Scheme 5.92).

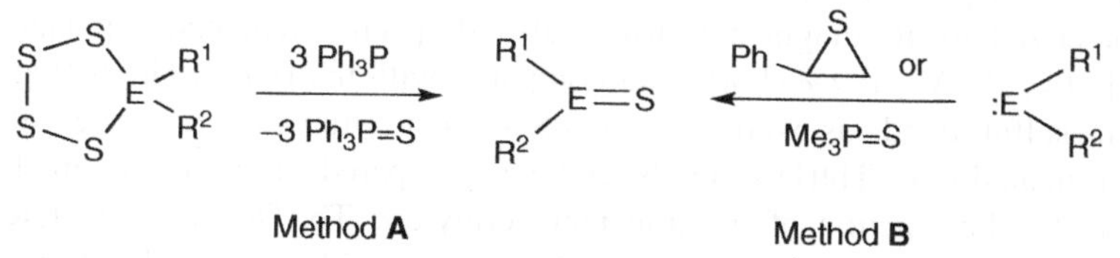

Scheme 5.92

Thus, the first isolable silanethione Tbt(Tip)Si=S **258** was prepared by Tokitoh *et al.* by method **A** (E = Si, R^1 = Tbt, R^2 = Tip).[226] The genuine nature of the electronically unsupported Si=S double bond in **258** was manifested in the characteristically strong deshielding of the Si center observed at 166.6 ppm, whose value well matched the calculated values of 158.0 ppm (for H_2Si=O) and 167.3 ppm (for Ph_2Si=S). The characteristic $n \rightarrow \pi^*$ electronic transition in **258** was observed at 396 nm. As was expected, the silanethione structural unit exhibited a trigonal-planar geometry at the silicon center (sum of the bond angles 359.7°/359.9° for the two independent molecules found in the unit cell), and remarkably short Si=S bonds of 1.948(4) Å/1.952(4) Å (ca. 9% shortening compared with standard Si–S single bonds).

Another stable silanethione **259** was available by the direct reaction of the stable silylene **260** with Me_3P=S according to method **B** (Scheme 5.93).[227]

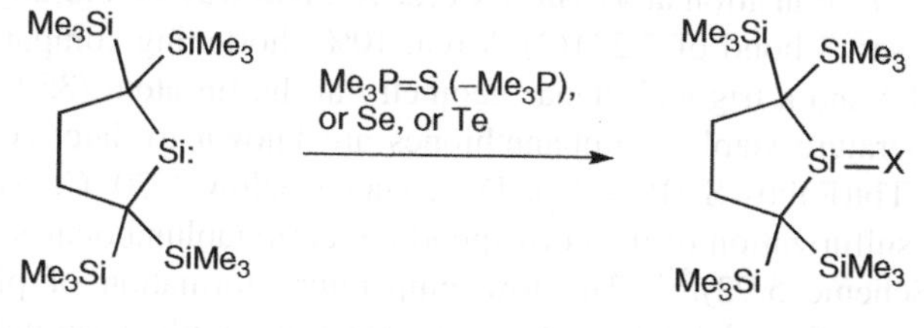

260

259 (X = S)
267 (X = Se)
272 (X = Te)

Scheme 5.93

Similarly to **258**, silaneketone **259** exhibited remarkable shortening [1.9575(7) Å] and planarity (360° at the doubly bonded silicon center) of the Si=S bond. However, compared with the diarylsilanethione **258**, dialkylsilanethione **259** revealed a blue shift of its UV band corresponding to the $n \rightarrow \pi^*$ electronic transition, 396 nm vs 336 nm, suggesting the presence of an interaction between the aryl substituents and the Si=S bond in **258**. Accordingly, the resonance of the sp^2-Si atom in **259** was observed at a notably lower field than that of **258**, 216.0 ppm vs 166.6 ppm, again indicating the less important electronic perturbation in the dialkylsilanethione **259**.

The first stable germanethione Tbt(Tip)Ge=S **261** was synthesized by Tokitoh *et al.* similarly to silanethione **258** according to method **A** (Scheme 5.92, E = Ge, R^1 = Tbt, R^2 = Tip).[228] As is typical for the heavy ketones, diarylgermanethione **261** was characterized by its diagnostic structural and spectral features: a planar (359.4°) and short [2.049(3) Å, ca. 9% shortening compared with the typical Ge–S single bonds] germanium–sulfur bond, as well as a Ge=S $n \rightarrow \pi^*$ transition observed at 450 nm. Another germanethione Tbt(Dis)Ge=S **262** was prepared similarly by method **A** or alternatively by the reaction of the transient germylene Tbt(Dis)Ge: (generated by the thermolysis of the corresponding germacyclopropene) with elemental sulfur according to method **B**.[228b]

As for the isolable stannanethiones, three compounds of such type are known. The first one, Tbt(Tip)Sn=S **263**, was prepared from the stable stannylene Tbt(Tip)Sn: by method **B** (Scheme 5.92) using styrene episulfide as the sulfur source.[229a,b] In the absence of X-ray structural and ^{119}Sn NMR spectral data, the identification of stannanethione **263** was based on its UV absorption at 473 nm corresponding to the Sn=S $n \rightarrow \pi^*$ electronic transition, as well as its characteristic reactivity toward carbon disulfide and phenyl isothiocyanate. Another diarylstannanethione Tbt(Ditp)Sn=S **264** [Ditp = 2,6-(2-i-Pr-C_6H_4)$_2$-C_6H_3] featuring the novel Ditp group (bulkier than Tip) was available by the dechalcogenation of the corresponding tetrathiastannolane according to method **A**.[229c] As was expected, stannanethione **264** exhibited a low-field ^{119}Sn NMR resonance at 531 ppm, as well as a 491 nm UV absorption attributable to the $n \rightarrow \pi^*$ (Sn=S) transition. Applying the same synthetic procedure and further increasing the steric bulk at the tin center, Tokitoh *et al.* were able to prepare diarylstannanethione Bbt(Titp)Sn=S **265** [Titp = 2,6-(2,4-i-Pr_2-C_6H_3)$_2$-C_6H_3].[229d] Novel stannanethione **265** manifested spectral features similar to those of **264**, namely a low-field resonance of the tin atom at 532 ppm and $n \rightarrow \pi^*$ transition at 475 nm. Moreover, **265** was structurally characterized, featuring a short Sn=S bond of 2.221(3) Å (ca. 10% shortening compared with typical Sn–S single bonds) and a trigonal-planar geometry at the tin atom (359.4°).[229d]

No room temperature stable plumbanethiones are known to date. A series of transient derivatives Tbt(R)Pb=S (R = Tip, Dis), stable below −20 °C, was generated at −78 °C by the desulfurization of the corresponding tetrathiaplumbolanes with $(Me_2N)_3P$ by method **A** (Scheme 5.92).[230] The low-temperature formation of plumbanethiones Tbt(R)Pb=S was confirmed by their trapping reactions with mesitonitrile oxide and phenyl isothiocyanate.

5.3.1.2.4.3 >E_{14}=Se Combinations

Similarly to the case of their sulfur analogs, all stable silane-, germane- and stannaneselones were prepared either by deselenation of the corresponding diaryltetraselenasilolanes/-germolanes/-stannolanes with Ph_3P or by the direct reaction of silylenes/germylenes/stannylenes with elemental selenium (see Scheme 5.92). Currently, two silaneselones >Si=Se, two germaneselones >Ge=Se, and one stannaneselone >Sn=Se have been reported as room temperature stable compounds. In line with the general trends for the heavy ketones, all compounds of this type featured diagnostically red-shifted $n \rightarrow \pi^*$ UV absorptions, low-field NMR resonances of the doubly bonded atoms, remarkably short E_{14}=Se double bonds and nearly planar geometry at their E_{14} centers [$n \rightarrow \pi^*$ absorption maximum, δ(NMR), E_{14}–Se bond distance, sum of the bond angles at the E_{14} center]: Tbt(Dip)Si=Se **266** [456 nm, $\delta(^{29}Si) = 174.0$ ppm, $\delta(^{77}Se) = 635.0$ ppm];[231] cyclic silaneselone **267** (Scheme 5.93) [383 nm, $\delta(^{29}Si) = 227.7$ ppm, 2.0963(5) Å, 360.0°];[227] Tbt(Tip)Ge=Se **268** [519 nm, $\delta(^{77}Se) = 940.6$ ppm, 2.180(2) Å, 359.3°];[228b,232] Tbt(Dis)Ge=Se **269** [492 nm, $\delta(^{77}Se) = 872.7$ ppm, 2.173(3) Å, 360.0°];[228b] Tbt(Ditp)Sn=Se (Ditp = 2,6-(2-*i*-Pr-C_6H_4)-C_6H_3) **270** [531 nm, $\delta(^{119}Sn) = 440.0$ ppm, 2.373(3) Å, 359.9°].[229c,233]

5.3.1.2.4.4 >E_{14}=Te Combinations

Stable silane-, germane- and stannanetellones were prepared by the reaction of silylenes/germylenes/stannylenes with either elemental tellurium or *n*-Bu_3P=Te. Two silanetellones >Si=Te, three germanetellones >Ge=Te and one stannanetellone >Sn=Te have been synthesized to date, characterized by their red-shifted $n \rightarrow \pi^*$ UV absorptions, low-field NMR resonances of their E_{14}=Te fragments, notably short E_{14}=Te double bonds and essentially planar geometry at the E_{14} center [$n \rightarrow \pi^*$ absorption maximum, δ(NMR), E_{14}–Te bond length, sum of the bond angles at the E_{14} center]: Tbt(Dip)Si=Te **271** [593 nm, $\delta(^{29}Si) = 171.0$ ppm, $\delta(^{125}Te) = 731.0$ ppm];[231] cyclic silanetellone **272** (Scheme 5.93) [476 nm, $\delta(^{29}Si) = 229.5$ ppm, 2.3210(6) Å, 360.0°];[227] Tbt(Tip)Ge=Te **273** [636 nm, 2.398(1) Å, 359.5°];[234] Tbt(Dis)Ge=Te **274** [599 nm, 2.384(2) Å, 360.0°];[234] Tbt(Dip)Ge=Te **275** [$\delta(^{125}Te) = 1\,194.0$ ppm];[231] Bbt(Titp)Sn=Te [Titp = 2,6-(2,4-*i*-Pr_2-C_6H_3)-C_6H_3] **276** [646 nm, $\delta(^{119}Sn) = 282.0$ ppm, $\delta(^{125}Te) = 1\,007.0$ ppm, 2.5705(6) Å, 359.7°].[235]

A series of isolable silanoic chalcogenoesters, featuring intramolecularly supported (N:→) Si=X (X = S, Se, Te) double bonds and isostructural with the above-described silanoic silylester **250** (see Scheme 5.90), was also recently reported by Driess *et al.*[236]

5.3.2 Heavy Analogs of 1,3-Dienes and Allenes

5.3.2.1 Heavy Analogs of 1,3-Dienes[237]

A remarkable hexakis(2,4,6-triisopropylphenyl)tetrasilabuta-1,3-diene Tip_2Si=Si(Tip)–Si(Tip)=SiTip_2 **277** was synthesized by Weidenbruch *et al.* in 1997 as the very first stable 1,3-diene of the heavy group 14 elements.[238] **277** was accessible by the reduction of tetraaryldisilene Tip_2Si=SiTip_2[24] with lithium followed by the addition of MesBr. Reaction was believed to proceed through the initial formation of disilenyllithium derivative Tip_2Si=Si(Tip)Li, followed by its subsequent reaction with MesBr to give Tip_2Si=Si(Tip)Br, which then underwent coupling with the remaining Tip_2Si=Si(Tip)Li forming the final tetrasilabuta-1,3-diene **277** (Scheme 5.94).

The terminal sp^2-Si atoms of the Si=Si–Si=Si 1,3-diene system in **277** resonated as expected at a higher field than the internal silicons: 52.3 ppm vs 89.5 ppm. Particularly

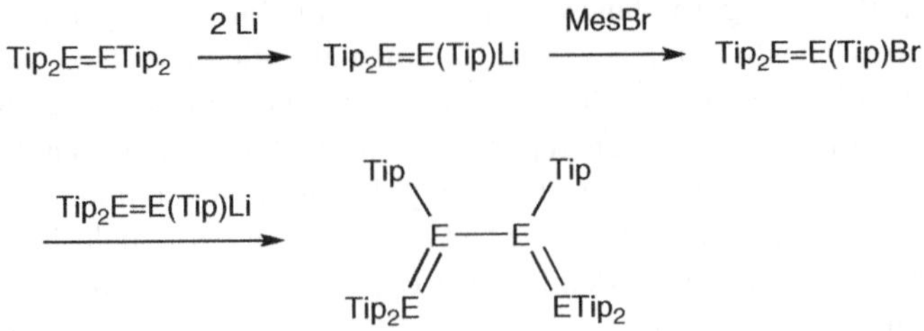

277: E = Si
278: E = Ge

Scheme 5.94

important was the diagnostic longest wavelength absorption in the electronic spectrum of **277** observed at 518 nm, whose value was nearly 100 nm bathochromically shifted in comparison with those of known tetraaryldisilenes, even exceeding the corresponding values for the isolable persilyldisilenes. This remarkable red shift in the UV spectrum of **277** was certainly indicative of the appreciable π-conjugation between the two Si=Si double bonds in solution. Crystallographic analysis of **277** corroborated the above conclusion: as expected for the conjugated system, the Si=Si bonds in **277** were slightly stretched to 2.175(2) Å, ca. 0.03 Å longer than those of tetraaryldisilenes, whereas the=Si–Si= single bond was compressed to 2.321(2) Å, shorter than standard Si–Si bonds of 2.34 Å (Figure 5.22). One should, however, take into account that the overall shortening of the central =Si–Si= single bond can also be interpreted in hybridization terms, namely, by the increase in s-character of the sp^2-hybrids used to form the Si–Si bond. Importantly, in the solid state **277** specifically featured an *s-cis* conformation with a dihedral angle between the two Si=Si units of 51.0°.

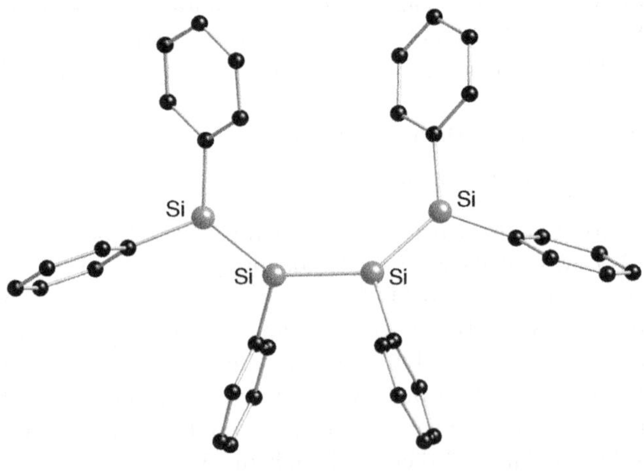

Figure 5.22 *Crystal structure of the tetrasilabuta-1,3-diene* ***277*** *(i-Pr groups of the Tip substituents and hydrogen atoms are not shown)*

Utilizing an analogous synthetic procedure, the same authors in due course prepared tetragermabuta-1,3-diene derivative **278** as the germanium analog of the above-described tetrasilabuta-1,3-diene **277** (Scheme 5.94).[98,239] Similarly to its silicon congener, **278** adopted an *s-cis* conformation in the crystalline form; however, the twisting between the two Ge=Ge fragments in **278** was notably less pronounced than that in **277**: 22.5° vs 51.0°. Both Ge=Ge bonds in **278** manifested distorted geometry around the sp^2-Ge centers: *trans*-bending and torsion (twist angles 21.3° and 22.4°). While the Ge=Ge double bonds in **278** of 2.3568(6) and 2.3439(5) Å were elongated compared with those of tetraaryldigermenes, the central Ge–Ge bond in **278** of 2.4581(5) Å was quite normal for a single bond. Although a possible conjugation between the Ge=Ge double bonds in **278** could not be unambiguously deduced based on its crystal structural data, this was evident from the remarkable longest wavelength absorption of **278** observed at 560 nm, which is ca. 140 nm red shifted compared with those of tetraaryldigermenes and nearly 100 nm red shifted compared with those of tetrasilyldigermenes.

Another representative of stable tetrasilabuta-1,3-dienes, *s-cis*-(t-$Bu_2MeSi)_2Si$=Si(Mes)–Si(Mes)=Si(SiMet-$Bu_2)_2$ **279**, was synthesized by a different approach: the reaction of 1,1-dilithiosilane (t-$Bu_2MeSi)_2SiLi_2$ and 1,1,2,2-tetrachlorodisilane $MesCl_2Si$–$SiCl_2Mes$ in a 2:1 ratio.[240] The doubly bonded silicon atoms in **279** are more deshielded than those of hexaaryltetrasilabuta-1,3-diene **277**, presumably because of the strong paramagnetic contribution caused by silyl substitution: 71.5 and 150.8 ppm (in **279**) vs 52.3 and 89.5 ppm (in **277**). The two Si=Si bonds in **279** are greatly twisted relative to each other with a Si=Si–Si=Si torsion angle of 72°, although planar geometry was maintained around all sp^2-Si atoms. The central Si–Si bond in **279** was quite normal [2.3376(11) Å], whereas the Si=Si bonds were stretched [2.1983(12) and 2.2003(12) Å] and longer than those in **277** [2.175(2) Å].[240] The twisting about the central Si–Si bond and elongation of the Si=Si bonds in **279** were attributed to the steric repulsion of the bulky silyl substituents. The longest UV absorption of **279** observed at 531 nm was red shifted compared with that of the related disilene (t-$Bu_2MeSi)_2Si$=$SiMes_2$ (437 nm) by ca. 100 nm. This bathochromic shift might be indicative of partial π-conjugation between the two Si=Si bonds, despite the great twisting of the buta-1,3-diene fragment in **279**.

The tetrasilabuta-1,3-diene **57**, in which the two terminal Si atoms of the 1,3-diene unit are incorporated into five-membered rings, was synthesized by the reduction of tribromodisilane **280** with sodium (Scheme 5.95).[54]

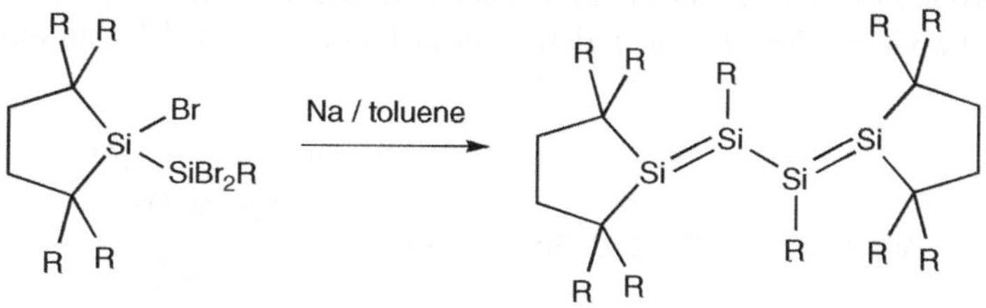

280 (R = $SiMe_3$) **57**

Scheme 5.95

In contrast to previously reported **277** and **279**, tetrasilabuta-1,3-diene **57** featured an *s-trans*-configuration of the two Si=Si bonds with a Si=Si–Si=Si torsion angle of 122.6°. The bond lengths of the 1,3-diene fragment in **57** were quite similar to those in **279**: 2.1980(16) and 2.2168(16) Å vs 2.1983(12) and 2.2003(12) Å (Si=Si); 2.3400(15) Å vs 2.3367(11) Å (Si–Si).[54] Because of the different substitution pattern, the resonances of the central and terminal sp^2-silicons in **57** were observed in largely distinctive regions: 9.3 and 210.2 ppm, respectively. The notable bathochromic shift in the UV-Vis spectrum of **57** (longest wavelength absorption at 510 nm) suggests the significant extent of the π–π conjugation between the two Si=Si bonds. As was mentioned in Section 5.3.1.1.1, both thermolysis and photolysis of the tetrasilabuta-1,3-diene **57** resulted in the formation of an unusual spirocyclic cyclotrisilene **56** (see Scheme 5.9).[54]

The above-described five-membered ring silagermene **191** (see Section 5.3.1.2.1.5, Scheme 5.60) revealed an *s-cis*-conformation of the Si=Ge–C=C diene unit caused by its cyclic composition.[165] Structural trends expected for the conjugated system, namely, stretching of the double bonds and shortening of the single bond, were not observed in **191** (despite the planarity of the five-membered ring): the lengths of the Si=Ge bond [2.250(1) Å], C=C bond [1.343(5) Å] and Ge–C bond [1.972(3) Å] were all within the normal ranges for these bonds. Accordingly, no notable bathochromic shift was observed in the UV-Vis spectrum of **191** compared with that of 1*H*-disilagermirene **186** featuring an isolated Si=Ge bond: 472 vs 467 nm. Overall, 1,2-disila-3-germacyclopenta-2,4-diene **191** represents a rather unusual cyclopentadiene system with two formally conjugated but actually isolated double bonds, of which one (Si=Ge) is much more reactive than the other (C=C). This may be rationalized by taking into account the remarkable mismatching in the size and energy levels of the Si=Ge and C=C π-orbitals, preventing their effective conjugative interaction. In contrast, for symmetrical 1,3-diene systems with two identical double bonds, such as 2,3-digermabuta-1,3-diene H_2C=GeH–GeH=CH_2, theory predicted about half the degree of conjugation compared with that of the parent buta-1,3-diene H_2C=CH–CH=CH_2.[241] Indeed, as shown above, all heavy group 14 element containing tetrasilabuta-1,3-dienes **277**,[238] **279**,[240] **57**[54] and tetragermabuta-1,3-diene **278**[98,239] revealed notable conjugation between their double bonds.

In some cases, the double bonds of the diene unit are incorporated into cyclic systems. The sole example of a stable cyclobutadiene containing heavy group 14 elements, 1,2-digermacyclobutadiene derivative **281**, was recently reported by Power *et al.*[242] **281** was prepared by the formal [2 + 2] cycloaddition of diphenylacetylene and the germanium analog of an alkyne ArGeGeAr [Ar = 2,6-(2,6-*i*-Pr_2-C_6H_3)$_2$-C_6H_3] (Scheme 5.96). The 1,3-diene carbons of **281** resonated as expected in the low field region, observed at 159.2 ppm. The Ge_2C_2 four-membered ring in **281** is represented by an almost planar

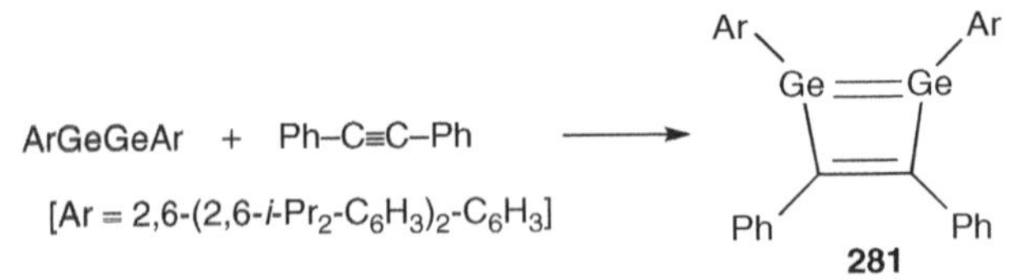

Scheme 5.96

trapezoid with remarkable pyramidalization at the germanium centers (sum of the bond angles of 317.3° and 318.0°). The cyclic bond lengths in **281** are consistent with the cyclobutadiene formulation: a C=C double bond [1.365(7) Å], two Ge–C single bonds [2.022(5) and 2.027(5) Å], and a 'weak' Ge=Ge double bond [2.4708(9) Å].[242] The weakness of the endocyclic Ge=Ge bond in **281** is indicated by its notable stretching (it is slightly longer than typical Ge–Ge single bonds of 2.44 Å) and significantly pyramidal Ge environments.

A remarkable spiropentasiladiene **282** was isolated as a side-product formed during the preparation of the above-described cyclotrisilene **49** (Section 5.3.1.1.1): reduction of $(t\text{-}Bu_2MeSi)_3Si–SiBr_2Cl$ with KC_8 (Scheme 5.97).[243] Isolation of **282** was facilitated by the effective steric protection by its very bulky R_3Si-groups, as well as by the important relaxation of spiropentasiladiene ring strain compared with that of the parent organic spiropentadiene: 61.1 vs 114.2 kcal/mol.

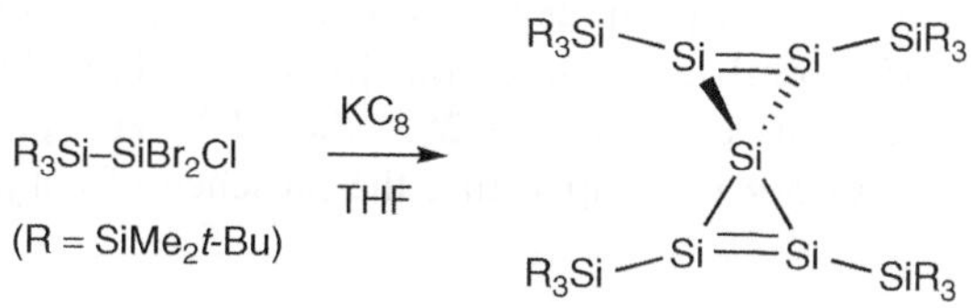

282

Scheme 5.97

The two spiro-connected cyclotrisilene units in **282** are twisted relative to each other by 78.3°, the endocyclic Si=Si bonds are also twisted by 30.0°, and the doubly bonded Si atoms featured insignificant pyramidalization with the sum of the bond angles around them amounting to 358.0° and 358.9°. The Si=Si bonds in **282** of 2.186(3) Å are longer than the typical double bonds in cyclotrisilenes, which was attributed to the effective π(Si=Si)–σ^*(Si–Si) orbital interaction (so-called σ^*-aromaticity: for discussion on this topic, see Section 5.3.1.1.1). A notable through-space interaction between π-orbitals (as well as between π^*-orbitals) in **282** was manifested in its longest wavelength absorption observed at 560 nm, which was remarkably red shifted compared with those of isolable cyclotrisilenes. Another indication of this pronounced through-space orbital interaction in **282** was the considerable deshielding of its sp^2-silicons, whose resonances were found at 154.0 ppm, which is in a remarkably lower-field region compared with those of other cyclotrisilenes.

An interesting tetrasiladiene **283** featuring two Si=Si bonds connected by a phenylene bridge was readily available by the reaction of disilenyllithium derivative $Tip_2Si=Si(Tip)Li$ (see Chapter 3, Section 3.5.2) with p-diiodobenzene (Scheme 5.98).[244]

The resonances of the doubly bonded Si atoms in **283** were found in the expected region of 70.7 and 56.8 ppm, whereas the UV-Vis spectrum of **283** revealed a significant bathochromic shift of the longest wavelength absorption observed at 508 nm (TD–DFT calculated value 564 nm).[244] This value was close to that of the conjugated tetrasilabuta-1,3-diene **277** (518 nm),[238] thus supporting the existence of π-conjugation in phenylene-bridged tetrasiladiene **283**, visualized in its most favorable $\pi_{C=C}$–$\pi^*_{Si=Si}$

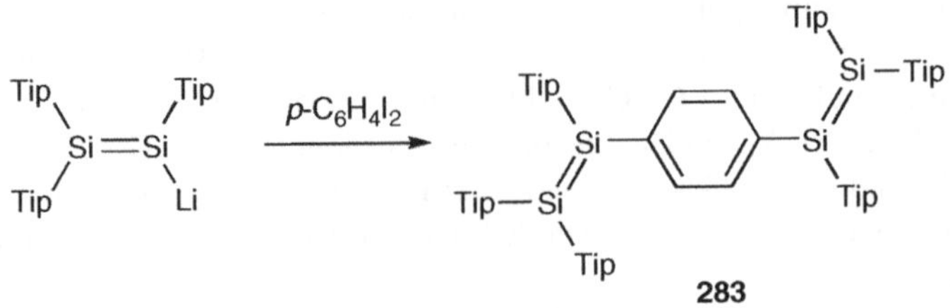

283

Scheme 5.98

orbital interaction. The Si=Si bonds of 2.1674(8) Å in **283** were usual for disilenes, and the Si=Si units were rotated out of the plane of the phenylene bridge to avoid steric repulsion.

By the reaction of disilenyllithium $Tip_2Si{=}Si(Tip)Li$ with iodobenzene, phenyl-substituted disilene $Tip_2Si{=}Si(Tip)Ph$ **284** was synthesized.[244] Although the spectral [$\delta(^{29}Si) = 71.8$ and 55.2 ppm] and structural [Si=Si: 2.1754(11) Å] features of **284** were similar to those of tetrasiladiene **283**, their UV-Vis spectral features were remarkably distinct, 439 vs 508 nm, supporting the presence of conjugation in **283** and its absence in **284**.

5.3.2.2 *Heavy Analogs of Allenes*[245]

The very first stable 1-silaallene **285** was synthesized by West *et al.* by the reduction of precursor **286** with 2 equiv. of *t*-BuLi (Scheme 5.99).[246] The resonances of the Si and central allenic C atoms were found at 48.4 and 225.7 ppm, respectively, values that were very close to those of Brook's silene **151** of 41.8 and 214.2 ppm (see Section 5.3.1.2.1.1). This phenomenon was ascribed to the 'reversed polarity' of the $Si^{\delta-}=C^{\delta+}$ bond in both cases. However, in contrast to Brook's silene **151**, which had a long Si=C bond of 1.764(3) Å, the Si=C bond in **285** of 1.704(4) Å was considerably shorter, and very close to that in Wiberg's silene **152** of 1.702(5) Å (see Section 5.3.1.2.1.1) and manifesting natural $Si^{\delta+}=C^{\delta-}$ bond polarization. The allenic moiety in **285** was slightly bent with a Si=C=C bond angle of 173.5°.

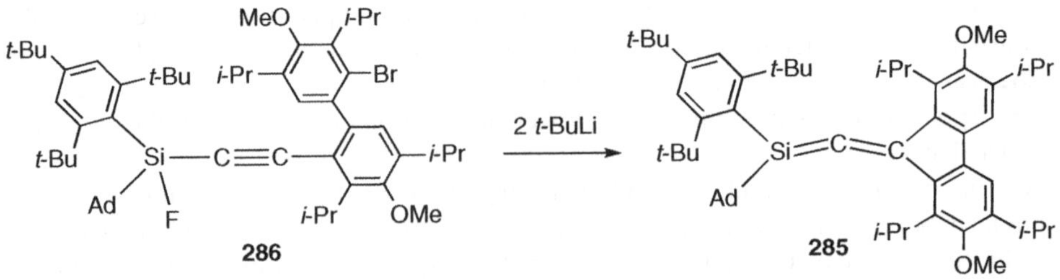

286

285

Scheme 5.99

In due course, West *et al.* prepared a series of other stable 1-silaallenes **287a-c** by a sequence of addition–elimination reactions starting from the corresponding (halosilyl)acetylenes (Scheme 5.100).[247]

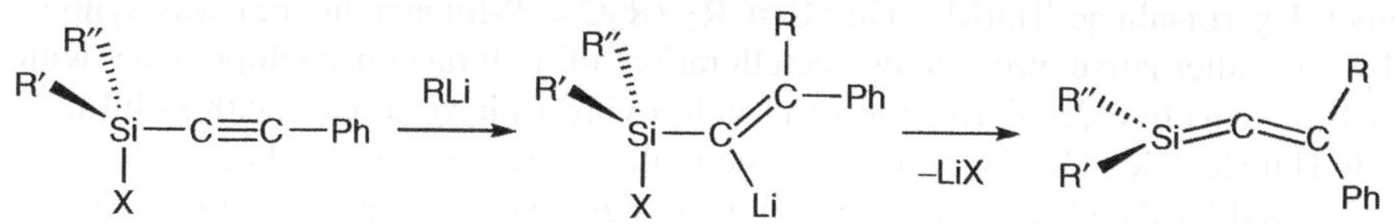

287a-c
a: R = *t*-Bu, R′ = R″ = Tip
b: R = R′ = *t*-Bu, R″ = Mes*
c: R = Ph, R′ = *t*-Bu, R″ = Mes*

Scheme 5.100

All 1-silaallenes **287** revealed characteristic signals for their central sp^2-C and sp^2-Si atoms, comparable with those in **285**: 223.6 and 13.1 ppm (for **287a**), 216.3 and 55.1 ppm (for **287b**), and 227.9 and 58.7 ppm (for **287c**). The geometry of the allenic Si=C=C skeleton in **287a** was also similar to that of **285**: the Si=C bond length of 1.693(3) Å, the C=C bond length of 1.325(4) Å, and the Si=C=C bent angle of 172.0°. The terminal carbon of the allene unit Si=C=C is essentially planar (359.9°), whereas silicon is insignificantly pyramidalized (357.2°).

Among other examples of 1-silaallenes, one can mention $(R_3Si)_2Si{=}C{=}C(SiMe_3)_2$ (R_3Si = Si*i*-Pr_3, SiMe*t*-Bu_2) photochemically generated upon the photolysis of the corresponding silacyclopropenes.[248] Thermally stable but photolytically labile 1-silaallenes $(R_3Si)_2Si{=}C{=}C(SiMe_3)_2$ were identified spectroscopically: −64.5 and −55.0 ppm (sp^2-Si), 268.3 and 267.8 ppm (central sp-C), 129.6 and 129.9 ppm (terminal sp^2-C). The appearance of resonances of the doubly bonded Si atoms at unusually high fields was explained in terms of the 'reversed polarity' of the $Si^{\delta-}{=}C^{\delta+}$ bond caused by substituent effects, which was supported by computational and reactivity studies.

Applying the procedure developed for 1-silaallenes, West *et al.* prepared the first stable 1-germaallene (Scheme 5.101).[249]

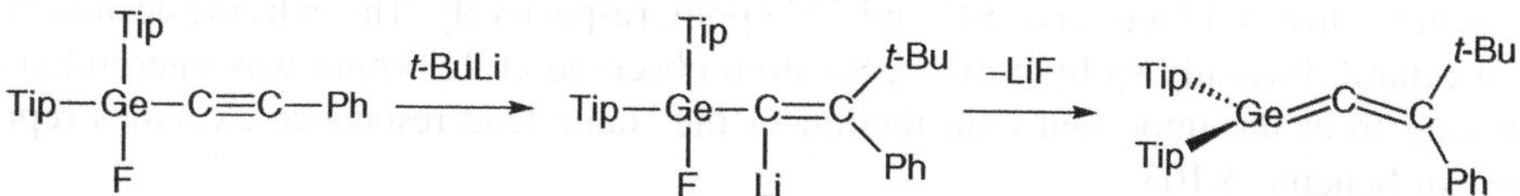

288

Scheme 5.101

The central allenic carbon in **288** resonated as expected in a very low field region at 235.1 ppm, a value that was comparable with that of the analogous 1-silaallene **287a** (223.6 ppm).[249a] The allenic fragment of **288** was remarkably bent with a Ge=C=C angle of 159.2° (cf. 172.0° in **287a**), and the Ge atom in **288** was more strongly pyramidalized than the Si atom in **287a**: 348.4 vs 357.2°. The Ge=C bond of 1.783(2) Å in 1-germaallene **288** was notably shorter than those of the structurally characterized germenes (see Section 5.3.1.2.1.2, Table 5.8).[249a] The C=C bond of the allenic unit in **288** was also slightly shortened to 1.314(2) Å.

Another 1-germaallene Tbt(Mes)Ge=C=CR_2 (R_2C = 9-fluorenylidene) was synthesized by two alternative procedures: detelluration of telluragermacyclopropane with $(Me_2N)_3P$ or reductive dechlorination of (1-chlorovinyl)chlorogermane with t-BuLi.[250] Although Tbt(Mes)Ge=C=CR_2 was not structurally characterized, its allenic composition was proved by the observation of a low-field signal at 243.6 ppm diagnostic for the central carbon of the propadiene system. Reactivity studies further corroborated the proposed allenic structure Tbt(Mes)Ge=C=CR_2, which, in the absence of trapping reagents, undergoes intramolecular cyclization forming the corresponding benzogermacyclobutene derivative.

No stable 1-stannaallenes >Sn=C=C<, 1-plumbaallenes >Pb=C=C<, or heavy allenes of the types >C=E=C<, >E=C=E′ < or >E=E′=C<(E, E′ = heavy group 14 elements) have been isolated to date.

The work in the field of all heavy group 14 element consisting allenes was pioneered by Wiberg *et al.*, who prepared the remarkable tristannaallene $(t\text{-}Bu_3Si)_2$Sn=Sn=Sn(Sit-$Bu_3)_2$ **289** as the first compound of such type in 1999.[108] Dark blue **289** was readily available by the simple reaction of Lappert's bis(amino)stannylene $[(Me_3Si)_2N]_2$Sn: with t-Bu_3SiNa (Scheme 5.102). Although the Sn=Sn=Sn fragment was sterically protected by four bulky t-Bu_3Si groups, **289** gradually isomerized in solution at room temperature into the above-described cyclotristannene **130** (see Section 5.3.1.1.3) with a half-life of 9.8 h at 25 °C (first-order reaction kinetics).

Scheme 5.102

As was expected, the terminal and central tins of the Sn=Sn=Sn fragment in **289** resonated in the low field region at 503 and 2233 ppm, respectively. The extreme deshielding of the central, formally sp-hybridized, Sn atom observed at 2233 ppm was interpreted as stemming from the important contribution of the stannylene resonance extremes represented in Scheme 5.103.

289 (R = Sit-Bu_3)

Scheme 5.103

The Sn=Sn bonds in tristannaallene **289** of 2.684(1) and 2.675(1) Å were considerably shorter than those of the majority of structurally characterized distannenes >Sn=Sn< (see Section 5.3.1.1.3, Table 5.5). As was found in other above-described heavy allenes and in contrast to undistorted organic allenes, the tristannaallene unit in **289** was markedly bent with a Sn=Sn=Sn bond angle of 156.0°, which was again realized by taking into account the stannylene resonance contributions shown in Scheme 5.103. The geometry around the terminal Sn atoms was pyramidal (again, in contrast to organic allenes) with the sum of the bond angles of 344.3° and 346.9°.

A family of isolable heavy analogs of allenes, representing all-silicon (trisilaallene), all-germanium (trigermaallene) and hybrid silicon/germanium (2-germadisilaallene and 1,3-digermasilaallene) versions, was recently reported by Kira *et al.*[251] The first trisilaallene **290** was prepared by the reductive dechlorination of the cyclic chloro(trichlorosilyl)silane **291** with potassium graphite (Scheme 5.104).[251a]

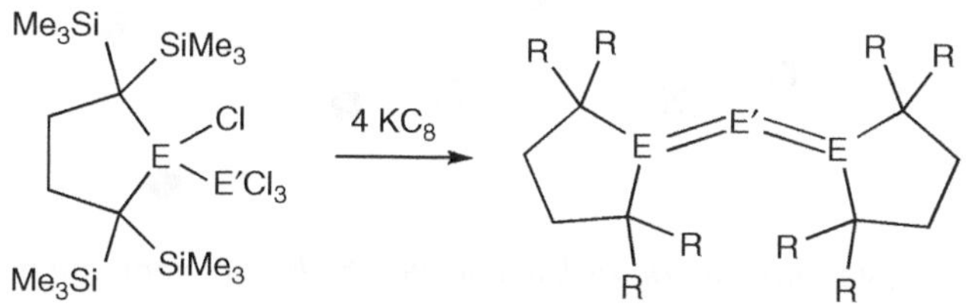

291 (E = E′ = Si)
295 (E = E′ = Ge)
297 (E = Ge, E′ = Si)

290 (E = E′ = Si; R = $SiMe_3$)
294 (E = E′ = Ge; R = $SiMe_3$)
296 (E = Ge, E′ = Si; R = $SiMe_3$)

Scheme 5.104

The resonances of the terminal and central silicons of the Si=Si=Si unit in **290** were observed at 196.9 and 157.0 ppm, respectively [calculated values were 212/221 and 144 ppm at the GIAO/6-311+G(2df,p)//B3LYP/6-31+G(d,p) level]. In line with the general trends for the heavy allenes, trisilaallene **290** revealed the following peculiar structural features: (1) a highly bent Si=Si=Si skeleton (136.5°); (2) Si=Si bond lengths of 2.177(1) and 2.188(1) Å that are typical for silicon–silicon double bonds; (3) a nearly perpendicular arrangement of the two five-membered rings; and (4) *trans*-bending and pyramidalization at the terminal Si atoms (354.1° and 354.9°) (Figure 5.23). Interestingly, the central allenic Si atom in the crystal structure of **290** was found in four different positions (from $Si2^{A}$ to $Si2^{D}$) at temperatures above −100 °C (Figure 5.23). These four structural isomers **A–D** were nearly isoenergetic (energy difference within 1 kcal/mol), thus suggesting the presence of a dynamic disorder because of the rotation of the Si2 atom about the Si1–Si3 axis.

In the UV-Vis spectrum, the trisilaallene **290** revealed its longest wavelength absorption at 584 nm, remarkably red shift compared, for example, with tetramethyldisilene (350 nm), which is indicative of the significant conjugation between the two cumulated Si=Si bonds.

Similar fast thermal motion of the central Ge atom of the allenic system was also observed in the crystalline state of 2-germadisilaallene **292**, prepared by the reduction of a mixture of isolable silylene **293** and dichlorogermylene–dioxane complex with potassium graphite (Scheme 5.105).[251b]

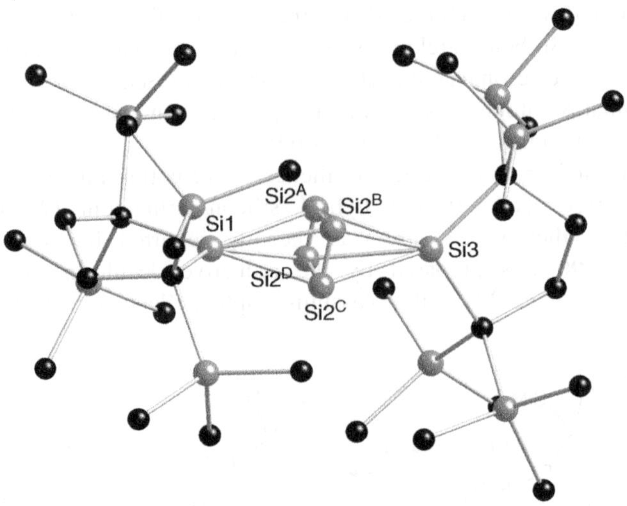

Figure 5.23 Crystal structure of trisilaallene **290** (hydrogen atoms are not shown)

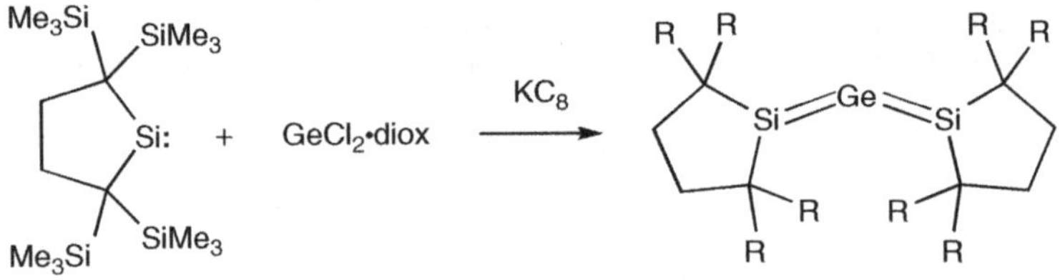

Scheme 5.105

Similarly to trisilaallene **290**, 2-germadisilaallene **292** showed a bent Si=Ge=Si skeleton (132.4°) and pyramidal geometry at the terminal Si atoms (353.9° and 354.0°). The Si=Ge bond lengths of 2.2366(7) and 2.2373(7) Å in **292** were slightly shorter than that of 1,2-disila-3-germacyclopenta-2,4-diene **191** (see Section 5.3.1.2.1.5) of 2.250(1) Å. The resonances of the terminal silicons in the allenic fragment of **292** were observed at low field (219.4 ppm), and a conjugation between the two Si=Ge bonds in **292** was manifested in the red shift of its longest wavelength UV-Vis absorption (599 nm).

Trigermaallene **294** and 1,3-digermasilaallene **296** were synthesized by the same procedure as that of trisilaallene **290** (Scheme 5.104).[251c] In contrast to trisilaallene **290** and 2-germadisilaallene **292**, neither **294** nor **296** exhibited a dynamic disorder, and accordingly the central Ge (in **294**) and Si (in **296**) atoms were found in only one of the four possible positions. This structural distinction between **290/292** and **294/296** was realized in terms of the different environmental features defining the spatial positions of their central allenic atoms.

The bending of the Ge=Ge=Ge (in **294**) and Ge=Si=Ge (in **296**) skeletons was notably more pronounced compared with those of **290** and **292**: 122.6° and 125.7° vs

136.5° and 132.4°. Likewise, the geometry at the terminal Ge atoms in **294** and **296** was markedly more distorted compared with that of **290** and **292**: 348.5°/348.6° and 349.3° vs 354.1°/354.9° and 353.9°/354.0°. This last observation is in line with the general trend of the greater pyramidalization at Ge in digermenes than at Si in disilenes. The Ge=Ge bonds in **294** [2.321(2) and 2.330(3) Å] were in the normal range for Ge=Ge bonds, whereas the Si=Ge bond in **296** [2.2694(8) Å] was slightly longer than that of 1,2-disila-3-germacyclopenta-2,4-diene **191** (see Section 5.3.1.2.1.5) of 2.250(1) Å. The central Si atom in 1,3-digermasilaallene **296** resonated at 236.6 ppm, that is, at a much lower field compared with that of trisilaallene **290** (157.0 ppm). Similarly to **290** and **292**, both trigermaallene **294** and 1,3-digermasilaallene **296** revealed considerable conjugation between the Ge=Ge (in **294**) and Si=Ge (in **296**) double bonds seen in the red shift of their longest wavelength UV-Vis absorptions: 630 nm (in **294**) and 612 nm (in **296**) (cf. the corresponding value for trisilaallene **290**, 584 nm).

As well as the allenes containing heavy group 14 elements, several stable allenic derivatives of the group 14–group 15 hybrids of the type >E=C=E′– (E = heavy group 14 element; E′ = group 15 element) have also been reported. They are briefly described below.

The first compound of such type, $Ar_2Sn{=}C{=}N{-}Mes$ [Ar = 2,4,6-$(CF_3)_3C_6H_2$] **298**, was synthesized by the stoichiometric reaction of stannylene Ar_2Sn: with mesitylisocyanide MesNC:.[252] Representing a formally allenic-type stannaketenimine structure, **298** revealed a highly bent Sn=C=N skeleton (153.9°) and distorted pyramidal geometry at the Sn atom (290.9°). The length of the Sn=C bond in **298** of 2.397(3) Å was much greater than those of the typical Sn=C double bonds in stannenes (see Section 5.3.1.2.1.3, Table 5.9), and even greater than those of standard Sn–C single bonds (2.15–2.20 Å). On the other hand, the C=N bond length of 1.158(3) Å was characteristic of isocyanides. Overall, the bonding between the Sn and C atoms in the >Sn=C=N– skeleton of **298** was described as a double donor–acceptor interaction between the starting Ar_2Sn: and :CNMes, similar to that in Lappert's distannene $Dis_2Sn{=}SnDis_2$ **121**. Given the inherent weakness of such dative interactions, it was not unexpected that **298** readily dissociated in solution into the starting stannylene and isocyanide. Accordingly, the reactivity of **298** toward 2,3-dimethylbuta-1,3-diene and *t*-BuOH was in fact that of the stannylene Ar_2Sn:.

The synthesis of the stable silicon congeners of **298**, that is, silaketenimines, was preceded by the report on their transient version *t*-$Bu_2Si{=}C{=}N{-}Ph$, prepared by the reaction of photolytically generated silylene *t*-Bu_2Si: with phenylisocyanide PhNC:.[253] Lacking sufficient kinetic stabilization, *t*-$Bu_2Si{=}C{=}N{-}Ph$ underwent head-to-tail dimerization across the Si=C bond forming the corresponding cyclic dimer.

The first isolable silaketenimines **299a,b** were prepared by the reaction of the stable dialkylsilylene **293** with isocyanides RNC: [R = Dip (2,6-*i*-Pr_2-C_6H_3), Ad] (Scheme 5.106).[254] Both **299a** and **299b** were stable in the solid state below 0 °C but readily dissociated into the silylene and isocyanide in solution even at low temperatures.

The allenic, rather than silylene–isocyanide donor–acceptor complex, composition of **299** was deduced from its structural features. Thus, the Si=C bonds of 1.794(3) Å (in **299a**) and 1.782(2) Å (in **299b**), although insignificantly longer than the typical Si=C double bonds in silenes (see Section 5.3.1.2.1.1, Table 5.7), were considerably shorter than standard Si–C single bonds (1.86 Å). In contrast, the C=N bonds of 1.203(3) Å

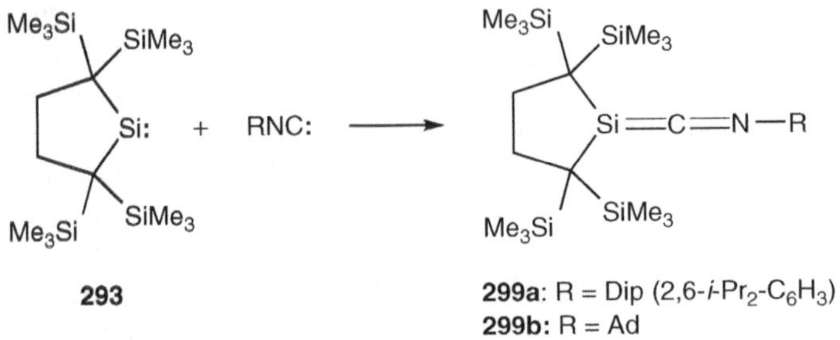

Scheme 5.106

(in **299a**) and 1.210(3) Å (in **299b**) were significantly stretched compared with the corresponding C≡N distances in aryl isocyanides (av. 1.16 Å). In accord with its allenic structure, the Si=C=N moiety was considerably bent (164.7° in **299a** and 163.1° in **299b**), whereas the terminal Si atoms were markedly pyramidalized (330.6° in **299a** and 331.6° in **299b**). NMR solution studies of **299** also corroborated its allenic structure formulation >Si=C=N–; the signals of the terminal silicon were observed at 15.9 ppm (in **299a**) and 23.9 ppm (in **299b**), whereas the central carbon was diagnostically deshielded and observed at very low fields of 214.6 ppm (in **299a**) and 221.3 ppm (in **299b**).

The reaction of the isolable disilyne with trimethylsilylcyanide provided a route to a bis(adduct) **300** (Scheme 5.107).[255]

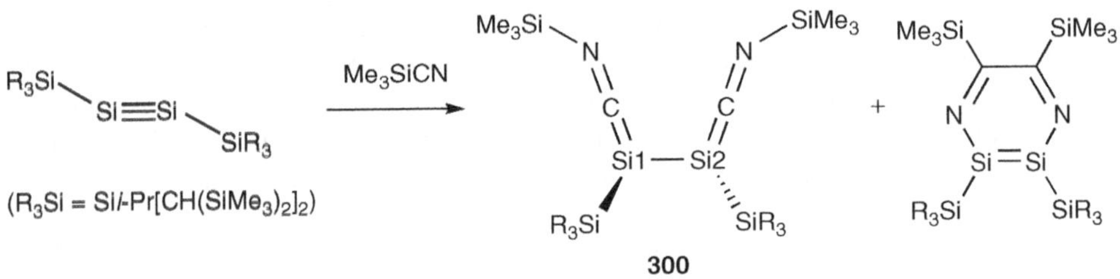

Scheme 5.107

The bis(adduct) **300** was classified as a bis(silaketenimine) based on its short silicon–carbon bonds of 1.826(3) and 1.830(2) Å, elongated (compared with C≡N bonds) carbon–nitrogen distances of 1.185(3) and 1.179(3) Å, and slightly bent Si=C=N fragments (171.1° and 170.8°). However, the extent of the silaketenimine character of **300** is definitely smaller than that of the above-described **299**. Given the large pyramidalization at the central Si1 and Si2 atoms (319.1° and 319.9°) and extreme shielding of these atoms [$\delta(^{29}Si)$ = −172.0 ppm], one should consider a zwitterionic form with a negative charge on either Si1 or Si2 as a strong contributor to the overall structure of **300**. Apparently, the best description of the bonding situation in

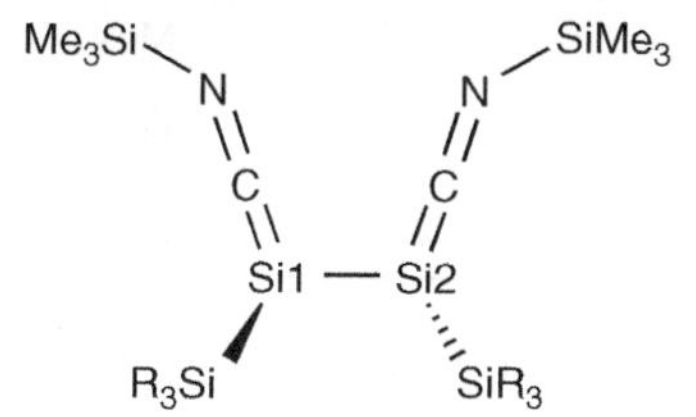

A

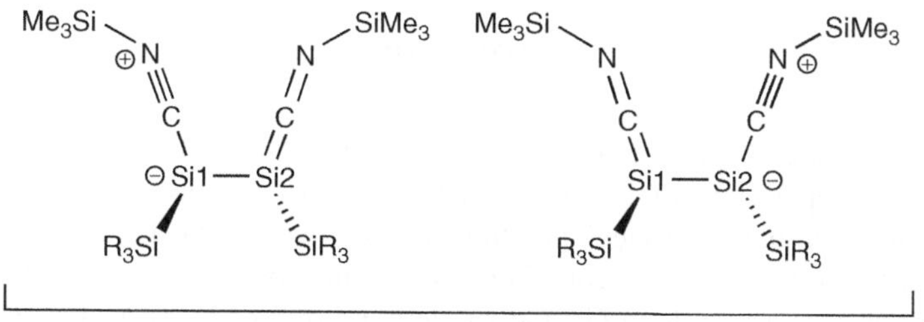

B

Scheme 5.108

300 would be a hybrid of the two resonance extremes, neutral **A** and charge-separated **B** (Scheme 5.108).

It is important to note that the bonding nature of silaketenimines (bent allene structure vs zwitterionic form) is predominantly governed by substituent effects: (*N*-aryl)diarylsilaketenimines are zwitterionic, (*N*-alkyl)dialkylsilaketenimines are allenic, and (*N*-aryl)dialkylsilaketenimines are either zwitterionic or allenic depending on the conformation of the *N*-aryl ring.[254] Thus, in contrast to the above-described silaketenimines **299** and **300**, the stable adducts of the thermally generated silylene Tbt(Mes)Si: and arylisocyanide ArNC: (Ar = Tip, Tbt, Mes*) were formulated as the silylene–isocyanide complexes Tbt(Mes)Si:←CNAr rather than the allenic derivatives Tbt(Mes)Si=C=N–Ar, based on their spectroscopic and reactivity studies and supported by theoretical calculations.[256]

The reaction of the stable digermyne ArGeGeAr with mesitylisocyanide MesNC: resulted in the formation of a bis(adduct), which was classified by Power *et al.* as the bis(germylene–isocyanide) complex **301** (in contrast to bis(silaketenimine) **300** obtained in the reaction of disilyne with trimethylsilylcyanide, see above) (Scheme 5.109).[257]

The formulation of **301** as a Lewis acid–Lewis base adduct, rather than as a bis(germaketenimine), was based on its structural characteristics. Both isocyanide molecules complex almost perpendicularly to the germanium coordination planes in the ArGeGeAr unit, and consequently the Ge centers in **301** are remarkably pyramidalized (293.2° and 292.5°). The germanium–carbon distances of 1.996(5)

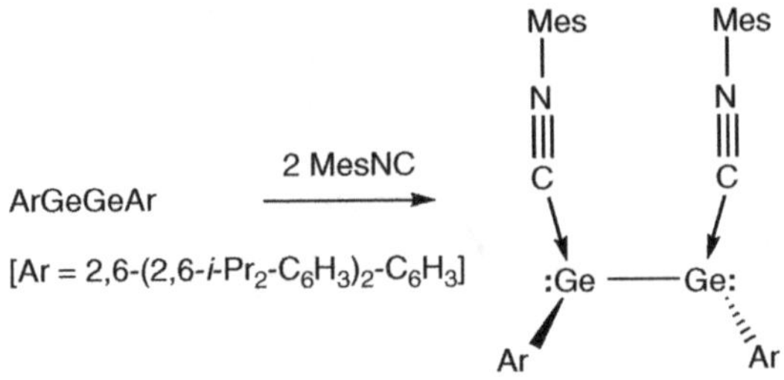

301

Scheme 5.109

and 2.026(5) Å are in the normal range for Ge–C single bonds, whereas the carbon–nitrogen separations of 1.145(6) and 1.157(6) Å are typical for aryl isocyanides. The germanium–carbon–nitrogen bending angles were 159.4 and 159.5°. Upon complexation to form **301**, the germanium–germanium bond of the starting digermyne was transformed into a very long Ge–Ge single bond of 2.6626(8) Å. This Ge–Ge bond length is much greater than that of typical digermanes (av. 2.44 Å), being ca. 0.38 Å longer than the germanium–germanium distance in the starting digermyne.

Representatives of the phosphorus analogs of sila-, germa- and stannaketenimines of the type >E=C=P– (E = heavy group 14 element) were prepared by Escudié *et al.* The transient phosphasilaallene **302** was generated by the reduction of a (*C*-chlorosilyl)phosphaalkene precursor **303** with *t*-BuLi (Scheme 5.110).[258] **302** was formed through the *β*-elimination of LiCl from the intermediate lithium derivative **304** taking place at −60 °C.

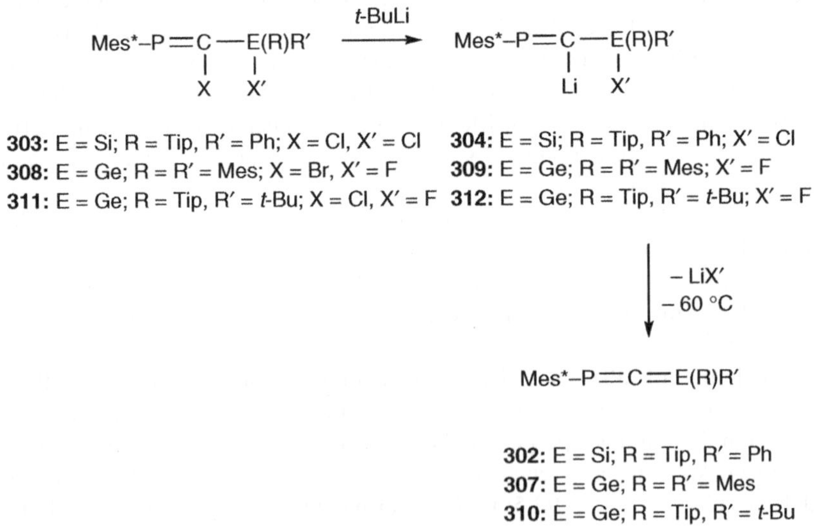

Scheme 5.110

The composition of **302** was proved by the diagnostic low-field chemical shifts of its allenic atoms observed by NMR spectroscopy at −60 °C: $\delta(^{31}P) = 288.7$ ppm, $\delta(^{13}C) = 269.1$ ppm [$^{1}J(^{13}C–^{31}P) = 45.8$ Hz], $\delta(^{29}Si) = 75.7$ ppm [$^{2}J(^{29}Si–^{31}P) = 18.3$ Hz]. The formulation of the transient **302** as an allenic derivative was also supported by its reaction with MeOH forming the expected adduct to a silicon–carbon double bond in accord with its normal polarity $Si^{\delta+} = C^{\delta-}$. Being thermally unstable, **302** underwent dimerization above −20 °C forming two types of cyclic dimers, the expected head-to-tail dimer **305** involving [2 + 2] cycloaddition of the two Si=C bonds, and the unsymmetrical dimer **306** formed upon the [2 + 2] cycloaddition of one Si=C and one P=C bond (Scheme 5.111).[258]

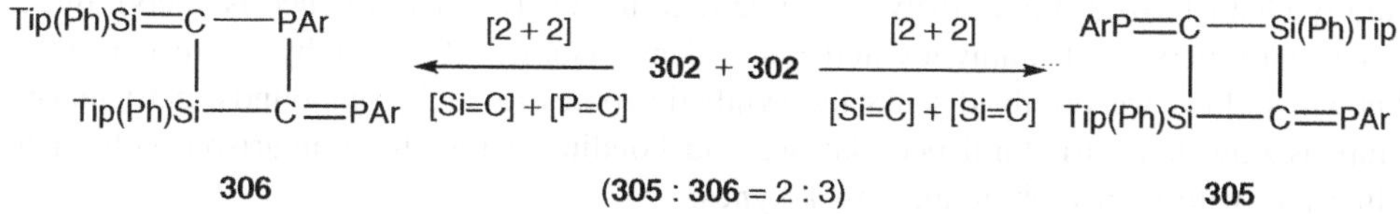

Scheme 5.111

The germanium analog of **302**, transient phosphagermaallene **307**, was generated by the same procedure: reduction of a (*C*-fluorogermyl)phosphaalkene precursor **308** with *t*-BuLi proceeding through the intermediate lithium salt **309** (Scheme 5.110).[259a] Formation of **307** was supported by the low-temperature observation of low-field resonances for the allenic P (240.0 ppm) and C [280.9 ppm, $^{1}J(^{13}C–^{31}P) = 54.3$ Hz] atoms, and additionally proved by its characteristic reactivity towards MeOH and MeLi. Similarly to phosphasilaallene **302** (Scheme 5.111), phosphagermaallene **307** dimerized at temperatures above −50 °C, giving analogous cyclic dimers of the same two types, a symmetrical head-to-tail dimer ([2 + 2] cycloaddition of the two Ge=C bonds) and an unsymmetrical dimer ([2 + 2] cycloaddition of one Ge=C and one P=C bond) formed in a 12:88 ratio.

Further increasing the steric bulk at the Ge center by replacing Mes substituents with Tip and *t*-Bu groups, Escudié *et al.* were able to isolate the room temperature stable phosphagermaallene **310** prepared by the same reaction procedure (Scheme 5.110).[259b] As is typical for >E=C=P– (E = heavy group 14 element) derivatives, **310** revealed diagnostic deshielding of its allenic phosphorus (249.9 ppm) and carbon [280.4 ppm, $^{1}J(^{13}C–^{31}P) = 60.9$ Hz] atoms. As well as from the spectroscopic observations, the allenic constitution of **310** was further supported by its reactions with water to give the 1,2-adduct across the Ge=C bond, and with carbonyl compounds forming the expected [2 + 2] cycloadducts (Ge=C + C=O).

5.3.3 Heavy Analogs of Alkynes: Disilynes, Digermynes, Distannynes, Diplumbynes, and their Valence Isomers[260]

The experimental success in isolation of the first stable silicon, germanium, tin and lead analogs of acetylenes was preceeded by important discoveries in the field of their transient congeners. Thus, in the 1990s, some parent compounds with triply bonded silicon atoms were identified by spectroscopic techniques: the silanitrile H–Si≡N was characterized by low-temperature (10 K) IR Ar-matrix isolation by Maier and Glatthaar,[261] whereas silynes H–C≡Si–F and H–C≡Si–Cl were detected in the gas phase by neutralization–reionization mass spectrometry by Schwarz *et al.*[262] In some cases, the existence of the transient heavy alkynes R–E≡E′–R′ (E, E′ = group 14 elements) was suggested based on their specific reactivity: Me–Si≡Si–Me trapped with anthracene,[263] Ar–Si≡Si–Ar [Ar = 2,6-(2,4,6-Me_3-C_6H_2)-C_6H_3] stabilized by intramolecular C–C bond insertion,[264] and Ar–Ge≡C–$SiMe_3$ [Ar = 2,6-(*i*-Pr_2NCH_2)$_2$-C_6H_3] trapped with alcohols.[265]

Regarding stable representatives, no heteronuclear derivatives of the type –E≡E′– have been reported to date despite the intensive efforts of many research groups. As for the homonuclear analogs –E≡E–, only a few of them have been isolated and structurally characterized: three disilynes, three digermynes (or their valence isomers), three valence isomers of distannynes and one valence isomer of diplumbyne are currently known. Below, we briefly discuss the synthetic routes to such compounds, particularly emphasizing their structural peculiarities and bonding nature in comparison with each other and with their carbon analogs, alkynes.

The chemistry of the stable derivatives featuring triple bonds between their heavy group 14 elements and transition metals was previously discussed by Power *et al.*[260a,266] and is not, therefore, covered in this section.

5.3.3.1 Disilynes RSiSiR

The relatively stable disilyne R_3Si–Si≡Si–SiR_3 [R_3Si = SiMe(Si*t*-Bu_3)$_2$] **313** was reported in 2004 by Wiberg *et al.* by the reductive dehalogenation of the 1,2-dichlorodisilene precursor (*E*)-R_3Si(X)Si=Si(SiR_3)X (X = Cl, Br) **22** (see Section 5.3.1.1.1) with lithium naphthalenide.[31,57,58,267] Because disilyne **313** has not been structurally characterized, its identification was based on the observation of the low-field resonance of the central silicon nuclei at 91.5 ppm, high resolution mass spectral data, and characteristic reactivity of **313** toward ethylene and buta-1,3-diene to form the corresponding [2 + 2] and [2 + 4] cycloadducts, respectively.[58] Although stable in the solid state, disilyne **313** underwent unavoidable isomerization in solution, accompanied by the 1,2-migration of one of the *t*-Bu_3Si-groups from the substituent Si to the central Si atom and subsequent cyclization forming unsymmetrical cyclotrisilene. In the absence of structural data, the length of the central silicon–silicon bond in **313** was computationally estimated to be as short as 2.072 Å with a Si–Si≡Si *trans*-bent angle of 148.0°, values that were in a good agreement with its Si≡Si triple bond formulation.[267b]

The first indefinitely stable and structurally characterized disilyne R_3Si–Si≡Si–SiR_3 (R_3Si = Si*i*-PrDis$_2$) **314** was also reported in 2004 by Sekiguchi *et al.*, being prepared by the reductive debromination of the 1,1,2,2-tetrabromodisilane derivative R_3Si–SiBr_2–SiBr_2–SiR_3 with 4 equiv. of potassium graphite.[268] The resonance of the triply bonded Si atoms in disilyne **314** was observed in the region similar to that of Wiberg's disilyne **313**: 89.9 ppm vs 91.5 ppm, respectively. The central Si1–Si1′ bond in **314** was extraordinarily short, 2.0622(9) Å, being ca. 4% shorter than typical Si=Si double bonds (2.14 Å) (Figure 5.24).[268a] Moreover, in sharp contrast to linear organic alkynes and in full agreement with theoretical predictions,[269] the geometry of the Si≡Si triple bond was found to be notably *trans*-bent with a Si2–Si1≡Si1′ bond angle of 137.4°.

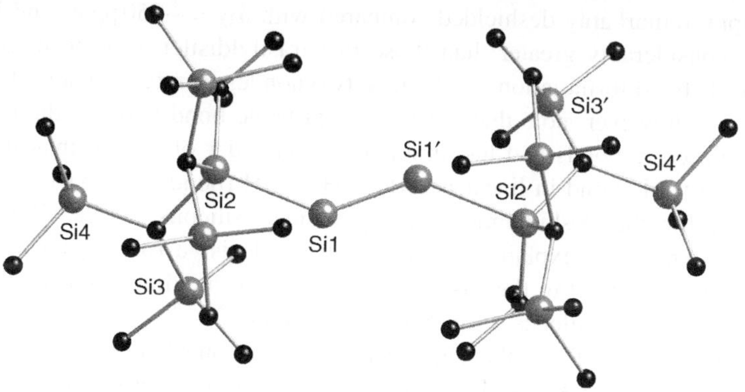

Figure 5.24 *Crystal structure of the disilyne* ***314*** *(hydrogen atoms are not shown)*

The results of NBO computations on **314** well corroborated its experimental Si≡Si triple bond formulation: WBI of the central Si1–Si1′ bond of 2.618 with the electron occupancy of the two $\pi_{Si\equiv Si}$ orbitals of 1.934 and 1.897 electrons.[268a,c] As was expected, upon *trans*-bending at the Si≡Si bond the degeneracies of both the HOMO and the LUMO diagnostic of the linear structure were broken, and the presence of the two nondegenerate HOMOs(π) and LUMOs(π^*) was manifested in the observation of the longest wavelength UV-Vis absorptions of **314** at 483 and 690 nm, corresponding to the π–π^* transitions, HOMO − 1 → LUMO/HOMO → LUMO + 1 and HOMO → LUMO, respectively.[268c]

The classification of **314** as a disilyne, featuring a genuine Si≡Si triple bond, found further support from its solid-state ^{29}Si NMR CPMAS studies, based on the general trend of the large chemical shift anisotropy (CSA) values indicative of the presence of π-bonds.[268b] For example, the large CSA of the parent acetylene HC≡CH stems from the two strongly deshielded chemical shift tensor (CST) contributors δ_{11} and δ_{22} that are orthogonal to the triple bond molecular axis (Figure 5.25, **A**). Accordingly, the measured values and orientations of the CST of disilyne **314** ($\delta_{11} = 364$ and

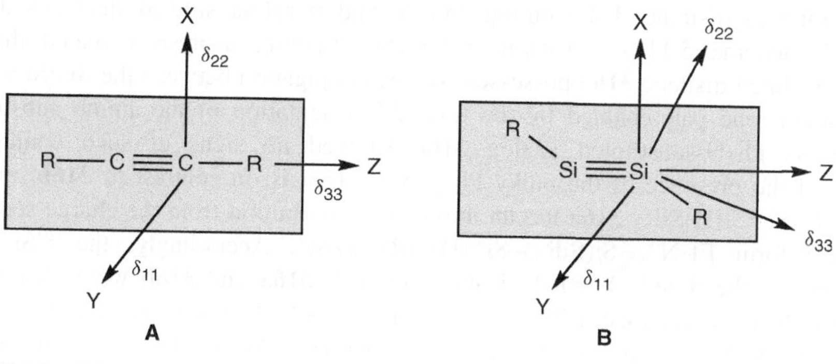

Figure 5.25 *Direction of the principal CST components in: (**A**) linear alkyne RC≡CR; (**B**) trans-bent disilyne RSi≡SiR*

$\delta_{22} = 221$ ppm, remarkably deshielded compared with $\delta_{33} = -350$ ppm) and large CSA (-643 ppm, considerably greater than those in tetrasilyldisilenes) were in accord with the Si≡Si triple bond formulation (σ–bond + two nondegenerate π-bonds) (Figure 5.25, **B**). One should, however, note that the *trans*-bent triple bond between the two silicons features weakened π-bonds and had a smaller bond order of ca. 2.6 than those of the classical strong triple bond in linear acetylene HC≡CH (bond order 3.0).

The reason for the deshielding of triply bonded silicons in the silyl-substituted *trans*-bent disilynes was explained in terms of the decisive role of the paramagnetic contribution, quite parallel to the case of the extreme deshielding of the doubly bonded Si atoms in tetrasilyldisilenes (see Section 5.3.1.1).[268b,270a] As in doubly bonded systems $(R_3Si)_2Si{=}Si(SiR_3)_2$, the large paramagnetic contributions σ^{p}_{11} and σ^{p}_{22} in bis(silyl)disilynes R_3Si–Si≡Si–SiR_3, stemming from the mixing between the occupied σ(Si–Si) and π^*_{in}/π^*_{out}(Si≡Si) orbitals, result from the decrease in the σ–π^*_{in}/π^*_{out} energy gaps because of the electropositive silyl substitution. Moreover, the increase in the *trans*-bending in disilynes reduces orbital overlap, raising the energy of the σ(Si–Si) orbital and decreasing the σ–π^* energy separation, which leads to an increase in the paramagnetic contribution to the chemical shielding, and finally to further deshielding of the triply bonded silicons compared with those in the hypothetical linear disilyne.

The crystal structure of diaryldisilyne Bbt–Si≡Si–Bbt (Bbt = 2,6-$[(Me_3Si)_2CH]_2$-4-$[(Me_3Si)_3C]$-C_6H_2) **315** was also recently reported by Tokitoh *et al.*[27] Disilyne **315** was prepared by the standard procedure: reductive debromination of 1,2-dibromodisilene (*E*)-Bbt(Br)Si=Si(Bbt)Br **16** (Section 5.3.1.1.1) with *t*-BuLi through initial lithium–bromine exchange followed by the β-elimination of LiBr from the transient Bbt(Li)Si=Si(Bbt)Br. Lacking the deshielding effects of σ-donating silyl substituents (see above), diaryldisilyne **315** revealed the resonance of its triply bonded Si atoms at a remarkably higher field (than disilynes **313** and **314**) at 18.7 ppm. In agreement with the theoretical expectations, aryl-substituted disilyne **315** exhibited a Si≡Si triple bond that was slightly longer than that in the silyl-substituted disilyne **314** [2.108(5) Å vs 2.0622(9) Å], with slightly greater *trans*-bending of the triple bond [133.0°(C_{Ar}–Si≡Si) vs 137.4°(Si–Si≡Si)].

Disilynes exhibited a rich reactivity towards a variety of reagents, forming acyclic and cyclic disilenes as 1,2-addition and cycloaddition products, respectively. Below, as a representative example of such reactivity, the addition of secondary amines and hydroborane to disilyne **314** forming amino- and boryl-substituted disilenes **316a,b** and **317** (Scheme 5.112) is shown.[271] Crystal structure analysis revealed that the Et_2N-substituted disilene **316a** possesses notable conjugation between the Si=Si π-bond and N atom lone pair enabled by the favorable orientation of the amino substituent, whereas its Ph_2N-substituted analog **316b** showed no signs of such conjugation because of the presence of the bulky Ph groups. That is, in contrast to **316b**, disilene $Et_2N(R_3Si)Si{=}Si(H)SiR_3$ **316a** has an important contribution from the charge-separated resonance form $Et_2N^+{=}Si(SiR_3)$–$Si^-(H)SiR_3$ **316a′**. Accordingly, the ^{29}Si NMR resonances of the doubly bonded Si atoms in both **316a** and **316b** were observed in markedly different regions: 170.5 (N–*Si*=Si) and −39.3 (Si=*Si*–H) ppm (for **316a**), and 136.4 (N–*Si*=Si) and 66.4 (Si=*Si*–H) ppm (for **316b**). The geometry of the boryl-substituted disilene $R_2B(R_3Si)Si{=}Si(H)SiR_3$ **317** also allowed for conjugation between the Si=Si π-bond and the vacant 2p-orbital on the B atom, seen in the

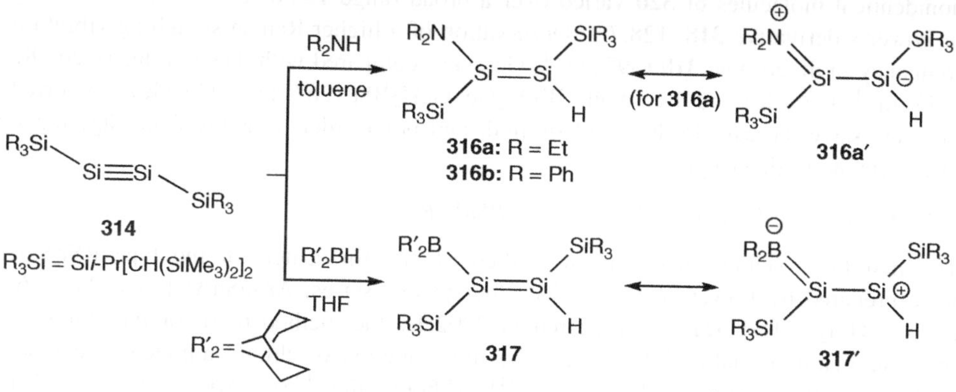

Scheme 5.112

substantial deshielding of the Si=*Si*–H center (150.5 ppm) due to the contribution of the resonance structure $R_2B^-{=}Si(SiR_3)–Si^+(H)SiR_3$ **317′**.

5.3.3.2 *Digermynes RGeGeR (and their Valence Isomers)*

Stable germanium analogs of alkynes have been known since 2002 when Power *et al.* reported the first representative ArGeGeAr [Ar = 2,6-(2,6-*i*-Pr_2-C_6H_3)$_2$-C_6H_3] **318**, prepared by the reduction of the chlorogermylene precursor Ar(Cl)Ge: with potassium in THF.[74,272] Although the germanium–germanium bond distance in **318** of 2.2850(6) Å was markedly shorter than standard Ge–Ge single bonds (ca. 2.44 Å) indicating its considerable multiple bond character, it was still in the range of the known digermenes (2.21–2.46 Å). It was therefore suggested that the bonding between the two germaniums in **318** is best described as a double bond with a bond order of 2 (digermene structure), rather than a real triple bond with a bond order of 3 (digermyne structure). Remarkable *trans*-bending of the germanium–germanium bond of 128.7°, because of the accumulation of nonbonding electron density at the Ge centers in the form of lone pairs, results in the decrease in the Ge–Ge bond order and stretching of the bond.

Another germanium analog of an alkyne, Ar*GeGeAr* [Ar* = 2, 6-(2,4,6-*i*-Pr_3-C_6H_2)$_2$-C_6H_3] **319**, was synthesized by an identical synthetic procedure: reduction of chlorogermylene Ar*(Cl)Ge: with potassium in THF; however, the crystal structure of **319** was not reported.[74]

The latest representative of RGeGeR derivatives, BbtGeGeBbt (Bbt = 2,6-[(Me_3Si)$_2$CH]$_2$-4-[(Me_3Si)$_3$C]-C_6H_2) **320**, was prepared by Tokitoh *et al.* by the reductive debromination of 1,2-dibromodigermene (*E*)-Bbt(Br)Ge=Ge(Bbt)Br **87** (Section 5.3.1.1.2) with potassium graphite.[273] The germanium–germanium bond in **320** was shorter than that in Power's compound **318**, 2.2060(8) Å and 2.2260(8) Å (for the two independent molecules in the unit cell) vs 2.2850(6) Å, thus indicating the greater extent of the triple bond/digermyne character in the former (the doublet–quartet energy gap of the BbtGe fragment of **320** was calculated to be ca. 4 kcal/mol smaller than that of the ArGe unit in **318**). The C_{Ar}–Ge≡Ge *trans*-bent angles in the two nonidentical molecules of **320** varied over a broad range 123.6–138.7° (*trans*-bending in Power's derivative **318**: 128.7°). Observation of a higher Raman stretching vibration frequency in digermyne **318** (397 cm^{-1}, Ge≡Ge), compared with those in digermene **87** (355 cm^{-1}, Ge=Ge) and digermane $BbtH_2Ge–GeH_2Bbt$ (267 cm^{-1}, Ge–Ge), supported the successive increase in the bond strength and bond order on going from digermane to digermene to digermyne.

5.3.3.3 *Valence Isomers of Distannynes RSnSnR*

The first three stable derivatives with dicoordinate tin centers of the type RSnSnR were prepared by Power *et al.*[74,274] The first in this series, ArSnSnAr [Ar = 2,6-(2,6-*i*-Pr_2-C_6H_3)$_2$-C_6H_3] **321** was prepared in 2002 by the method used for the synthesis of its germanium analog **318**, namely, by the reduction of chlorostannylene precursor Ar(Cl)Sn: with potassium in benzene.[74,274a]. Although the ^{119}Sn NMR resonance of **321** was not detected, probably because of line broadening resulting from the large CSA, its composition was established by X-ray crystallography to show a *trans*-bent ArSnSnAr skeleton characteristic for the heavy alkyne analogs (125.2°). The tin–tin distance in **319** of 2.6675(4) Å was markedly shorter than those of typical Sn–Sn single bonds (2.81 Å), and virtually identical to the value of 2.6683(10) Å measured for the tetrasilyldistannene (*t*-Bu_2MeSi)$_2$Sn=Sn(SiMe*t*-Bu_2)$_2$ **126** (Section 5.3.1.1.3) that features the shortest Sn=Sn double bond among the acyclic distannenes. Accordingly, similarly to the case of its germanium analog **318**, the tin–tin bonding interaction in **321** was described as a double rather than a triple bond with the bond order estimated as intermediate between 1.5 and 2.0.[74] Moreover, the diagnostic *trans*-bent geometry of **321** was indicative of the developing lone pair character at each tin center.

The second stable tin analog of an alkyne, Ar*SnSnAr* [Ar* = 2, 6-(2,4,6-*i*-Pr_3-C_6H_2)$_2$-C_6H_3] **322**, was similarly available by the reduction of chlorostannylene Ar*(Cl)Sn: with potassium in toluene, but this compound was not crystallographically characterized.[74]

Interestingly, modification of the bulky ligand in **321** by the introduction of a Me_3Si-group at the 4-position of the central aryl ring caused great changes to the structural parameters in the resulting distannyne analog **323** Ar′SnSnAr′ [Ar′ = 2,6-(2,6-*i*-Pr_2-C_6H_3)$_2$-4-$SiMe_3$-C_6H_2].[274b] The new tin alkyne analog **323** was synthesized by the reduction of a stannylene Ar′(Cl)Sn: with potassium in diethyl ether. The tin–tin bond length of 3.066(1) Å and *trans*-bent angle of 99.3° in **323** were dramatically different from those in the original **321**, 2.6675(4) Å and 125.2°, respectively. Such extraordinary stretching of the tin–tin bond and narrowing of the bending angle in **323** compared with **321** makes the former closer to the previously reported lead analog Ar*PbPbAr* [Ar* = 2, 6-(2,4,6-*i*-Pr_3-C_6H_2)$_2$-C_6H_3] **324**,[275] for which single bonding between the two plumbylene units Ar*–$\ddot{\text{Pb}}$–$\ddot{\text{Pb}}$–Ar* was established. Likewise, the bond linking the two Ar′Sn fragments in **323** was described as a single bond rather than a triple or even double bond.[274b]

Recently, another organotin derivative Ar″SnSnAr″ [Ar″ = 2,6-(Me_2NCH_2)$_2$-C_6H_3] **325**, prepared by reduction of the corresponding chlorostannylene derivative, was classified as a distannyne [δ(^{119}Sn NMR) = 612 ppm].[276] However, given its long tin–tin separation of 2.9712(12) Å and acute *trans*-bent angle of 94.3°, which closely resemble those of the distannyne analog **323**, the tin–tin bonding in **325** should be recognized as a single bond, and this was supported by NBO computations (central Sn–Sn bond order = 0.91). Overall, **325** represents a valence isomer of a distannyne in the form of bis(stannylene) Ar″–S̈n–S̈n–Ar″, in which the vacant p-orbital at each tin center is intramolecularly stabilized by nitrogen lone pair electron donation (N→Sn) coming from a pair of chelating Me_2NCH_2-ligands.

5.3.3.4 *Valence Isomers of Diplumbynes RPbPbR*

The lead analog of an alkyne Ar*PbPbAr* [Ar* = 2,6-(2,4,6-*i*-Pr_3-C_6H_2)$_2$-C_6H_3] **324** was the first in the series of group 14 organometallics of the type REER (E = Si–Pb). Diplumbyne valence isomer **324** was prepared by the reduction of the corresponding bromoplumbylene Ar*(Br)Pb: with $LiAlH_4$ in diethyl ether (no ^{207}Pb NMR signal could be detected in solution).[275] Compared with other above-described heavy alkyne analogs, **324** exhibited extreme stretching of the lead–lead bond and an extraordinary degree of *trans*-bending of the Ar*PbPbAr* skeleton. The observed Pb–Pb bond distance of 3.1881(1) Å was even longer than the typical values for singly bonded diplumbanes such as Ph_3Pb–$PbPh_3$ [2.844(4) Å], whereas the C_{Ar*}–Pb–Pb *trans*-bent angle of 94.3° (close to the right angle of 90°) indicated little hybridization of the Pb valence orbitals and the presence of a lone pair at each lead center. In hybridization terms, this implies that each Pb atom in **324** uses two 6p-orbitals for bonding to a substituent and the other lead partner, that the remaining 6p-orbital is vacant, and that the most stable 6s-orbital is occupied by a lone pair of electrons. Thus, the extreme lengthening of the Pb–Pb single bond may be accounted for by the head-to-head overlap of p-orbitals (instead of sp^n-hybrids) to form this bond, as well as by the presence of lone pairs at the neighboring lead centers. The geometrical features of **324** (long Pb–Pb bond, strong *trans*-bending) are indicative of its existing in the unique bis(plumbylene) form **B** instead of the usual linear acetylene-like diplumbyne form **A** (Scheme 5.113). However, the preference of the nonclassical resonance structure **B** over the classical **A** is not due to the inherent weakness of the lead–lead multiple bonds in **A**, but instead should be attributed to the increased reluctance for s,p-hybridization in heavier group 14 elements, which in the case of lead is further reinforced by relativistic effects.

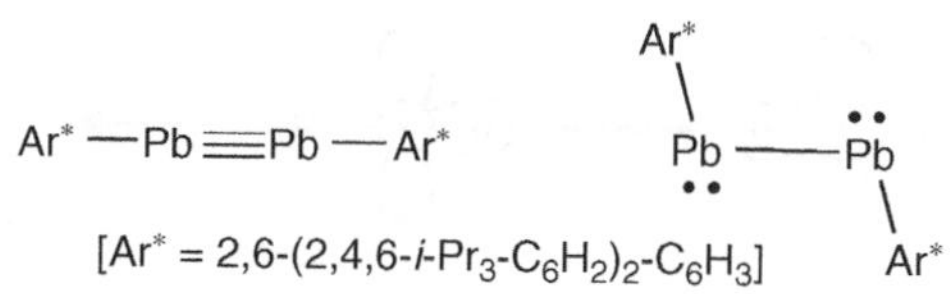

324: A **324: B**

Scheme 5.113

5.3.3.5 *Heavy Analogs of Alkynes: Structure and Bonding*

Quite similarly to the case of heavy alkenes (see Section 5.3.1.1), the most fundamental structural deformation of the heavy analogs of alkynes R–E≡E–R (E = Si–Pb) is a remarkable *trans*-bending/pyramidalization at the E centers, which strikingly contrasts with the linear geometry of organic alkynes R–C≡C–R. The unusual *trans*-bent geometry of heavy alkynes has been explained by two alternative MO models.[260,269a,b]

According to the first approximation, the triple bond between the two silicons in the disilyne H–Si≡Si–H is considered the result of the bonding interaction between two HSi monomers, which are known to possess a doublet ground state ($^2\Pi$) that is much more stable (by 42.6 kcal/mol) than the quartet state ($^4\Sigma$).[269a,b] By contrast, in the case of organic HC monomers, the destabilization of the quartet compared with the doublet is significantly diminished to only 16.7 kcal/mol. Accordingly, the interaction between the singly occupied MOs of the two quartet HC monomers results in the three covalent interactions (one σ and two π) of the classical linear C≡C triple bond (Figure 5.26, **A**: E = C, R= H).

This bonding mode (Figure 5.26, **A**: E = Si–Pb) is, however, disfavored for the heavy alkyne analogs R–E≡E–R, because of their large doublet–quartet energy separation, leading to a preferential bonding interaction between the two doublet R–E monomers. Because an important Pauli repulsion between the two electron pairs is envisaged in the linear alkyne-type structure (Figure 5.26, **B**), the structure may gain additional stabilization through the stretching of the E–E bond and, most importantly, *trans*-bending at the E centers. The resulting E–E triple bond (for E = Si–Sn) may therefore be regarded as a combination of one π-bond (perpendicular to the REER plane) and two donor–acceptor bonds (Figure 5.26, **C**).[269a,b] According to this approach, upon the extreme *trans*-bending to 90° in the diplumbyne analog RPbPbR, the two donor–acceptor bonds transform into the nonbonding electron pairs, whereas the π-bond is converted to a σ-bond, resulting in overall single bonding between the two lead atoms (Scheme 5.113, **B**). One can note that this bonding model is highly reminiscent of that of the *trans*-bent heavy alkene analogs featuring nonclassical donor–acceptor double bonds (see Figure 5.1).

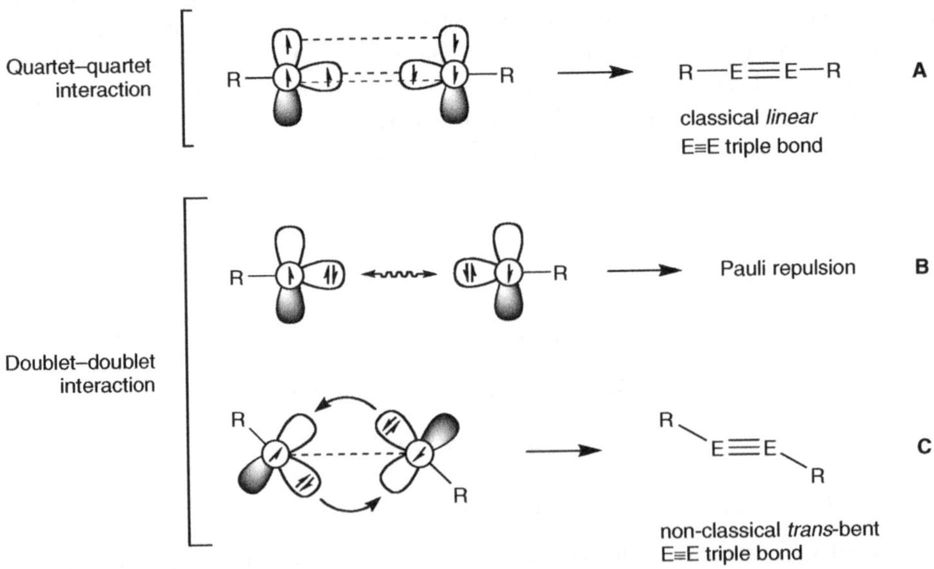

Figure 5.26

Alternatively, the *trans*-bending at E in REER derivatives may be explained in terms of a second-order Jahn–Teller effect by the mixing of the E–E $\pi_{\text{in-plane}}$ and σ^*-orbitals, facilitated in the case of heavy group 14 elements because of the remarkable decrease in the energy separation between the bonding and antibonding levels on descending group 14.[74,260,266,277] Mixing of the E–E σ- and $\pi^*_{\text{in-plane}}$-orbitals, whose energy separation is markedly greater, also contributes to the overall *trans*-bending at the E centers but to a much lesser extent. A diagram of the selected orbital interactions in linear (left) and *trans*-bent (right) REER molecules is shown in Figure 5.27.

Because of the increasing closeness of the E–E π- and σ^*-orbitals energy levels on descending group 14, the mixing of these orbitals becomes more prominent in the same direction, resulting in an increase in *trans*-bending and decrease in bond shortening on going from silicon to lead. One of the hybrid MOs generated by the π–σ^* interactions is essentially nonbonding, and this accumulation of the lone pair character at E occurs at the expense of electron density in the bonding region between the two E centers. Accordingly, the nature of the E–E bond in the heavy alkyne analogs REER considerably differs from that of the classical C≡C triple bond (bond order = 3); namely, the triple bond character in E≡E is essentially retained in disilynes (E = Si, bond order

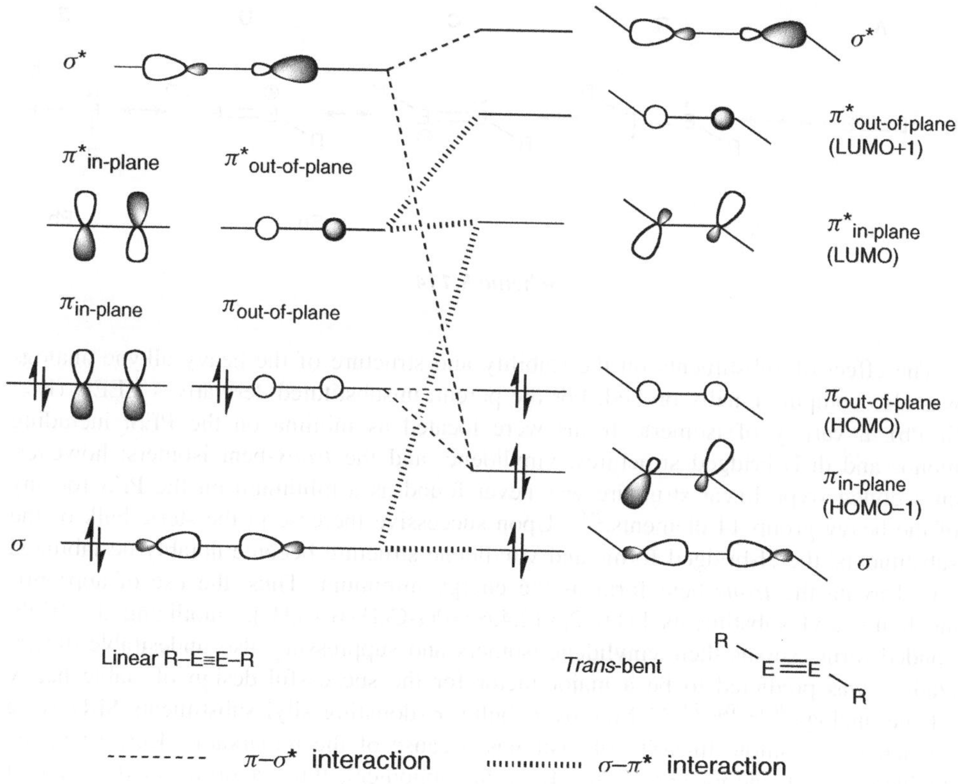

Figure 5.27

greater than 2), digermynes and distannynes feature an approximate E=E double bond (E = Ge and Sn, bond order approximately or greater than 2 for Ge and approximately or less than 2 for Sn), whereas the lead–lead bonding in diplumbynes is essentially a single E–E bond (E = Pb, bond order = 1).[260] In valence bond terms, this carbon to lead progression down group 14 can be illustrated by the sequence of canonical structures in accord with the decrease in the bond order on going from linear alkynes (**A**) to *trans*-bent disilynes/digermynes/distannynes in the form of either biradicals (**B**) or zwitterions (**C, D**), and finally, to the most distorted lead analogs (PbPb) represented by diplumbylene resonance form (**E**) (Scheme 5.114).[91,260,277] The importance of the singlet biradical contribution **D** for disilynes and digermynes (in contrast to distannynes and diplumbynes) was supported by recent computations to reveal ca. 17% and 15% biradical character in the bonding $\pi_{\text{out-of-plane}}$ and nonbonding $\pi_{\text{in-plane}}$ orbitals of MeSiSiMe and MeGeGeMe, respectively (in contrast to ca. 4% biradical character in distannyne MeSnSnMe and diplumbyne MePbPbMe).[91,260b,266,278] The significance of the singlet biradical contribution to the structure of stable germanium alkyne analog ArGeGeAr **316** was well corroborated by the experimental results showing its remarkably higher reactivity toward unsaturated compounds compared with that of isostructural distannyne ArSnSnAr **319**.[91]

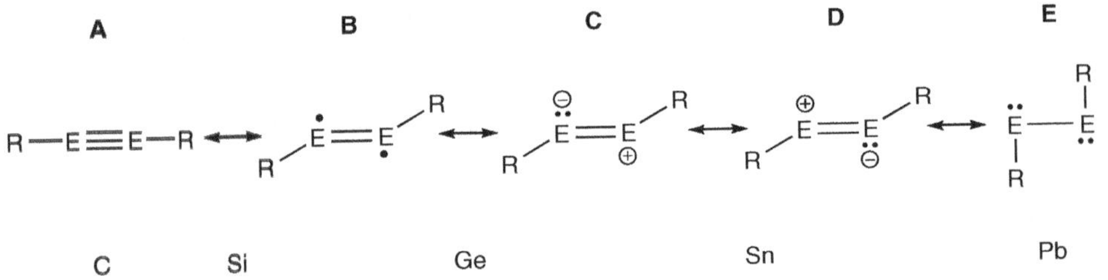

Scheme 5.114

The effect of substituents on the stability and structure of the heavy alkyne analogs was also computationally probed. For the parent, unsubstituted derivatives HEEH (E = Si–Pb), a variety of isomeric forms were located as minima on the PES, including mono- and di-H-bridged structures, vinylidene, and the *trans*-bent isomers; however, an acetylene-type linear structure was never found as a minimum on the PES for any of the heavy group 14 elements.[260b] Upon successive increase in the steric bulk of the substituents, the H-bridged forms and vinylidene structure become notably destabilized thus leaving the *trans*-bent form as the energy minimum. Thus, the use of appropriate bulky aryl substituents [Tbt, 2,6-(2,4,6-*i*-Pr_3-C_6H_2)$_2$-C_6H_3], stabilizing the triply bonded forms versus their vinylidene isomers and suppressing also undesirable dimerization, was predicted to be a major factor for the successful design of stable heavy alkyne analogs.[269,270a,279,280] Moreover, bulky σ-donating silyl substituents SiR_3 were predicted to promote linearity of disilynes because of the remarkable decrease in the doublet–quartet energy gap of the R_3Si–Si monomers, thus favoring quartet–quartet interaction mode **A** (Figure 5.26) and leading to stabilization of the linear disilyne form.[269a,b] The advantage of the electron-donating silyl groups was attributed to the increase in the negative charge at the triply bonded Si centers, which diminishes the size and energy difference between valence s- and p-orbitals, thus making their hybridization more effective.

5.4 Summary and Outlook

The new field of stable multiply bonded derivatives of the heavy group 14 elements was established in the 1970–1980s by the pioneering syntheses of the first compounds featuring double bonds between tins (Sn=Sn, 1973), between silicons (Si=Si, 1981), and between silicon and carbon (Si=C, 1981). Since then, a number of other stable representatives (both homo- and heteronuclear) of the heavy alkenes >E=E′<, heavy 1,3-dienes >E′ = E″–E=E<, heavy allenes >E=E′ = E″<, and heavy alkynes – E≡E– have been prepared, and in most cases structurally characterized. Now, these compounds are no longer fleeting intermediates, but are readily available as preparative scale reagents very useful as starting materials for the synthesis of a great variety of linear and cyclic organometallic derivatives, that are not available in any other way. The important structural distinctions between the heavy alkenes >E=E< (E = Si–Pb) and their organic counterparts >C=C< (*trans*-bent and/or twisted vs planar configuration), as well as between the heavy alkynes –E≡E– and organic acetylenes –C≡C– (*trans*-bent vs linear geometry), are now firmly experimentally ascertained, and the reasons for these distinctions are well understood. The fundamental issue of the E=E and E≡E bonding situations, compared with those in the prototypical alkenes and alkynes, was comprehensively addressed in recent computational studies to disclose the principal differences in the bonding types: classical covalent in the planar alkenes and linear alkynes versus nonclassical donor–acceptor in their *trans*-bent heavy analogs.

As future prospects, one can anticipate further developments in the synthesis of novel types of double bonding between different elements of the types R_2E=E′–R and R_2E=E″ (E = heavy group 14 element, E′ = group 13/15 element, E″ = group 16 element). Compounds involving the heaviest elements of these groups, as well as doubly bonded derivatives with oxygen (silanones, germanones, etc.), are among the most challenging of such types. Triply bonded species of the types R–E≡E′–R and R–E≡E″ (E, E′ = heavy group 14 elements, E″ = group 15 element) are the other prominent candidates for experimental pursuits. Apart from synthetic developments, the application of heavy alkenes and alkynes in the field of material science (such as the large-scale use of alkenes for the multiton production of polymers) is expected for development in the near future. The use of heavy alkenes and alkynes as novel ligands for transition metal complexes is the other promising direction for their practical utilization.

5.5 References

1. (a) F. S. Kipping, J. E. Sands, *J. Chem. Soc. Trans.* **1921**, *119*, 830; (b) F. S. Kipping, J. E. Sands, *J. Chem. Soc. Trans.* **1921**, *119*, 848; (c) F. S. Kipping, *J. Chem. Soc.* **1927**, 2719.
2. D. N. Roark, G. J. D. Peddle, *J. Am. Chem. Soc.* **1972**, *94*, 5837.

3. Leading reviews on the transient disilenes: (a) L. E. Gusel'nikov, N. S. Nametkin, V. M. Vdovin, *Acc. Chem. Res.* **1975**, *8*, 18; (b) L. E. Gusel'nikov, N. S. Nametkin, *Chem. Rev.* **1979**, *79*, 529; (c) R. West, *Angew. Chem., Int. Ed. Engl.* **1987**, *26*, 1201; (d) G. Raabe, J. Michl, *Chem. Rev.* **1985**, *85*, 419; (e) G. Raabe, J. Michl, in *The Chemistry of Organic Silicon Compounds* (Eds S. Patai, Z. Rappoport), John Wiley & Sons, Ltd, Chichester, **1989**, Chapter 17; (f) T. Tsumuraya, S. A. Batcheller, S. Masamune, *Angew. Chem., Int. Ed. Engl.* **1991**, *30*, 902.
4. Leading reviews on the transient digermenes: (a) J. Satgé, *Adv. Organomet. Chem.* **1982**, *21*, 241; (b) J. Satgé, *Pure Appl. Chem.* **1984**, *56*, 137; (c) J. Satgé, *Rev. Silicon Germanium TinLead Compd.* **1985**, *8*, 291; (d) J. Barrau, J. Escudié, J. Satgé, *Chem. Rev.* **1990**, *90*, 283; (e) J. Satgé, *J. Organomet. Chem.* **1990**, *400*, 121.
5. The structure and bonding nature of the alkene analogs of the heavy group 14 elements is discussed in a vast number of computational and experimental studies, and we will be unable to cover all of them in the present chapter. Instead, we refer our readers to recent excellent reviews (and literature cited therein), in which this issue is comprehensively addressed: (a) P. P. Power, *Chem. Rev.* **1999**, *99*, 3463; (b) M. Karni, Y. Apeloig, J. Kapp, P. von R. Schleyer, in *The Chemistry of Organic Silicon Compounds, Vol. 3* (Eds Z. Rappoport, Y. Apeloig), John Wiley & Sons, Ltd, Chichester, **2001**, Chapter 1; (c) K. W. Klinkhammer, in *The Chemistry of Organic Germanium, Tin and Lead Compounds, Vol. 2, Part 1* (Ed. Z. Rappoport), John Wiley & Sons, Ltd, Chichester, **2002**, Chapter 4.
6. (a) D. E. Goldberg, D. H. Harris, M. F. Lappert, K. M. Thomas, *J. Chem. Soc., Chem. Commun.* **1976**, 261; (b) P. J. Davidson, D. H. Harris, M. F. Lappert, *J. Chem. Soc., Dalton Trans.* **1976**, 2268; (c) J. D. Cotton, P. J. Davidson, M. F. Lappert, J. D. Donaldson, J. Silver, *J. Chem. Soc., Dalton Trans.* **1976**, 2286.
7. (a) E. A. Carter, W. A. Goddard III, *J. Phys. Chem.* **1986**, *90*, 998; (b) G. Trinquier, J.-P. Malrieu, P. Rivière, *J. Am. Chem. Soc.* **1982**, *104*, 4529; (c) G. Trinquier, J.-P. Malrieu, *J. Am. Chem. Soc.* **1987**, *109*, 5303; (d) J.-P. Malrieu, G. Trinquier, *J. Am. Chem. Soc.* **1989**, *111*, 5916; (e) G. Trinquier, J.-P. Malrieu, in *Supplement A: The Chemistry of Double-Bonded Functional Groups, Vol. 2, Part 1* (Ed. S. Patai), John Wiley & Sons, Ltd, Chichester, **1989**, Chapter 1; (f) G. Trinquier, J.-P. Malrieu, *J. Phys. Chem.* **1990**, *94*, 6184; (g) G. Trinquier, *J. Am. Chem. Soc.* **1990**, *112*, 2130; (h) G. Trinquier, J.-C. Barthelat, *J. Am. Chem. Soc.* **1990**, *112*, 9121; (i) G. Trinquier, *J. Am. Chem. Soc.* **1991**, *113*, 144.
8. R. West, J. D. Cavalieri, J. J. Buffy, C. Fry, K. W. Zilm, J. C. Duchamp, M. Kira, T. Iwamoto, T. Müller, Y. Apeloig, *J. Am. Chem. Soc.* **1997**, *119*, 4972.
9. N. F. Ramsey, *Phys. Rev.* **1950**, *78*, 699.
10. M. Ichinohe, Y. Arai, A. Sekiguchi, N. Takagi, S. Nagase, *Organometallics* **2001**, *20*, 4141.
11. D. Auer, C. Strohmann, A. V. Arbuznikov, M. Kaupp, *Organometallics* **2003**, *22*, 2442.
12. Leading reviews on the stable disilenes: (a) R. Okazaki, R. West, *Adv. Organomet. Chem.* **1996**, *39*, 231; (b) M. Weidenbruch, in *The Chemistry of Organic Silicon Compounds, Vol. 3* (Eds Z. Rappoport, Y. Apeloig), John Wiley & Sons, Ltd, Chichester, **2001**, Chapter 5; (c) M. Kira, T. Iwamoto, *Adv. Organomet. Chem.* **2006**, *54*, 73.

13. (a) R. West, M. J. Fink, J. Michl, *Science* **1981**, *214*, 1343; (b) M. J. Fink, M. J. Michalczyk, K. J. Haller, R. West, J. Michl, *J. Chem. Soc., Chem. Commun.* **1983**, 1010; (c) R. West, *Pure Appl. Chem.* **1984**, *56*, 163; (d) M. J. Fink, M. J. Michalczyk, K. J. Haller, R. West, J. Michl, *Organometallics* **1984**, *3*, 793; (e) B. D. Shepherd, C. F. Campana, R. West, *Heteroatom Chem.* **1990**, *1*, 1; (f) M. Wind, D. R. Powell, R. West, *Organometallics* **1996**, *15*, 5772.
14. M. J. Michalczyk, R. West, J. Michl, *J. Am. Chem. Soc.* **1984**, *106*, 821.
15. B. D. Shepherd, D. R. Powell, R. West, *Organometallics* **1989**, *8*, 2664.
16. R. S. Archibald, Y. van den Winkel, A. J. Millevolte, J. M. Desper, R. West, *Organometallics* **1992**, *11*, 3276.
17. R. S. Archibald, Y. van den Winkel, D. R. Powell, R. West, *J. Organomet. Chem.* **1993**, *446*, 67.
18. M. J. Michalczyk, R. West, J. Michl, *Organometallics* **1985**, *4*, 826.
19. H. B. Yokelson, D. A. Siegel, A. J. Millevolte, J. Maxka, R. West, *Organometallics* **1990**, *9*, 1005.
20. (a) S. Masamune, Y. Hanzawa, S. Murakami, T. Bally, J. F. Blount, *J. Am. Chem. Soc.* **1982**, *104*, 1150; (b) S. Masamune, S. Murakami, J. T. Snow, H. Tobita, D. J. Williams, *Organometallics* **1984**, *3*, 333.
21. S. Masamune, Y. Eriyama, T. Kawase, *Angew. Chem., Int. Ed. Engl.* **1987**, *26*, 584.
22. R. Tanaka, T. Iwamoto, M. Kira, *Angew. Chem., Int. Ed.* **2006**, *45*, 6371.
23. P. Boudjouk, B.-H. Han, K. R. Anderson, *J. Am. Chem. Soc.* **1982**, *104*, 4992.
24. H. Watanabe, K. Takeuchi, N. Fukawa, M. Kato, M. Goto, Y. Nagai, *Chem. Lett.* **1987**, 1341.
25. S. A. Batcheller, T. Tsumuraya, O. Tempkin, W. M. Davis, S. Masamune, *J. Am. Chem. Soc.* **1990**, *112*, 9394.
26. (a) N. Tokitoh, H. Suzuki, R. Okazaki, K. Ogawa, *J. Am. Chem. Soc.* **1993**, *115*, 10428; (b) H. Suzuki, N. Tokitoh, R. Okazaki, J. Harada, K. Ogawa, S. Tomoda, M. Goto, *Organometallics* **1995**, *14*, 1016.
27. T. Sasamori, K. Hironaka, Y. Sugiyama, N. Takagi, S. Nagase, Y. Hosoi, Y. Furukawa, N. Tokitoh, *J. Am. Chem. Soc.* **2008**, *130*, 13856.
28. T. Sasamori, A. Yuasa, Y. Hosoi, Y. Furukawa, N. Tokitoh, *Organometallics* **2008**, *27*, 3325.
29. M. Kira, T. Maruyama, C. Kabuto, K. Ebata, H. Sakurai, *Angew. Chem., Int. Ed. Engl.* **1994**, *33*, 1489.
30. M. Kira, S. Ohya, T. Iwamoto, M. Ichinohe, C. Kabuto, *Organometallics* **2000**, *19*, 1817.
31. N. Wiberg, W. Niedermayer, G. Fischer, H. Nöth, M. Suter, *Eur. J. Inorg. Chem.* **2002**, 1066.
32. M. Ichinohe, R. Kinjo, A. Sekiguchi, *Organometallics* **2003**, *22*, 4621.
33. A. Sekiguchi, S. Inoue, M. Ichinohe, Y. Arai, *J. Am. Chem. Soc.* **2004**, *126*, 9626.
34. H. Kobayashi, T. Iwamoto, M. Kira, *J. Am. Chem. Soc.* **2005**, *127*, 15376.
35. S. Tsutsui, K. Sakamoto, M. Kira, *J. Am. Chem. Soc.* **1998**, *120*, 9955.
36. T. A. Schmedake, M. Haaf, Y. Apeloig, T. Müller, S. Bukalov, R. West, *J. Am. Chem. Soc.* **1999**, *121*, 9479.
37. M. Weidenbruch, A. Schäfer, K. L. Thom, *Z. Naturforsch.* **1983**, 38b, 1695.

38. M. Weidenbruch, A. Pellmann, Y. Pan, S. Pohl, W. Saak, H. Marsmann, *J. Organomet. Chem.* **1993**, *450*, 67.
39. (a) N. Wiberg, W. Niedermayer, K. Polborn, *Z. Anorg. Allg. Chem.* **2002**, *628*, 1045; (b) N. Wiberg, W. Niedermayer, K. Polborn, P. Mayer, *Chem. Eur. J.* **2002**, *8*, 2730.
40. T. Iwamoto, J. Okita, C. Kabuto, M. Kira, *J. Organomet. Chem.* **2003**, *686*, 105.
41. M. Zirngast, M. Flock, J. Baumgartner, C. Marschner, *J. Am. Chem. Soc.* **2008**, *130*, 17460.
42. S. Boomgaarden, W. Saak, H. Marsmann, M. Weidenbruch, *Z. Anorg. Allg. Chem.* **2002**, *628*, 1745.
43. (a) P. Jutzi, A. Mix, B. Rummel, W. W. Schoeller, B. Neumann, H.-G. Stammler, *Science* **2004**, *305*, 849; (b) P. Jutzi, A. Mix, B. Neumann, B. Rummel, W. W. Shoeller, H.-G. Stammler, A. B. Rozhenko, *J. Am. Chem. Soc.* **2009**, *131*, 12137.
44. Y. Wang, Y. Xie, P. Wei, R. B. King, H. F. Schaefer III, P. von R. Schleyer, G. H. Robinson, *Science* **2008**, *321*, 1069.
45. S. Inoue, M. Ichinohe, A. Sekiguchi, *Chem. Lett.* **2008**, *37*, 1044. Recently, an interesting zirconium complex $Cp_2(Cl)Zr\{\eta^1\text{-}[Si(Tip){=}SiTip_2]\}$, featuring monohaptocoordinated disilenide ligand, was also prepared by the reaction of Cp_2ZrCl_2 with the lithium disilenide $LiSi(Tip){=}SiTip_2$: T. Nguyen, D. Scheschkewitz, *J. Am. Chem. Soc.* **2005**, *127*, 10174.
46. (a) S. Matsumoto, S. Tsutsui, E. Kwon, K. Sakamoto, *Angew. Chem., Int. Ed.* **2004**, *43*, 4610; (b) S. Tsutsui, E. Kwon, H. Tanaka, S. Matsumoto, K. Sakamoto, *Organometallics* **2005**, *24*, 4629.
47. Reviews on the cyclopropene analogs of the heavy group 14 elements: (a) V. Ya. Lee, A. Sekiguchi, in *The Chemistry of Organic Germanium, Tin and Lead Compounds, Vol. 2, Part 1* (Ed. Z. Rappoport), John Wiley & Sons, Ltd, Chichester, **2002**, Chapter 14; (b) A. Sekiguchi, V. Ya. Lee, in *Organosilicon Chemistry V: From Molecules to Materials* (Eds N. Auner, J. Weis), Wiley-VCH, Weinheim, **2003**, p. 92; (c) A. Sekiguchi, V. Ya. Lee, *Chem. Rev.* **2003**, *103*, 1429.
48. π–σ^* interaction (σ^*-aromaticity) concept: (a) A. Göller, H. Heydt, T. Clark, *J. Org. Chem.* **1996**, *61*, 5840; (b) Y. Naruse, J. Ma, S. Inagaki, *Tetrahedron Lett.* **2001**, *42*, 6553.
49. (a) T. Iwamoto, C. Kabuto, M. Kira, *J. Am. Chem. Soc.* **1999**, *121*, 886; (b) T. Iwamoto, M. Tamura, C. Kabuto, M. Kira, *Organometallics* **2003**, *22*, 2342.
50. M. Ichinohe, T. Matsuno, A. Sekiguchi, *Angew. Chem., Int. Ed.* **1999**, *38*, 2194.
51. (a) V. Ya. Lee, M. Ichinohe, A. Sekiguchi, N. Takagi, S. Nagase, *J. Am. Chem. Soc.* **2000**, *122*, 9034; (b) V. Ya. Lee, M. Ichinohe, A. Sekiguchi, *J. Organomet. Chem.* **2003**, *685*, 168.
52. V. Ya. Lee, H. Yasuda, A. Sekiguchi, *J. Am. Chem. Soc.* **2007**, *129*, 2436.
53. M. Ichinohe, M. Igarashi, K. Sanuki, A. Sekiguchi, *J. Am. Chem. Soc.* **2005**, *127*, 9978.
54. K. Uchiyama, S. Nagendran, S. Ishida, T. Iwamoto, M. Kira, *J. Am. Chem. Soc.* **2007**, *129*, 10638.
55. M. Kira, T. Iwamoto, C. Kabuto, *J. Am. Chem. Soc.* **1996**, *118*, 10303.
56. N. Wiberg, H. Auer, H. Nöth, J. Knizek, K. Polborn, *Angew. Chem., Int. Ed.* **1998**, *37*, 2869.

57. N. Wiberg, W. Niedermayer, H. Nöth, M. Warchhold, *Z. Anorg. Allg. Chem.* **2001**, *627*, 1717.
58. N. Wiberg, S. K. Vasisht, G. Fischer, P. Mayer, *Z. Anorg. Allg. Chem.* **2004**, *630*, 1823.
59. R. Kinjo, M. Ichinohe, A. Sekiguchi, N. Takagi, M. Sumimoto, S. Nagase, *J. Am. Chem. Soc.* **2007**, *129*, 7766.
60. K. Takanashi, V. Ya. Lee, M. Ichinohe, A. Sekiguchi, *Angew. Chem., Int. Ed.* **2006**, *45*, 3269.
61. V. Ya. Lee, K. Takanashi, T. Matsuno, M. Ichinohe, A. Sekiguchi, *J. Am. Chem. Soc.* **2004**, *126*, 4758.
62. A. Sekiguchi, T. Matsuno, M. Ichinohe, *J. Am. Chem. Soc.* **2001**, *123*, 12436.
63. V. Ya. Lee, S. Miyazaki, H. Yasuda, A. Sekiguchi, *J. Am. Chem. Soc.* **2008**, *130*, 2758.
64. A. Grybat, S. Boomgaarden, W. Saak, H. Marsmann, M. Weidenbruch, *Angew. Chem., Int. Ed.* **1999**, *38*, 2010.
65. (a) E. K. Pham, R. West, *J. Am. Chem. Soc.* **1989**, *111*, 7667; (b) E. K. Pham, R. West, *Organometallics* **1990**, *9*, 1517.
66. (a) D. H. Berry, J. H. Chey, H. S. Zipin, P. J. Carroll, *J. Am. Chem. Soc.* **1990**, *112*, 452; (b) D. H. Berry, J. H. Chey, H. S. Zipin, P. J. Carroll, *Polyhedron* **1991**, *10*, 1189; (c) P. Hong, N. H. Damrauer, P. J. Carroll, D. H. Berry, *Organometallics* **1993**, *12*, 3698.
67. (a) E. A. Zarate, C. A. Tessier-Youngs, W. J. Youngs, *J. Am. Chem. Soc.* **1988**, *110*, 4068; (b) E. A. Zarate, C. A. Tessier-Youngs, W. J. Youngs, *J. Chem. Soc., Chem. Commun.* **1989**, 577.
68. (a) H. Hashimoto, Y. Sekiguchi, T. Iwamoto, C. Kabuto, M. Kira, *Organometallics* **2002**, *21*, 454; (b) H. Hashimoto, Y. Sekiguchi, Y. Sekiguchi, T. Iwamoto, C. Kabuto, M. Kira, *Can. J. Chem.* **2003**, *81*, 1241; (c) M. Kira, Y. Sekiguchi, T. Iwamoto, C. Kabuto, *J. Am. Chem. Soc.* **2004**, *126*, 12778; (d) H. Hashimoto, K. Suzuki, W. Setaka, C. Kabuto, M. Kira, *J. Am. Chem. Soc.* **2004**, *126*, 13628; (e) T. Iwamoto, Y. Sekiguchi, N. Yoshida, C. Kabuto, M. Kira, *Dalton Trans.* **2006**, 177.
69. R. Fischer, M. Zirngast, M. Flock, J. Baumgartner, C. Marschner, *J. Am. Chem. Soc.* **2005**, *127*, 70.
70. (a) M. J. S. Dewar, *Bull. Soc. Chim. Fr.* **1951**, C71; (b) J. Chatt, L. A. Duncanson, *J. Chem. Soc.* **1953**, 2939.
71. Leading reviews on the stable digermenes: (a) J. Escudié, C. Couret, H. Ranaivonjatovo, J. Satgé, *Coord. Chem. Rev.* **1994**, *130*, 427; (b) K. M. Baines, W. G. Stibbs, *Adv. Organomet. Chem.* **1996**, *39*, 275; (c) J. Escudié, H. Ranaivonjatovo, *Adv. Organomet. Chem.* **1999**, *44*, 113; (d) N. Tokitoh, R. Okazaki, in *The Chemistry of Organic Germanium, Tin and Lead Compounds, Vol. 2, Part 1* (Ed. Z. Rappoport), John Wiley & Sons, Ltd, Chichester, **2002**, Chapter 13.
72. (a) P. B. Hitchcock, M. F. Lappert, S. J. Miles, A. J. Thorne, *J. Chem. Soc., Chem. Commun.* **1984**, 480; (b) D. E. Goldberg, P. B. Hitchcock, M. F. Lappert, K. M. Thomas, A. J. Thorne, T. Fjeldberg, A. Haaland, B. E. R. Schilling, *J. Chem. Soc., Dalton Trans.* **1986**, 2387.

73. R. S. Simons, L. Pu, M. M. Olmstead, P. P. Power, *Organometallics* **1997**, *16*, 1920.
74. L. Pu, A. D. Phillips, A. F. Richards, M. Stender, R. S. Simons, M. M. Olmstead, P. P. Power, *J. Am. Chem. Soc.* **2003**, *125*, 11626.
75. M. Stender, L. Pu, P. P. Power, *Organometallics* **2001**, *20*, 1820.
76. M. Weidenbruch, M. Stürmann, H. Kilian, S. Pohl, W. Saak, *Chem. Ber.* **1997**, *130*, 735.
77. A. Schäfer, W. Saak, M. Weidenbruch, H. Marsmann, G. Henkel, *Chem. Ber.* **1997**, *130*, 1733.
78. A. Schäfer, W. Saak, M. Weidenbruch, *Z. Anorg. Allg. Chem.* **1998**, *624*, 1405.
79. H. Schäfer, W. Saak, M. Weidenbruch, *Organometallics* **1999**, *18*, 3159.
80. G. Ramaker, W. Saak, D. Haase, M. Weidenbruch, *Organometallics* **2003**, *22*, 5212.
81. B. Pampuch, W. Saak, M. Weidenbruch, *J. Organomet. Chem.* **2006**, *691*, 3540.
82. T. Sasamori, Y. Sugiyama, N. Takeda, N. Tokitoh, *Organometallics* **2005**, *24*, 3309.
83. S. Masamune, Y. Hanzawa, D. J. Williams, *J. Am. Chem. Soc.* **1982**, *104*, 6136.
84. J. T. Snow, S. Murakami, S. Masamune, D. J. Williams, *Tetrahedron Lett.* **1984**, *25*, 4191.
85. J. Park, S. A. Batcheller, S. Masamune, *J. Organomet. Chem.* **1989**, *367*, 39.
86. S. A. Batcheller, T. Tsumuraya, O. Tempkin, W. M. Davis, S. Masamune, *J. Am. Chem. Soc.* **1990**, *112*, 9394.
87. M. Kira, T. Iwamoto, T. Maruyama, C. Kabuto, H. Sakurai, *Organometallics* **1996**, *15*, 3767.
88. (a) K. Kishikawa, N. Tokitoh, R. Okazaki, *Chem. Lett.* **1998**, 239; (b) N. Tokitoh, K. Kishikawa, R. Okazaki, T. Sasamori, N. Nakata, N. Takeda, *Polyhedron* **2002**, *21*, 563.
89. (a) G. H. Spikes, J. C. Fettinger, P. P. Power, *J. Am. Chem. Soc.* **2005**, *127*, 12232; (b) A. F. Richards, A. D. Phillips, M. M. Olmstead, P. P. Power, *J. Am. Chem. Soc.* **2003**, *125*, 3204.
90. C. Cui, M. M. Olmstead, P. P. Power, *J. Am. Chem. Soc.* **2004**, *126*, 5062.
91. C. Cui, M. M. Olmstead, J. C. Fettinger, G. H. Spikes, P. P. Power, *J. Am. Chem. Soc.* **2005**, *127*, 17530.
92. A. Sekiguchi, H. Yamazaki, C. Kabuto, H. Sakurai, *J. Am. Chem. Soc.* **1995**, *117*, 8025.
93. A. Sekiguchi, N. Fukaya, M. Ichinohe, N. Takagi, S. Nagase, *J. Am. Chem. Soc.* **1999**, *121*, 11587.
94. A. Sekiguchi, Y. Ishida, N. Fukaya, M. Ichinohe, N. Takagi, S. Nagase, *J. Am. Chem. Soc.* **2002**, *124*, 1158.
95. (a) V. Ya. Lee, H. Yasuda, M. Ichinohe, A. Sekiguchi, *Angew. Chem., Int. Ed.* **2005**, *44*, 6378; (b) V. Ya. Lee, H. Yasuda, M. Ichinohe, A. Sekiguchi, *J. Organomet. Chem.* **2007**, *692*, 10.
96. V. Ya. Lee, H. Yasuda, A. Sekiguchi, *J. Am. Chem. Soc.* **2007**, *129*, 2436.
97. V. Ya. Lee, K. Takanashi, M. Ichinohe, A. Sekiguchi, *J. Am. Chem. Soc.* **2003**, *125*, 6012.

98. G. Ramaker, A. Schäfer, W. Saak, M. Weidenbruch, *Organometallics* **2003**, *22*, 1302.
99. (a) J. D. Cotton, P. J. Davidson, M. F. Lappert, J. D. Donaldson, J. Silver, *J. Chem. Soc., Dalton Trans.* **1976**, 2286; (b) T. Fjeldberg, A. Haaland, M. F. Lappert, B. E. R. Schilling, R. Seip, A. J. Thorne, *J. Chem. Soc., Chem. Commun.* **1982**, 1407.
100. C. Stanciu, A. F. Richards, P. P. Power, *J. Am. Chem. Soc.* **2004**, *126*, 4106.
101. (a) M. Weidenbruch, H. Kilian, K. Peters, H. G. von Schnering, H. Marsmann, *Chem. Ber.* **1995**, *128*, 983; (b) M. A. Della Bona, M. C. Cassani, J. M. Keates, G. A. Lawless, M. F. Lappert, M. Stürmann, M. Weidenbruch, *J. Chem. Soc., Dalton Trans.* **1998**, 1187.
102. (a) K. W. Klinkhammer, W. Schwarz, *Angew. Chem., Int. Ed. Engl.* **1995**, *34*, 1334; (b) K. Klinkhammer, *Polyhedron* **2002**, *21*, 587; (c) J. Klett, K. W. Klinkhammer, M. Niemeyer, *Chem. Eur. J.* **1999**, *5*, 2531.
103. (a) T. Fukawa, V. Ya. Lee, M. Nakamoto, A. Sekiguchi, *J. Am. Chem. Soc.* **2004**, *126*, 11758; (b) V. Ya. Lee, T. Fukawa, M. Nakamoto, A. Sekiguchi, B. L. Tumanskii, M. Karni, Y. Apeloig, *J. Am. Chem. Soc.* **2006**, *128*, 11643.
104. S. Masamune, L. R. Sita, *J. Am. Chem. Soc.* **1985**, *107*, 6390.
105. (a) A. Schäfer, M. Weidenbruch, W. Saak, S. Pohl, H. Marsmann, *Angew. Chem., Int. Ed. Engl.* **1991**, *30*, 834; (b) A. Schäfer, M. Weidenbruch, W. Saak, S. Pohl, H. Marsmann, *Angew. Chem., Int. Ed. Engl.* **1991**, *30*, 962; (c) M. Weidenbruch, A. Schäfer, H. Kilian, S. Pohl, W. Saak, H. Marsmann, *Chem. Ber.* **1992**, *125*, 563.
106. K. W. Klinkhammer, T. F. Fässler, H. Grützmacher, *Angew. Chem., Int. Ed.* **1998**, *37*, 124.
107. M. Stürmann, W. Saak, K. W. Klinkhammer, M. Weidenbruch, *Z. Anorg. Allg. Chem.* **1999**, *625*, 1955.
108. N. Wiberg, H.-W. Lerner, S.-K. Vasisht, S. Wagner, K. Karaghiosoff, H. Nöth, W. Ponikwar, *Eur. J. Inorg. Chem.* **1999**, 1211.
109. U. Lay, H. Pritzkow, H. Grützmacher, *J. Chem. Soc., Chem. Commun.* **1992**, 260.
110. M. Weidenbruch, A. Stilter, H. Marsmann, K. Peters, H. G. von Schnering, *Eur. J. Inorg. Chem.* **1998**, 1333.
111. C. Drost, P. B. Hitchcock, M. F. Lappert, *Angew. Chem., Int. Ed.* **1999**, *38*, 1113.
112. W.-P. Leung, W.-H. Kwok, F. Xue, T. C. W. Mak, *J. Am. Chem. Soc.* **1997**, *119*, 1145.
113. W.-P. Leung, H. Cheng, R.-B. Huang, Q.-C. Yang, T. C. W. Mak, *Chem. Commun.* **2000**, 451.
114. B. E. Eichler, P. P. Power, *Inorg. Chem.* **2000**, *39*, 5444.
115. A. D. Phillips, S. Hino, P. P. Power, *J. Am. Chem. Soc.* **2003**, *125*, 7520.
116. B. E. Eichler, P. P. Power, *J. Am. Chem. Soc.* **2000**, *122*, 8785.
117. F. Breher, H. Rüegger, *Angew. Chem., Int. Ed.* **2005**, *44*, 473.
118. M. Stürmann, W. Saak, H. Marsmann, M. Weidenbruch, *Angew. Chem., Int. Ed.* **1999**, *38*, 187.
119. M. Stürmann, W. Saak, M. Weidenbruch, K. W. Klinkhammer, *Eur. J. Inorg. Chem.* **1999**, 579.
120. S. Hino, M. Olmstead, A. D. Phillips, R. J. Wright, P. P. Power, *Inorg. Chem.* **2004**, *43*, 7346.

121. M. Stürmann, M. Weidenbruch, K. W. Klinkhammer, F. Lissner, H. Marsmann, *Organometallics* **1998**, *17*, 4425.
122. M. Stürmann, W. Saak, M. Weidenbruch, *Z. Anorg. Allg. Chem.* **1999**, *625*, 705.
123. S. Hino, M. M. Olmstead, P. P. Power, *Organometallics* **2005**, *24*, 5484.
124. Reviews on silenes: (a) H. F. Schaefer III, *Acc. Chem. Res.* **1982**, *15*, 283; (b) A. G. Brook, K. M. Baines, *Adv. Organomet. Chem.* **1986**, *25*, 1; (c) R. S. Grev, *Adv. Organomet. Chem.* **1991**, *33*, 125; (d) A. G. Brook, M. A. Brook, *Adv. Organomet. Chem.* **1996**, *39*, 71; (e) T. Müller, W. Ziche, N. Auner, in *The Chemistry of Organic Silicon Compounds, Vol. 2* (Eds Z. Rappoport, Y. Apeloig), John Wiley & Sons, Ltd, Chichester, **1998**, Chapter 16; (f) J. Escudié, C. Couret, H. Ranaivonjatovo, *Coord. Chem. Rev.* **1998**, 178–180, 565; (g) T. L. Morkin, W. J. Leigh, *Acc. Chem. Res.* **2001**, *34*, 129; (h) L. E. Gusel'nikov, *Coord. Chem. Rev.* **2003**, *244*, 149; (i) V. Ya. Lee, A. Sekiguchi, *Organometallics* **2004**, *23*, 2822; (j) H. Ottosson, A. M. Eklöf, *Coord. Chem. Rev.* **2008**, *252*, 1287.
125. L. E. Gusel'nikov, M. C. Flowers, *J. Chem. Soc., Chem. Commun.* **1967**, 864.
126. (a) O. L. Chapman, C. C. Chang, J. Kole, M. E. Jung, J. A. Lowe, T. J. Barton, M. L. Tumey, *J. Am. Chem. Soc.* **1976**, *98*, 7844; (b) M. R. Chedekel, M. Skoglund, R. L. Kreeger, H. Shechter, *J. Am. Chem. Soc.* **1976**, *98*, 7846; (c) A. K. Mal'tsev, V. N. Khabashesku, O. M. Nefedov, *Izv. Akad. Nauk SSSR, Ser. Khim.* **1976**, 1193.
127. (a) A. G. Brook, F. Abdesaken, B. Gutekunst, G. Gutekunst, R. K. Kallury, *J. Chem. Soc., Chem. Commun.* **1981**, 191; (b) A. G. Brook, S. C. Nyberg, F. Abdesaken, B. Gutekunst, G. Gutekunst, R. K. M. R. Kallury, Y. C. Poon, Y.-M. Chang, W. Wong-Ng, *J. Am. Chem. Soc.* **1982**, *104*, 5667; (c) S. C. Nyberg, A. G. Brook, F. Abdesaken, G. Gutekunst, W. Wong-Ng, *Acta Cryst.* **1985**, *C41*, 1632; (d) K. M. Baines, A. G. Brook, R. R. Ford, P. D. Lickiss, A. K. Saxena, W. J. Chatterton, J. F. Sawyer, B. A. Behman, *Organometallics* **1989**, *8*, 693; (e) A. G. Brook, A. Baumegger, A. J. Lough, *Organometallics* **1992**, *11*, 3088.
128. Y. Apeloig, M. Karni, *J. Am. Chem. Soc.* **1984**, *106*, 6676.
129. (a) N. Wiberg, G. Wagner, *Angew. Chem., Int. Ed. Engl.* **1983**, *22*, 1005; (b) N. Wiberg, G. Wagner, G. Müller, *Angew. Chem., Int. Ed. Engl.* **1985**, *24*, 229; (c) N. Wiberg, G. Wagner, J. Riede, G. Müller, *Organometallics* **1987**, *6*, 32; (d) N. Wiberg, G. Wagner, G. Müller, J. Riede, *J. Organomet. Chem.* **1984**, *271*, 381; (e) N. Wiberg, G. Wagner, G. Reber, J. Riede, G. Müller, *Organometallics* **1987**, *6*, 35; (f) N. Wiberg, K.-S. Joo, K. Polborn, *Chem. Ber.* **1993**, *126*, 67.
130. Y. Apeloig, M. Bendikov, M. Yuzefovich, M. Nakash, D. Bravo-Zhivotovskii, D. Bläser, R. Boese, *J. Am. Chem. Soc.* **1996**, *118*, 12228.
131. D. Bravo-Zhivotovskii, R. Dobroversky, D. Nemirovsky, V. Molev, M. Bendikov, G. Molev, M. Botoshansky, Y. Apeloig, *Angew. Chem., Int. Ed.* **2008**, *47*, 4343.
132. K. Sakamoto, J. Ogasawara, Y. Kon, T. Sunagawa, C. Kabuto, M. Kira, *Angew. Chem., Int. Ed.* **2002**, *41*, 1402.
133. (a) S. Inoue, M. Ichinohe, A. Sekiguchi, *Angew. Chem. Int. Ed.* **2007**, *46*, 3346; (b) In contrast to **159**, structurally similar silene $[R_2Si{\cdots}Si(OR){\cdots}C(H)CAr_3]^{-}\bullet[Li^{+}(thf)]$ ($R = SiMe\textit{t}\text{-}Bu_2$, $Ar = 3,5\text{-}\textit{t}\text{-}Bu_2\text{-}C_6H_3$) revealed an η^3-coordination of the Li cation, which, being direcly bound to the sp^3-*Si*R_2, interacts also with both the sp^2-*Si*(OR) and sp^2-C(H)Ar_3 atoms of the Si=C fragment. [K. Tanaka, S. Inoue, M. Ichinohe, A. Sekiguchi,

Organometallics **2009**, *28*, 6625]. This η^3-disilaallyl coordination mode is retained in toluene solution, as was evidenced by NMR spectroscopy. However, given the peculiar structural (unremarkable stretching of the Si=C and shortening of the Si–Si bonds, nearly planar geometries of the sp^2-Si/sp^2-C and notable pyramidalization at the sp^3-Si atoms) and spectral (high-field signal of the sp^3-Si and low-field signal of the sp^2-Si atoms) features, one should recognize the rather small extent of η^3-π-allyl delocalization with the predominant contribution from the $R_2Si(Li)$–Si(OR)=C(H)CAr_3 localized resonance form. Indeed, this silene-substituted η^1-silyllithium structure was exclusively observed in polar THF.

134. I. Bejan, D. Güclü, S. Inoue, M. Ichinohe, A. Sekiguchi, D. Scheschkewitz, *Angew. Chem,. Int. Ed.* **2007**, *46*, 3349.
135. I. Bejan, S. Inoue, M. Ichinohe, A. Sekiguchi, D. Scheschkewitz, *Chem. Eur. J.* **2008**, *14*, 7119.
136. M. Igarashi, M. Ichinohe, A. Sekiguchi, *J. Am. Chem. Soc.* **2007**, *129*, 12660.
137. (a) G. Delpon-Lacaze, C. Couret, *J. Organomet. Chem.* **1994**, *480*, C14; (b) G. Delpon-Lacaze, C. de Battisti, C. Couret, *J. Organomet. Chem.* **1996**, *514*, 59; (c) M. Kira, S. Ishida, T. Iwamoto, C. Kabuto, *J. Am. Chem. Soc.* **1999**, *121*, 9722; (d) D. Bravo-Zhivotovskii, G. Korogodsky, Y. Apeloig, *J. Organomet. Chem.* **2003**, *686*, 58; (e) I. S. Toulokhonova, I. A. Guzei, R. West, *J. Am. Chem. Soc.* **2004**, *126*, 5336; (f) S. Ozaki, T. Sasamori, N. Tokitoh, *Organometallics* **2008**, *27*, 2163.
138. Generation of 2-silenolates: (a) J. Ohshita, Y. Masaoka, S. Masaoka, M. Ishikawa, A. Tachibana, T. Yano, T. Yamabe, *J. Organomet. Chem.* **1994**, *473*, 15; (b) J. Ohshita, S. Masaoka, Y. Masaoka, H. Hasebe, M. Ishikawa, A. Tachibana, T. Yano, T. Yamabe, *Organometallics* **1996**, *15*, 3136; (c) J. Ohshita, S. Masaoka, M. Ishikawa, *Organometallics* **1996**, *15*, 2198; (d) J. Ohshita, S. Masaoka, Y. Morimoto, M. Ishikawa, *Organometallics* **1997**, *16*, 910; (e) J. Ohshita, S. Masaoka, Y. Morimoto, M. Sano, M. Ishikawa, *Organometallics* **1997**, *16*, 1123; (f) J. Ohshita, H. Sakurai, Y. Tokunaga, A. Kunai, *Organometallics* **1999**, *18*, 4545; (g) J. Ohshita, Y. Tokunaga, H. Sakurai, A. Kunai, *J. Am. Chem. Soc.* **1999**, *121*, 6080; (h) J. Ohshita, M. Tamai, H. Sakurai, A. Kunai, *Organometallics* **2001**, *20*, 1065; (i) J. Ohshita, H. Sakurai, S. Masaoka, M. Tamai, A. Kunai, M. Ishikawa, *J. Organomet. Chem.* **2001**, *633*, 131.
139. (a) T. Guliashvili, I. El-Sayed, A. Fischer, H. Ottosson, *Angew. Chem., Int. Ed.* **2003**, *42*, 1640; (b) I. El-Sayed, T. Guliashvili, R. Hazell, A. Gogoll, H. Ottosson, *Org. Lett.* **2002**, *4*, 1915; (c) H. Ottosson, *Chem. Eur. J.* **2003**, *9*, 4144; (d) H. Ottosson, P. G. Steel, *Chem. Eur. J.* **2006**, *12*, 1577; (e) A. M. Eklöf, T. Guliashvili, H. Ottosson, *Organometallics* **2008**, *27*, 5203.
140. (a) M. Mikoleit, K. Schmohl, R. Kempe, H. Oehme, *Angew. Chem., Int. Ed.* **2000**, *39*, 1610; (b) M. Mikoleit, R. Kempe, H. Oehme, *Chem. Eur. J.* **2001**, *7*, 987; (c) K. Schmohl, H. Reinke, H. Oehme, *Eur. J. Inorg. Chem.* **2001**, 481; (d) M. Pötter, U. Bäumer, M. Mickoleit, R. Kempe, H. Oehme, *J. Organomet. Chem.* **2001**, *621*, 261; (e) M. Mickoleit, K. Schmohl, M. Michalik, H. Oehme, *Eur. J. Inorg. Chem.* **2004**, 1538; (f) U. Bäumer, H. Reinke, H. Oehme, *J. Organomet. Chem.* **2006**, *691*, 229.

141. (a) B. K. Campion, R. H. Heyn, T. D. Tilley, *J. Am. Chem. Soc.* **1988**, *110*, 7558; (b) B. K. Campion, R. H. Heyn, T. D. Tilley, *J. Am. Chem. Soc.* **1990**, *112*, 4079; (c) T. S. Koloski, P. J. Carroll, D. H. Berry, *J. Am. Chem. Soc.* **1990**, *112*, 6405; (d) B. K. Campion, R. H. Heyn, T. D. Tilley, A. L. Rheingold, *J. Am. Chem. Soc.* **1993**, *115*, 5527; (e) D. Bravo-Zhivotovskii, H. Peleg-Vasserman, M. Kosa, G. Molev, M. Botoshanskii, Y. Apeloig, *Angew. Chem., Int. Ed.* **2004**, *43*, 745.
142. (a) N. Wiberg, C.-K. Kim, *Chem. Ber.* **1986**, *119*, 2966; (b) D. Bravo-Zhivotovskii, I. Zharov, M. Kapon, Y. Apeloig, *J. Chem. Soc., Chem. Commun.* **1995**, 1625.
143. H. Meyer, G. Baum, W. Massa, A. Berndt, *Angew. Chem., Int. Ed. Engl.* **1987**, *26*, 798.
144. (a) C. Couret, J. Escudié, J. Satgé, M. Lazraq, *J. Am. Chem. Soc.* **1987**, *109*, 4411; (b) M. Lazraq, J. Escudié, C. Couret, J. Satgé, M. Dräger, R. Dammel, *Angew. Chem., Int. Ed. Engl.* **1988**, *27*, 828; (c) M. Lazraq, C. Couret, J. Escudié, J. Satgé, M. Soufiaoui, *Polyhedron* **1991**, *10*, 1153; (d) G. Anselme, J. Escudié, C. Couret, J. Satgé, *J. Organomet. Chem.* **1991**, *403*, 93.
145. M.-A. Chaubon, J. Escudié, H. Ranaivonjatovo, J. Satgé, *J. Chem. Soc., Dalton Trans.* **1996**, 893.
146. C. Couret, J. Escudié, G. Delpon-Lacaze, J. Satgé, *Organometallics* **1992**, *11*, 3176.
147. M. Weidenbruch, M. Stürmann, H. Kilian, S. Pohl, W. Saak, *Chem. Ber.* **1997**, *130*, 735.
148. (a) M. Stürmann, W. Saak, M. Weidenbruch, A. Berndt, D. Scheschkewitz, *Heteroatom Chem.* **1999**, *10*, 554; (b) F. Meiners, W. Saak, M. Weidenbruch, *Chem. Commun.* **2001**, 215; (c) B. Pampuch, W. Saak, M. Weidenbruch, *J. Organomet. Chem.* **2006**, *691*, 3540; (d) F. Meiners, W. Saak, M. Weidenbruch, *Organometallics* **2000**, *19*, 2835; (e) F. Meiners, D. Haase, R. Koch, W. Saak, M. Weidenbruch, *Organometallics* **2002**, *21*, 3990.
149. N. Tokitoh, K. Kishikawa, R. Okazaki, *J. Chem. Soc., Chem. Commun.* **1995**, 1425.
150. V. Ya. Lee, M. Ichinohe, A. Sekiguchi, *J. Am. Chem. Soc.* **2002**, *124*, 9962.
151. H. Schumann, M. Glanz, F. Girgsdies, F. E. Hahn, M. Tamm, A. Grzegorzewski, *Angew. Chem., Int. Ed. Engl.* **1997**, *36*, 2232.
152. A. J. Arduengo III, H. V. R. Dias, J. C. Calabrese, F. Davidson, *Inorg. Chem.* **1993**, *32*, 1541.
153. (a) W.-P. Leung, Z.-X. Wang, H.-W. Li, T. C. W. Mak, *Angew. Chem., Int. Ed.* **2001**, *40*, 2501; (b) W.-P. Leung, C.-W. So, Z.-X. Wang, J.-Z. Wang, T. C. W. Mak, *Organometallics* **2003**, *22*, 4305; (c) W.-P. Leung, C.-W. So, K.-W. Kan, H.-S. Chan, T. C. W. Mak, *Organometallics* **2005**, *24*, 5033; (d) W.-P. Leung, C.-W. So, K.-W. Kan, H.-S. Chan, T. C. W. Mak, *Inorg. Chem.* **2005**, *44*, 7286.
154. N. Wiberg, S.-K. Vasisht, *Angew. Chem., Int. Ed. Engl.* **1991**, *30*, 93.
155. (a) H. Meyer, G. Baum, W. Massa, S. Berger, A. Berndt, *Angew. Chem., Int. Ed. Engl.* **1987**, *26*, 546; (b) A. Berndt, H. Meyer, G. Baum, W. Massa, S. Berger, *Pure Appl. Chem.* **1987**, *59*, 1011.
156. (a) K. D. Dobbs, W. J. Hehre, *Organometallics* **1986**, *5*, 2057; (b) T. L. Windus, M. S. Gordon, *J. Am. Chem. Soc.* **1992**, *114*, 9559; (c) H. Jacobsen, T. Ziegler, *J. Am. Chem. Soc.* **1994**, *116*, 3667.

157. M. Weidenbruch, H. Kilian, M. Stürmann, S. Pohl, W. Saak, H. Marsmann, D. Steiner, A. Berndt, *J. Organomet. Chem.* **1997**, *530*, 255.
158. (a) G. Anselme, C. Couret, J. Escudié, S. Richelme, J. Satgé, *J. Organomet. Chem.* **1991**, *418*, 321; (b) G. Anselme, H. Ranaivonjatovo, J. Escudié, C. Couret, J. Satgé, *Organometallics* **1992**, *11*, 2748; (c) G. Anselme, J.-P. Declercq, A. Dubourg, H. Ranaivonjatovo, J. Escudié, C. Couret, *J. Organomet. Chem.* **1993**, *458*, 49; (d) A. Fatah, R. El Ayoubi, H. Gornitzka, H. Ranaivonjatovo, J. Escudié, *Eur. J. Inorg. Chem.* **2008**, 2007.
159. Y. Mizuhata, N. Takeda, T. Sasamori, N. Tokitoh, *Chem. Commun.* **2005**, 5876.
160. A. Schäfer, M. Weidenbruch, W. Saak, S. Pohl, *J. Chem. Soc., Chem. Commun.* **1995**, 1157.
161. N. Kuhn, T. Kratz, D. Bläser, R. Boese, *Chem. Ber.* **1995**, *128*, 245.
162. (a) K. M. Baines, J. A. Cooke, *Organometallics* **1991**, *10*, 3419; (b) K. M. Baines, J. A. Cooke, *Organometallics* **1992**, *11*, 3487.
163. R. S. Grev, H. F. Schaefer III, K. M. Baines, *J. Am. Chem. Soc.* **1990**, *112*, 9458.
164. (a) V. Ya. Lee, M. Ichinohe, A. Sekiguchi, N. Takagi, S. Nagase, *J. Am. Chem. Soc.* **2000**, *122*, 9034; (b) V. Ya. Lee, M. Ichinohe, A. Sekiguchi, *J. Organomet. Chem.* **2003**, *685*, 168. Review on Silagermenes Si=Ge (and other heteronuclear doubly bonded compounds E=E′, E=heavy group 14 element, E′=group 13-16 element): V. Ya. Lee, A. Sekiguchi, J. Escudie', H. Ranaivonjatovo, *chem. Lett*, **2010**, *39*, 312.
165. (a) V. Ya. Lee, M. Ichinohe, A. Sekiguchi, *J. Am. Chem. Soc.* **2000**, *122*, 12604; (b) V. Ya. Lee, M. Ichinohe, A. Sekiguchi, *J. Organomet. Chem.* **2001**, *636*, 41.
166. M. Ichinohe, Y. Arai, A. Sekiguchi, N. Takagi, S. Nagase, *Organometallics* **2001**, *20*, 4141.
167. M. Igarashi, M. Ichinohe, A. Sekiguchi, *Heteroatom Chem.* **2008**, *19*, 649.
168. C. Drost, B. Gehrhus, P. B. Hitchcock, M. F. Lappert, *Chem. Commun.* **1997**, 1845.
169. A. Sekiguchi, R. Izumi, V. Ya. Lee, M. Ichinohe, *J. Am. Chem. Soc.* **2002**, *124*, 14822.
170. M.-A. Chaubon, J. Escudié, H. Ranaivonjatovo, J. Satgé, *Chem. Commun.* **1996**, 2621.
171. A. Schäfer, W. Saak, M. Weidenbruch, *Organometallics* **2003**, *22*, 215.
172. A. Sekiguchi, R. Izumi, V. Ya. Lee, M. Ichinohe, *Organometallics* **2003**, *22*, 1483.
173. (a) V. Ya. Lee, K. Takanashi, M. Ichinohe, A. Sekiguchi, *J. Am. Chem. Soc.* **2003**, *125*, 6012; (b) V. Ya. Lee, K. Takanashi, M. Nakamoto, A. Sekiguchi, *Russ. Chem. Bull., Int. Ed.* **2004**, *53*, 1102.
174. P. von R. Schleyer, D. Kost, *J. Am. Chem. Soc.* **1988**, *110*, 2105.
175. (a) N. Nakata, R. Izumi, V. Ya. Lee, M. Ichinohe, A. Sekiguchi, *J. Am. Chem. Soc.* **2004**, *126*, 5058; (b) Theoretical treatment: J. Hsiao, M.-D. Su, *Organometallics* **2007**, *26*, 4432.
176. N. Nakata, R. Izumi, V. Ya. Lee, M. Ichinohe, A. Sekiguchi, *Chem. Lett.* **2008**, *37*, 1146.
177. N. Nakata, R. Izumi, V. Ya. Lee, M. Ichinohe, A. Sekiguchi, *Chem. Lett.* **2005**, *34*, 582.
178. N. Nakata, A. Sekiguchi, *J. Am. Chem. Soc.* **2006**, *128*, 422.

179. Reviews on silaimines: (a) S. Walter, U. Klingebiel, *Coord. Chem. Rev.* **1994**, *130*, 481; (b) I. Hemme, U. Klingebiel, *Adv. Organomet. Chem.* **1996**, *39*, 159; (c) P. Neugebauer, B. Jaschke, U. Klingebiel, in *The Chemistry of Organic Silicon Compounds, Vol. 3* (Eds Z. Rappoport, Y. Apeloig), John Wiley & Sons, Ltd, Chichester, **2001**, Chapter 6; (d) U. Klingebiel, C. Matthes, *J. Organomet. Chem.* **2007**, *692*, 2633.
180. M. Hesse, U. Klingebiel, *Angew. Chem., Int. Ed. Engl.* **1986**, *25*, 649.
181. (a) N. Wiberg, K. Schurz, G. Reber, G. Müller, *J. Chem. Soc., Chem. Commun.* **1986**, 591; (b) N. Wiberg, K. Schurz, G. Müller, J. Riede, *Angew. Chem., Int. Ed. Engl.* **1988**, *27*, 935; (c) G. Reber, J. Riede, N. Wiberg, K. Schurz, G. Müller, *Z. Naturforsch.* **1989**, 44b, 786.
182. P. von R. Schleyer, P. D. Stout, *J. Chem. Soc., Chem. Commun.* **1986**, 1373.
183. D. Stalke, N. Keweloh, U. Klingebiel, M. Noltemeyer, G. M. Sheldrick, *Z. Naturforsch.* **1987**, 42b, 1237.
184. J. Niesmann, U. Klingebiel, M. Schäfer, R. Boese, *Organometallics* **1998**, *17*, 947.
185. J. Niesmann, U. Klingebiel, M. Noltemeyer, R. Boese, *Chem. Commun.* **1997**, 365.
186. H.-W. Lerner, S. Scholz, M. Bolte, *Z. Anorg. Allg. Chem.* **2001**, *627*, 1638.
187. S. Walter, U. Klingebiel, D. Schmidt-Bäse, *J. Organomet. Chem.* **1991**, *412*, 319.
188. M. Denk, R. K. Hayashi, R. West, *J. Am. Chem. Soc.* **1994**, *116*, 10813.
189. (a) R. Boese, U. Klingebiel, *J. Organomet. Chem.* **1986**, *315*, C17; (b) D. Stalke, U. Pieper, S. Vollbrecht, U. Klingebiel, *Z. Naturforsch.* **1990**, 45b, 1513.
190. (a) H.-W. Lerner, N. Wiberg, J. W. Bats, *J. Organomet. Chem.* **2005**, *690*, 3898; (b) H.-W. Lerner, M. Bolte, K. Schurz, N. Wiberg, G. Baum, D. Fenske, J. W. Bats, M. Wagner, *Eur. J. Inorg. Chem.* **2006**, 4998.
191. Reviews on germaimines: (a) A. Kandri Rodi, H. Ranaivonjatovo, J. Escudié, A. Kerbal, *Main Group Metal Chem.* **1996**, *19*, 199; (b) J. Barrau, G. Rima, *Coord. Chem. Rev.* **1998**, 178–180, 593.
192. T. Tsumuraya, W. Ando, *Chem. Lett.* **1989**, 1043.
193. (a) C. Glidewell, D. Lloyd, K. W. Lumbard, *J. Chem. Soc., Dalton Trans.* **1987**, 501; (b) C. Glidewell, D. Lloyd, K. W. Lumbard, J. S. McKechnie, *Tetrahedron Lett.* **1987**, *28*, 343.
194. (a) G. Ossig, A. Meller, S. Freitag, O. Müller, H. Gornitzka, R. Herbst-Irmer, *Organometallics* **1996**, *15*, 408; (b) G. Ossig, A. Meller, O. Müller, R. Herbst-Irmer, *Organometallics* **1996**, *15*, 5060.
195. J. Pfeiffer, W. Maringgele, M. Noltemeyer, A. Meller, *Chem. Ber.* **1989**, *122*, 245.
196. M. Rivière-Baudet, J. Satgé, A. Morère, *J. Organomet. Chem.* **1990**, *386*, C7.
197. H. G. Ang, F. K. Lee, *J. Chem. Soc., Chem. Commun.* **1989**, 310.
198. A. Meller, G. Ossig, W. Maringgele, D. Stalke, R. Herbst-Irmer, S. Freitag, G. M. Sheldrick, *J. Chem. Soc., Chem. Commun.* **1991**, 1123.
199. G. Trinquier, J.-C. Barthelat, J. Satgé, *J. Am. Chem. Soc.* **1982**, *104*, 5931.
200. (a) W. Ando, T. Ohtaki, Y. Kabe, *Organometallics* **1994**, *13*, 434; (b) T. Ohtaki, Y. Kabe, W. Ando, *Heteroatom Chem.* **1994**, *5*, 313.
201. (a) M. Veith, S. Becker, V. Huch, *Angew. Chem., Int. Ed. Engl.* **1990**, *29*, 216; (b) M. Rivière-Baudet, A. Morère, *J. Organomet. Chem.* **1992**, *431*, 17; (c) M. Rivière-Baudet, A. Khallaayoun, J. Satgé, *Organometallics* **1993**, *12*, 1003;

(d) M. Rivière-Baudet, A. Khallaayoun, J. Satgé, *J. Organomet. Chem.* **1993**, *462*, 89; (e) M. Rivière-Baudet, J. Satgé, F. El Baz, *J. Chem. Soc., Chem. Commun.* **1995**, 1687; (f) M. Veith, A. Rammo, *Z. Anorg. Allg. Chem.* **1997**, *623*, 861.

202. (a) G. Ossig, A. Meller, S. Freitag, R. Herbst-Irmer, *J. Chem. Soc., Chem. Commun.* **1993**, 497; (b) G. Ossig, A. Meller, S. Freitag, R. Herbst-Irmer, G. M. Sheldrick, *Chem. Ber.* **1993**, *126*, 2247.
203. Reviews on phospasilenes: (a) M. Driess, *Coord. Chem. Rev.* **1995**, *145*, 1; (b) M. Driess, *Adv. Organomet Chem.* **1996**, *39*, 193.
204. J.-G. Lee, J. E. Boggs, A. H. Cowley, *J. Chem. Soc., Chem. Commun.* **1985**, 773.
205. (a) C. N. Smit, F. M. Lock, F. Bickelhaupt, *Tetrahedron Lett.* **1984**, *25*, 3011; (b) C. N. Smit, F. Bickelhaupt, *Organometallics* **1987**, *6*, 1156; (c) Y. van den Winkel, H. M. M. Bastiaans, F. Bickelhaupt, *J. Organomet. Chem.* **1991**, *405*, 183.
206. (a) E. Niecke, E. Klein, M. Nieger, *Angew. Chem., Int. Ed. Engl.* **1989**, *28*, 751; (b) H. R. G. Bender, E. Niecke, M. Nieger, *J. Am. Chem. Soc.* **1993**, *115*, 3314.
207. (a) M. Driess, *Angew. Chem., Int. Ed. Engl.* **1991**, *30*, 1022; (b) M. Driess, R. Janoschek, *J. Mol. Struct. (Theochem)* **1994**, *313*, 129; (c) M. Driess, S. Rell, H. Pritzkow, *J. Chem. Soc., Chem. Commun.* **1995**, 253; (d) M. Driess, H. Pritzkow, S. Rell, U. Winkler, *Organometallics* **1996**, *15*, 1845; (e) M. Driess, S. Rell, K. Merz, *Z. Anorg. Allg. Chem.* **1999**, *625*, 1119.
208. M. Driess, S. Rell, H. Pritzkow, R. Janoschek, *Angew. Chem., Int. Ed. Engl.* **1997**, *36*, 1326.
209. M. Driess, S. Block, M. Brym, M. T. Gamer, *Angew. Chem., Int. Ed.* **2006**, *45*, 2293.
210. M. Driess, H. Pritzkow, U. Winkler, *J. Organomet. Chem.* **1997**, *529*, 313.
211. S. Yao, S. Block, M. Brym, M. Driess, *Chem. Commun.* **2007**, 3844.
212. V. Ya. Lee, M. Kawai, A. Sekiguchi, H. Ranaivonjatovo, J. Escudié, *Organometallics* **2009**, *28*, 4262.
213. B. Li, T. Matsuo, D. Hashizume, H. Fueno, K. Tanaka, K. Tamao, *J. Am. Chem. Soc.* **2009**, *131*, 13222.
214. (a) J. Escudié, C. Couret, J. Satgé, M. Andrianarison, J.-D. Andriamizaka, *J. Am. Chem. Soc.* **1985**, *107*, 3378; (b) M. Dräger, J. Escudié, C. Couret, H. Ranaivonjatovo, J. Satgé, *Organometallics* **1988**, *7*, 1010; (c) H. Ranaivonjatovo, J. Escudié, C. Couret, J. Satgé, M. Dräger, *New J. Chem.* **1989**, *13*, 389.
215. (a) H. Ranaivonjatovo, J. Escudié, C. Couret, J. Satgé, *J. Organomet. Chem.* **1991**, *415*, 327; (b) H. Ranaivonjatovo, J. Escudié, C. Couret, A. Kandri Rodi, J. Satgé, *Phosphorus Sulfur Silicon* **1993**, *76*, 61.
216. (a) C. Couret, J. Escudié, J. Satgé, A. Raharinirina, J.-D. Andriamizaka, *J. Am. Chem. Soc.* **1985**, *107*, 8280; (b) H. Ranaivonjatovo, J. Escudié, C. Couret, J. Satgé, *J. Chem. Soc., Chem. Commun.* **1992**, 1047.
217. Reviews on arsasilenes: (a) M. Driess, *Coord. Chem. Rev.* **1995**, *145*, 1; (b) M. Driess, *Adv. Organomet Chem.* **1996**, *39*, 193.
218. (a) M. Driess, H. Pritzkow, *Angew. Chem., Int. Ed. Engl.* **1992**, *31*, 316; (b) M. Driess, H. Pritzkow, M. Sander, *Angew. Chem., Int. Ed. Engl.* **1993**, *32*, 283.
219. (a) N. Tokitoh, T. Matsumoto, R. Okazaki, *Bull. Chem. Soc. Jpn.* **1999**, *72*, 1665; (b) R. Okazaki, N. Tokitoh, *Acc. Chem. Res.* **2000**, *33*, 625; (c) N. Tokitoh, R. Okazaki, *Adv. Organomet. Chem.* **2001**, *47*, 121; (d) N. Takeda, N. Tokitoh, R.

Okazaki, in *Handbook of Chalcogen Chemistry* (Ed. F. A. Devillanova), Royal Society of Chemistry, Cambridge, **2007**, Chapter 3.
220. P. Arya, J. Boyer, F. Carré, R. Corriu, G. Lanneau, J. Lapasset, M. Perrot, C. Priou, *Angew. Chem., Int. Ed. Engl.* **1989**, *28*, 1016.
221. (a) I. M. T. Davidson, J. F. Thompson, *J. Chem. Soc., Chem. Commun.* **1971**, 251; (b) V. N. Khabashesku, Z. A. Kerzina, E. G. Baskir, A. K. Maltsev, O. M. Nefedov, *J. Organomet. Chem.* **1988**, *347*, 277; (c) V. N. Khabashesku, S. E. Boganov, O. M. Nefedov, *Izv. Akad. Nauk SSSR, Ser. Khim.* **1990**, 1199; (d) S. E. Boganov, V. I. Faustov, V. N. Khabashesku, Z. A. Kerzina, N. D. Kagramanov, A. A. Kutin, O. M. Nefedov, P. Mazerolles, G. Manuel, *Izv. Akad. Nauk, Ser. Khim.* **1993**, 701; (e) V. N. Khabashesku, S. E. Boganov, K. N. Kudin, J. L. Margrave, O. M. Nefedov, *Organometallics* **1998**, *17*, 5041.
222. (a) S. Yao, M. Brym, C. van Wüllen, M. Driess, *Angew. Chem., Int. Ed.* **2007**, *46*, 4159; (b) S. Yao, Y. Xiong, M. Brym, M. Driess, *J. Am. Chem. Soc.* **2007**, *129*, 7268; (c) Y. Xiong, S. Yao, M. Driess, *J. Am. Chem. Soc.* **2009**, *131*, 7562; (d) S. Yao, Y. Xiong, M. Driess, *Chem. Commun.* **2009**, 6466.
223. N. Tokitoh, T. Matsumoto, R. Okazaki, *Chem. Lett.* **1995**, 1087.
224. E. Bonnefille, S. Mazières, C. Bibal, N. Saffon, H. Gornitzka, C. Couret, *Eur. J. Inorg. Chem.* **2008**, 4242.
225. T. Kudo, S. Nagase, *Organometallics* **1986**, *5*, 1207.
226. (a) H. Suzuki, N. Tokitoh, S. Nagase, R. Okazaki, *J. Am. Chem. Soc.* **1994**, *116*, 11578; (b) H. Suzuki, N. Tokitoh, R. Okazaki, S. Nagase, M. Goto, *J. Am. Chem. Soc.* **1998**, *120*, 11096.
227. T. Iwamoto, K. Sato, S. Ishida, C. Kabuto, M. Kira, *J. Am. Chem. Soc.* **2006**, *128*, 16914.
228. (a) N. Tokitoh, T. Matsumoto, K. Manmaru, R. Okazaki, *J. Am. Chem. Soc.* **1993**, *115*, 8855; (b) T. Matsumoto, N. Tokitoh, R. Okazaki, *J. Am. Chem. Soc.* **1999**, *121*, 8811.
229. (a) N. Tokitoh, M. Saito, R. Okazaki, *J. Am. Chem. Soc.* **1993**, *115*, 2065; (b) M. Saito, N. Tokitoh, R. Okazaki, *Organometallics* **1996**, *15*, 4531; (c) M. Saito, N. Tokitoh, R. Okazaki, *J. Am. Chem. Soc.* **2004**, *126*, 15572; (d) T. Tajima, T. Sasamori, N. Takeda, N. Tokitoh, *Bull. Chem. Soc. Jpn.* **2007**, *80*, 1202.
230. N. Kano, N. Tokitoh, R. Okazaki, *Chem. Lett.* **1997**, 277.
231. N. Tokitoh, T. Sadahiro, K. Hatano, T. Sasaki, N. Takeda, R. Okazaki, *Chem. Lett.* **2002**, 34.
232. T. Matsumoto, N. Tokitoh, R. Okazaki, *Angew. Chem., Int. Ed. Engl.* **1994**, *33*, 2316.
233. M. Saito, N. Tokitoh, R. Okazaki, *J. Am. Chem. Soc.* **1997**, *119*, 11124.
234. N. Tokitoh, T. Matsumoto, R. Okazaki, *J. Am. Chem. Soc.* **1997**, *119*, 2337.
235. T. Tajima, N. Takeda, T. Sasamori, N. Tokitoh, *Organometallics* **2006**, *25*, 3552.
236. S. Yao, Y. Xiong, M. Brym, M. Driess, *Chem. Asian J.* **2008**, *3*, 113.
237. Reviews on 1,3-dienes: (a) S. Willms, M. Weidenbruch, in *Organosilicon Chemistry IV: From Molecules to Materials* (Eds N. Auner, J. Weis), Wiley-VCH, Weinheim, **2000**, p. 117; (b) M. Weidenbruch, *J. Organomet. Chem.* **2002**, *646*, 39; (c) M. Weidenbruch, in *Silicon Chemistry* (Eds P. Jutzi, U. Schubert), Wiley-VCH, Weinheim, **2003**, p. 100; (d) M. Weidenbruch, in *Organosilicon Chemistry V: From*

Molecules to Materials (Eds N. Auner, J. Weis), Wiley-VCH, Weinheim, **2003**, p. 114.
238. M. Weidenbruch, S. Willms, W. Saak, G. Henkel, *Angew. Chem., Int. Ed. Engl.* **1997**, *36*, 2503.
239. H. Schäfer, W. Saak, M. Weidenbruch, *Angew. Chem., Int. Ed.* **2000**, *39*, 3703;
240. (a) M. Ichinohe, K. Sanuki, S. Inoue, A. Sekiguchi, *Organometallics* **2004**, *23*, 3088; (b) M. Ichinohe, K. Sanuki, S. Inoue, A. Sekiguchi, *Silicon Chemistry* **2005**, *3*, 111.
241. C. Jouany, S. Mathieu, M.-A. Chaubon-Deredempt, G. Trinquier, *J. Am. Chem. Soc.* **1994**, *116*, 3973.
242. C. Cui, M. M. Olmstead, P. P. Power, *J. Am. Chem. Soc.* **2004**, *126*, 5062.
243. T. Iwamoto, M. Tamura, C. Kabuto, M. Kira, *Science* **2000**, *290*, 504.
244. I. Bejan, D. Scheschkewitz, *Angew. Chem., Int. Ed.* **2007**, *46*, 5783.
245. Reviews: (a) J. Escudié, H. Ranaivonjatovo, L. Rigon, *Chem. Rev.* **2000**, *100*, 3639; (b) B. Eichler, R. West, *Adv. Organomet. Chem.* **2001**, *46*, 1; (c) J. Escudié, H. Ranaivonjatovo, *Organometallics* **2007**, *26*, 1542.
246. G. E. Miracle, J. L. Ball, D. R. Powell, R. West, *J. Am. Chem. Soc.* **1993**, *115*, 11598.
247. M. Trommer, G. E. Miracle, B. E. Eichler, D. R. Powell, R. West, *Organometallics* **1997**, *16*, 5737.
248. M. Ichinohe, T. Tanaka, A. Sekiguchi, *Chem. Lett.* **2001**, 1074.
249. (a) B. E. Eichler, D. R. Powell, R. West, *Organometallics* **1998**, *17*, 2147; (b) B. E. Eichler, D. R. Powell, R. West, *Organometallics* **1999**, *18*, 540.
250. N. Tokitoh, K. Kishikawa, R. Okazaki, *Chem. Lett.* **1998**, 811.
251. (a) S. Ishida, T. Iwamoto, C. Kabuto, M. Kira, *Nature* **2003**, *421*, 725; (b) T. Iwamoto, T. Abe, C. Kabuto, M. Kira, *Chem. Commun.* **2005**, 5190; (c) T. Iwamoto, H. Masuda, C. Kabuto, M. Kira, *Organometallics* **2005**, *24*, 197; (d) T. Iwamoto, T. Abe, S. Ishida, C. Kabuto, M. Kira, *J. Organomet. Chem.* **2007**, *692*, 263.
252. H. Grützmacher, S. Freitag, R. Herbst-Irmer, G. S. Sheldrick, *Angew. Chem., Int. Ed. Engl.* **1992**, *31*, 437.
253. M. Weidenbruch, B. Brand-Roth, S. Pohl, W. Saak, *Angew. Chem., Int. Ed. Engl.* **1990**, *29*, 90.
254. T. Abe, T. Iwamoto, C. Kabuto, M. Kira, *J. Am. Chem. Soc.* **2006**, *128*, 4228.
255. K. Takeuchi, M. Ichinohe, A. Sekiguchi, *J. Am. Chem. Soc.* **2008**, *130*, 16848.
256. (a) N. Takeda, H. Suzuki, N. Tokitoh, R. Okazaki, S. Nagase, *J. Am. Chem. Soc.* **1997**, *119*, 1456; (b) N. Takeda, T. Kajiwara, H. Suzuki, R. Okazaki, N. Tokitoh, *Chem. Eur. J.* **2003**, *9*, 3530.
257. G. H. Spikes, P. P. Power, *Chem. Commun.* **2007**, 85.
258. L. Rigon, H. Ranaivonjatovo, J. Escudié, A. Dubourg, J.-P. Declercq, *Chem. Eur. J.* **1999**, *5*, 774.
259. (a) H. Ramdane, H. Ranaivonjatovo, J. Escudié, S. Mathieu, N. Knouzi, *Organometallics* **1996**, *15*, 3070; (b) Y. El Harouch, H. Gornitzka, H. Ranaivonjatovo, J. Escudié, *J. Organomet. Chem.* **2002**, 643–644, 202.
260. Reviews: (a) P. P. Power, *Chem. Commun.* **2003**, 2091; (b) P. P. Power, *Organometallics* **2007**, *26*, 4362. (c) A. Sekiguchi, *Pure Appl. Chem.* **2008**, *80*, 447.

261. G. Maier, J. Glatthaar, *Angew. Chem., Int. Ed. Engl.* **1994**, *33*, 473.
262. M. Karni, Y. Apeloig, D. Schröder, W. Zummack, R. Rabezzana, H. Schwarz, *Angew. Chem., Int. Ed.* **1999**, *38*, 332.
263. (a) A. Sekiguchi, S.S. Zigler, K.J. Haller, R. West, *Recueil des Travux Chimiques des Pays-Bas* **1998**, *107*, 197. (b) A. Sekiguchi, S. S. Zigler, R. West, J. Michl, *J. Am. Chem. Soc.* **1986**, *108*, 4241.
264. R. Pietschnig, R. West, D. R. Powell, *Organometallics* **2000**, *19*, 2724.
265. C. Bibal, S. Mazières, H. Gornitzka, C. Couret, *Angew. Chem., Int. Ed.* **2001**, *40*, 952.
266. E. Rivard, P. P. Power, *Inorg. Chem.* **2007**, *46*, 10047.
267. (a) N. Wiberg, in *Organosilicon Chemistry V: From Molecules to Materials* (Eds N. Auner, J. Weis), Wiley-VCH, Weinheim, **2003**, p. 101; (b) N. Takagi, S. Nagase, *Eur. J. Inorg. Chem.* **2002**, 2775.
268. (a) A. Sekiguchi, R. Kinjo, M. Ichinohe, *Science* **2004**, *305*, 1755; (b) V. Kravchenko, R. Kinjo, A. Sekiguchi, M. Ichinohe, R. West, Y. S. Balazs, A. Schmidt, M. Karni, Y. Apeloig, *J. Am. Chem. Soc.* **2006**, *128*, 14472; (c) A. Sekiguchi, M. Ichinohe, R. Kinjo, *Bull. Chem. Soc. Jpn.* **2006**, *79*, 825.
269. (a) K. Kobayashi, S. Nagase, *Organometallics* **1997**, *16*, 2489; (b) S. Nagase, K. Kobayashi, N. Takagi, *J. Organomet. Chem.* **2000**, *611*, 264; (c) K. Kobayashi, N. Takagi, S. Nagase, *Organometallics* **2001**, *20*, 234.
270. (a) M. Karni, Y. Apeloig, N. Takagi, S. Nagase, *Organometallics* **2005**, *24*, 6319; (b) D. Auer, M. Kaupp, C. Strohmann, *Organometallics* **2005**, *24*, 6331.
271. K. Takeuchi, M. Ikoshi, M. Ichinohe, A. Sekiguchi, *J. Am. Chem. Soc.* **2010**, *132*, 930.
272. M. Stender, A. D. Phillips, R. J. Wright, P. P. Power, *Angew. Chem., Int. Ed.* **2002**, *41*, 1785.
273. Y. Sugiyama, T. Sasamori, Y. Hosoi, Y. Furukawa, N. Takagi, S. Nagase, N. Tokitoh, *J. Am. Chem. Soc.* **2006**, *128*, 1023.
274. (a) A. D. Phillips, R. J. Wright, M. M. Olmstead, P. P. Power, *J. Am. Chem. Soc.* **2002**, *124*, 5930; (b) R. C. Fischer, L. Pu, J. C. Fettinger, M. A. Brynda, P. P. Power, *J. Am. Chem. Soc.* **2006**, *128*, 11366.
275. L. Pu, B. Twamley, P. P. Power, *J. Am. Chem. Soc.* **2000**, *122*, 3524.
276. R. Jambor, B. Kasna, K. N. Kirschner, M. Schürmann, K. Jurkschat, *Angew. Chem., Int. Ed.* **2008**, *47*, 1650.
277. P. P. Power, *Appl. Organomet. Chem.* **2005**, *19*, 488.
278. Y. Jung, M. Brynda, P. P. Power, M. Head-Gordon, *J. Am. Chem. Soc.* **2006**, *128*, 7185.
279. N. Takagi, S. Nagase, *Organometallics* **2001**, *20*, 5498.
280. M. Karni, Y. Apeloig, *Silicon Chemistry* **2002**, *1*, 61.

6

Heavy Analogs of Aromatic Compounds

6.1 Introduction

Aromaticity is one of the milestone chemical concepts widely applicable in both organic and inorganic chemistry. Aromatic compounds have been known for nearly two centuries, ever since the discovery of the first aromatic hydrocarbon, 'bicarburet of hydrogen' C_6H_6 (which is now called benzene), by Faraday in 1825.[1] A remarkable cyclic structure of benzene with alternating C–C single and C=C double bonds was later suggested by Kekulé in 1866.[2] A vast number of aromatic compounds have been synthesized and reported in the scientific literature since then, growing exponentially from year to year: a literature search generated an enormous number of references directly related to the topic of aromaticity and aromatic compounds, totaling more than 300 000 entries for the last 30 years. Since the pioneering discovery of Faraday, the field of aromatic compound chemistry has undergone a spectacular evolution, which is clearly reflected in the synthesis of novel classes of aromatic derivatives including charged species, nonclassical (Möbius, three-dimensional, homo-, metalla-) aromatics and fullerenes. However, the theory of aromaticity lagged behind such rapidly progressing experimental achievements, and not much was known about the electronic structure and bonding nature of aromatic compounds until the development of molecular orbital (MO) theory. Thus, in fact, the seemingly simple question 'How can one define aromaticity, and to what extent is such a definition of aromaticity valid?' was, and still is, not easy to answer simply. A general simplified textbook definition of aromatic compounds, derived from Hückel's fundamental MO theory, states: 'Planar monocyclic completely conjugated hydrocarbons will be aromatic when the ring contains $(4n + 2)$ π-electrons'.[3] This statement, better known as the famous Hückel's rule (1931),[4] served well for a long time as the

Organometallic Compounds of Low-Coordinate Si, Ge, Sn and Pb: From Phantom Species to Stable Compounds
Vladimir Ya. Lee and Akira Sekiguchi

principal criterion for the assignment of organic/inorganic derivatives to the class of aromatic or nonaromatic compounds. However, the above-mentioned groundbreaking advances in experimental organic chemistry, primarily associated with the synthesis of new classes of compounds, which did not satisfy all of the requirements of Hückel's rule (for example, fullerenes, *trans*-annulenes, Möbius aromatics, which are nonplanar), featuring nonetheless aromatic properties (so-called nonclassical aromatics), undoubtedly required the development of a more unified and general approach to the definition of aromaticity as a chemical phenomenon. It quickly became apparent that simple electron counting is too simplified an approach for the recognition of aromaticity, whose real nature turned out to be far more complicated and versatile than initially thought. It is not surprising, therefore, that the theory of aromaticity has undergone a significant change (particularly during the last two decades), first of all regarding the most fundamental physicochemical criteria based on which compounds could be reliably classified as aromatic, nonaromatic or antiaromatic. Various criteria were proposed as more or less effective aromaticity evaluation tools; however, only three of them have been finally recognized to be of fundamental importance: energetic, structural and magnetic.[5] The first criterion, the energetic criterion of aromaticity, based on a quantum chemical calculation of the overall stabilization gained through the cyclic electron delocalization [the so-called resonance energy (RE)], plays a major role in evaluation of the degree of aromaticity: the most commonly used estimate of RE is the aromatic stabilization energy (ASE) calculated employing either isodesmic or homodesmotic reactions. The structural criterion of aromaticity deals with the geometrical features of the molecule: planarity of the ring and cyclic carbon–carbon bond equalization are considered as the default prerequisites necessary for optimum overlap of the carbon p_π-orbitals and hence aromaticity. The magnetic criterion of aromaticity is based on the phenomenon of the diamagnetic ring current caused by the circulation of the π-electrons and diamagnetic susceptibility exaltation (Λ). This can also be estimated by quantum chemical calculations, whereas the experimental assessment of aromaticity can be achieved by NMR spectroscopic measurements. In particular, deshielding of the nuclei lying in the plane of the aromatic ring and shielding of those lying inside the anisotropy cone of the aromatic ring are typically considered as direct evidence for induced ring current effects and, consequently, aromaticity of the compound. Such tendencies in the chemical shift changes can be quantified on the basis of its calculation at the center of the aromatic ring, the so-called *nucleus independent chemical shift* (NICS),[6] which can be calculated directly at the center of the ring meanplane [NICS(0)], or at any other distance from this center [for example, at 1 Å above this ring center, NICS(1)]. Accordingly, the classical aromatic compounds display negative NICS values; for example, benzene has a NICS(1) value of -10.6, whereas antiaromatic species exhibit positive values, cyclobutadiene has NICS(1) $= +12.7$, and nonaromatic compounds have values near zero.[7] In many cases, the principal energetic criterion of aromaticity reasonably correlates with the structural and magnetic criteria; however, in a number of cases, compounds that would be classified as aromatic according to two of these fundamental aspects (for example, energetic and magnetic) do not fulfill the third requirement (for example, planarity, bond equalization) of aromaticity. Such conflict situations clearly reflect the necessity for further more sophisticated developments in the classification of aromatic compounds.

Striking progress in the development of the theory of aromaticity has been achieved in recent years, and well documented in a number of excellent reviews and monographs published in the past two decades,[5,8] of which the monograph by Minkin *et al.*[5] and the two thematic issues of *Chemical Reviews*, devoted to the problems of aromaticity (2001)[8b] and π- and σ-delocalization (2005),[8c] are of particular importance. All-metal aromatics, which bridge the fields of organic and inorganic chemistry, have also been recently reviewed.[8c–e] Many aspects of the contemporary chemistry of aromatic compounds (homoaromaticity, metallaaromaticity, spherical and three-dimensional aromaticity, Möbius aromaticity, structural and magnetic aspects of aromaticity, antiaromaticity, etc.) were comprehensively covered by the leading experts in the field of theoretical chemistry. Therefore, in this chapter, we are not going to discuss the essentials of the theory of aromaticity again; instead, we would like to present a viewpoint of synthetic chemists on this problem: how the novel aromatic compounds of the heavy group 14 elements can be synthesized, what the most reliable estimates for the degree of their aromaticity (or nonaromaticity) are, and how such new members of the aromatic compounds family can be utilized in organometallic chemistry.

In contrast to the well-developed field of organic aromatics, the chemistry of their heavier congeners of Si, Ge, Sn and Pb is still rather 'uncultivated', leaving most of the synthetic and mechanistic problems to be solved and thus representing an important challenge for both experimental and theoretical organometallic chemists. However, it should be noted that the synthesis of the sila-, germa-, stanna- and plumbaaromatic compounds is quite a nontrivial problem, because the classical synthetic methods, widely and successfully utilized in organic chemistry for the preparation of aromatic derivatives, turned out to be of very limited applicability for the synthesis of heavy aromatics. Such striking synthetic difficulties in the chemistry of the aromatic compounds of heavy group 14 elements can be explained from the general viewpoint of the fundamental differences between carbon and its lower neighbors (Si, Ge, Sn, Pb): electronegativity, size and polarizability. This sharp distinction, along with the well-recognized tendency for the heavy main group elements to be more and more reluctant toward s,p-hybridization because of the progressively increasing difference in the size and energy of the s- and p-orbitals descending the group, are responsible for the great problems of thermodynamic stabilization of the heavy aromatics. Moreover, even after being synthesized, such species often manifest a greatly enhanced reactivity compared with the ordinary organic aromatics; therefore, kinetic stabilization by bulky protective groups is a necessary prerequisite for their successful isolation. It is therefore not surprising that the synthesis of the stable Si-, Ge-, Sn- and Pb-containing aromatic compounds required development of novel original synthetic strategies, sharply distinct from those of organic aromatic derivatives.

The problem of aromaticity of the heavy group 14 elements received a deep theoretical insight in a series of reviews published by Apeloig *et al.*,[9] and the most important experimental achievements in this field have also been recently reviewed.[10,11] In this chapter, we are going to discuss the current state of affairs in the chemistry of the group 14 heavy aromatics, particularly emphasizing the latest experimental achievements, as well as future prospects in the field. Both neutral and charged 2π- and 6π-electron species, especially those that could be isolated as room temperature stable derivatives, will be the focus of our discussion.

6.2 Early Studies

6.2.1 Transient Species: Generation, Identification and Trapping

The most important experimental achievements at the early stage of research in such an intriguing field dealt with the generation and identification (by either direct spectroscopic studies or trapping experiments) of the heavy group 14 elements containing benzenoids; that is, sila- and germaarenes.[9b] Particular attention has been paid to the most fundamental prototypical silabenzene, the Me-derivative of which, **1b**, (Scheme 6.1) was generated for the first time by Barton *et al.* and whose intermediate formation was convincingly proved by its dimerization product or trapping reactions with MeOH and alkynes (HC≡CH, $CF_3C{\equiv}CCF_3$).[12] Direct spectroscopic detection of the extremely reactive silabenzene **1a** and methylsilabenzene **1b** was accomplished during the 1980s by the efforts of the research groups of Bock (photoelectron spectroscopy),[13] Maier (IR and UV matrix isolation technique),[14] Chapman and Barton (IR and UV matrix isolation technique),[15] and Nefedov (mass spectrometry).[16] Some other interesting examples of transient substituted silabenzenes, whose intermediacy was proved by either trapping reactions or direct spectroscopic observation in a low-temperature matrix, were reported in due course (Scheme 6.1): 1,4-di-*tert*-butyl-1-silabenzene **2** (formation of a [2 + 2] head-to-tail dimer, [4 + 2] cycloaddition with 1,3-dienes),[17] 1-methyl-2,3,4,5-tetraphenyl-1-silabenzene **3** (1,2- and 1,4-addition of MeOH and [4 + 2] cycloaddition with 2,3-dimethyl-1,3-butadiene),[18] 4-*tert*-butyl-1-methoxy-2,6-bis(trimethylsilyl)-1-silabenzene **4c** (1,2-addition of MeOH),[19] 1-*tert*-butyl-2,4,6-tris(trimethylsilyl)-1-silabenzene **4a** and 1-*tert*-butyl-2,6-bis(isopropyldimethylsilyl)-4-trimethylsilyl-1-silabenzene **4b** (matrix IR and UV spectroscopy, stable up to 90 K even without an Ar matrix).[20]

Several examples of 1,4-disilabenzenes were also postulated to be formed as reactive intermediates (Scheme 6.1); namely, hexamethyl-1,4-disilabenzene **6** (1,4-addition of MeOH, [4 + 2] cycloaddition with alkynes, [4 + 1] cycloaddition with molecular oxygen),[21] and the parent 1,4-disilabenzene **5** (Ar matrix UV spectroscopy at 10 K).[22]

The sole example of a transient unsubstituted 1,3,5-trisilabenzene **7** was believed to be generated in the gas phase mass spectrometrically by collision-induced dissociation experiments;[23] however, this interesting result definitely needs to be verified by other independent studies; for example, trapping experiments.

Several Si-containing polynuclear aromatic hydrocarbons were also generated as reactive intermediates. Thus, the thermally generated transient 2-methyl-2-silanaphthalene **8a**, 2-methoxy-2-silanaphthalene **8b** and 2-allyl-2-silanaphthalene **8c** were identified by their 1,2-adducts with MeOH,[24] whereas the intermediate formation of 9-silaanthracenes **9** was convincingly proved by the isolation of their adducts with MeOH and [4 + 2] cycloadducts with alkynes.[25] The transient 9-silaanthracene derivatives **9** were also stabilized and directly detected in an argon matrix at low temperatures by UV and fluorescence spectroscopy, and in the gas phase of a mass spectrometer by the neutralization-reionization mass spectrometry technique.[26]

The record of thermal stability of silabenzene derivatives was demonstrated by Märkl and Schlosser, who reported the synthesis of 1,4-di-*tert*-butyl-2,6-bis(trimethylsilyl)-1-silabenzene **4d** (Scheme 6.1), whose intermediate formation was shown by its trapping reaction with MeOH producing the expected 1,2-adduct.[27] In the absence of trapping reagents, **4d** was stable in solution up to ca. −100 °C, irreversibly decomposing at higher temperatures to form unidentified products. The ^{29}Si NMR spectrum of **4d** measured in a reaction mixture at −100 °C showed the signal of the cyclic Si atom at +26.8 ppm, a value that is markedly shifted to higher field compared with the normal low-field resonances expected for doubly bonded Si atoms. This definitely points to the existence of silabenzene **4d** in solution in the form of its solvated complex, in which the THF solvent acts as a Lewis base donor toward the Lewis acidic sp^2-Si atom.

The heavier congeners of arenes, that is, germaarenes, were studied rather scantily. One can only mention the work of Märkl's group on the generation of 4-*tert*-butyl-1-R-1-germabenzene derivatives **10** (R=H, Me, *t*-Bu) (Scheme 6.1) identified by their trapping reactions (formation of a [2 + 2] head-to-tail dimer, [4 + 2] cycloaddition with 1,3-dienes)[28] and detection in the gas phase by means of photoelectron spectroscopy.[29]

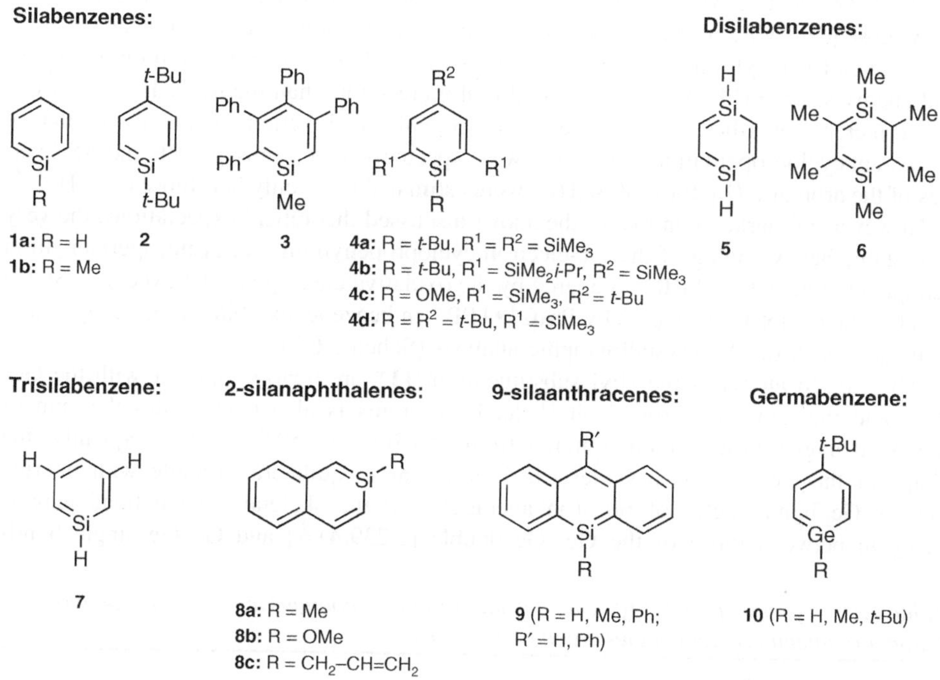

Scheme 6.1

6.3 Stable Compounds (Synthesis and Structure)

6.3.1 2π-Electron Species

The most fundamental 2π-electron aromatic compounds are the derivatives of the singly charged cyclopropenylium ion $R_3C_3^+$ and doubly charged cyclobutadiene dication $R_4C_4^{2+}$. The properties and stabilities of these two classes of π-delocalized cyclic systems are strikingly different: whereas several of the stable $R_3C_3^+$-type compounds have been synthesized since the pioneering work of Breslow,[30] no stable derivatives of the $R_4C_4^{2+}$ type have been reported to date. Below, the most important advances in the chemistry of the heavy analogs of the 2π-electron cationic species of the types $R_3E_3^+$ and $R_4E_4^{2+}$ (E = Si, Ge, Sn, Pb) will be discussed.

6.3.1.1 Cyclopropenylium Ion Derivatives

The synthesis of stable Si and Ge versions of the cyclopropenylium ion was preceded by calculations by Schleyer *et al.*, who predicted that the classical planar cyclopropenylium ion with D_{3h} symmetry is favored over the nonplanar triply H-bridged C_{3v} isomeric structure only in the case of the $Si_3H_3^+$ ion by 23.7 kcal/mol at the B3LYP/6-311++G(2d,2p) level, whereas for the heavier derivatives of the types $Ge_3H_3^+$, $Sn_3H_3^+$ and $Pb_3H_3^+$, the planar D_{3h} conformation was largely destabilized in favor of the bridged C_{3v} conformer by 17.4, 32.4 and 63.3 kcal/mol, respectively, at the B3LYP/TZ2P+ level (Table 6.1).[31]

A subsequent MP2/6-31G* study by the same authors showed that the planar D_{3h} form of the cyclotrisilenylium ion $Si_3H_3^+$ represents, indeed, a global minimum on the potential energy surface (PES), being 42.1 kcal/mol more stable than the triply H-bridged C_{3v} conformer.[32] Nevertheless, the degree of its aromatic stabilization compared with that of the parent cyclopropenylium ion $C_3H_3^+$ was expectedly lower; thus, the RE and ASE values of the aromatic D_{3h} form of $Si_3H_3^+$ were estimated to be only half those of $C_3H_3^+$.[32]

However, in marked contrast to the above-discussed theoretical expectations, the very first stable heavy analog of the 2π-electron cyclopropenylium ion, cyclotrigermenylium tetraphenylborate **11$^+$•TPB$^-$**, prepared by the oxidative cleavage of the exocyclic Si–Ge bond of the cyclotrigermene **12** by Ph_3C^+•TPB^- in benzene, exhibited a perfectly planar structure, as shown by crystallographic analysis (Scheme 6.2).[33]

All three Si atoms of the silyl substituents in **11$^+$** are nearly coplanar with the Ge_3 ring, and the geometry around all skeletal Ge atoms is also planar with the sum of the bond angles around them ranging from 359.7(2) to 359.9(2)°. Consequently, the shape of the Ge_3 ring is best described as a nearly equilateral triangle with internal Ge–Ge–Ge bond angles of 60° (on average), and the skeletal Ge–Ge bond lengths being in between those of the Ge=Ge double [2.239(4) Å] and Ge–Ge single bonds

Table 6.1 *Stability of the $E_3H_3^+$ cyclic isomers (relative energies are given in kcal/mol, number of imaginary frequencies are in parentheses)*

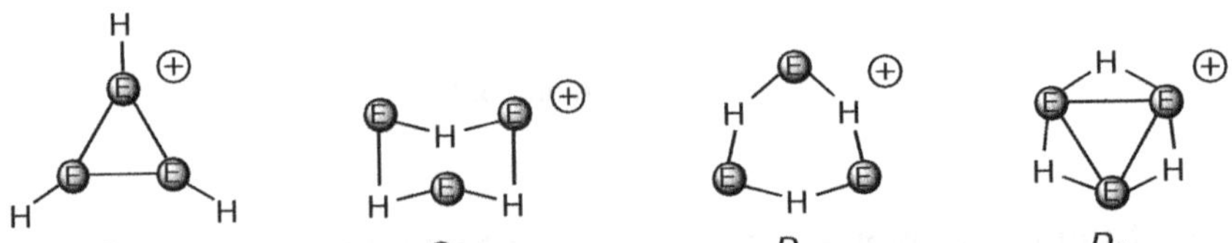

	D_{3h}	C_{3v}	D_{3h}	D_{3h}
C	0.0(0)	–	–	194.8(3)
Si	0.0(0)	23.7(0)	55.6(3)	53.5(3)
Ge	0.0(0)	−17.4(0)	5.5(3)	–
Sn	0.0(1)	−32.4(0)	−15.7(3)	–
Pb	0.0(3)	−63.3(0)	−8.1(3)	–

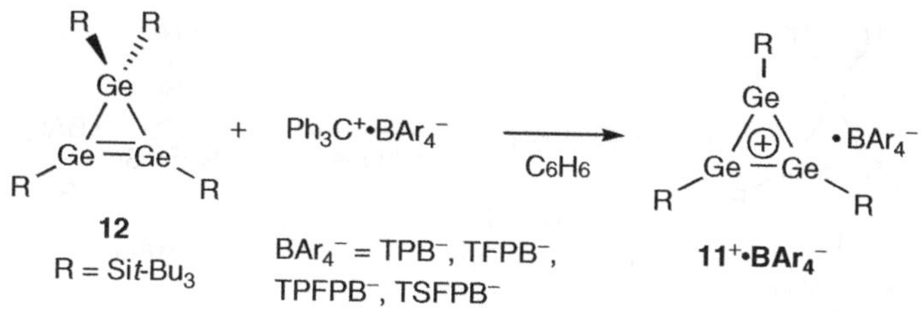

Scheme 6.2

[2.522(4) Å] of precursor **2** [2.321(4)–2.326(4) Å]. Such special geometrical features of $\mathbf{11}^+$ were interpreted as direct evidence for the even delocalization of the positive charge over the three skeletal Ge atoms, thus producing a planar 2π-electron aromatic structure closely resembling that of the cyclopropenylium ion.[33a] The closest distances between the cationic Ge atoms and TPB^- counterion were long enough to rule out the possibility of any noticeable cation–anion interactions, implying that $\mathbf{11}^+$ is a 'free' cation in the crystalline state. The striking discrepancy between the calculated nonplanar H-bridged C_{3v} structure[31] and experimentally observed planar aromatic D_{3h} structure[33a] is evidently because of the different substitution pattern: small and easily mobile H atoms on the calculated models versus very bulky t-Bu_3Si groups in the real compound, which are very reluctant to move and whose migratory and bridging aptitudes are almost negligible. As well as the decisive kinetic protection of the highly reactive cationic centers, the electropositive t-Bu_3Si groups play another key role as the 'extinguisher' of the positive charge on the highly electrophilic Ge atoms, greatly contributing to the overall thermodynamic stabilization of the compound. This was demonstrated by the appreciable low-field shift of the resonances of the Si substituents in $\mathbf{11}^+$ (64.0 ppm in CD_2Cl_2) and supported by the HF/6-31G* calculations: the Mulliken charge of the skeletal Ge atoms was −0.07, whereas that of the substituent Si atoms was +0.64.[34b,c] Employing the same synthetic procedure, cyclotrigermenylium ion $\mathbf{11}^+$ was prepared with other counterions, such as $B[3,5\text{-}(CF_3)_2C_6H_3]_4^-$ (TFPB), $B(C_6F_5)_4^-$ (TPFPB) and $B[4\text{-}(t\text{-}BuMe_2Si)C_6F_4]_4^-$ (TSFPB) (Scheme 6.2).[34] All of these derivatives appeared to be significantly more stable in solution than the original cyclotrigermenylium ion $\mathbf{11}^+$ with the BPh_4^- counterion, which opened the possibility of studying the structure of $\mathbf{11}^+$ in the condensed phase. Particularly important is the fact that $\mathbf{11}^+$, which was found to be a 'free' cation in the crystalline form, retained its freedom in solution as well, as was clearly demonstrated by the ^{29}Si NMR spectra of $\mathbf{11^+{\bullet}BAr_4^-}$ being virtually independent of both counterions (TFPB, TPFPB, TSFPB) and solvents, being almost identical in CD_2Cl_2, $CDCl_3$, toluene-d_8 and Et_2O: 64.0, 64.2, 64.4 and 64.5 ppm, respectively.[34]

In good agreement with the theoretical predictions,[32] the all-Si version of the cyclopropenylium ion, cyclotrisilenylium ion $\mathbf{13}^+$ bearing voluminous t-Bu_2MeSi and t-Bu_3Si groups, is remarkably planar, quite similar to the above case of the cyclotrigermenylium ion $\mathbf{11}^+$. $\mathbf{13}^+$ was prepared by the same synthetic procedure as that for the Ge_3^+ ion;[33a] that is, by the one-electron oxidation of the cyclotrisilene **14** with $Ph_3C^+{\bullet}BAr_4^-$ (BAr_4^- = TPFPB, TSFPB, $B[2,3,5,6\text{-}F_4C_6H]_4^-$ (TTFPB)) (Scheme 6.3).[35] The absence

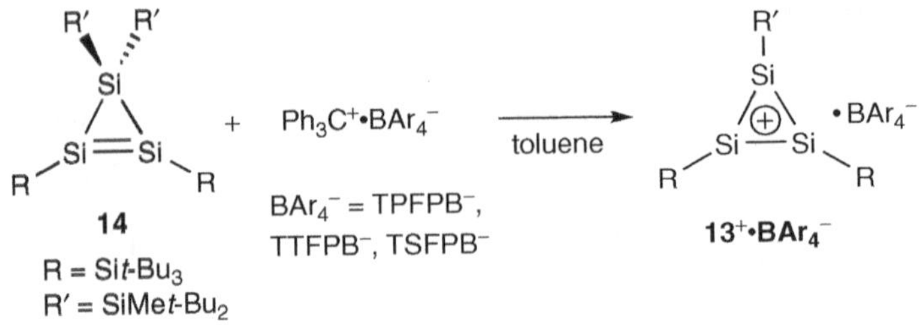

Scheme 6.3

of any observable coordination of external nucleophiles (either counterion or solvent) to **13+** in both the crystalline form or in solution was demonstrated by X-ray crystallographic analysis and NMR spectroscopy; for example, the latter revealed the independence of the ^{29}Si NMR chemical shifts of **13+** of both counterion (TPFPB, TSFPB, TTFPB) and solvent (benzene, toluene, chlorobenzene). The manifestation of the positive charges developed on the skeletal Si atoms was seen in their diagnostic low-field shifted resonances at 284.6 and 288.1 ppm, in the highly deshielded region typical for noncoordinated silyl cations [calculated value for the model $(H_3Si)_3Si^+$ was 300.9 ppm]. The cyclopropenylium-type aromatic delocalization of the 2π-electrons over the entire three-membered Si_3 ring was also confirmed by the particular structural features of **13+**; namely, its equilateral triangle shape with internal Si–Si–Si bond angles of nearly 60° (on average) and endocyclic Si–Si bond lengths of 2.217(3) Å (on average), which are intermediate between those of the Si=Si double [2.1612(8) Å] and Si–Si single bonds [av. 2.3728(8) Å] of the precursor **14** (Figure 6.1).[35]

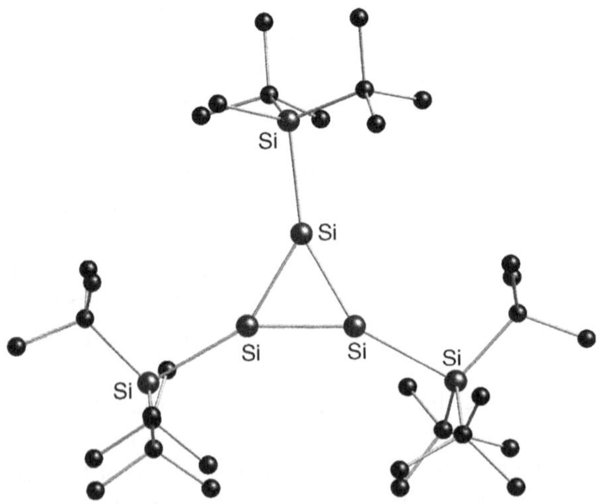

Figure 6.1 *Crystal structure of cyclotrisilenylium ion derivative* **13+** *(TSFPB− counterion and hydrogen atoms are not shown)*

6.3.1.2 Cyclobutadiene Dication Derivatives

Neither organic cyclobutadiene dication derivatives nor their congeners of the heavy group 14 elements have been isolated as stable individual compounds to date. The first organic species of this type, tetramethylcyclobutadiene dication, was generated and observed spectroscopically by the ionization of 3,4-dichloro-1,2,3,4-tetramethylcyclobutene in SbF_5–SO_2 solution at −75 °C by Olah *et al.* (Scheme 6.4: E = C, R = Me),[36] followed by the generation of tetraphenyl-,[37] 1,2-difluoro-3,4-diphenyl-[38] and 1,2-diphenylcyclobutadiene[38] dications under stable ion conditions by the same research group. However, as was mentioned above, the synthesis of cyclobutadiene dications that are stable at room temperature was not achieved; therefore, the structural information necessary for a discussion of the bonding nature of such intriguing species is still lacking.

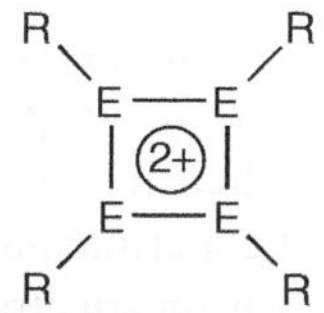

E = C, Si, Ge, Sn, Pb

Scheme 6.4

According to the results of *ab initio* calculations on the $Me_4C_4^{2+}$ species (Scheme 6.4: E = C, R = Me), the planar D_{4h} structure represents a transition state for the inversion of the folded D_{2d} structure (folding angle 39°) with an inversion barrier of ca. 7.0 kcal/mol.[39a] The destabilization of the planar D_{4h} structure compared with the puckered D_{2d} structure was attributed to the unfavorable 1,3-antibonding interactions in the former.[39] Moreover, the experimentally observed ^{13}C NMR chemical shifts of $Me_4C_4^{2+}$ are in better agreement with those calculated for the D_{2d} than for the D_{4h} structure.[38,39a,40,41] It was thus concluded that in small doubly charged four-membered ring systems, such as the cyclobutadiene dication and cyclobutadiene dianion, the repulsive Coulomb interaction between the two charges becomes more important than the advantage gained by cyclic π-delocalization and may therefore outbalance the effects of Hückel aromatic stabilization.[39a,42,43] The latest computations by Schleyer *et al.* agree well with the earlier conclusions, confirming the preference of the folded structure for the cyclobutadiene dication and its alkyl derivatives (Scheme 6.4: E = C, R = H, alkyl), which nevertheless were considered as aromatic according to NICS(0) calculations: −9.0 (unsubstituted cyclobutadiene dication, D_{2d} minimum structure), −17.2 (tetra-*tert*-butylcyclobutadiene dication, C_2 minimum structure).[44]

For the heavy analogs of the cyclobutadiene dication derivatives, no experimental information on stable compounds of the $R_4E_4^{2+}$ type (E = Si, Ge, Sn, Pb) is available at the present time (Scheme 6.4: E = Si–Pb). Calculations on the parent tetrasilacyclobutadiene dication (Scheme 6.4: E = Si, R = H) revealed that the most stable isomer on the $Si_4H_4^{2+}$ PES has a slightly puckered (9.0°) D_{2d} structure, similar to that of its carbon counterpart, the cyclobutadiene dication (Scheme 6.4: E = C, R = H), although the latter is considerably more puckered (32.5°) than the former.[45,46] However, despite its nonplanar geometry, the tetrasilacyclobutadiene dication features equivalent cyclic Si–Si bonds of 2.268 Å, which points to some extent of electron delocalization and, consequently, 2π-electron aromaticity of the species.[45,46] Considering the perspectives for the experimental realization of the preparation of tetrasilacyclobutadiene dication and its heavier analogs, one can imagine either two-electron oxidation of tetrasilacyclobutadiene or oxidative dehalogenation of dihalotetrasilacyclobutene among the most feasible synthetic strategies.

6.3.2 6π-Electron Species

These compounds can be classified into two main groups: neutral (benzene and its homologs) and charged {anionic (cyclopentadienide ion, cyclobutadiene dianion) and cationic [cycloheptatrienylium (tropylium) ion, cyclooctatetraene dication]}. The stability of the neutral 6π-electron aromatic compounds is generally associated with their chemical inertness, whereas the charged aromatic species, even being thermodynamically stabilized by π-delocalization, preserve their inherent high reactivity. In this section, we will deal with the most significant recent experimental accomplishments in the field of the stable heavy analogs of such well-known 6π-electron aromatics as benzene, cyclopentadienide ion, cyclobutadiene dianion and tropylium ion.

6.3.2.1 Benzene (and its Homologs) Derivatives

6.3.2.1.1 Theoretical Studies

The experimental accomplishments in the field of the heavy arenes were preceded by numerous computational studies, which were recently comprehensively covered by Apeloig and Karni,[9b] the most important of which are discussed below.

The properties of silabenzene and the degree of its aromaticity compared with that of benzene have been extensively studied by many research groups; all of them agreed that silabenzene is markedly aromatic. It was found to possess a planar six-membered ring with an Si–C bond length of 1.760 Å (at 3-21G*)[47] and 1.771 Å (at B3LYP/6-311+G**),[48a] being just intermediate between those of the normal Si–C and Si=C bonds. These diagnostic structural features were taken as a direct manifestation of the π-delocalized system, thus confirming the aromaticity of silabenzene, at least to some observable extent. Energetic considerations also pointed to a significant aromaticity of silabenzene of ca. 70–85% of that of benzene, depending on the reference compounds and computational level.[47,49] The magnetic criterion also supported the aromaticity of silabenzene, in which the NICS value of −9.1 was virtually identical to that of benzene (−9.7).[48a] The isomeric sila(Dewar)benzene was destabilized by 38.4 kcal/mol compared with silabenzene, although these early calculations employed the relatively low HF/3-21G*//HF/STO-3G level of theory.[50] One can compare this value with the experimentally found 60 kcal/mol for the energy difference between benzene and (Dewar)benzene.[9b] Several substituted silabenzenes (substituents F, Cl, OH, SH, OCH_3, NH_2, PH_2, CH_3, SiH_3, CN, NO_2, COOH at the *ipso, ortho, meta* and *para* positions) were studied by Baldridge and Gordon, who have also made a comparison between them and the analogous substituted benzenes from the viewpoint of relative energies, degree of aromaticity and electron density distributions.[51] Subsequently, Baldridge *et al.* reported another comprehensive computational study of the electronic structures and properties (including aromaticity) of the series of silabenzenes of the type $Si_nC_{6-n}H_6$ ($n = 1-6$).[52] The aromatic stabilization of germabenzene was predicted to be comparable with that of silabenzene, 17.97 vs 16.75 kcal/mol, whereas such stabilization sharply dropped on going from germabenzene to stannabenzene, 16.75 vs 12.20 kcal/mol.[47]

All three possible isomers of disilabenzene, i.e. 1,2-, 1,3- and 1,4-disilabenzenes, were found to have planar geometries at the B3LYP/3-21G*[9b] and HF/STO-2G[53] levels. The most stable isomer was 1,3-disilabenzene, while the other two, 1,2- and 1,4-disilabenzenes, were destabilized by 6.1 and 11.1 kcal/mol, respectively.[9b] ASEs for 1,3-, 1,4- and 1,2-disilabenzenes were calculated as 47.2, 36.4 and 22.5 kcal/mol (3-21G basis set); these values, compared with that of benzene of 59.0 kcal/mol, correspond to 80, 62 and 38% of the aromaticity of benzene, respectively.[53] Accordingly, all cyclic bond lengths were intermediate between the corresponding single and double bond lengths, thus reflecting the π-electrons' delocalization.[9b]

The planar D_{3h} geometry was found as a minimum on the PES for 1,3,5-trisilabenzene studied by Gordon *et al.* at the RHF and MP2 levels [6-31G(d) and 6-311G(d,p) basis sets],[54] Vacek *et al.* at the HF/TZ2P and CCSD(T)/DZP//CISD/DZP,[55] and Apeloig and Karni at the MP2/6-31G**[9b] levels of theory. The resonance energies of 1,3,5-trisilabenzene and benzene were calculated as 53.1 and 63.8 kcal/mol, respectively, from which it was concluded that 1,3,5-trisilabenzene possesses ca. 83% of the aromaticity of benzene.[55] Indeed, the optimized Si–C bond lengths in the 1,3,5-trisilabenzene molecule of 1.754–1.759 Å were intermediate between those of the typical Si–C single and Si=C double bond lengths, and the C–Si–C and Si–C–Si bond angles were found to be 120.5° and 119.5°, respectively, pointing to aromaticity of the system. Relative to its valence isomers, 1,3,5-trisilylene and trisilaprismane, 1,3,5-trisilabenzene was the most favored by 35.0[55] and 35.3[54b] kcal/mol, respectively.

Perhaps one of the most comprehensively studied heavy arenes is the full silicon analog of benzene; that is, hexasilabenzene Si_6H_6 (as well as its derivatives), in which all cyclic carbon atoms are replaced with silicons. The early HF/3-21G level calculations found Si_6H_6 to possess D_{6h} symmetry with equivalent Si–Si bond lengths of 2.220 Å, which are longer than normal Si=Si double bonds and shorter than Si–Si single bonds.[56] This was also corroborated by the work of other groups who obtained essentially the same results.[57] However, subsequent calculations employing higher levels of theory disproved such statements, pointing to the existence of a chair-like puckered structure of D_{3d} symmetry for Si_6H_6 as the true minimum on the PES.[54b,58] Importantly, at all levels of theory the Si–Si bond distances in both D_{6h} and D_{3d} structures were found to be very similar, being intermediate between the typical Si=Si (2.15 Å) and Si–Si (2.34 Å) bond lengths, thus implying the presence of cyclic electron delocalization. The ASE value of the planar D_{6h} hexasilabenzene, depending on the theoretical method used, was reported to be ca. 50–80% that of benzene.[54a,56,57a,58e] The ASE of the nonplanar D_{3d} structure of hexasilabenzene is 2.3 kcal/mol larger than that for the planar D_{6h} isomer, reflecting the reduced ring strain in the former.[58e] The aromaticity of hexasilabenzene, suggested on the basis of energetic considerations, was further supported by the evaluation of magnetic criteria. Thus, the NICS(π, 0.5 Å above the ring center) value of Si_6H_6 of

−14.1 was just barely smaller than the respective C_6H_6 value of −16.8, indicating the relatively large π-electron delocalization in hexasilabenzene.[58e] A study on the stability of the Si_6H_6 structural isomers surprisingly showed that the most stable isomer on the PES is hexasilaprismane, which is more favorable than the planar D_{6h} hexasilabenzene, puckered D_{3d} hexasilabenzene, hexasila(Dewar)benzene and hexasilabenzvalene by 11.9, 7.7, 10.5 and 5.9 kcal/mol, respectively.[58f,59] This situation is totally reversed in relation to that of organic C_6H_6 isomers, for which benzene is by far the most stable isomer followed by prismane (+117.5 kcal/mol), (Dewar)benzene (+81.0 kcal/mol) and benzvalene (+74.8 kcal/mol).[59] The earlier calculations have drawn similar conclusions concerning the energetic preference of hexasilaprismane over hexasilabenzene.[56,57a,60] A study of the influence of the R groups on the structure and ASE of R_6Si_6 isomers revealed that the substituent effects are quite parallel to those of heavy alkenes; namely, strongly σ-accepting groups, such as F, impose the largest deviation from planarity and stretching of the Si–Si bonds, whereas σ-donating (SiH_3) and π-accepting (BH_2) substituents exhibited the opposite effect, resulting in an overall aromatic stabilization and flattening of the R_6Si_6 molecule.[58d] These results suggest that the hexasilyl-substituted hexasilabenzene $(R_3Si)_6Si_6$ with bulky R groups, required for kinetic stabilization, can be reasonably considered as a feasible candidate for the synthesis of the planar delocalized hexasilabenzene, although such an attractive synthetic target has not yet been achieved. The departure of the heavier congeners of hexasilabenzene of the type E_6H_6 (E = Ge, Sn, Pb) from planarity becomes progressively greater descending group 14: the puckering angles for Ge_6H_6, Sn_6H_6 and Pb_6H_6 are 38.0, 50.8 and 58.0°, respectively. In the same direction, the energy difference between the planar D_{6h} and puckered D_{3d} structures of E_6H_6 also increase: D_{6h} is destabilized compared with D_{3d} by 9.1 (Ge_6H_6), 23.1 (Sn_6H_6) and 63.3 (Pb_6H_6) kcal/mol.[58c] The above-described regularities were explained in terms of the increased tendency for π-electron localization going down group 14, a trend that is consistent with the progressive preference for pyramidalization on going from germyl to stannyl and plumbyl radicals.[58c,d] The subsequent calculations showed that the D_{3d} structure of hexagermabenzene is 14.5 kcal/mol more stable than its planar D_{6h} isomer, and that the aromaticity of Ge_6H_6 is comparable with that of Si_6H_6 judging from their similar values of the NICS(π, 0.5 Å above the ring center) computed for the planar D_{6h} forms: −14.4 vs −14.1 (−16.8 for C_6H_6).[58e] Similarly to the case of Si_6H_6 valence isomers, the prismane structures were found to be preferable for all heavier E_6H_6 systems, which was explained by the reluctance of the heavy group 14 elements to form double bonds, as well as their capability to form less strained polycyclic compounds.[58c,f]

6.3.2.1.2 *Experimental Accomplishments*

Silabenzene, as well as its heavier analogs, has been a long-standing synthetic attraction for many research groups, yet the isolation of the first silaaromatics was accomplished only a decade ago (see Section 6.2.1). The first room temperature stable representative of the silaaromatics, 2-silanaphthalene **15**, was prepared by Tokitoh *et al.* in 1997.[48] **15** was synthesized by the dehydrohalogenative reduction of the bromosilane precursor with *t*-BuLi in hexane, utilizing as a crucial factor the kinetic stabilization of the highly reactive molecule by a very bulky substituent, the 2,4,6-tris[bis(trimethylsilyl)methyl]phenyl group (Tbt) (Scheme 6.5).

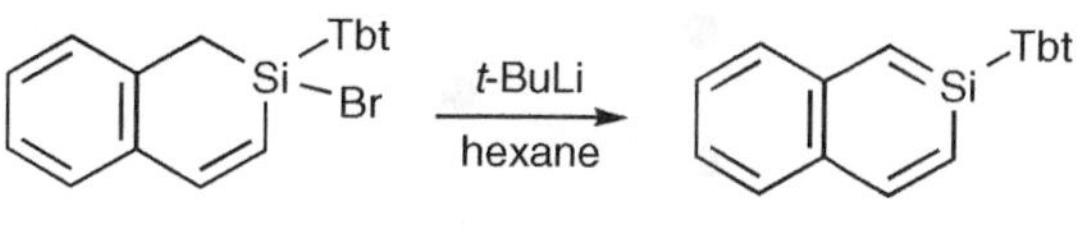

15

Scheme 6.5

Applying the same synthetic procedure and taking advantage of the protective power of Tbt and Bbt (2,6-bis[bis(trimethylsilyl)methyl]-4-[tris(trimethylsilyl)methyl]phenyl) groups, the same authors prepared the most fundamental silabenzene **16**[61] (Scheme 6.6), as well as other stable silaaromatics, such as 1-silanaphthalene **17**,[62] 9-silaanthracene **18**[63] and 9-silaphenanthrene **19**[64] (Scheme 6.6).

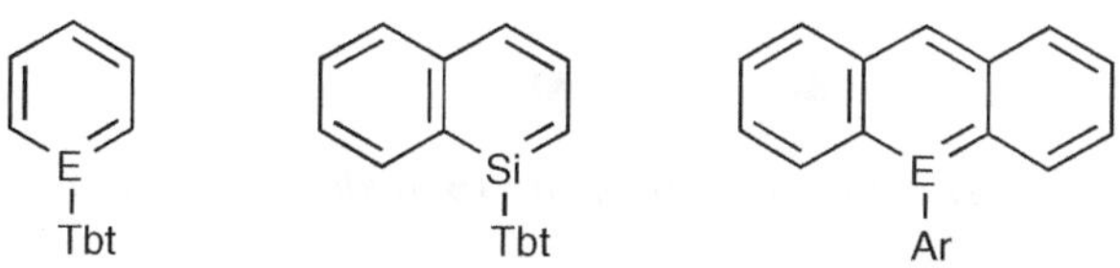

16 (M = Si)
20 (M = Ge)

17

18 (E = Si, Ar = Tbt)
22 (E = Ge, Ar = Tbt, Bbt)

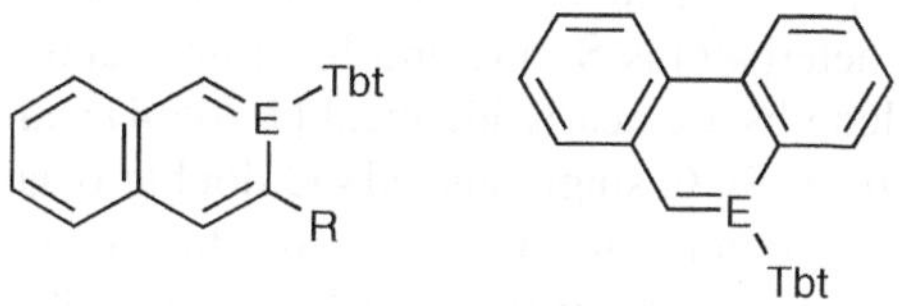

21 (E = Ge, R = H)
24 (E = Sn, R = *t*-Bu)

19 (E = Si, Ar = Tbt)
23 (E = Ge, Ar = Tbt, Bbt)

Scheme 6.6

Descending group 14, the problem of kinetic stabilization of heavy aromatics becomes more important, reflecting the general reluctance of the heavier elements (particularly Sn and Pb) for s,p-hybridization and formation of multiple bonds. Therefore, while several germaaromatic derivatives have been synthesized as stable compounds, including germabenzene **20**,[65] 2-germanaphthalene **21**,[66] 9-germaanthracene **22**[67] and 9-germaphenanthrene **23**[67] (Scheme 6.6), the number of corresponding stable stannaaromatic compounds is limited to the lone example of the 2-stannanaphthalene derivative **24**[68] (Scheme 6.6). Synthesis of a stable stannabenzene derivative has not been accomplished yet, whereas the attempted synthesis of 9-stannaphenanthrene resulted in the formation of its head-to-tail dimer, although the intermediacy of 9-stannaphenanthrene at −78 °C was confirmed by trapping reactions.[69] No stable representatives of aromatic compounds incorporating Pb atoms in the ring have been reported thus far.

The most fundamental question in the chemistry of sila/germabenzene and their homologs relates to their aromaticity, compared with that of their carbon counterparts: is

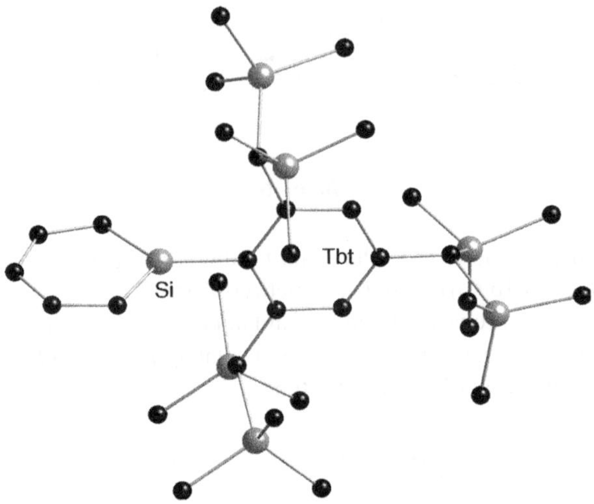

Figure 6.2 *Crystal structure of silabenzene **16** (hydrogen atoms are not shown)*

this a real, observable and measurable phenomenon? The requirements of the structural aromaticity criterion are satisfied for the sila-, germa- and stannaaromatic compounds discussed above: all of them possess almost planar six-membered rings with a trigonal-planar geometry around the heteroatoms Si, Ge and Sn. Importantly, in silabenzene **16**, the two adjacent Si–C bond lengths are nearly identical [1.765(4) and 1.770(4) Å], values that are between those of typical Si–C single and Si=C double bonds (Figure 6.2).

Moreover, the cyclic C–C bonds in **16** are also almost of the same lengths [1.381(6)–1.399(6) Å], and very similar to those in benzene (1.39–1.40 Å). The same tendency in the skeletal E–C and C–C bond lengths is also operative for other heavy arenes, including polynuclear derivatives (except for the 2-sila- and 2-germanaphthalene derivatives **15** and **21**, for which a small C–C bond alternation was observed similar to that of the parent naphthalene) (Scheme 6.6). The calculated geometries for these compounds were in fairly good agreement with the experimental observations. The heavy group 14 element atoms were diagnostically deshielded, indicating their sp^2-hybridization, as was seen in their ^{29}Si NMR [93.6 (silabenzene **16**),[61a] 91.7 (1-silanaphthalene **17**),[62a] 87.3 (2-silanaphthalene **15**),[48a] 87.2 (9-silaanthracene **18**),[63] 86.9 (9-silaphenanthrene **19**) ppm],[64] and ^{119}Sn NMR [264 ppm (2-stannanaphthalene **24**)][68] spectra (Scheme 6.6). Accordingly, the resonances of the ring carbon atoms were observed in the region diagnostic of aromatic compounds. The ring protons were also characteristically deshielded, indicating the presence of a diamagnetic ring current, which was taken as another manifestation of aromaticity.[11] The experimental observations were well supported by the theoretical considerations. Thus, the NICS(1) calculations for model unsubstituted compounds provided negative values for all heavy aromatics (Table 6.2).[11]

In view of the magnetic criterion, such negative NICS(1) values point to the presence of a diamagnetic ring current, which was also demonstrated by the diamagnetic susceptibility exaltation calculations providing large negative Λ values for

Table 6.2 *NICS(1) values calculated for the model unsubstituted heavy aromatics (values for the parent organic derivatives are given in parentheses)*

Heavy aromatics (organic aromatics)	NICS(1)
Silabenzene **16**	−9.1
(benzene)	(−11.1)
Germabenzene **20**	−9.0
(benzene)	(−11.1)
1-silanaphthalene **17** (Si–containing ring)	−9.1
(naphthalene)	(−11.5)
2-silanaphthalene **15** (Si–containing ring)	−9.4
(naphthalene)	(−11.5)
2-germanaphthalene **21** (Ge–containing ring)	−9.3
(naphthalene)	(−11.5)
9-silaanthracene **18** (Si–containing ring)	−11.0
(anthracene)	(−13.7)

germabenzene and 2-germanaphthalene (−19.5 and −27.9), reasonably comparable with those for the parent benzene and naphthalene (−22.7 and −30.4).[11b] The values of ASE for sila- and germabenzene were estimated to be almost identical to those of the parent benzene: 34.0 and 33.8 kcal/mol vs 34.1 kcal/mol.[11a] Finally, the authors convincingly concluded that benzene, naphthalene, anthracene and phenanthrene analogs of the heavy group 14 elements (Si, Ge, Sn) are aromatic to a large extent; that is, the introduction of the heavy elements in the six-membered rings does not interrupt the aromatic cyclic π-delocalization.[11] As expected, the heavy arenes feature red shifts in their electronic spectra compared with those of organic congeners, reflecting the decrease in the HOMO–LUMO energy gaps: 331 (silabenzene **16**) and 326 (germabenzene **20**) nm vs 256 nm (benzene); 378 (1-silanaphthalene **17**), 387 (2-silanaphthalene **15**) and 386 (2-germanaphthalene **21**) nm vs 312 nm (naphthalene); 503 (9-silaanthracene **18**) and 520 (9-germaanthracene **22**) nm vs 375 nm (anthracene) (Scheme 6.6).[11,67]

The number of stable representatives of benzene analogs where two cyclic carbon atoms are replaced with their heavy group 14 element congeners is limited to only three examples, all of them being synthesized starting from heavy alkyne analogs. The first compound in this series, 1,4-diaza-2,3-digermabenzene **25**, reported by Power *et al.*, was smoothly prepared by the reaction of the digermyne analog Ar′GeGeAr′ [Ar′ = 2,6-(2,6-i-Pr_2-C_6H_3)$_2$-C_6H_3] with PhCN at room temperature in hexane (Scheme 6.7).[70a]

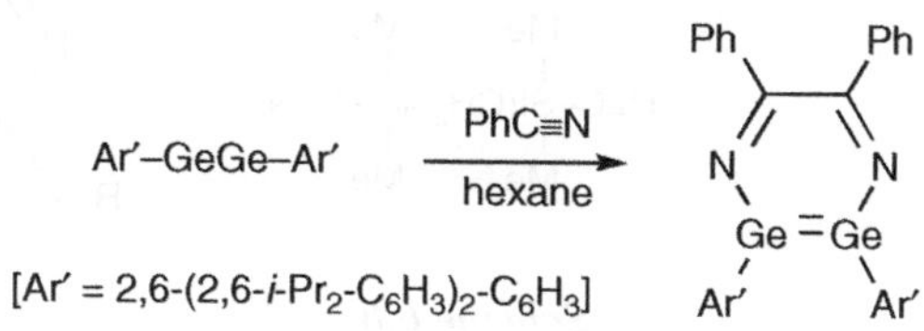

25

Scheme 6.7

The heavy pyrazine analog **25**, incorporating two Ge atoms in the ring, formally obeys Hückel's 6π-electron rule; however, in fact it lacks cyclic delocalization of its π-electrons. This was clearly seen in its peculiar structural features; namely, puckering of the six-membered ring and highly pronounced trigonal pyramidal geometry of both Ge atoms (the sum of the bond angles around the Ge atoms = ca. 326°). Moreover, the Ge–Ge bond length of 2.4963(4) Å in **25** was slightly longer than that in elemental Ge (2.44 Å), whereas Ge–N [average 1.803(3) Å] were best described as single bonds, N–C [average 1.281(3) Å] as double bonds and C–C [1.517(3) Å] as a single bond. That is, the Ge–Ge bond in **25**, which has a bond order less than one, is actually a weakened single bond rather than a double bond, which is reasonable taking into account the fact that the bond between the two Ge atoms in the starting Ar′GeGeAr′ is not a real triple bond but an approximately double bond instead.

The silicon analogs of **25**, 1,4-diaza-2,3-disilabenzene derivatives **26a,b**, were similarly synthesized by the reaction of the isolable disilyne Dis_2*i*-PrSi–Si≡Si–Si*i*-PrDis$_2$ (see Chapter 5, Section 5.3.3) with silylcyanides (Scheme 6.8).[70b,c] Although the $Si_2N_2C_2$ six-membered ring in **26b** is remarkably puckered and the geometry of the skeletal Si atoms is not trigonal-planar (the sum of the bond angles around the Si atoms = 353.0°/354.0°), the skeletal bond lengths were found to be intermediate between those of the typical single and double bonds: Si–Si = 2.2531(6) Å, Si–N = 1.678(4)/1.679(4) Å, N–C = 1.320(5)/1.341(5) Å, C–C = 1.473(6) Å.[70c] This indicates some (but by no means full) extent of 6π-electron aromatic delocalization in the 1,4-diaza-2,3-disilabenzene ring of **26b**, a conclusion that was further supported by the NICS calculations: NICS(1) = −10.0 for the Me_3Si-substituted model at the B3LYP/6-31G(d) level. The resonances of the skeletal Si and C atoms in **26** were observed at 40.2 and 164.5 ppm (for **26a**), and at 54.0 and 166.4 ppm (for **26b**).

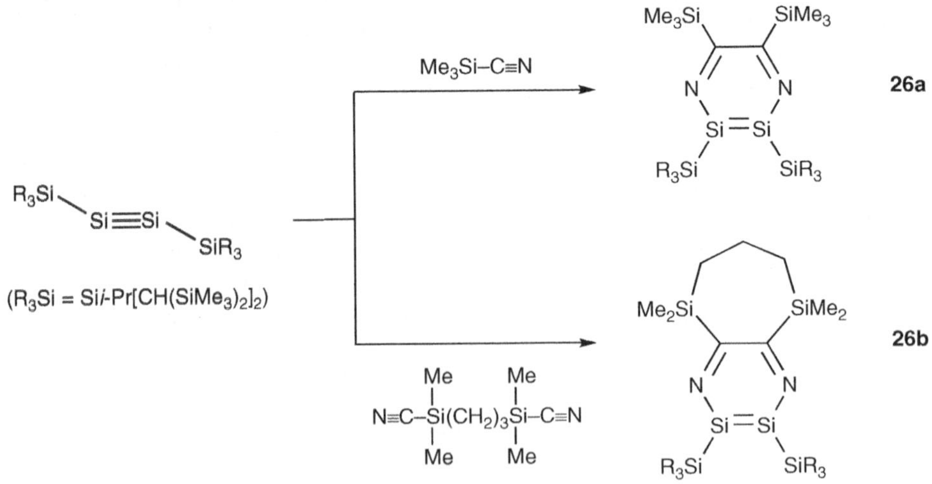

Scheme 6.8

The reaction pathways to form 1,4-diaza-2,3-disilabenzene derivatives **26** were computationally verified based on the reaction of the H_3Si–Si≡Si–SiH_3 model with

$Me_3Si–C≡N$.[70c] As the most probable mechanistic scenario, the initial σ-coordination of the nitrogen lone pair of the silylcyanide to the LUMO ($\pi^*_{in\text{-}plane}$) of disilyne to form a disilyne–silylcyanide intermediate, followed by the σ-coordination of a second silylcyanide molecule to another silicon center giving a disilyne–silylcyanide bis-adduct intermediate, was suggested. The final intramolecular C–C bond formation in the bis-adduct complex produces a 1,4-diaza-2,3-disilabenzene derivative.

Compared with the above-described 1,4-diaza-2,3-digermabenzene **25** and 1,4-diaza-2,3-disilabenzenes **26**, 1,2-disilabenzene **27** was found to be even more aromatic. It was prepared as a mixture of two regioisomers (**27a**:**27b** = 2:3) by the reaction of the disilyne Dis_2i-PrSi–Si≡Si–Sii-PrDis$_2$ [Dis = $CH(SiMe_3)_2$] with PhC≡CH in hexane at room temperature (Scheme 6.9).[71]

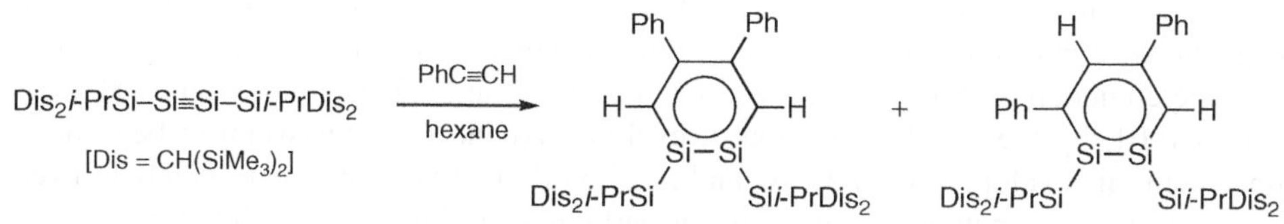

27a **27b**

(**27a**:**27b** = 2:3)

Scheme 6.9

Typically for silaaromatic derivatives, the resonances of the cyclic Si atoms were observed in low-field at 99.2 ppm for **27a** and 99.4 and 106.8 ppm for **27b**, whereas those of the ring proton and carbon atoms were found in the characteristic aromatic region. As expected, the disilabenzene ring in the isomer **27a** was almost flat, and the geometries around the cyclic Si atoms were nearly trigonal planar (the sum of the bond angles around the Si atoms = ca. 360°). The Si–Si bond length of 2.2018(18) Å was intermediate between the Si–Si single and Si=Si double bond lengths. The endocyclic Si–C bond lengths were nearly identical [1.804(4) and 1.799(5) Å], being also intermediate between those of typical single and double bonds. The two opposite cyclic C–C bonds were almost the same length [1.389(6) and 1.386(6) Å], whereas the central C–C bond was longer, 1.452(6) Å. Evidently, even upon the introduction of the two heavy group 14 elements (Si) into the six-membered ring, 1,2-disilabenzene **27a** is still able to benefit (at least, in part) from the 6π-electron aromatic delocalization, similar to the prototypical benzene and silabenzene. Accordingly, NICS(1) values for the model **27a** and **27b** (Me_3Si instead of Dis_2i-PrSi groups) were calculated to be negative: −8.0 (**27a**) and −8.1 (**27b**). According to theoretical calculations, the first step in the reaction of disilyne and phenylacetylene may involve the formation of a 1,2-disilacyclobutadiene-type intermediate, followed by its Diels–Alder reaction with a second molecule of phenylacetylene with the subsequent valence isomerization of the resulting 1,2-disila(Dewar)benzene into the final 1,2-disilabenzene.

One should also mention a transient 1,4-dimethyl-2,3,5,6-tetrakis(trimethylsilyl)-1,4-disilabenzene, which was believed to be formed upon the photolysis of its valence isomer 1,4-dimethyl-2,3,5,6-tetrakis(trimethylsilyl)-1,4-disila(Dewar)benzene, and whose intermediate formation was proved by [4 + 2] cycloaddition trapping reactions with either 2,3-dimethylbuta-1,3-diene or C_6D_6.[72]

6.3.2.2 Cyclopentadienide Ion Derivatives

The cyclopentadienide ion, $C_5H_5^-$, which obeys Hückel's $(4n + 2)$ π-electron rule, should be classified as an aromatic system and is indeed aromatic.[5] It can be easily prepared by the highly thermodynamically favorable deprotonation of cyclopentadiene, which is known to be one of the most acidic hydrocarbons with a pK_a value of 16.0,[73] to produce the cyclopentadienide ion, greatly stabilized by aromatic delocalization with all six π-electrons occupying bonding orbitals (Figure 6.3).

The spectral (NMR, IR, Raman) and crystal structure data (X-ray) point to a planar delocalized D_{5h} geometry for the cyclopentadienide ion, the degree of aromaticity being smaller but still comparable with that of benzene: ~80–90% estimated on the basis of magnetic criteria of aromaticity.[74] The cyclopentadienide ion is one of the oldest known aromatic compounds, being synthesized for the first time by Thiele more than 100 years ago, in 1901, by the simple reduction of cyclopentadiene with potassium in benzene.[75] Since then, the cyclopentadienide ion and its alkyl derivatives have been widely utilized as very useful organometallic reagents; in particular, as ligands for a huge number of transition metal complexes.

Systematic investigation of analogs of the cyclopentadienide ion, incorporating one heavy group 14 element (Si, Ge, Sn) in the ring, dates back to 1958 when Gilman and Gorsich reduced bis(dibenzosilole) with lithium to generate the first silacyclopentadienide ion derivative (Scheme 6.10).[76]

Since then, the chemistry of such silacyclopentadienide ions (and their germa- and stanna-congeners) has been greatly developed;[77,78] however, the fundamental question of aromaticity of such 6π-electron anionic species, which formally satisfy the requirements of Hückel's rule, was almost unexplored until the early 1990s. One can mention only two theoretical papers addressing this problem published in the mid 1980s and these

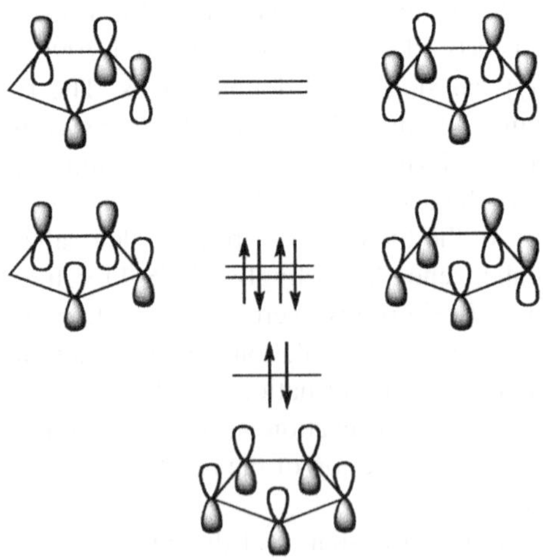

Figure 6.3 *Frontier molecular orbitals diagram of the cyclopentadienide ion*

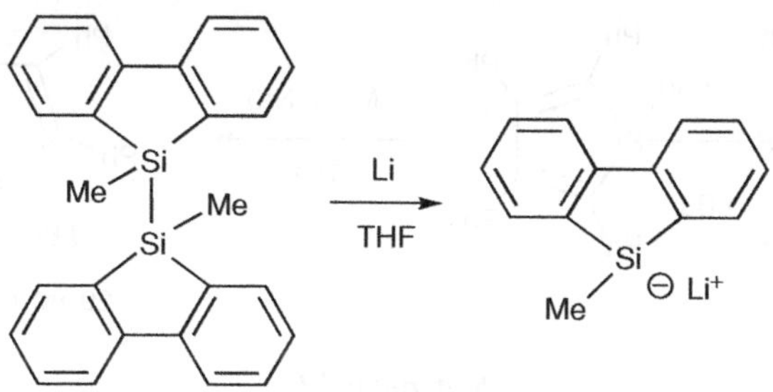

Scheme 6.10

led to contradictory conclusions. The first paper by Gordon *et al.* proposed a planar C_{2v} structure for the unsubstituted silacyclopentadienide ion $C_4SiH_5^-$, which was claimed to possess ca. 25% of the aromaticity of the parent $C_5H_5^-$. However, the evident drawback of this work was the low HF/STO-2G level utilized in the calculations.[49b] Higher HF/6-31G* level calculations by Damewood showed a pyramidal C_s conformation as the true minimum structure for the $C_4SiH_5^-$ anion, whereas the planar C_{2v} structure was found to be a transition state for the pyramidal inversion at silicon.[79] Based on such pyramidal geometry, Damewood finally concluded that the degree of aromaticity of silacyclopentadienide ion is only ca. 3% that of $C_5H_5^-$.[79]

The story of the silacyclopentadienide ion derivatives experienced a true renaissance in the 1990s, beginning with a paper published by Joo *et al.* on the generation of the 1,1-disodio-2,3,4,5-tetraphenyl-1-silacyclopentadienediide **28^{2-}•2Na$^+$** by the reductive dehalogenation of 1,1-dichloro-2,3,4,5-tetraphenyl-1-silacyclopentadiene with metallic sodium in dioxane (Scheme 6.11: M = Na, solvent = 1,4-dioxane).[80]

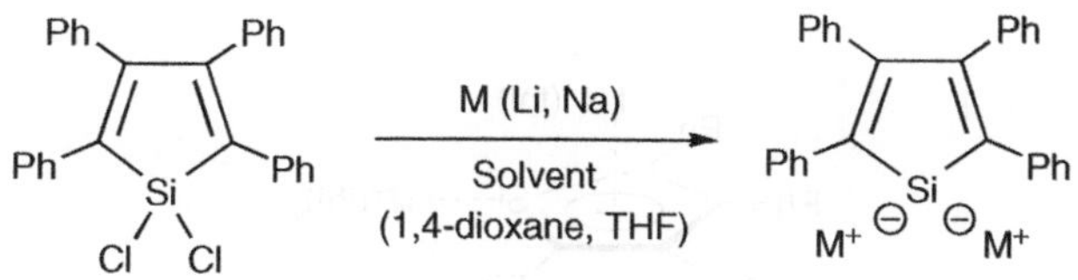

28^{2-}•2M$^+$

Scheme 6.11

The next important contribution was made by Boudjouk *et al.*, who synthesized a lithium salt of 1-*tert*-butyl-2,3,4,5-tetraphenyl-1-silacyclopentadienide **29$^-$•M$^+$** (M = Li, Na) by cleavage of the exocyclic Si–Si bond of bis(1-*tert*-butyl-2,3,4,5-tetraphenyl-1-silacyclopentadienyl) with either lithium or sodium (Scheme 6.12).[81] Based on the deshielding of the skeletal Si atom observed upon the reduction, Boudjouk *et al.* proposed a significant delocalization of the negative charge over the five-membered ring and, consequently, its 6π-electron aromaticity.[81]

Reducing 1,1-dichloro-2,3,4,5-tetraphenyl-1-silacyclopentadiene with lithium upon sonication, Boudjouk *et al.* also prepared its dianionic derivative, dilithio-2,3,4,5-tetraphenyl-1-silacyclopentadienediide **28^{2-}•2Li$^+$**, whose aromaticity was deduced from

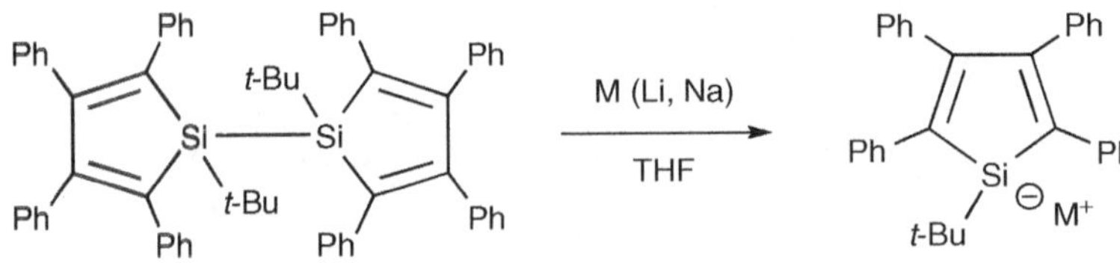

29^{-}•M^{+}(M = Li, Na)

Scheme 6.12

its characteristic NMR spectral data: a large deshielding of the Si atom ($\Delta\delta = 61.7$ ppm) and shielding of the skeletal C atoms, taking place upon reduction (Scheme 6.11: M = Li, solvent = THF).[82] Later, the crystal structure of the same compound, **28^{2-}•2Li^{+}**, was reported by West *et al.*, who described the essentially unfolded five-membered ring with almost planar geometry around the Si atom and nearly equal C–C skeletal bond lengths (1.426–1.448 Å), implying cyclic aromatic π-delocalization.[83] The most peculiar structural feature of this dianion was the coordination mode of the Li countercations: whereas the Li1 ion was pentahaptocoordinated to the five-membered ring (as would be expected for a delocalized system), the Li2 ion was bound exclusively to the Si atom (Scheme 6.13). The accompanying calculations showed two dilithio derivatives as the minimum structures on the PES: the first one closely resembling the experimental η^5-Li, η^1-Li structure, whereas the second one with C_{2v} symmetry featured the two Li ions above and below the ring plane (η^5-Li, η^5-Li structure) representing an 'inverse sandwich' structure. The latter structure was more stable than the former by 21 kcal/mol, although the solvation and crystal packing forces effects were not considered in the computations.

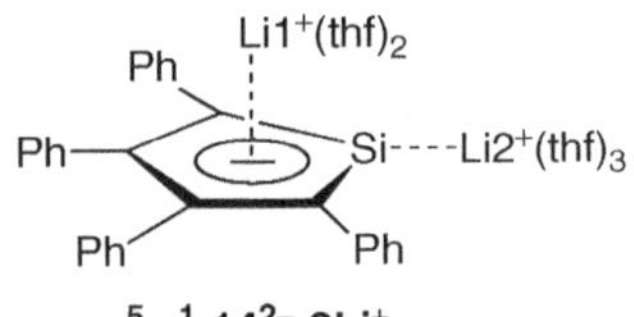

η^5,η^1-**14^{2-}•2Li^{+}**

Scheme 6.13

The experimental realization of both theoretically predicted η^5, η^1- and η^5, η^5-structures was achieved by the synthesis of the germanium version of **28^{2-}•2Li^{+}**; namely, dilithio-2,3,4,5-tetraphenyl-1-germacyclopentadienediide **30^{2-}•2Li^{+}**, prepared by the reduction of 1,1-dichloro-2,3,4,5-tetraphenyl-1-germacyclopentadiene with lithium (Scheme 6.14).[84]

The dianionic derivative **30^{2-}•2Li^{+}** crystallized from dioxane in two structurally distinct forms depending on the crystallization temperature. Thus, crystallization at −20 °C produced an 'inverse sandwich' η^5, η^5 structure of **30^{2-}•2Li^{+}** (Scheme 6.15, **A**) with the two Li ions situated above and below the five-membered ring. In contrast,

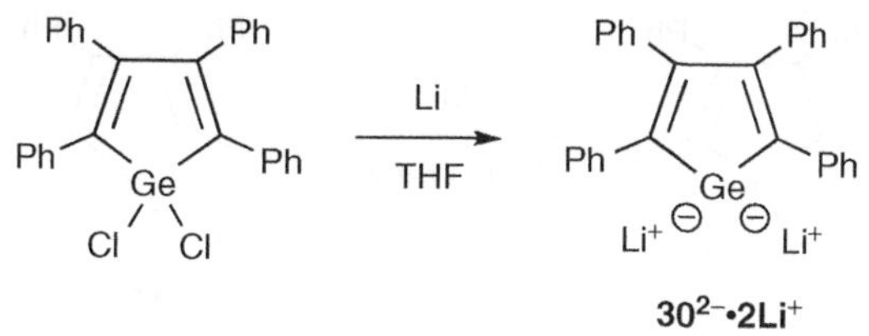

Scheme 6.14

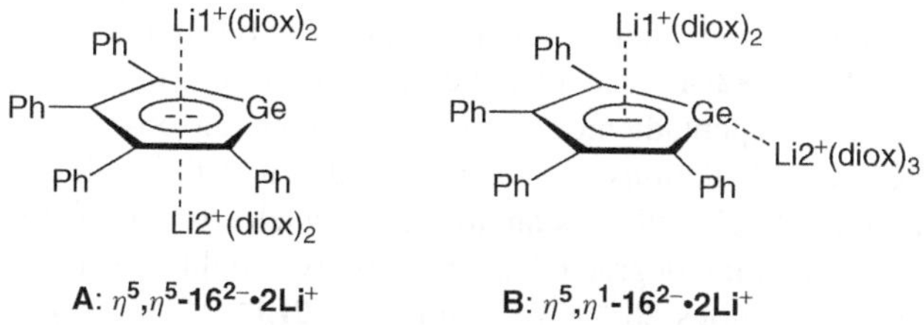

Scheme 6.15

crystallization at room temperature provided the η^5, η^1-structure of $\mathbf{30^{2-}{\bullet}2Li^+}$ (Scheme 6.15, **B**) closely resembling that of the above-discussed $\mathbf{28^{2-}{\bullet}2Li^+}$ with the only difference between them being the geometry around the element center: pyramidal at Ge for $\mathbf{30^{2-}}$[84] and nearly planar at Si for $\mathbf{28^{2-}}$.[83] The symmetrical η^5, η^5-structure of the calculated unsubstituted model of $\mathbf{30^{2-}{\bullet}2Li^+}$ was found to be 25 kcal/mol more stable than the unsymmetrical η^5, η^1-structure, similar to the above case of the silacyclopentadienediide ion.

Experimentally, both η^5, η^1- and η^5, η^5-structures of $\mathbf{30^{2-}{\bullet}2Li^+}$ featured almost identical skeletal C–C bonds, which was also supported by calculations: this observation led the authors to conclude there was a high degree of aromaticity in the germacyclopentadienediide ion, and indeed, a value of 13.0 kcal/mol was calculated from the isodesmic reaction for the ASE of the model unsubstituted $H_4C_4Ge^{2-}$.[84] The same conclusion concerning the cyclic π-delocalization in $\mathbf{30^{2-}{\bullet}2Li^+}$ was drawn earlier by Boudjouk *et al.* on the basis of ^{13}C NMR studies.[85]

The heaviest analog of the cyclopentadienide ion, that is, dilithio-2,3,4,5-tetraphenyl-1-stannacyclopentadienediide $\mathbf{31^{2-}{\bullet}2Li^+}$, was recently prepared by the reduction of either bis(1,2,3,4,5-pentaphenyl-1-stannacyclopentadienyl) or hexaphenyl-1-stannacyclopentadiene with lithium (Scheme 6.16).[86]

In contrast to the previous case of germanium derivative $\mathbf{30^{2-}{\bullet}2Li^+}$, the dianionic $\mathbf{31^{2-}{\bullet}2Li^+}$ exists in a single structural modification of η^5, η^5 type with a planar five-membered ring and nearly equivalent skeletal C–C bonds of 1.422(6)–1.446(6) Å, thus providing evidence for the delocalized aromatic nature of $\mathbf{31^{2-}{\bullet}2Li^+}$.[86c] Calculations on $\mathbf{31^{2-}{\bullet}2Li^+}$ showed the existence of two minimum structures: η^5, η^1 of C_1 symmetry and η^5, η^5 of C_2 symmetry, of which the latter agrees well with the experimental structure, being more favorable than the former by 20 kcal/mol. The delocalization of the negative

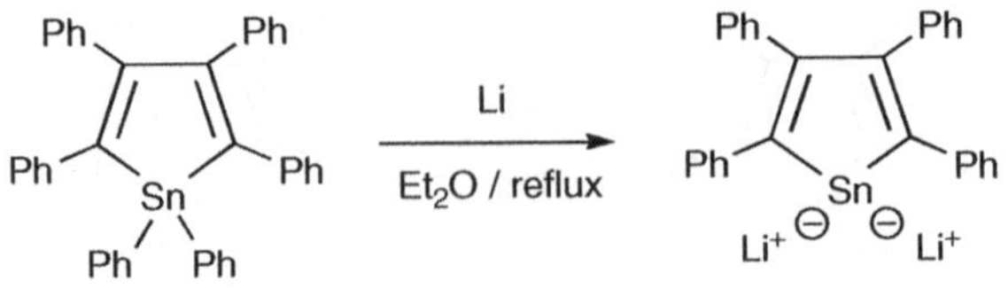

31^{2-}•2Li$^+$

Scheme 6.16

charge into the five-membered ring was also manifested in the appreciable deshielding of the Sn atom taking place upon reduction: $\Delta\delta = 251.3$ ppm.[86c] Very importantly, the ^{7}Li resonance of **31^{2-}•2Li$^+$** was observed at a high field of −4.36 ppm, typical of the classical lithium cyclopentadienide aromatic systems, an observation that was also supported by NICS(1) calculations, which provided a negative value of −5.96. Thus, **31^{2-}•2Li$^+$** can be reliably classified as an aromatic 6π-electron derivative, although one can reasonably expect that the degree of its aromaticity should be smaller than that of its silicon and germanium analogs **28^{2-}•2Li$^+$** and **30^{2-}•2Li$^+$**.[86c] The same group has also recently reported an interesting example of the aromatic dilithium dibenzopentalenediide bearing i-Pr$_3$Si substituents on the five-membered rings.[87]

Since the above-mentioned pioneering works, some other interesting examples of anionic heterocyclic systems, incorporating sila-, germa- or stannacyclopentadiene fragments, have been reported possessing a greater or lesser extent of 6π-electron aromaticity. These examples include exclusively dianionic species listed in Table 6.3.[88–97]

In sharp contrast, the monoanions of the heavy cyclopentadienide derivatives are represented exclusively by localized (nonaromatic) structures, featuring pyramidal geometries at the element centers, pronounced alternation of the cyclic C–C bonds and specific η^1-coordination of the countercations to the elements (Table 6.4).[88,98–100] However, upon the coordination of such monoanionic heavy cyclopentadienide ions to transition

Table 6.3 *Aromatic dianionic derivatives including a sila-, germa- or stannacyclopentadiene fragment*

Compound	Reference
Dipotassio-2,3,4,5-tetramethyl-1-silacyclopentadienediide	88
Dipotassio-2,3,4,5-tetramethyl-1-germacyclopentadienediide	88b
Dilithio-3-*n*-butyl-2-phenyl-1-silaindenediide	89
Disodio-3-*n*-butyl-2-phenyl-1-silaindenediide	89
Dipotassio-1-silafluorenediide	90, 91
Dilithio-2,3,4,5-tetraethyl-1-germacyclopentadienediide	92
(its related dianionic trimer)	(93)
Dilithio-2,3-diphenyl-1-germaindenediide	94
Disodio-2,3-diphenyl-1-germaindenediide	94
Dipotassio-9-germafluorenediide	95
Dilithio-3-*n*-butyl-2-phenyl-1-stannaindenediide	96
Dipotassio-9-stannafluorenediide	97

Table 6.4 *Nonaromatic monoanionic derivatives including a sila- or germacyclopentadiene fragment*

Compound	Reference
1-lithio-1-methyl-1-silafluorenide	98
1,2-bis(1-lithio-3-*n*-butyl-2-phenyl-1-silaindenyl)ethane	99
Lithio-1-tris(trimethylsilyl)silyl-2,3,4,5-tetramethyl-1-germacyclopentadienide	88b, 100
Lithio-1-(2,4,6-trimethylphenyl)-2,3,4,5-tetramethyl-1-germacyclopentadienide	88b
Sodio-1,2,3,4,5-pentamethyl-1-germacyclopentadienide	88
Potassio-1-trimethylsilyl-2,3,4,5-tetramethyl-1-silacyclopentadienide	88b

metal fragments, a considerable π-delocalization and aromatization of the five-membered rings was observed, manifested in a planarization at the heavy group 14 element and equalization of the C–C bond lengths. This was described for the first time by Tilley *et al.*, who prepared a variety of transition metal complexes with either sila- or germacyclopentadienes as cyclic polyene ligands.[101]

The groundbreaking experimental achievements in the synthesis of the heavy analogs of the cyclopentadienide ion revived a great deal of theoretical interest in the issue of aromaticity of these species.[102,103] The most important contribution to this problem was made by Schleyer *et al.* in a series of papers published in the mid 1990s. Using high-level computations, they systematically studied the analogs of the cyclopentadienide ion $C_4H_4EH^-$ ($E = C$, Si, Ge, Sn, Pb) and cyclopentadienediide ion $C_4H_4E^{2-}$ (as well as their lithium derivatives) and found an appreciable aromaticity for these compounds.[102]

This conclusion largely disagreed with those of the earlier low-level calculations, which predicted either a small[49b] or almost negligible[79] degree of aromaticity for the silacyclopentadienide ion. The latest calculations by Schleyer *et al.* showed also that the magnitudes of ASE and Λ of the silacyclopentadienide ion $C_4H_4SiH^-$ were about half the values for the parent aromatic cyclopentadienide ion, $C_5H_5^-$.[102a] Moreover, the η^5-coordination of the Li^+ countercation to the silacyclopentadienide ring markedly increases the aromaticity of the latter (compared with that of the 'naked' $C_4H_4SiH^-$), as demonstrated by planarization at the Si atom, appreciable equalization of the cyclic C–C bonds, increase in the ASE of η^5-Li-silacyclopentadienide up to 80% (compared with that of the η^5-Li-cyclopentadienide) and increase in the Λ value of η^5-Li-silacyclopentadienide to be as large as that of η^5-Li-cyclopentadienide (-14.1 ppm vs -14.5 ppm).[102b] That is, a thorough consideration of the geometric, energetic and magnetic features together led to the conclusion that the η^5-Li-silacyclopentadienide is highly aromatic, even more so than the 'naked' silacyclopentadienide ion itself. The dianion $H_4C_4Si^{2-}$, as well as its alkali metal salts $H_4C_4SiLi_2$, $H_4C_4SiNa_2$ and $H_4C_4SiK_2$, were also found to be highly aromatic.[102c] Thus, for example, the most stable configurations of the salts $C_4H_4SiLi_2$, $C_4H_4SiNa_2$ and $C_4H_4SiK_2$ were represented by η^5, η^5 'inverse sandwiches' with equalized C–C bond lengths diagnostic of their delocalized structures, and markedly negative values of Λ pointing to large aromatic ring currents and responsible for the shielding of the Li nucleus in $C_4H_4SiLi_2$ (calculated δ value of the ^7Li NMR resonance is -7.7 ppm).[102c] The influence of the nature of the heavy group 14 element on the overall degree of aromaticity of mono- and dianions was also systematically studied by Goldfuss and Schleyer.[102d] Thus, the aromaticity of C_s monoanions $C_4H_4EH^-$ (E = C, Si, Ge, Sn, Pb) notably dropped descending group 14 because of the increasing pyramidality at E and inversion barriers on going down from C to Pb.

Compared with the pyramidal C_s geometries, the planar C_{2v} configurations of monoanions $C_4H_4EH^-$ were found to be significantly more aromatic; however, they represented only the transition states for pyramidal inversion. In contrast, the degree of aromaticity of dianions $C_4H_4E^{2-}$ and their dilithium derivatives $C_4H_4ELi_2$, which do not suffer from the problem of pyramidalization at E, was remarkably constant for all derivatives regardless of the element E.[102d] This made the dilithium salt derivatives $C_4H_4ELi_2$ the most aromatic among all calculated structures, even more aromatic than the free dianions $C_4H_4E^{2-}$.

The particularly exciting example of a cyclopentadienide ion analog consisting only of heavy group 14 elements, namely, pentasilacyclopentadienide ion $H_5Si_5^-$, was computationally studied by Schleyer *et al.*[45] Two nonplanar minimum structures of C_s and C_2 symmetry with equal energies were located on the PES, whereas the planar delocalized D_{5h} structure was found to be 8.3 kcal/mol higher in energy. The value of ASE for the C_s form of $Si_5H_5^-$ was calculated from the isodesmic equation as 52.8 kcal/mol, being much smaller than the value of 84.7 kcal/mol calculated for its carbon analog, $C_5H_5^-$, pointing to a significantly lower degree of aromaticity of $Si_5H_5^-$.[45] Nevertheless, such an ASE value (52.8 kcal/mol) was still much greater than that for the silacyclopentadienide ion $C_4H_4SiH^-$ (2.2 kcal/mol), calculated on the basis of an analogous isodesmic reaction.[79] However, experimentally the highly challenging persilacyclopentadienide derivatives $R_5Si_5^-$, as well as their heavier analogs $R_5E_5^-$ (R = Ge, Sn, Pb), have not been realized, remaining as a very attractive target for organometallic chemists.

At present, the closest approach to the above-mentioned problem was achieved by Lee *et al.*, who reported the first analog of the cyclopentadienide ion incorporating *three* heavy group 14 elements in the ring: two Si and one Ge atoms.[104] This novel heavy cyclopentadienide derivative, lithium 1,2-disila-3-germacyclopenta-2,4-dienide **32$^-$•[Li$^+$(thf)]**, was prepared by the reduction of the most electronically perturbed metallacyclopentadiene **33** featuring the Si=Ge–C=C diene system[105] with KC_8, followed by exchange of the countercation from K^+ to Li^+ by treatment with LiBr (Scheme 6.17).[104]

Because the LUMO of **33** is largely localized on the Si=Ge π*-orbital, the first step of the reaction was proposed to be the two-electron reduction of the Si=Ge double bond followed by the β-elimination of *t*-Bu_2MeSiK. The driving force for the last step would be the release of the highly destabilizing electrostatic repulsion between the two negative charges in **34^{2-}•2K$^+$** and the final aromatization of the cyclic system. Consideration of the structural peculiarities of **32$^-$•[Li$^+$(thf)]** definitely points to its delocalized structure (Figure 6.4). Because the skeletal atoms in **32$^-$** are different (two C, two Si and one Ge), a discussion in terms of equalization of the cyclic bonds is meaningless in such a case. Instead, one should take into account the general tendencies in the change of these skeletal bonds taking place upon the reduction. Indeed, all double bonds of the starting **33**[105] (Si=Ge and C=C) were *elongated* to form the final lithium heavy cyclopentadienide **32$^-$•[Li$^+$(thf)]**[104] [Si=Ge: 2.250(1) vs 2.3220(5) Å; C=C: 1.343(5) vs 1.402(2) Å], whereas all single bonds of **33** (Ge–C, Si–Si, and Si–C) were *shortened* [Ge–C: 1.972(3) vs 1.9303(17) Å; Si–Si: 2.364(1) vs 2.2403(7) Å; Si–C: 1.888(3) vs

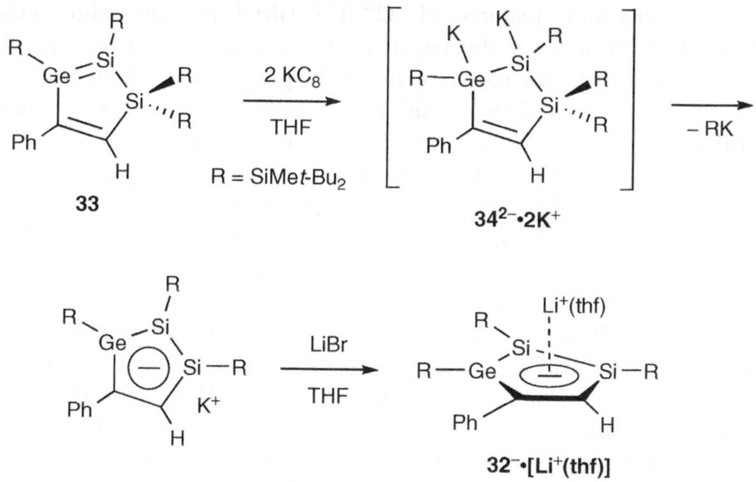

Scheme 6.17

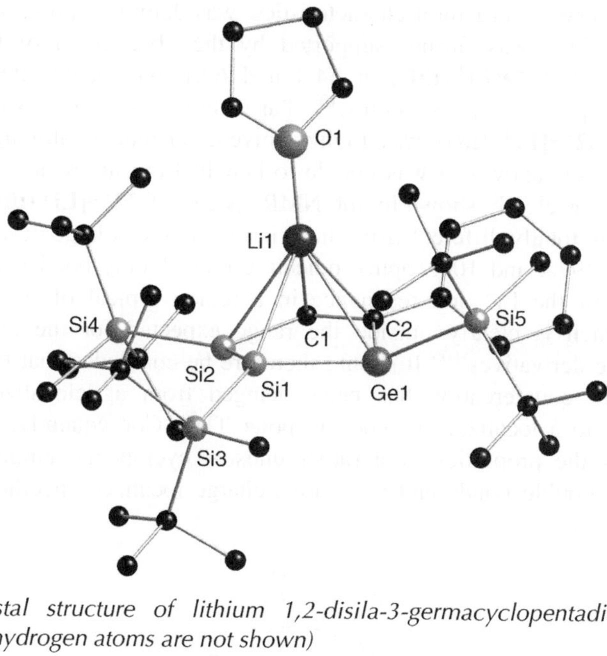

Figure 6.4 *Crystal structure of lithium 1,2-disila-3-germacyclopentadienide derivative* ***32⁻•[Li⁺(thf)]*** *(hydrogen atoms are not shown)*

1.8269(18) Å]. Importantly, the lengths of all skeletal bonds in **32⁻** were intermediate between those of the typical single and double bonds. Moreover, the planarity of the C_2Si_2Ge ring of the starting **33** was almost unaffected by the reduction to form anionic **32⁻**.[104] Furthermore, the Li^+ cation in **32⁻• [Li⁺(thf)]** was diagnostically pentahapto-coordinated to the anionic ring, resulting in the formation of a classical 'half-sandwich' complex typical of the aromatic cyclopentadienide derivatives of the C_5R_5Li type.

Overall, these structural features of **32^{-}•[Li^{+}(thf)]** provide solid evidence for the significant delocalization of the negative charge electron density over the entire five-membered ring; that is, the requirements of the geometrical criterion of aromaticity are definitely satisfied for **32^{-}•[Li^{+}(thf)]**. However, the extent of aromaticity of **32^{-}•[Li^{+}(thf)]** is certainly smaller than that of the prototypical lithium cyclopentadienide: in contrast to C_5H_5Li, where the negative charge is expected to be equally distributed between all carbon atoms, one cannot assume such an even distribution of electron density over all skeletal atoms in **32^{-}•[Li^{+}(thf)]** simply because of their different electronegativities. Instead, a major part of the negative charge is expected to be accumulated on the Ge atom, whose pyramidalization degree was the most appreciable of all the skeletal atoms. With respect to the magnetic criterion of aromaticity, **32^{-}•[Li^{+}(thf)]** should also qualify as an aromatic compound. Thus, the NICS(1) value for the model compound **32^{-}•Li^{+}** (H_3Si instead of *t*-Bu_2MeSi substituents) was negative, −8.4, being comparable with (albeit smaller than) those of the classical 6π-electron aromatic derivatives: C_5H_5Li [NICS(1) = −10.3] and C_6H_6 [NICS(1) = −11.2].[104b] Accordingly, the ^7Li NMR resonance of **32^{-}•[Li^{+}(thf)]** was observed in toluene at a high field, −5.4 ppm, characteristic of the aromatic lithium cyclopentadienide derivatives.[106] That is, the aromaticity of **32^{-}•[Li^{+}(thf)]**, initially deduced from its crystal structural characteristics, was definitely preserved in nonpolar toluene solution. This was further supported by the observation of the skeletal Si atom resonances of **32^{-}•[Li^{+}(thf)]** at 54.4 and 69.1 ppm, values that are between those of typical sp^2- and sp^3-Si atoms.[107] Particularly interesting was the fact that the structure of **32^{-}•[Li^{+}(thf)]** was highly solvent-dependent: although aromatic in nonpolar toluene (see above), it was unable to benefit from aromatic delocalization in polar THF. This was clearly shown by the NMR spectra of **32^{-}•[Li^{+}(thf)]** measured in THF-d_8, which are totally different from those in toluene-d_8: a large deshielding of both skeletal Si atoms (97.4 and 104.9 ppm) indicating their doubly bonded character,[107,108] and observation of the Li^{+} ion resonance in a region typical of η^1-germyllithiums, −0.6 ppm,[109] which is clearly outside the range expected for the aromatic lithium cyclopentadienide derivatives.[106] It should therefore be concluded that the coordination mode of the Li^{+} countercation has been changed from a delocalized η^5-mode in nonpolar toluene to a localized η^1-mode in polar THF. Consequently, **32^{-}•[Li^{+}(thf)]** in THF acquires the properties of a rather unusual cyclopentadienide ion, featuring Si=Si and C=C double bonds and a negative charge localized specifically on the Ge

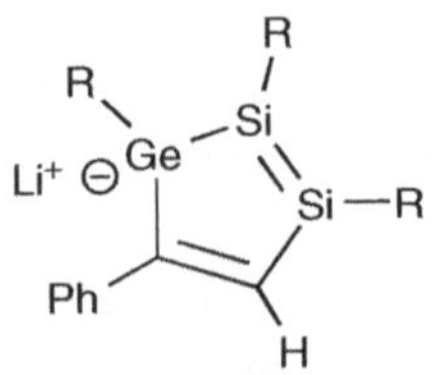

R = SiMe*t*-Bu$_2$

32^{-}

Scheme 6.18

atom (Scheme 6.18). That is, the heavy cyclopentadienide derivative $\mathbf{32^{-}\bullet[Li^{+}(thf)]}$ should be classified as an aromatic compound in the solid state as well as in nonpolar solvents, such as toluene and benzene, whereas it should be considered as a nonaromatic species in polar THF. One should also note that $\mathbf{32^{-}\bullet[Li^{+}(thf)]}$ appears to be the first monoanionic heavy cyclopentadienide derivative whose aromaticity was reliably established experimentally and verified computationally,[104] thus representing an experimental realization of Schleyer's theoretical prediction about the significant aromaticity of lithium silacyclopentadienide C_4H_4SiHLi.[102b]

As a related system isoelectronic with the cyclopentadienide ions, one should also mention N-heterocyclic silylenes, germylenes, stannylenes and plumbylenes, whose 6π-electron aromaticity stems from the cyclic p_π–p_π delocalization involving C=C bond π-electrons and endocyclic N lone pairs (Scheme 6.19).[110] Such accomplishments in the field of heavy group 14 elements were preceded by the pioneering work of Arduengo *et al.* on the synthesis and crystallographic determination of the first stable singlet N-heterocyclic carbenes **35a–e** (Scheme 6.19).[111] Shortly afterwards, the first representative of the heavy analogs of such five-membered ring carbenes, namely, N-heterocyclic germylene **36a**, was prepared by Herrmann's group,[112] followed by the synthesis of the stable silylenes **38** by Denk *et al.*[113] and **39a,b** by Lappert *et al.*[114] Since then, several other remarkable derivatives of this class have been prepared,

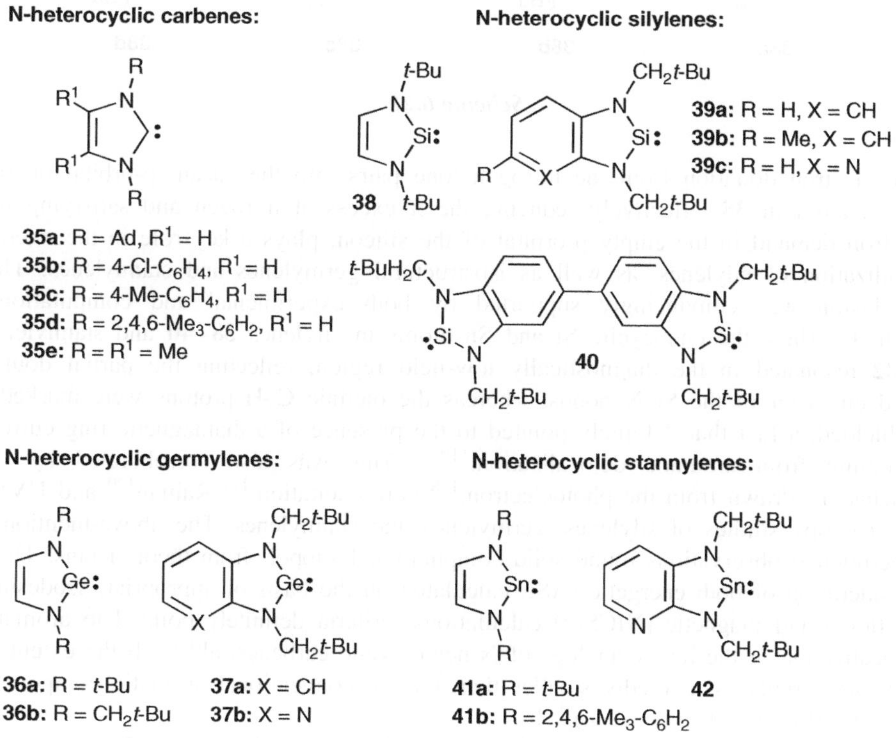

Scheme 6.19

including pyrido-fused silylene **39c**, germylene **37b** and stannylene **42**,[115] germylenes **36b** and **37a**,[116] and stannylenes **41a,b**.[117] Mainly, these compounds were prepared by the reductive dehalogenation of the corresponding dichlorides (in the case of silylenes) or by the direct cyclization of dilithio-1,4-diamides with either $GeCl_2$•diox or $SnCl_2$ (in the case of germylenes and stannylenes).

The most peculiar feature of all these five-membered ring compounds was their remarkable thermal stability – for example, silylene **38** indefinitely survives heating at 150 °C in toluene, being decomposed only at its melting point of 220 °C – and chemical inertness: **38** was amazingly unreactive compared with transient silylenes, being inert toward Et_3SiH and PhC≡CPh (effective silylene scavengers) and pyridine, Me_3P, Et_3N and THF (Lewis bases).[113] Such unusual properties of N-heterocyclic silylenes, as well as their heavier congeners, germylenes and stannylenes, were explained in terms of appreciable $(4n + 2)$ π-electron cyclic delocalization leading to aromatic stabilization of the whole system (Scheme 6.20).

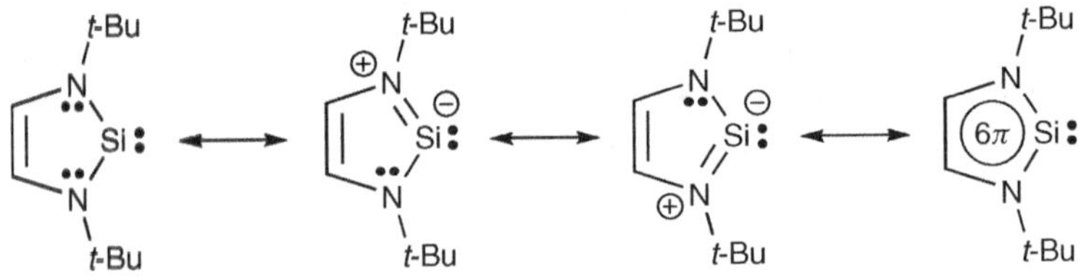

Scheme 6.20

π-Electron donation from the nitrogen lone pairs into the vacant p-orbital on the silicon atom in **38**, effectively reducing the π-excess at nitrogen and satisfying the electron demand in the empty p-orbital of the silicon, plays a key role in the overall stabilization of silylenes, as well as isostructural germylenes and stannylenes. This conclusion was convincingly supported by both experimental and computational methods. Thus, the endocyclic Si and Sn atoms in silylenes **38–40** and stannylenes **41,42** resonated in the diagnostically low-field region, reflecting the partial double bond character of the Si–N bonds, whereas the olefinic C–H protons were markedly deshielded, a fact that definitely pointed to the presence of a diamagnetic ring current stemming from aromatic delocalization.[110a,b] This was also corroborated by the conclusions drawn from the photoelectron,[118] core excitation,[119] Raman[120] and UV[117] spectroscopy studies of silylenes, germylenes and stannylenes. The above-mentioned experimental observations found solid computational support from theoreticians. Thus, consideration of both energetic (ASE calculated on the basis of appropriate isodesmic reactions) and magnetic [NICS(2) calculations] criteria definitely pointed to aromatic delocalization in the heavy analogs of N-heterocyclic carbenes, although the extent of such aromaticity is markedly smaller than that in conventional aromatic compounds (ca. 50–60% of that in benzene).[110,121]

In distinct contrast, the bonding picture in the N-heterocyclic silylene **38** and germylene **36a** was viewed by Arduengo *et al.* in terms of σ-donation from a diimine to a neutral silicon (or germanium) atom, based on the experimental data of solid-state NMR and photoelectron spectroscopy.[122] This bonding model, implying the major contribution of the neutral resonance form featuring localized C=C double bond and nitrogen p_π-lone pairs, did not recognize a dominant role for π-delocalization in the stabilization of the N-heterocyclic carbenes. Such a viewpoint was opposed by groups of theoreticians, who pointed to a significant cyclic electron delocalization (aromaticity) of N-heterocyclic carbenes (as well as isostructural silylenes and germylenes), deduced from energetic, magnetic and structural considerations.[121a,123] This was evidenced by, for example, a significant electronic population of the vacant p_π-orbital of the carbene center (55.8% of a perfect π-delocalization).[123] The extent of such cyclic delocalization in silylenes was found to be smaller than that in isostructural carbenes, in which it was in turn markedly smaller compared with prototype aromatics, such as the imidazolium ion and benzene.

6.3.2.3 Cyclobutadiene Dianion Derivatives

As was discussed in Section 6.3.1.2, the behavior of small doubly charged systems, such as the 2π-electron cyclobutadiene dication, is greatly dictated by the severe Coulomb repulsive interaction between the two charges, which can outbalance the advantage gained through cyclic π-delocalization and aromatic stabilization.[39a,42,43] The same trends are evidently operative in the case of the doubly negatively charged 6π-electron cyclobutadiene dianion species (Scheme 6.21: E = C), whose story is sharply distinct from that of its singly negatively charged 6π-electron analog, the cyclopentadienide ion (see Section 6.3.2.2).

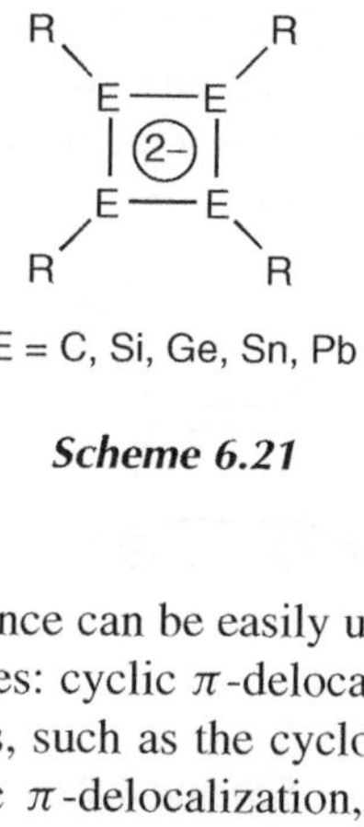

Scheme 6.21

The origin of such a great difference can be easily understood in terms of the interplay between the two principal tendencies: cyclic π-delocalization and thermodynamic stabilization. If the monoanionic species, such as the cyclopentadienide ion, can (and indeed do) greatly benefit from the cyclic π-delocalization, resulting in the overall aromatic stabilization of the molecule, the same delocalization in the doubly charged systems, such as the cyclobutadiene dianion, has the opposite, destabilizing, effect because of the unavoidable mixing of the two extra electrons, which cannot be efficiently separated from each other, resulting in a strong Coulomb repulsion. As in the case of the cyclobutadiene dication, such a Coulomb repulsion can overwhelm the delocalization stability, resulting in a highly pronounced trend for electron localization and separation, which may finally cause a significant departure of the cyclobutadiene dianion ring from planarity. Another important reason, greatly contributing to the overall destabilization

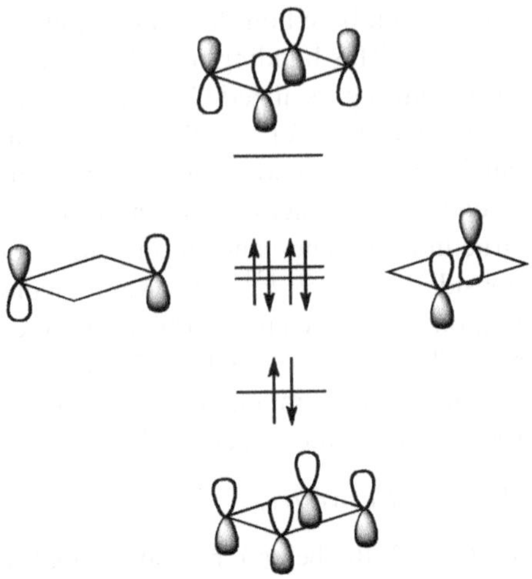

Figure 6.5 *Frontier molecular orbitals diagram of the cyclobutadiene dianion*

of the cyclobutadiene dianion molecule, is the fact that in the planar delocalized D_{4h} structure, four (out of the total of six) π-electrons are forced to occupy nonbonding orbitals to form a pair of doubly degenerate HOMOs (Figure 6.5), thus giving rise to strongly repulsive 1,3-antibonding interactions responsible for the overall destabilization and high reactivity of such species.

Indeed, an early calculation by Hess *et al.* described the structure of the cyclobutadiene dianion $H_4C_4^{2-}$ as a puckered four-membered ring featuring one negative charge localized on a cyclic carbon atom and the other allylic-type charge delocalized over the remaining three skeletal carbon atoms, thus implying a small degree of cyclic conjugation of the six π-electrons (Scheme 6.22, **A**).[42]

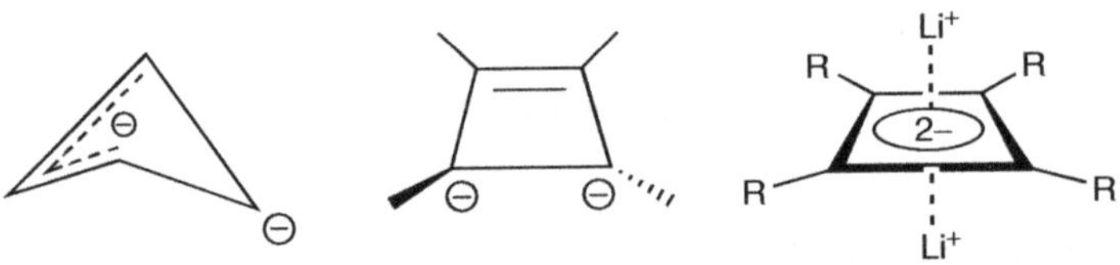

A **B** **C**

Scheme 6.22

More elaborate calculation by van Zandwijk *et al.* showed that the above-described folded structure actually possesses an imaginary frequency, being therefore a saddle point on the PES, whereas the true minimum structure of the cyclobutadiene dianion was found to be one with a C=C double bond and the two negative charges located on the other two adjacent tetrahedral carbon atoms, corresponding to an overall trapezoid shape of the four-membered ring (Scheme 6.22, **B**).[124] However, as was convincingly demonstrated by Schleyer *et al.*, the introduction of positive lithium ions effectively counteracts the destabilizing tendency of the Coulomb repulsion, markedly diminishing the degree of mixing of the extra electrons, which finally results in a much greater extent of cyclic π-delocalization and, in effect, in the overall aromatic stabilization of the dianionic system.[125] Thus, the dilithium cyclobutadiene dianion $[C_4H_4]^{2-}\bullet 2Li^+$ was calculated to have a planar 'inverse sandwich' D_{4h} structure (Scheme 6.22, **C**; R = H).[125] The experimental realization of such a theoretical prediction was achieved recently by the group of Sekiguchi, who prepared the dilithium salt of the tetrakis(trimethylsilyl)cyclobutadiene dianion $[(Me_3Si)_4C_4]^{2-}\bullet 2Li^+$, thermodynamically stabilized by the four π-accepting trimethylsilyl groups, and showed that its structure possessed a planar C_4 ring with a delocalized aromatic system of the 6π-electrons (Scheme 6.22, **C**; R = $SiMe_3$).[126] However, the problem of the synthesis of heavy analogs of the cyclobutadiene dianion, in which all skeletal C atoms are replaced with heavy group 14 elements, still unresolved, and no experimental studies on such heavy analogs of the cyclobutadiene dianion have been reported in the literature prior to our investigations. The only theoretical paper on the tetrasilacyclobutadiene dianion predicted that the most favorable structure of the unsubstituted compound $[Si_4H_4]^{2-}$ is tetrasilabicyclo[1.1.0]butane-2,4-diide of C_s symmetry, which is more stable than the planar delocalized D_{4h} configuration of the tetrasilacyclobutadiene dianion by 29.9 kcal/mol.[46]

Because one of the most convenient methods for making >E=E'< (E, E' = group 14 element) double bonds is the dehalogenative reduction of 1,2-dihalides of the type >E(X)–E'(X)< with alkali metals, at first, the appropriate precursors for the preparation of the heavy analogs of cyclobutadiene dianions, *trans*-1,2-dichloro-$^3\Delta$-1,2,3,4-disiladigermetene **43**[127] and *trans,trans,trans*-1,2,3,4-tetrabromo-1,2,3,4-tetrasiletane **44**,[104b] were designed. The formation of the target dipotassium salts of 1,2-disila-3,4-digerma- and 1,2,3,4-tetrasilacyclobutadiene dianions $\mathbf{45^{2-}\bullet 2[K^+(thf)_2]}$ and $\mathbf{46^{2-}\bullet 2[K^+(thf)_2]}$ was smoothly achieved by the reductive dehalogenation of either **43** or **44** with KC_8 in THF (Scheme 6.23).[128a]

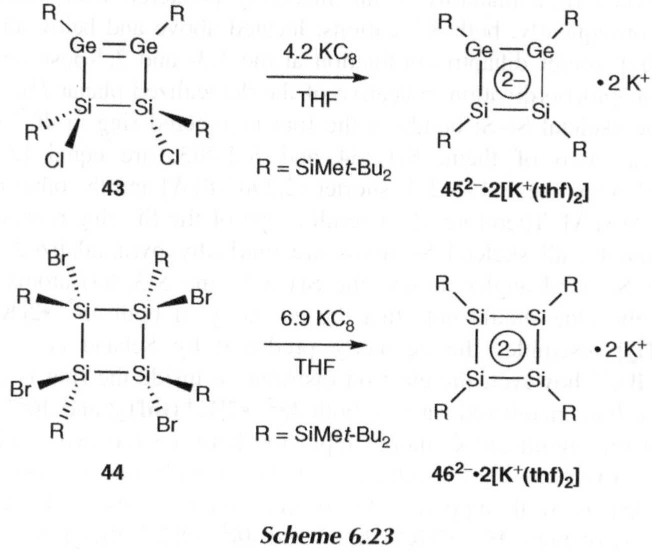

Scheme 6.23

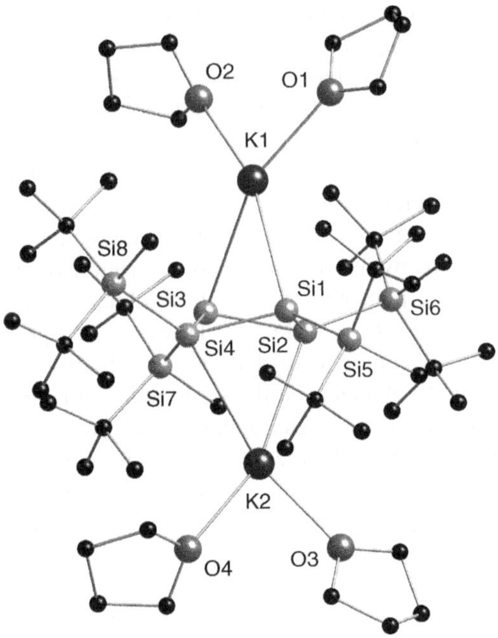

Figure 6.6 *Crystal structure of the dipotassium tetrasilacyclobutadiene dianion derivative* $\mathbf{46^{2-}\bullet 2[K^+(thf)_2]}$ *(hydrogen atoms are not shown)*

The crystal structures of both $\mathbf{45^{2-}\bullet 2[K^+(thf)_2]}$ and $\mathbf{46^{2-}\bullet 2[K^+(thf)_2]}$[104b,128a] (Figure 6.6) are very similar to each other and distinctly different from that of the above-mentioned D_{4h} aromatic carbon analog $[(Me_3Si)_4C_4]^{2-}\bullet 2Li^+$.[126a] First of all, both four-membered Si_2Ge_2 and Si_4 rings of $\mathbf{45^{2-}\bullet 2[K^+(thf)_2]}$ and $\mathbf{46^{2-}\bullet 2[K^+(thf)_2]}$ significantly deviate from planarity, being markedly puckered with folding angles of 43° and 34°. Consequently, both K^+ cations, located above and below the Si_4 ring in $\mathbf{46^{2-}\bullet 2[K^+(thf)_2]}$, prefer dihaptocoordination at the 1,3- and 2,4-positions of the ring (instead of tetrahaptocoordination indicative of the delocalized planar D_{4h} structure).

Moreover, the skeletal Si–Si bonds of the four-membered ring in $\mathbf{46^{2-}\bullet 2[K^+(thf)_2]}$ are not identical: two of them, Si1–Si4 and Si2–Si3, are equal [2.3301(8) and 2.3300(8) Å], whereas one, Si1–Si2, is shorter [2.2989(8) Å] and the other one, Si3–Si4, is longer [2.3576(8) Å]. Therefore, the overall shape of the Si_4 ring represents a folded trapeze. Remarkably, all skeletal Si atoms are markedly pyramidalized: the sums of the three Si–Si–Si bond angles around the Si1(Si2) and Si3(Si4) atoms are 341 and 326°, respectively. One should note that the geometry of both $\mathbf{45^{2-}\bullet 2[K^+(thf)_2]}$ and $\mathbf{46^{2-}\bullet 2[K^+(thf)_2]}$ resembles the geometry predicted by Schaad *et al.* for $C_4H_4^{2-}$ (Scheme 6.22, **B**);[42] however, the electron distribution inside the ring is different. The puckering of the four-membered rings in both $\mathbf{45^{2-}\bullet 2[K^+(thf)_2]}$ and $\mathbf{46^{2-}\bullet 2[K^+(thf)_2]}$ originates from the significant Coulomb repulsion between the two negative charges, which cannot be overcome by the influence of the spatially remote potassium cations, a conclusion that is well supported by recent computations.[129] Undoubtedly, the structural features of both $\mathbf{45^{2-}\bullet 2[K^+(thf)_2]}$ and $\mathbf{46^{2-}\bullet 2[K^+(thf)_2]}$ derivatives fail to satisfy the geometrical requirements of aromaticity in their classical sense; i.e. ring planarity, cyclic bond equalization (anticipated for homonuclear $\mathbf{46^{2-}\bullet 2[K^{+}(thf)_{2}]}$), and tetrahaptocoordination expected for the aromatic cyclobutadiene dianion in its D_{4h} conformation. The conditions of the magnetic criteria of aromaticity[6] are also not fulfilled: NICS(1) values for the models of $\mathbf{45^{2-}\bullet 2[K^{+}(thf)_{2}]}$ and $\mathbf{46^{2-}\bullet 2[K^{+}(thf)_{2}]}$ (Me_3Si instead of t-Bu_2MeSi substituents) were calculated to be positive, +4.3 and +6.1, respectively, pointing to the absence of an aromatic ring current. Thus, it was concluded that the 6π-electron heavy cyclobutadiene dianion derivatives $\mathbf{45^{2-}\bullet 2[K^{+}(thf)_{2}]}$ and $\mathbf{46^{2-}\bullet 2[K^{+}(thf)_{2}]}$ should be classified as nonaromatic, at least in the solid state. However, the extent of π-delocalization and, accordingly, aromaticity of the heavy cyclobutadiene dianions markedly depends on the nature of the counterions: smaller cations could approach the cyclobutadiene dianion ring more closely and thus more effectively counteract the Coulomb repulsion of the two negative charges to stabilize the system. For example, both dilithium and disodium derivatives of tetrasilacyclobutadiene dianion $\mathbf{46^{2-}\bullet 2[M^{+}(thf)_{2}]}$ **(M = Li, Na)** revealed the diagnostic tetrahaptocoordination of both M atoms to the Si_4 ring, which was notably more planar than that in the dipotassium derivative $\mathbf{46^{2-}\bullet 2[K^{+}(thf)_{2}]}$.[104b] Interestingly, when the reduction of **43** was carried out with alkaline earth metals Mg, Ca, Sr (instead of the alkali metals Li, Na, K), different products, alkaline earth metal salts of 1,3-disila-2,4-digermabicyclo[1.1.0]butane-2,4-diide $\mathbf{47^{2-}\bullet [M^{2+}(thf)_{n}]}$ (M = Mg, $n = 3$; M = Ca, $n = 4$; M = Sr, $n = 4$), were exclusively formed,[130] whose structures were very similar to that of the tetrasilabicyclo[1.1.0]butane-2,4-diide predicted to be a minimum structure for $[Si_4H_4]^{2-}$ (Scheme 6.24).[46] In effect, the above-described reaction clearly demonstrated the decisive influence of the nature of the reducing reagent, namely, alkali vs alkaline earth metals, on the structure of the final product.

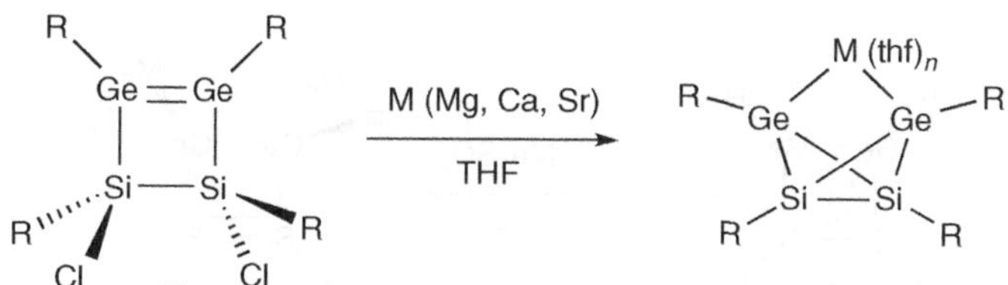

43

R = SiMe*t*-Bu$_2$

$\mathbf{47^{2-}\bullet [M^{2+}(thf)_{n}]}$
M = Mg, $n = 3$
M = Ca, $n = 4$
M = Sr, $n = 4$

Scheme 6.24

Despite the close similarity of the solid-state structures of both $\mathbf{45^{2-}\bullet 2[K^{+}(thf)_{2}]}$ and $\mathbf{46^{2-}\bullet 2[K^{+}(thf)_{2}]}$ (Scheme 6.25, **A**), their solution structures were dramatically different, which was clearly manifested by the resonances of their skeletal silicon atoms.[104b,128a] Thus, the signals of the endocyclic Si atoms in heteronuclear $\mathbf{45^{2-}\bullet 2K^{+}}$ were observed at 113.7 ppm, that is, at a low-field characteristic of doubly bonded Si atoms,[107] indicating that the most important contribution is from the resonance structure **B** (Scheme 6.25) with

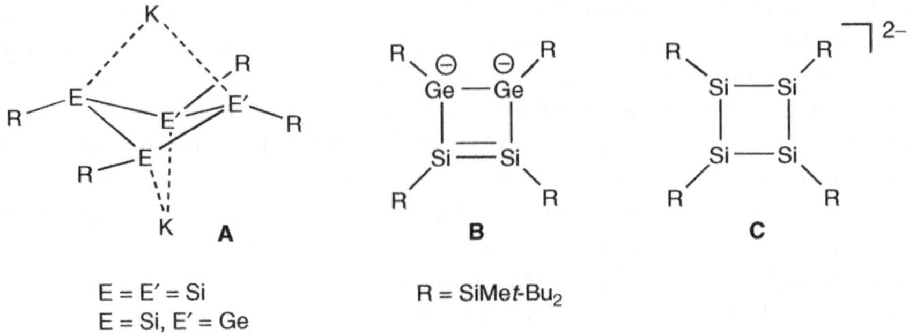

Scheme 6.25

both negative charges being situated on the more electronegative Ge atoms. Accordingly, the methylation of **45^{2-}•2[K$^+$(thf)$_2$]** with Me_2SO_4 proceeded exclusively at the Ge atoms to form *trans*-3,4-dimethyl-$^1\Delta$-1,2,3,4-disiladigermetene **48** as a single product (Scheme 6.26).[104b,128a] This observation agrees well with the general postulate that the electronegativity difference favors electron localization, overcoming the trend of cyclic π-delocalization and, consequently, making the molecule nonaromatic.[124] In the homonuclear all-Si version **46^{2-}•2[K$^+$(thf)$_2$]**, the resonances of the skeletal Si atoms were observed at 17.0 ppm, evidently outside the range expected for doubly bonded Si atoms, thus indicating a more important degree of π-delocalization of negative charges in **46^{2-}•2[K$^+$(thf)$_2$]** corresponding to the resonance contribution **C** (Scheme 6.25). Neither of the solution structures of the heavy cyclobutadiene dianion salts **45^{2-}•2[K$^+$(thf)$_2$]** and **46^{2-}•2[K$^+$(thf)$_2$]** depended on the polarity of the liquid medium, being virtually identical in both polar THF and nonpolar toluene.

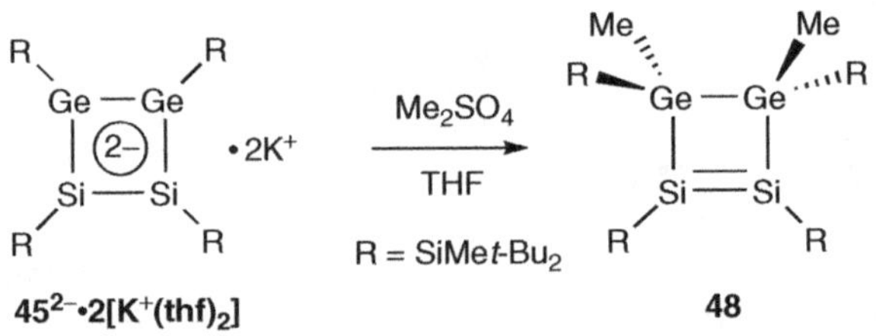

Scheme 6.26

Comparing the experimentally observed structures with those predicted theoretically, one should note that the resonance contribution **B** for **45^{2-}•2[K$^+$(thf)$_2$]** (Scheme 6.25) closely resembles the localized cyclobutene-1,2-diide structure **B** (Scheme 6.22) found by van Zandwijk *et al.*,[124] whereas the resonance form **C** for **46^{2-}•2[K$^+$(thf)$_2$]** (Scheme 6.25) more closely approaches the planar delocalized D_{4h} structure **C** (Scheme 6.22) calculated for $[C_4H_4]^{2-}$•2Li$^+$ by Schleyer *et al.*[125] Regarding the possibility of synthesizing the heavy analogs of planar aromatic cyclobutadiene dianion derivatives, one can bear in mind a couple of points: first, the idea of preparing homonuclear derivatives of identical group 14 elements may help to avoid the undesirable electron localization due to the electronegativity differences; secondly, utilization of small countercations (preferably Li^+), which are more capable of forming covalent bonds to the skeletal atoms, may outbalance the influence of the destabilizing Coulomb repulsion.

6.3.2.4 *Cycloheptatrienylium (Tropylium) Ion Derivatives*

The synthesis of the first stable cycloheptatrienylium derivative, tropylium bromide, by Doering and Knox in 1954, clearly demonstrated the predictive power of Hückel's $(4n + 2)$ π-electron rule.[131] According to this rule, the very stable tropylium ion should be recognized as a 6π-electron aromatic compound, and it was indeed aromatic, capable of being prepared by a variety of experimental procedures.[132] Furthermore, the tropylium ion was a very familiar species for organic mass spectrometry specialists, who often observed the abundant formation of tropylium ions upon the ionization of toluene.[133] Both X-ray crystallography[134] and theoretical[135] studies reliably concluded that the tropylium ion in its ground state possesses a planar D_{7h} structure with equal endocyclic C–C bond lengths of 1.400(2) Å (X-ray)[134b] and 1.405 Å (calculations),[135a] which pointed to the cyclic delocalization of its π-electrons and, consequently, its aromaticity.

The synthesis of stable analogs of the tropylium ion incorporating the heavy group 14 elements in the ring still has not been achieved. On the basis of early gas-phase experiments, Beauchamp *et al.* stated that of the two experimentally observed distinct $C_6SiH_7^+$ isomers, namely, the silatropylium ion and the hypothetical silabenzene cation, the former was thermodynamically more favorable than the latter, similar to the case of their carbon analogs.[136] However, in a subsequent study, Jarek and Shin criticized and disproved their statement, indicating that the most stable $C_6SiH_7^+$ isomer is η^4-C_6H_6•SiH^+, rather than the silatropylium ion.[137] This contradiction was rationalized by the high-level computations of Radom *et al.*, who demonstrated that the silatropylium ion is destabilized relative to the silabenzyl cation by 9.1 kcal/mol, this tendency being opposite to that in carbon chemistry, where among the $C_7H_7^+$ isomers the tropylium ion is more stable than the benzyl cation by 6.9 kcal/mol.[138] The optimized structure of the planar C_{2v} silatropylium ion exhibited cyclic C–C bond alternation; however, the degree of electron delocalization or aromaticity of the silatropylium ion was not discussed.[138] Experimentally, the closest approach to the synthesis of silatropylium ion derivatives was demonstrated by Komatsu and his group, who employed a modification of the seven-membered ring of the silatropylium ion with a rigid σ-framework by annelation with bicyclo[2.2.2]octene units.[139] This metastable derivative of silatropylium ion **49**$^+$•**TPFPB**$^-$ bearing a bulky mesityl group on the Si atom was prepared by the 'hydride-transfer reaction' of the silepin **50** and Ph_3C^+•TPFPB$^-$ in CD_2Cl_2 at $-50\,°C$, utilizing a general tendency of the stronger C–H vs Si–H bonds (Scheme 6.27).[139a]

The 6π-electron aromaticity of the silatropylium ion **49**$^+$ was manifested by the low-field resonance of its Si atom observed at 142.9 ppm, as well as by appreciable deshielding of all skeletal carbon atoms, pointing to a significant delocalization of the positive charge over the entire seven-membered ring. The requirements of the magnetic criteria of aromaticity were also satisfied for **49**$^+$: NICS(1) for the unsubstituted model $H_7C_6Si^+$ was calculated to be -7.3 [for comparison, NICS(1) for the parent tropylium ion

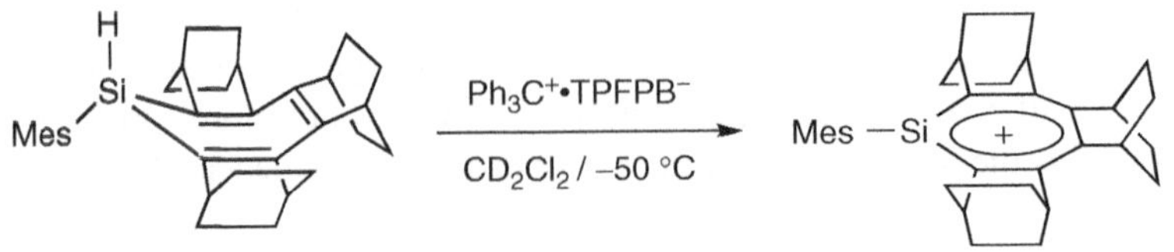

50

49+•TPFPB−

Scheme 6.27

$H_7C_7^+$ was estimated as −10.7]. However, the silatropylium ion **49+** was not thermally stable: once generated at low temperature, it can survive only below −50 °C, it underwent an unavoidable reaction with solvent through a chlorine abstraction reaction at higher temperatures. No other examples of silatropylium or any other heavy group 14 elements containing tropylium ions, both stable and metastable, have been reported thus far; therefore, this synthetic problem still presents a great deal of interest for organometallic chemists, particularly in respect of the structural information of these intriguing species.

6.3.3 Homoaromaticity

The term *homoaromaticity* is generally used to describe those cyclic compounds in which the conjugated system is formed by bypassing a saturated unit (typically, a CH_2 group).[3] Despite the interrupted chain of the cyclic conjugation, such systems can still benefit from the effects of aromaticity; however, their overall stabilization is reduced, compared with normal aromatics, because of the poorer orbital overlap. The term 'homoaromatic' was introduced for the first time by Winstein in 1959.[140] Since then, the concept of homoaromaticity has been widely accepted in organic chemistry,[5,141] being well recognized in cationic systems where delocalization of the positive charge was considered as an important driving force for homoaromaticity.[141d] However, this was not so evident in anionic and neutral systems, where the application of homoaromaticity remains highly controversial.[141d] The key compound among the cationic homoaromatic systems is definitely the homotropylium ion **51+**, synthesized for the first time by Pettit *et al.* in 1962 by the treatment of cyclooctatetraene with $SbCl_5$/HCl in nitromethane (Scheme 6.28, **A**).[142] The homotropylium ion, comprehensively studied both experimentally and computationally, is considered the prototypical homoaromatic system, serving as a reference species relative to which the properties of other homoaromatic compounds can be estimated.[5]

Cyclobutenylium ion **52+**, the homoaromatic analog of the very stable aromatic cyclopropenylium ion, is the simplest homoaromatic system, one of the best studied homoconjugated compounds (Scheme 6.28, **B**). This was prepared by the treatment of 3-acetoxycyclobutene with $HOSO_2F$ under superacidic conditions (SbF_5–SO_2ClF/−78 °C) by Olah *et al.*[143] The ^{13}C NMR spectra of **52+** were consistent with its homoaromatic structure,[143] which further found theoretical support that well reproduced the structural and spectral features of **52+**.[144] However, the overall stabilization of the cyclobutenylium ion **52+** is reduced compared with that of the homotropylium ion **51+**: the homoaromatic stabilization energy of **52+** is ca. 40% that of **51+**.[5]

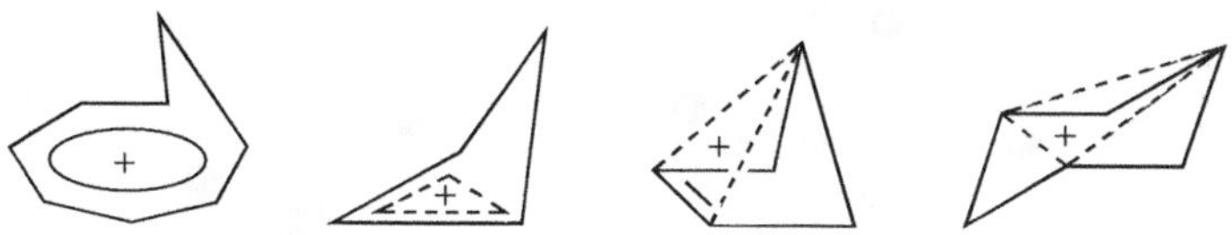

A: 51^+ B: 52^+ C: 53^+ D: 54^+

Scheme 6.28

The all-Si version of the cyclobutenylium derivative, homocyclotrisilenylium ion **55^+•TPFPB$^-$**, was recently prepared by Sekiguchi *et al.*[145] Its synthesis was smoothly achieved by the oxidative demethylation of cyclotrisilene **56** with Et_3Si^+ cation in benzene (Scheme 6.29).

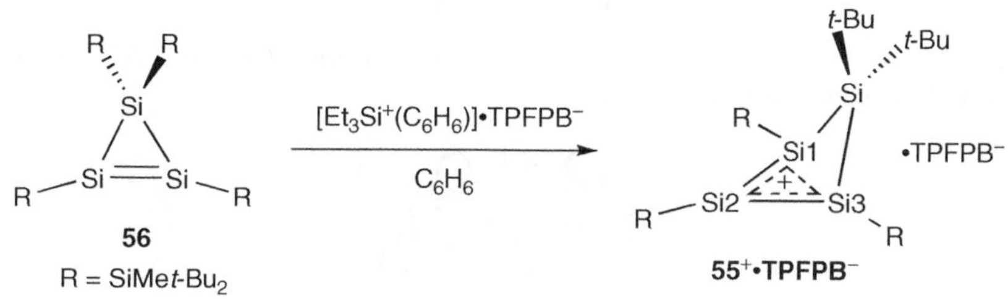

Scheme 6.29

Control experiments with deuterium-labeled cyclotrisilene **56-d_6** showed that the Me group was abstracted from one of the *t*-Bu_2MeSi substituents attached to the skeletal sp^3-Si atom, accompanied by the formation of an intermediate silyl cation, and followed by its isomerization to homocyclotrisilenylium ion **55^+•TPFPB$^-$**. In the crystalline form, **55^+** lacks any observable interactions with either counterion or solvent molecules, thus being a truly free cation in the solid state (Figure 6.7).

Similar to its carbon analog, the homocyclopropenylium ion,[141d] homocyclotrisilenylium ion **55^+** adopted a significantly puckered (46.6°) structure with a short nonbonding Si1---Si3 interatomic distance of 2.692(2) Å, reasonably consistent with its homoaromaticity (Figure 6.8).[145]

As is typical for homoaromatic systems, an important amount of the positive charge was built up at the central Si2 atom of the homoaromatic system, which was manifested in its greatly low-field shifted resonance at +315.7 ppm (for comparison, the chemical shifts of the terminal Si1 and Si3 atoms of the homoaromatic system were observed at the much higher field of +77.3 ppm). In full agreement with the homoaromatic composition of **55^+**, the geometry around the three cationic sp^2-Si atoms (Si1, Si2 and Si3) was perfectly planar, whereas the environment around the bypassed sp^3-Si4 atom was markedly pyramidal (Figure 6.8). Furthermore, the Si–Si bond lengths of the cationic portion of **55^+** were intermediate between those of the typical Si–Si single and Si=Si double bonds: Si1–Si2 = 2.240(2) Å and Si2–Si3 = 2.244(2) Å.[145]

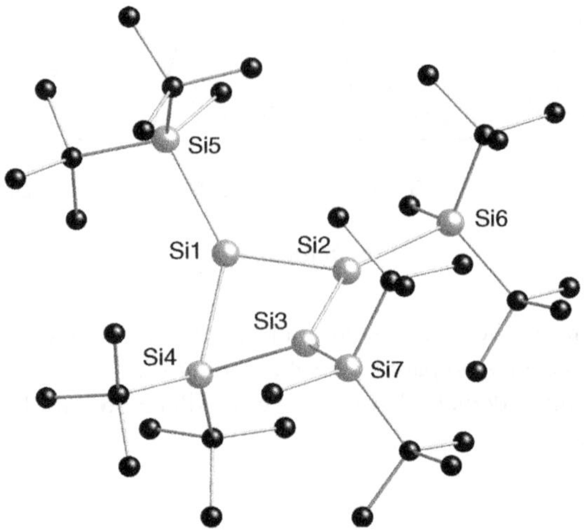

Figure 6.7 *Crystal structure of homocyclotrisilenylium ion derivative **55**$^+$ (counterion TPFPB$^-$ and hydrogen atoms are not shown)*

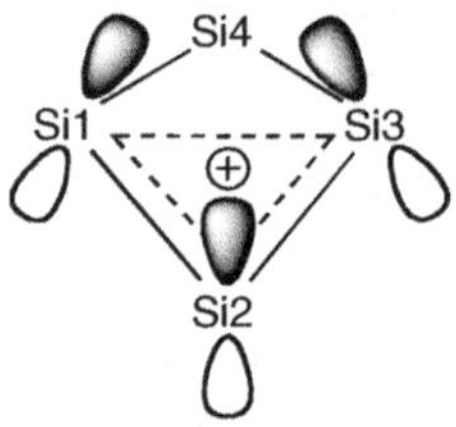

Figure 6.8 *Schematic representation of HOMO of the homocyclotrisilenylium ion **55**$^+$ calculated for the $[(H_3SiSi)_3SiMe_2]^+$ model at the B3LYP/6-31G(d) level (Me and H_3Si substituents are not shown)*

The homoconjugated bishomocyclopropenylium ion **53**$^+$, folded by ca. 90° (Scheme 6.28, **C**), was calculated to be more stable than its planar less strained conformer by 6 kcal/mol,[146] nevertheless, the parent bishomocyclopropenylium ion remains experimentally unknown.[141d] In contrast, its numerous 3,5-bridged derivatives have been widely studied both experimentally and computationally. For example, the recently reported crystal structures of such compounds showed an elongation of the C=C double bond (ca. 1.41 Å) and a remarkable tilting of the bridging carbon atom toward the C=C bond, resulting in a significant shortening of the bridging sp^3-C••• doubly bonded sp^2-C interatomic distances (ca. 1.75–1.88 Å).[147] Such structural features strongly support the homoaromaticity of these systems. The number of stable analogs of bishomocyclopropenylium systems of heavy group 14 elements is limited to only one example.[148] This compound, a derivative of 1,4,5-trigermabicyclo[2.1.0]pent-2-en-5-ylium **57**$^+$**•TPFPB**$^-$, was prepared by the oxidative debromination of the precursor **58** with the Et_3Si^+ cation in benzene (Scheme 6.30).

The positive charge, developed on the bridging sp^3-Ge atom upon its debromination, was effectively stabilized by intramolecular through-space interaction with the endocyclic C=C double bond, thus forming a bishomocyclopropenylium system (Figure 6.9).

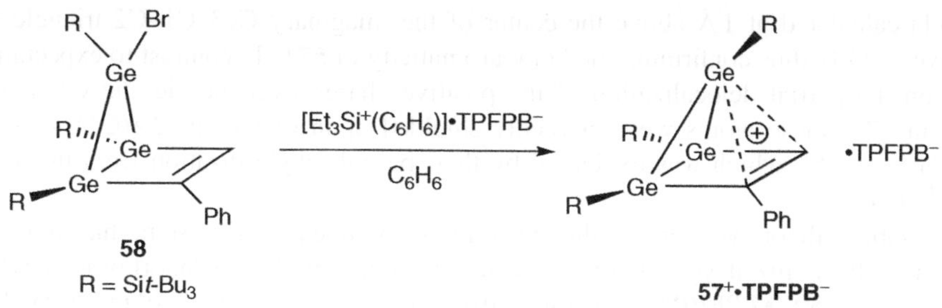

Scheme 6.30

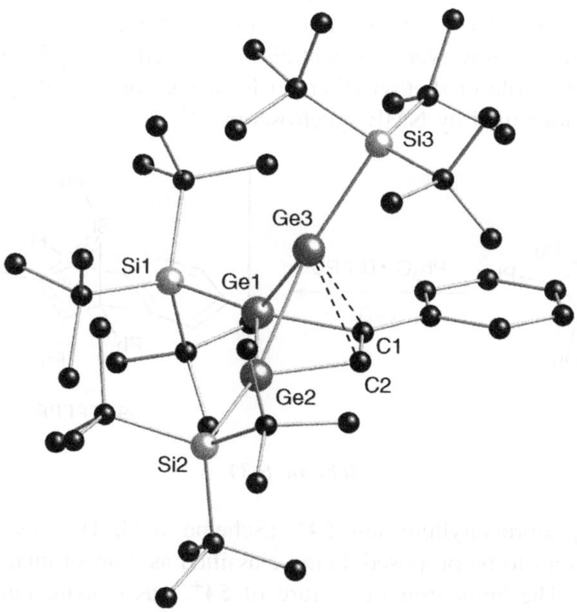

Figure 6.9 *Crystal structure of bishomocyclopropenylium ion derivative **57**$^+$ (counterion TPFPB$^-$ and hydrogen atoms are not shown)*

Quite parallel to the above-mentioned cases of organic bishomocyclopropenylium derivatives, the Ge3–C1 and Ge3–C2 interatomic distances were rather short [2.415(7) and 2.254(7) Å], whereas the C1=C2 double bond was markedly stretched [1.411(9) Å], unambiguously indicating the through-space π(C=C) $\rightarrow$ 4p(Ge) homoconjugation. As a result, the Ge3 atom was significantly tilted toward the C=C moiety resulting in the formation of a highly folded bicyclic structure with an interplanar angle of 67°. The formation of such a three-center two-electron bond was further supported by NBO calculations, showing the notable electron deficiency of the doubly occupied π(C=C) orbital (1.28 electrons) and appreciable electron occupancy of the vacant 4p(Ge) orbital (0.42 electrons).[148] Consequently, the bond orders (Wiberg bond indices, WBI) of Ge3–C1, Ge3–C2 and C1–C2 were computed as 0.37, 0.46 and 1.39, respectively. Furthermore,

NICS(1) calculated at 1 Å above the center of the imaginary Ge3–C1–C2 triangle was negative, −11.0, thus confirming the homoaromaticity of **57$^+$**. In contrast to expectations based on the partial delocalization of the positive charge over the olefinic C1 and C2 atoms in **57$^+$**, these atoms were markedly shielded upon oxidation [$\Delta\delta$(C1) = −29.8, $\Delta\delta$(C2) = −54.5], which was explained by their partial rehybridization from the sp^2 to the sp^3 state.

The stable silicon version of the bishomocyclopropenylium system has not been experimentally realized yet, although the intermediacy of the 7-silanorbornadienylium ion derivative **59$^+$•TPFPB$^-$**, generated from the corresponding 7-silanorbornadiene **60** by oxidation with Ph_3C^+•TPFPB$^-$, was recently discussed by Müller *et al.* (Scheme 6.31).[149] In sharp contrast to the well-known stable 7-norbornadienylium ion, **59$^+$•TPFPB$^-$** could not be detected even at temperatures as low as −60 °C, irreversibly rearranging at room temperature into the π-type complex ($PhSi^+$/1,2,3,4-tetraphenylnaphthalene). However, transient **59$^+$** can be effectively intercepted by Lewis bases, such as acetonitrile or diethyl ether, to form the corresponding silylnitrilium or silyloxonium ions identified by NMR spectroscopy.[149]

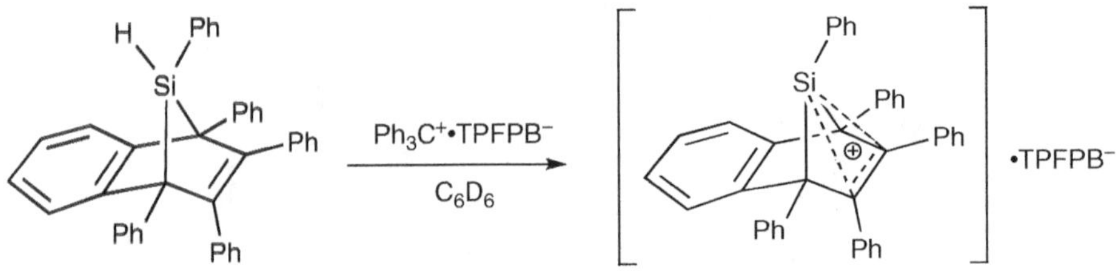

60 **59$^+$•TPFPB$^-$**

Scheme 6.31

The trishomocyclopropenylium ion **54$^+$** (Scheme 6.28, **D**) was one of the first homoaromatic species to be proposed, being classified as homoaromatic by Winstein as early as 1959.[140a] The homoaromatic nature of **54$^+$** was convincingly proved by the investigations of Cremer *et al.*, who pointed out the equality of the homoconjugative interatomic C–C nonbonding distances of 1.82 Å and the even distribution of the positive charge over all three homoconjugative C atoms.[146] The closest approach to the stable trishomocyclopropenylium system of the heavy group 14 elements was reported by Sekiguchi *et al.* in 2002.[150] This trishomoaromatic cationic Ge_{10}-cluster compound **61$^+$•TTFPB$^-$** was unexpectedly prepared by the thermolysis of the iodide **62** in the presence of excess KI in toluene at 50 °C for 1 week (Scheme 6.32).

61$^+$•TTFPB$^-$ represents a metal-rich cluster, in which six (out of a total of 10) Ge atoms carry silyl groups, one Ge atom has an iodo substituent, and the three remaining Ge atoms are tricoordinate substituent-free ‘naked’ atoms. The interatomic nonbonding distances between these three ‘naked’ Ge atoms were virtually identical [3.2642(15), 3.2622(17) and 3.2542(15) Å], being significantly longer than the normal Ge–Ge single bond length but still in the range of the metallic Ge–Ge bond lengths in Zintl anions of the type Ge_9^{3-}.[151] Such structural features definitely pointed to a homoconjugative interaction between the three central Ge atoms, thus making a trishomoaromatic cationic system.

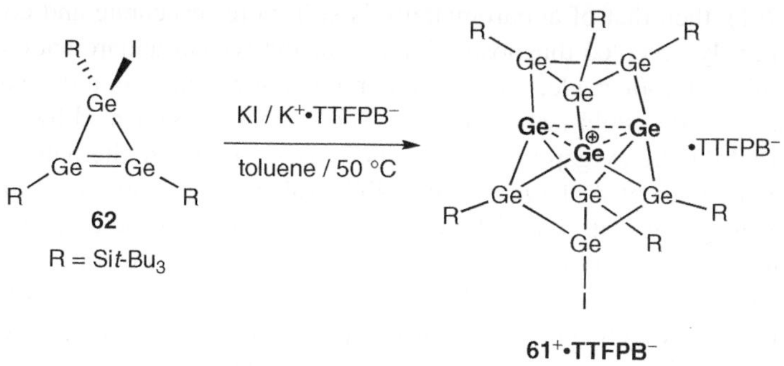

Scheme 6.32

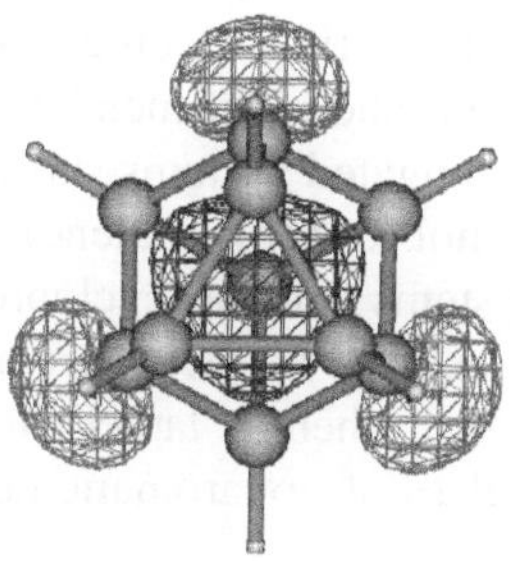

Figure 6.10 *HOMO of the model cationic cluster $Ge_{10}H_7{}^+$ calculated at the MP2/6-31G(d)//B3LYP/6-31G(d) level*

This conclusion was firmly supported by subsequent theoretical considerations. Thus, the trishomocyclotrigermenylium model $Ge_{10}H_7{}^+$ represented a global minimum on the PES, featuring three-center two-electron bonding inside the central core of the cluster between the three 'naked' Ge atoms with an even distribution of the positive charge over all of them. Such a homoconjugative Ge_3 bond was clearly seen in the HOMO of $Ge_{10}H_7{}^+$ manifesting the in-phase interaction of the p-orbitals of the Ge atoms (Figure 6.10).

The Ge–Ge interatomic distances were calculated to be 3.423 Å with Ge–Ge bond orders (WBI) of 0.288. The cluster **61⁺•TTFPB⁻** also satisfied the requirements of energetic and magnetic criteria of aromaticity: the ASE value of $Ge_{10}H_7{}^+$ was highly exothermic (−19.2 kcal/mol) and the NICS(0) was significantly negative (−26.4), consistent with the appreciable homoaromatic stabilization of $Ge_{10}H_7{}^+$.[150]

6.3.4 Antiaromaticity

The term *antiaromaticity* was introduced for the first time by Breslow as early as 1965[152] to describe adequately the cyclic conjugated 4*n* π-electron systems, whose '. . . π-electron energy is higher than that of a suitable reference compound which is not cyclically delocalized'.[153] Thus, the phenomenon of antiaromaticity represents a particular aspect of antibonding interactions; likewise, aromaticity reflects specific bonding relations. If the definition of the phenomenon of aromaticity is somewhat ambiguous (see Section 6.1), then that of antiaromaticity is still more inaccurate and controversial. The most generally accepted diagnostic features of the typical antiaromatic compounds include the following: $4n$ π-electron cyclic conjugation resulting in a decrease of delocalization and overall destabilization of the cyclic system accompanied by a decrease in the HOMO–LUMO energy gap (energetic criterion), cyclic bond alternation (geometric criterion), small values of Λ, large positive NICS values and paratropic shifts in the ^1H NMR spectra (magnetic criteria).[154] However, $4n$ π-electron cyclic conjugated systems could be either antiaromatic or nonaromatic. How can one distinguish between these two types of molecules; in other words, which $4n$ π-electron rings should be classified as antiaromatic and which ones as nonaromatic species? Such a distinction is rather subtle and based mainly on energetic considerations: small $4n$ π-electron compounds, such as three-, four- and five-membered rings, are highly destabilized if they are planar (+40, +35 and +29 kcal/mol, respectively).[155] However, departure from planarity markedly decreases their overall antiaromatic destabilization (+6 kcal/mol for the nonplanar three-membered ring). When the ring becomes larger, the antiaromaticity notably decreases, being small for the seven- and eight-membered rings (ca. 0 and +4 kcal/mol, respectively). The larger cyclic conjugated systems show very little (if any) antiaromatic character (for example, ca. 0 kcal/mol for [16]annulene).[155] Therefore, one should recognize the small ring 4π-electron systems (such as cyclopropenyl anion **62**, cyclobutadiene **63**, cyclopentadienylium ion **64**, as well as their derivatives) as the most fundamental antiaromatic species (Scheme 6.33), whereas larger $4n$ π-electron annulenes (such as cyclooctatetraene) belong to the class of nonaromatic derivatives.

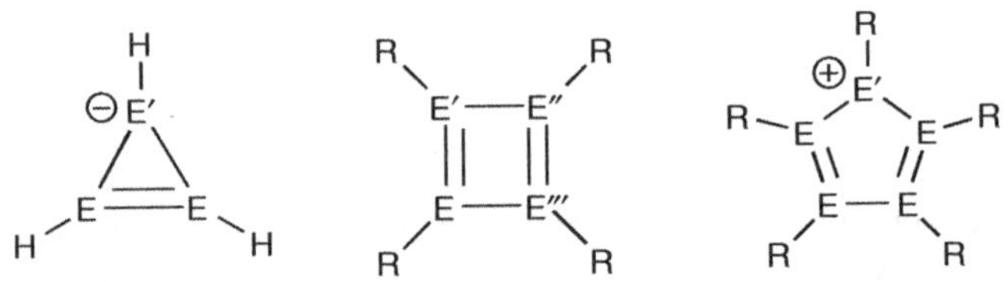

Cyclopropenyl anion
62

Cyclobutadiene
63

Cyclopentadienylium ion
64

Scheme 6.33

Studies in the field of antiaromatic compounds, incorporating group 14 elements heavier than carbon in the ring, are rather scarce. It should be noted that such chemistry is mainly developed computationally, whereas the experimental accomplishments in preparing stable heavy analogs of antiaromatic compounds are limited to the sole example of the isolable 1,2-digermacyclobutadiene derivative (see below). Below we will briefly discuss the current situation for each class of the 4π-electron antiaromatic compounds; namely, the group 14 heavy analogs of neutral cyclobutadiene, the cyclopropenyl anion and the cyclopentadienylium ion.

6.3.4.1 Cyclobutadiene Derivatives

At first, for the sake of comparison, one should mention the highly pronounced antiaromatic destabilization of cyclobutadiene itself, because this compound serves as a benchmark for the classical manifestation of antiaromaticity, being a highly desirable target for both experimental and theoretical pursuits. The early calculations by Hess and Shaad showed a large value of the RE (85.4 kcal/mol employing the 6-31G* basis set) for cyclobutadiene **63** ($E = E' = E'' = E''' = C$, $R = H$), thus pointing to its significant cyclic destabilization and high degree of antiaromaticity.[156] A subsequent more elaborate investigation by Glukhovtsev and Pross produced a value of 40.6 ± 1.7 kcal/mol calculated at the G2 level of theory for the antiaromatic destabilization energy of **63** ($E = E' = E'' = E''' = C$, $R = H$).[157] Recent photoacoustic calorimetry measurements provided direct experimental estimates of the antiaromaticity of the parent **63** ($E = E' = E'' = E''' = C$, $R = H$), which was dramatically destabilized by its 4π-electron delocalization.[158] Thus, relative to a hypothetical conjugated diene reference, **63** is destabilized by 87 kcal/mol, of which 32 kcal/mol are due to its ring strain and 55 kcal/mol can be attributed to the antiaromaticity of **63**. Relative to a reference with isolated double bonds, **63** showed an antiaromatic contribution of 48 kcal/mol.[158] Such highly pronounced antiaromaticity of cyclobutadiene **63**, reflected in its extreme reactivity, hampered synthesis of the parent compound, which is indeed still unknown as a stable species. By contrast, several heavily substituted and therefore kinetically stabilized cyclobutadiene derivatives have been synthesized and isolated, the latest example **63** ($E = E' = E'' = E''' = C$, $R = SiMe_3$) being prepared by the two-electron oxidation of the corresponding cyclobutadiene dianion derivative.[159]

Among the heavy group 14 element analogs of the antiaromatic cyclobutadiene, the silacyclobutadiene system has been most extensively studied by both theoreticians and experimentalists. The early computations by Gordon performed at the HF/6-31G*//HF/3-21G level showed that the unsubstituted silacyclobutadiene **63** ($E = E' = E''' = C$, $E'' = Si$, $R = H$) is highly antiaromatic by 53.3 kcal/mol (cyclobutadiene itself was found to be more antiaromatic than silacyclobutadiene by 14.6 kcal/mol at the same level of calculations), being nevertheless more stable than the isomeric silatetrahedrane by 33.3 kcal/mol.[160] The more extensive HF/DZ+d calculations by Colvin and Schaefer showed that the singlet silacyclobutadiene has a planar (with alternating single and double bonds) closed-shell ground state, whereas the lowest triplet state of silacyclobutadiene (with the much more 'square-like' geometry) lies ca. 5 kcal/mol above the ground state (for comparison: 23 kcal/mol for the singlet–triplet energy separation in cyclobutadiene).[161] The 'antiaromatic' destabilization of the singlet silacyclobutadiene was estimated as 49.1 kcal/mol, a value that was ca. 17 kcal/mol smaller than the analogous destabilization of cyclobutadiene.[161] A comprehensive HF/6-31G* study on the relative stability of 18 H_4C_3Si valence isomers located 2-methylsilacycloprop-2-en-1-ylidene, stabilized by the 2π-electron aromatic contribution due to interaction of the $\pi(C{=}C) \rightarrow p(Si)$ orbitals, as the global minimum structure on the PES, whereas silacyclobutadiene was destabilized by 59.5 kcal/mol, being on the other hand more favorable than silatetrahedrane by

32.0 kcal/mol.[162] The subsequent calculation by Maier *et al.* at the BLYP/6-31G* level determined the destabilization energy of silacyclobutadiene compared with the global minimum 2-methylsilacycloprop-2-en-1-ylidene as 45.3 kcal/mol.[163] The most recent computations by Lawson *et al.* at RHF, B3LYP and BLYP levels employing large 311++G** and TZV basis sets convincingly demonstrated the antiaromatic nature of the parent silacyclobutadiene **63** ($E = E' = E''' = C$, $E'' = Si$, $R = H$) deduced from the consideration of its particular geometric, energetic and magnetic properties. Thus, the optimized C_s structure of silacyclobutadiene diagnostically featured all different cyclic bond lengths, two of them corresponding to single bonds (Si–C = 1.92 Å and C–C = 1.54 Å) and the other two to double bonds (Si=C = 1.71 Å and C=C = 1.34 Å). This definitely pointed to a localization of both π-bonds (Si=C and C=C) and hence to the antiaromaticity of the whole system.[164] Such a conclusion agreed well with that made on the basis of energetic considerations: the value of the destabilization homodesmotic RE calculated at the B3LYP level for silacyclobutadiene was 58.0 kcal/mol (77.1 kcal/mol for cyclobutadiene), whereas that of the isodesmic RE calculated at the same level was 38.7 kcal/mol (47.7 kcal/mol for cyclobutadiene).[164] Consideration of the magnetic criterion also pointed to an appreciable antiaromaticity of the silacyclobutadiene: the values of NICS(0) and NICS(2) calculated at the B3LYP level were positive, +19.0 and +6.6, respectively (+27.3 and +4.3 for cyclobutadiene).[164]

The thorough HF/3-21G theoretical studies of the 13 cyclic $H_4C_2Si_2$ valence isomers by Gordon *et al.* showed 2-H_3Si-silacycloprop-2-en-1-ylidene as a global minimum on the PES (quite similar to the above-described situation with the H_4C_3Si valence isomers).[165] Neither possible rectangular-planar 1,2-disilacyclobutadienes **63** ($E = E''' = C$, $E' = E'' = Si$, $R = H$) nor **63** ($E = E' = C$, $E'' = E''' = Si$, $R = H$) were located as minima structures, whereas rectangular-planar 1,3-disilacyclobutadiene **63** ($E = E'' = C$, $E' = E''' = Si$, $R = H$), albeit representing a minimum on the PES, exhibited significant biradical character and thus cannot be adequately described at the given levels of calculations (RHF, UHF–NO CI and FORS MCSCF). The latter structure is disfavored compared with the most stable 2-H_3Si-silacycloprop-2-en-1-ylidene by 49.7 kcal/mol, which should be ascribed to antiaromatic destabilization effects. On the other hand, 1,3-disilacyclobutadiene **63** ($E = E'' = C$, $E' = E''' = Si$, $R = H$) was more favorable by 37.7 kcal/mol than the highly strained isomeric disilatetrahedrane, which was the least stable $H_4C_2Si_2$ isomer.[165] In accordance with these results, the later calculations by Maier *et al.* also showed a nearly planar 1,3-disilacyclobutadiene **63** ($E = E'' = C$, $E' = E''' = Si$, $R = H$) as the minimum on the $H_4C_2Si_2$ PES, whereas planar 1,2-disilacyclobutadienes **63** ($E = E''' = C$, $E' = E'' = Si$, $R = H$) and **63** ($E = E' = C$, $E'' = E''' = Si$, $R = H$) did not represent minimum structures.[166] 1,3-Disilacyclobutadiene derivatives have often been discussed as feasible dimerization products of the simple silynes R–Si≡C–R′ having small substituents R and R′; however, in no case have these dimers been either isolated or characterized. To prevent the highly exothermic dimerization of silynes and consequently to stabilize kinetically the Si≡C triple bond, the use of extremely bulky substituents, such as *t*-Bu_3Si, Tbt and 2,6-bis[bis(trimethylsilyl)methyl]phenyl groups, was suggested by Karni and Apeloig.[167]

The PES of tetrasilacyclobutadiene H_4Si_4 **63** ($E = E' = E'' = E''' = Si$, $R = H$), which represents a full silicon analog of cyclobutadiene H_4C_4, is rather complicated. The early study by Sax and Kalcher performed at the SCF, TC–SCF and MC–SCF levels surprisingly found a cyclic C_{2v} bis(silanediyl) (–H_2Si–Si–)$_2$ as the lowest energy H_4Si_4 isomer.[168] The planar D_{2h} tetrasilacyclobutadiene was not located as the minimum on the PES; however, its symmetry reduction to a C_{2h} structure by *trans*-bending of the hydrogen atoms led to the overall stabilization of the system by 4.8 kcal/mol, although an even more favorable C_{2h} isomer is still highly destabilized compared with bis(silanediyl) by 28.9 kcal/mol.[168] Schaefer *et al.* showed the three nonplanar four-membered ring structures of C_s, D_2 and D_{2d} symmetries as minima on the singlet tetrasilacyclobutadiene PES calculated at the HF/DZP level.[169] The most stable is a puckered D_{2d} structure with equal Si–Si bond lengths and an axial arrangement of the hydrogen atoms, which was more favorable than the isomeric tetrasilatetrahedrane by 28.3 kcal/mol.[169b] A subsequent study by the same authors at the SCF/DZP level including electron correlation discovered several other minima structures on the PES of H_4Si_4, of which the highest energy isomer was represented by tetrasilatetrahedrane, and, what was most puzzling, the lowest energy isomer was predicted to have a rather experimentally confusing structure of the puckered C_s tetrasilacyclobuta-1,2-diene.[170] The energy separation between these two most (tetrasilacyclobuta-1,2-diene) and least (tetrasilatetrahedrane) favorable structures was estimated to be as great as 50.6 kcal/mol.[170] Nagase and Nakano found that the symmetry-forbidden rearrangement of tetrasilatetrahedrane to tetrasilacyclobutadiene is ca. 8 kcal/mol exothermic, although the planar D_{2h} tetrasilacyclobutadiene is not a minimum structure, being transformed into the more stable nonplanar C_s conformer featuring one axial and three equatorial hydrogen atoms.[171] The semiempirical calculations by Glukhovtsev *et al.* confirmed the instability of the planar D_{2h} configuration of tetrasilacyclobutadiene toward its departure from planarity, resulting in the formation of the more stable folded structures of C_{2h} (pyramidalization at the silicon atoms) or C_s (puckering of the Si_4 skeleton) symmetry.[46] Tetrasilacyclobutadiene derivatives are the expected products of the dimerization of disilynes R–Si≡Si–R; however, such expectations have never been realized experimentally, presumably because of the intrinsically high reactivity of the resulting tetrasilacyclobutadiene species. It has been predicted by Nagase *et al.* that the utilization of extremely bulky aryl [2,6-bis(2,4,6-*i*-Pr_3-C_6H_2)C_6H_3][172a] or silyl [SiMe(Si*t*-Bu_3)$_2$][172b] substituents may very efficiently protect the Si≡Si triple bond and thus stabilize disilyne against its dimerization. Indeed, isolation of the first room temperature stable disilyne was finally achieved taking advantage of the kinetic protection of the highly reactive Si≡Si bond by the very voluminous Si*i*-Pr[CH($SiMe_3$)$_2$]$_2$ substituent (see Chapter 5).[173]

The initial experimental accomplishments in the field of cyclobutadiene derivatives containing the heavy group 14 elements are limited to a couple of examples, in which the intermediacy of the title compounds was evidenced by trapping reactions. The first compound in this series, 2,3,4-tri-*tert*-butyl-1-(2,4,6-trimethylphenyl)-1-silacyclobutadiene **65**, was generated upon the photochemical rearrangement (>400 nm) of the transient (2,4,6-trimethylphenyl)(cyclopropenyl)silylene **66** at 77 K in a 3-methylpentane glass matrix (Scheme 6.34).[174]

Albeit thermally unstable, **65** was smoothly trapped with either EtOH or $MeOSiMe_3$ at low temperature to produce the corresponding 1,2-adducts across the Si=C double bond.

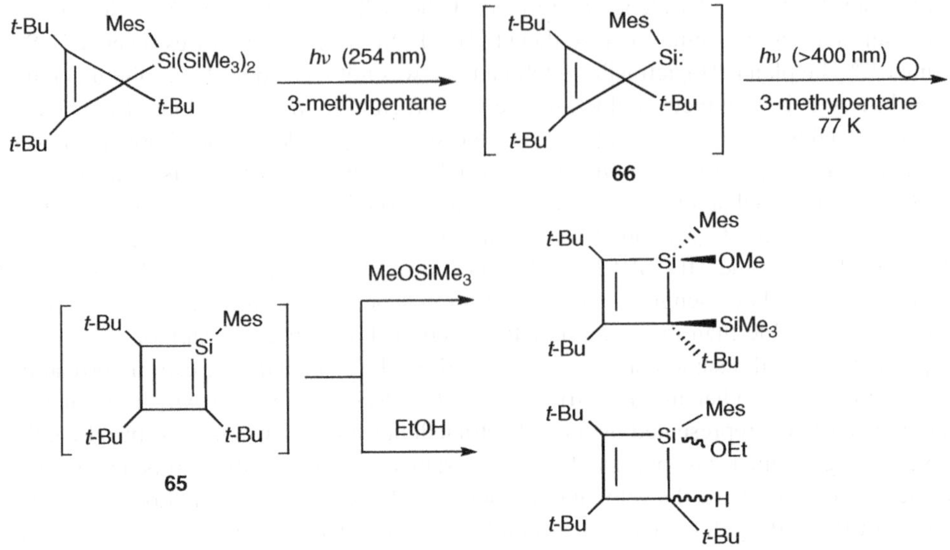

Scheme 6.34

Interestingly, in the absence of trapping reagents, **65** underwent an unusual dimerization to form polycyclic derivative **67** through an initial [1,5] sigmatropic migration and subsequent Diels–Alder reaction of the resulting *o*-silaxylylene intermediate **68** and starting **65** (Scheme 6.35).[175]

Another representative of the transient silacyclobutadiene family, namely, 2,3,4-tri-*tert*-butyl-1-(2,4,6-triisopropylphenyl)-1-silacyclobutadiene **69**, was similarly generated and trapped with EtOH by the photochemical rearrangement (>400 nm) of (2,4,6-triisopropylphenyl)(cyclopropenyl)silylene **70** at 77 K in a 3-methylpentane glass matrix (Scheme 6.36).[176] Without trapping reagents, silacyclobutadiene **69** thermally reverted to the starting silylene **70**, a fact that pointed to the greater thermodynamic stability of the latter species.

Annealing of a 3-methylpentane glass matrix containing silacyclobutadienes **65** or **69** produced species manifesting strong well-resolved triplets in the EPR spectra, which were ascribed to the triplet biradicals of the silacyclobutadiene dimers **71** (Scheme 6.37).[177] The presence of the triplet species **71** was convincingly proved by the observation of the very diagnostic 'half-field' signals in their EPR spectra. The surprising persistence of triplet biradicals **71** was attributed to allylic delocalization of the unpaired electrons, as well as effective steric protection of the carbon radical centers by the bulky *tert*-butyl groups.[177]

Another example of a chemically proven transient silacyclobutadiene derivative was reported by Kira *et al.* This 2,3-di-*tert*-butyl-1,4-bis(trimethylsilyl)-1-silacyclobutadiene **72** was formed upon the facile isomerization of the transient silatriafulvene derivative **73**, which was generated by the sila-Peterson reaction of di-*tert*-butylcyclopropenone with $(Me_3Si)_3SiLi$ in refluxing benzene (Scheme 6.38).[178] The intermediate formation of the silacyclobutadiene **72** was confirmed by its trapping

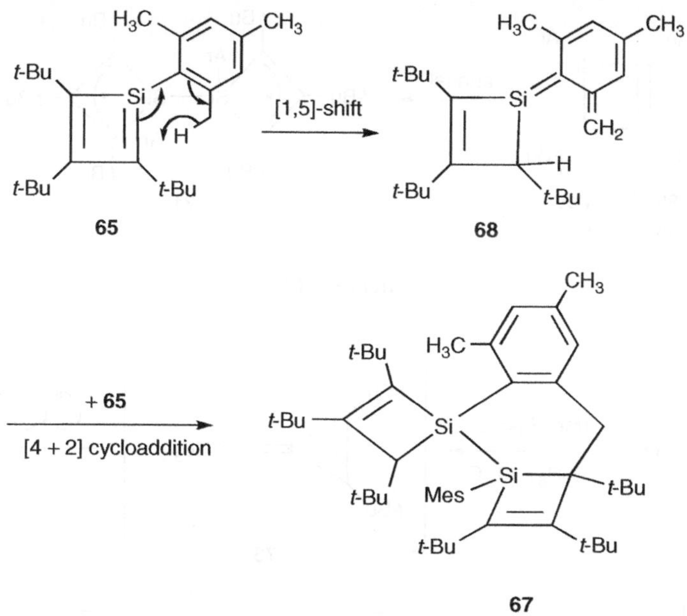

Scheme 6.35

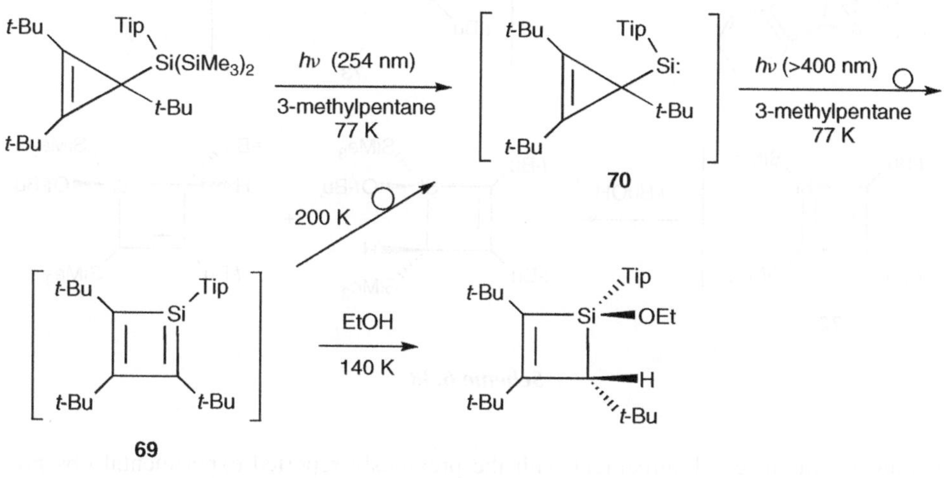

Scheme 6.36

with *tert*-butanol to produce the corresponding alcohol 1,2- and 1,4-addition products. The mechanism of such an intramolecular transformation of silatriafulvene **73′** to silacyclobutadiene **72′** may involve the formation of the cyclopropenylsilylene **74′** as a key reactive intermediate, which isomerizes to **72′** via a pyramidal transition state with a relatively low activation barrier of 5.9 kcal/mol (Scheme 6.39).[179] This theoretical

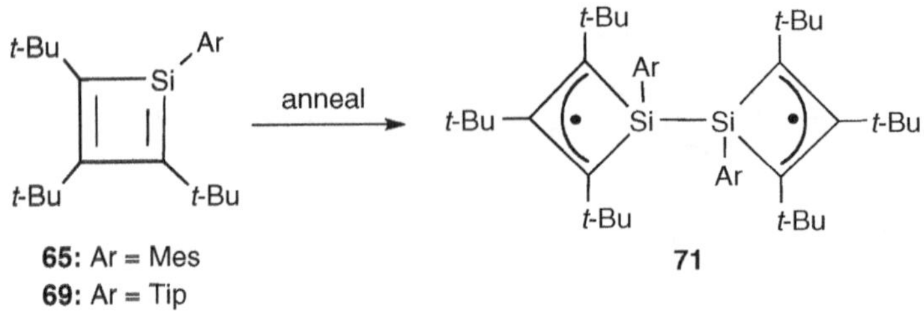

Scheme 6.37

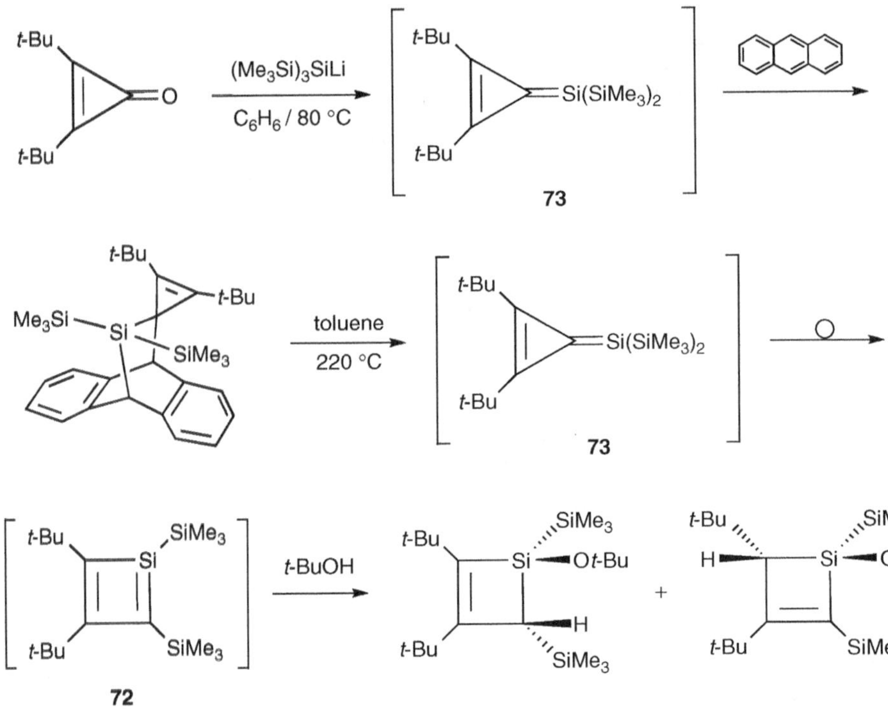

Scheme 6.38

conclusion was in good agreement with the previously reported experimental observations by Fink *et al.* pointing to a rearrangement of the transient (cyclopropenyl)silylenes **66** and **70** to the corresponding silacyclobutadienes **65** and **69**.[174,176]

The other transient silacyclobutadiene bearing bulkier t-BuMe$_2$Si substituents was generated similarly by the isomerization of silatriafulvene **75**, stabilized by reaction with CpCo(CO)$_2$ and isolated in the form of its (η^5-cyclopentadienyl)(η^4-silacyclobutadiene)cobalt complex **76** (Scheme 6.40).[180]

The only example of an isolable representative of the heavy cyclobutadienes, namely, 1,2-digermacyclobutadiene derivative **77**, was recently synthesized by

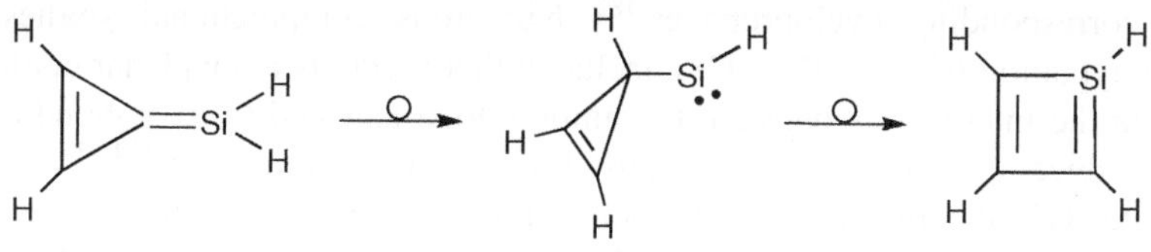

Scheme 6.39

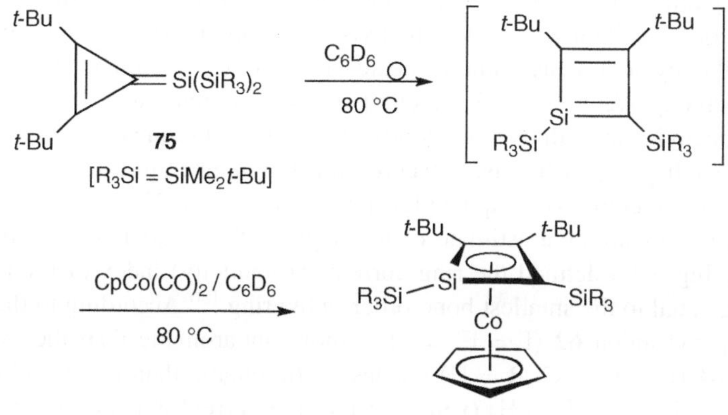

Scheme 6.40

Power *et al.* by the formal [2 + 2] cycloaddition of the stable digermyne analog ArGeGeAr [Ar = 2,6-(2,6-i-Pr_2-C_6H_3)$_2$-C_6H_3] and diphenylacetylene (Scheme 6.41).[70,181] The structural and bonding characteristics of **77** are discussed in Chapter 5 (Section 5.3.2).

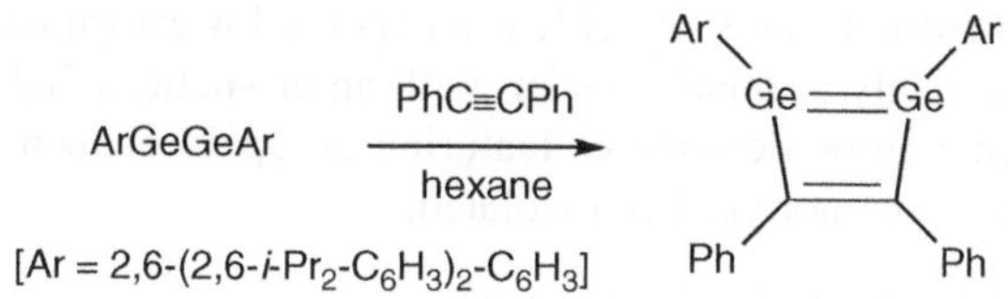

Scheme 6.41

6.3.4.2 *Cyclopropenyl Anion Derivatives*

Antiaromaticity of the cyclopropenyl anion was first noted in the early works of Breslow *et al.*, who pointed to the exceptional difficulty of proton abstraction from the sp^3-C atom of substituted cyclopropene derivatives, which were much less acidic than the corresponding cyclopropanes.[182] Numerous computational studies of the cyclopropenyl anion **62** (E = E′ = C) concluded that it adopts a nonplanar configuration to minimize the interactions between the anionic lone pair and C=C π-bond electrons, and thus reduce the destabilization caused by cyclic conjugation.[154,155] According to elaborate G2 theoretical investigations, the cyclopropenyl anion is somewhat destabilized by 4.1 kcal/mol.[183] The fully antiaromatic cyclopropenyl anion **62** (E = E′ = C) was calculated to be planar; however, it was not located as a minimum structure, possessing three imaginary frequencies and being 34 kcal/mol higher in energy than its nonplanar conformer. Kass *et al.* reported for the first time the gas-phase formation of a substituted cyclopropenyl anion by the fluoride ion promoted desilylation of 3-carbomethoxy-3-trimethylsilylcyclopropene.[184] The existence in solution of the cyclopropenyl anion species derived by a similar fluorodesilylation of 3-carbomethoxy-1,2-diphenyl-3-trimethylsilylcyclopropene and 3-carbomethoxy-1-*tert*-butyl-2-phenyl-3-trimethylsilylcyclopropene was subsequently demonstrated by the same research group, who also observed the low-temperature UV spectrum of the long-lived 3-carbomethoxy-1,2-diphenylcyclopropenyl anion.[185]

An alternative quantitative estimate of the degree of aromaticity/antiaromaticity was suggested by Jug, who defined the ring current (aromaticity) index of monocyclic systems as being equal to the smallest bond order in the ring.[186] According to this approach, the cyclopropenyl anion **62** (E = E′ = C) is more antiaromatic than the cyclopentadienylium ion **64** (E = E′ = C, R = H) but less antiaromatic than the cyclobutadiene **63** (E = E′ = E″ = E‴ = C, R = H) (values of the ring current indices are given in parentheses): $C_5H_5^+(1.042) < C_3H_3^-(1.016) < C_4H_4(0.980)$. Based on such a definition, it was suggested that the silacyclopropenyl anion $H_3C_2Si^-$ **62** (E = C, E′ = Si) is nonaromatic in its singlet ground state: the Si anionic center is highly pyramidal, whereas the skeletal Si–C bond order (and ring current index) was calculated to be 1.21, thus reflecting their single bond character (CI/SINDO1 calculations).[187] However, upon excitation the degree of antiaromaticity of the silacyclopropenyl anion markedly increases; thus, the lowest energy triplet state is highly antiaromatic with a ring current index of 1.03. The situation with the parent cyclopropenyl anion **62** (E = E′ = C) is greatly different: in its singlet ground state, it should be best described as a localized moderately antiaromatic structure with a ring current index of 1.14, whereas the lowest triplet has the properties of a delocalized moderately aromatic system with an aromaticity index of 1.46.[187]

Stable cyclopropenyl anion derivatives featuring group 14 elements heavier than carbon in their skeletons have not yet been isolated.

6.3.4.3 Cyclopentadienylium Ion Derivatives

Early attempts to generate pentaphenylcyclopentadienylium ion $Ph_5C_5^+$ **64** (E = E′ = C, R = Ph) by the reaction of pentaphenylcyclopentadienol with concentrated H_2SO_4 were made by Ziegler and Schnell.[188] Later studies by Breslow *et al.* showed that $Ph_5C_5^+$ was indeed generated under such conditions and was long-lived below −40 °C, being characterized by UV and EPR spectroscopy as a low-lying excited triplet.[189] On the other hand, the cyclopentadienyl halides were remarkably resistant toward solvolysis; thus, cyclopentadienyl iodide was surprisingly inert toward ionization with silver perchlorate in propionic acid.[190] Subsequently, the solvolytic generation of the cyclopentadienylium ion derivatives has been reported, albeit at very low reaction rates;[191] for example, the reaction of pentamethylcyclopentadienyl bromide with silver tetrafluoroborate in CH_2Cl_2 resulted in the initial formation of the transient pentamethylcyclopentadienylium ion $Me_5C_5^+$ **64** (E = E′ = C, R = Me) trapped with nucleophiles (MeOH, $MeNH_2$).[191a] In a limited number of cases, thermodynamically stabilized cyclopentadienylium ions have been obtained as stable derivatives.[192] The pentachlorocyclopentadienylium ion $Cl_5C_5^+$ **64** (E = E′ = C, R = Cl) was also generated and shown to be a ground state triplet,[189b] whose IR spectrum was measured in an SbF_5 matrix and whose structure was calculated to be of D_{5h} symmetry.[193] The parent cyclopentadienylium ion $H_5C_5^+$ **64** (E = E′ = C, R = H) was also generated by the reaction of bromocyclopentadiene with SbF_5 in a low-temperature matrix at 78 K and observed by EPR spectroscopy, implying its triplet ground state.[194] Such experimental observations[194] are in good agreement with theoretical expectations,[195] which predict $H_5C_5^+$ to possess a triplet ground state with a highly symmetrical delocalized D_{5h} geometry. Of all possible $C_5H_5^+$ isomers, the early calculations predicted the vinylcyclopropenylium ion to be the most favorable structure,[196] although subsequent computations[195f] emphasized that the triplet D_{5h} cyclopentadienylium ion and the singlet C_s vinylcyclopropenylium ion isomers are very close in energy and therefore the estimation of their energy difference is very sensitive to computational levels. Thus, the energetic preference of the vinylcyclopropenylium ion calculated at the G2 and G2(B3LYP/MP2/CC) levels was very small, 2.8 and 3.1 kcal/mol, respectively. However, utilization of other computational levels {MP2/6-311G(d,p)//MP2/6-31G(d), MP4/6-311+G(d,p)//MP2/6-31G(d), CCSD(T)/6-311G(d,p)//B3LYP/6-31G(d)+ZPE [B3LYP/6-31G(d)]} may significantly decrease this energy gap and even reverse the stability order, thus making the triplet cyclopentadienylium ion the most stable structure.[195f]

The experimental estimates of the antiaromaticity of the cyclopentadienylium ion were made based on thermodynamic studies of the gas-phase ionization potentials of the cyclopentenyl and cyclopentadienyl radicals, showing a very large destabilization energy for the cyclopentadienylium ion compared with the cyclopentenylium ion of 31.6 kcal/mol.[197] Magnetic criteria of aromaticity also indicated the strong antiaromatic character of the singlet cyclopentadienylium ion (B3LYP/6-31G* level): $\Lambda = 32.6$ and NICS(0) = 54.1.[198] Comparative calculations at the HF/6-31+G* level gave similar values of Λ and NICS: 30.5 and 49.2 (for singlet); −3.3 and −4.5 (for triplet), thus confirming the antiaromaticity of the C_{2v} singlet and aromaticity of the D_{5h} triplet states of the cyclopentadienylium ion.[199] According to this computation, the triplet is more stable than the singlet by 10.5 [B3LYP/6-311+G(d,p)] and 7.6 [CCSD(T)/cc-pVDZ//B3LYP/6-311+G(d,p)] kcal/mol.[199]

The stable heavy group 14 element analogs of the cyclopentadienylium ion are still unknown; however, several theoretical studies related to this problem have been published by Schleyer *et al.* Thus, the antiaromaticity of the silacyclopentadienylium ion $H_5C_4Si^+$ **64** (E = C, E′ = Si, R = H) was acknowledged for the first time in the course of the MP2/6-31G* *ab initio* computational studies on a series of five-membered C_4H_4X ring systems.[102a] The antiaromatic destabilization energy of the C_{2v} silacyclopentadienylium ion was calculated on the basis of a homodesmotic reaction to be 24.1 kcal/mol, whereas the magnetic susceptibility exaltation value Λ was found to be positive +13.2 indicating paramagnetic ring current effects. Moreover, the cyclic C–C bond lengths in $H_5C_4Si^+$ of 1.357 and 1.518 Å clearly manifest highly pronounced C–C bond alternations, typical for a localized antiaromatic system.[102a] A subsequent HF/6-31+G* investigation showed a positive NICS value (+12.8) for $H_5C_4Si^+$.[6] Overall, all energetic, magnetic and structural characteristics of the silacyclopentadienylium ion **64** (E = C, E′ = Si, R = H) definitely point to its antiaromatic character. However, the degree of antiaromaticity of the silacyclopentadienylium ion is markedly reduced compared with that of its carbon analog, the cyclopentadienylium ion. Thus, a systematic study showed the general trend of a sharp drop in antiaromaticity extent within group 14 on going from C to Si and its rather insignificant decrease on going from Si to Pb.[102d] For example, computational analysis [B3LYP/6-31+G*(for C and H), /LanL2DZdp (for E)] of the optimized geometries of the $H_5C_4E^+$ (E = C–Pb) species **64** (E = C, E′ = Si–Pb, R = H) showed the following trend in the changes of =**C–C**= cyclic bond lengths (in Å): 1.575 (E′ = C), 1.526 (E′ = Si), 1.522 (E′ = Ge), 1.514 (E′ = Sn), 1.512 (E′ = Pb), hence manifesting the decrease in C–C bond alternation descending group 14. The magnitudes of the antiaromatic isodesmic destabilization energy also follow the same tendency: 13.4 (E′ = C), 5.8 (E′ = Si), 5.8 (E′ = Ge), 5.5 (E′ = Sn), 5.6 (E′ = Pb), emphasizing the largest antiaromatic destabilization for the carbon system.[102d] Magnetic criteria of aromaticity are also in line with the above conclusion. Thus, the Λ value is highly positive for the cyclopentadienylium ion $H_5C_5^+$, whereas they are much smaller (or even negative) for the sila-, germa-, stanna- and plumbacyclopentadienylium ions **64** (E = C, E′ = Si–Pb, R = H): +28.2 (E′ = C), +6.4 (E′ = Si), +3.4 (E′ = Ge), −0.5 (E′ = Sn), −3.8 (E′ = Pb). The NICS value computed at the ring center is exceptionally large for the cyclopentadienylium ion, thus confirming its pronounced antiaromaticity, whereas those of the heavy analogs of the cyclopentadienylium ion are much smaller: +49.2 (E′ = C), +18.6 (E′ = Si), +13.7 (E′ = Ge), +9.3 (E′ = Sn), +5.2 (E′ = Pb).

A detailed investigation of the PES of the $H_5Si_5^+$ valence isomers at the MP2/6-31G* *ab initio* level established quite remarkable differences compared with the case of its carbon analog $H_5C_5^+$ (see above).[200] First of all, the most stable $H_5Si_5^+$ C_{2v} isomer turned out to be pentasila[1.1.1]propellanyl cation **78**, which has no analogs in carbon chemistry (Scheme 6.42). The next lowest energy minimum structure was found to be the distorted C_s pyramid-like **79**, which was only 1.8 kcal/mol less stable than **78**. The planar delocalized triplet D_{5h} isomer **80** was not an energy minimum (two imaginary frequencies) on the PES, being 24.1 kcal/mol destabilized compared with **78**.[200] This situation is totally different from that of the parent cyclopentadienylium ion, whose triplet D_{5h} isomer was one of the two most stable structures.[195f] Neither of the planar localized singlet C_{2v} isomers **81** (alkene form, one imaginary frequency) and **82** (diene form, three imaginary frequencies) were energy minima either, being 34.6 and 34.7 kcal/mol higher in energy than **78** (Scheme 6.42). However, when the symmetry constraints of both **81** and **82** were reduced to the C_s group, they transformed into the same folded conformer **83** (local minimum), which was stabilized by 13.6–13.7 kcal/mol compared with **81** and **82**, but was still destabilized by 21.0 kcal/mol compared with the global minimum **78**.[200] One should remember that in carbon chemistry, the localized C_{2v} allylic cation singlet, even being destabilized by 9.1 kcal/mol compared with the delocalized D_{5h} triplet, was still located as a minimum structure at the HF level. On the other hand, the singlet C_s vinylcyclopropenylium ion $H_5Si_5^+$ isomer **84** (local minimum) is greatly

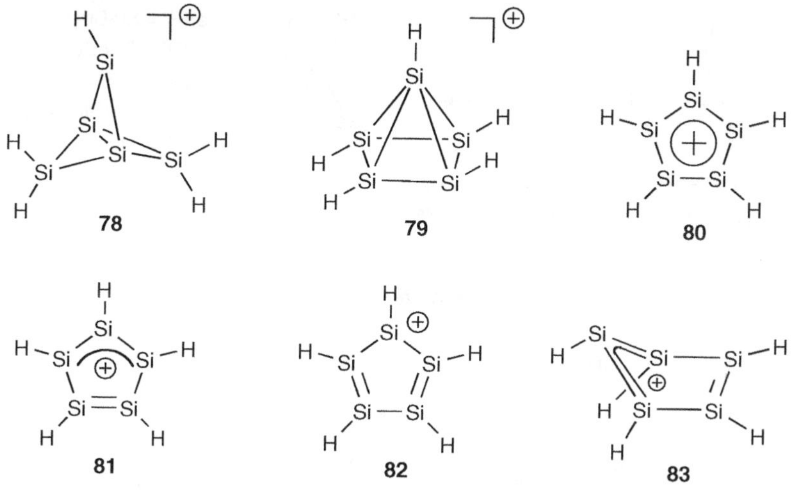

Scheme 6.42

destabilized compared with the global minimum **78** by 35.4 kcal/mol,[200] whereas the analogous vinylcyclopropenylium ion was found to be one of the two most favorable structures among all $H_5C_5^+$ isomers.[195f,196]

6.3.5 Reactivity of Heavy Aromatics: Complexation to Transition Metals

The chemistry of the heavy aromatics of group 14, similar to their aromatic carbon congeners, is rich and diverse, in many respects markedly exceeding that of the classical arenes because of the enhanced reactivity of the E=E (E = heavy group 14 element) bonds incorporated into the aromatic rings. A variety of reactions with nucleophiles (for cationic heavy aromatics) and electrophiles (for anionic heavy aromatics), as well as addition and cycloaddition reactions (for neutral heavy aromatics), have been reported in recent years. Below we will briefly describe the most important examples of such reactivity, particularly emphasizing the reactions of the heavy aromatics with transition metals.

One interesting example of the reactivity of 2π-electron heavy aromatics is the electrophilic interaction of cyclotrigermenylium derivative **11$^+$•TFPB$^-$** with a number of anionic derivatives of group 14 elements, such as t-Bu_3SiNa, t-Bu_3GeNa, $(Me_3Si)_3SiLi$, $(Me_3Si)_3GeLi$ and MesLi, providing a simple and effective approach for the synthesis of unsymmetrically substituted cyclotrigermenes **85a–e** (Scheme 6.43).[201] Interestingly, the $(Me_3Si)_3Si$-substituted cyclotrigermene **85c** prepared by this method featured an unprecedented example of *cis*-bent geometry around the Ge=Ge double bond[202] as a consequence of both steric and electronic effects of the very bulky electropositive $(Me_3Si)_3Si$ substituent.

On the other hand, the reaction of **11$^+$•TTFPB$^-$** with potassium halides KX (X = Cl, Br, I) in diethyl ether resulted in the formation of halogen-substituted cyclotrigermenes **86a–c** (Scheme 6.44), in which the halogen atoms underwent unusual room-temperature migration over the entire Ge_3 skeleton, as revealed by dynamic ^1H NMR spectroscopy

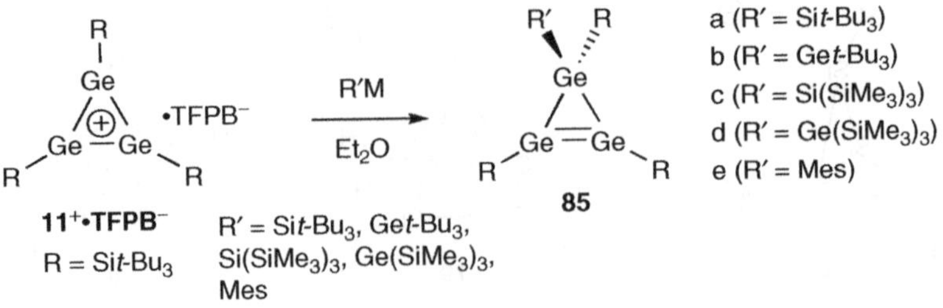

Scheme 6.43

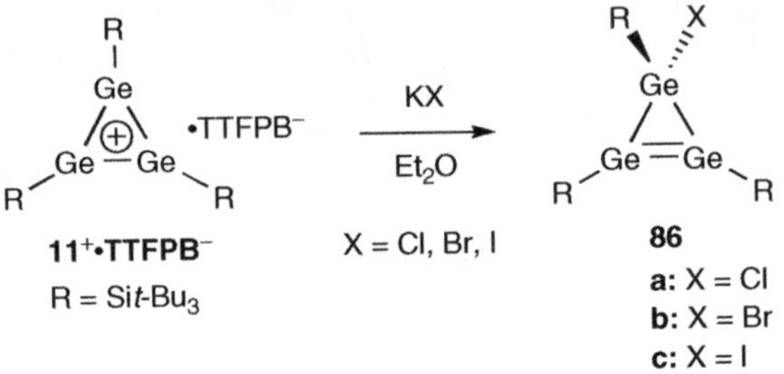

Scheme 6.44

(Scheme 6.45).[203] The Br- and I-substituted cyclotrigermenes **86b** and **86c** featured *cis*-bent geometry around the Ge=Ge bond, which was explained in terms of the contribution of π(Ge=Ge) → σ^*(Ge–X) through-space orbital interaction (Figure 6.11).

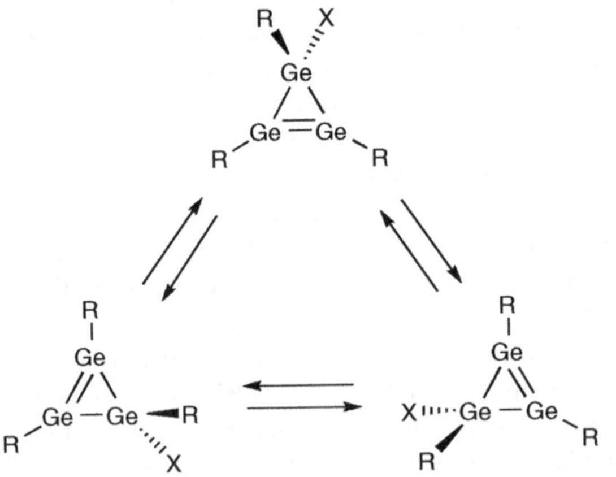

Scheme 6.45

The reactivity of the neutral heavy 6π-electron aromatics, that is, benzene and its analogs, includes a variety of addition and cycloaddition reactions leading to novel cyclic and polycyclic compounds, as well as thermal and photochemical transformations

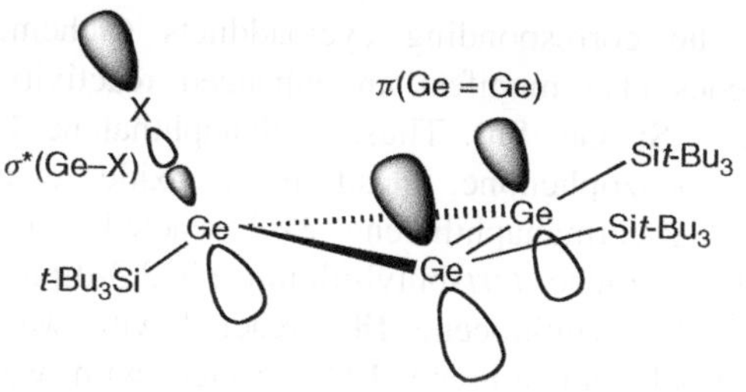

Figure 6.11 *Schematic representation of the π(Ge=Ge) → σ*(Ge–X) orbital interaction in halogen-substituted cyclotrigermenes*

resulting in the formation of valence isomers. Several examples of such reactions will be briefly discussed below.

Thus, silabenzene **16**,[61] being thermally stable up to 100 °C, on standing in C_6D_6 solution at room temperature gradually forms a [4 + 2] dimer, which can be cycloreverted back to the starting **16**.[11] 1-Silanaphthalene **17**[62] also undergoes [4 + 2] cycloaddition, slowly at room temperature and faster at 100 °C, to form a stable dimer that is not prone to retrocycloaddition.[62] 9-Silaanthracene **18**[63] can also be dimerized at 110 °C/C_6D_6/15 days or at 180 °C/solid state/1 h to give a very thermally stable [4 + 4] cycloaddition product.[204] Upon photochemical irradiation, silaaromatics can be transformed into their valence isomers. For example, silabenzene **16**,[61] when photolysed with 290–350 nm light in C_6D_6, exclusively produced the corresponding silabenzvalene,[61b] whereas the irradiation of 9-silaanthracene **18**[63] (300–500 nm, C_6D_6) resulted in the quantitative formation of 9,10-Dewar-9-silaanthracene, which readily undergoes thermal reversion to the starting **18** even at −80 °C (Scheme 6.46).[204]

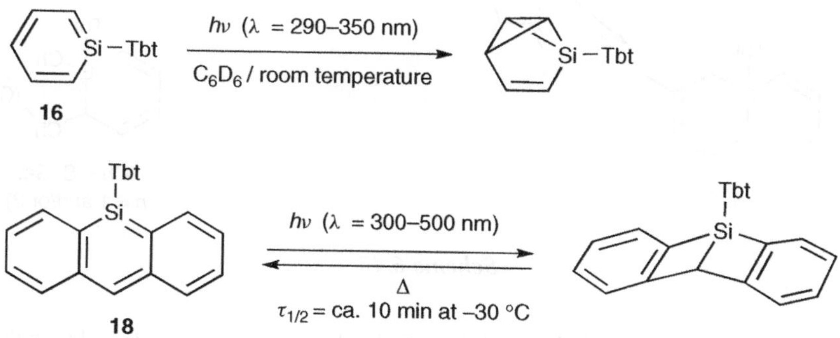

Scheme 6.46

All neutral heavy 6π-electron aromatics, even being kinetically stabilized by the extremely bulky Tbt group, undergo a variety of addition and cycloaddition reactions, forming novel cyclic and polycyclic derivatives. Thus, silabenzene **16**[61] and germabenzene **20**[65] readily react with water, methanol, styrene, phenylacetylene, benzophenone, mesitonitrile oxide, 2,3-dimethylbuta-1,3-diene, elemental sulfur and selenium to produce the corresponding cycloadducts (Scheme 6.47).[11] Likewise, polynuclear heavy arenes also manifest an enhanced reactivity of their endocyclic E=C double bonds (E = Si, Ge, Sn). Thus, 1-silanaphthalene **17**[62] smoothly reacted with water, methanol, benzophenone, mesitonitrile oxide, elemental selenium and carbon tetrachloride;[62b] 2-germanaphthalene **21**[66] reacted with water, mesitonitrile oxide, 2,3-dimethylbuta-1,3-diene, *tert*-butyllithium, *cis*-2,3-diphenylthiirane, elemental sulfur and selenium;[66b] 9-silaanthracene **18**[63] reacted with water, methanol, styrene and benzophenone;[63] 9-silaphenanthrene **19**[64] reacted with water and mesitonitrile oxide;[64] and 9-stannaphenanthrene[69] reacted with methanol, mesitonitrile oxide and 2,3-dimethylbuta-1,3-diene.[69]

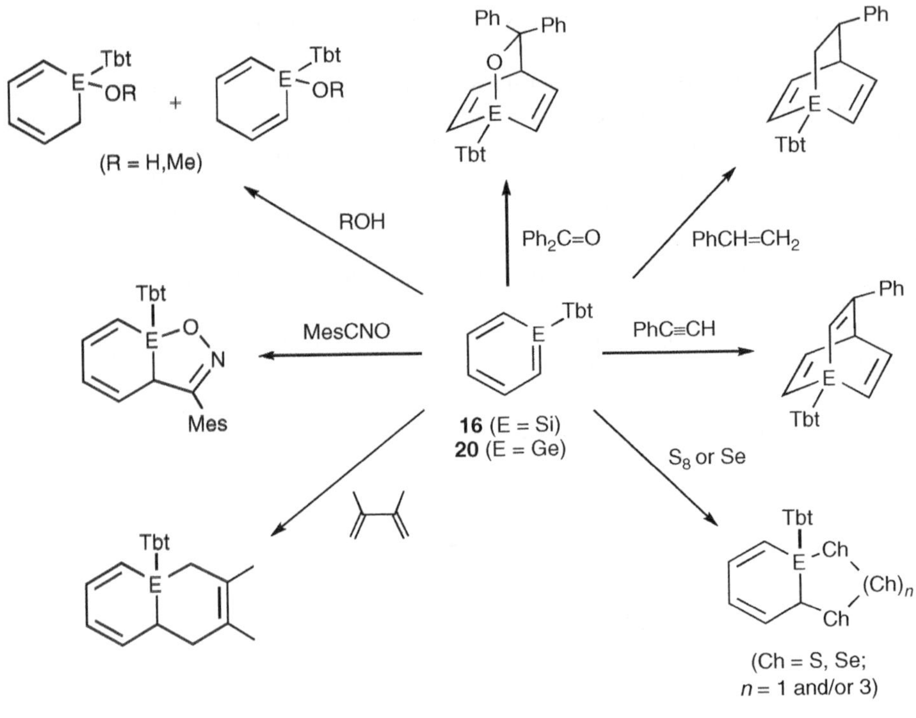

Scheme 6.47

The most thoroughly studied reactivity of the heavy cyclopentadienide analogs is that of dilithio-2,3,4,5-tetraphenyl-1-silacyclopentadienediide $\mathbf{28}^{2-}\bullet\mathbf{2Li}^{+}$, which was reacted with 1,1-dichloro-2,3-diphenylcyclopropene to give a stable biradical,[205] with 2-adamantanone to produce a novel silene,[206] with diphenylcyclopropenone to form the dilithium salt of 2,3,5,6-tetraphenyl-1,4-hydroquinone,[207] and with 1,3-dienes (2,3-dimethylbuta-1,3-diene and cyclohexa-1,3-diene) to yield spirosilacyclopentadienes and elemental lithium.[208] The dimerization of 1-methyl-2,3,4,5-tetraphenyl-1-silacyclopentadienide in THF solution at room temperature produced a rather unusual tricyclic diallylic dianion, isolated in the form of its dilithium salt,[209] whereas the reaction of dilithio-2,3,4,5-tetramethyl-1-silacyclopentadienediide with diphenylcyclopropenone resulted in the formation of a silaspirotropylidene derivative.[210]

The synthetic utilization of the homoaromatic derivatives of heavy group 14 elements is limited to a couple of interesting examples. They concern the reduction of the homocyclotrisilenylium ion **55**$^+$ described in Section 6.3.3. Thus, the one-electron reduction of **55**$^+$ with an equivalent amount of either t-Bu$_3$SiNa or KC$_8$ in diethyl ether produced the stable cyclotetrasilenyl radical **87**, the first one in the series of isolable silicon-centered free radicals (Scheme 6.48).[211] Further reduction of **55**$^+$ with an excess of metallic Li, Na or KC$_8$ yielded the lithium, sodium or potassium salts of cyclotetrasilenide ion **88**$^-$•**M**$^+$ (M = Li, Na, K) (Scheme 6.48).[212] The X-ray diffraction analysis of the lithium derivative **88**$^-$•**[Li**$^+$**(thf)]** disclosed a cyclotetrasilene structure, in which the Li cation was intramolecularly coordinated to the skeletal Si=Si double bond.

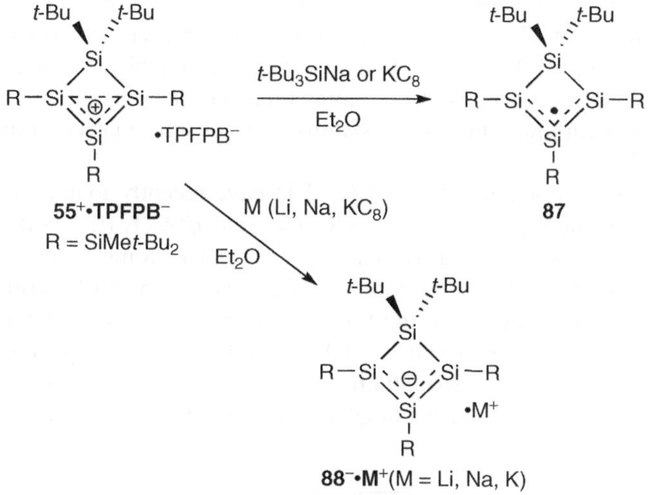

Scheme 6.48

Undoubtedly, the most promising application of the heavy group 14 aromatics is their utilization as novel cyclic polyene ligands for new generation transition metal complexes. In contemporary organometallic chemistry, carbocyclic polyene ligands *cyclo*-C_nR_n ($n = 3$–8), such as cyclopropenyls η^3-C_3R_3, cyclobutadienes η^4-C_4R_4, cyclopentadienyls η^5-C_5R_5, arenes η^6-C_6R_6, cycloheptatrienyls η^7-C_7R_7, and cyclooctatetraenes η^8-C_8R_8, have found widespread utilization in the synthesis of a vast number of transition metal complexes.[213] As the most famous examples, one should mention such 18-electron complexes as (cyclobutadiene)tricarbonyliron (η^4-H_4C_4)Fe(CO)$_3$,[214] bis(cyclopentadienyl)iron(II), ferrocene, (η^5-H_5C_5)$_2$Fe[215] and bis(benzene)chromium (η^6-H_6C_6)$_2$Cr.[216] Such transition metal complexes are widely applied in the field of material science because of the high technological demands for the production of new advanced materials. Thus, metallocenes can be utilized in catalysis (chiral ferrocene as a ligand scaffold), supramolecular chemistry (macroheterocycles for recognition of charged and neutral species), medicine (anticancer drugs), nanotechnology (carbon nanotubes), polymer science (vinyl ferrocene as a ferrocenyl version of polystyrene), the car industry (fuel antiknock agents), molecular magnets (decamethylferrocenium ion-radical as a donor in charge-transfer complexes), etc.[217] The heavy analogs of the carbocyclic polyene ligands, in which the skeletal carbons are fully (or partially) replaced with Si, Ge, Sn and Pb atoms, were unknown until the mid 1990s, when Tilley *et al.* reported the synthesis of the first transition metal complexes featuring sila- and germacyclopentadienyl ligands.[101] Since then and to date, there has been a permanently growing interest in this prospective field of organometallic chemistry, resulting in the synthesis of approximately two dozen hybrid transition metal–main group element complexes.

Transition metal complexes with the heavy cyclopropenyls η^3-E_3R_3 (E = Si, Ge, Sn, Pb) (Section 6.3.1.1) as ligands are still unknown, although such complexes were theoretically predicted to be feasible synthetic targets. Thus, several complexes of group 9 (Co, Rh, Ir) and 10 (Ni) metals having η^3-E_3H_3 (E = Si, Ge) π-ligands were found to be minimum structures, in which an important ligand-to-metal electron donation made η^3-E_3H_3 cationic ligands.[218] The η^3-Si_3H_3 ligand was even found to be capable of forming sandwich-type complexes with the core atoms represented by the main group elements, such as boron and carbon, which were stabilized by charge transfer from the ligand to the core atom.[219]

The neutral heavy aromatics (Section 6.3.2.1) have recently found an application as π-arene ligands of the type η^6-C_5E (E = Si, Ge) and η^6-C_9E (E = Sn) for complexes of group 6 (Cr, Mo, W) and 8 (Ru) metals.[68,220–222] The complexation smoothly proceeded upon the ligand exchange reaction of stable silabenzene **16**,[61] germabenzene **20**[65] and 2-stannanaphthalene **24**[68] with the transition metal complexes $[M(CH_3CN)_3(CO)_3]$ (M = Cr, Mo, W) in either benzene or THF to produce the corresponding neutral [η^6-silabenzene]$M(CO)_3$ (M = Cr, Mo) **89a,b**,[220] [η^6-germabenzene]$M(CO)_3$ (M = Cr, Mo, W) **90a–c**[221,222] and [η^6-2-stannanaphthalene]$Cr(CO)_3$ **91**[68] complexes (Scheme 6.49).

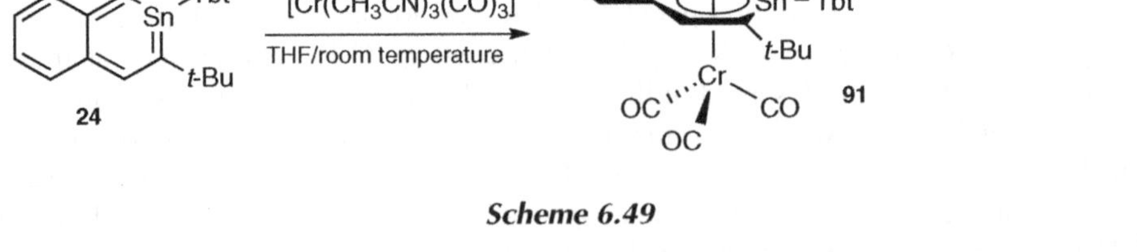

Scheme 6.49

All of these complexes featured nearly planar heavy arene rings, which were η^6-coordinated to transition metals. The spectral data of complexes **90a–c** suggest that germabenzene can efficiently serve as the π-arene ligand for transition metal complexes, like the parent benzene.[222] The reaction of germabenzene **20**[65] with the

Ru derivative [(η^5-C_5Me_5)RuCl]$_4$ in THF afforded the novel complex [(η^5-C_5Me_5)Ru{(η^5-C_5H_5)GeTbt(Cl)}] **92a** with a reduced hapticity of coordination, in which the germacyclohexadienyl ligand was pentahaptocoordinated to the Ru center (Scheme 6.50).[222]

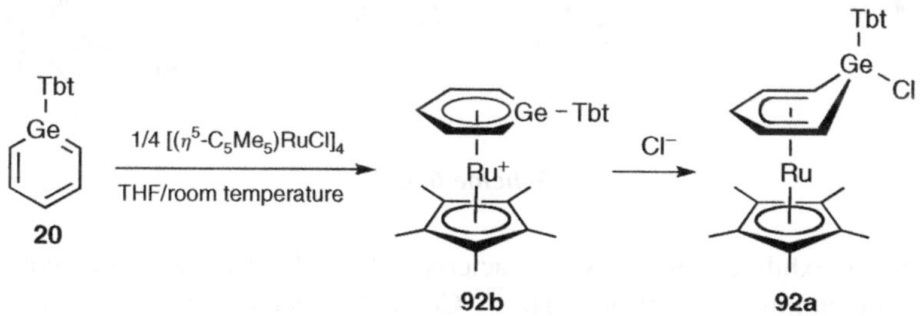

Scheme 6.50

The formation of this complex was achieved through the initial formation of cationic η^6-complex **92b**, followed by nucleophilic attack of the chloride ion on the Ge atom accompanied by cleavage of the Ge–Ru bond (Scheme 6.50). In complex **92b**, the η^5-C_5Me_5 ligand and η^5-germacyclohexadienyl fragment, both of which are essentially flat, are nearly coplanar, which is a characteristic feature of sandwich complexes.[222]

One should also mention the interesting sila- and 1,4-disilabenzene complexes of ruthenium recently synthesized by Tilley *et al.*[223] Thus, the first transition metal silabenzene complex [(η^5-C_5Me_5)Ru(η^6-C_5H_5Si*t*-Bu)]$^+$•[BH(C_6F_5)$_3$]$^-$ **93$^+$**•[BH(C_6F_5)$_3$]$^-$ was readily prepared by oxidation of the silacyclohexadienyl complex (η^5-C_5Me_5)Ru[(η^5-C_5H_5SiH*t*-Bu)] with B(C_6F_5)$_3$ (Scheme 6.51). The characterization of **93$^+$**•[BH(C_6F_5)$_3$]$^-$ was based on spectroscopic methods, which suggested that, as well as the delocalized η^6-type bonding of silabenzene to the Ru atom, there is some important mixed η^5, η^1-type bonding contribution with limited delocalization in the silabenzene ligand.

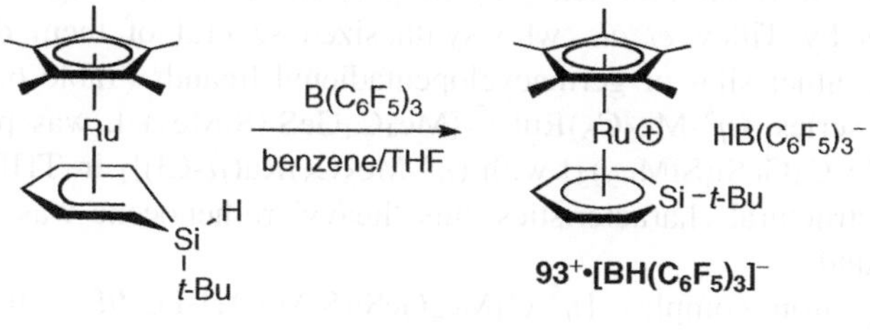

Scheme 6.51

On the other hand, heating the Ru complex (η^5-C_5Me_4Et)(PMe_3)$_2$RuCH$_2$SiMe$_3$ with the *trans*-1,4-dihydrohexamethyl-1,4-disilacyclohexa-2,5-diene in toluene at 90 °C unexpectedly produced the 1,4-disilabenzene complex (η^5-C_5Me_4Et)(PMe_3)RuH(η^2-hexamethyl-1,4-disilabenzene) **94** (Scheme 6.52).

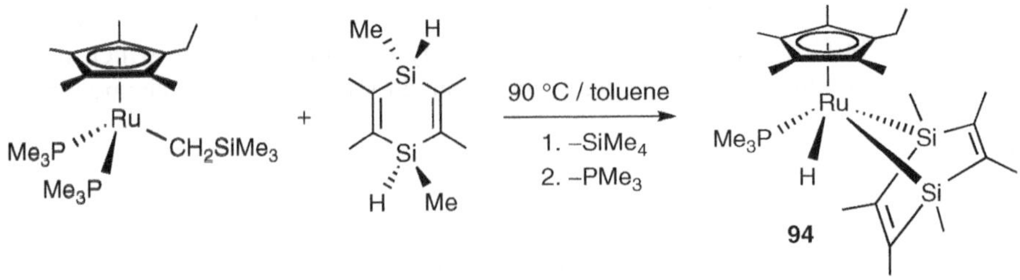

Scheme 6.52

Complex **94** exhibited the very characteristic hydride peak at −12.6 ppm with a typical $^{2}J_{P-H}$ coupling constant of 34 Hz.[223] Computational studies showed that the 1,4-disilabenzene–Ru atom bonding in **94** should be best described as being intermediate between the two extreme resonance forms **94a** and **94b** (Scheme 6.53), in which the metallodisilanorbornadiene structure possesses some degree of transannular Si–Si σ-bonding that can weakly interact with the Ru center.[223]

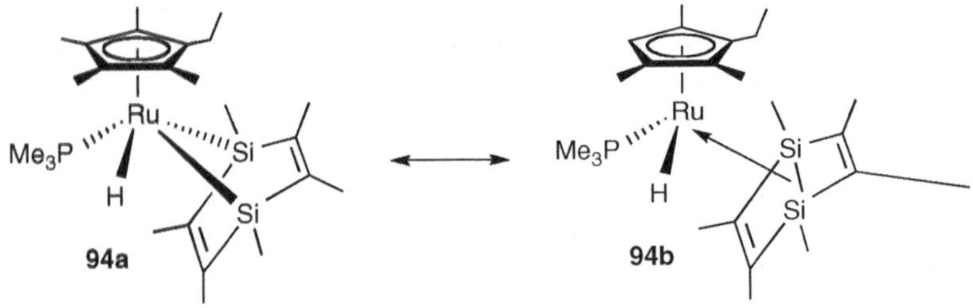

Scheme 6.53

As mentioned above (Section 6.3.2.2), the heavy η^5-C_4E and η^4-C_4E cyclopentadienyls (E = Si, Ge) were the first heavy cyclic polyenes utilized as ligands for transition metal complexes by Tilley *et al.*, who synthesized several of them during the past decade featuring either sila- or germacyclopentadienyl ligands (Table 6.5).[101] The first complex in this series, (η^5-Me_5C_5)Ru[η^5-($Me_4C_4GeSi(SiMe_3)_3$], was prepared by the reaction of Li[$Me_4C_4GeSi(SiMe_3)_3$] with (η^5-Me_5C_5)Ru(μ-Cl)]$_4$ in THF.[101a] Based on its spectral and structural characteristics, this 'heavy' ruthenocene was classified as an aromatic compound.

An interesting iron complex [η^5-$C_4Me_4GeSi(SiMe_3)_3$]$_2$Fe **95** with the two germacyclopentadienyl ligands sandwiching the central Fe atom was recently prepared by Tilley *et al.* by the coupling of the lithium germacyclopentadienide Li^+[$C_4Me_4GeSi(SiMe_3)_3$]$^-$ with $FeCl_2(tht)_{1.5}$ in THF at low temperature (Scheme 6.54).[224] The reaction of this complex **95** with MeLi surprisingly produced a novel ferrocenophane-type *ansa*-$(Me_3Si)_2Si$(η^5-GeC_4Me_4)$_2$Fe complex **96** (Scheme 6.54).

The ferrocene derivatives consisting entirely of heavy group 14 elements are the most intriguing, representing an inorganic version of classical metallocenes. Thus, the

Table 6.5 *Transition metal complexes with sila- or germacyclopentadienyl ligands*

Complex	Reference
$(\eta^5\text{-}C_5Me_5)Ru[\eta^5\text{-}C_4Me_4GeSi(SiMe_3)_3]$	101a
$(\eta^5\text{-}C_5Me_5)(\eta^5\text{-}C_4Me_4SiSiMe_3)HfCl_2$	101c,d
$(\eta^5\text{-}C_5Me_5)(\eta^5\text{-}C_4Me_4GeSiMe_3)HfCl_2$	101c
$[(\eta^5\text{-}C_5Me_5)(\eta^5\text{-}C_4Me_4Ge)HfMe_2Li(thf)]_2$	101d
$(\eta^5\text{-}C_5Me_5)(\eta^5\text{-}C_4Me_4GeSiMe_3)HfMe_2$	101d
$(\eta^5\text{-}C_5Me_5)(\eta^4\text{-}C_4Me_4GeMeEt)HfMe$	101d
$(\eta^5\text{-}C_5Me_5)(\eta^4\text{-}C_4Me_4GeMe\textit{t}Bu)HfMe$	101d
$(\eta^5\text{-}C_5Me_5)(\eta^4\text{-}C_4Me_4GeMePh)HfMe$	101d
$(\eta^5\text{-}C_5Me_5)(\eta^5\text{-}C_4Me_4SiSiMe_3)ZrCl_2$	101d
ansa-$\{\eta^4\text{-}C_4Me_4Ge[Rh(PMe_3)_3H](\eta^5\text{-}C_5Me_4CH_2)\}HfMe_2$	101e
$(\eta^5\text{-}C_5Me_5)(\eta^5\text{-}C_4Me_4Ge[Rh(dmpe)_2])HfMe_2$	101e
$(\eta^5\text{-}C_5Me_5)(\eta^5\text{-}C_4Me_4Ge[Ir(dmpe)_2])HfMe_2$	101e
$[(\eta^4\text{-}C_4Me_4GeMeCl)TaCl_3(Et_2O)x]_2(x = 0.5\text{–}1.0)$	101f
$(\eta^4\text{-}C_4Me_4GeMeCl)TaCl_3(PPh_3)$	101f
$(\eta^4\text{-}C_4Me_4GeMeCl)TaCl_3(CNXyl)$(Xyl = 2,6-dimethylphenyl)	101f
$(\eta^4\text{-}C_4Me_4GeMeCl)TaCl_3(CNXyl)_2$	101f
$(\eta^4\text{-}C_4Me_4GeMeCl)TaCl_3(pyridine)_2$	101f
$(\eta^5\text{-}C_5H_5)(\eta^4\text{-}C_4Me_4GeMeCl)TaCl_2$	101f

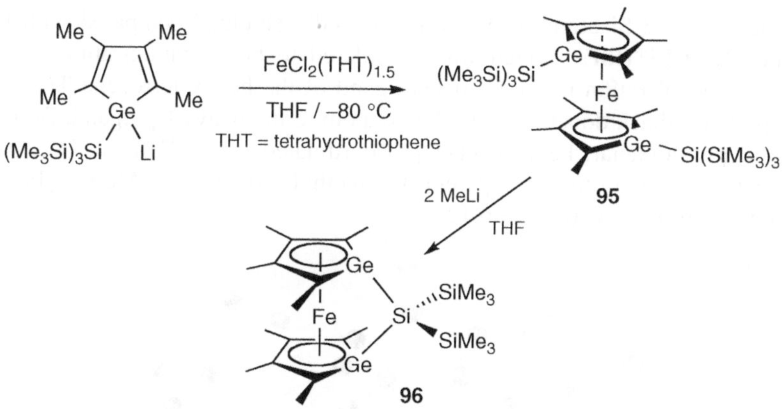

Scheme 6.54

persilaferrocene $[(\eta^5\text{-}H_5Si_5)_2Fe]$ was studied computationally at the HF level without electron correlation functions to show the symmetrical D_{5d} structure [similar to the parent ferrocene $(\eta^5\text{-}H_5C_5)_2Fe]$, in which the Fe–$(\eta^5\text{-}H_5Si_5)$-ligand binding energy was smaller than that in the parent ferrocene: 113.6 vs 144.1 kcal/mol.[225] Because of the shallow PES, transformation of the D_{5d} configuration to others of lower symmetry was expected to be facile. The positive charge in $(\eta^5\text{-}H_5Si_5)_2Fe$ was diminished compared with that in $(\eta^5\text{-}H_5C_5)_2Fe$, which was given as the reason for the weaker Fe-to-ligand interaction in the former. The HOMO–LUMO energy gap was smaller in persilaferrocene than in the parent ferrocene, thus reflecting the smaller HOMO–LUMO separation in the pentasilacyclopentadienyl ligand compared with that of the cyclopentadienyl ligand.

However, the synthesis of such highly attractive heavy ferrocenes (η^5-$R_5E_5)_2Fe$ (E = Si, Ge, Sn, Pb) has not yet been accomplished. As the closest approach to this problem, one can mention a recent study of the ferrocene derivative incorporating *three* heavy group 14 elements (two Si and one Ge) in one of the cyclopentadienyl rings.[226] The target complex (η^5-Me_5C_5)Fe[η^5-(CPh)(CH)Si_2Ge(SiMe*t*-$Bu_2)_3$] **97** was synthesized by the reaction of the heavy lithium cyclopentadienide **32$^-$•[Li$^+$(thf)]** with the *in situ* generated (η^5-Me_5C_5)Fe(acac) complex[227] (acac = 2,4-pentanedionate) in THF at −78 °C (Scheme 6.55).

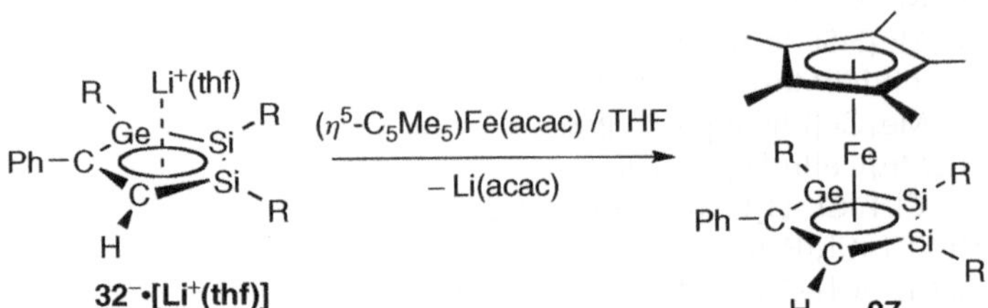

32$^-$•[Li$^+$(thf)]
R = SiMe*t*-Bu_2

97

Scheme 6.55

All skeletal C and Si atoms in **97** were markedly shielded compared with those of the starting **32$^-$•[Li$^+$(thf)]**, which was considered to be a manifestation of the high degree of π-delocalization in the C_2Si_2Ge ligand of the heavy ferrocene **97**, reinforced by the important δ-back donation from the Fe atom to the heavy Cp ligand. Both ligands in **97** are nearly coplanar, being pentahaptocoordinated to the Fe atom in a staggered manner relative to each other [the parent decamethylferrocene (η^5-$Me_5C_5)_2Fe$ also has a staggered conformation] (Figure 6.12).

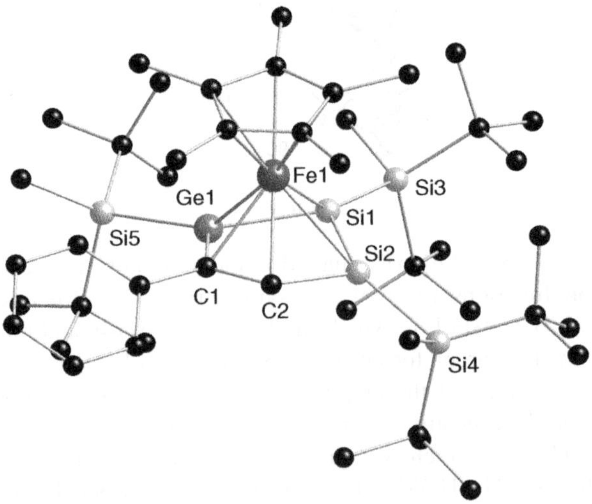

Figure 6.12 *Crystal structure of the heavy ferrocene* ***97*** *(hydrogen atoms are not shown)*

The structural manifestation of the cyclic π-delocalization in the C_2Si_2Ge ligand was clearly seen in the skeletal bond lengths of **97**, which were intermediate between the typical single and double bonds, being very similar to those of aromatic **32$^-$•[Li$^+$(thf)]**. Other evidence of effective cyclic π-delocalization in **97** includes the striking flattening of the C_2Si_2Ge ring and planarization around all skeletal atoms taking place upon complexation. The cyclic voltammetry (CV) measurement of **97** (Ag/Ag^+, CH_2Cl_2, 0.1 M $n Bu_4NClO_4$) revealed two irreversible oxidation waves, of which the first one at $E_p^{ox} = -0.53$ V most likely corresponds to a heavy ferrocene cation-radical. Notably, this oxidation process proceeded at significantly more negative potentials than the corresponding one-electron oxidation of the parent decamethylferrocene[226] (−0.32 V, reversible) and even bis(germacyclopentadienyl)ferrocene $[\eta^5\text{-}C_4Me_4GeSi(SiMe_3)_3]_2Fe$ **95**[224] (−0.45 V, irreversible). This suggests that the heavy cyclopentadienyl ligand $[\eta^5\text{-}(CPh)(CH)Si_2Ge(SiMe\textit{t}\text{-}Bu_2)_3]$ in **97** is a more powerful electron donor to the Fe center than cyclopentadienyl $\eta^5\text{-}Me_5C_5$ and even germacyclopentadienyl $\eta^5\text{-}C_4Me_4GeSi(SiMe_3)_3$ ligands. Detailed computational studies on the model complex $(\eta^5\text{-}H_5C_5)Fe[\eta^5\text{-}(CPh)(CH)Si_2Ge(SiH_3)_3]$ showed that the most important contribution to the overall bonding in **97** results from the strong π-donation from the doubly occupied heavy cyclopentadienyl ligand $[\eta^5\text{-}(CPh)(CH)Si_2Ge(SiH_3)_3]$ π-orbitals to the vacant doubly degenerate d-orbitals of the $(\eta^5\text{-}H_5C_5)Fe^+$ unit.[226] On the other hand, mixing of σ- and δ-type orbitals of the $(\eta^5\text{-}H_5C_5)Fe^+$ and heavy cyclopentadienyl fragments is much less important. The theoretical studies also demonstrated that the electrostatic bonding (52.4%) in **97** dominates over the covalent contribution (47.6%).

As the latest accomplishments in the field of transition metal complexes featuring heavy cyclopentadienyl ligands, Rh half-sandwich **98** and Ru sandwich (ruthenocene) **99** with the η^5-1,2,3-trisilacyclopentadienyl ligand were synthesized by the reaction of the lithium trisilacyclopentadienide derivative with $Rh(CO)_2(acac)$ and $[(\eta^5\text{-}C_5Me_5)RuCl]_4$, respectively (Scheme 6.56).[228]

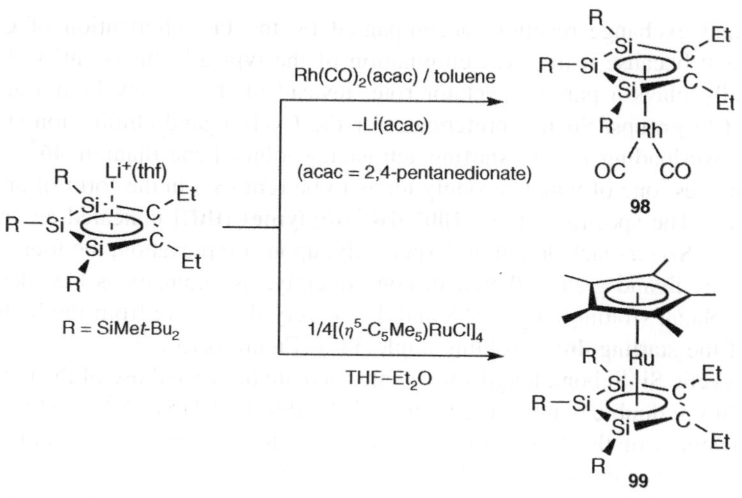

Scheme 6.56

Structural and spectroscopic features of both **98** and **99** pointed to an appreciable delocalization of the 6π-electrons over the Si_3C_2-ligand, and hence, aromaticity of these complexes.

The heavy η^4-E_4R_4 cyclobutadienes (E = Si, Ge) as cyclic polyene ligands for transition metal complexes became synthetically accessible quite recently, since the preparation of the heavy cyclobutadiene dianion derivatives **45^{2-}•2[K$^+$(thf)$_2$]** and **46^{2-}•2[K$^+$(thf)$_2$]**.[104b,128a] This is in marked contrast to the well-established chemistry of η^4-C_4R_4 transition metal complexes.[229] Only a few examples of complexes featuring tetrasilacyclobutadiene ligands have been synthesized thus far:[230] an anionic cobalt dicarbonyl complex [η^4-(*t*-Bu$_2$MeSi)$_4$Si$_4$]Co(CO)$_2$$^-$•[K$^+$(diglyme)$_2$(thf)] **$100^-$•[K$^+$(diglyme)$_2$(thf)]**,[231] an iron tricarbonyl complex [η^4-(*t*-Bu$_2$MeSi)$_4$Si$_4$]Fe(CO)$_3$ **101**,[232] a ruthenium tricarbonyl complex [η^4-(*t*-Bu$_2$MeSi)$_4$Si$_4$]Ru(CO)$_3$ **102**,[233] and a sandwich cobalt cyclopentadienyl complexes [η^4-(*t*-Bu$_2$MeSi)$_4$Si$_3$E]Co(η^5-C_5H_5) (E = Si, Ge) **103a,b**.[234] The 18-electron [η^4-tetrasilacyclobutadiene]dicarbonylcobalt complex **100^-•[K$^+$(diglyme)$_2$(thf)]** was prepared by the reaction of tetrasilacyclobutadiene dianion dipotassium salt **46^{2-}•2[K$^+$(thf)$_2$]**[104b,128a] with an excess of (C_5H_5)Co(CO)$_2$ in THF and isolated in the form of its diglyme complex (Scheme 6.57).[231]

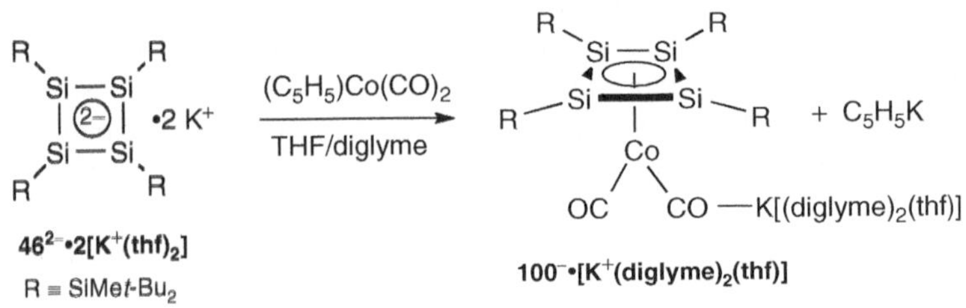

Scheme 6.57

This ligand exchange reaction, accompanied by the fast elimination of C_5H_5K, is rather unusual because it involves elimination of the typically 'innocent' C_5H_5 ligand, which usually plays a purely spectator role, instead of the loosely bound and readily removable CO groups. Such a preference for the C_5H_5 ligand elimination stems from the severe overloading of the starting tetrasilacyclobutadiene dianion **46^{2-}** with two negative charges, one of which strongly tends to be removed in the form of an aromatic $C_5H_5^-$ anion. The spectral data of **100^-•[K$^+$(diglyme)$_2$(thf)]** indicated an appreciable 3d(Co) → π*(Si$_4$) π-back donation. Expectedly, upon complexation, the four-membered ring of the Si_4 ligand greatly flattened; consequently, its geometry is best described as rectangular-planar (folding angles 2.8 and 4.8°), very distinctive from the highly folded structure of the starting **46^{2-}** (folding angle 34°) (Figure 6.13).[231]

All four cyclic Si–Si bond lengths were intermediate between those of the typical Si–Si single and Si=Si double bonds, ranging from 2.2570(9) to 2.3154(8) Å. All these peculiar structural features of the Si_4 fragment, together with its diagnostic η^4-coordination to the Co center, are indicative of cyclic π-delocalization.

The tetrasilacyclobutadiene iron complex **101**, a silicon version of the famous Pettit's (η^4-H_4C_4)Fe(CO)$_3$,[214] was obtained by the treatment of 3,4-dibromo-1,2,3,4-tetrakis

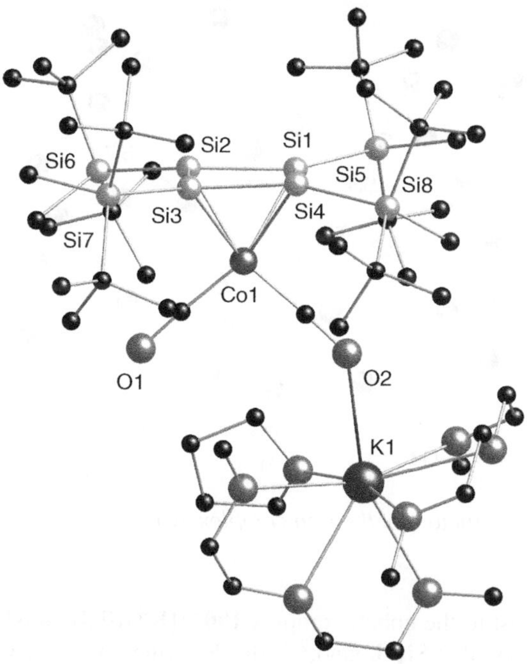

Figure 6.13 *Crystal structure of the anionic cobalt complex* ***100⁻•[K⁺(diglyme)₂(thf)]*** *(hydrogen atoms are not shown)*

(di-*tert*-butylmethylsilyl)cyclotetrasilene **104** with an excess of $Na_2[Fe(CO)_4]$ in THF at room temperature (Scheme 6.58).[232]

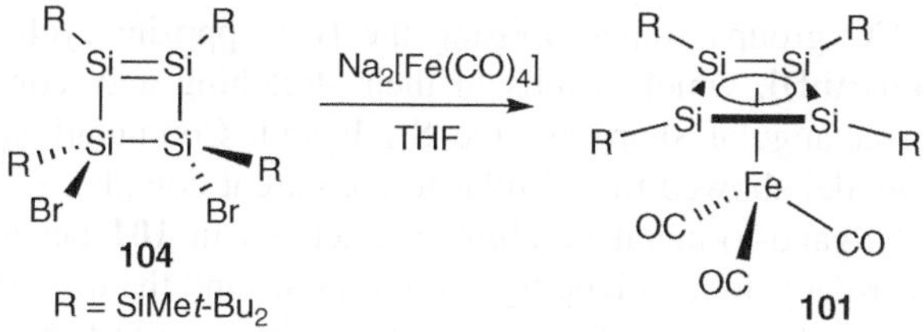

Scheme 6.58

An appreciable 3d(Fe) → π*(Si_4) π-back donation was also operative in this case, as was shown by the high-field shifted resonance of the skeletal Si atoms at −15.8 ppm. Similar to the case of the Co complex **100^-•[K^+(diglyme)$_2$(thf)]**, the tetrasilacyclobutadiene ring in **101** is tetrahaptocoordinated to iron and nearly planar (folding angle 1.8°), with the skeletal Si–Si bond distances [2.2610(7)–2.2802(7) Å] fitting well within the range between the typical Si–Si single (av. 2.34 Å) and Si=Si double (av. 2.20 Å) bond lengths (Figure 6.14).

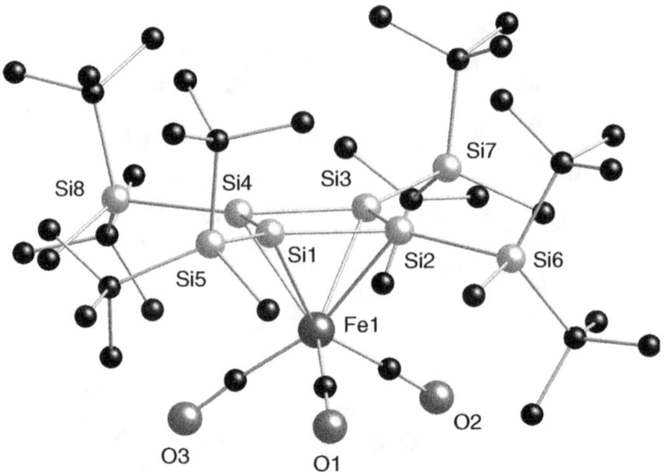

Figure 6.14 *Crystal structure of the iron complex **101** (hydrogen atoms are not shown)*

However, in contrast to the cobalt complex **100⁻•[K⁺(diglyme)₂(thf)]** which features a rectangular shape of the Si_4 ligand,[231] the Si_4 ring in the iron complex **101** has a nearly regular square shape,[232] characteristic of the cyclobutadiene–transition metal complexes.[229a,c] Such a striking difference in the geometry of the tetrasilacyclobutadiene ligand in **100⁻•[K⁺(diglyme)₂(thf)]** and **101** was attributed to the partial distribution of the negative charge over the Si_4 ligand in **100⁻•[K⁺(diglyme)₂(thf)]**, seen in the lower CO stretching vibration frequencies of **100⁻•[K⁺(diglyme)₂(thf)]** (1870 cm^{-1}) compared with those of **101** (1922 and 1973 cm^{-1}), implying stronger π-back donation from the transition metal to CO groups in the case of the Co complex **100⁻•[K⁺(diglyme)₂(thf)]**. Another important reason affecting the overall geometry of the Si_4 ligand is the spatial orientation of the CO groups, superimposing the two opposite cyclic Si–Si bonds in **100⁻•[K⁺(diglyme)₂(thf)]**, which results in their stretching and, consequently, in the preference for the rectangular shape for the Si_4 ligand. Computational studies on the Me_3Si-substituted model showed that similar to the parent complex (η^4-H_4C_4)$Fe(CO)_3$, the most important ligand-to-metal bonding interactions in **101** occurred between the singly occupied tetrasilacyclobutadiene ligand π-orbitals and the d_{xz}-, d_{yz}-orbitals of the Fe atom, resulting in the formation of two pairs of doubly degenerate bonding (HOMO–1, HOMO) and antibonding (LUMO, LUMO+1) hybrid orbitals (Figure 6.15).

The HOMO and LUMO energy levels of the model complex $[(Me_3Si)_4Si_4]Fe(CO)_3$ were higher and lower, respectively, than the corresponding levels of the hypothetical carbon analog $[(Me_3Si)_4C_4]Fe(CO)_3$, thus giving rise to a markedly smaller HOMO–LUMO energy gap for the silicon version $[(Me_3Si)_4Si_4]Fe(CO)_3$ (4.61 eV) compared with that of its carbon analog $[(Me_3Si)_4C_4]Fe(CO)_3$ (5.64 eV). Particularly notable is the appreciable increase in the tetrasilacyclobutadiene ligand HOMO energy level responsible for its stronger π-donating ability (compared with that of the C_4 ligand in the cyclobutadiene complex) and, accordingly, enhanced π-back donation from the Fe atom to the CO ligands. The potential aromaticity of **101** was verified by the NICS(1) calculation

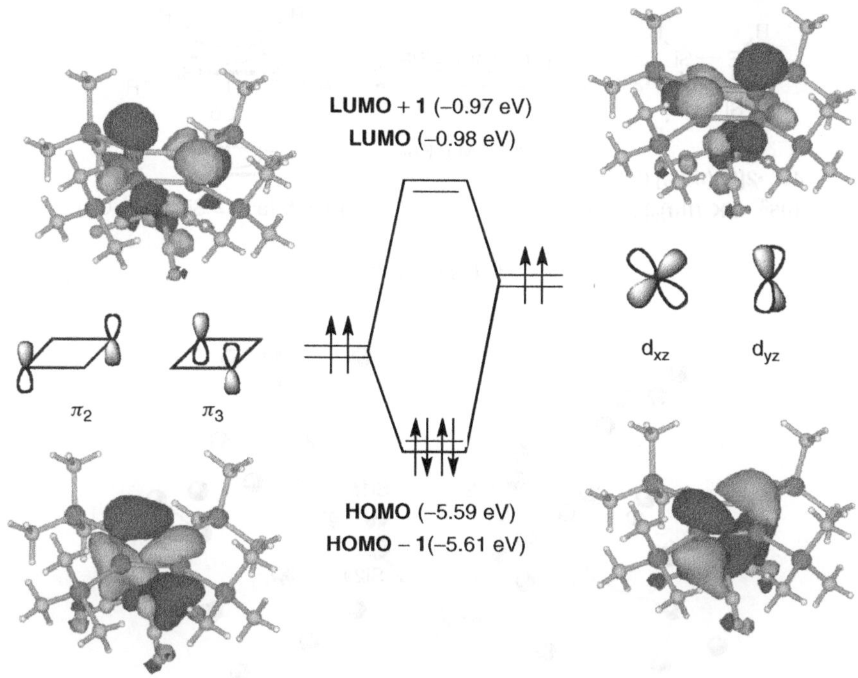

*Figure 6.15 Frontier MOs of **101** [calculated for the model complex with Me_3Si-groups instead of t-Bu_2MeSi substituents at the B3LYP/6-31G(d) for H, C, O, Si and LANL2DZ for Fe levels]*

on the Me_3Si-substituted model providing a negative value of −12.5, which was taken as evidence for the classification of this complex as an organometallic aromatic, similar to the parent complex $(H_4C_4)Fe(CO)_3$.

Similarly to the iron complex **101**, its ruthenium analog $[\eta^4$-$(t$-$Bu_2MeSi)_4Si_4]Ru(CO)_3$ **102** was also prepared by the reaction of the tetrasilacyclobutadiene dianion derivative **46^{2-}•2[K^+(thf)$_2$]** with $[Ru(CO)_3Cl_2]_2$.[233] Like other transition metal complexes featuring the tetrasilacyclobutadiene ligand, **102** was classified as an organometallic aromatic.

The tetrasila- and trisilagermacyclobutadiene cobalt cyclopentadienyl complexes $[\eta^4$-$(t$-$Bu_2MeSi)_4Si_3E]Co(\eta^5$-$C_5H_5)$ (E = Si, Ge) **103a,b**, as the first sandwich complexes featuring cyclic polyene ligands consisting entirely of heavy group 14 elements, were prepared by the reaction of dipotassium salts of either the tetrasilacyclobutadiene dianion **46^{2-}•2[K^+(thf)$_2$]**[104b,128a] or trisilagermacyclobutadiene dianion **105^{2-}•2[K^+(thf)$_2$]**[234] with the $(\eta^5$-$H_5C_5)CoI_2(PPh_3)$ complex in THF (Scheme 6.59).[234] Trisilagermacyclobutadiene Co complex **103b** could also be synthesized using $[(\eta^5$-$H_5C_5)_2Co]^+$•$[PF_6]^-$ [instead of $(\eta^5$-$H_5C_5)CoI_2(PPh_3)$], accompanied by the elimination of $(H_5C_5)K$.

Quite similar to the case of other tetrasilacyclobutadiene complexes,[231–233] sandwich complexes **103** exhibited an important extent of 3d(Co) → π*(Si_4 or Si_3Ge) π-back donation, manifested in the appreciable shielding of the skeletal Si atoms (0.4 ppm for

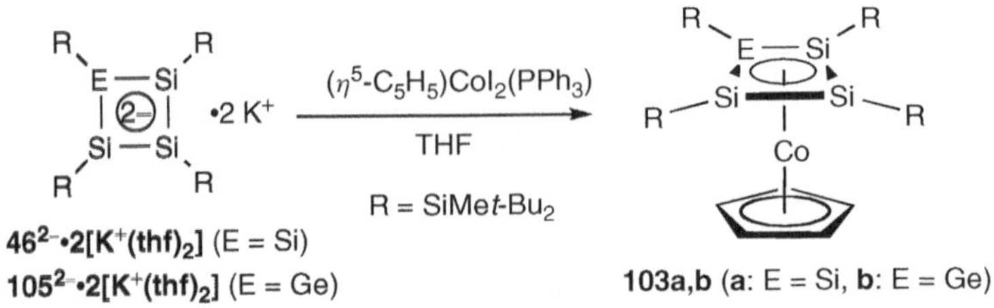

Scheme 6.59

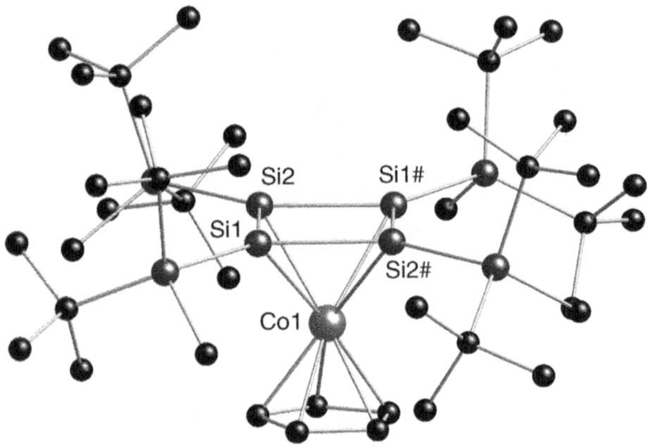

Figure 6.16 *Crystal structure of the sandwich cobalt complex **103a** (hydrogen atoms are not shown)*

103a, GIAO calculation value −4.7 ppm for the model complex $[\eta^4\text{-}(Me_3Si)_4Si_4]Co(\eta^5\text{-}C_5H_5))$.[234] The geometry of the Si_4 ring in **103a** is highly reminiscent of that of the iron complex **101**[232]: an almost regular square-planar shape with the endocyclic Si–Si bond lengths spreading over the very narrow range of 2.2718(10)–2.2725(9) Å and being between those of the usual Si–Si single (av. 2.34 Å) and Si=Si double (av. 2.20 Å) bond lengths (Figure 6.16).

6.4 Summary and Outlook

Aromaticity is one of the oldest and most fundamental concepts in organic chemistry. The study of aromatic derivatives covers almost two centuries, since the milestone discovery of the first aromatic compound, benzene, by Faraday. The number of experimental accomplishments in this field is vast, and a modern theory of aromaticity with the invaluable support of sophisticated computational methods provides a rather satisfactory insight into the nature of aromaticity. Particularly important are the recent developments in the definition of aromaticity as a peculiar phenomenon from the viewpoint of various criteria, which have allowed for a more reliable assignment of organic compounds to the classes of aromatic, nonaromatic or antiaromatic derivatives. In contrast, the field of the heavy congeners of aromatic compounds, containing group 14 elements heavier than carbon, is still waiting to be comprehensively investigated. Being considered as fleeting, not stable (and even not detectable) intermediates just a few decades ago, the heavy aromatic species have now turned into a class of real compounds that can be isolated, stored for an indefinite period, and even structurally characterized. In other words, the heavy aromatic derivatives are becoming readily available and synthetically very useful organometallic reagents. However, their chemistry is very far from completion, and there is still much that remains to be done. For example, stable derivatives of the heavy analogs of the 2π-electron cyclobutadiene dication, as well as those of the 6π-electron cycloheptatrienylium (tropylium) ion and cyclooctatetraene dication, are still experimentally unknown. Moreover, aromatic compounds consisting of all heavy group 14 elements, representing a particular interest from the viewpoint of the aromaticity of inorganic rings, have been prepared only in the case of the cyclopropenylium ion and cyclobutadiene dianion species. Undoubtedly, the tetrasilacyclobutadiene dication, pentasilacyclopentadienide ion, hexasilabenzene and heptasilatropylium ion (as well as their heavier homologs of Ge, Sn and, possibly, Pb atoms) are among the most desirable synthetic targets in the near future. Apart from these synthetic challenges, one can also expect major breakthroughs in the theory of heavy (organometallic) aromaticity, which is markedly distinct from its organic counterpart. To date, there are very few theoretical studies related to this topic; therefore, new theoretical insights based on state-of-the-art high-level computations using appropriate model compounds would be very welcome. As a culmination, we would also predict a permanently increasing interest in the chemistry of transition metal complexes with heavy aromatic compounds as the cyclic polyene ligands of the new generation. Apart from its undoubted fundamental interest, the chemistry of such transition metal complexes is also expected to be a highly promising attraction in applied science for the design of new advanced materials.

6.5 References

1. M. Faraday, *Phil. Trans. R. Soc*. **1825**, 440.
2. A. Kekulé, *Justus Liebigs Ann. Chem*. **1866**, *137*, 129.
3. F. A. Carey, R. J. Sundberg, *Advanced Organic Chemistry, Part A: Structure and Mechanisms*, 4th Edn, Kluwer Academic/Plenum Publishers, New York, **2000**.
4. E. Hückel, *Z. Phys*. **1931**, *70*, 204.
5. V. I. Minkin, M. N. Glukhovtsev, B. A. Simkin, *Aromaticity and Antiaromaticity: Electronic and Structural Aspects*, John Wiley & Sons, Ltd, New York, **1994**.
6. P. von R. Schleyer, C. Maerker, A. Dransfeld, H. Jiao, N. J. R. van E. Hommes, *J. Am. Chem. Soc*. **1996**, *118*, 6317.
7. P. von R. Schleyer, M. Manoharan, Z.-X. Wang, B. Kiran, H. Jiao, R. Puchta, N. J. R. v E. Hommes, *Organic Lett*. **2001**, *3*, 2465.
8. (a) P. J. Garrat, *Aromaticity*, John Wiley & Sons, Ltd, New York, **1986**; (b) P. von R. Schleyer (Ed.), thematic issue 'Aromaticity': *Chem. Rev*. **2001**, *101*, issue 5; (c) P. von R. Schleyer (Ed.), thematic issue 'Delocalization – π and σ': *Chem. Rev*. **2005**, *105*, issue 10; (d) C. A. Tsipis, *Coord. Chem. Rev*. **2005**, *249*, 2740; (e) Y. Wang, G. H. Robinson, *Organometallics* **2007**, *26*, 2.

9. (a) Y. Apeloig, in *The Chemistry of Organic Silicon Compounds, Part 1* (Eds S. Patai, Z. Rappoport), John Wiley & Sons, Ltd, Chichester, **1989**, Chapter 2; (b) Y. Apeloig, M. Karni, in *The Chemistry of Organic Silicon Compounds, Vol. 2, Part 1* (Eds Z. Rappoport, Y. Apeloig), John Wiley & Sons, Ltd, Chichester, **1998**, Chapter 1.
10. (a) V. Ya. Lee, A. Sekiguchi, *Angew. Chem., Int. Ed.* **2007**, *46*, 6596; (b) V. Ya. Lee, A. Sekiguchi, *Chem. Soc. Rev.* **2008**, *37*, 1652.
11. (a) N. Tokitoh, *Acc. Chem. Res.* **2004**, *37*, 86; (b) N. Tokitoh, *Bull. Chem. Soc. Jpn.* **2004**, *77*, 429.
12. (a) T. J. Barton, D. S. Banasiak, *J. Am. Chem. Soc.* **1977**, *99*, 5199; (b) T. J. Barton, G. T. Burns, *J. Am. Chem. Soc.* **1978**, *100*, 5246; (c) T. J. Barton, M. Vuper, *J. Am. Chem. Soc.* **1981**, *103*, 6788.
13. (a) B. Solouki, P. Rosmus, H. Bock, G. Maier, *Angew. Chem., Int. Ed. Engl.* **1980**, *19*, 51; (b) H. Bock, R. A. Bowling, B. Solouki, T. J. Barton, G. T. Burns, *J. Am. Chem. Soc.* **1980**, *102*, 429; (c) H. Bock, P. Rosmus, B. Solouki, G. Maier, *J. Organomet. Chem.* **1984**, *271*, 145.
14. (a) G. Maier, G. Mihm, H. P. Reisenauer, *Angew. Chem., Int. Ed. Engl.* **1980**, *19*, 52; (b) G. Maier, G. Mihm, H. P. Reisenauer, *Chem. Ber.* **1982**, *115*, 801; (c) G. Maier, G. Mihm, R. O. W. Baumgärtner, H. P. Reisenauer, *Chem. Ber.* **1984**, *117*, 2337.
15. C. L. Kreil, O. L. Chapman, G. T. Burns, T. J. Barton, *J. Am. Chem. Soc.* **1980**, *102*, 841.
16. A. K. Mal'tsev, N. D. Kagramanov, I. O. Bragilevskii, N. G. Komalenkova, S. A. Bashkirova, D. A. Zagorets, M. Ya. Kel'man, E. A. Chernyshev, O. M. Nefedov, *Izv. Akad. Nauk SSSR, Ser. Khim.* **1989**, 1051 (English translation: *Bull Acad. Sci. USSR, Div. Chem. Sci.* **1989**, 950).
17. G. Märkl, P. Hofmeister, *Angew. Chem., Int. Ed. Engl.* **1979**, *18*, 789.
18. (a) W. Ando, H. Tanikawa, A. Sekiguchi, *Tetrahedron Lett.* **1983**, *24*, 4245; (b) A. Sekiguchi, H. Tanikawa, W. Ando, *Organometallics* **1985**, *4*, 584.
19. G. Märkl, W. Schlosser, W. S. Sheldrick, *Tetrahedron Lett.* **1988**, *29*, 467.
20. P. Jutzi, M. Meyer, H. P. Reisenauer, G. Maier, *Chem. Ber.* **1989**, *122*, 1227.
21. (a) J. D. Rich, R. West, *J. Am. Chem. Soc.* **1982**, *104*, 6884; (b) K. M. Welsh, J. D. Rich, R. West, J. Michl, *J. Organomet. Chem.* **1987**, *325*, 105; (c) A. Sekiguchi, G. R. Gillette, R. West, *Organometallics* **1988**, *7*, 1226.
22. G. Maier, K. Schöttler, H. P. Reisenauer, *Tetrahedron Lett.* **1985**, *26*, 4079.
23. A. Bjarnason, I. Arnason, *Angew. Chem., Int. Ed. Engl.* **1992**, *31*, 1633.
24. (a) Y.-W. Kwak, J.-B. Lee, K.-K. Lee, S.-S. Kim, B. H. Boo, *Bull. Korean Chem. Soc.* **1994**, *15*, 410; (b) Y.-W. Kwak, K.-K. Lee, S.-D. Yoh, *Bull. Korean Chem. Soc.* **1997**, *18*, 552.
25. M. Oba, Y. Watanabe, K. Nishiyama, *Organometallics* **2002**, *21*, 3667.
26. (a) Y. van den Winkel, B. L. M. van Baar, F. Bickelhaupt, W. Kulik, C. Sierakowski, G. Maier, *Chem. Ber.* **1991**, *124*, 185; (b) H. Hiratsuka, M. Tanaka, T. Okutsu, M. Oba, K. Nishiyama, *J. Chem. Soc., Chem. Commun.* **1995**, 215; (c) H. Hiratsuka, M. Tanaka, H. Horiuchi, Naris, T. Yoshinaga, M. Oba, K. Nishiyama, *J. Organomet. Chem.* **2000**, *611*, 71.
27. G. Märkl, W. Schlosser, *Angew. Chem., Int. Ed. Engl.* **1988**, *27*, 963.

28. G. Märkl, D. Rudnick, *Tetrahedron Lett.* **1980**, *21*, 1405.
29. G. Märkl, D. Rudnick, R. Schulz, A. Schweig, *Angew. Chem., Int. Ed. Engl.* **1982**, *21*, 221.
30. R. Breslow, *J. Am. Chem. Soc.* **1957**, *79*, 5318.
31. E. D. Jemmis, G. N. Srinivas, J. Leszczynski, J. Kapp, A. A. Korkin, P. von R. Schleyer, *J. Am. Chem. Soc.* **1995**, *117*, 11361.
32. G. N. Srinivas, E. D. Jemmis, A. A. Korkin, P. von R. Schleyer, *J. Phys. Chem. A* **1999**, *103*, 11034.
33. (a) A. Sekiguchi, M. Tsukamoto, M. Ichinohe, *Science*, **1997**, *275*, 60; (b) A. Sekiguchi, M. Tsukamoto, M. Ichinohe, N. Fukaya, *Phosphorus Sulfur Silicon Relat. Elem.* **1997**, 124–125, 323.
34. (a) M. Ichinohe, N. Fukaya, A. Sekiguchi, *Chem. Lett.* **1998**, 1045; (b) A. Sekiguchi, N. Fukaya, M. Ichinohe, *Phosphorus Sulfur Silicon* **1999**, 150-151, 59; (c) A. Sekiguchi, N. Fukaya, M. Ichinohe, Y. Ishida, *Eur. J. Inorg. Chem.* **2000**, 1155.
35. (a) M. Ichinohe, M. Igarashi, A. Sekiguchi, *J. Am. Chem. Soc.* **2005**, *127*, 9978; (b) A closely related Si_2C-hybrid disilacyclopropenylium ion derivative was recently prepared as a 2π-electron aromatic system, featuring diagnostically deshielded cationic cyclic Si and C atoms (+208.2 ppm for the ^{29}Si nuclei and +253.7 ppm for the ^{13}C nucleus): M. Igarashi, M. Ichinohe, A. Sekiguchi, *J. Am. Chem. Soc.* **2007**, *129*, 12660.
36. G. A. Olah, J. M. Bollinger, A. M. White, *J. Am. Chem. Soc.* **1969**, *91*, 3667.
37. G. A. Olah, G. D. Mateescu, *J. Am. Chem. Soc.* **1970**, *92*, 1430.
38. G. A. Olah, J. S. Staral, *J. Am. Chem. Soc.* **1976**, *98*, 6290.
39. (a) M. Bremer, P. von R. Schleyer, U. Fleischer, *J. Am. Chem. Soc.* **1989**, *111*, 1147; (b) K. Krogh-Jespersen, P. von R. Schleyer, J. A. Pople, D. Cremer, *J. Am. Chem. Soc.* **1978**, *100*, 4301.
40. G. A. Olah, G. Liang, *J. Am. Chem. Soc.* **1976**, *98*, 3033.
41. G. A. Olah, G. Liang, *J. Am. Chem. Soc.* **1977**, *99*, 6045.
42. B. A. Hess Jr, C. S. Ewig, L. J. Schaad, *J. Org. Chem.* **1985**, *50*, 5869.
43. T. Clark, D. Wilhelm, P. von R. Schleyer, *Tetrahedron Lett.* **1982**, *23*, 3547.
44. M. Balci, M. L. McKee, P. von R. Schleyer, *J. Phys. Chem. A* **2000**, *104*, 1246.
45. A. Korkin, M. Glukhovtsev, P. von R. Schleyer, *Int. J. Quantum Chem.* **1993**, *46*, 137.
46. M. N. Glukhovtsev, B. Ya. Simkin, V. I. Minkin, *Zh. Strukt. Khim.* **1989**, *30*, 23.
47. K. K. Baldridge, M. S. Gordon, *J. Am. Chem. Soc.* **1988**, *110*, 4204.
48. (a) N. Tokitoh, K. Wakita, R. Okazaki, S. Nagase, P. von R. Schleyer, *J. Am. Chem. Soc.* **1997**, *119*, 6951; (b) K. Wakita, N. Tokitoh, R. Okazaki, S. Nagase, P. von R. Schleyer, H. Jiao, *J. Am. Chem. Soc.* **1999**, *121*, 11336; (c) K. Wakita, N. Tokitoh, R. Okazaki, *Bull. Chem. Soc. Jpn.* **2000**, *73*, 2157.
49. (a) H. B. Schlegel, B. Coleman, M. Jones Jr, *J. Am. Chem. Soc.* **1978**, *100*, 6499; (b) M. S. Gordon, P. Boudjouk, F. Anwari, *J. Am. Chem. Soc.* **1983**, *105*, 4972; (c) P. George, C. W. Bock, M. Trachtman, *Theor. Chim. Acta* **1987**, *71*, 289.
50. J. Chandrasekhar, P. von R. Schleyer, R. O. W. Baumgärtner, M. T. Reetz, *J. Org. Chem.* **1983**, *48*, 3453.
51. K. K. Baldridge, M. S. Gordon, *Organometallics* **1988**, *7*, 144.

52. K. K. Baldridge, O. Uzan, J. M. L. Martin, *Organometallics* **2000**, *19*, 1477.
53. K. K. Baldridge, M. S. Gordon, *J. Organomet. Chem.* **1984**, *271*, 369.
54. (a) N. Matsunaga, T. R. Cundari, M. W. Schmidt, M. S. Gordon, *Theor. Chim. Acta* **1992**, *83*, 57; (b) N. Matsunaga, M. S. Gordon, *J. Am. Chem. Soc.* **1994**, *116*, 11407.
55. R. A. King, G. Vacek, H. F. Schaefer III, *J. Mol. Struct. (Theochem)* **1995**, *358*, 1.
56. S. Nagase, T. Kudo, M. Aoki, *J. Chem. Soc. Chem. Commun.* **1985**, 1121.
57. (a) A. Sax, R. Janoschek, *Angew. Chem., Int. Ed. Engl.* **1986**, *25*, 651; (b) D. A. Clabo Jr, H. F. Schaefer III, *J. Chem. Phys.* **1986**, *84*, 1664.
58. (a) S. Nagase, H. Teramae, T. Kudo, *J. Chem. Phys.* **1987**, *86*, 4513; (b) A. F. Sax, J. Kalcher, R. Janoschek, *J. Comput. Chem.* **1988**, *9*, 564; (c) S. Nagase, *Polyhedron* **1991**, *10*, 1299; (d) S. Nagase, *Pure Appl. Chem.* **1993**, *65*, 675; (e) P. von R. Schleyer, H. Jiao, N. J. R. van Eikema Hommes, V. G. Malkin, O. L. Malkina, *J. Am. Chem. Soc.* **1997**, *119*, 12669; (f) Stable prismanes with bulky substituents were reported: hexasilaprismane R_6Si_6 (R = 2,6-*i*-Pr_2-C_6H_3) and its low-temperature photochemical isomerization to hexasila(Dewar)benzene: A. Sekiguchi, T. Yatabe, C. Kabuto, H. Sakurai, *J. Am. Chem. Soc.* **1993**, *115*, 5853; hexagermaprismane R_6Ge_6 [R = $CH(SiMe_3)_2$]: A. Sekiguchi, C. Kabuto, H. Sakurai, *Angew. Chem., Int. Ed. Engl.* **1989**, *28*, 55; hexastannaprismane R_6Sn_6 (R = Si*t*-Bu_3): N. Wiberg, H.-W. Lerner, H. Nöth, W. Ponikwar, *Angew. Chem., Int. Ed.* **1999**, *38*, 1103; (g) A tricyclic aromatic isomer of hexasilabenzene was very recently reported: K. Abersfelder, A.J.P. White, H.S. Rzepa, D. Schesckewitz, *Science* **2010**, *327*, 564.
59. M. Zhao, B. M. Gimarc, *Inorg. Chem.* **1996**, *35*, 5378.
60. R. Janoschek, *J. Inorg. Organomet. Polymers* **1995**, *5*, 155.
61. (a) K. Wakita, N. Tokitoh, R. Okazaki, S. Nagase, *Angew. Chem.* **2000**, *112*, 648; *Angew. Chem., Int. Ed.* **2000**, *39*, 634; (b) K. Wakita, N. Tokitoh, R. Okazaki, N. Takagi, S. Nagase, *J. Am. Chem. Soc.* **2000**, *122*, 5648.
62. (a) N. Takeda, A. Shinohara, N. Tokitoh, *Organometallics* **2002**, *21*, 4024; (b) A. Shinohara, N. Takeda, T. Sasamori, N. Tokitoh, *Bull. Chem. Soc. Jpn.* **2005**, *78*, 977.
63. N. Takeda, A. Shinohara, N. Tokitoh, *Organometallics* **2002**, *21*, 256.
64. N. Tokitoh, A. Shinohara, T. Matsumoto, T. Sasamori, N. Takeda, Y. Furukawa, *Organometallics* **2007**, *26*, 4048.
65. N. Nakata, N. Takeda, N. Tokitoh, *J. Am. Chem. Soc.* **2002**, *124*, 6914.
66. (a) N. Nakata, N. Takeda, N. Tokitoh, *Organometallics* **2001**, *20*, 5507; (b) N. Nakata, N. Takeda, N. Tokitoh, *Organometallics* **2003**, *22*, 481.
67. T. Sasamori, K. Inamura, W. Hoshino, N. Nakata, Y. Mizuhata, Y. Watanabe, Y. Furukawa, N. Tokitoh, *Organometallics* **2006**, *25*, 3533.
68. (a) Y. Mizuhata, T. Sasamori, N. Takeda, N. Tokitoh, *J. Am. Chem. Soc.* **2006**, *128*, 1050; (b) Y. Mizuhata, T. Sasamori, N. Nagahora, Y. Watanabe, Y. Furukawa, N. Tokitoh, *Dalton Trans.* **2008**, 4409.
69. Y. Mizuhata, N. Takeda, T. Sasamori, N. Tokitoh, *Chem. Lett.* **2005**, *34*, 1088.
70. (a) C. Cui, M. M. Olmstead, J. C. Fettinger, G. H. Spikes, P. P. Power, *J. Am. Chem. Soc.* **2005**, *127*, 17530; (b) K. Takeuchi, M. Ichinohe, A. Sekiguchi, *J. Am.*

Chem. Soc. **2008**, *130*, 16848; (c) K. Takeuchi, M. Ichinohe, A. Sekiguchi, J.-D. Guo, S. Nagase, *J. Phys. Org. Chem.* **2010**, *23*, 390.
71. R. Kinjo, M. Ichinohe, A. Sekiguchi, N. Takagi, M. Sumimoto, S. Nagase, *J. Am. Chem. Soc.* **2007**, *129*, 7766.
72. Y. Kabe, K. Ohkubo, H. Ishikawa, W. Ando, *J. Am. Chem. Soc.* **2000**, *122*, 3775.
73. A. Streitwieser Jr, L. L. Nebenzahl, *J. Am. Chem. Soc.* **1976**, *98*, 2188.
74. P. W. Fowler, E. Steiner, *J. Phys. Chem. A* **1997**, *101*, 1409.
75. J. Thiele, *Chem. Ber.* **1901**, *34*, 68.
76. H. Gilman, R. D. Gorsich, *J. Am. Chem. Soc.* **1958**, *80*, 3243.
77. Reviews: (a) K. Tamao, A. Kawachi, *Adv. Organomet. Chem.* **1995**, *38*, 1; (b) P. D. Lickiss, C. M. Smith, *Coord. Chem. Rev.* **1995**, *145*, 75; (c) J. Belzner, U. Dehnert, in *The Chemistry of Organic Silicon Compounds*, *Vol. 2, Part 1* (Eds Z. Rappoport, Y. Apeloig), John Wiley & Sons, Ltd, Chichester, **1998**, Chapter 14; (d) J. Dubac, C. Guérin, P. Meunier, in *The Chemistry of Organic Silicon Compounds*, *Vol. 2, Part 3* (Eds Z. Rappoport, Y. Apeloig), John Wiley & Sons, Ltd, Chichester, **1998**, Chapter 34; (e) A. Sekiguchi, V. Ya. Lee, M. Nanjo, *Coord. Chem. Rev.* **2000**, *210*, 11; (f) P. Riviere, A. Castel, M. Riviere-Baudet, in *The Chemistry of Organic Germanium, Tin and Lead Compounds*, *Vol. 2, Part 1* (Ed. Z. Rappoport), John Wiley & Sons, Ltd, Chichester, **2002**, Chapter 11; (g) M. Saito, M. Yoshioka, *Coord. Chem. Rev.* **2005**, *249*, 765.
78. E. Colomer, R. J. P. Corriu, M. Lheureux, *Chem. Rev.* **1990**, *90*, 265.
79. J. R. Damewood Jr, *J. Org. Chem.* **1986**, *51*, 5028.
80. W.-C. Joo, J.-H. Hong, S.-B. Choi, H.-E. Son, *J. Organomet. Chem.* **1990**, *391*, 27.
81. J.-H. Hong, P. Boudjouk, *J. Am. Chem. Soc.* **1993**, *115*, 5883.
82. J.-H. Hong, P. Boudjouk, S. Castelino, *Organometallics* **1994**, *13*, 3387.
83. R. West, H. Sohn, U. Bankwitz, J. Calabrese, Y. Apeloig, T. Mueller, *J. Am. Chem. Soc.* **1995**, *117*, 11608.
84. R. West, H. Sohn, D. R. Powell, T. Müller, Y. Apeloig, *Angew. Chem.* **1996**, *108*, 1095; *Angew. Chem., Int. Ed. Engl.* **1996**, *35*, 1002.
85. J.-H. Hong, P. Boudjouk, *Bull. Soc. Chim. Fr.* **1995**, *132*, 495.
86. (a) M. Saito, R. Haga, M. Yoshioka, *Chem. Commun.* **2002**, 1002; (b) M. Saito, R. Haga, M. Yoshioka, *Chem. Lett.* **2003**, *32*, 912; (c) M. Saito, R. Haga, M. Yoshioka, K. Ishimura, S. Nagase, *Angew. Chem.* **2005**, *117*, 6711; *Angew. Chem., Int. Ed.* **2005**, *44*, 6553; (d) R. Haga, M. Saito, M. Yoshioka, *J. Am. Chem. Soc.* **2006**, *128*, 4934; (e) R. Haga, M. Saito, M. Yoshioka, *Eur. J. Inorg. Chem.* **2007**, 1297; (f) A related dilithio - 2,3,4,5-tetraphenyl-1-plumbacyclopentadienediide was very recently reported: M. Saito, M. Sekiguchi, T. Tajima, K. Ishimuar, S. Nagase, M. Hada, *Science* **2010**, *328*, 339.
87. M. Saito, M. Nakamura, T. Tajima, M. Yoshioka, *Angew. Chem.* **2007**, *119*, 1526; *Angew. Chem., Int. Ed.* **2007**, *46*, 1504.
88. (a) W. P. Freeman, T. Don Tilley, G. P. A. Yap, A. L. Rheingold, *Angew. Chem.* **1996**, *108*, 960; *Angew. Chem., Int. Ed. Engl.* **1996**, *35*, 882; (b) W. P. Freeman, T. Don Tilley, L. M. Liable-Sands, A. L. Rheingold, *J. Am. Chem. Soc.* **1996**, *118*, 10457.
89. S.-B. Choi, P. Boudjouk, P. Wei, *J. Am. Chem. Soc.* **1998**, *120*, 5814.

90. Y. Liu, T. C. Stringfellow, D. Ballweg, I. A. Guzei, R. West, *J. Am. Chem. Soc.* **2002**, *124*, 49.
91. In marked contrast, the dilithio-1-silafluorenediide was earlier reported to possess a localized nonaromatic structure on the basis of its ^{29}Si NMR spectra; however, this conclusion suffered from the lack of crystal structure: S.-B. Choi, P. Boudjouk, *Tetrahedron Lett*. **2000**, *41*, 6685.
92. S.-B. Choi, P. Boudjouk, J.-H. Hong, *Organometallics* **1999**, *18*, 2919.
93. J.-H. Hong, Y. Pan, P. Boudjouk, *Angew. Chem*. **1996**, *108*, 213; *Angew. Chem., Int. Ed. Engl*. **1996**, *35*, 186.
94. S.-B. Choi, P. Boudjouk, K. Qin, *Organometallics* **2000**, *19*, 1806.
95. Y. Liu, D. Ballweg, T. Müller, I. A. Guzei, R. W. Clark, R. West, *J. Am. Chem. Soc*. **2002**, *124*, 12174.
96. M. Saito, M. Shimosawa, M. Yoshioka, K. Ishimura, S. Nagase, *Organometallics* **2006**, *25*, 2967.
97. M. Saito, M. Shimosawa, M. Yoshioka, K. Ishimura, S. Nagase, *Chem. Lett*. **2006**, *35*, 940.
98. J.-H. Hong, P. Boudjouk, I. Stoenescu, *Organometallics* **1996**, *15*, 2179.
99. S.-B. Choi, P. Boudjouk, *J. Chem. Soc., Dalton Trans*. **2000**, 841.
100. W. P. Freeman, T. D. Tilley, F. P. Arnold, A. L. Rheingold, P. K. Gantzel, *Angew. Chem*. **1995**, *107*, 2029; *Angew. Chem., Int. Ed. Engl*. **1995**, *34*, 1887.
101. (a) W. P. Freeman, T. D. Tilley, A. L. Rheingold, R. L. Ostrander, *Angew. Chem*. **1993**, *105*, 1841; *Angew. Chem., Int. Ed. Engl*. **1993**, *32*, 1744; (b) W. P. Freeman, T. D. Tilley, A. L. Rheingold, *J. Am. Chem. Soc*. **1994**, *116*, 8428; (c) J. M. Dysard, T. D. Tilley, *J. Am. Chem. Soc*. **1998**, *120*, 8245; (d) J. M. Dysard, T. D. Tilley, *J. Am. Chem. Soc*. **2000**, *122*, 3097; (e) J. M. Dysard, T. D. Tilley, *Organometallics* **2000**, *19*, 2671; (f) J. M. Dysard, T. D. Tilley, *Organometallics* **2000**, *19*, 4720.
102. (a) P. von R. Schleyer, P. K. Freeman, H. Jiao, B. Goldfuss, *Angew. Chem*. **1995**, *107*, 332; *Angew. Chem., Int. Ed. Engl*. **1995**, *34*, 337; (b) B. Goldfuss, P. von R. Schleyer, *Organometallics* **1995**, *14*, 1553; (c) B. Goldfuss, P. von R. Schleyer, F. Hampel, *Organometallics* **1996**, *15*, 1755; (d) B. Goldfuss, P. von R. Schleyer, *Organometallics* **1997**, *16*, 1543.
103. T. Müller, Y. Apeloig, H. Sohn, R. West, in *Organosilicon Chemistry III: From Molecules to Materials* (Eds N. Auner, J. Weis), Wiley-VCH, Weinheim, **1998**, p. 144.
104. (a) V. Ya. Lee, R. Kato, M. Ichinohe, A. Sekiguchi, *J. Am. Chem. Soc*. **2005**, *127*, 13142; (b) V. Ya. Lee, K. Takanashi, R. Kato, T. Matsuno, M. Ichinohe, A. Sekiguchi, *J. Organomet. Chem*. **2007**, *692*, 2800; (c) Very recently, a novel 6π-electron aromatic lithium cyclopentadienide derivative *cyclo*-$\{(EtC)_2[Si(SiMe\textit{t}\text{-}Bu_2)]_3\}^{-}\bullet[Li^{+}(L)]$ (L=thf, *t*-$Bu_2C{=}O$) was synthesized by the reduction of the heavy cyclopentadiene precursor *cyclo*-$\{(EtC)_2[Si(SiMe\textit{t}\text{-}Bu_2)]_2[Si(SiMe\textit{t}\text{-}Bu_2)CH_2SiMe\textit{t}\text{-}Bu_2]\}$ by KC_8: H. Yasuda, V. Ya. Lee, A. Sekiguchi, *J. Am. Chem. Soc*. **2009**, *131*, 6352.
105. V. Ya. Lee, M. Ichinohe, A. Sekiguchi, *J. Am. Chem. Soc*. **2000**, *122*, 12604.
106. The aromatic lithium cyclopentadienide derivatives show very diagnostic high field signals in their ^{7}Li NMR spectra because of ring current effects: (a) −8.60 ppm

for C_5H_5Li: C. Elschenbroich, *Organometallics*, 3rd Edn, Wiley-VCH, Weinheim, **2006**; (b) −7.51 ppm for $(HMe_2Si)_5C_5Li$: A. Sekiguchi, Y. Sugai, K. Ebata, C. Kabuto, H. Sakurai, *J. Am. Chem. Soc.* **1993**, *115*, 1144; (c) −5.63 ppm for $[\eta^5\text{-Li•TMEDA}][\eta^5\text{-Li}]_{1/2}[\eta^1\text{-Li}]_{1/2}[Et_4C_4Ge]$: see ref. 92.

107. For the chemical shifts of disilenes >Si=Si<, see the following reviews: (a) R. Okazaki, R. West, *Adv. Organomet. Chem.* **1996**, *39*, 231; (b) M. Weidenbruch, in *The Chemistry of Organic Silicon Compounds, Vol. 3* (Eds Z. Rappoport, Y. Apeloig), John Wiley & Sons, Ltd, Chichester, **2001**, Chapter 5.
108. The resonances of the cyclic doubly bonded Si atoms in the five-membered rings are typically observed in the range 86.0–124.2 ppm: A. Grybat, S. Boomgaarden, W. Saak, H. Marsmann, M. Weidenbruch, *Angew. Chem.* **1999**, *111*, 2161; *Angew. Chem., Int. Ed.* **1999**, *38*, 2010.
109. The ^7Li NMR resonance of η^1-Li σ-bonded to a Ge atom in trisgermole dianion is −0.93 ppm in THF-d_8: see ref. 93.
110. (a) R. West, M. Denk, *Pure Appl. Chem.* **1996**, *68*, 785; (b) M. Haaf, T. A. Schmedake, R. West, *Acc. Chem. Res.* **2000**, *33*, 704; (c) N. J. Hill, R. West, *J. Organomet. Chem.* **2004**, *689*, 4165; (d) A related aromatic derivative of the N-heterocyclic silylene, 2-silaimidazolium ion, was generated and observed by ^{29}Si NMR spectroscopy at temperatures below −10°C: S. Ishida, T. Nishinaga, R. West, K. Komatsu, *Chem. Commun.* **2005**, 778.
111. (a) A. J. Arduengo III, R. L. Harlow, M. Kline, *J. Am. Chem. Soc.* **1991**, *113*, 361; (b) A. J. Arduengo III, H. V. R. Dias, R. L. Harlow, M. Kline, *J. Am. Chem. Soc.* **1992**, *114*, 5530.
112. W. A. Herrmann, M. Denk, J. Behm, W. Scherer, F.-R. Klingan, H. Bock, B. Solouki, M. Wagner, *Angew. Chem., Int. Ed. Engl.* **1992**, *31*, 1485.
113. (a) M. Denk, R. Lennon, R. Hayashi, R. West, A. V. Belyakov, H. P. Verne, A. Haaland, M. Wagner, N. Metzler, *J. Am. Chem. Soc.* **1994**, *116*, 2691; (b) M. Haaf, A. Schmiedl, T. A. Schmedake, D. R. Powell, A. J. Millevolte, M. Denk, R. West, *J. Am. Chem. Soc.* **1998**, *120*, 12714.
114. (a) B. Gehrhus, M. F. Lappert, J. Heinicke, R. Boese, D. Bläser, *J. Chem. Soc., Chem. Commun.* **1995**, 1931; (b) B. Gehrhus, P. B. Hitchcock, M. F. Lappert, J. Heinicke, R. Boese, D. Bläser, *J. Organomet. Chem.* **1996**, *521*, 211.
115. J. Heinicke, A. Oprea, M. K. Kindermann, T. Karpati, L. Nyulászi, T. Veszprémi, *Chem. Eur. J.* **1998**, *4*, 541.
116. O. Kühl, P. Lönnecke, J. Heinicke, *Polyhedron* **2001**, *20*, 2215.
117. T. Gans-Eichler, D. Gudat, M. Nieger, *Angew. Chem., Int. Ed.* **2002**, *41*, 1888.
118. M. Denk, J. C. Green, N. Metzler, M. Wagner, *J. Chem. Soc., Dalton Trans.* **1994**, 2405.
119. J. F. Lehmann, S. G. Urquhart, L. E. Ennis, A. P. Hitchcock, K. Hatano, S. Gupta, M. K. Denk, *Organometallics* **1999**, *18*, 1862.
120. (a) L. A. Leites, S. S. Bukalov, M. Denk, R. West, M. Haaf, *J. Mol. Struct.* **2000**, *550*, 329; (b) L. A. Leites, S. S. Bukalov, A. V. Zabula, I. A. Garbuzova, D. F. Moser, R. West, *J. Am. Chem. Soc.* **2004**, *126*, 4114.
121. (a) C. Heinemann, T. Müller, Y. Apeloig, H. Schwarz, *J. Am. Chem. Soc.* **1996**, *118*, 2023; (b) R. West, J. J. Buffy, M. Haaf, T. Müller, B. Gehrhus, M. F. Lappert,

Y. Apeloig, *J. Am. Chem. Soc*. **1998**, *120*, 1639; (c) J. Oláh, T. Veszprémi, *J. Organomet. Chem*. **2003**, *686*, 112.
122. (a) A. J. Arduengo III, D. A. Dixon, K. K. Kumashiro, C. Lee, W. P. Power, K. W. Zilm, *J. Am. Chem. Soc*. **1994**, *116*, 6361; (b) A. J. Arduengo III, H. Bock, H. Chen, M. Denk, D. A. Dixon, J. C. Green, W. A. Herrmann, N. L. Jones, M. Wagner, R. West, *J. Am. Chem. Soc*. **1994**, *116*, 6641; (c) A. J. Arduengo III, H. V. R. Dias, D. A. Dixon, R. L. Harlow, W. T. Klooster, T. F. Koetzle, *J. Am. Chem. Soc*. **1994**, *116*, 6812.
123. C. Boehme, G. Frenking, *J. Am. Chem. Soc*. **1996**, *118*, 2039.
124. G. van Zandwijk, R. A. J. Janssen, H. M. Buck, *J. Am. Chem. Soc*. **1990**, *112*, 4155.
125. (a) A. J. Kos, P. von R. Schleyer, *J. Am. Chem. Soc*. **1980**, *102*, 7928; (b) M. Balci, M. L. McKee, P. von R. Schleyer, *J. Phys. Chem. A* **2000**, *104*, 1246; (c) Y. Jung, T. Heine, P. von R. Schleyer, *J. Am. Chem. Soc*. **2004**, *126*, 3132.
126. (a) A. Sekiguchi, T. Matsuo, H. Watanabe, *J. Am. Chem. Soc*. **2000**, *122*, 5652; (b) K. Ishii, N. Kobayashi, T. Matsuo, M. Tanaka, A. Sekiguchi, *J. Am. Chem. Soc*. **2001**, *123*, 5356; (c) A. Sekiguchi, T. Matsuo, M. Tanaka, *Organometallics* **2002**, *21*, 1072; (d) B. A. Shainyan, A. Sekiguchi, *Theochem*. **2005**, *728*, 1; (e) A. Sekiguchi, T. Matsuo, *Synlett* **2006**, 2683.
127. V. Ya. Lee, K. Takanashi, M. Ichinohe, A. Sekiguchi, *J. Am. Chem. Soc*. **2003**, *125*, 6012.
128. (a) V. Ya. Lee, K. Takanashi, T. Matsuno, M. Ichinohe, A. Sekiguchi, *J. Am. Chem. Soc*. **2004**, *126*, 4758; (b) Very recently, dipotassium derivative of polycyclic compound featuring a planar aromatic tetragermacyclobutadiene dianion fragment was reported: W. Wang, S. Yao, C. van Wüllen, M. Driess, *J. Am. Chem. Soc*. **2008**, *130*, 9640. In the same paper, another 6π-electron species, potassium salt of the cyclogermylidenide derivative, whose aromaticity was deduced from crystallographic and NMR spectra data, was also described.
129. (a) T. Sommerfeld, *J. Am. Chem. Soc*. **2002**, *124*, 1119; (b) S. Feuerbacher, L. S. Cederbaum, *J. Am. Chem. Soc*. **2003**, *125*, 9531.
130. V. Ya. Lee, K. Takanashi, M. Ichinohe, A. Sekiguchi, *Angew. Chem*. **2004**, *116*, 6871; *Angew. Chem., Int. Ed*. **2004**, *43*, 6703.
131. W. v. E. Doering, L. H. Knox, *J. Am. Chem. Soc*. **1954**, *76*, 3203.
132. (a) W. v. E. Doering, H. Krauch, *Angew. Chem*. **1956**, *68*, 661; (b) K. M. Harmon, in *Carbonium Ions, Vol. IV* (Ed. G. A. Olah), John Wiley & Sons, Ltd, New York, **1973**, Chapter 2.
133. C. Lifshitz, *Acc. Chem. Res*. **1994**, *27*, 138.
134. (a) E. S. Gould, *Acta Cryst*. **1955**, *8*, 657; (b) G. R. Clark, G. J. Palenik, *J. Organomet. Chem*. **1973**, *50*, 185.
135. (a) C. Cone, M. J. S. Dewar, D. Landman, *J. Am. Chem. Soc*. **1977**, *99*, 372; (b) C. Glidwell, D. Lloyd, *Tetrahedron* **1984**, *40*, 4455.
136. (a) S. Murthy, Y. Nagano, J. L. Beauchamp, *J. Am. Chem. Soc*. **1992**, *114*, 3573; (b) Y. Nagano, S. Murthy, J. L. Beauchamp, *J. Am. Chem. Soc*. **1993**, *115*, 10805.
137. R. L. Jarek, S. K. Shin, *J. Am. Chem. Soc*. **1997**, *119*, 6376.

138. (a) A. Nicolaides, L. Radom, *J. Am. Chem. Soc.* **1994**, *116*, 9769; (b) A. Nicolaides, L. Radom, *J. Am. Chem. Soc.* **1996**, *118*, 10561; (c) A. Nicolaides, L. Radom, *J. Am. Chem. Soc.* **1997**, *119*, 11933.
139. (a) T. Nishinaga, Y. Izukawa, K. Komatsu, *J. Am. Chem. Soc.* **2000**, *122*, 9312; (b) T. Nishinaga, Y. Izukawa, K. Komatsu, *Tetrahedron* **2001**, *57*, 3645; (c) K. Komatsu, *Bull. Chem. Soc. Jpn.* **2001**, *74*, 407.
140. (a) S. Winstein, *J. Am. Chem. Soc.* **1959**, *81*, 6524; (b) S. Winstein, J. Sonnenberg, L. de Vries, *J. Am. Chem. Soc.* **1959**, *81*, 6523.
141. Reviews on the concept of homoaromaticity: (a) S. Winstein, *Q. Rev., Chem. Soc.* **1969**, *23*, 141; (b) L. A. Paquette, *Angew. Chem.* **1978**, *90*, 114; *Angew. Chem., Int. Ed. Engl.* **1978**, *17*, 106; (c) R. V. Williams, H. A. Kurtz, *Adv. Phys. Org. Chem.* **1994**, *29*, 273; (d) R. V. Williams, *Chem. Rev.* **2001**, *101*, 1185.
142. J. L. v. Rosenberg Jr, J. E. Mahler, R. Pettit, *J. Am. Chem. Soc.* **1962**, *84*, 2842.
143. G. A. Olah, J. S. Staral, R. J. Spear, G. Liang, *J. Am. Chem. Soc.* **1975**, *97*, 5489.
144. (a) R. C. Haddon, K. Raghavachari, *J. Am. Chem. Soc.* **1983**, *105*, 118; (b) M. Schindler, *J. Am. Chem. Soc.* **1987**, *109*, 1020; (c) S. Sieber, P. von R. Schleyer, A. H. Otto, J. Gauss, F. Reichel, D. Cremer, *J. Phys. Org. Chem.* **1993**, *6*, 445.
145. A. Sekiguchi, T. Matsuno, M. Ichinohe, *J. Am. Chem. Soc.* **2000**, *122*, 11250.
146. K. J. Szabo, E. Kraka, D. Cremer, *J. Org. Chem.* **1996**, *61*, 2783.
147. (a) T. Laube, *J. Am. Chem. Soc.* **1989**, *111*, 9224; (b) T. Laube, C. Lohse, *J. Am. Chem. Soc.* **1994**, *116*, 9001; (c) T. Laube, *Acc. Chem. Res.* **1995**, *28*, 399; (d) W. J. Evans, K. J. Forrestal, J. W. Ziller, *J. Am. Chem. Soc.* **1995**, *117*, 12635.
148. Y. Ishida, A. Sekiguchi, Y. Kabe, *J. Am. Chem. Soc.* **2003**, *125*, 11468.
149. J. Schupann, B. Herrschaft, T. Müller, *Organometallics* **2001**, *20*, 4584.
150. A. Sekiguchi, Y. Ishida, Y. Kabe, M. Ichinohe, *J. Am. Chem. Soc.* **2002**, *124*, 8776.
151. (a) T. F. Fässler, M. Hunziker, *Inorg. Chem.* **1994**, *33*, 5380; (b) J. D. Corbett, *Angew. Chem.* **2000**, *112*, 682; *Angew. Chem., Int. Ed.* **2000**, *39*, 671.
152. R. Breslow, *Chem. Eng. News* **1965**, *43*, 90.
153. R. Breslow, *Acc. Chem. Res.* **1973**, *6*, 393
154. A. D. Allen, T. T. Tidwell, *Chem. Rev.* **2001**, *101*, 1333
155. K. B. Wiberg, *Chem. Rev.* **2001**, *101*, 1317
156. B. A. Hess Jr, L. J. Schaad, *J. Am. Chem. Soc.* **1983**, *105*, 7500
157. M. N. Glukhovtsev, S. Laiter, A. Pross, *J. Phys. Chem.* **1995**, *99*, 6828
158. A. A. Deniz, K. S. Peters, G. J. Snyder, *Science* **1999**, *286*, 1119
159. A. Sekiguchi, M. Tanaka, T. Matsuo, H. Watanabe, *Angew. Chem., Int. Ed.* **2001**, *40*, 1675 (see also references 6 and 7 cited therein).
160. M. S. Gordon, *J. Chem. Soc., Chem. Commun.* **1980**, 1131
161. M. E. Colvin, H. F. Schaefer III, *Faraday Symp. Chem. Soc.* **1984**, *19*, 39.
162. G. W. Schriver, M. J. Fink, M. S. Gordon, *Organometallics* **1987**, *6*, 1977.
163. G. Maier, H. P. Reisenauer, J. Jung, H. Pacl, H. Egenolf, *Eur. J. Org. Chem.* **1998**, 1297.
164. A. Lugo, J. Fischer, D. B. Lawson, *J. Mol. Struct. (Theochem)* **2004**, *674*, 139.
165. T. A. Holme, M. S. Gordon, S. Yabushita, M. W. Schmidt, *Organometallics* **1984**, *3*, 583.

166. G. Maier, H. P. Reisenauer, A. Meudt, *Eur. J. Org. Chem*. **1998**, 1291.
167. M. Karni, Y. Apeloig, *Silicon Chem*. **2002**, *1*, 61.
168. (a) A. F. Sax, J. Kalcher, *J. Chem. Soc., Chem. Commun*. **1987**, 809; (b) A. F. Sax, J. Kalcher, *J. Comput. Chem*. **1989**, *10*, 309.
169. (a) D. A. Clabo Jr, H. F. Schaefer III, *J. Am. Chem. Soc*. **1986**, *108*, 4344; (b) B. F. Yates, D. A. Clabo Jr, H. F. Schaefer III, *Chem. Phys. Lett*. **1988**, *143*, 421.
170. B. F. Yates, H. F. Schaeffer III, *Chem. Phys. Lett*. **1989**, *155*, 563.
171. S. Nagase, M. Nakano, *Angew. Chem., Int. Ed. Engl*. **1988**, *27*, 1081.
172. (a) N. Takagi, S. Nagase, *Chem. Lett*. **2001**, 966; (b) N. Takagi, S. Nagase, *Eur. J. Inorg. Chem*. **2002**, 2775.
173. A. Sekiguchi, R. Kinjo, M. Ichinohe, *Science* **2004**, *305*, 1755.
174. M. J. Fink, D. B. Puranik, M. P. Johnson, *J. Am. Chem. Soc*. **1988**, *110*, 1315.
175. D. B. Puranik, M. P. Johnson, M. J. Fink, *J. Chem. Soc., Chem. Commun*. **1989**, 706.
176. D. B. Puranik, M. J. Fink, *J. Am. Chem. Soc*. **1989**, *111*, 5951.
177. J. R. Gee, W. A. Howard, G. L. McPherson, M. J. Fink, *J. Am. Chem. Soc*. **1991**, *113*, 5461.
178. K. Sakamoto, J. Ogasawara, H. Sakurai, M. Kira, *J. Am. Chem. Soc*. **1997**, *119*, 3405.
179. T. Veszpremi, M. Takahashi, J. Ogasawara, K. Sakamoto, M. Kira, *J. Am. Chem. Soc*. **1998**, *120*, 2408.
180. Y. Kon, K. Sakamoto, C. Kabuto, M. Kira, *Organometallics* **2005**, *24*, 1407.
181. C. Cui, M. M. Olmstead, P. P. Power, *J. Am. Chem. Soc*. **2004**, *126*, 5062.
182. (a) R. Breslow, J. Brown, J. J. Gajewski, *J. Am. Chem. Soc*. **1967**, *89*, 4383; (b) R. Breslow, D. A. Cortes, B. Jaun, R. D. Mitchell, *Tetrahedron Lett*. **1982**, *23*, 795; (c) J. Klicic, Y. Rubin, R. Breslow, *Tetrahedron* **1997**, *53*, 4129.
183. M. N. Glukhovtsev, S. Laiter, A. Pross, *J. Phys. Chem*. **1996**, *100*, 17801.
184. R. K. Sachs, S. R. Kass, *J. Am. Chem. Soc*. **1994**, *116*, 783.
185. T. L. Arrowood, S. R. Kass, *J. Am. Chem. Soc*. **1999**, *121*, 7272.
186. K. Jug, *J. Org. Chem*. **1983**, *48*, 1344.
187. E. J. P. Malar, *Tetrahedron* **1996**, *52*, 4709.
188. K. Ziegler, B. Schnell, *Liebigs Ann*. **1925**, *445*, 266.
189. (a) R. Breslow, H. W. Chang, *J. Am. Chem. Soc*. **1961**, *83*, 3727; (b) R. Breslow, H. W. Chang, R. Hill, E. Wasserman, *J. Am. Chem. Soc*. **1967**, *89*, 1112.
190. R. Breslow, J. M. Hoffman Jr, *J. Am. Chem. Soc*. **1972**, *94*, 2110.
191. (a) P. Jutzi, A. Mix, *Chem. Ber*. **1992**, *125*, 951; (b) A. D. Allen, M. Sumonja, T. T. Tidwell, *J. Am. Chem. Soc*. **1997**, *119*, 2371.
192. R. Gompper, H. Glöckner, *Angew. Chem., Int. Ed. Engl*. **1984**, *23*, 53.
193. (a) H. Vancik, I. Novak, D. Kidemet, *J. Phys. Chem. A* **1997**, *101*, 1523; (b) H. Vancik, I. Novak, D. Kidemet, *J. Phys. Chem. A* **1998**, *102*, 8437.
194. (a) M. Saunders, R. Berger, A. Jaffe, J. M. McBride, J. O'Neill, R. Breslow, J. M. Hoffman Jr, C. Perchonock, E. Wasserman, R. S. Hutton, V. J. Kuck, *J. Am. Chem. Soc*. **1973**, *95*, 3017; (b) E. Wasserman, R. S. Hutton, *Acc. Chem. Res*. **1977**, *10*, 27.
195. (a) W. J. Hehre, P. von R. Schleyer, *J. Am. Chem. Soc*. **1973**, *95*, 5837; (b) M. J. S. Dewar, R. C. Haddon, *J. Am. Chem. Soc*. **1974**, *96*, 255; (c) W. T. Borden, E. R. Davidson, *J. Am. Chem. Soc*. **1979**, *101*, 3771; (d) J. Fabian, A. Mehlhorn,

N. Tyutyulkov, *J. Mol. Struct. (Theochem)* **1987**, *151*, 355; (e) M. N. Glukhovtsev, B. Reindl, P. von R. Schleyer, *Mendeleev Commun.* **1993**, 100; (f) M. N. Glukhovtsev, R. D. Bach, S. Laiter, *J. Phys. Chem.* **1996**, *100*, 10952; (g) E. P. F. Lee, T. G. Wright, *Phys. Chem. Chem. Phys.* **1999**, *1*, 219.
196. J. Feng, J. Leszczynski, B. Weiner, M. C. Zerner, *J. Am. Chem. Soc.* **1989**, *111*, 4648.
197. F. P. Lossing, J. C. Traeger, *J. Am. Chem. Soc.* **1975**, *97*, 1579.
198. H. Jiao, P. von R. Schleyer, Y. Mo, M. A. McAllister, T. T. Tidwell, *J. Am. Chem. Soc.* **1997**, *119*, 7075.
199. V. Gogonea, P. von R. Schleyer, P. R. Schreiner, *Angew. Chem., Int. Ed.* **1998**, *37*, 1945.
200. A. A. Korkin, V. V. Murashov, J. Leszczynski, P. von R. Schleyer, *J. Phys. Chem.* **1995**, *99*, 17742.
201. A. Sekiguchi, N. Fukaya, M. Ichinohe, N. Takagi, S. Nagase, *J. Am. Chem. Soc.* **1999**, *121*, 11587.
202. The Ge=Ge double bonds are typically *trans*-bent: P. P. Power, *Chem. Rev.* **1999**, *99*, 3463.
203. A. Sekiguchi, Y. Ishida, N. Fukaya, M. Ichinohe, *J. Am. Chem. Soc.* **2002**, *124*, 1158.
204. A. Shinohara, N. Takeda, N. Tokitoh, *J. Am. Chem. Soc.* **2003**, *125*, 10804.
205. I. S. Toulokhonova, T. C. Stringfellow, S. I. Ivanov, A. Masunov, R. West, *J. Am. Chem. Soc.* **2003**, *125*, 5767.
206. I. S. Toulokhonova, I. A. Guzei, R. West, *J. Am. Chem. Soc.* **2004**, *126*, 5336.
207. V. I. Timokhin, I. A. Guzei, R. West, *Silicon Chem.* **2006**, *3*, 75.
208. I. S. Toulokhonova, D. R. Friedrichsen, N. J. Hill, T. Müller, R. West, *Angew. Chem.* **2006**, *118*, 2640; *Angew. Chem., Int. Ed.* **2006**, *45*, 2578.
209. H. Sohn, D. R. Powell, R. West, J.-H. Hong, W.-C. Joo, *Organometallics* **1997**, *16*, 2770.
210. H. Sohn, J. Merritt, D. R. Powell, R. West, *Organometallics* **1997**, *16*, 5133.
211. A. Sekiguchi, T. Matsuno, M. Ichinohe, *J. Am. Chem. Soc.* **2001**, *123*, 12436.
212. T. Matsuno, M. Ichinohe, A. Sekiguchi, *Angew. Chem.* **2002**, *114*, 1645; *Angew. Chem., Int. Ed.* **2002**, *41*, 1575.
213. A. F. Hill, *Organotransition Metal Chemistry*, Wiley-Interscience, New York, **2002**.
214. J. D. Fitzpatrick, L. Watts, G. F. Emerson, R. Pettit, *J. Am. Chem. Soc.* **1965**, *87*, 3254.
215. T. J. Kealy, P. L. Pauson, *Nature* **1951**, *168*, 1039.
216. E. O. Fischer, W. Hafner, *Z. Naturforsch.* **1955**, 10b, 665.
217. C. Elschenbroich, *Organometallics*, 3rd Edn, Wiley-VCH, Weinheim, **2006**.
218. G. N. Srinivas, L. Yu, M. Schwartz, *Organometallics* **2001**, *20*, 5200.
219. G. N. Srinivas, T. P. Hamilton, E. D. Jemmis, M. L. McKee, K. Lammertsma, *J. Am. Chem. Soc.* **2000**, *122*, 1725.
220. (a) A. Shinohara, N. Takeda, T. Sasamori, T. Matsumoto, N. Tokitoh, *Organometallics* **2005**, *24*, 6141; (b) Y. Tanabe, Y. Mizuhata, N. Tokitoh, *Chem. Lett.* **2008**, *37*, 720.

221. N. Nakata, N. Takeda, N. Tokitoh, *Angew. Chem.* **2003**, *115*, 119; *Angew. Chem., Int. Ed.* **2003**, *42*, 115.
222. N. Tokitoh, N. Nakata, A. Shinohara, N. Takeda, T. Sasamori, *Chem. Eur. J.* **2007**, *13*, 1856.
223. J. M. Dysard, T. D. Tilley, T. K. Woo, *Organometallics* **2001**, *20*, 1195.
224. W. P. Freeman, J. M. Dysard, T. D. Tilley, A. L. Rheingold, *Organometallics* **2002**, *21*, 1734.
225. T. Kudo, S. Nagase, *J. Mol. Struct. (Theochem)* **1994**, *311*, 111.
226. V. Ya. Lee, R. Kato, A. Sekiguchi, A. Krapp, G. Frenking, *J. Am. Chem. Soc.* **2007**, *129*, 10340.
227. (a) E. E. Bunel, L. Valle, J. M. Manriquez, *Organometallics* **1985**, *4*, 1680; (b) J. M. Manriquez, E. E. Bunel, B. Oelckers, *Inorg. Synth.* **1997**, *31*, 214.
228. H. Yasuda, V. Ya. Lee, A. Sekiguchi, *J. Am. Chem. Soc.* **2009**, *131*, 9902.
229. (a) A. Efraty, *Chem. Rev.* **1977**, *77*, 691; (b) P. K. Baker, H. Silgram, *Trends Organomet. Chem.* **1999**, *3*, 21; (c) D. Seyferth, *Organometallics* **2003**, *22*, 2.
230. A related (η^5-cyclopentadienyl)(η^4-silacyclobutadiene)cobalt complex **75**, prepared by the thermal reaction of 4-silatriafulvene with (η^5-H_5C_5)$Co(CO)_2$, was recently reported: see ref. 180.
231. K. Takanashi, V. Ya. Lee, T. Matsuno, M. Ichinohe, A. Sekiguchi, *J. Am. Chem. Soc.* **2005**, *127*, 5768.
232. K. Takanashi, V. Ya. Lee, M. Ichinohe, A. Sekiguchi, *Angew. Chem.* **2006**, *118*, 3347; *Angew. Chem., Int. Ed.* **2006**, *45*, 3269.
233. K. Takanashi, V. Ya. Lee, A. Sekiguchi, *Organometallics* **2009**, *28*, 1248.
234. K. Takanashi, V. Ya. Lee, M. Ichinohe, A. Sekiguchi, *Eur. J. Inorg. Chem.* **2007**, 5471.

Index

Page references in **bold** refer to tables. Those in *italics* refer to figures.

Organometallic Compounds of Low-Coordinate Si, Ge, Sn and Pb: From Phantom Species to Stable Compounds
Vladimir Ya. Lee and Akira Sekiguchi

Printed and bound in the UK by
CPI Antony Rowe, Eastbourne

Printed and bound by CPI Group (UK) Ltd, Croydon, CR0 4YY
16/09/2026
14963473-0002